Thinking About Evolution

Historical, Philosophical, and Political Perspectives

This is the second of two volumes published by Cambridge University Press in honor of Richard Lewontin. The first volume, *Evolutionary Genetics from Molecules to Morphology*, honors Lewontin's more technical contributions to genetics and evolutionary biology. In this second volume of essays, philosophical, historical, and political dimensions of his work are honored.

Given the range of Lewontin's own contributions, it is fitting that the volume covers such a wide range of perspectives on modern biology. He is not only a very successful practitioner of evolutionary genetics but also a rigorous critic of the practices of genetics and evolutionary biology, and an articulate analyst of the social, political, and economic contexts and consequences of genetic and evolutionary research. The volume begins with an essay by Lewontin titled "Natural History and Formalism in Evolutionary Genetics." Chapter 2 is an extended interview with Lewontin covering the history of evolutionary genetics as seen from his perspective and as exemplified by his career. The remaining chapters, contributed by former students, postdoctoral fellows, colleagues, and collaborators, cover issues ranging from the history and conceptual foundations of evolutionary biology and genetics, to the implications of human genetic diversity, to the political economy of agriculture and public health.

Rama S. Singh is Professor in the Department of Biology at McMaster University.

Costas B. Krimbas is Professor of Philosophy and History of Science at the University of Athens.

Diane B. Paul is Professor of Political Science at the University of Massachusetts at Boston.

John Beatty is Professor of Ecology, Evolution, and Behavior at the University of Minnesota.

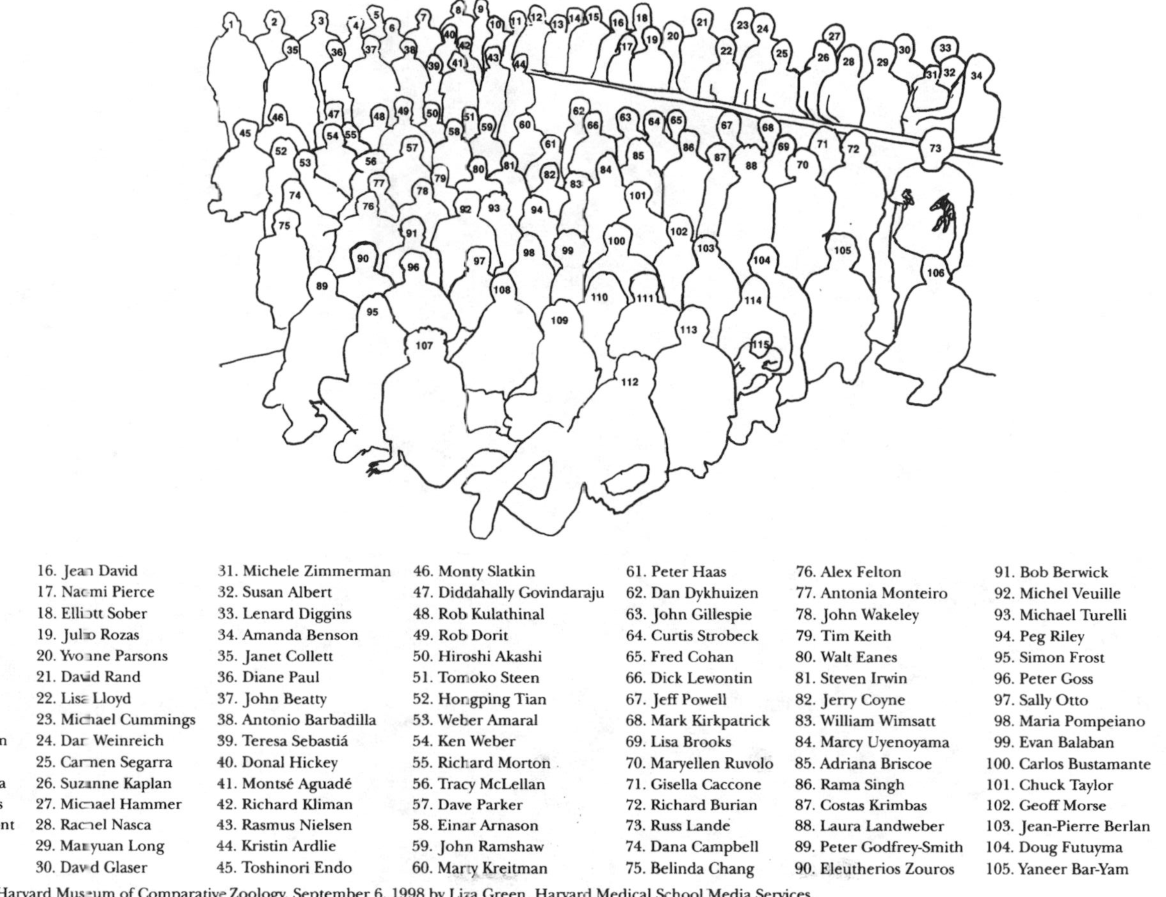

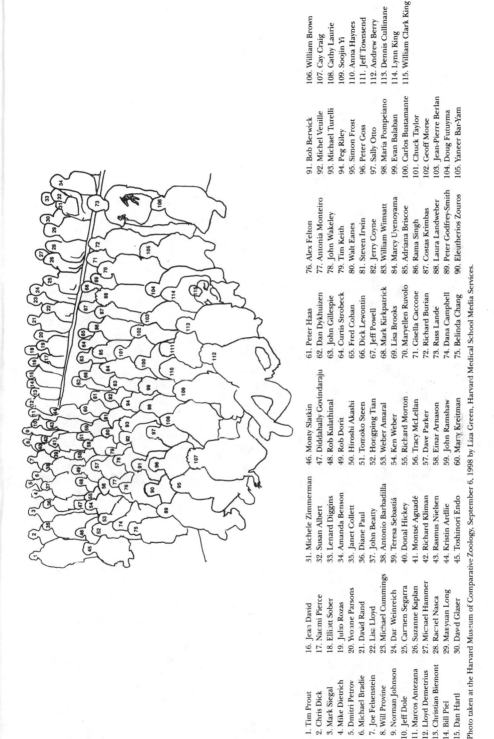

1. Tim Prout
2. Chris Dick
3. Mark Siegal
4. Mike Dietrich
5. Dmitri Petrov
6. Michael Bradie
7. Joe Felsenstein
8. Will Provine
9. Norman Johnson
10. Jeff Dole
11. Marcos Antezana
12. Lloyd Demetrius
13. Christian Biemont
14. Bill Piel
15. Dan Hartl
16. Jean David
17. Naomi Pierce
18. Elliott Sober
19. Julio Rozas
20. Yvonne Parsons
21. David Rand
22. Lisa Lloyd
23. Michael Cummings
24. Dan Weinreich
25. Carmen Segarra
26. Suzanne Kaplan
27. Michael Hammer
28. Rachel Nasca
29. Manyuan Long
30. David Glaser
31. Michele Zimmerman
32. Susan Albert
33. Lenard Diggins
34. Amanda Benson
35. Janet Collett
36. Diane Paul
37. John Beatty
38. Antonio Barbadilla
39. Teresa Sebastiá
40. Donal Hickey
41. Montsé Aguadé
42. Richard Kliman
43. Rasmus Nielsen
44. Kristin Ardlie
45. Toshinori Endo
46. Monty Slatkin
47. Diddahally Govindaraju
48. Rob Kulathinal
49. Rob Dorit
50. Hiroshi Akashi
51. Tomoko Steen
52. Hongping Tian
53. Weber Amaral
54. Ken Weber
55. Richard Morton
56. Tracy McLellan
57. Dave Parker
58. Einar Arnason
59. John Ramshaw
60. Marty Kreitman
61. Peter Haas
62. Dan Dykhuizen
63. John Gillespie
64. Curtis Strobeck
65. Fred Cohan
66. Dick Lewontin
67. Jeff Powell
68. Mark Kirkpatrick
69. Lisa Brooks
70. Maryellen Ruvolo
71. Gisella Caccone
72. Richard Burian
73. Russ Lande
74. Dana Campbell
75. Belinda Chang
76. Alex Felton
77. Antonia Monteiro
78. John Wakeley
79. Tim Keith
80. Walt Eanes
81. Steven Irwin
82. Jerry Coyne
83. William Wimsatt
84. Marcy Uyenoyama
85. Adriana Briscoe
86. Rama Singh
87. Costas Krimbas
88. Laura Landweber
89. Peter Godfrey-Smith
90. Eleutherios Zouros
91. Bob Berwick
92. Michel Veuille
93. Michael Turelli
94. Peg Riley
95. Simon Frost
96. Peter Goss
97. Sally Otto
98. Maria Pompeiano
99. Evan Balaban
100. Carlos Bustamante
101. Chuck Taylor
102. Geoff Morse
103. Jean-Pierre Berlan
104. Doug Futuyma
105. Yaneer Bar-Yam
106. William Brown
107. Cay Craig
108. Cathy Laurie
109. Soojin Yi
110. Anna Haynes
111. Jeff Townsend
112. Andrew Berry
113. Dennis Cullinane
114. Lynn King
115. William Clark King

Photo taken at the Harvard Museum of Comparative Zoology, September 6, 1998 by Liza Green, Harvard Medical School Media Services.

Thinking About Evolution

Historical, Philosophical, and Political Perspectives

VOLUME TWO

Edited by

RAMA S. SINGH
McMaster University

COSTAS B. KRIMBAS
University of Athens

DIANE B. PAUL
University of Massachusetts at Boston

JOHN BEATTY
University of Minnesota

CAMBRIDGE
UNIVERSITY PRESS

CAMBRIDGE UNIVERSITY PRESS
Cambridge, New York, Melbourne, Madrid, Cape Town,
Singapore, São Paulo, Delhi, Tokyo, Mexico City

Cambridge University Press
The Edinburgh Building, Cambridge CB2 8RU, UK

Published in the United States of America by Cambridge University Press, New York

www.cambridge.org
Information on this title: www.cambridge.org/9780521178310

First published 2001
First paperback edition 2011

A catalogue record for this publication is available from the British Library

Library of Congress Cataloguing in Publication data

Thinking about evolution : historical, philosophical, and political perspectives /
edited by Rama S. Singh ... [et al.].

p. cm.

ISBN 0-521-62070-8

1. Evolution (Biology) I. Singh, Rama S. (Rama Shankar), 1945–

QH366.2.T53 2000
576.8 – dc21

00-037822

ISBN 978-0-521-62070-3 Hardback
ISBN 978-0-521-17831-0 Paperback

Contents

Section B: Philosophy of Evolutionary Biology

Section C: The Politics of Evolutionary Biology

List of Contributors

BALABAN, E., The Neurosciences Institute, San Diego, California 92121

BEATTY, J., Department of Ecology, Evolution and Behavior, University of Minnesota, St. Paul, Minnesota 55108

BERLAN, J.-P., Economie et Sociologie Rurales, Institut National de la-Recherche Agronomique, 34060 Montpellier, France

BRANDON, R. N., Departments of Philosophy and Zoology, Duke University, Durham, NC 27708

FALK, R., Department of Genetics, The Hebrew University, 91904 Jerusalem, Israel

GAYON, J., Université Paris 7-Denis Diderot, UFR GHSS, cc7001, 75251 Paris, France

GODFREY-SMITH, P., Philosophy Department, Stanford University, Stanford, CA 94305-2155

GOULD, S. J., Museum of Comparative Zoology, Harvard University, Cambridge, MA 02138

GRAY, R. D., Department of Psychology, University of Auckland, Auckland 92019, New Zealand

HAYNES, R. H., Department of Biology, York University, Toronto, Ontario, M3J 1P3, Canada

HUBBARD, R., Department of Biology, Harvard University, Cambridge, MA 02138

KITCHER, P., Department of Philosophy, Columbia University, New York, NY 10027

KRIMBAS, C. B., Department of Philosophy and History of Science, The National University of Athens, Greece

LEWONTIN, R. C., Museum of Comparative Zoology, Harvard University, Cambridge, MA 02138

LEVINS, R., Department of Population Sciences, Harvard University, Boston, MA 02115

LLOYD, E. A., Department of History and Philosophy of Science, Indiana University, Bloomington, IN 47405

PAUL, D. B., Department of Political Science, University of Massachusetts Boston, Boston, MA 02125

PIATTELLI-PALMARINI, M., Cognitive Sciences Program, University of Arizona, Tucson, AZ 85721–0025

PROCTOR, R. N., Department of History, Pennsylvania State University, University Park, PA 16802

ROSE, S. P. R., Brain and Behavioral Research Group, The Open University, Milton Keynes, MK7 6AA UK

RUVOLO, M., Department of Anthropology, Harvard University, Peabody Museum, Cambridge, MA, 02138

SARKAR, S., Faculty of Philosophy, University of Texas at Austin, Austin, TX 78712-1180

SCHANK, J. C., Department of Psychology, University of California, Davis, CA 95616

SEIELSTAD, M., Program for Population Genetics, Harvard School of Public Health, Boston, MA 02115

SINGH, R. S., Department of Biology, McMaster University, Hamilton, Ontario, Canada, L8S 4K1

SOBER, E., Department of Philosophy, University of Wisconsin, Madison, WI 53706

SPENCER, H. G., Department of Zoology, University of Otago, Dunedin, New Zealand

TAYLOR, P. J., Program on Critical and Creative Thinking, Graduate College of Education, University of Massachusetts Boston, Boston, MA 02125

VANDERMEER, J., Department of Biology, University of Michigan, Ann Arbor, MI 48109

VEUILLE, M., Ecole Pratique des Hautes Etudes, Laboratoire d'Ecologie, cc237, Université Pierre et Marie Curie, 75252 Paris cedex 05, France

WIMSATT, W. C., Department of Philosophy, Committee on Evolutionary Biology, The University of Chicago, Chicago, IL 60637

Preface

Scientists earn their reputation by making special contributions in a variety of ways. Some become known for a discovery that revolutionizes their science. Others are respected as intellectual leaders for significant contributions leading to sustained progress in their field. Still others become known for providing guidance, opportunity, and uniquely inspiring rapport to a large number of graduate students, writers, and research colleagues. A rare few do all the above and remarkably enough still find time to deal with the broader issues of epistemology, philosophy, history, and sociology of science. Richard Lewontin is one of these rare scientists.

If we are to attach a major discovery or a conceptual breakthrough to Lewontin's name (such as Haldane's cost of natural selection, Fisher's fundamental theorem of natural selection, Wright's shifting balance theory, or Maynard Smith's game theory applications) then the successful completion of the genetic variation research program of the Chetverikov–Dobzhansky School will be known as the outstanding highlight of Lewontin's career. Dobzhansky and his students and collaborators pursued the twin problems of the amount and the adaptive role of genetic variation for nearly 25 years without a satisfactory solution. All estimates of genetic variation were indirect or inadequate, for there was no reductionist research program that could allow the study of genetic variation at the level of the gene. Lewontin's pioneering success in the application of protein electrophoresis to the problem of genetic variation changed the scene radically. The estimation of electrophoretic variation was direct and more useful than any one had expected. The technique also removed the experimental limitations imposed by genetic incompatibility between species and allowed reliable comparisons of genetic variation between populations and species without any need to make genetic crosses. The impact and the anticipation of the avalanche of future results from the use of electrophoresis were discussed in his well-known book *The Genetic Basis of Evolutionary Change* (1974). This book sets out the problem of population genetics in a rationally constructed historical context and is required reading for all aspiring population geneticists.

Evolutionary research requires broad interest and versatility in modeling, experimental design, statistics, field biology, and much more. Such breadth allowed Lewontin to be successful, time and again, in designing new experimental systems or suggesting key concepts to answer old questions or pursue new ones. Lewontin became interested in the uniqueness of the phenotype and genotype–environment interactions inspired mainly by the Russian biologist I. Schmalhausen's book *Factors of Evolution.* Lewontin's doctoral thesis studied fitness as a function of genotype frequency and density and showed that "viability of a genotype is a function of the other genotypes which coexist with it, the result of any particular combination not being predictable on the basis of the viabilities of the coexisting genotypes when tested in isolation." This was followed by studies of interlocus epistatic interactions in fitnesses and the evolution of naturally occurring inversion polymorphism in *Drosophila.* His mathematical work on linkage disequilibrium provided a new direction for research, and results from a series of papers on multilocus fitness effects anticipated discussion on the units of selection. His experimental work on norms of reaction in *Drosophila* was exemplary in exposing the problem of genetic determination and led to a new appreciation of genotype–environment interaction and phenotypic plasticity. He pointed out the importance of developmental time in fitness, which is something usually forgotten when describing fitness components. His 1972 article on the "Apportionment of Human Diversity" pointing out that any genetic differences between races has to be compared with genetic variation within population and races, is a landmark in human genetics and evolution. More recently, his laboratory has been a major center for studies of DNA sequence variation. Lewontin has provided training and guidance to a large number of graduate students and postdoctoral fellows. The number is well over one hundred! Many more have worked in Lewontin's laboratory but have not necessarily coauthored publications with him.

But what makes Lewontin known more in the wider circle of evolutionary biology and in science in general is his role as a critic of how science is done on the one hand and his passionate engagements with the issues of science and society on the other. He has made important contributions and has influenced research workers in the history and philosophy of science and in the areas of science and society such as agriculture, social health problems, bioethics, and genetics and IQ. If you drop Lewontin's name in any group of biologists, an animated discussion is sure to follow! These discussions are not about science but about its relevance and applications to human affairs. His concern about social issues springs directly from his unique perspective of evolutionary biology. Lewontin's research program may be reductionist, but he is not. He has encouraged and challenged evolutionary biologists to find the most desirable combination of Platonic and Aristotelian traditions in studying nature. Accordingly, the mathematical rigor of early population biology must be extended to accommodate interactive, hierarchical, probabilistic, and historical factors as learned empirically in the field. To him, "*context and interaction are of the essence*" (Lewontin 1974, p. 318) whether one is talking about interactions between hierarchical levels, between organisms and the environment, or between causes and effects. A reductionist approach to science does not necessitate a reductionist view of the world. No level of analysis is specially privileged for a general understanding of causality. Genetic and environmental effects are interdependent, and the phenotypic variance cannot be partitioned into fixed components. Organisms do not fit in preexisting ecological niches but create their own niches. History and contingencies are so important in evolution that looking for adaptive explanations for all organismic traits undermines the role of natural history. These ideas essentially follow from his belief that relationships between organisms and their environments, and likewise, those between groups and hierarchical levels, are governed by forces so weak that the outcomes are neither fixed nor predetermined.

John Maynard Smith has written (first volume, pages 628–640) that "Richard Lewontin has contributed to science not only by his own work on evolution theory and molecular variation, and by his influence on the many young scientists who have worked with him but also by asking us to think about the relationships between the science we do and the world we do it in." Although one may not agree with Lewontin on all issues (he would be surprised if you did!), one thing is sure – Lewontin has been a colorful personality who has made evolutionary biology rigorous and interesting at the same time. We affectionately dedicate this volume to him.

This volume has been long in preparation, and we thank the authors for their patience. We are extremely grateful to many colleagues who provided help as reviewers. At the Cambridge University Press we express our sincere thanks to Robin Smith for his early enthusiasm and help in the preparation of this book, and to Ellen Carlin for supervising its completion. Eleanor Umali and her associates at Techbooks have done a superb job of production. Finally we would like to thank Kathy McIntosh of McMaster University for her enthusiasm and the enormous amount of work that she has put in, from communication with authors to preparation of final manuscripts of this book.

Rama Singh
Hamilton, Ontario, Canada

Costas Krimbas
Athens, Greece

Introduction

This is the second of two volumes published in honor of Richard Lewontin. The first volume, *Evolutionary Genetics from Molecules to Morphology*, honors his more technical contributions to genetics and evolutionary biology. In this second volume of essays, we honor the philosophical, historical, and political dimensions of his work. It is not our intention to separate science from philosophy, history, and politics; indeed, the essays in this volume concern broader dimensions of technical scientific developments. But we acknowledge that the structure of the volumes to some extent reinforces the conventional divide between the scientific and social that Lewontin himself has fought so hard to dissolve.

If one is looking for central threads in Lewontin's wide-ranging work, then surely his epistemological concerns are a good place to start. One can follow his general skepticism to several other major concerns and interests. Lewontin is well known – even infamous – for persistently questioning whether geneticists and evolutionary biologists can possibly know what they want to know and often claim to know. His epistemological critiques of adaptationist thinking (including the justly famous "Spandrels" article coauthored with Stephen Gould), hereditarian thought, and sociobiology are especially well known. But just as, or perhaps even more, indicative of his epistemological concerns are the volleys that fall closer to home in his own field of evolutionary genetics. Skepticism is the motif of his classic *Genetic Basis of Evolutionary Change* in which he set out *not* to resolve the major debates of evolutionary genetics of the time but to articulate clearly just how difficult it would be to answer the questions at issue. He managed to remain fairly independent of the major competing schools of thought: the "balance" school associated most prominently with his mentor, Theodosius Dobzhansky, and the "classical"–"neoclassical" school associated most prominently with Dobzhansky's archrival H. J. Muller and with James Crow and Motoo Kimura. Dobzhansky certainly benefitted from Lewontin's critiques of the Muller camp but felt somewhat persecuted when the same intense skepticism was aimed in his direction (by his own "son"):

Dear Dick:

Unless my memory is faulty, there has not been a case when anything that I said, wrote, or done met with your approval. I suppose this phenomenon may have some Freudian explanation, and for that reason or otherwise, I am so used to your disapproval that it no longer hurts me as much as it used to....

Anyway, since when do you consider the Wisconsin–Indiana [Crow–Muller] line so sacred that it should not even be questioned? I thought that you like defiance of authority for defiance sake, and that a rebellion appeals to you as something heroic regardless of anything so small as whether the rebellion is justified....

In his periodic reviews of the state of population genetics, Lewontin continues to point out the limitations of existing techniques and methods – tools that he himself has played a key role in developing and putting to use. The title of one such review well illustrates his discontent with methods of his own device: "Electrophoresis in the Development of Evolutionary Genetics: Milestone or Millstone?" The *Chronicle of Higher Education* carried a story on Lewontin's dissatisfactions as expressed at a conference for leaders of the field and young scientists; the article was titled "An Influential Scientist's Scathing Critique of His Field." Some conference attendees expressed concern that Lewontin's skepticism might demoralize the young investigators who were present. Probably the opposite was true.

Also indicative of his epistemological concerns are influential articles on a variety of special topics, including the deeply "historical" (where "historical" does not just mean evolutionary) nature of the biological world, on organism–environment interactions, and on units of selection. All of these issues have to do with ways in which evolutionary outcomes are even more difficult to anticipate and understand than had previously been thought. These contributions, along with Lewontin's critiques of adaptationist and hereditarian thought, also exemplify other general philosophical themes that run throughout his work – especially concerns about causation versus correlation and reductionism and emergentism.

Through these concerns, Lewontin has had an enormous influence on philosophy of biology. This influence has been extended not only through his writings but through personal contact. In addition to participating in formal philosophical seminars, he has also made office space and a considerable amount of his time available to the many philosophers from around the world who have spent all or part of a sabbatical or a summer in his laboratory. (This is probably a good place to acknowledge the role that his own grad ate students have played over the years hosting all those visitors, bringing them up-to-speed on one or another technical issue, participating in Socratic dialogues about concepts and methods, and of course enduring lots of not-fully-baked philosophical esoterica.)

Like his colleagues Ernst Mayr and Stephen Gould, who have also had so much influence on the philosophy of biology, Lewontin has been concerned not to bemoan but rather to appreciate the ways in which biology differs from the physical sciences. He is well known among philosophers for once publically disassembling a prominent philosopher of science (not a philosopher of biology) whose reach far exceeded his grasp of his topic – evolutionary biology. As Lewontin himself recalls,

I was once asked to be the commentator at a Philosophy of Science Association meeting on a paper by a famous epistemologist. At first I was worried that I would not understand his paper; but after I read it, what worried me was that I *did* understand it.

Biology has been an embarrassment to the philosophers of science and so, for a very long time, most of them pretended it did not exist, or denied that it was a real science at all.

However, as Lewontin proceeds to explain, the situation in philosophy of science was quickly changing as more and more philosophers of science turned to philosophy of biology (Lewontin himself has played no insignificant role in this development).

Lewontin's philosophy of science should not be analyzed independently of his Marxism. His general skepticism derives in part from his conviction that alternative scientific methods and theories often reflect alternative ideologies (i.e., to some extent hidden): "science is its own era reflected in thought." Thus, Lewontin's analyses of scientific controversies often involve exposes of previously unrecognized sociopolitical dimensions of the positions in question. For example, in *The Genetic Basis of Evolutionary Change*, he situated the classical and balance schools within the context of ideological differences concerning the significance of genetic diversity. His numerous critiques of evolutionary equilibrium analyses quite often begin or end with an account of their ideological context (a "preoccupation with stability" – a reaction to the revolutions of Darwin and Marx – "change had to be tamed in science as it was in society"). But Lewontin is well aware of the self-reflexivity of such contextual concerns:

It is sure to be pointed out that the dialectical approach is no less contingent historically and socially than the viewpoints we criticize, and that the dialectic must itself be analyzed dialectically. This is no embarassment; rather, it is a necessary awareness for self-criticism.

Lewontin's epistemological skepticism is also a reflection of his dialectical materialism, which plays a cautionary, heuristic role in his work. This is especially well expressed in the conclution to his searching account of Lysenkoism,

The error of the Lysenkoist claim arises from attempting to apply a dialectical analysis of physical problems from the wrong end. Dialectical materialism is not, and never has been, a programmatic method for solving particular physical problems. Rather, dialectical analysis provides an overview and a set of warning signs against particular forms of dogmatism and narrowness of thought. It tells us, "Remember that history may leave an important trace. Remember that being and becoming are dual aspects of nature. Remember that conditions change and that the conditions necessary to the initiation of some process may be destroyed by the process itself. Remember to pay attention to real objects in time and space and not lose them in utterly idealized abstractions. Remember that qualitative effects of context and interaction may be lost when phenomena are isolated." And above all else, "Remember that all the other caveats are only reminders and warning signs whose application to different circumstances of the real world is contingent."

But the purpose of Lewontin's philosophy and science is not just to understand, but to change the world. He has done much more than just to ponder social change. His political activities range from his various contacts with the Black Panther Party, the Socialist Workers Party and the Communist party, and the New World Agriculture Group. We think it is especially telling that his political concens extend to the most local details – for example, the organization of his own laboratory. The laboratory's physical structure – with its large central space and small private offices and with the most desirable space allocated to those who spend the most time in the laboratory (particularly the technicians) – quite self-consciously reflects egalitarian ideals. This egalitarian commitment was taken a step further – indeed, a step too far – in the 1970s. Perhaps inspired by Marx's dictum that "one basis for science and another for life is a priori a lie," he decreed that his own workplace would henceforth be democratically and collectively organized. Thus, tasks such as stockkeeping, dishwashing, and general laboratory maintenance were to be shared rather than delegated to technical workers, grant applications were to be written by the group as a whole, and the group would also decide which prospective students to admit. Even the research direction of the laboratory was to be discussed collectively. But within a year, the structure collapsed. In Lewontin's own words:

> There was no political agreement and the "collectivity" had been decreed by the Professor. The students went along, humouring him, but resented having to do manual and administrative work that others should have been doing. I, on the other hand, increasingly failed to do my share of the chores because of the press of other work, and, secretly, other priorities.

And he further acknowledged that,

> the entire arrangement was in contradiction to the objective situation of power, namely that the Professor had the real power over the money, space, work, etc., and only on whim had created the collective attempt. A change in whim could create its opposite. The objective reality of power involves all such attempts at collective work within hierarchical institutions in indissoluble contradictions.

Lewontin's political critiques of science have had a wide audience. In his popular writing (such as his frequent essays for the *New York Review of Books*), lectures, television and radio appearances, and testimony as an expert witness in various court cases, he has employed his scientific expertise, institutional position, and prestige to undermine genetic determinism, racism, and the existing system of publicly funded agricultural research. He has testified in opposition to the University of California's agricultural mechanization projects, arguing that they harmed the interests of small family farmers, laborers, consumers, and the rural population, thus violating the provisions of the Hatch Act. He has challenged efforts to measure the heritability of human mental traits and to establish a genetic basis for racial differences in IQ scores. He has tangled with enthusiasts of the Human Genome Project and with the FBI in connection with forensic uses of DNA. Indeed, virtually anywhere that there is a controversy about genetics and social issues, Lewontin can be found at or near its center.

In these disputes, Lewontin's strategy has generally been to expose error. His opponents tend to be characterized as bad scientists whose data are shoddy, understanding of population genetics inadequate, and generalizations overly broad. In this polemical mode, Lewontin wears the mantle of a scientist upholding the conventions of his field. His opponents are judged and found wanting by criteria generally agreed-upon by population geneticists.

His Marxist approach to science leads Lewontin away from the mainstream, whereas his efforts to expose the scientific mistakes of biological determinists and other opponents tug him back toward the center. There is a serious tension here. The authority of scientists in matters political ultimately derives from their standing with their peers. Political efficacy depends on being seen to speak for a community. It would be futile to argue, for example, that Arthur Jensen's work fails from a Marxist standpoint. The point is to show that it fails by standards shared by population geneticists, whatever their politics. Lewontin has never accused Jensen (or Richard Herrnstein, or E. O. Wilson) of being insufficiently dialectical – a charge that would have meaning only for other Marxists – but rather with doing bad science, which is a claim that can only be made by a geneticist who can speak credibly for the field. To move too far from its mainstream would be to destroy the credibility that is a sine qua non of political effectiveness. That Lewontin has handled this contradiction as well as he has is testimony to his particular scientific, political, and personal skills.

For us, one of Lewontin's most admirable traits is an almost ruthless intellectual honesty. He is aware that we all have ideological commitments and that it is not only our opponents who may be blinded by them. He is always prepared to acknowledge a really good argument, however unappealing its content and whatever its source. A story told to Costas Krimbas nicely illustrates this facet of his intellectual style:

> He (the philosopher of biology Peter Godfrey-Smith) and I went to a seminar [on epistemological issues in evolutionary biology] offered by the Philosophy Department, by Bob Nozick and the philosopher and economist Amartya Sen.... I know Bob Nozick from a long time ago. He is a very good philosopher, but I disagree with almost everything he writes. So Peter and I went to the seminar and thought we would set them straight. They will be confused, they will make all the stupid mistakes in the world, and we will fix them. But it turned out that they were not confused at all. They really understood what the issues were, perfectly. They really were excellent.... Bob and I have a completely different view of the theory of justice, about everything... so for that reason I felt that this is not a guy who will set me straight in evolutionary biology, but he did!

Lewontin does not demand deference from his student or colleagues. His own approach is no-holds-barred, and he expects the same treatment. Although the essays in this volume concern subjects of interest to Lewontin and are written by his former graduate students, postdoctoral fellows, colleagues, and collaborators, some of whom are also friends, they do not necessarily reflect his views. Indeed, some are explicitly critical, reflecting the fact that Richard Lewontin has never surrounded himself with epigones. A hagiography would not be the appropriate way to honor his work.

The Editors

CHAPTER ONE

Natural History and Formalism in Evolutionary Genetics

R. C. LEWONTIN

In a famous synecdoche, Dobzhansky (1951) once defined evolution as "a change in the genetic composition of populations" (p. 16), an epigram that should not be mistaken for the claim that everything worth saying about evolution is contained in statements about genes, although from reading Dobzhansky's own work, it is a mistake that might be easy to make. It is the changes in biological properties of organisms that we are trying to understand, and our interest in genes arises because individual organisms are mortal. As a consequence, changes in the composition of populations can be propagated through time only to the extent that characteristics are heritable. Of course, because DNA sequences are part of the biological apparatus of the organism, we are concerned in their evolution as molecular biological characteristics just as we are interested in the evolution of any other part of the machinery of the cell. But the substitution of genes for biological properties influenced by those genes fails to come to grips with the original problematic of evolutionary biology in two ways, one more subtle than the other.

The first consequence of the substitution of genes for the biological properties that they influence is at the descriptive level and is well known. In its most extreme form, gene language can dispense entirely with other properties of organisms. There is nothing to prevent us, at least for extant species, from describing the organisms as DNA sequences and then recounting their variation, evolution, and speciation by specifications entirely in the space of DNA descriptors. We reject such an extreme program not because of any formal incompleteness but because, in the absence of bridging statements between genes and other characters, it does not allow us to engage even the descriptive issues that motivated us in the first place such as the evolution of size and shape or of sex. Of course, we would also be barred from using any evidence from paleontology, or indeed even thinking about the mammal-like reptiles. (Thinking about mammal-like reptiles is not the same as thinking about DNA sequences that may have been common ancestors in the gene tree that links certain extant DNA sequences with each other.) Moreover, most of the causal explanations of evolutionary change involve assertions about organisms. Although it is possible formally to assign fitnesses to genotypes and to carry out the complete projection of the evolutionary trajectory of a population of DNA molecules (Godfrey-Smith and Lewontin 1993), the actual assignment of such fitnesses in any particular case demands information about material properties of organisms and a set of bridging statements between those properties and DNA sequences. Even gene-centered descriptions of evolution, such as selfish gene theories in sociobiology, do not dispense with such bridging statements. On the contrary, they use rich descriptions of organismal properties but make the particular metaphysical claim that organisms are merely instrumentalities by which genes propagate themselves. That organismic properties are seen as subserving the ends of genes does not eliminate those properties from the explanatory program but, on the contrary, *is* their claimed explanation. The reason that evolutionary genetic theory does not operate entirely in the space of DNA sequences is that it leaves out all the rest of biology. To paraphrase Hobbes, genes are biologists' counters; they do but reckon with them, but they are the money of computer modelers.

In practice, we link DNA to biology by connecting DNA sequences with organismic characters using statements of the form, "the genes for. . . ." If such statements are taken as shorthand for "genes that are causally antecedent in development and physiology, either directly or indirectly, to. . . ," then this seems to allow us to do all the work we need. A set of mappings between genes and organismic characters provides a flexibility of descriptors that can, in principle, accommodate all the causal statements we want to make about the evolution of organisms. The statements connecting gene and character must, of course, be sophisticated enough to include the developmental contingencies that arise from interaction with environment. It might be supposed that all the real work has been done by introducing the phenotype of the organisms and that the reference to the genes is really superfluous beyond the implication that the characters of interest have some heritability. That is, we might, in diametrical opposition to selfish gene theorists, attempt to carry out evolutionary explanation entirely in the phenotypic space above the genotypic level. But that is not possible. One reason is that there are causal phenomena, influencing the evolution of phenotypic characters, that are a consequence of the properties of the system of inheritance. So, the genes "for" a character may increase or decrease in a species entirely because they are tightly linked chromosomally to genes "for" other characters that are themselves under the influence of natural selection. A great deal of the change in bacterial populations is a consequence of such linkages because the entire genome of asexually reproducing organisms is completely linked in a single hereditary unit. It seems that in sexual organisms as well, genetic "hitchhiking" can sweep out genetic variation at many loci as the result of selection at a single locus. Moreover, the probabilistic nature of gene segregation at meiosis results in an irreducible stochastic element in evolution. Another reason is that the molecular properties of DNA enter in an unavoidable way into explanation of changes in phenotypic characters. The possibility of new variation by mutation is contingent on the actual molecular configuration of the DNA in particular organisms. Thus, Hall (1978) has shown that if two bacterial strains with the same biochemical phenotype owe that phenotype to slightly different mutations, one may evolve further by added mutations, whereas the other is in an evolutionary blind alley because the required further mutations are not accessible by single mutational steps in that strain.

The problematic of evolutionary biology is the explanation of changes and diversity in the characters of organisms, given the particular material nature of the transmission of developmental information across generational lines, the genetic system. The theoretical apparatus is in the form of a dynamical theory that takes the composition of a population in a given generation and transforms it into the composition of the next generation by taking into account a variety of causal forces such as mating patterns, mutation, natural selection, and the rules of inheritance. There is a peculiarity of the dynamical system describing organic evolution that makes evolutionary explanation particularly difficult and unsatisfactory – a peculiarity that does not appear in simpler physical systems.

In general, a dynamical theory of the change of a system in time consists of two elements. First it requires the specification of a state space within which the system's movement can be calculated. That is, the system being transformed must be described by a set of variables that characterize the system in such a way that the changes over time in these variables suffice to generate the temporal evolution in which we are interested. Formally, the state space is a multidimensional one whose dimensions are the state variables that are changing, and the state of the system at time t, E_t can be represented as a point in that space. Second, there must be a set of laws derived from the causal forces that suffice to take the state at time t, E_t and transform it into the new state, E_{t+1} at the next instant of time. These laws may be deterministic, unambiguously transforming E, or they may be probabilistic, providing only a probability distribution for E_{t+1} given E_t, but in either case they generate a prediction of the dynamical behavior of the system through time only if the state variables and the dynamical laws have been correctly chosen with respect to each other. One cannot choose any arbitrary set of state variables and demand that there be laws of their transformation. It may be that we have no way of connecting the variables to transformation laws, or that the variables do not form a sufficient set in themselves, but that yet further variables must be considered before it is possible to follow the dynamical changes in the variables of interest. It might be, for example, that we are interested only in changes in the position of a space probe in three-dimensional space, but there is no set of physical laws that can transform position alone into position. We need, in addition, at least the three dimensions of changing velocity to do the work of prediction. The physical laws transform six variables into six new values, but then we extract only the three of interest to describe the history of the system.

There is another possibility for the relation of variables of interest to the dynamical laws of transformation – one that is sometimes believed to apply to evolutionary biology. The variables we care about may not be state variables

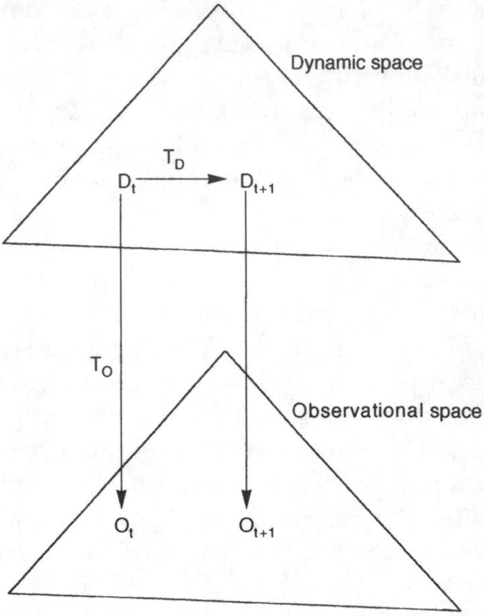

Figure 1.1. Relation between successive values of an epiphenomenal observed state description and successive values of the dynamic states from which they are calculated. See text for a detailed description.

that appear in any dynamical system of laws at all, yet they may be completely specified by such a system. That is, there is a dynamical state space in which some set of dynamical variables is transformed by some set of dynamical laws, but then we map or calculate the successive positions in the dynamical space into a new space of description that itself has no dynamical laws. This is illustrated in Figure 1.1. The system evolves in the dynamical space D according to rules of dynamical transformation, T_D but is seen in an observational space O using some rules of transformation T_O from one space into the other. If the nature of the transformation T_O is such that it can be unambiguously inverted, then, in principle, we can make laws of transformation in the O space. But if there is no possibility of such an inversion, then we must always operate in the D space, accepting our variables of interest as merely epiphenomenal.

The application of these abstract considerations to evolutionary explanation is straightforward. What is to be explained in evolution is the change and diversification of living organisms, of their shape, metabolism, and behavior. But microevolutionary dynamics has no laws that transform in evolutionary time the shapes of leaves or the songs of birds. These variables do not appear in the equations of population genetics, nor would we know how to make them do so. An unspoken assumption of populations genetics is that those variables in the O space, can, if desired, be calculated from genetic variables in the D space that are dynamically sufficient to allow evolutionary trajectories to be predicted and then mapped onto the space of interesting phenotypes. Of course, the possibility of carrying out the mapping from the genotypic dynamical space into the phenotypic observational space requires an immense knowledge of the mechanism of development, including its contingency on environment. Unfortunately, the developmental transformations T_O cannot be inverted because there are many genotypes that can produce the same phenotype and many phenotypes that can correspond to the same genotype, but that is not an in-principle difficulty for evolutionary prediction and explanation. At least in principle, the array of genotypes corresponding to a particular phenotypic variation in a particular population could be discovered by gene mapping techniques and a study of norms of reaction. Then the business of population genetics could be carried out in the dynamical genetic space, and the trajectory of the epiphenomenal phenotypic characters could be calculated by means of the transformation T_O. In fact, this characterization of evolutionary dynamics misses most of the action that we are trying to understand.

The machinery of microevolutionary transformation from one generation to the next contains two sorts of causal forces operating on two sorts of entities. First, there are the forces operating on organisms to change the pattern of organismic variation within a population. These include the migration into the population of individuals with a different distribution of characteristics than is present in the recipient group, the assortment of organisms within the population into pairs for reproduction, and the differential schedule of offspring production by different mated pairs. All three of these causal processes are occurring in real finite populations of organisms and cannot be completely described only by some deterministic rules. They are, of necessity, stochastic, and thus the evolutionary race may not be to the swift nor the battle to the strong, for time and chance happeneth to all. This seemingly trite observation will come back to haunt us. Second, there are the causes of the generation of heritable variation and its passage between generations, including gene mutations, horizontal transfer of DNA, the recombination of genes, and their final segregation into effective gametes. Like the causal mechanisms operating on organisms, these processes too are stochastic. The basic mechanism of Mendelian segregation in which one and only one of two parental types is included in a gamete, one and only one of which will fuse with another gamete to produce a zygote, guarantees a stochastic element in the production of future generations. The most important point is that this taxonomy of mechanisms is also a taxonomy of the entities to which these mechanisms apply. The first category is a list of causal processes involving the behavior, morphology, and physiology of *organisms* in relation to each other and influenced by the external physical and biotic world in which they are embedded. The second category contains those processes that apply to *genomes* as subcellular molecular entities with their own phenomena of replication and assortment into gametes and that are generally robust to variations in the external circumstances of the organisms. Thus, we have two separate state spaces: (1) a genotypic space spanned by genotypic state variables with dynamical laws that apply to these entities, and (2) an organismic state space with its own state variables and laws of transformation. But the topological relationship between these spaces is very different from the parallel worlds of genotype and phenotype that are the subject of the usual confrontation of gene-centered and organism-centered descriptions of evolution. They are not alternative equivalent dynamical spaces, each of which is dynamically sufficient in itself, with bridging statements that will allow the translation of one into the other (see the discussions in Sterelny and Kitcher 1988 and Godfrey-Smith and Lewontin 1993). Rather, they are serially related as sections of a single linear dynamical transformation in which it is necessary to move from one space to the other and then back again to complete the evolutionary transformation in time. To make this movement possible, we must somehow add to the dynamical laws operating within each of the subspaces a set of laws that carry us from one space to the other. These laws, however, are not simply laws of calculation that map one equivalent description into another. They must be, at least in part, causal laws that move the evolving ensemble from one dynamical subspace to the other. The relation between the spaces and their laws of transformation is illustrated in Figure 1.2, taken essentially from a similar figure in Lewontin (1974), but our emphasis here is somewhat different from the original description.

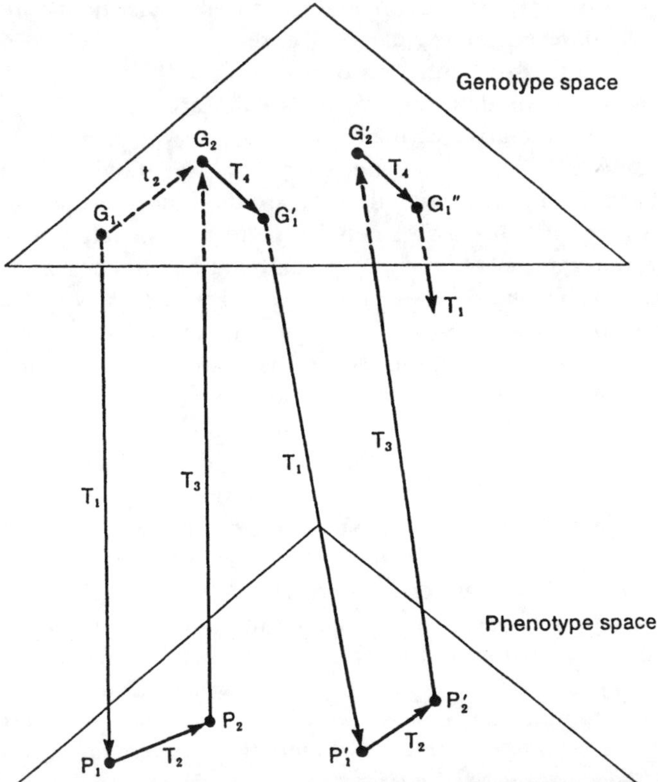

Figure 1.2. Relations of transformations in the genotypic and phenotypic spaces producing successive states in the evolution of a population. See text for a detailed description.

In the figure, G_1 and G_2 are genotypic state variables at two different points in the transformation of the population during a single generation and, similarly, P_1 and P_2 are different phenotypic states within a generation. The state variables G'_1, G'_2, P'_1, and P'_2 are the equivalent states in the next generation, and so on. The laws of transformation are

T_1: a set of epigenetic relations that give the distribution of phenotypes resulting from the development of a set of genotypes in a given array of environments and take into account developmental noise;

T_2: the patterns of mating-pair formation, migration, and reproductive schedules of different phenotypes that transform the organismic array in a population *within* the course of a generation;

T_3: epigenetic equivalences that give the array of genotypes carried by the phenotypic array at P_2 (because of the many–many relation between genotype and phenotype, the transformation T_3 is not simply the inverse of the transformation T_1);

T_4: genetic phenomena of mutation (*sensu lato*), recombination, segregation, and effective gamete formation and fertilization that generate the array of genotypes G'_1 in the next generation. The two dynamical transformations, T_4 and T_2, have asymmetrical properties with respect to the entire sequence. Transformation T_4 operates purely on genetic entities and has no effect on the distribution of organisms in the population, but T_2, which changes the distribution of organisms, necessarily simultaneously changes the distribution of genes contained in those organisms. The changes in genotypic composition as a result of T_2 are indicated in the figure by the broken line, labeled t_2, connecting G_2 and G_1. These changes in genetic composition are epiphenomenal with respect to the forces operating in the organismal space, but nevertheless they are intrinsic to the entire dynamic sequence because they are the basis on which the next genetic transformation, T_4, is built. As we will see, the existence of the shadow transformation t_2 has profound effects on the structure of inference in evolutionary genetics.

It is a relatively simple matter to construct a deductive machinery that can be used to predict the changes that will occur in the forward direction, as shown in Figure 1.2. A model of any degree of detailed specification can be made of the transformations in both the organismic and genic spaces. Sometimes predictions of successive values of state variables for these models can be expressed in closed mathematical form, but for the more complex models, especially those that take random events into account, the forward projection must be made by numerical calculations or by computer simulations. Explanations then take the form of showing how the observed history of population composition can be generated by the particular combination of genetic and organismal forces. Usually the "observed history" is not an observation of temporal changes but of temporal stability over a small number of generations as compared with the lifetime of a species. The virtually static composition of a particular population is then explained as a long-term equilibrium arising from a balance of the forces of mutation, natural selection, and breeding structure. Variation among contemporaneous populations can be derived from some combination of local variation in selective forces or can be explained as a stochastic steady-state distribution produced by random genetic drift in finite populations and a regular flow of migrants between populations. Such explanations assume, of course, that the values of the forces have remained reasonably constant over a period long enough to establish a steady state.

At first sight, it would appear that explanations in evolutionary genetics involve observations of processes and entities at two biological levels and rules of mapping between the levels to provide a complete causal story of the dynamical process in Figure 1.2. At the genic level we would require information on the kinds and rates of mutations of genes, of the linkage relations among different loci, and about any variations that might occur on the normal Mendelian rules of gene segregation and effective gamete formation. That is, we require information about cellular processes. At the organismic level explanations involve observations about mating patterns, about migration among populations, and about the schedules of births and deaths of various classes of individuals, all of which are contingent on the environments experienced and created by the organisms. That is, we require information about natural history. We also need to be able to convert genotypes into phenotypes and infer genotypes from phenotypes (transformations T_1 and T_3) to complete the dynamical pathway, but the requirement for developmental information is only a secondary consequence of the necessity of the transition to the organismic space, and if we could eliminate the need for the dynamical transformations in that space we could dispense with developmental information.

There are two possibilities for completing the program of population genetic explanation. One is to observe the dynamic processes directly and measure their rates, thus producing *physical* estimates of the forces. There is no difficulty, in principle or practice, in observing the processes that occur in the genotypic space. Mutation, recombination, segregation, and effective gamete formation are cellular processes that are not contingent on unique external circumstances or on the complete behavior and physiology of the organism, and thus they can be measured in controlled circumstances and extrapolated to nature. The rate of nucleotide mutation in *Drosophila melanogaster* or the degree of segregation distortion at the T locus in *Mus musculus* can be established in the laboratory. The real problem for evolutionary genetics is the practical impossibility of making the necessary natural historical observations that the transformation T_2 in the organismic space demands. Anyone who has worked with real organisms in real, natural historical circumstances knows that it is, in practice, impossible to recover complete demographic information that includes the geographical structuring of the pool of potential mates, the numbers of males and females, the actual patterns of mating by phenotypes, and the complete age-specific mortality and fertility of mated individuals, all of which enter into a correct modeling of migration, mating, natural selection, and random events that are part of the dynamics. In unusually favorable cases such as that of the live bearing fish *Zoarces* (Christianson, Frydenberg, and Simonsen 1973), it is possible to obtain most of this information, but not, for example, migration patterns between local concentrations of animals, nor, for that matter, any clear notion of what would be meant by a "population." But even if we could obtain total information of the required sort for a species, that information would only be for extant populations and would not tell us about historical changes in demographic variables. Obviously the past cannot be directly observed. How then, does population genetics go about its business?

The actual operation of inference in population genetics is a form of the usual method of statistics. A concrete model of the population process is constructed involving causal processes and entities at the genetic and organismic levels that cannot be observed directly but will have predictable consequences for the intergenerational observations of genetic and organismic state variables. In constructing the model, a distinction is made between structural features that are fixed parts of the model and a variety of parameters associated with these structural features, which may vary from case to case. Thus, for example, the model might postulate that there are discrete populations occupying islands of territory but that some number of individuals migrate from each local population to immediately neighboring islands (the Kimura and Weiss 1964 "stepping stone" model of migration). The parameters are the local population size and the proportion of migrants. The object of the enterprise is to estimate the values of the parameters for the unobservable processes that would lead to the observed changes in composition of the population between generations or the distribution of differences among an ensemble of populations that are assumed to be at a stochastic steady state, including the possibility that the parameter associated with a particular causal pathway is small enough to be ignored. The method of estimation is the usual one of searching through the parameter space for that combination of parameter values that best predicts the observed state variables in successive generations when the forward transformations shown in Figure 1.2 are applied. In a famous example, Dobzhansky and Wright (1941) explained the frequency and allelism of lethal mutations in natural populations of *Drosophila pseudoobscura* by postulating a stochastic equilibrium of local "neighborhoods" of interbreeding flies with some migration between neighborhoods. They used the estimated variance in lethal frequency among sampling localities to estimate effective population size and migration rate between populations. Thus, population genetics begins as a deductive scheme intended to convert information about causal processes into predictions of evolutionary outcomes but operates as an inductive scheme that estimates parameters of causal processes from evolutionary results. This sounds like what goes on in much of science and seems harmless enough until it is realized that the inferential process is severely underdetermined.

Any attempt to construct a model of population processes that includes both the cellular processes in the genetic space and the migration, mating patterns, and differential offspring production of individuals in the organismic space will be rich enough to ensure that some set of values of the inferred parameters will be more than sufficient to generate the observations of population evolution. Conversely, then, the observed evolution or presumed stationary distribution of an ensemble of populations does not offer a unique inference about causative forces. No observations of the changes or stability of the frequency distribution of some inherited phenotype from generation to generation can distinguish between a model that postulates differential survival of offspring from birth to adulthood and a model that postulates equal survival but unequal fertility of different mating pairs. It is only a question of degrees of freedom. For a character completely determined by, say, three alleles at a locus there are six possible phenotypes whose frequencies add to unity. Thus, there are only five independent frequency changes that can occur across generational lines. These changes depend only on relative probabilities of survival and reproduction of the phenotypes. But there are 5 relative survivorships of the 6 phenotypes and 35 relative fertilities of the 36 different mating types, and thus either a survivorship or a fertility model would be sufficient to predict the observed frequency changes; unfortunately there is no way to choose between them from the observation of those changes. Worse still, if the fertility model is postulated, there is far too little information in phenotypic frequencies to even begin to estimate the 35 fertility parameters. Population structure offers another example. One model of structure and migration is of an extremely simple discrete structure in which there are islands, each of a fixed population size, with migrants into each population from a common pool. Another is a continuously distributed large population within which there is limited movement in the physical neighborhood. Between these models is a vast array of more complex possibilities. All of these models predict essentially the same compositional outcome, which is a negative correlation of frequency with distance between points and an excess of homozygotes in the collection of populations at large over what would be expected from a completely panmictic assemblage. Any attempt to estimate parameters of a more a complete model, including migration, population structure, and natural selection compounds the underdetermination problem.

The response of population genetics to the problem of inductive underdetermination is to make use of a few minimal standard models that are just sufficient to deduce the observed intergenerational or interpopulational state variables and then to estimate the parameters of those models. These standard models begin their existence as concrete biological models with clearly defined parameters. The features of the standard models are as follows:

1. Discrete generations. It is assumed that the organism is semelparous and thus that generations begin and end for all individuals simultaneously. There are deductive models of continuous, iteroparous reproduction (for example, Charlesworth and Giesel 1972), but these have too many parameters to allow the inferential inversion.
2. Discrete population structure. The population is a well-defined, clearly bounded collection of a fixed population size N within which random mating takes place.

There are one or more such entities, and migrants enter each from a common pool with a particular frequency structure. This pool may be the average of all the local populations.

3. Viability selection. Each genotype has a probability of survivorship from conception to adulthood, its viability, but all survivors have an equal expected fertility.
4. Mating structure. There are $N/2$ males and $N/2$ females after selection. A male and a female are chosen at random with replacement to produce an offspring, and this process is repeated until a new generation of $N/2$ males and $N/2$ females after viability selection are produced again. This is equivalent to forming $N/2$ mating couples and having them produce a Poisson distribution of offspring with an average of $2/V$ offspring per couple, where V is the average viability of all offspring at birth.
5. A genetic apparatus. There is mutation among allelic forms and phenomena of segregation and recombination that may include deviations from the normal Mendelian rules.
6. Stochastic steady state. For stochastic models, in which the observations are not of change between generations but of the distribution of genetic structures either among populations or among genes that have undergone the same population history, it is assumed that the evolution has gone on long enough under constant conditions to reach a stochastic steady state. Thus, unique historical events have left no trace.

No one takes the standard model to be a realistic model of any real species. Its purpose is to introduce the phenomena that are regarded as the causes of evolutionary change and to parameterize them with a number of degrees of freedom that will allow the parameters to be estimated from intergenerational and interpopulation data. But the estimated parameters are the properties of the particular biological structure that has been used to make the standard model – even though that structure may not apply to the case being investigated. Thus, what is produced, in contrast to the physical measurements of forces in a specific biological model, are *formal* estimates of forces. These estimates are only the observations themselves inverted by the hypothetical and probably fictitious physical structure of causation in the standard model. They in no way test that structure, and if the structure does not correspond to the actual physical processes, then they are not estimates of anything concrete. They are, instead, *als ob* quantities, which are values that would be sufficient to explain the frequency data if the structural model had been correct in the first place. Moreover, the use of the standard model in this way effectively removes all uniquely organismic information and thus allows the entire set of dynamic transformations to occur in the genotypic space. The dynamic transformation T_2 in the organismal space has been replaced by its shadow transformation t_2 in the genotypic space. Because what appeared first as a concrete biological model now appears as a purely formal *als ob* model, there is no reason not to substitute the genotypic equivalents for phenotypes. Details about organisms are doing no dynamic work because the parameters are merely formal re-representations of the genotypic frequency data from which they have been estimated.

One well-known example of the substitution of a formal, as-if quantity for a concrete biological one is effective population size N_e. Effective population size, or deme size, or panmictic unit size, are all derived from the operation of random genetic drift in a standard model population. If we have a well-defined, bounded collection of $N/2$ males and $N/2$ females mating to produce offspring at random with a Poisson distribution of offspring number in a diploid organism, there will be a random fluctuation in the frequency of genotypes from generation to generation. This random fluctuation can be observed temporally within a population, or it may be seen as a variance in genotype frequencies among replicated populations. The standard model provides a prediction of the rate of random variation as an exact function of the effective size N_e. Thus, given an observed temporal or spatial fluctuation in genotype frequencies we can estimate how many couples are in the populations under investigation. But suppose there are unequal numbers of males and females, or couples do not have an equal probability of offspring production and thus that there is not a Poisson distribution in offspring number, or population numbers vary in time, or there is no well-defined, bounded population but, instead, a continuously distributed species with some mating of neighbors? What do we mean by "effective population size"? What relationship does this size have to the actual physical number of indviduals in the population? The answer is that the effective population size in these cases is that number N_e, which, if substituted in the standard model, would produce the same rate of divergence in gene frequencies as has been observed to occur in the actual, nonstandard population. Therefore, in a continuously distributed population, the deme size or panmictic unit is that number of organisms which, under the standard model of isolated islands of population, would produce the observed divergence among sample sites. It is a pure fiction, not a count of actual individuals. But, if it is simply a formal redescription of the observed genotypic frequency divergence, then what has been explained?

Every population geneticist is conscious of the fictional status of N_e, but he or she is rather more mystified by another such fiction, *fitness*. The fitness of a genotype or phenotype is meant to be a single quantity that expresses the intensity with which natural selection favors or discriminates against the type. It is also meant to predict the change in frequency of the type during the course of evolution. If we invoke the standard model, the concept of fitness is unproblematical. It is the relative viability of the genotype, and these fitnesses can be estimated from successive changes in genotype frequencies. For example, if there are two alleles at a locus, there are three genotypes and two independent fitnesses, and these can be estimated from two intergenerational changes in genotype frequencies. If the frequencies are not changing, we cannot distinguish between an absence of selection and a stable, selective polymorphism, but an estimate of fitnesses can be made from successive changes in genotypic composition within a generation. When we depart from the standard model, however, it is unclear what we mean by the fitness of a genotype even in the deductive direction. How are differential fertility and viability to be combined into a single measure? The problem is even more difficult in an iteroparous species in which not only the total fertility but its age distribution are relevant to the outcome of selection. A great deal of paper and ink has been devoted to this problem, and even R. A. Fisher produced a solution, the intrinsic rate of increase or Malthusian parameter *m* that only applied to asexual species and only if the population was stationary in size, although he failed to realize it. One might think, on the model of effective population size, that the fitness of a genotype can be defined as the number necessary to produce a change in genotype frequencies in a standard model equal to the change actually observed in the more complicated case. But this is unsatisfactory for two reasons. First, it does not help when genotype frequencies are not changing. Second, it has a curious consequence for populations with different growth rates. For a fixed-age schedule of births and deaths, the same genotype may increase or decrease in a population depending upon whether the population is growing or shrinking in size. In a growing population, genotypes that have their fertility skewed toward early ages will increase in frequency because an offspring born early in life is born when the total population is smaller than it will be in the future, and thus each early birth is worth more than a later birth. In a shrinking population, an early birth occurs when the total population is larger than it will be in the future, and thus later births are worth more (Charlesworth and Giesel 1972). Do we really want the fitness of a genotype to be reversible in this way? It would rupture the relationship of relative fitness with relative physiological properties of individuals. Moreover it completes the tautology. If fitness is defined tautologically by the evolutionary result, then cause and effect are completely confounded and nothing is explained by invoking fitness. It makes more sense to abolish the concept of fitness as a single summary value that is to have explanatory value and return to the more complex information needed to project evolutionary changes. Of course, that means a return to natural history.

To say that the estimates of population genetic parameters are only formal calculations from a fictitious model is not the same as saying that they are useless for explanatory purposes. The issue is how the fictitious parameters map onto the parameters of the realistic process. A widely used statistical test of whether natural selection is influencing the variation of nucleotide sequences within a species makes use of the standard stochastic model (Hudson, Kreitman, and Aguade 1987). The judgment that natural selection does or does not operate in this test is, strictly speaking, a judgment that viability selection does or does not exist in stochastically stationary populations. But it might very well be that any other scheme of selection, including age-dependent fertility differences, would still necessarily map onto a nonzero value in the fictitious standard model. Thus, any numerical value estimated for the formal selection coefficient might be quite wrong, but the judgment that selection is occurring would be correct. The converse is less likely to be true. That is, if no selection were really operating, we might easily judge there to be some selection from the test because some other aspect of the standard model was not fulfilled in nature. The assumption of stationarity is certainly critical. If the genetic structure of the populations still bore the imprint of a past unique historical event, then no inferences can be trusted because we have no idea of what effect that event had on the population. This problem of the lingering effects of history on the present genetic status of populations is widespread. It is fairly clear that most of the systematic causal forces acting on the evolution of populations are weak. Mutation rates are of the order of 10^{-6} to 10^{-5} per gene, and it now seems likely that most selection, especially on nucleotide variation, is of no greater order than 10^{-3}. Although some local populations are small, if we take into account even low rates of migration, random differentiation among populations will go on quite slowly. As a result we cannot observe temporal changes directly and, instead, have come more and more to depend for our inferences on the differences between populations that have accumulated over long periods. But the weakness of the dynamic forces also means that the time for return to a steady state from occasional perturbations, the "relaxation time," is also very long, and so populations are likely not to be in the steady state. Once again, history matters.

REFERENCES

Charlesworth, B., and Giesel, J. T. 1972. Selection in populations with overlapping generations II. Relations between gene frequency and demographic variables. *American Naturalist* 106:388–401.

Christiansen, F. B., Frydenberg, O., and Simonsen, V. 1973. Genetics of *Zoarces* populations III. Selection component analysis of an esterase polymorphism using population samples including mother offspring combinations. *Hereditas* 73:291–304.

Dobzhansky, Th. 1951. *Genetics and the Origin of Species.* New York: Columbia University Press.

Dobzhansky, Th. and Wright (1941). Genetics of natural populations. V. Relations between mutation rate and accumulation of lethals in populations of *Drosophila pseudoobscura. Genetics* 26:23–51.

Godfrey-Smith, P., and Lewontin, R. C. 1993. The dimensions of selection. *Philosophy of Science* 60:373–95.

Hall, B. 1978. Experimental evolution of a new enzymatic function. II. Evolution of multiple functions for EBG enzyme in *E. coli. Genetics* 89:453–65.

Hudson, R. R., Kreitman, M., and Aguade, M. 1987. A test of neutral molecular evolution based on nucleotide data. *Genetics* 116:153–9.

Kimura, M., and Weiss, G. H. 1964. The stepping stone model of population structure and the decrease of genetic correlation with distance. *Genetics* 49:561–76

Lewontin, R. C. 1974. *The Genetic Basis of Evolutionary Change.* New York: Columbia University Press.

Sterelny, K., and Kitcher, P. S. 1988. The return of the gene. *Journal of Philosophy* 85:339–61.

SECTION A

HISTORY OF AND IN EVOLUTIONARY BIOLOGY

Introductory Remarks

The following discussion is a composite of five lengthy interviews recorded over a 3-year period (1996–98). We have selected and arranged the material with three principles in mind. The first is to focus, not on personal aspects of Dick's life (at his request), but on the history of population genetics from his point of view and as exemplified by his career.

The second is to move beyond the published record. We anticipate a varied audience (scientists, philosophers, historians) with diverse prior knowledge of Dick's work. Some of the interview material will be familiar to one segment of the audience and new to another. But we chose not to incorporate material on topics about which Dick's work or perspective would already be known to most readers and about which the interviews did not divulge anything substantially new. For this reason, and also because Dick himself focused on the history of population genetics, there is relatively little, for example, on his critique of sociobiology. However, we have included substantial segments on his work on the political economy of agriculture, which is material that is not nearly as well known.

Third, we aimed to present a coherent narrative (within the page constraints of a chapter in an anthology).

It was our pleasure.

CHAPTER TWO

Interview of R. C. Lewontin

R. C. LEWONTIN, DIANE PAUL, JOHN BEATTY, AND COSTAS B. KRIMBAS

2.1 Introduction

RCL: I would like to begin by saying something for John's benefit that I said to Diane. I was somewhat alarmed when I read the list of questions, beginning as they do with personal issues, parents, upbringing, political life, college, and so on. The question is not whether I have to answer such questions but what the posing of them reveals about what we are engaged in here. I assumed that we were engaged in a piece of oral history in which an interview with a participant in population genetics during an era when it was being founded, in a sense, by Dobzhansky and others, and when it underwent major changes would be interesting because there would be first-hand recollections of events, people, and ideas that were not available through the written record. One could get nuanced, if biased views of a whole variety of substantive topics. I take it that is the purpose of oral history, but none of that has anything to do with the personal history of the actor.

There obviously are very extraordinary circumstances in which the life history of the actor may have some relevance. For example, if you believe as I do that Dobzhansky essentially single-handedly created the problematic of modern population genetics, namely, the description and explanation of patterns of variation in natural populations, then, of course, the question of where he got the idea is significant, and the fact that he spent early years collecting beetles in Central Asia and was impressed with the diversity of those beetles is directly relevant. That is really an extraordinary situation and certainly does not apply to me. When it comes down to it, after all, I have spent my life working out the problematic created by Dobzhansky, and I do not see how anything personal is relevant. What bothers me is that the posing of those questions suggests that you have a very different project in mind than I do. Somehow, you are engaged in writing the career history of an individual, and I do not know why you want to do that. I understand that these interviews are supposed have something to do with these volumes that are being produced on the excuse of my Festschrift, but my understanding was that this was simply an excuse and that the volumes were, in fact, to take up substantive issues in sociology and philosophy of science, and the science itself. I am very worried that the article that is prepared here will somehow be the life of RCL. Who the hell cares? Obviously, I care, but I have already made up a complete story about my own life, and I do not see why I should inflict it on anybody else. So, I hope you will not get off on the wrong foot. I hope we will do a piece of oral history in which personal experiences can illuminate or, obviously, distort from a personal perspective the development of our field. That is the project I am interested in anyway. So, let me go on to the substantive questions: Columbia, life in Dobzhansky's lab, and so forth. This seems to me relevant precisely because Dobzhansky's laboratory was really a school, in the old sense of school, and the intellectual atmosphere there and the way life was carried on have had a very strong influence on the development of population genetics.

2.2 Life in Dobzhansky's Laboratory and Dobzhansky's Problematic

Let me say a word about Dobzhansky's personal influence that relates to the remarks I made before. Here's an opportunity to test a number of hypotheses about the role that individuals can play in creating and molding a field, and it's interesting in the case of Dobzhansky because he was a person of such immense strength of personality and drive. If it can't be shown that Dobzhansky had such an effect, then it would be impossible to show it for anybody else. In this connection, I think the hypothesis of personal influence fails in the following sense:

Dobzhansky had an immense missionary zeal for population genetics, and he went to Brazil, to Israel, to Egypt, to other parts of the Middle East, to other South American countries, spreading the gospel and recruiting people into his vision of population genetics. He goes to a country that is a scientific backwater, like Brazil, in which Rockefeller was putting money. He goes around Brazil, puts his arms up, and says in his loud voice, "now, my friend, you must work on the problem of genetic variation in *Drosophila*." That's what I mean by charisma. Everybody, unless you're repelled by that, as some people are – if you're not repelled by it, you immediately become taken by him; he *takes* you, and he moves you; he moves you physically. I remember an occasion with Dobzhansky when I wanted to go downtown and he wanted to go uptown, and I engaged in a physical struggle with him in the middle of Broadway and 115th Street about whether I was going to go uptown with him. He wouldn't take no for an answer. So he goes to a country like Brazil, and he meets Dreyfus, who was a very important Brazilian scientist. And Dreyfus takes him around, introduces him to all kinds of people. And he recruits people to come to the United States on Rockefeller money; I think they all came to his laboratory. That's what I mean; he had personal charisma, but he also operated in conditions in which people also saw an opportunity to do something: Egypt, Brazil, Colombia. We're not talking about countries with big scientific establishments. He didn't recruit anybody from England. Englishmen came, but only with a certain authority. Phil Sheppard came to the laboratory, but Dobzhansky's charisma didn't work on Phil Sheppard.

He was quite successful in recruiting people, but the fact of the matter is that none of that actually took in the long term. There are no active schools of population genetics of any size in Brazil, Germany, France, Yugoslavia, Egypt – all places where Dobzhansky had done very successful recruiting and had sent back people from his lab to begin programs. But none of the programs ever took, and population genetics is today an essentially Anglo–Saxon, if not totally American, operation. (I think of Japan as an extension of the American school through Kimura, and in any case, Dobzhansky didn't recruit there). So despite Dobzhansky's driving personality, despite his recruitment of people, despite his almost charismatic appearance, he could not create a worldwide science, and this raises very interesting questions.

I bring this matter up because it was characteristic of Columbia in Dobzhansky's day that it was filled with people from all over the world. While I was a graduate student there, there passed through the lab Charles Birch from Australia; [Antonio] Da Cunha, [A. G. Lagden] Cavalcanti, [O.] Frota-Pessoa, Vicky [Warwick Estevam] Kerr, Antonio Cordeiro, and Waldo Pavan, all from Brazil; Danko Brncic from Chile; Lalith Sanghvi from India and Azim Tantawy from Egypt – all recruited by Dobzhansky (Sanghvi possibly by Dunn) to come to the laboratory.

This was in addition to people like Phil Sheppard who came just on sabbatical but who were already part of the Fisher–Ford school in Britain. These people all passed through the lab while I was associated with it in one way or another as a graduate student, or perhaps one or two of them reappeared a little later when I used to spend periods in the lab while teaching on weekends at Columbia after I was a professor at Rochester. Later on, there were Germans, Frenchmen, and Yugoslavs, but I did not overlap with them. But the point is that the laboratory was always an international center. In addition, of course, there were emigré Russians and Ukrainians. Mikhail Vetukhiv, who did important experiments on coadaptation, was a postdoctoral fellow and also the President of the Ukrainian Academy of Sciences in exile in New York. The two Spasskys, Natasha and Boris, were Dobzhansky's chief assistants, and later added to that was Olga Pavlovsky, who made the slides for Dobzhansky and was regarded by everyone as Dobzhansky's favorite. Then there were individuals who appeared in the lab for strange reasons: Dr. Arkhimovich, who was a former beet breeder in the Ukraine, and a woman who for a long time did bottle washing in the kitchen. It was a whole crew of exiled Russians and Ukrainians who had not been recruited by Dobzhansky in the usual way, of course, since he was not allowed into the Soviet Union, but who hung out in the general atmosphere of the Upper West Side and who, in one way or another, came into the laboratory.

Nearly all of the people who came into the lab in this way, postdoctoral fellows of all kinds, with the exception of Phil Sheppard, who was already a very accomplished evolutionary geneticist, and Charles Birch, who was an established ecologist, acted essentially as Dobzhansky's laboratory assistants. That is to say, by his driving personality, Dobzhansky would essentially tell each person what experiment really needed to be done. The person might or might not participate in the planning of the experiment, but usually Dobzhansky planned the experiment, picked out the flies to be used – sometimes in consultation with Howard Levene, who was his statistical adviser – and then the person did the experiment. The data were then brought to Dobzhansky, and he analyzed them or had Howard analyze them. Then Dobzhansky wrote the paper, and it was published under the name of the person who had done the physical work, with or without Dobzhansky's name as a coauthor. This included Krimbas, for example. Nobody seemed to be able to escape this mill. There was a great deal of falsification of history in the process, because many papers were published in the laboratory that didn't even have Dobzhansky's name on them, like the Vetukhiv papers, and yet for which he had planned the whole thing, designed the experiment, analyzed the experiment, and written the paper.[1] He often said that a good professor has written many more papers than he has his name on, but I regard his inversion of Robert Merton's Matthew principle as a serious falsification of history.[2] Even a strong-minded, established, young professional like Waldo Pavan from Brazil, who was already a successful cytogeneticist, was pressed into doing work that later on he would not continue. Dobzhansky was virtually irresistible in this respect.

Q: How were you recruited into Dobzhansky's lab?
RCL: That wasn't straightforward. I was an undergraduate at Harvard. L.C. Dunn had come to Harvard as Visiting Professor for one semester. We had no geneticists, so he was going to give the genetics course. I took the course and got interested, and then he offered that if anybody would work with him in the mouse colony – he brought some mice with him – they wouldn't have to do the regular labs. And I thought, ah, I'd be glad to get rid of that, so I went to work with him doing mouse crosses. And we actually worked side-by-side for 5 or 6 months. At the same time, his assistant in the genetics course was Eliot Spiess. Spiess was a registered graduate student at Harvard but worked with Dobzhansky in absentia for his thesis since Harvard had no geneticists. After doing his thesis research, Eliot came back here as the chief instructor in genetics. So after finishing with Dunn, I became Spiess's research assistant, and I used X-rays trying to produce mutations in *Drosophila persimilis*. In my senior year, I was working with Spiess. Dobzhansky came to see Spiess. (Dobzhansky, as part of his empire, would go out and visit his former students and make sure everything was okay). Spiess introduced me to Dobzhansky.

Then, when time came for graduate school, I either wrote or spoke to Dunn saying that I would like to get into Columbia but I have not had many of the courses that are required, like embryology, and so on. Dunn said, "we'll take care of that." So I came to Columbia on the invitation of Dunn and Dobzhansky together. What I knew about population genetics was zero.

The thing Dobzhansky did when I arrived was to hand me Schmalhausen's book.[3] We had a seminar there on evolutionary genetics, run by Dobzhansky. It was a non-systematic seminar, on different topics, and he said to me, "in 3 weeks I want you to give a seminar on chapters 2 and 3 or 1 and 3 or 2 and 4 of Schmalhausen; the business where he lays out the notion of *morphoses*." And so I started to read Schmalhausen, and I was immensely impressed; as they say now, it blew my mind.

I would go around to Tim Prout and explain Schmalhausen's ideas and how wonderful they were. And of course Schmalhausen never left me. So that was the first influence Dobzhansky had on me – to push my nose into Schmalhausen. I began some experiments; they must have been Dobzhansky's ideas, I don't know.

Q: Did Dobzhansky dictate your graduate research?
RCL: I resisted him, but by a subterfuge. He assigned me a thesis experiment and then took off for one of his long periods in Brazil. I was totally uninterested in the experiment he'd assigned to me and simply didn't do it and began work that I found interesting myself. When he returned, I just told him that I began the experiment he suggested, which was a lie, and I couldn't get it to work so I did something else. I may have been the only person in the lab who did that.

Q: What was your thesis research?
RCL: It was an experimental demonstration of the importance of frequency- and density-dependent selection in *Drosophila*. From a general consideration of ecology and from a few papers in the literature, it seemed obvious to me that the relative viability of different genotypes should change with both population density and with the frequencies of the genotypes in the population. This turned out to be true for *Drosophila melanogaster*, for which I found facilitation effects. That means the presence of larvae of competing genotypes actually increased the viability of some strains, and there was an intermediate density at which the probability of survival was the highest. I maintained an interest in this subject for several years and did another slightly more complex experiment with Yoshiro Matsuo showing very strong effects of this kind in another species of *Drosophila*. Also, I did a general theoretical treatment of frequency-dependent selection.

There's an interesting story about this theoretical paper. I sent it to *Heredity* and got it back with a note from R.A. Fisher saying that no doubt I would want to publish it somewhere, but not in *his* journal. I then sent it to *Genetics*, where it was published with no problem. I never did find out what Sir Ron didn't like. Perhaps he was right. There was a brief flurry of interest in the subject. For example, Bob Sokal tried my experiment in house flies and found the same results, but after a while interest died and has not been revived, and the theoretical paper has only occasionally been cited. I still think it's essential to our understanding of how selection really works, but it makes things more complicated, and people, especially theoreticians, hate complications because they can't put out grand-sounding general theorems like the maximization of mean fitness.

To return to Dobzhansky's lab, Tim Prout, Don Cooper, and Dan Marien all did experimental things that were on Dobzhansky's list of things to be done. It's of more than incidental interest perhaps that we shared the floor with L.C. Dunn and his students, and the students were all on very close terms with each other. I don't think Dunn's students cared too much for Dobzhansky, and in at least one instance, Reba Mirsky, who was Alfred Mirsky's daughter – therefore, a family friend of the Dobzhanskys – had an active war with Dobzhansky. Dobzhansky and Dunn, on the other hand, got on extremely well. They taught together and they had similar political view, although Dunn was much more of a fellow traveler of the Communist party and its movements than Dobzhansky ever dreamed of being. But they were both left-liberals, and they certainly were antieugenicists and antiracist. Dobzhansky actually had an unusually large number of left-Red students, including several members of the Communist party. I don't know how that happened. I'm pretty sure he was aware of it, but that didn't seem to bother him. He claimed to have been a strong supporter of the February Revolution but not of the October Revolution, and I think he would certainly be described as a social democrat.

Q: Did the politics of the lab appeal to you?
RCL: The only thing I thought about was population genetics. I have to tell you the truth; you're not going to believe me, but it's the truth. The beauty of the subject was that it was such a backwater of science; that and so irrelevant to political and social issues. So you could see what I knew about population genetics! I thought that I could do my thing, and I would be left alone; I would be free of all the pressures from the different sides. I really thought the subject had nothing to do with human welfare, and that's why I went into it. We are talking about a 21-year-old who knew nothing.

When I was first in college, I was very radical. I went to the meetings of the John Reed Society; I was a Communist sympathizer. I was a radical in high school. Mary Jane and I founded a movement in high school for progressive politics called "World We Want" (W.W.W.). I came to college. People used to say, "Here he comes again, telling us they can make a Ford car for $330; you know, it's all profits." I was considered the bad Commie, and I loved it. My father told me when I went to college that I was not to get involved in politics because it's dangerous. Then I failed all my courses and got thrown out of college. At the same time, Mary Jane and I were married; then not long after I was thrown out, we had a kid. I didn't know what the hell I was going to do with my life. I had a job, but I could have worked in the textile industry for the rest of my life. And I had the opportunity to go back to college. I put my nose to the grindstone. I thought, *oh-oh, I've got to make up for this catastrophe in my life – I've got a family to support.* So I turned my back on any political or other extracurricular interest; I deliberately turned my back on all that and just became very careerist. And that included when I arrived in Columbia. That was an abreaction, so to speak. It was very deliberate. Most of my friends were radicals. When the workers went out on strike at Columbia, they wouldn't cross the picket line. I crossed the picket line. This was serious business. So that's the answer. I chose population genetics precisely because I thought it was apolitical.

Q: What happened to bring you back to politics?
RCL: My career program worked. That pressure was off me.

Q: What about the politics of the other students?
RCL: Dobzhansky had a lot of students who were in the YCL [Young Communist League] or in the Communist party. For some reason, the graduate students at Columbia were pretty radical. They were mostly New Yorkers for one thing. And we used to fight a lot on issues and have big discussions about what you should and shouldn't do. So there was a political atmosphere in the laboratory. Dobzhansky didn't publicly participate, but he didn't discourage it either. He knew who his students were. I thought it was a great mystery.

Q: Did you ever talk with him about it?
RCL: No, I never did. Given his relationship to the Bolsheviks, I could never understand how such people came to be in the lab. Mikhail Vetukhiv was a real sweetheart of a man, very nice, spoke in a low tone of voice. But what was the history of Mikhail Vetukhiv? When the Germans invaded the Ukraine and then were kicked out of the Ukraine, he retreated with the Germans. He had to choose between the Bolsheviks and the Germans; like many Ukrainians, he chose the Germans. I think he was probably somewhere to the right of Attila the Hun as the president of the Ukrainian Academy of Sciences in exile. But he was very nice to me during his whole life. I haven't doped that out yet – in one sense, having deep political commitments; on the other hand, not taking them seriously on some level.

Q: What were Dobzhansky's social views, on eugenics and race, for instance?
RCL: I don't recall that we ever had a discussion of eugenics or racism in the laboratory. Everybody in the lab was antiracist, including Dobzhansky. Everybody knew that. If we thought anything about Dobzhansky, it was what we learned, what I learned, from Fred Osborn. I never heard of Fred Osborn until I was introduced to him by Dobzhansky. What did Fred Osborn say to me about Dobzhansky? "Dobzhansky was my savior. Dobzhansky is the one who taught me the error of my ways and showed me that my views about eugenics were racist and were all wrong; the Eugenics Society was wrong." And he's the one who turned the Society from being a eugenicists' society to being one that studied human variation. And Fred Osborn, when I knew him, was a very antiracist liberal guy, and he claimed that Dobzhansky did that to him. I was on the Board of Directors of the American Eugenics Society. They changed their name after a while; they changed the name of the journal to *Social Biology*. It published a paper, for example, by Carl Bajema showing that people with very low IQs didn't have more children than people with average IQs.[4] It was really attacking all that eugenics stuff. Dobzhansky did that. And I became conscious of it maybe in my late graduate years, or just after I left.

Dobzhansky was a very strong biological determinist, and that, I think, has some real intellectual interest. On the one hand, Dobzhansky popularized the notion of the norm of reaction that he got from Schmalhausen and kept insisting that one had to concern oneself with the norm of reaction. He taught the Clausen, Keck, and Hiesey papers[5] and so on. Yet, at the same time, he thought that populations with the highest genetic variation would have a great advantage over those which didn't because each person could be specialized in one thing. He thought that human populations depended on the fact there were many different genetic talents in them. What made him different from the eugenicists was his attitude toward homozygotes and heterozygotes. He thought heterozygotes to be superior to homozygotes and, therefore, eugenics would not work.

Now, you might ask, How can he on the one hand be a biological determinist and then on the other hand keep insisting on the importance of norm of reaction, that is, the reaction of genotypes in development to environment? The answer is that his particular form of biological determinism declared that the norm of reaction of heterozygotes was much flatter than the norm of reaction of homozygotes. That was his idea about homeostasis. So, although there was, in principle, a variable response of any genotype to the environment, what was characteristic of heterozygotes was that they were on the average superior to homozygotes and also much flatter in their environmental response. In a sense, you can say, he was a biological determinist in that heterozygotes were biologically determined in their phenotypes, but homozygotes he thought were narrow specialists whose phenotypes were very sensitive to the environment. So, if you picture the norms of reaction, there would be a very jagged norm of reaction with a peak, perhaps, or a strong slope for homozygotes which might, for a few homozygotes in a particular environment actually project up above the norm of reaction for the heterozygote, but generally speaking the heterozygote lines would lie above those of the homozygotes, and they would be flat. So he could have his cake and eat it too. In fact, he spent a lot of his time measuring the norms of reaction for chromosomal homozygotes from nature that did have these strong response curves. But for him, heterozygotes were all much of a muchness, and they were all on the average better. He called homozygotes narrow specialists and heterozygotes generalists, but they were not only generalists but superior generalists.

He was not always particularly logical about this point of view. For example, I kept taxing him with his exclusive use of variance as a measure of homeostasis. I kept pointing out to him that you couldn't use just the variability of response to measure homeostasis because that would make an unconditional lethal the most homeostatic organism there was, since it was lethal in all environments. Dobzhansky thought this was ludicrous, but he didn't see that it was an attempt to reduce to absurdity his insistence on using only the variance to measure homeostasis rather than a low variance and a high average performance together.

This brings up a point for the people trained by Dobzhansky. While they were all wowed by his personality, and he was pretty much of a steamroller, a significant fraction of them felt that his intellectual sharpness and especially his ability to do logical thinking was not all it could be. Maybe graduate students all think that about their professors, but we certainly thought that about Dobzhansky. Tim Prout and I, especially, because we were knowledgeable about mathematics and statistics, thought it ludicrous that Dobzhansky could barely add 2 + 2. It is definitely true that Dobzhansky was not strong on logical consistency; he was a very intuitive person.

But having pointed out these limitations, I want to acknowledge that my own problematic is the problematic of my professor. For me, Dobzhansky was a fulcrum. He had a lot of admirable qualities and a lot of qualities that I'm glad I don't have. But one thing he was, was a person with a strong intellectual program against which you could struggle and with which you could struggle. He was a fulcrum! You need a fulcrum to use a lever to pick something up, but you have to put pressure on it. Dobzhansky was different things to different students. For me, he was a really hard rock. I recognize that everything I do in science I get in one way or another from that program that he initiated.

Q: Back to life in the lab . . .

RCL: Life in the laboratory was highly social. There were a large number of small rooms shared by two or sometimes three graduate students and postdoctoral fellows who were mixed in with each other. There was a large lab in which the Spasskys worked and a small room off that in which Mrs. Pavlovsky worked, and Dobzhansky had a teeny office in which he sat at his desk and wrote papers or some correspondence. He had no secretary but shared in the typing pool of the department. Then, he spent a lot of his time in the big lab where he sat at the microscope reading chromosome slides and holding court. Typically, someone would come and sit down in the chair next to him and have a discussion with him about something of interest while Dobzhansky looked through the microscope and scored chromosomes. Dobzhansky would also do egg-counting for egg-laying experiments, but again, he would hold court while doing these things. There was a coffee club that met twice a day to which everybody came. We each chipped in a few pennies a week to buy the coffee, and when any visitor came to the lab, say Sewall Wright or E.B. Ford, or anybody of any importance, that person would be dragged along to the coffee club, either by Dobzhansky or by one of us who went to grab him away from Dobzhansky, and the graduate students got to chat with this person. The coffee club included both Dunn's students and Dobzhansky's students. We had many arguments and discussions at those coffee clubs, and they were a very important source of education for all of us.

I have no insight into the way things went in the field, as I never went to the field with Dobzhansky. The one time that he and I collected together, John Moore and Tim Prout and he and I went up to Palm Canyon to collect some flies and visit an old locality. That was years afterwards when Dobzhansky was already retired and in California. I don't remember that Dobzhansky was particularly knowledgeable about the field, although no doubt he was. We collected a few flies and walked back down again. Almost the entire walk up and down from Palm Canyon, about an hour or more each way, over stony ground, was occupied with a violent argument between me and Dobzhansky on biological determinism. I remember it very well because I sprained my ankle because I wasn't watching where I was going.

This brings me to the question of his vision of population genetics. At the very center of Dobzhansky's whole intellectual structure was the importance of genetic variability. Darwin built his theory of evolution on selection among genetic variants and Dobzhansky, therefore, felt that population genetics really was a science of description of the variability and an attempt to sort out the forces operating and especially to demonstrate (a) that natural selection was operating but (b) that the outcome of natural selection might be contingent on chance events. That is to say, he was a non-Fisherian strongly influenced by Sewall Wright.

By 1950, when I first met Dobzhansky and began to work with him, he'd abandoned his original infatuation with a purely drift neutral theory and was trying to bring together both drift and selection, always insisting that both were operating simultaneously. What interested him was the differentiation in space and time of populations through selection and drift that would then introduce the early stages of speciation that would then result in the formation of species.

As I wrote recently in *Genetics*,[6] all graduate students had to be able to reproduce on their qualifying examination – which was a very serious written qualifying examination, that people could actually be thrown out of Columbia for failing – the mantra that populations became isolated from one another and differentiated in space through isolation and local selection in combination with drift; that these divergences became greater and greater until finally, when the populations came back together again, they were so divergent they couldn't mate anymore very successfully, and those who did try didn't leave any offspring; therefore, there was reinforcing selection, and, finally, sexual, isolation as a consequence of the reinforcement, and species were formed. So, you began with population differentiation, geographic race formation, and then species formation.

Dobzhansky's view of population genetics was that this process is what we were supposed to work out. That was always informed essentially by a deep-seated selectionism on his part. That is to say, he did agree that random events were important, but for him the most interesting aspect of random events was that you might find yourself with different sets of genotypes and that the result was that the direction you actually went changed, you might go up a different adaptive peak. But his primary interest, by the time I got to know him, really was in the selective process itself, and he regarded population genetics as having to work out the origin of variation and the selection of that variation. So, he thought recombination was a very important process, and he developed a whole program to study the origin of fitness variation through recombination, the so-called "synthetic lethals." He studied the chromosomal polymorphisms because he thought they were a model of how natural selection operated – not because he was particularly interested in inversions but because they showed the power of coadaptation and the effect of selection winnowing out those alleles which nicked in heterozygous condition. As inversions holding the genes together, they exaggerated the effect, but it was really the idea that populations maintained those alleles that were superior in heterozygous condition and rejected those that were not.

The whole theory of coadaptation was at the very heart of much of his view of the selective process. So, for Dobzhansky, the plan for population genetics was (a) to show that there was a lot of variation of all kinds in populations at the genic level, at the chromosomal level, and so on, (b) to try to measure the rates of introduction of that variation through mutation and recombination, and (c) to show how selection would operate to winnow out that variation largely, as I said, by purifying out those mutations and variants that were not superior in heterozygous condition and, therefore, would disappear and would not be part of the variation, leaving behind variation that was superior in heterozygous condition and, therefore, balanced.

It was explicitly not part of his plan to give ecological and physiological explanations of this superiority except as an armchair exercise once in a while. No part of his experimental program or, indeed, of what he tried to get people interested in doing, involved real experimental work in the field or real attempts to measure physiological properties that would serve fitness. In this sense, Dobzhansky's vision of population genetics was a formal one. It really did not intersect very seriously with the British view that the object of the study of natural populations was to give ecological and behavioral explanations of the selection that was operating. For Dobzhansky, these would be particular circumstances in each case and in themselves not all that interesting, except as teaching devices. After all, he did not expect that the gene contents of Arrowhead or Chiricahua subserved any particular physiological function but rather large numbers of different physiological functions.

He was a generalizer and regarded the task of population genetics to provide generalizations about the evolutionary process through exemplification. Heterozygotes were generally more fit. Heterozygotes were generally more homeostatic.

Populations generally consisted of loci that were coadapted. He wouldn't have dreamed of spending a large amount of time studying one particular behavioral or physiological mechanism in one particular organism the way the E.B. Ford school did. Only a tiny handful of his papers was generally biological in this sense; for example, his work with yeast to try to show that the yeasts that flies really ate were also the ones on which they had higher viability (which didn't work out), and one paper in which he claimed that maybe changes in chromosomal frequencies in California were a consequence of the use of DDT. But, again, he never actually tried it.

I emphasize all this because I believe that attitude of Dobzhansky has had a profound effect on the current shape of population and evolutionary genetics. The school of ecological genetics that we characterize as the English school – Ford, Fisher, Phil Sheppard, Arthur Cain, and so on – just never made any real inroads into the bulk of work in the United States. An important exception was the work of Bill Heed on desert *Drosophila*, but that work was not really appreciated by most population geneticists. In the United States, population genetics adheres to that formal structure that Dobzhansky laid on it. Certainly, that's what all of electrophoresis was, that's what all of the work on DNA polymorphism is, all of the theoretical work being done on coalescence theory, and so on. It's all meant to provide observations and tools to make generalized statements about the importance of selection, the importance of drift, how strong are our selection intensities, and so on, but as general statements.

2.3 North Carolina State: Quantitative Genetics

Q: How did you end up going to North Carolina State?
RCL: As far as I know, I was hired at North Carolina State precisely because it was felt by the quantitative geneticists there that Mendelian population genetics needed to be on the scene and to interact with quantitative genetics. That was very farseeing of them, and I was brought there precisely to do that. Originally, they had wanted Bruce Wallace, but he decided to stay in Cold Spring Harbor, so I was second choice.

There had been (and remained) no quantitative genetics at Columbia. One couldn't even learn it. Columbia graduate students never heard of additive genetic variance, and I do not know that Dobzhansky ever heard of it. There were certainly selection experiments going on, even for Ph.D. theses, but these were never analyzed in terms of heritabilities or additive genetic variance or anything of that kind. They were simply made to illustrate that there was a lot of genetic variation for traits in natural populations and that the same selection might reach different results because of sampling of the initial genomes or because of some drift during the course of the experiment. At North Carolina State, the situation was exactly the reverse. There was no population genetics there, but there was quantitative genetics, largely in the Department of Experimental Statistics with Ralph Comstock, Clark Cockerham, and Harold Robinson, Robinson being largely an experimentalist, whereas Cockerham and Comstock were theoreticians planning experiments and analyzing them in corn, but also in pigs, and generally providing consultations with all kinds of agricultural people.

I was not in statistics but in the four-person Department of Genetics headed by a very well-known geneticist of cotton, Stanley Stephens, who worked on speciation in cotton, and with Dan Grosch, a person who worked on radiation sensitivity in wasps, and Ben Smith, who worked on cytogenetics in speciation in plants. The formalization that Dobzhansky had pushed for population genetics plus my own predilections for statistics and mathematics put me in strong contact with the statistical people, and they, of course, as I said, had hired me because they wanted somebody like that.

What each side discovered was that the other side was thinking of some of the very same problems and making the same models using different letters for the parameters and the variables but coming out with the same conclusion. My own work was supported by funds from Research and Marketing and other USDA funds, and we planned experiments together with the theoretical statisticians Comstock and Cockerham. They had been concerned with the possible role that restricted recombination might play in the estimates of the dominance for yield in corn, and I got kind of interested in recombination as I began to teach population genetics, which is what my task was. And so Ken Kojima and I started to work on the formalism of recombination and selection together, which had previously not been done. Indeed, in 1957, I asked Sewall Wright whether he didn't think we ought to treat selection at two loci with recombination built into it specifically, and he said no, he didn't think that was worthwhile because recombination was such a strong force that it would randomize the genes with respect to each other, and we didn't have to worry about it. That's not what I found. It is not what Ken Kojima found, and Ken Kojima's thesis on the optimum model actually was the first complete work on this subject.

I had worked out the general approach, the problem, and some solutions, but I wasn't very happy with how much you could do analytically, so I put it aside and then Motoo Kimura published a paper showing the results for a very specific model in butterflies, and that impelled me to get back to work on the selection–recombination problem. But all this was in the context of the interest that the quantitative geneticists had in the role of recombination in biasing their estimates of overdominance. After years of experiments they were able to show that there was no average overdominance for yield in maize and that all the apparent overdominance was a linkage bias effect.

It's probably important to realize that putting together classical population genetics with quantitative genetics was an impetus from the quantitative genetics side and not from the population genetics side. The American school of Mendelian population genetics really had very little interest in these issues, again, I think, under the influence of Dobzhansky. Quantitative geneticists, on the other hand, understood that all of their models were really based on Mendelizing factors, and they really somehow felt that the experimental and theoretical point of view that Mendelian population geneticists had would inform what they were doing; thus they took the active steps to bring them together. For example, Alan Robertson was a visiting professor at North Carolina when I first started there, and it was he who first suggested that computer simulation of genetics of populations would be useful to quantitative genetics people. He pushed the first programming of simulations on the old IBM 650. Everett Dempster came as a visiting professor, and Everett was, again, somebody with one foot in quantitative genetics of poultry and one foot in Mendelian population genetics. Jim Crow was a frequent visitor, and he again had started life interested in overdominance in corn but was, in fact, a Mendelian population geneticist of the Sewall Wright and Fisher schools. Somehow, he managed to combine the two schools. All of these people made long-term or regular appearances in Raleigh. Indeed, when I gave my job seminar, they were sitting in the audience. When I started to teach population genetics, Alan Robertson used to sit in the back of the room and listen.

It's interesting that, although Mendelian population geneticists did lots of selection experiments in *Drosophila* and mice, they never incorporated the quantitative genetics point of view. Only Mather did that in Britain. What's even more curious is that Sewall Wright himself began as a quantitative geneticist working on inbreeding in guinea pigs at the USDA but nevertheless just didn't give much attention to the quantitative genetics formulation of selection problems. He was a gene-frequency monger as Dobzhansky was and as the whole school of population genetics was.

Of course there was a potential link between the two schools with Michael Lerner, who was a very close friend of Dobzhansky's and a very influential person. Dobzhansky had immense admiration for Lerner's intellect. Lerner knew quantitative genetics. He was an animal breeder, but he also was steeped in Mendelian population genetics. And if anyone from the side of Mendelian population genetics could be said to be doing any bridging, it was Michael, but I am never quite clear from which side he was really coming originally. At any rate, he never succeeded in getting Dobzhansky to take a really quantitative genetics viewpoint largely because Dobzhansky could not do the math and statistics. In the later generation, Jim Crow was also a bridging person between the two fields since he was an adept at Fisher's quantitative genetic theory and also a close follower of Sewall Wright's. Of all of the generations that essentially followed Dobzhansky, Crow would be the chief bridging person. But again, most of Crow's active research work was on the Mendelian side.

2.4 Rochester and Chicago: Origins of Electrophoretic Methods and Population Biology

When I left North Carolina State and went to Rochester, population genetics was still pretty much in the state it had been when I was a graduate student. The problematic was the same. The Dobzhansky school of population genetics still characterized the field. One was interested in selection for inversions in characterizing the variation that existed in populations, and so on. Indeed, I would say that in the 6 years I was in Rochester, population genetics was probably at its lowest ebb. The old problematic was still with us. No real progress was being made. Dobzhansky's crop of students and postdoctoral fellows really were not quite sure what to do next. There was a kind of scattering of a bit of this and a bit of that, which was also reflected in my own work. I pursued the theoretical work – largely on recombination and selection – more actively than experimental work. The work that was done on house mice, again, used computers.

What people were doing was a kind of unfocused collection of bits and pieces of the Dobzhansky program. What hung over all of our heads was the classic problem of whether there is a lot of heterozygosity or not at the genic level and whether the selection is purifying or not. The experiments were the same old experiments of making chromosomal homozygotes and heterozygotes and then having big arguments about how to interpret the data. I'm not saying that there wasn't a lot of work being done. What I am saying is that it did not seem to be going anywhere. There were B/A ratios, D/L ratios, and so on.

The one promising direction was Bruce Wallace's very original work on the effect of newly induced heterozygous mutations on an otherwise homozygous background. This experiment was the first really new idea in the field. And when Bruce, in his huge experiment, showed that there was some average selective advantage to the newly arisen heterozygous mutations, Dobzhansky went wild. This was the thing he had been looking for, and Bruce became his absolute hero. But unfortunately, as time went on, the evidence got weaker, and it didn't work out as Dobzhansky had hoped. The feeling of a thickening impasse pervaded the whole field. There was the Macy conference on genetic load. There were papers in the *American Naturalist.* We could go on and on and on.

The late 1950s and early 1960s were really the time when the field was beginning to collapse of its own internal contradictions. It seemed pretty obvious that we needed to find some way out that would enable us to actually observe what the frequencies of different variants at different loci were so we could settle once and for all how much heterozygosity on a per-locus basis was out there and what the frequency of the alleles was and, if possible, even try to then go on and determine the fitness relations. I chewed on this problem a lot while I was in Rochester and even developed an experimental protocol involving a special technique for detecting isoalleles at a locus whose only effect was to modify the phenotype of a known mutant at another locus, but to make this into a general program would have meant finding many, many such modifier loci and studying each one separately, and it would have been a real pain. I also had the idea that one might use the fourth chromosome of *Drosophila melanogaster* as a model for the rest of the genome, and when Alice Kenyon did that experiment, she found essentially no variation on the fourth chromosome, and we really didn't know what to make of it. It's only with Andrew Berry's demonstration, more recently, that there's really no variation on the fourth chromosome at the DNA level, together with our present understanding of the effect of selective sweeps, that the old fourth chromosome data could be made sense of.

At any rate, we were all pretty discouraged. I kept formulating in my mind what the requirements of an experimental program would be, but I didn't have the faintest idea how to carry it out. Then, I went on a visit to the University of Chicago to give a talk, and I met Jack Hubby there. And Jack showed me these mind-blowing observations on protein variation in *Drosophila.* What Jack was doing was taking strains of *Drosophila,* grinding them up, extracting protein wholesale by alcohol and salt precipitations, and then running these proteins on electrophoretic gels and characterizing the different *Drosophilas* according to their pattern of proteins, equating each band on the gel with some protein coded for by some unknown gene. There were no enzymes in the problem – just these gross proteins – and Jack's interest was essentially systematic rather than population genetic.

Jack came from the school of evolutionary genetics at Texas where there was no real population genetics in our sense going on but a tremendous interest in the genetics of speciation and sexual isolation. Wilson Stone, who was the head of that program, and his students wrote a lot of stuff on species isolation, on sexual isolaton, and on the differentiation in chromosomes between species, and they are responsible for having done the major work on chromosomal evolution in the genus *Drosophila* and on doing all of the systematic work as well, based on morphology. So when Jack came to Chicago, he had this system of comparing proteins as a way of comparing species. He was working with Lynn Throckmorton on phylogenetic problems in *Drosophila*. He also realized that one could find differences even between strains in the same species. It must be emphasized that Jack was not a population geneticist and was not participating in the struggles in population genetics, nor was he even aware of what the problematic in the field was. And here I was, puzzling over the problematic of the field and seeing a technique that could clearly be applied directly. It was a case of a problem without a technique meeting a technique without a problem, at least in population genetics. It would really be interesting to know how many cases of experiments that have had some influence in a field are a result of a similar process for which a technique is developed for one purpose and it seems suddenly by accident to fit the purposes of some other question.

When I went to work in Chicago, Jack and I immediately started the electrophoresis work. By this time, Jack and Sumiko Narise was looking at enzyme proteins that could be visualized on gels from single flies rather than having to grind up huge numbers and produce proteins of unknown function. We at once started work together on the population genetic problem and within a very short time produced results that seemed very exciting to us. Most of that story is in my 1974 book; at least my point of view is in that book.

At the same time, Chicago had a strong program in ecology, which continued. Tom Park was a very strong experimentalist in *Tribolium* demography in ecology. Alfred Emerson, whose position I took, had of course been very powerful in ecology. While I was there, the department hired Monty Lloyd and Dan Janzen and a number of plant ecologists and, in general, ecology was pretty important. The result was that we had an interaction between a group interested in ecology and a program in population genetics that began originally with me but then came to include Jack. The outcome of that was a fusion of interests in genetics and ecology.

Being entrepreneurial, I got together with Tom Park, and we applied to the Ford Foundation (I guess Tom had the *entree* into that) for a large training program in what we called "population biology," which was to include evolutionary genetics and population genetics and demography and other forms of population ecology. We got the money, some millions of dollars actually, for, I think, a 5-year program chiefly to pay stipends to bring postdoctoral fellows and graduate students into this program. We had a field day. We could appoint anybody

time, guilty of a lot of hubris. But there was a struggle going on in the general scene that was also going on at Harvard, between organismic biologists and the claims of the Watsons, the Cricks, the Meselsons, the Brenners, and the whole Cal-Tech group around Max Delbrück, who really claimed that they were going to discover everything that was interesting about biology, and this really pissed off the evolutionary biologists. It came to a head at Harvard with the demand by the developmental cellular and molecular biologists to have a separate department. Already Matt Meselson and Jim Watson had been appointed to a separate Department of Biochemistry and Molecular Biology because the biologists wouldn't have them, the biologists having been dominated by old fashioned biology. And it *was* old fashioned. When I was a student in the early 1950s, it was a bastion of 19th-century biology.

Later, when Ernst Mayr decided to step down as Director of the Museum of Comparative Zoology, Nate Pusey [President of Harvard] called me in Chicago and asked if I would be the new director. I was Ernst's designated successor. That was due to Ernst. And Ernst knew what he was doing; he knew what the future was. Anyway, I declined. Then I was on a committee called by Pusey from outside the university to advise him on whether the Biology Department should be broken up into two sections because, by that time, the late 1960s, enough modern biology had infiltrated into that department that there was real tension. Jim and Matt were still not in that department; they were in Biochemistry and Molecular Biology. The Biology Department had become a mixed bag and was having growing pains. I advised Pusey against breaking the department apart for purely intellectual reasons. Wisely, they did break the department apart into two subsections – for political reasons. Cellular and developmental biology formed one group, and evolutionary biology formed the other, and the two barely spoke to each other.

At the same time, Dobzhansky, who was near retirement from Columbia, had gone to Rockefeller University because Alfred Mirsky had pulled strings to get him there. But he was a complete fish out of water. Nobody else at Rockefeller was slightly interested in his world view, and he really felt out of it. They were all hostile to what they regarded as a kind of soft biology done by evolutionary biologists. The electrophoresis work had an important role to play in all this because it seemed to bring together molecular biology and evolutionary biology, and the consequence was that I was on good terms with a variety of the hard core molecular biologists. Max Delbrück invited me to come to Cal-Tech to give a talk to his group. Norton Zinder was a close person to me at Rockefeller. I was the only evolutionary biologist he trusted. In general, I was regarded as a sensible evolutionary biologist who understood that molecular biology was the savior of biology. I spent a fair amount of time on NSF and NIH grant panels and, in general, built up a circle of acquaintances among the molecular biologists in those groups. There was a definite feeling that the two sides could be brought together.

At the time, this was pretty one-sided. That is to say, population geneticists in pursuit of their own problematic were becoming molecular and interested in molecular biology at various levels. But molecular biologists, by and large, were not much interested in evolution. Then, this began to change, and more and more molecular biologists have been showing – and it is still increasing – an interest in evolutionary problems. It is not entirely clear why that reciprocal movement has happened. It would be tempting to think that the introduction of molecular methods into population genetics and evolutionary genetics was instrumental, but I can hardly believe that is true. At the same time, there was, of course, the accumulation of information on protein sequences from a variety of organisms that eventually became the basic data for the neutral theory of protein evolution. Emmanuel Margoliash, for example, was studying cytochrome C sequences in a whole variety of organisms, and thus it became clear that there was evolution of proteins. But I don't even think that can fully explain the growing interest of molecular biologists in evolutionary questions. I think that it has to do simply, if I may put it this way, with the maturation of molecular biology, that is to say, the taking of molecular biology into biology as a whole.

You have to remember that molecular biology began largely as the work of biological outsiders (biochemists, biophysicists, physicists, and so on), who, as I said, were really quite ignorant of even the elementary facts of genetics. They had, for example, a cockamamie theory of recombination that involved nothing but copy-choice with no actual breakage of chromosome strands, and that obviously cannot be true. As molecular biology has become the core of biology as a science as a whole, cell and molecular biologists have become real biologists. Molecular biology is biology, and they have more and more become interested in the evolution of their phenomena. For example, the penetration of molecular biology into developmental genetics cannot be done unless you have some evolutionary perspective on organisms. So, I think that is what happened.

But that was a later event. The earlier event, as I say, was the taking into evolutionary genetics of a molecular biological approach, originally through electrophoresis. That meant that there was a small group of us who were kind of bridges between the two halves of biology. That surely was the reason why I was offered a job at Harvard despite the split between the Mayr–Dobzhansky evolutionary biologists and the molecular biologists. Both Mayr and Dobzhansky understood the necessity of incorporation of molecular biology into population genetics. Dobzhansky, for example, while I was still a student, spent time in Texas in Roger Williams' lab learning how to do amino acid chromatography from urine and created with L.C. Dunn a whole Institute for the Study of Human Variation that was looking at patterns of urinary excretion in human urine of amino acids in the hopes, again, of finding single locus polymorphisms to feed the problematic of the field. That did not work out, but he was quite ready to take molecular biology in. As soon as electrophoresis appeared, I went down to Rockefeller and gave a little in-house talk to his group before we ever published our work, and immediately they seized on it and set up shop to do some more work. I think Francisco Ayala, who was then an assistant professor in Dobzhansky's group, immediately set up a lab.

Mayr, despite his commitment to whole organisms and systematics, really did have an understanding of the importance that molecular biology would play, and he was responsible, surely, for Harvard's deciding that they had to have a population geneticist like me working in molecular biology. Ernst clearly wanted me hired in order to molecularize evolutionary stuff, and he certainly saw me as being a bridge with the people here. But by being a bridge you also protect the evolutionary biologists from the accusations that they're old-fashioned and behind.

2.6 Continuing with the Problematic: *The Genetic Basis of Evolutionary Change*

These were the heady days of electrophoresis, with every laboratory dropping whatever it was doing in population genetics and suddenly fastening onto electrophoresis as the answer to a maiden's prayer. At the same time that was happening, there was obviously trouble on the horizon. For example, the first attempt to actually measure fitness differences associated electromorphs that was really a proper attempt, not artifactual, was Yamazaki's study of the esterases when he was a graduate student in our group in Chicago, and that attempt failed. That is to say, he was not able to measure any fitness differences. It began to dawn on all of us that if there were fitness differences associated with allozymes, they were probably pretty small. There was an industry for a while of measuring fitness components of the ADH fast–slow polymorphism, and there appeared to be some real results there. But then it slowly dawned on people that the amount of alcohol you had to put onto the flies to produce these results was vastly more than the amount of alcohol they ever saw in nature, even in wineries. So that kind of faded out. Slowly but surely, it began to dawn on people that selection coefficients for individual loci were probably very small. A gloom began to settle again. Here we had this huge amount of information about variation that got even bigger after Rama and I came to Cambridge and started the work on sequential gel electrophoresis demonstrating that there were even much greater amounts of polymorphism than anybody had yet measured. At the same time, nobody seemed to be able to figure out how to measure the fitness relations. The problematic had remained unchanged. A big advance had been made at the observational level. At least we knew what kind of variation there was, and we could follow gene frequencies in populations both in space and time, but we did not know how to measure the forces operating.

At the same time, Motoo was telling us that there probably were no forces operating except random drift, and even if there were, the selection coefficients were so small that the product of population size, N, and selection intensity, s, was less than 1, and there was no effective natural selection. Harry Harris and Hopkinson, in London, were measuring real differences in kinetic parameters for electrophoretic variants in humans, but again they were not able to tie that to any fitness differences. And nobody was able to show that there was any balancing selection for any of these polymorphisms, which was sort of required by the observation of what appeared to be long-term stable polymorphism. So, things slowly began to return to the atmosphere of before 1966. That is to say, we did not know how to detect the forces operating, even though we now knew what the variation was. It was at that time in the late 1960s and early 1970s that the pattern was set for how to deal with this problem, and that was to copy the pattern of the earlier – late 1950s and early 1960s – period. That is, you try to develop a theory that makes predictions about the distribution of genetic variation – how many alleles there will be, what their frequency will be, and so on, and then compare the observations about standing variation to see whether it meets that null theory or not. If it does, you say there is no selection. If it does not, you say there is selection. In the old days of B/A and D/L ratios, the null hypothesis was purifying selection, and the deviations would be regarded as evidence of balanced selection. In the new situation, the null hypothesis was no selection, and the significant deviations from the null prediction would say there was selection. But it turned out that these predictions were not good enough. The Watterson Test and Ewens Test turn out to have no power to detect selection or to figure out in what direction it was, and as rich as the electrophoretic data were, they were not rich enough to distinguish between identical selection in different populations and a little migration without selection. And once again, I think people got pretty gloomy. I certainly did. Notice that the problematic has remained unchanged.

Q: Those epistemological problems are presented in your book, *Genetic Basis of Evolutionary Change*, in which you describe the major contentions in population genetics, not just in terms of theories and techniques but also in terms of political or, more broadly ideological, conflicts. But when you describe the history of the field on the tape, there's virtually no reference to those elements. How do you explain the apparent inconsistencies? The history that you told us was an almost pure history of the logic of ideas of the field.

RCL: I'm not sure. Perhaps it's got to do with my state of mind when I did the tape – what I thought would be important. Partly, I guess, because in doing the tape I was always under the influence of my initial remarks, I wanted to make it as nonpersonal as possible and talk about the field.

Biographical information can be revealing; for example, Dobzhansky certainly had a whole ideological view of the world, which was that human welfare depended on diversity. Dobzhansky would have said, and I inherited that notion from him, that a population you made essentially homozygous by directional selection would lose the possibility of adaptation in changed circumstances. That's the general theory of population genetics. Both Muller and Dobzhansky might have wanted to improve the human species, but whereas Dobzhansky would say, "maximum diversity is the best thing you could have," Muller would say, "no, we ought to get the best genotype, and we ought to select for it." Dobzhansky's response, I think, would have been, and my response would be, if you really believe that genetic diversity (and I don't believe that) is the basis for adaptability, then, although you might get a short-term advantage out of selecting for the best genotypes you can create, you're really going to screw up the works again because Dobzhansky believed, remember, that homozygotes were narrow specialists. So you could select the homozygote that was better for the present situation, but it would have no ability to protect you against future demands and changes. Since the world is changing, you've got to have genetic diversity. Dobzhansky would have been one of the chief screamers about all the uniform corn plants; he'd have loved that. But all that's based on the notion that the issue of adaptability, and continued historical change of population, depends on the genetic diversity. You only believe that if you're a biological determinist, which he was.

What I guess I'm saying is I find a certain consistency between general world view, political ideology, and attitudes about scientific questions. I'm not saying they're determinate. I'm not saying there aren't variants. My problem, as a person trying to talk about the history of ideas, is that I'm confused. I don't have a doctrinaire position on the relative importance of individuals and their predilections, and how that comes together as social movement. That's a problem we haven't solved, any of us, and I haven't solved it in my own mind. To say that no individual matters is clearly wrong, but to write the history of science as a history of individual careers is also clearly wrong.

I think it helps to distinguish between output and input. The effect of the output in the field has immense political implications quite aside from what the people in the field intended in the first place. And there, I think, I'm not so confused; there, I think, the lines are clearer. The issue of how the content of a science reaches the outside, and the statements that are made, and the implications made, about what the press says and what people take from it, and whether they give support to the eugenics movement or not – that I think we can talk about without a reference to individual people at all.

Q: Could you say in what particular ways the claims in the book reflected your politics?

RCL: The book is not primarily a political book. But it has remarks in it about the relevance of these issues to ideology, and the ideological implications. And those were thrown in, if I could be very uncharitable about it, because I wished to do some propaganda. How do you feel about that explanation? That is to say that I would regard it as politically indefensible to write a book about all those things without connecting them to political issues. And it was also a golden opportunity to do a little politics. Please remember that we're talking about a book written in 1971 at a time when political issues were really hot.

Q: You say that you saw the book as an opportunity for propaganda, but I'm wondering how. You have a few remarks in the beginning, and a few toward the end. How does it really bear on the politics of the day?

RCL: Okay. I mean, since I'm not a biological determinist, I don't think that the ultimate issues depend on how much diversity exists. I don't think that the discovery of large amounts of polymorphism within populations has an important political implication except for one, and that is the issue of the relative diversity between populations as opposed to within populations. I beat that one over and over and over again. There's a political implication. And if it did turn out that 95% of human genetic differences were between blacks and whites and yellows and browns and reds, I'd have been very unhappy. What can I say?

Q: You've been writing about the nature–nurture issue, in one form or another, since the early 1970s. In a 1972 critique of Arthur Jensen, you accepted a substantial heritability of IQ and also wrote that social scientists should pay more attention to biology. What did you have in mind?
RCL: I can't recall what I had in mind when I said that. On the issue of putting various heritabilities between .6 and .8 and not quarreling with that – there had been no careful analysis of the data. Leo Kamin's careful analysis came later.[8] So I simply said, yeah, if that's what all these studies show, I'll accept it. This was ignorance. I just accepted what was in the literature. That's a simple answer to that question. Let me go further and say, however, that over the years my position has moved to the following one, which is not very different. Namely, I don't think the heritability studies are very good, and I don't trust the estimates, but my position is consistently that there is no performance, behavior, test, or whatever that you can give humans that wouldn't be heritable. I think it is crystal clear that every variation among humans has some influence of genes just because every variation of humans is a manifestation of a biological organism. Genetic differences between individuals cannot be totally irrelevant to anything. The important issue however is mediation, not whether genes have some influence.

An example I give in classes is that genes must be relevant to performance on IQ tests because you sit in a classroom, the sun's coming in the window, the kids are shuffling their feet, they haven't had a bath, and you're supposed to answer all these contextless questions; what you need most is concentration and to not be distracted by all the sensory inputs. I think it would help a lot if you were slightly deaf and didn't smell too well; let's say, if your general level of sensory input sensitivity were lowered so you could concentrate on the job. I can see that genes must have an effect. But that's not a very interesting mediation from the standpoint of IQ. Again, the question is always, who cares about heritability or what work that does for one, as opposed to what the mediation story is, and what work that does for one?

2.7 New Attacks on the Problematic

In about 1981 or so, Marty Kreitman got the idea that it would be interesting to look at nucleotide variation. He was the first one, to my knowledge, to have the idea that one could learn something new about population genetics by sequencing DNA. It does say something about population genetics. Some restriction enzyme polymorphisms had already been observed, but that did not seem to help anything. Marty thought that maybe we could learn something from nucleotide sequences. In particular, it was clear we could learn whether an electrophoretic polymorphism was a one-amino acid difference or a multi-amino acid difference and whether an electrophoretic class that did not break up by tricks of electrophoresis, nevertheless would turn out to be heterogeneous and would have many different amino acid compositions contained within it. What I do not remember is whether Marty also realized at that moment when he first started the work that there is a fundamental property of nucleotide sequences that could put a whole new twist on the solution to our problems. The reason I say I don't remember whether he thought that or not is that he is not very explicit about it in his 1983 paper in *Nature*,[9] although he certainly uses that kind of reasoning.

The reasoning runs as follows: Since nucleotide sequences do not have a one-to-one correspondence with amino acid sequences because there are synonymous positions, then it is entirely conceivable that a locus that was monomorphic for some gene would nevertheless turn out to be polymorphic at its synonymous positions and monomorphic only for amino acids. If that were true, then, that would rule out the monomorphism as simply a result of common ancestry, that is, genetic drift, and say that the monomorphism of amino acids in face of the polymorphism of third positions could only be explained by selection against all of the amino acid polymorphisms, that is, by purifying selection. So, if one did a nucleotide sequence and found that a locus was totally monomorphic for both nucleotides and amino acids, then it is not quite clear what you would conclude, but you might easily conclude that it was a result of drift. But if you found it was monomorphic for amino acids but polymorphic for synonymous positions, you would have to conclude that selection had been operating, and that is the conclusion Marty came to when he wrote his paper because ADH has 6% nucleotide polymorphism, and it is all synonymous. There is some of it in the introns, and there is a lot of it in third positions. So, what he showed was that you could not avoid the conclusion that purifying selection was operating. This left open the question of why there was a fast–slow polymorphism in ADH, and his data did not speak to that point. It could be that this was the one position that had escaped selection and was purely a random drift polymorphism, or it could be that there was balancing selection at that position. Later on, more observations were made, which certainly lead you toward the latter conclusion, but that was not clear at the beginning. I really do want to emphasize that nucleotide sequencing is not just looking at another level. It really is a major methodological break because of the lack of one-to-one correspondence between nucleotide sequences and amino acid sequences.

Marty himself was reluctant to make as strong a conclusion from his data as I was. I kept pointing out in talks that I gave about it and in discussions with Marty that this ADH protein was 25% leucine, isoleucine, and valine. Simple-minded biochemistry told us that it ought not make any difference to the protein, if there were these substitutions among leucine, valine, and isoleucine, yet all leucine, isoleucine, and valine mutational substitutions had been cleaned out. There was no variation. Then Steve Schaeffer came to the lab and proposed to do the same task in *Drosophila pseudoobscura*, which did not even have the fast–slow polymorphism, and eventually did 99 sequences of *pseudoobscura* and found absolutely no amino acid variation with the exception of one strain that had an isoleucine–valine substitution. The point was made even more clearly, and that is that natural selection operates on very subtle differences, is very fine in its operation, and can detect every amino acid substitution in ADH.

I think that is an excessively strong result, but I do not know what else you can conclude. I mean, it is hard to believe that it makes an average difference to survival and reproduction of fruit flies when they have a single isoleucine–valine substitution in some damn spot in their ADH. It's a long way between that and differential reproduction. Sure, the selection would not tend to be very strong but *Ns* has to be reasonably large to clean out all of that variation. So, we are left with a very interesting physiological mystery that nobody seems to be working on.

Nobody outside of population genetics understood the importance of this first experiment. *Nature* actually returned the manuscript without review originally, saying that it was of no interest, and it was only when I insisted that they send it on to Mike Ashburner for his opinion that they finally published it. Molecular biologists who heard that Marty wanted to sequence the same gene over and over again for different individuals thought he was out of his mind, although Wally Gilbert was immensely helpful in providing him with the instruction, the materials, and the lab space to do the work in. But, of course, these observations dropped like a bomb on population genetics because now it seemed once again that the impasse could be broken, that one could distinguish between purely random drift explanations of homozygosity from selection explanations of homozygosity. It still left the issue of how you would explain heterozygosity, but at least you could explain homozygosity, and so the entire field dropped electrophoresis almost immediately and shifted over to DNA sequencing. You could also study introns; you could study flanking sequences. The kinds of inferences you could make were tremendously increased, and the major problematic, which was (a) how much variation there is, but (b) how you explain it, seemed like it could be dealt with using the argument that I just made.

Of course, as you might expect, things did not work out that way. The distribution of nucleotide differences with the occasional amino acid differences is as difficult to interpret as ever. First of all, it has become absolutely clear from the last 15 years of work that there are no such things as unselected, unconstrained nucleotide sequences, at least by category. Third positions are under selection because we have the maintenance of codon bias. Flanking sequences are under some kind of selection because we have some incredible conservation of flanking sequences. Introns are under some selection because the variation is not homogeneous throughout the introns, and so on, and so on. So, now we observe a particular amino acid polymorphism in a species – is that amino acid polymorphism maintained by selection or is it neutral? There are differences in amino acid sequences between two species. Has selection driven that differentiation or is that random drift? How do you find out? Well, you could try to measure the selection, but we already said you cannot succeed. So, we are back to square one again, namely, making null models, comparing the pattern of variation of the null models, and seeing if those null models will allow us to detect some kind of cumulative selection effect over long periods of time even though it is small. Now the null model becomes the random drift model – the no selection model – just as it was in the Ewens and Watterson cases, but now it is based on a more articulated random drift theory of nucleotides and a lot more theoretical work since Ewens' original work. Now, you compare nucleotide heterozygosity within and between species and within and between loci and amino acid heterozygosity within and between species and loci and you try to manipulate those small numbers of numbers to test whether the footprints of selection are there or not. The answer is that sometimes you get the footprints of selection – in fact, quite often. Unfortunately, we do not know whether the tests are really being significant because of selection or because of other assumptions built into the test theories – for example, that the alleles are at stochastic steady state, or that there is no effective migration, or that there is either a lot of recombination or no recombination depending on the theory, and so on, and so on. We seem to be better off in dealing with the problematic because we have the richest possible kind of data set, and we have the contrasts within and between species and synonymous and nonsynonymous variation, but it is not clear yet whether the predictions of the null theory are sharp enough and whether the assumptions of those predictions are unimportant enough that deviations from the null predictions really are detecting selection. The important point I want to make is that all of the work done beginning with that first nucleotide sequencing work and all of the laboratory work done like that since that time all over the country is a repeat of where we were in 1950 in Dobzhansky's laboratory. The problematic has not changed, and the problems of measurement have not changed. What has changed is that we now have not only a gene-by-gene but a nucleotide-by-nucleotide view of exactly what the variation looks like. That cannot be bad, but I am not sure it is good enough.

The Genetic Basis of Evolutionary Change simply makes this point, but as of 1974, when we had no nucleotide data. I could write the book all over again. It would contain all of the nucleotide data. It would contain a lot of interesting new stuff, but the bottom line would still be the same: that we as yet do not know how to turn static data into a firm understanding of processes, although we have suspicions, and sometimes pretty good suspicions, and we as yet do not know how to measure fitness. So, in that sense, population genetics has not changed.

2.8 Funding

Q: Can you say something about the sort of funding that's been available for population genetics during the period in which you've been active? We note that you've had considerable support from the Atomic Energy Commission (and the agencies that succeeded it, the Energy Research and Development Administration and the Department of Energy), the National Institutes of Health, and the National Science Foundation.

RCL: The Atomic Energy Commission had a Division of Biology and Medicine, which had as one of its programs, genetics. They obviously had real interest in population genetics because they were supposed to make policy about the effect of radiation on human populations. My relationship began in connection with my early quantitative genetic simulation studies. I went to them to ask for money to do theoretical work on computer simulation of populations. I titled it "The Effect of Mutations on Populations," but I wasn't really interested in that; I was interested in general population genetic modeling. When I was still in North Carolina, a group of us would meet at Oak Ridge to work on the big, high-speed computer there, simulating Mendelian population genetics. I went there three or four times while I was at Raleigh. We had free access to the computer and began programs of simulation. Then, when I moved to Rochester, Dick Levins came, and he and I began a very large program of computer simulation of the selection recombination problem and the varying environment problem. Dick was hired as a research associate on the AEC grant. We shared a room and talked about computer models and the division of labor!

Slowly that kind of work spread both from North Carolina and from what I was doing in Rochester. As I say, it was supported continuously by the AEC, which then turned into ERDA, which then turned into the DOE. I had a contract with them for 20 years or so. Indeed, they kept pumping in the money even when I didn't ask for it. They were anxious to have this stuff done. I would get letters from them saying that "we haven't had your renewal application yet." (It was an annual contract.) "Please hurry up and send it in because we put the money aside already and we have to have the documents." It wasn't big money, but it was plenty to support that kind of stuff.

Now, what's interesting about the AEC was that the people who were in charge of the genetics program, a succession of people, were all people I knew anyway. I don't know whether it's well understood that people who go into these government programs as executive secretaries, as program officers, and so on, have a certain social homogeneity about them. They are all people who, of course, are professionals and Ph.D.s in the subject. It means that many of them are already part of the club. When Charlie Eddington was the program officer, I knew Charlie from Oak Ridge. He was a pal of Bill Baker's and Larry Sandler's, so Charlie wasn't exactly a stranger. Part of your relationship to these people was based on your professional acquaintance anyway. Let me say the only agency I've ever dealt with, as an alienated, strange, bureaucratic agency, was the Office of Naval Research (ONR). They came to me asking if I would like some money to do computer studies because they'd heard that I was doing them. I said, why is the ONR interested? They said, well, we're just interested in computer studies. Some guy showed up in my office. I said, sure, send me the money. So they sent me the money. But that only lasted a few years.

The other cases were all part of a network. At NSF, the program director for genetic biology for years and years and years was Herman Lewis. The program director for genetic biology at NSF was not like the executive secretary at NIH. At NIH they have numerical ratings, and the executive secretary more or less has to go along with the numerical score. The program director at an NSF program has the complete power to decide who gets the money. And the panel isn't a study section – it's an advisory council. We used to meet three times a year in Washington with Herman, and we would go through proposals. Charlie Yanofsky was on the panel, and it was a club. I was in that club for several years. And we would read all the proposals; we never paid any attention to mail reviews. We'd read the proposals and we'd talk about them around the table. And then Herman had a little card and he'd say, "Okay you guys, do I put this one before or after?" And he'd go through the cards that we'd already done and say this is better than this person, and he would put it wherever we suggested. But then he would go home and take the cards and do whatever he wanted to do. He took our advice, but he wasn't bound by it.

There was no numerical rating system. Herman really sat at the center and did what he did. So you had this very strange business, which I actually wrote a long piece about with Loren Baritz. (Loren and I were codirectors of a program of study of "Research and the Universities" organized by the American Academy of Arts and Sciences, but we never published the huge report that we produced.) People who are part of the club but who are not high research-output people – they don't regard themselves as separate bureaucrats but as part of the whole culture. They have to be pals with the leading researchers and be on good terms with them because that's where their adhesion to the social group comes in. At the same time they have a certain power that people need to court. So there's kind of mutual courting that goes on, which is extremely interesting.

In those days there was a lot of money, and I was able to say to Mary Jane that the day I can't get a research grant from either the NIH or NSF, I know I'd better retire. They'll deny it furiously, but when I was on the NSF panel, my research proposal would come to the panel on which I was sitting. Of course I would leave the room, but you're not going to tell me that my pals in the room were going to give me a bad recommendation so that when I got home I would be told that Charlie and all these guys had rejected the proposal. No way. So the best news of all would be that your research proposal was going to be reviewed by a panel of which you were a member. Now they've changed the procedures. Now proposals of panel members are reviewed by special sections.

The way that AEC contracts worked was that you'd get a 5-year or 3-year contract and you'd get renewals of the contract. The whole contract would be rewritten after a major outside review; then, the contract would be renewed and you'd have annual reviews. I'd get a letter from someone at the AEC saying, "Where the hell is your renewal proposal? The money is put aside already and I'm waiting for the renewal proposal. If you don't send me the proposal, I don't know what I'll do because somebody's already accounted for it." The program officer'relationship to the AEC was clearly one of considerable looseness. They did what they damned pleased with the money. That changed in one respect when I got out of the AEC funding business. A guy I saw at a meeting in Washington said, "well, you know, your contract has expired. Look, we're not going to fund what you're doing now. The word has come down. Our policy now is that we're going to make a big play for the genome, for sequencing the genome." And he said, would you please change what you're doing and write a proposal, if you want to keep getting money, that says something about the genome." I said, "not interested." So a policy change did come down, and the program officer for genetics had to reorganize his program according to the general policy direction. But my impression was that, before that time, program officers had complete freedom.

Q: Weren't your politics suspect?
RCL: As I wrote in a recent essay on universities and the Cold War,[10] no political questions were ever asked by the AEC. They continued to support me without question all the time that I was very active with the antiwar movement. I was in

contact with the Socialist Workers Party; I was involved with the Communist party and the Panthers. No, no, no; that issue never arose. But as I wrote in that piece, that was also true before that. L. C. Dunn was supported by the AEC for years when he was a noted "fellow traveler." As far as I know, Dunn was never a member of the Party, but he was the president or founder of every Party-sponsored Soviet–American friendship this and that and the other thing. He was a Red. And that never stopped the AEC from supporting him. It stopped him from getting something he wanted though. He wanted to be the scientific attaché in the Paris embassy, but they wouldn't clear him. My tutor at Harvard got that job instead.

Oh, but the ONR. I had ONR money for a couple of years just after I got my degree and I was still fleeing as far away from politics as I could get. I didn't even think about political issues in 1954, and when I did, I got in big fights with my friends in the Party. So the ONR didn't bother me. And then I never thought of political issues in connection with the AEC because I thought it was a cash cow. I didn't think I was doing them any favors. I felt that what I was doing was totally independent.

I want to talk about that. I think that people in science programs, the extramural science programs of the AEC, had the view that this was nothing other than another excuse for getting taxpayer money out to the scientific community. I know that in the big Dobzhansky–Muller fight a lot was made about where the money came from and where Dobzhansky had bent, Bruce Wallace had bent – not the truth but their interpretation of it – to suit their patrons. I wasn't aware of that issue until much later, and the issue certainly never arose for me because of my work. I was working on a two-locus theory, which had nothing directly to do with mutations or anything. The claimed relevance was a fake is what I'm trying to say. And it led me to a very cynical attitude that carried over. Let's look at the NIH. It used to be in the old days, you were supposed to tell what the health-related importance of your work was in the NIH proposal. I don't think you have to any more, but you used to. And everybody understood that was boilerplate. Nobody in the study section ever paid the slightest attention to it. The executive secretary never paid the slightest attention to it, and the council never paid the slightest attention to it. That was all for the consumption of the fools in Congress. People were very, very cynical about that. So no, the AEC wouldn't have bothered me in the least.

Q: Can you think of one proposal, or one contract, which contributed to the mission of the AEC concerning the effects of radiation?

RCL: I have no recollection of any such case in my own work. When my 5-year contract renewal came up, I would have a paragraph in there saying of course we need to understand the effects of mutations in populations because radiation produces mutations, but that would have been a piece of boilerplate. I'm not sure people understand the social relationship that existed between the research community – at least that faction that I knew – and its patrons. I'm telling you, it's a very cynical attitude, which is shared by all my colleagues. The Cold War and all of that made it possible for us to get our hands on public money. We'd tell them a lot of nonsense about relevance.

Now I'm not saying that's true of everybody. There was some muttering about Bill Russell's program, which was an immense on-site factory to measure the rates of mutation in mice caused by ionizing radiation. A lot of people sort of felt that he was doing the work of the AEC; this was not science, and it was not intellectually interesting. He was making a single measure, which is, What is the doubling dose for visible mutations? It's clearly work the AEC would use in some way.

Let me press a little further. I actually worked physically once in a while for two or three days at a time at Oak Ridge, when Alex Hollaender was the head of the Division of Biology and Medicine. And Oak Ridge at that time was the employer of many people who later on went to be professors of genetics: Larry Sandler, Eddie Novitski, Bill Baker, and so on. They all worked there right after the War, and Alex Hollaender was the director of genetics. Everybody laughed at Union Carbide, which ran the outfit. They thought the whole thing was a joke. Nobody took it seriously. Now, I'm not saying they were right to not take it seriously; that's another issue. Maybe they were using us. We thought we were using them.

Q: You mentioned the NIH. Why were they interested in population genetics?
RCL: The paper justification is obvious. Since NIH is interested in human disease, and since there's an epidemiology of genetic disease, you could make claims, for example, that we need to understand how the genes that cause disease change their frequencies in populations and whether there is really heterosis. You know, predicting the course of genetic disease in human populations, that would be the argument, to try to understand why certain diseases are polymorphic. Sickle-cell anemia was a gift of God in this respect.

But that's not why the NIH *was* interested. Remember, the Division of Research Grants does the judging of proposals, but there were separate institutes that actually funded the grants, and they're orthogonal to the Division of Research Grants. All the money that went into population genetics came from one unit called General Medical Sciences (GMS). What the hell is General Medical Sciences? It was a euphemism for basic biological research that you couldn't really demonstrate had any direct use for health, and so they funded all kinds of stuff – they funded population genetics, they funded ecology. Because the people who ran the NIH below the bureaucratic level, the actual operating level, belonged to our culture. They were us, and they were the conduit by which money passed from the general federal purse to the universities. Now, how much conscious policy there was even at the congressional level to help support higher education in universities by this ploy, I don't know. It may very well be that even at the highest level this was understood. But I can sure as hell tell you that the Institute of General Medical Sciences was there to support higher education and research.

Q: Can you say more about NIH funding? What part of your work did it support?
RCL: NIH funded my molecular population genetics. When we began that work, Jack and I, it became clear that NSF would not have enough money to support us, and so we went to NIH. NIH supported work on molecular population genetics of the kind I had been talking about – "find them and grind them" and then nucleotide sequencing – beginning in 1966, I guess, until almost 1996, close to 30 years, first at Chicago and then at Harvard. There was never any question of renewal, and toward the end of that period, the laboratory was spending about $300,000 or $400,000 a year – $300,000 in direct costs from a combination of two NIH grants. That's now run out of steam. Because the problematic of population genetics has not

moved on to a solution, NIH is getting more picky. They no longer want to support purely observational work. They want very well formulated hypotheses that can then be tested by very precise testing of statistical measurements – things which I do not believe in. We no longer have any research money because the NIH is not interested in this continued kind of exploration of patterns of variation and constraint, which is what I believe is all we could do. There is still interesting work to be done, as far as I can see. We need to look more at the DNA binding regions of the homeobox genes, of which *dpp* is one. We only finished the coding region of *dpp*. There's a lot more interesting work to be done in *dpp*, but there's no money for it because it doesn't have a well-defined hypothesis.

The whole hypothesis business has become very important with the diminution in the amount of money available. When I say the diminishing amount of money, the total amount of money available, of course, is as much as ever, but the research gets more expensive every year and the number of workers gets greater, and so the competition for funds gets greater and greater and greater. There have been complaints by population geneticists that population genetics has not been as well funded as other fields, but I have not reached that conclusion from a survey over time. There is a slightly lower rate of success of population genetics proposals than of, say, developmental genetics proposals, but not much less. I don't think it can be shown that populations genetics has been starved out by developmental genetics. Population genetics has benefitted from federal funding in exactly the same way as all science has benefitted. The Cold War was the making of academic science, and we are still riding on that. But people are finding it increasingly difficult to get the funds they need.

On the other hand, journal publication of population genetics is flourishing. There's quite a bit of it in *Genetics.* We have our whole new journal, *Molecular Biology and Evolution.* We also have *Molecular Phylogeny and Evolution* and the *Journal of Molecular Evolution,* which has a good deal of molecular population genetics in it. I don't think that population geneticists have any complaint about their ability to publish their stuff, or even a very strong complaint about their ability to get it funded – although, I guess if I had to choose a field based on funding in genetics, I probably would choose developmental genetics. The competition there, however, is cutthroat, whereas, there isn't much competition in population genetics. I have noticed when I have gone back to the NIH panels as an ad hoc member that criteria have changed. One is now asked the question: Well, here are these four different labs all asking essentially the same question. Which one will get it done most efficiently? It used to be that much more consideration was given to developing the laboratory as a place where training could go on, a place where a long-term program could be produced. The question of who is going to get the answer first and most efficiently did not arise. It now is a very important issue.

2.9 Life in Lewontin's Lab

Q: Earlier, we discussed "life in Dobzhansky's lab." What about life in Lewontin's lab? What sort of lab environment have you tried to create?

RCL: I don't know what you have in mind, but let me say what I think is important.

The image of how a laboratory should be organized comes to each of us, presumably, from our personal experiences as a student and postdoctoral fellow. My own view is that Dobzhansky's laboratory was, in part, organized in a very bad way, namely, that every person who came into it was under extreme pressure to fill part of Dobzhansky's intellectual program. But on the other hand, it had its very good sides, namely, Dobzhansky's reluctance to grab credit from his junior colleagues or his postdoctoral fellows and put his name on their papers. Also, people felt quite free to fight with Dobzhansky on intellectual issues in a very strong give and take. Of course, his ego was never threatened. My own practice has been somewhat different from Dobzhansky's in some respects. I have taken seriously his admonition not to put your name on other people's work, since I believe strongly in the correctness of Merton's Matthew Effect paper, and in fact, discover that one has to work very hard to make sure that the head of the laboratory is not given credit for things that the head of the laboratory has not done.

On the other hand, I disagree strongly with Dobzhansky's attempt to enforce his immediate program. Our laboratory has always been exactly the reverse. Everyone is welcome so long as there is space and money and everyone is welcome to do what they want to do. In part, this was the vision that Ernst Mayr had when our laboratory was first set up, namely, that we would provide instruction in electrophoretic methods and other developing methods to people in systematics and other parts of the department. That is why we have a whole floor. The original lab was built with large numbers of places to do electrophoresis, and people from botanical and zoological departments of the museums and herbaria would come up and learn how to run stuff. But that soon developed into a much more general system, which is that any person with any biological problem of any kind who felt that they would be at home in this atmosphere was welcome to work here on any kind of organism provided there was space and money for materials. Sometimes that was provided by the people themselves but more often than not it came out of our general slush fund.

We had more support than we actually needed. The work planned in the grant applications was always a small part of what was done. Indeed, the work reported in those progress reports that got us the renewals often had very little resemblance to what was claimed was going to be done in the previous research grant. The renewal gave progress reports of all the work that was actually done in the laboratory by the great variety of people who were graduate students and postdoctoral fellows, and then a plan would be made for a follow-up of some particular part of it, but then again, a large number of new things would be done. The result was that the lab supported work on the fish of Lake Victoria, on human Y chromosomes, on fungi, on plants, on snails, on fruit flies, and so on, and so on. The principle, if there was one, was that any investigations in population genetics and species differentiation would contribute to the general solution of the problems of the field.

The professor was available for consultation, if someone wanted to talk with him, but I never had the habit of going to people's rooms once a week and asking to see what they were doing and how things were going. My attitude has always been that we are a collection of independent persons who know what they're doing, and if they do not and they are too dumb to ask somebody, then too bad for them. Our seminar series essentially follows along those lines. Anyone who wants to give a talk in our lab about anything is welcome to do so. I only wish there were more talks about philosophy, sociology, and history.

I don't particularly enjoy the role of mentor. I like the company of adults. I have taken many graduate students, but if I'd had only postdoctoral fellows, that would have been ok. I don't really have a lot to do with graduate students when they first arrive – only after they're folded in. It really has worked very well, and it has demonstrated that, aside from doing the administration of getting grants and keeping the lab doors open, the professor is barely necessary.

This gets on to the whole issue of how scientific work is organized, ought to be organized, quite aside from the problematic of population genetics. Our students have generally been somewhat older than the average, and many of them have come from outside of biology. Many of them came from computer work. What a laboratory should do is to provide the infrastructure and a varied intellectual atmosphere and let people do what they can. If you run such an organization, letting into it only people who seem a bit more mature than the just-graduated senior from college, it works very well. In fact, many of our just-graduated seniors from college are 5 or 10 years older than the usual just-graduated senior, many of them having dropped out of one college after another until they found out what they wanted to do. One graduate student seemed to have failed out of every state college in the California system before he arrived here, but it did not stop him from being a first-class scientist.

There is another issue in the sociology of laboratory life that is an interesting general one, and that is, how one handles the objective asymmetry of power relations between the professor in the laboratory and everybody else. In many labs, there is some attempt at camaraderie, beer drinking, and so on, but this seems to me to be a veil that is drawn over the real social relations, which are a reflection of the asymmetry of power. The professors cannot be, in fact, really friends to their students, because only peers can be friends. They can be nonexploitative. They can be considerate. They can go out of their way to help or to stay out of the way. They can do all of those things, but they cannot create a peer relationship where one objectively does not exist. Obviously, that is not the same as saying that professors cannot be instructed by their students. Of course, they not only can but are, but that simply says that people who are in a lower status and lower power relationship may nevertheless be more perceptive and more intelligent than the people above them. I am really not much satisfied with what has been written on the social structure of laboratory life, and I think it would be worth a whole new careful look.

2.10 The Political Economy of Agriculture

Q: You've described the Ford Foundation project on "The Political Economy of Agricultural Research" as perhaps the most satisfying project you've undertaken. What was it, and why was the Ford Foundation interested in supporting such work?

RCL: The question being addressed was, What are the forces that influence the directions taken by agricultural research – especially public agricultural research? Why does the agricultural research community do this instead of that? My point of view was that research is a factor of production, and the usual standard analysis of such a factor of production is that there is a production possibility frontier; what technological change does is to push the frontier outward, and the solution taken is at the farthest-out frontier. But the question, How come the frontier went that way instead of that way? was never asked. So our initial question was, What is it that directs what agricultural research has done? The desire to ask that question came out of a political issue, which was what was going on in agriculture in the Third World.

The Ford Foundation reasoned: We're not sure that the kind of agricultural research being supported by the Foundation is really accomplishing what we need. What we need is completely out of left-field – a literally left-field view of this – so we're going to take these two radicals, Levins and Lewontin who have a good track record of doing science and analysis, whose politics are rather odd, and we'll agree that they should look at the problem and maybe we'll learn something. It was quite clever on the part of the Foundation; the idea that you should ask someone wacko. So they gave us a lot of money through the Harvard School of Public Health.

Dick Levins and I then went out and recruited people of various kinds, including Sam Baker, who got his Ph.D. in economics at Harvard and is now a professor at the University of South Carolina. We recruited Flavio Valente, a Brazilian nutritionist; Dick Franke and Barbara Chasin, two sociologists who wrote a book about the Sahel; and Najwa Makhoul, a Palestinian radical sociologist with a degree from M.I.T. She did field work for us on the Israeli system, which is one of the most interesting systems in support of agricultural research.

We organized ourselves into field teams. My idea was that we should actually go and interview people who had done agricultural research at various institutions around the country and around the world. The study consisted of going to agricultural research stations in the South, the Northeast, the Midwest, and Far West and included a special group of Black agricultural research institutions and a study of the national agricultural research projects in France (which I did together with Jean-Pierre Berlan) in Brazil, in Israel, and the Sahel project.

We sat in this office, and we made an outline of how you would do a political–economic study of agricultural research. We were very interested in labor relations, in power relations within the working units, and where the money came from. We had this complete articulated outline to make a set of questions that we would probe and ask people. We talked to deans, we talked to agricultural research workers, we talked to research assistants, we talked with people filling different roles and asked them their view of the matter, where the money came from, why they did what they did, who the boss was, and how much supervision there really was, to develop a whole picture of the situation. We also made a complete study of the history of agricultural research supported in the United States, and then we took certain specialty things, like hybrid corn and chickens. I think finally we lit on hybrid corn and chickens because they tested certain hypotheses about the reproduction of capital.

Finally, when we got through all of that and we'd asked all our questions, we found ourselves in a new ballpark. We were no longer primarily concerned with how agricultural research decisions were made. It became a study of the penetration of capital into a sphere of production, which, up until that time, had been petty production. The problematic became, How can we explain in capitalism, in world capitalism, the existence of a huge sphere of production in the hands of millions of petty producers? What the hell ever happened to capital? And we found out what happened to capital.

We have a story, the penetration of capital, which is right, which really makes the thing work, and every prediction we've made has come true. Farmers are becoming more and more proletarianized. More and more they are working as the providers of labor. Our analysis finally was that what agriculture had become was the putting-out system, where you owned some of the means of production, namely the sewing machine on your shoulder, and when the boss called you to work, you went with the sewing machine, and when he was finished, he sent you home and you didn't work for weeks. Tough luck. But he provided some of the inputs, and you provided the labor plus a small piece of machinery. It's become increasingly the case that farmers provide the labor and the land, and that's all. Everything else is a capital input, which is purchased from the oligopolistic providers, and the output is purchased usually now on contract. Now farmers are just laborers who happen to own the land, if they own it – if they have title. And so the project became a study of the penetration of capital into a highly dispersed sector of production.

It was also a study of the way in which explicit political ends can shape an agricultural research program. In Israel, everything had to meet several political demands having to do with relations with Arabs, as contrasted with the anarchy of research in the United States, where you have a chronic crisis of overproduction in agriculture since the middle of the nineteenth century, which is relieved only occasionally by a good war or export of a lot of stuff. There's an increasing contradiction in a research system that has as a constant goal improving agricultural productivity, which has as an outcome exaggerating the crisis of overproduction, which results in falling prices, which in turn increases pressure on farmers to produce.... It really shows the consequence of an anarchic system.

Let me give you a wonderful example, predicted by the whole analysis. In Vermont the farmers got together 2 years ago and decided not to use bovine growth hormone because the price of milk on the market was already so low that they were frightened to death of an increase of milk supply. So they refused to use bovine growth hormone, the idea being to keep the supply down a little so they could keep the price up. But what they didn't realize was that most milk isn't used for fresh milk. Most of the milk they were providing was used for cheese. So the Vermont cheese manufacturers simply went to New York, where the farmers were using bovine growth hormone and strangling themselves by increasing their productivity. But what choice does an individual farm producer have? If everybody else is going to do it, then I have to do it. There's no cooperative equilibrium, and therefore every farmer has to do what kills him. That's been the story of U.S. agricultural research: increasing the penetration of capital, decreasing the margin of profit to the individual farmer, increasing the rate of bankruptcy, increasing the rate of farm failure. And that's not just the last 10 years. That's a continuous process.

My own involvement derives from two sources. One is a political motivation to try to use what methodologies and substantive knowledge I had to understand the workings of capitalism. The other is a more academic interest in historical and political issues quite aside from any ideology. The actual study itself has produced extremely interesting results about the way in which capital has penetrated the last major section of production, which seems to be out of historical synchrony with the development of capitalism. To the extent that we are historians interested in how social relations and relations of production develop over time, this is a very important sector of inquiry, and it is bound to be deeply affected in the inquiry by ideological presuppositions of the investigator.

A lot of what has been written about agriculture and its development seems to me to make little or no sense. Some of it is very romantic, and some of it is apologetic, but virtually none of it takes a purely analytic approach beginning with an explicitly asserted analysis of the whole way in which capital operates. Of course, a Marxist approach is conducive to that latter kind of view since it tends to begin with certain a priori assertions about the nature of the relationships and tries to follow through on them. What I have intended to do in the analysis of agricultural research is to use a specifically Marxist economic framework of an understanding of capitalism and the penetration of capital and the nature of commodities, and so forth, to understand how research has been an element in the subjection of agricultural production to the general dominion of capital investment.

In some ways, I feel that the work on agricultural research is the most important intellectual work that I myself have participated in, and in many ways the most intellectually rewarding. On the other hand, I recognize the very great importance that an a priori standpoint makes in the analysis of such a system and, therefore, the lack of a sort of cleanness of the result that you get because it can never be freed of a generally ideologically formed stance about the way that the economy works. It would be interesting to discuss whether any form of evaluation could be carried out. For example, we make certain predictions about the future direction of plant breeding (predictions which we claim are coming true) having to do with the nature of contracts, genome control, and so on, which we say would not be predicted by the more usual view of the development of agriculture and its research. For example, a very large amount of effort is put by Pioneer Hybrid Seed Company into what is called genome control, that is, the identification by DNA tests of the genomes of plants. This is then used not as a method of plant breeding but to ensure property rights in the varieties that farmers actually plant in their fields. When this is combined, as Monsanto has done, with a contract form of farming, which gives Monsanto complete control over all the inputs and the harvesting and allows Monsanto to decide whether the crop will be harvested or whether it will be left to the producer to do with it what he pleases, I would claim that these developments are only explainable in the terms that have been projected by our Ford Foundation project.

The problem for plant breeders has always been how to acquire property rights in the product of their work. In the old days, there was no way for a breeder to acquire property rights because once the farmer got the seed he could propagate it, and you were out in the cold. Hybrid corn was a deliberate move to secure property rights for the producers of the seed; it was designed consciously to be a method of copy protection. Again a prediction has been coming true – namely, that there is now a change in the method of plant breeding from the madness for hybrids, which appeared in tomatoes and chickens, into a totally new method of capturing property rights. And it works.

I'll tell you how I know this. I went on a new field trip in the Midwest to update my information on biotechnology before starting to write the book. And when the people at Pioneer Hybrid, the seed company, heard that I was coming to Minnesota, they all came to see me, including the current and former directors of research. And I said: "What are you guys doing in biotech"? They said, "Oh, our big pressure is for genome control. Steve here is in charge of the genome-control program." What's that? Genome control is a program of identifying by DNA fingerprints the genome of the material released by different companies. I ask: "Why do you want to do that"? The answer is as follows: You release a genetically engineered variety – whether it's hybrid doesn't matter now you don't have to go to the trouble of making a hybrid – these genetically engineered varieties of soybeans, which are not hybrids; potatoes, which are not hybrids. At the same time, when we sell the seed to the farmer, the farmer signs a contract. The contract says, "I promise I will not give seed of this to anybody else; I'll only use it myself. In addition, I promise not to use it for my next year's crop." If you're a potato producer, how do you make potatoes? You cut up potatoes to produce pieces; they're called seed potatoes. If you buy my variety, you're not entitled to use that for seed potatoes next year.

What's important is genome control. You send enforcement people into the field next year. You just walk by and pick up a potato – that's all you have to do – and bring it back to the lab; they do a little genome fingerprinting on it, and if it has your fingerprint, you sue the hell out of the farmer, and they did. Monsanto has the grip on potatoes for potato chips – one of the biggest uses of potatoes in America. They use a special variety of potato, with high carbohydrates that resists sopping up grease, makes a light potato chip. It's all genetically engineered. All those potatoes are sold on contract. And the contract – talk about proletarianization – the contract normally stipulates that you're not allowed to use the potatoes for seed the next year. It includes the fact that they provide all the inputs in addition to the seed, they provide the chemicals and so on, and they do they the harvesting. They have big potato harvesters. And the understanding of the contract is, if they don't come and harvest your potatoes because the factory already has all the potatoes it needs – because they like to control the supply in order to control the price – if they don't harvest your potatoes, they're yours. We didn't harvest them; they belong to you. I don't know what you're going to do with them. Now, this was our prediction of what would happen in one form or another, and it's happening. So hybrids are no longer the way you protect property rights. It's genome control plus contracts, and those contracts are upheld in the courts.

If any potato farmer could succeed in getting away with it and getting a potato from this year's crop to use for next year's crop, Monsanto would be totally out of the picture – and they are not out of the picture. So, one of the satisfying things about this work has been that it has actually enabled us not just to make an analysis, but to make predictions about what's going to happen that turned out to be true.

Q: Was there a link between this project and your involvement in the 1979 California Rural Legal Assistance "mechanization" lawsuit against the University of California?

RCL: The first time I ever was an expert in court was as a result of this agricultural work. The California Agrarian Action Project, a public interest group, put together an alliance of farm workers, small farmers, and consumers to sue the University of California claiming that the research done by the State Agricultural Experiment Station at the University of California was not fulfilling their legal mandate, which was explicitly to benefit small farmers, consumers, and farmworkers. Their claim was that all the research being done by the University of California was antifarm-worker, objectively against the interests of small farmers and consumers, and all in favor of the biggest-capital farmers, and that the University of California trustees included some of the big-capital farmers, and that was no accident. So there was a trial in Oakland county. The California Rural Legal Assistance Program, through its legal counsel, Public Advocates, got together a group of several experts to testify on behalf of the plaintiffs. That included Bill Friedlander from Santa Cruz, me as director of this project – that was my credential in a court of law, that I was director of the project and a professor at Harvard – and a few other people in a loose confederation called the New World Agriculture Group.

I learned a lot. Testifying was the most psychically stressful thing I'd ever done in my life. Of course I was prepped in advance by Public Advocates and deposed for a whole day by the lawyers for the defendants. That deposition was offered in court, and I testified for 2 days. Well, what happens if I say something on the stand that's contradicted by the deposition? They'll throw me in jail for perjury! That's why it was so stressful. For the first time in my life I had to appear in a public situation speaking words such that every word, as far as I was concerned, counted: no hand waving and approximations and saying later, "oh, I didn't mean that"; it had to be right. That and the fact that I had been prepped not to volunteer any information – exactly the opposite of what my tendency would be.

But it was a great revelation. One amazing thing was that I wasn't Red-baited. Also, I discovered the immense importance of status in that circumstance – greater than any other circumstance I've ever been in. An hour went by to qualify me as an expert. Not only my curriculum vitae but all kinds of other stuff [was presented] to prove to the jury that I was an expert so that my words could have credibility. I explained the structure of the program and then explained that Professor Baker had done certain interviews and that our analysis was based on his field notes. The first question I'm asked on cross is, "is Professor Baker an economist"? I say "yes." "Is he on the faculty of Harvard? Where is he a professor"? I say, "University of South Carolina." And you could see the smile on their faces. The fact that I was a named professor at Harvard meant a lot. On the other hand, I had to be qualified because I don't have a degree in economics from Harvard. So I had to explain that I worked for 4 years, in an agricultural research institution. And finally I was qualified without objection. We had 2 days of direct, cross, and redirect.

In particular I had been asked to go through a large numbers of proposals made to the University of California by the Extension Research Division, and then I was asked about them; for example, what do you find in these research proposals that supports the contentions of the plaintiffs? I said: "Now here's an interesting research proposal. It says, we wish to do research on dwarf fruit trees. When I look at the research proposal it says, reason for doing research: in order to be able to utilize a different quality of labor." Or spraying schools. (You're obliged by the USDA to have a spraying school if you spray insecticide; you're supposed to have a program to educate workers that insecticides are poisonous.) And in the report it says, well, we had a lot of trouble with these spray schools because the workers were constantly bringing up real or imaginary complaints. The smoking gun was right in the documents because these people never imagined that anybody else would read this stuff. The antilabor bias of farmers is transferred ideologically to research workers, many of whom are the children of farmers who wish they were farmers but can't afford to farm. That's what our research brought out, that they identified, not with the worker, not with the consumer, but with the farmers who were employers of labor. And the whole thing was just laid out. So we won the case.

Then the hard part came. As you know, the hardest thing isn't winning the case. It's designing the remedies. We were cautioned over and over again by the attorneys that we shouldn't get into issues that might sound like we were interfering with academic freedom. What are the remedies that will not interfere with academic freedom but would guarantee, or at least push the bias in some other direction? Not easy to do. (By the way, there were many amicus curiae briefs filed, including by other state experiment stations – all on the side of the defendants, of course.)

We tried to imagine, for example, having a group of people representing the different constituencies who would read the research proposals and not rank the proposal on the basis of its particular scientific merits – because that would interfere with academic freedom – but would certify to the court either that they found a continuing bias or they didn't. The trouble with that remedy is, How do you choose the representative of the constituencies? Who is to represent the farm workers? Who is to represent the consumers? Who is to represent the farmers?

2.10.1 Interviewers' Note on the Outcome of the Case

The remedy that was ultimately approved created (1) five constituency advisory boards (representing small farmers, family farmers, farm labor, rural residents, and consumers) to advise the director of the local experiment station, (2) a program for agricultural research and methodology assessment on each of the three agricultural campuses, and (3) a university-wide faculty committee to oversee implementation. However, the University of California appealed the ruling, and in 1989 it was reversed. Thus, the remedy was never implemented.[11]

Notes

1. Vetukhiv, M. A. 1953. Viability of hybrids between local populations of *Drosophila pseudoobscura*. *PNAS* 39:30–34.
2. After Matthew 25:29–30: "For unto every one that hath shall be given, and he shall have abundance: but from him that hath not shall be taken away even that which he hath."
3. Schmalhausen, I. I. 1949. *Factors of Evolution*. Philadelphia: Blackiston.
4. Osborn, F., and Bajema, C. J. 1972. The eugenic hypothesis. *Social Biology* 19:337–49.
5. Clausen, J., Keck, D. D., and Hiesey, W. M. 1940. *Experimental Studies on the Nature of Species* I. Carnegie Institute of Washington Publication 520:1–452.
6. Lewontin, R. C. 1997. Dobzhansky's *Genetics and the Origin of Species*: Is it still relevant? *Genetics* 47:351–5.

7. Hubby, J. L., and Lewontin, R. C. 1966. A molecular approach to the study of genic heterozygosity in natural populations. I. The number of alleles at different loci in *Drosophila pseudoobscura. Genetics* 54:577–94; Lewontin, R. C., and Hubby, J. L. 1966. A molecular approach to the study of genic heterozygosity in natural populations. II. Amount of variation and degree of heterozygosity in natural populations of *Drosophila pseudoobscura. Genetics* 54:595–609.
8. Kamin, L. 1974. *The Science and Politics of IQ.* Potomac, Maryland: Lawrence Erlbaum.
9. Kreitman, M. 1983. Nucleotide polymorphism at the alcohol dehydrogenase locus of *Drosophila melanogaster. Nature* 304:412–417.
10. Lewontin. R. C. 1997. The cold war and the transformation of the academy. In *The Cold War and the University: Toward an Intellectual History of the Postwar Years.* New York: The New Press, 1–34.
11. Brabo, L. 1998. The mechanical tomato harvester and the CRLA suit against the University of California. Unpublished manuscript.

CHAPTER THREE

Hannah Arendt and Karl Popper

Darwinism, Historical Determinism, and Totalitarianism

JOHN BEATTY

> Just as Darwin discovered the law of evolution in organic nature, so Marx discovered the law of evolution in human history.
>
> Friedrich Engels at the graveside of Marx

> Underlying the Nazi's belief in race laws is Darwin's idea of man as the product of a natural development . . . , just as under the Bolsheviks' belief in class-struggle as the expression of the law of history lies Marx's notion of society as the product of a gigantical historical movement.
>
> Hannah Arendt, *Origins of Totalitarianism*

3.1 Introduction

Darwin's theory of evolution and Marx's theory of history have been compared and contrasted many times, in many respects, approvingly and disapprovingly. These diverse assessments reflect, in part, equally diverse interpretations of Darwinism and Marxism.

Take Darwinism. There is today considerable discussion of the indeterminism and nondirectionality of Darwinian evolution. On Gould's influential account, if the history of life could be replayed, natural selection would produce different results every time (Gould 1989). And as Lewontin has long argued, this is not just a matter of what might have been; evolution by natural selection *does* lead to very different outcomes under very similar conditions (Lewontin 1966, 1969, 1974, 1990).

This notion has a distinguished past dating back to Darwin himself (1862, pp. 282–93; 1872, v. 1, pp. 241–4). But evolutionary indeterminism has co-existed fitfully with a more deterministic view of evolution: namely, that natural selection inevitably leads to directional change, even progress (Ruse 1996). On this view, progress in the organic world is as inevitable as increase in entropy in the physical world (Fisher 1930, p. 37).

Similarly, there has been considerable controversy as to whether, or to what degree, Marxism implies a deterministic, directional, and progressive view of history. This is in part related to the distinction between Marxism as a lawlike theory of history and Marxism as a call for change. The question is often posed, Why exhort the proletariat to struggle against the bourgeoisie if the success of the former is historically guaranteed – if societies are moving in that progressive direction anyway?

Without consensus concerning the determinism or indeterminism of Darwinism and without consensus concerning the determinism or indeterminism of Marxism, we are not likely to agree on whether to compare or contrast Darwinism and Marxism in this respect.

Nor is this just an abstract philosophical issue. Different interpretations of Darwinism, Marxism, and their relationships also reflect the different historical contexts of their interpreters. For example, Marxists of the Frankfurt School (and other "critical theorists") stressed the *noninevitability* of emancipatory revolution and hence the need for voluntary initiative. This was in part a reaction to the rise of Nazism and Stalinism and the Hitler–Stalin pact, all of which suggested that the sort of socialist state these Marxists desired was hardly imminent (Held 1980, pp. 16–23). These Marxists generally downplayed connections with Darwinism, which some of them identified with a deterministic view of history that leaves little room for volition (e.g., Horkheimer [1947] 1974, p. 125).

Indeed, *anti-Marxists* found it convenient to link Marxism with a deterministic interpretation of Darwinism. This link was increasingly promoted, beginning in the late 1930s, in response to the rise of fascism and communism. At bottom, it was argued, fascism and communism were species of the same genus: totalitarianism. What totalitarian ideologies supposedly had in common was a sinister historical determinism that paraded as progress and that lulled people into acceptance of their dismal fates by convincing them that they were powerless to avoid the inevitable. Several prominent critics cited evolutionary thought, and Darwinism in particular, as the source of, or at least reinforcement for, the historical determinism underlying totalitarianism.

The question of the totalitarian connection between Darwinism and Marxism is the central topic of this chapter. I will focus on two influential proponents of the connection: Hannah Arendt, especially in her book, *The Origins of Totalitarianism* ([1951] 1968), and Karl Popper in his antitotalitarian works, *The Poverty of Historicism* ([1944–45] 1957) and *The Open Society and Its Enemies* ([1945] 1966).

Arendt's and Popper's works fit into a genre of interwar and post-World War II literature linking science, and especially evolutionary biology, to the rise of totalitarianism. Some authors stressed the materialistic implications of evolutionary biology (Barzun [1941] 1958, Hayes 1941). Arendt, Popper, and others emphasized the historically fatalist implications.

I am pleased to offer this small contribution to the festschrift honoring the Marxist Darwinian, Richard Lewontin, who has an important place in the changing interpretations of Darwinism and Marxism. Lewontin acknowledges that nineteenth-century Darwinism and Marxism were directional, progressive theories of change. He himself has worked hard to explicate and support a different view of evolution that is in the tradition of Darwin, Marx, and Engels, but that is nondirectional and nonprogressive. His view is relatively nondeterministic in several important respects. For example, he criticizes the assumption – implicit in evolutionary equilibrium analyses – that there are evolutionary states that will be converged upon regardless of differences in starting points or historical pathways. He also emphasizes that there are no timeless laws of biology; all supposed "laws" of biology have evolutionary histories, including the "laws" of evolution themselves (e.g., Fisher's "fundamental theorem" of natural selection and presumably all other maximization and minimization principles underlying evolutionary equilibrium analyses). More generally, "What characterizes the dialectical world, in all its aspects . . . is that it is constantly in motion. Constants become variables, causes become effects, and systems develop destroying the conditions that gave rise to them" (Levins and Lewontin 1985, p. 279). So there are no certain evolutionary outcomes, and no universal rules of directional (much less progressive) evolutionary change (Lewontin 1966, 1969, 1974, pp. 269–71; [1977] 1985, 1991).

In turn, Lewontin's relatively nondeterministic interpretations of Darwinism have been used to support nondeterministic explications of Marx's view of history (e.g., Miller 1984, p. 234 invoking Lewontin 1974).

Lewontin's Darwinism could hardly be more different from the version that I am considering here: the historically deterministic Darwinism of the critics of totalitarianism. This version of Darwinism fit exceedingly well with concurrent materialist interpretations in helping to make sense of the fatalist and materialist interpretations of totalitarianism. An appropriately odious Darwinism helped to explain the odiousness of totalitarianism.

3.2 Arendt and Popper: Influence of and Influences on

The Foreign Affairs 50-Year Bibliography (Dexter 1972, covering the years 1920–1970) had this to say about the context and significance of Arendt's totalitarianism:

> This has probably been the most influential single book on the theme of totalitarianism. It appeared at that excruciating juncture in the twentieth century when the task of absorbing and comprehending the horrors of the barely vanquished Nazi régime was compounded by the dark prospect of forthcoming conflict with Stalinist Russia [quoting Arendt]: "This moment of anticipation is like the calm that settles after all hopes have died." Linking the Nazi and Stalinist phenomena as essentially identical and as transcending all traditional concepts of "left" and "right," Miss Arendt was instrumental . . . in preparing the way for a whole series of studies of totalitarianism." (p. 45)

The sense of despair referred to (that "all hopes have died"), and to which Arendt was reacting, was attributed by her to the "historical necessity" associated with totalitarianism (Arendt [1951] 1968, p. viii). That is, it was supposedly useless to hope for a future other than the fate prophesied by totalitarian ideology. At best, one could choose between the different "chains of fatality" (p. 345) associated with different versions of totalitarianism.

Popper is also recognized as having been a very influential theorist and critic of totalitarianism. For example:

> Karl Popper employed a radical critique of Plato, Hegel and Marx to indict what he called "utopian social engineering" in modern "totalitarian" regimes in Germany and Russia – a procedure reproduced in recent times by André Glucksman, Bernard Henri-Lévy and other European "new philosophers" disappointed in Marxism's twentieth-century incarnations in the Soviet Union and elsewhere. Popper, in fact, set the tone from the outset for the ideological use of the idea of totalitarianism, which would become a significant weapon in the West's arsenal of political rhetoric. (Benjamin R. Barber, "Totalitarianism," in *The Blackwell Encyclopedia of Political Thought,* 1987)

Even more so than Arendt, Popper associated totalitarianism primarily with historical determinism. For instance, he dedicated *The Poverty of Historicism* to "the countless men and women of all creeds or nations or races who fell victims to the facist and communist belief in Inexorable Laws of Historical Destiny."

As I mentioned, and as I will discuss in more detail, both Arendt and Popper associated this historical determinism in no small part with Darwinian evolutionary thought. To the best of my knowledge, this aspect of their work has received very little attention. For example, the most important and extensive analysis of Arendt's political philosophy by Margaret Canovan (1992) includes no discussion of her references to evolutionary thought. John Watkins barely mentions, and in the process seriously underestimates, the political context of Popper's concerns about Darwinism (1995, p. 193).

Arendt's and Popper's views on the connections among totalitarianism, fatalism, and evolutionary thought stand out more when compared. I will address the similarities between their views in the next section. For now it is worth noting some common influences. Not surprisingly for a Prussian/German and an Austrian (respectively), Arendt and Popper were both well aware of the nineteenth-century German naturalist and popularizer of Darwinian evolutionary ideas (and promoter of pan-Germanism), Ernst Haeckel. In the German-speaking world, in the late nineteenth and early twentieth centuries, "Darwinismus" was truly a worldview, and Darwinismus was Haeckel's version of Darwinism (Gasman 1971). Both Arendt and Popper identified Haeckel with an influential if "crude" "form" or "brand" of Darwinism (Popper [1945] 1966, v. 1, p. 61; Arendt [1951] 1968, p. 159).

Haeckel was particularly well known, or infamous, for his historical determinist thesis about the implications of Darwinism for human history: namely, that human history was governed by the "iron law" of natural selection (Haeckel [1899] 1900, pp. 270–1; Gasman 1971, pp. 34–5). Thus, one might link Darwinian evolutionary thought and historical determinism.

To get a glimpse of how Arendt and Popper might have further linked Darwinian historical determinism, through Haeckel, to fascism on the one hand and communism on the other, consider two more facts about Haeckel's reputation. First, for his evolutionary views, he was frequently eulogized in the Third Reich as a forerunner of National Socialism (Gasman 1971, pp. 170 ff.; Arendt [1951] 1968, p. 159, n. 5).[1]

Second, consider that Haeckel, along with Darwin, had a prominent place in the writings of Friedrich Engels. To be sure, Haeckel was often Engels's target, but he was just as often Engels's authority on evolutionary issues and was certainly considered by Engels to be a representative Darwinian (e.g., Engels [1885] 1939, pp. 75–85; [1927] 1940). Moreover, although Engels had objections to Darwinian evolutionary theory, he nonetheless believed that there was an important connection between Darwin's and Marx's achievements. As he remarked in his famous eulogy at the graveside of Marx: "Just as Darwin discovered the law of evolution in organic nature, so Marx discovered the law of evolution in human history" ([1883] in Marx and Engels 1978, p. 16; Arendt [1951] 1968, p. 463). Indeed, as George Lichtheim suggested in connection with Engels's influence, " 'From Hegel to Haeckel' might serve as a summary of the evolution of Marxist thinking between the 1840s and the 1880s" ([1961] 1965, p. 244).[2]

Add all that together and you have a preview of the (supposed) link between evolutionary historical determinism and totalitarianism.

That link was not only perceived by Arendt and Popper but also by the Austrian economist Friedrich von Hayek, whose analysis influenced Arendt and Popper. They both reacted to his analysis of "scientism" – "the slavish imitation of the method and language of science" – and what he called "historism," an outgrowth of scientism devoted to the pursuit of laws of history (Hayek 1942–44, part 1, p. 269; part 2, pp. 50–63; also 1941). Hayek complained that "the idea of evolution" was often invoked as "a kind of incantation" in support of historism (Hayek 1941, p. 312; Arendt [1951] 1968, pp. 345–47; Popper 1957 and [1945] 1966, passim).

Connections between totalitarianism, historical determinism, and evolutionary thought were, as I mentioned, also being pursued by other public intellectuals during the same period, perhaps most prominently by Jacques Barzun in his book *Darwin, Marx, Wagner* ([1941] 1958) although Barzun, together with his mentor Carleton Hayes (*The Age of Materialism*, 1941), put more emphasis on materialism than historical determinism as the principal connection between evolutionary thought and totalitarianism.[3] Isaiah Berlin also linked Darwinism to historical determinism, and the latter to the rise of totalitarianism (e.g., [1953] 1996, p. 11; and more implicitly in [1954] 1969, pp. 109–10).

Interestingly, especially considering the similarities to be explored here, Arendt and Popper never referred to each other – in any connection. The histories of their publications and revisions overlap, and thus mutual reference was possible. It seems most likely that they independently perceived and elaborated the connections between evolutionary thought, historical determinism, and totalitarianism.

3.3 Connecting Fascism and Communism through Evolutionary Determinism

Arendt and Popper both emphasized that evolutionary thought had been invoked in support of the idea that there are inescapable, scientific laws of history – a position that Popper called "historicism," following Hayek's use of the term "historism."[4] According to Arendt and Popper, evolutionarily inspired historicism was a central assumption of all versions of totalitarianism, not only those versions concerned with the rise and fall of races, but also those concerned with the rise and fall of classes.

Here is an example of the point in Arendt's own words:

> [A totalitarian government] claims to obey strictly and unequivocally laws of Nature or of History....
>
> Underlying the Nazi's belief in race laws [concerning the inevitable domination of particular races] is Darwin's idea of man as the product of a natural development..., just as under the Bolsheviks' belief in class-struggle as the expression of the law of history lies Marx's notion of society as the product of a gigantical historical movement. (Arendt [1951] 1968, pp. 461, 463; my emphasis)

(It is worth noting the structural similarity between Arendt's point and Engels's eulogy of Marx – quoted above – of which Arendt was well aware.)

Popper makes basically the same connection, as seen in the following passages:

> Thus the formula of the fascist brew is in all countries the same: Hegel plus a dash of nineteenth-century materialism (especially Darwinism in the somewhat crude form given to it by Haeckel). The "scientific" element in racialism can be traced back to Haeckel, who was responsible, in 1900, for a prize-competition whose subject was: "What can we learn from the principles of Darwinism in respect of the internal and political development of a state?" The first prize was allotted to a voluminous racialist work by W. Schallmeyer, who thus became the grandfather of racial biology. [The basic idea of fascism–racialism] is that degeneration, particularly of the upper classes, is at the root of political decay ([1945] 1966, v. 2, p. 61)

> Historicism... the doctrine that history is controlled by specific historical or evolutionary laws... can be well illustrated by... the doctrine of the chosen people.
>
> [The value of this doctrine] can be seen from the fact that its chief characteristics are shared by the two most important modern versions of historicism... the historical philosophy of racialism or facism on the one (the right) hand and the Marxian historical philosophy on the other (the left). For the chosen people racialism substitutes the chosen race..., selected as the instrument of destiny, ultimately to inherit the earth. Marx's historical philosophy substitutes for it the chosen class, the instrument for the creation of the classless society, and at the same time the class destined to inherit the earth. ([1945] 1966, v. 1, pp. 8–9)

These passages may suggest that evolutionary thought plays a merely analogical role in linking fascism and communism: Darwinism or some version of it is the historical science underlying fascism, just as Marxism is the historical science underlying communism. But Arendt and Popper both argued for the existence of a more substantial connection between evolutionarily inspired fascism and Marxist-inspired communism.

Arendt seems to have relied heavily on Engels's understanding of the relationship between Darwin's and Marx's thought (e.g., see the following quotation; see also Arendt [1953] 1993, p. 377). In any case, her reading of Marx and (or through) Engels, suggested to her that Darwin's and Marx's views were instances of a more general, supposedly scientific theory of history:

> The difference between Marx's historical and Darwin's naturalistic approach has frequently been pointed out, usually and rightly in favor of Marx. This has led us to forget the great and positive interest Marx took in Darwin's theories; Engels could not think of a greater compliment to Marx's scholarly achievements than to call him the "Darwin of history." If one considers, not the actual achievement, but the basic philosophies of both men, it turns out that ultimately the movement of history and the movement of nature are one and the same. Darwin's introduction of the concept of development into nature . . . means in fact that nature is, as it were, being swept into history, that natural life is considered to be historical. The "natural" law of the survival of the fittest is just as much a historical law and could be used as such by racism as Marx's law of the survival of the most progressive class . . . . Engels saw the affinity between the basic convictions of the two men very clearly because he understood the decisive role which the concept of development played in both theories. (Arendt [1951] 1968, pp. 463–4)

For Popper, as for Arendt, the connections between evolutionarily inspired fascism and Marxist-inspired communism are more than just analogical. But for Popper the connections are not quite as substantial. For example, Popper sometimes distinguishes between the "natural laws" underlying fascism–racialism and Marx's "economic laws" underlying communism (Popper [1945] 1966 v. 1, pp. 9–10). Nonetheless, he more often chastises Marx for pursuing and claiming to find – in Marx's own words – a "natural law" of society that governs the "natural phases of its evolution" (Marx [1867] 1967, p. 20; discussed in Popper [1944–45] 1957, p. 51; [1945] 1966, v. 2, pp. 86, 202).

For Popper, Marxism was not an extension of Darwinism, but rather Marxism and Darwinism were both extensions of the sort of positivistic, lawlike approaches to history promoted in the nineteenth century by Auguste Comte and John Stuart Mill. By lending credence to the positivists' program, Popper believed that Darwinism also lent support to Marxism:

> The belief . . . that it is the task of the social sciences to lay bare the law of the evolution of society in order to foretell its future . . . might be perhaps described as the central historicist doctrine . . . . [It] is the same view that gives rise to the pro-naturalistic – and scientistic – belief in so-called "natural laws of succession"; a belief which, in the days of Comte and Mill, could claim to be supported by the long-term predictions of astronomy, and more recently, by Darwinism. Indeed, the recent vogue of historicism might be regarded as merely part of the vogue of evolutionism. (Popper [1944–45] 1957, pp. 105–6)

Thus, although Marxism is not just a variant on Darwinism, it rode the coattails of Darwinism.[5]

3.4 What Stalinist Purges and Eugenics Have in Common

Another respect in which Arendt and Popper associated totalitarianism, historical determinism, and evolutionary thought is with regard to the means by which the predictions of historicist theories are guaranteed. Totalitarian ideologies, according to Arendt and Popper, claim to be based on laws of history but are not; there are no historical laws but only claims that parade as such. Predictions based on those claims come true only to the extent that their subjects can be compelled to cooperate in securing the supposedly inevitable outcomes. Quoting Arendt, "totalitarian government can be safe only to the extent that it can mobilize man's own will power in order to force him into that gigantic movement of History or Nature" described by the totalitarian ideology (Arendt [1951] 1968, p. 473). This compulsion or mobilization often takes the form of terror, which "executes on the spot the death sentences which Nature is supposed to have pronounced on races or individuals who are 'unfit to live,' or History on 'dying classes,' without waiting for the slower and less efficient processes of nature or history themselves" (p. 466).

> In both instances the same objective is accomplished: the liquidation is fitted into a historical process in which man only does or suffers what, according to immutable laws, is bound to happen anyway. As soon as the execution of the victims has been carried out, the "prophecy" becomes a retrospective alibi: nothing has happened but what had already been predicted. It does not matter whether the "laws of history" spell the "doom" of the classes and their representatives, or whether the "laws of nature . . . exterminate" all those elements – democracies, Jews, Eastern subhumans (*Untermenschen*), or the incurably sick – that are not "fit to live" anyway. (Arendt [1951] 1968, p. 350)

Thus, Stalinist purges and eugenics take over where Marxism and evolutionism fall short.

> Stalin, in the great speech before the Central Committee of the Communist Party in 1930 in which he prepared the physical liquidation of intraparty right and left deviationists, described them as representatives of "dying classes." This definition not only gave the argument its specific sharpness but also announced, in totalitarian style, the physical destruction of those whose "dying out" had just been prophesied. (Arendt [1951] 1968, p. 350)

Similarly, where Darwinian prophesies had failed, eugenic action was required to ensure the survival of the fittest:

> Eugenics promised to overcome the troublesome uncertainties of the survival doctrine according to which it was impossible either to predict who would turn out to be the fittest or to provide the means for the nations to develop everlasting fitness. . . . The process of selection had only to be changed from a natural

necessity which worked behind the backs of men into an "artificial," consciously applied physical tool.... Ernst Haeckel's early remark that mercy-death would save "useless expenses for family and state" is quite characteristic. (Arendt [1951] 1968, pp. 178–9)

Popper had similar concerns about large-scale social engineering, or "social midwifery" in the service of predicted historical outcomes. The midwife–birthing language is borrowed from a quotation from Marx that Popper makes much of: "When a society has discovered the natural law that determines its own movement, even then it can neither overleap the natural phases of its evolution, nor shuffle them out of the world by the stroke of the pen. But this much it can do: it can shorten and lessen the birth-pangs" (preface to *Capital*; Popper [1944–45] 1957, p. 51).

Totalitarian social engineering is guided by what Popper called "historicist moral theory," which equates the predictions of supposedly scientific theories of history with what is good and desirable. What will happen anyway thus becomes the goal to be strived for: "A historicist sociology can... be interpreted as a kind of technology which may help (as Marx puts it) to 'shorten and lessen the birth-pangs' of a new historical period" (Popper [1944–45] 1957, p. 71; also pp. 49–76; and [1945] 1966, v. 2, pp. 204–11).

Popper considered evolutionary ethics to be a totalitarian-style, historicist moral theory, and eugenics as totalitarian-style social engineering. His example of an evolutionary ethicist was C. H. Waddington, who had argued that "a social system [is] something which essentially involves motion along an evolutionary path," and who claimed that "we must accept the direction of evolution [of a social system] as good simply because it *is* good" (Waddington 1942, pp. 17–18; Popper [1944–45] 1957, pp. 106–7, n. 2).

For Waddington, it was the proper job of the scientist – in this case the evolutionary biologist – to discover the natural course of evolution and hence to determine what is good and what should serve as the goal for further evolutionary change (Waddington 1942, pp. 18–19, 135–6). This sounded to Popper like eugenics ([1944–45] 1957, p. 159 and n. 1). (And indeed Waddington supported eugenics, even if he was not its most avid proponent.) Eugenics, Popper complained, was promoted whenever human evolution did not seem to be going in the direction that it naturally should, thus supposedly leading to degenerative changes requiring genetic ameliorative measures ([1945] 1966, v. 1, pp. 317–8, n. 71).

Popper's suspicions about biologists like Waddington were reinforced by Waddington's own acknowledgment that his views entailed "cosmic [if not local] fatalism" (1942, p. 18). Even worse, Waddington had argued that "totalitarianism... does seem to be inevitable; the whole of recent history is towards it" (Waddington [1941] 1948, p. 22). And this is on the whole good. For Waddington foresaw the day when fascist and communist versions of totalitarianism would fall by the wayside as more scientifically controlled versions of totalitarianism emerged ([1941] 1948 passim, and especially pp. 70–71).[6]

What frustrated Popper so much about the way in which Darwinism legitimized the search for laws of history is that Darwinian evolutionary theory itself only pretended to be lawlike; it actually failed in this respect, and as a result fell short of the standards of science. There could be no lawlike, hence scientific, theory of evolution. There could be no lawlike, hence scientific, theory of history – whether natural history or social.

I will not rehearse at length Popper's critique of evolutionary theory. But I want to point out that his critique is usually presented by philosophers as a stand-alone argument with no mention of the political dimensions. Personally, I have always been sympathetic with his more narrowly philosophical arguments against the existence of evolutionary laws.[7] But it seems clear that Popper also set his sights on evolutionary biology because of its supporting role in historicist thought and hence, he believed, in the rise of totalitarianism. The similarity of his views to Arendt's, Hayek's, Berlin's, and to a lesser extent Barzun's, suggest that this sort of concern was not idiosyncratic.

Popper's political concerns about Darwinian evolutionary thought are invisible in his later works, in which he was somewhat more positive about the virtues of Darwinism, and in which he even promoted a Darwinian model of theory-change in science. But of course he only elevated the status of Darwinism to a "metaphysical research program." And it should not be forgotten that his first critique of evolutionary biology was part of a project that he dedicated to "the countless men and women of all creeds or nations or races who fell victims to the facist and communist belief in Inexorable Laws of Historical Destiny."[8]

(There is much more that could be said in connection with Arendt's and Popper's concerns about historical determinism in general and Darwinism in particular. For instance, it is important to view Arendt and Popper not only in the context of the rise of totalitarianism but also as part of a tradition of German philosophers responding to, and reacting against, positivistic, lawlike conceptions of history.)

3.5 Concluding Remarks

Times changed. By the 1970s and 1980s, Darwinism and Marxism could be compared without sullying each other's reputations. For example, "analytical" Marxists (in the sense of "analytical philosophy") could quite unapologetically view Darwinian and Marxist accounts as instances of a legitimately scientific form of "functional" explanation (e.g., Cohen 1978), or view Darwinian and Marxist theories as instances of a legitimately scientific form of theory – albeit nonlawlike and nondeterministic in contrast to Newtonian mechanics (e.g., Miller 1984).

Through the many comparisons and contrasts over the years, Darwinism and Marxism have become partly constitutive of each other: "Darwinism, like Marxism . . . ," "Darwinism, unlike Marxism . . . ," "Marxism, like Darwinism . . . ," "Marxism, unlike Darwinism . . . . " Their perceived virtues and vices reinforce each other or stand out in contrast to one another. They are inseparable even when perceived to differ – indeed precisely in order to demonstrate the difference.

Questions about the relationship between Marxism and Darwinism so often lead back to Engels. With somewhat mixed appreciation, Lewontin and Richard Levins dedicated *The Dialectical Biologist* to Engels, "who got it wrong a lot of the time, but who got it right where it counted." Presumably, what Engels got right was his general dialectical approach to nature and history, which Lewontin and Levins interpret as a precautionary heuristic (see the Introduction to this volume, and also Levins and Lewontin 1985, pp. 191–2). And what Engels got wrong was the relationship between Darwin's and Marx's views of history. At any rate, somehow – not incomprehensibly, and surely owing in part to Engels – that relationship came to be viewed by several anti-Marxists in the late 1930s, 1940s, and 1950s in terms of historical determinism. Two children of that generation, Lewontin and Levins, Marxists and Darwinians both, came to view the relationship very differently.

3.6 Acknowledgments

My thanks to Diane Paul, James Farr, Lisa Disch, and Garland Allen for help with this chapter. And thanks especially to Dick for his inspiration and support over many years.

Notes

1. There is considerable disagreement as to whether, or to what extent, Haeckel should be viewed as a proto-Nazi (see e.g., Gasman 1971, Kelly 1981, Mocek 1991). My concern here, though, is how Arendt and Popper viewed Haeckel (and how they viewed Darwinism through Haeckel).
2. The connection between Marxism and Darwinism – and in particular the question of whether they both involve deterministic views of history – is controversial and often involves distinctions between Marx's Marxism and Engels's Marxism (and also between early Marx and later Marx). Engels is commonly blamed for the view that Marxism and Darwinism are both lawlike views of history (Lichtheim [1961] 1965; Berlin [1953] 1996, p. 90; Ball 1979; but see also Paul 1979).

   My concern in this chapter is with Arendt's and Popper's views – which are certainly not my own – of the relationship between Marxism and Darwinism in this respect. It is clear that Marx and Darwin raised profound issues concerning historical contingency and necessity. But it is just as clear (to me!) that they did not resolve those issues, even to their own satisfaction. Thus, one can cite passages from both Marx and Darwin that suggest the inherent directionality and progressiveness of change in their respective domains (the social and the biological). One can cite Marx complaining that Darwin's theory leaves progress to chance (Marx to Engels, 7 August 1866, in *Marx and Engels* 1956–1990, v. 31; discussed in Paul 1979, pp. 123, 132–3). And one can as well cite passages from both that emphasize historical contingency (Darwin 1862, pp. 282–93; 1872, v. 1, pp. 241–44; Marx to the Editorial Board of *Otechestvenniye Zapiski*, Nov. 1877 in *Marx and Engels* 1975, pp. 291–4, esp.

p. 294; see the discussion of Darwin in Beatty 1995, and the discussion of Marx in Miller 1984, ch. 7).

3. Garland Allen is currently writing a book on Barzun's views about the connections between Darwinism and totalitarianism.
4. The point has often been made that what Popper calls and criticizes as "historicism" is a far cry from what historians and most philosophers of history refer to and defend as "historicism," which is closer to being the sort of outlook or approach that *differentiates* history from science (rather than being a scientistic imitation of science). More specifically, "historicism," properly labeled, is antithetical to the idea that there are laws of history. So Popper himself is a historicist!

   Again, Popper's use of the term follows Hayek's. Why Hayek, and following him Popper, used the term in that way is not my concern here. I am concerned instead with the ideas and approaches that Popper associated with that term and not the appropriateness of the term. For criticisms of Popper's use of the term "historicism," see Lee and Beck 1953–54, Rand 1964, Keaney 1997.
5. Connections have often been drawn between Darwinism and Marxism through positivism. Thus, for example, Isaiah Berlin refers to the "half-positivist, half-Darwinian interpretation of Marx's thought [about history] which we owe mainly to Kautsky, Plekanhov, and above all to Engels." (Berlin 1978, p. 90). See Farr (1984) and Miller (1984) for thorough critiques of the idea that Marx himself was a positivist.
6. Interestingly, Popper seems to have found Nazi eugenics so loathsome that he refused to refer to any Nazi eugenicists by name ([1945] 1966, v. 1, p. 339). Indeed, as far as I can tell he never referred to any modern fascists in *Poverty of Historicism* or *Open Society* except to implicate Haeckel. Plato was his stand-in fascist–racialist totalitarian who exemplified just how despicable that position could be. Popper subjected Plato's eugenics to considerable analysis and scorn throughout the first volume of *Open Society*. Of particular interest is Popper's discussion of how Plato, in *The Republic*, urges statesmen to institute rigged marriage lotteries so that inferior people will blame fate when they are not chosen to reproduce (*Republic* 460a; Popper [1945] 1966, v. 1, pp. 150, 339).
7. Although I disagree that the lack of laws renders evolutionary theory un- or subscientific. And I think the argument against the existence of laws needs to be substantially elaborated along the lines of the Lewontin and Gould references cited earlier. My own contribution to that argument is Beatty 1995.
8. Reflecting on his change of heart concerning Darwinism, Popper later acknowledged

   > ... when I was younger I used to say very contemptuous things about evolutionary philosophies. When twenty two years ago Canon Charles E. Raven, in his *Science, Religion and the Future*, described the Darwinian controversy as a "storm in a Victorian teacup," I agreed, but criticized him for paying too much attention "to the vapors still emerging from the cup," by which I meant *the hot air of the evolutionary philosophies (especially those which told us that there were inexorable laws of evolution)*. But now I have to confess that this cup of tea has become, after all, *my* cup of tea; and with it I have to eat humble pie. (Popper [1966] 1972a, p. 241; the long emphasis is mine)

   At the same time, Popper made clear that his change of heart had its limits:

   > Darwin's discovery of the theory of natural selection has often been compared to Newton's discovery of the theory of gravitation. This is a mistake. Newton formulated a set of universal

laws intended to describe the interaction, and consequent behavior, of the physical universe. Darwin's theory of evolution proposed no such universal laws. There are no Darwinian laws of evolution. (Popper [1966] 1972b, p. 267)

REFERENCES

Arendt, Hannah. [1951] 1968. *The Origins of Totalitarianism.* New York: Harcourt Brace.

Arendt, Hannah. [1953] 1993. *Essays in Understanding.* New York: Harcourt Brace.

Ball, Terence. 1979. Marx and Darwin: A reconsideration. *Political Theory* 7:469–83.

Barber, Benjamin R. 1987. Totalitarianism. *The Blackwell Encyclopedia of Political Thought.* Oxford: Blackwell.

Barzun, Jacques. [1941] 1958. *Darwin, Marx, and Wagner.* New York: Anchor.

Beatty, John. 1995. The evolutionary contingency thesis. In *Concepts, Theories, and Rationality in the Biological Sciences,* eds. Gereon Wolters and James G. Lennox. Pittsburgh: University of Pittsburgh Press.

Berlin, Isaiah. [1953] 1996. The sense of reality. In *The Sense of Reality.* New York: Farrar, Straus and Giroux.

Berlin, Isaiah. [1954] 1969. Historical inevitability. In *Four Essays on Liberty.* Oxford: Oxford University Press.

Berlin, Isaiah. 1978. *Karl Marx,* 4th ed. Oxford, UK: Oxford University Press.

Canovan, Margaret. 1992. *Hannah Arendt: A Reinterpretation of Her Political Thought.* Cambridge, UK: Cambridge University Press.

Cohen, G. A. 1978. *Karl Marx's Theory of History: A Defence.* Princeton, NJ: Princeton University Press.

Darwin, Charles. 1862. *On the Various Contrivances by which British and Foreign Orchids Are Fertilised by Insects, and on the Good Effects of Intercrossing.* London: Murray.

Darwin, Charles. 1872. *On the Origin of Species by Means of Natural Selection,* 6th ed. London: Murray.

Dexter, Bryan, ed. 1972. *The Foreign Affairs 50-Year Bibliography.* New York: Bowker.

Engles, Friederich. 1883. Speech at the graveside of Marx. In Marx, Karl, and Friedrich Engels. *The Marx–Engels Reader,* ed. Robert C. Tucker. Princeton, NJ: Princeton University Press.

Engels, Friedrich. [1885] 1939. *Herr Eugen Dühring's Revolution in Science (Anti-Dühring).* New York: International Publishers.

Engels, Friedrich. [1927] 1940. *Dialectics of Nature.* New York: International Publishers.

Farr, James. 1984. Marx and positivism. In *After Marx,* eds. Terence Ball and James Farr. Cambridge, UK: Cambridge University Press.

Fisher, Ronald A. 1930. *The Genetical Theory of Natural Selection.* Oxford, UK: Oxford University Press.

Gasman, Daniel. 1971. *The Scientific Origins of National Socialism.* New York: American Elsevier.

Gould, Stephen J. 1989. *Wonderful Life.* New York: Norton.

Haeckel, Ernst. [1899] 1900. *Riddle of the Universe.* New York: Harper.

Hayek, F. A. von. 1941. The counter-revolution of science. *Economica* 8:9–36, 119–41, 281–320.

Hayek, F. A. von. 1942. Scientism and the study of society. *Economica* 9:267–91; 1943, 10:34–63; 1944, 11:27–39.

Hayes, Carleton J. H. 1941. *A Generation of Materialism, 1871–1900.* New York: Harper.

Held, David. 1980. *Introduction to Critical Theory: Horkheimer to Habermas.* Berkeley, CA: University of California Press.

Horkheimer, Max. [1947] 1974. *Eclipse of Reason.* New York: Continuum.

Keaney, Michael. 1997. The poverty of rhetoricism: Popper, Mises and the riches of historicism. *History of the Human Sciences* 10:1–22.

Kelly, Alfred. 1981. *The Descent of Darwin: The Popularization of Darwinism in Germany, 1860–1914.* Chapel Hill, NC: University of North Carolina Press.

Lee, Dwight E., and Beck, Robert N. 1953–54. The meaning of historicism. *American Historical Review* 59:568–77.

Levins, Richard, and Lewontin, Richard. 1985. *The Dialectical Biologist.* Cambridge, MA: Harvard University Press.

Lewontin, Richard C. 1966. The principle of historicity in evolution. In *Mathematical Challenges to the Neo-Darwinian Interpretation of Evolution,* eds. Paul S. Moorhead and Martin M. Kaplan. Philadelphia: Wistar Institute Press.

Lewontin, Richard C. 1969. The bases of conflict in biological explanation. *Journal of the History of Biology* 2:35–45.

Lewontin, Richard C. 1974. *The Genetic Basis of Evolutionary Change.* New York: Columbia University Press.

Lewontin, Richard C. [1977] 1985. Evolution as theory and ideology. In Levins and Lewontin, *The Dialectical Biologist.* Cambridge, MA: Harvard University Press.

Lewontin, Richard C. 1990. Fallen angels. *New York Review of Books* 14 June:3–5.

Lewontin, Richard C. 1991. Facts and the factitious in natural sciences. *Critical Inquiry* 18:140–53.

Lichtheim, George. [1961] 1965. *Marxism: An Historical and Critical Study.* London: Routledge and Kegan Paul.

Marx, Karl. [1867] 1967. *Capital,* Vol. 1. New York: International Publishers.

Marx, Karl, and Engels, Friedrich. 1975. *Selected Correspondence.* Moscow: Progress Publishers.

Marx, Karl, and Engels, Friedrich. 1956–1990. *Werke.* Berlin: Dietz.

Marx, Karl, and Engels, Friedrich. 1978. *The Marx-Engles Reader,* ed. Robert C. Tucker. Princeton, NJ: Princeton University Press.

Miller, Richard W. 1984. *Analyzing Marx: Morality, Power and History.* Princeton, NJ: Princeton University Press.

Mocek, Reinhard. 1991. Two faces of biologism: Some reflections on a difficult period in the history of biology in Germany. In *World Views and Scientific Discipline Formation,* eds. William R. Woodward and Robert S. Cohen. Dordrecht, The Netherlands: Kluwer.

Paul, Diane B. 1979. Marxism, Darwinism, and the theory of two sciences. *Marxist Perspectives* 2:116–43.

Paul, Diane B. 1992. Fitness: Historical perspectives. In *Keywords in Evolutionary Biology,* eds. Evelyn Fox Keller and Elisabeth A. Lloyd. Cambridge, MA: Harvard University Press.

Popper, Karl. [1944–45] 1957. *The Poverty of Historicism.* New York: Harper.

Popper, Karl. [1945] 1966. *The Open Society and Its Enemies.* Princeton, NJ: Princeton University Press.

Popper, Karl. [1966] 1972a. Of clouds and clocks: An approach to the problem of rationality and the freedom of man. In *Objective Knowledge.* Oxford, UK: Oxford University Press.

Popper, Karl. [1966] 1972b. Evolution and the tree of knowledge. In *Objective Knowledge.* Oxford, UK: Oxford University Press.

Rand, Calvin G. 1964. Two meanings of historicism in the writings of Dilthey, Troeltsch, and Meinecke. *Journal of the History of Ideas* 25:503–18.

Ruse, Michael. 1996. *Monad to Man: The Concept of Progress in Evolutionary Biology.* Cambridge, MA: Harvard University Press.

Waddington, Conrad H. [1941] 1948. The *Scientific Attitude.* West Drayton, Middlesex: Penguin.

Waddington, Conrad H. 1942. *Science and Ethics.* London: Allen and Unwin.

Watkins, John. 1995. Popper and Darwinism. *Philosophy* 70:191–207.

Wood, Allen. 1981. *Karl Marx.* London: Routledge and Kegan Paul.

CHAPTER FOUR

The Genetics of Experimental Populations

L'Héritier and Teissier's Population Cages

JEAN GAYON AND MICHEL VEUILLE

4.1 Introduction

France has the reputation of being "the only major scientific nation that did not contribute significantly to the evolutionary synthesis" (Mayr 1980, p. 320) because of the contempt shown by French zoologists for Darwinism. This is not entirely true. In the 1930s, an influential school of mathematically trained population geneticists developed in Paris. It flourished in a peculiar and privileged part of the French university system, *l'Ecole Normale Supérieure* (ENS), and was led by Philippe L'Héritier and Georges Teissier. The contribution of this school was acknowledged by Wright (1977), who, in the third volume of his book *Evolution and the Genetics of Natural Populations*, devoted virtually the whole chapter on "natural selection in the laboratory" (one tenth of the volume) to their results.

That L'Héritier and Teissier's group developed at the same time as Dobzhansky and Wright's influence was flourishing in the United States may provide a way of analyzing the respective roles of biologists and of biological facts in the rise of scientific ideas. Both schools began their work in the 1930s with the same aim of evaluating the predictions of theoretical population genetics, which had recently been elaborated by Haldane, Fisher, and Wright. After years of effort, both groups reached a completely unexpected conclusion, namely, that natural selection maintains polymorphism in populations. For many years, the two schools were unaware of each other, partly because they followed different approaches. L'Héritier and Teissier set up experimental populations of *Drosophila melanogaster*, a species that Dobzhansky (1965) later described as a "weed species." At this time, Dobzhansky's work was focused on natural populations of *D. pseudoobscura*. Shortly before the outbreak of World War II, the two groups swapped roles by adopting each other's approach. In 1937, Georges Teissier commissioned Wright to write a short book, *Statistical Genetics in Relation to Evolution*, which was published in Paris in 1939. According to Provine (1986), this represented "the clearest, most accessible, and comprehensive statement" of Wright's thought at the time and was to be seminal in the later turn of the ENS school to the study of natural populations of land snails and marine isopods. Conversely, population cages – the experimental device that had been developed by the French – were used by Wright and Dobzhansky for their joint contribution to the twelfth article in the "Genetics of Natural Population" series (GNP XII), which was a critical step in their collaboration.

In his introduction to *Dobzhansky's Genetics of Natural Populations*, Lewontin (1981) pointed out that there was a major change in Dobzhansky's research agenda around 1950. Two papers published in the 1940s seem to have been critical for this. In GNP IX (1943), which was devoted to seasonal fluctuations in inversion frequencies, Dobzhansky "ruled out genetic drift as an explanation and suggested that the cause was seasonally varying natural selection.... This hypothesis was confirmed in one of the key papers of the series, GNP XII [1946], in which repeatable changes in inversion frequencies were obtained in population cages" (Lewontin 1981, p. 109). Only an experimental test of the new hypothesis could convince Dobzhansky that natural selection, rather than genetic drift, was the key factor maintaining inversion polymorphism. This led him to abandon the project he had been pursuing for 10 years of applying Wright's equations to chromosome inversions for measuring population parameters. After 1950, Dobzhansky increasingly used population cages to study the evolution of polymorphism at the gene level. By allowing Dobzhansky to adopt a more experimental (laboratory) approach to population genetics, the population cages played an important role in the development of his research program. They probably made it possible for him to (partially) break with his former approach of subjecting field observations to subtle statistical analyses with the hope of determining the parameters necessary for applying Sewall Wright's view of evolution to Mendelian populations.

Historians have paid little attention to the origins of these "population boxes." In the methods section of GNP XII, Wright and Dobzhansky pay credit to L'Héritier and Teissier for what may have simply seemed to be a clever invention. But just as there is no machine without a designer, there is no experimental system without a theoretical purpose. In GNP XII Wright and Dobzhansky considered, among other hypotheses, that the *ST–CH* polymorphism was maintained in their boxes by heterozygote advantage. The same apparatus had been used in Paris to show the indefinite coexistence of mutants under what L'Héritier and Teissier then interpreted as heterosis. However, heterosis was not Wright and Dobzhansky's main concern, for at the time they were trying to explain seasonal frequency cycles. They may have seen no reason to discuss the results of the two French geneticists, whose interests in larval competition, the dynamics of allele replacement, heterosis, and frequency-dependent selection may have appeared of little use to them.

The purpose of this study is to characterize the contribution of L'Héritier and Teissier to the rise of experimental population genetics in the 1930s and 1940s. We will not provide a detailed description of the school that developed from the mid-1940s under the specific responsibility of Teissier. We will focus primariliy on the origins of this school. This will lead us to propose answers to the following questions: How was it possible for this school to develop given the unfavorable national context? Why and to what extent did it remain intellectually isolated from its American counterparts? How did the two schools, from their different perspectives, come to such similar conclusions? We will assume that readers are familiar with the history of population genetics in the United States.[1]

4.2 Philippe L'Héritier (1907–1994) and the "Démomètre"

Population cages were designed by Philippe L'Héritier in 1932 just 2 years after the publication of Fisher's (1930) *Genetical Theory of Natural Selection* and 1 year after Wright's (1931) *Evolution in Mendelian Populations.* In 1932, both Wright and Fisher attended the Sixth International Congress of Genetics in Ithaca, where Wright presented his "adaptive surfaces" for the first time. L'Héritier, then aged 25, attended the meeting and was enthused by what he heard.

L'Héritier had specialized in mathematics at high school before entering the Ecole Normale Superieure (ENS Paris) in 1926 where he began working with Georges Teissier. Seven years older, Teissier was also a mathematician with an interest in natural history. Teissier's main field was biometry, and he tried to influence L'Héritier in this direction. However, L'Héritier was advised by André Mayer, a professor of physiology at the Collège de France (Paris) to become a geneticist. Mayer strongly resented the absence of genetics in France and thought that this young man, trained in both mathematics and biology, would be the right person to send to the United States to work in a genetics laboratory. L'Héritier was perhaps the first French biologist to benefit from a Rockefeller grant long before Boris Ephrussi and Jacques Monod. He left France in 1931 with "the task of learning genetics and of finding a research project which he could continue in France after returning from the U.S." (L'Héritier 1981, p. 335). He studied genetics at Iowa State College under Professor Lindstrom, who worked on tomatoes.

> During this stay in the U.S., I discovered the existence of population genetics, which I believe I had never heard of before. I read R. A. Fisher's book, which had just appeared. I also discovered Sewall Wright's work. I met and heard these men and heard them speak at one of the first International Congresses of Genetics held at Cornell in July 1932. I also met other founding fathers of modern genetics and evolutionary biology, Theodosius Dobzhansky and H. J. Muller . . . I discovered the famous little fly, *Drosophila*. I had never seen one before. (L'Héritier 1981, pp. 335–6)

The idea of population cages soon occured to him. Before returning to Europe, he paid a visit to Woods Hole:

> One day, while walking on an American beach, I realized it would be possible to breed the fly, not in small bottles where only one generation could be observed, but in boxes in which food would be periodically renewed. This was the origin of the famous population cages, or "démomètres." This word, "the measure of populations," corresponded to my initial idea of comparing the ability of flies from different origins or strains to establish themselves, demographically speaking, in a given milieu. (L'Héritier 1981, p. 337; see also Mayr 1980).

In October 1932, L'Héritier returned to France with this idea in mind and some wildtype strains of *Drosophila* in his luggage. He intended to use them for a Ph.D. dissertation on quantitative variation in *Drosophila melanogaster.* Shortly after his return, he published his first article on the genetics of resistance of five wildtype strains to toxic food (L'Héritier 1932), which had been carried out during his stay in Lindstrom's laboratory. Pesticides did not exist at this time, but a series of chemicals had already been tested for killing insects, including sodium arseniate, which L'Héritier used in his experiments. In a perfect diallel cross, he demonstrated the advantage of hybrids for most strains, the higher resistance of females over males, and a significant *X*-linked effect in one strain. The American influence appears clearly in this paper, which looks more "modern" than any of L'Héritier and Teissier's later publications. It is the only one to show the primary data in a table and to include standard deviations. Systematic statistical testing would not appear in their articles until the 1950s. At this point, there was no question of testing evolutionary hypotheses, and the history of population cages had not yet begun. However, survival on toxic medium was thought to provide "a measure of survival" of the strains.

The history of experimental population genetics really began with L'Héritier's experiments in Paris. His research had three parts that corresponded to the three chapters of his doctoral thesis (L'Héritier 1937): (1) the measurement of several quantitative differences between strains, (2) the experimental estimation of genetic factors involved in quantitative inheritance, and (3) the study of the growth of these strains in population cages. L'Héritier wanted to see whether under experimental conditions each strain would reach a stable demographic equilibrium and if these equilibria would differ.

L'Héritier carried out his experiments with four wildtype strains of *Drosophila.* Three of them had been given to him in the United States by Lindstrom (Ames), Muller (Austin), and Morgan (Pasadena). The fourth strain had been collected by Guyénot around 1920 and kept in Paris by Caullery. All these strains were probably highly inbred after generations spent in the laboratory.

It was L'Héritier himself who built the "population cages."[2] Raised in the country, he was a skilled woodworker, and throughout his life enjoyed making his own furniture. Before 1945, the population cages were made of wood. A short description is given by L'Héritier and Teissier (1933): the box was maintained at a constant temperature and contained 20 food vials. Every day, a vial of fresh food was introduced to replace the oldest one. The 20 vials or days thus corresponded to the development of two generations of *Drosophila* at 26°C (there were 21 vials in later versions of the cage to accommodate a 3-week schedule). Adults were anesthesized using CO_2 and sprinkled in the dark onto a sheet of light-sensitive paper that was then exposed to a brief pulse of light. This produced a photograph on which the entire imaginal population could be counted. Between 1933 and 1937, L'Héritier counted 500,000 flies for his Ph.D. research (L'Héritier 1937). Besides carrying out this study of population stability, competition, and survival in the demometer, L'Héritier carried out population genetics experiments with Teissier. The first version of the cage was intended to follow genetically homogenous populations and was inappropriate for the study of polymorphic populations containing several genotypes. It meant that L'Héritier and Teissier had to count the phenotypes of the anesthesized flies. They later added sampling vials to the demometer to collect the eggs laid over a 24-hour cycle. These eggs were reared under noncompetitive conditions to provide a reliable estimate of the zygotes formed at a given time.

At the beginning of their collaboration, population cages were only one of the experimental methods they employed. These cages contained pure strains that laid their eggs in vials in which competition took place. Demometers became more and more important owing to technical improvements and to a better understanding of what was going on in the boxes. L'Héritier and Teissier's laboratory notebooks are still in existence and provide a precious source of information. They show that in 1936 no less than 32 population cages were followed simultaneously in the laboratory.

Although L'Héritier played a crucial role in setting up the population cage program, his contribution only lasted 5 years. In 1937, an unexpected discovery pulled him in another direction: he found that some *Drosophila* lines did not survive CO_2 anaesthesis. Five papers published on this subject (L'Héritier and Teissier 1937d; 1938a,b; 1940; 1944) showed that CO_2 sensitivity was a non-Mendelian hereditary trait, which was only transmitted in the female line. L'Héritier later showed that the effect was caused by the *Sigma* rhabdovirus. In 1938, L'Héritier was appointed as professor in Strasbourg, 300 miles east of Paris, near the German border. The two geneticists agreed that Teissier would continue the population cage project, and L'Héritier would concentrate on CO_2 sensitivity.

At the outbreak of the war in 1939, L'Héritier was conscripted until the French defeat in 1940. He was then transferred to Clermont–Ferrand (central France) along with the rest of the Strasbourg University staff. He lost his strains and zoology teaching notes[3] and was unable to do any research until 1943. After the end of the war, he participated in the foundation of the CNRS genetics laboratories at Gif-sur-Yvette south of Paris (Burian et al. 1988) and later went back to Clermont–Ferrand, where he and his students discovered the I–R hybrid dysgenesis system (Picard and L'Héritier 1971), thus participating in the discovery of transposable elements in *Drosophila* (see Kidwell 1979, Bregliano and Kidwell 1983).

Although L'Héritier's work on population genetics was limited to his doctoral research, he made a considerable contribution to the founding of the ENS school. He introduced population genetics to France and designed an experimental system for testing theoretical models. He published a short book on the mathematical theory of genetical selection inspired by the work of Fisher and Wright (L'Héritier 1934). He later published a two-volume textbook of population genetics (1954) and a textbook on biological statistics (1949). Apart from an elegant presentation of Fisher's and Wright's conceptions, the 1934 book contains reservations that help one understand L'Héritier and Teissier's experimental work at the time. These will be discussed further.

4.3 Georges Teissier (1900–1972) and Allometry

Before meeting L'Héritier, Teissier studied the systematics and development of Hydrozoa and biometry.[4] The latter in fact involved two fields: the growth of the organism and the growth of populations. For Teissier, this was not an eclectic choice: his whole aim was to integrate mathematics and biology. In the early 1930s, he spent a great deal of time trying to establish rigorous mathematical growth laws. He wrote several papers and syntheses, including one with Huxley on the growth curve. This article, which was published simultaneously in *Nature* and *Naturwissenschaften,* coined the word "allometry," which has been used ever since to describe such studies. Teissier apparently did not think that the growth of individuals and the growth of populations were formally different questions; he collaborated with Boris Ephrussi in physiological mathematics and with Jacques Monod (at the time his student) in population biology. As the editor of a collection of scientific books published in Paris by Hermann, Teissier oversaw the production of monographs in "Biometry and Biological Statistics," written by major contemporary figures, such as Lotka, Volterra, D'Ancona, and Gause. The last to be published, just before the outbreak of the war, was Wright's *Statistical Genetics in Relation to Evolution,* a title that summarized the spirit of the collection.

Teissier probably saw a mathematical similarity between development and the dynamics of competition in population biology. He once said that the replacement of one allele by another in a finite population was like an equilibrium between two exponential rates of increase. This is akin to his conception of allometry, as expressed by the ratio of the exponential growth of two organs, then measured by the "equilibrium constant," the other name of the allometry coefficient. It was because of Teissier's unitary view of mathematics and biology that the demometer was sometimes used to study competition between two species (*Drosophila melanogaster* and *D. funebris*) and sometimes to study competition between two alleles. The estimator used by L'Héritier and Teissier in their first population genetic study is illuminating in this respect. The change in the frequency of two competing alleles, p and q, was measured as

$$\lambda = \mathrm{abs}(\Delta p/p - \Delta q/q).$$

This formula, which is reminiscent of the exponential term of the logistic population growth curve, is not generally used in population genetics.[5] It is a rather poor estimator from a Darwinian point of view because it describes the rise of a mutant by its effects rather than by a well-identified cause, as in a selection coefficient. Moreover, it is more appropriate for comparing competition between two species, for which there are two independent classes of organisms, than for describing the competition between genes, which involves three genotypes that generally have different adaptive values. It was 11 years after the population genetics project had begun before Teissier used his mathematical skills to derive population genetics equations in a classical way. His study on "the equilibrium of lethal genes in stationary panmictic populations" (1944) might have been inspired by reading Wright and Haldane. This paper was accompanied by data from an experiment on the maintenance of lethal recessive mutants in demometers that he had been running for several years. It should be noted that such experiments involved the competition of only two genotype classes – the wild type and the heterozygous mutant. Although artificial, these experiments constituted a completely controlled competitive system involving two forms.

4.4 The Genetics of Experimental Populations Papers (1932–37 and 1942–54)

The heroic period, during which genuine inventions were made in a strictly local context, extended from 1932 to 1954. During this period, L'Héritier, Teissier, or both published 20 "notes" reporting experimental work on *Drosophila* (L'Héritier and Teissier 1933; 1934a,b; 1935; 1936; 1937a,b,c; L'Héritier, Neefs, and Teissier 1937; Teissier 1942a,b; 1943; 1944; 1947a,b; 1952; 1953; 1954a,b). To this list can be added the third part of L'Héritier's thesis on quantitative variation in *Drosophila melanogaster* (L'Héritier 1937), which is devoted to larval competition in several wildtype lines of *Drosophila* and is related to L'Héritier and Teissier's joint work on population cages. This provides technical details about the population cages not given elsewhere (which probably explains why the thesis itself was cited by Wright and Dobzhansky in GNP XII). These papers each report a different experiment.

The nine papers published before L'Héritier's move to Strasbourg (1938) were written jointly by L'Héritier and Teissier. The following 11, written between 1942 and 1954, were by Teissier alone.

During the intervening 4-year period (from 1938 to 1942), Teissier did not publish anything in population genetics, although he wrote papers in his two other areas of interest, biometry and invertebrate morphology. Their correspondence in this period shows that Teissier insistently tried to bring L'Héritier back to Paris to continue their collaboration. Indeed, Teissier succeeded in finding a position for him, which he refused for personal reasons.

4.5 Population Dynamics and the First Experiments

The population cages were not specifically designed for the study of competition between genes. L'Héritier (1981, p. 338) later admitted that this idea first occurred to Teissier. In its early days, however, the project was exclusively L'Héritier's. Before 1931, Teissier had communicated his strong interest in biometry to L'Héritier, and they had already collaborated on Teissier's major theoretical concern, the "laws of growth." Thus, the technical invention was certainly L'Héritier's, but it came about in the context of a common interest in demographic studies.

This point deserves further discussion. In his biometrical papers, before 1933 and after, Teissier often emphasized Pearl's pioneering experimental work on population growth. In the mid-1920s, Pearl had recognized the technical difficulties of experimental studies on *Drosophila* population growth. Geneticists used to breed fruitflies in bottles containing approximately 300 individuals, which were changed every generation. This technique was not appropriate for experiments on population dynamics. Pearl considered three alternative solutions, all of which in their way were unsatisfactory: (1) food was provided once, in a limited amount; (2) food was renewed in suboptimal amounts by adding fresh food to the bottle; (3) unlimited food was provided by serially transferring the flies to new bottles. In case (1), flies were left to die; it was only possible to measure the maximum size of the population, the effect of crowding on population dynamics, and mortality as a function of time. Situation (3) meant transferring all flies to new bottles. After a few generations, this required a huge number of bottles, shelves, technicians, and students with only the possibility of measuring the intrinsic growth rate under a nonlimiting food supply. Situation (2) was probably the closest to realistic conditions of competition in the wild. However, Pearl noted that it did not permit population size to be measured: "It has not been possible to devise any method of holding these populations in a steady state at the level of the asymptote [of the theoretical logistic curve], when there is at all times an abundance of fresh food. The population simply waves up and down the average size" (Pearl 1927, p. 544; see also Pearl 1925 and 1928).

L'Héritier and Teissier's first published study using population cages ("Study of a *Drosophila* Population at Equilibrium," 1933) was a direct response to the problem addressed by Pearl, as clearly stated at the beginning of the paper:

> Following many experiments on *Drosophila*, Pearl and collaborators [1927] have been able to analyze precisely some of the factors controlling population growth. But it does not seem that anyone has ever been able to maintain a population in a steady state with such a method. Using a relatively simple technique that involves keeping the breeding conditions constant, we have been able to maintain the equilibrium of a *Drosophila* population and to measure its main demographic characteristics. (L'Héritier and Teissier 1933, p. 1765)

This first experiment, using the Ames wildtype stock, lasted 65 days and was run in parallel in two cages. The two populations stabilized at an equilibrium of approximately 3,200 flies, which was studied for a month (about three generations). The sex ratio stabilized at the level of 34% males and 66% females. The authors stated that competition occurred at the larval stage (survival from egg to adult was only 1–2%). The three main results of this paper were confirmed by all subsequent experiments. Note that it was necessary to know the sex ratio to estimate the effective size of the population.

At this stage, there was as yet no population genetics, although natural selection was not far from the authors' thoughts, as suggested by their closing sentences: "It is not surprising that the rigorous selection resulting from our breeding conditions is advantageous to females [supposed to be more vigorous]. The same process of natural selection, which maintains the population in a state of equilibrium, results in a numerical predominance of females." (L'Héritier and Teissier 1933, p. 1767).

Of course, like the first aeroplanes, the experiment did not go very far: the fragile equilibrium was only followed for a few generations. But the experiment had been successful beyond their wildest hopes, and they were soon able to follow a population cage for several years.

4.6 Larval Competition and Population Equilibrium

In a paper published in 1934 ("On Some Factors of Success in Larval Competition in *Drosophila melanogaster*," 1934a), L'Héritier and Teissier used wildtype and *white* mutants to mark strains of competing larvae. Food vials were successively introduced into demometers, each containing a single strain, to allow females to lay eggs. The vials were then removed, and the emerging adults were observed. The authors showed that, whatever their origin, eggs that were laid first had a definite competitive advantage, and those that were laid beyond 2 days were unlikely to produce adults. They also showed that wildtype flies had a higher "selective value" than *white* flies. This experiment did not use population cages as a competition site, and, in fact, the design of the first demometers meant they were not appropriate for such experiments. The experiment was, however, supposed to reflect conditions of larval competition indirectly within the demometer.

The first two papers (1933–34) may look like hesitant steps towards real population genetics because they addressed "struggle-for-life" questions rather than "differential-survival" questions. In fact they proved critical to the development of L'Héritier and Teissier's approach by helping them assess the key features of an experimental system that they were later to use repeatedly. The experiments showed that a very stable population of around 3000–4000 flies was established, that its sex ratio was biased, and that the equilibrium was maintained at the cost of overpopulation because only 3% of zygotes survived to the reproductive stage. The kind of natural selection that dominated in population cages was larval competition.

By using population cages, L'Héritier and Teissier were clearly pursuing three objectives. First, *Drosophila* experimental populations reached an equilibrium state, and their size (including their effective size) was controlled as a fixed parameter. Second, demographic pressure induced a reproducible force of natural selection in "experimental" conditions. Third, this device allowed a reliable control of physical parameters. This made the demometer an especially suitable system to study the biotic component of natural selection, which was L'Héritier and Teissier's main interest. A reliable experimental system is one in which several parameters are separated; it is also one in which results are reproducible. Population cages met these requirements for studying natural selection, and L'Héritier and Teissier were certainly conscious of this from the very beginning.

Two years later, they published a paper with approximately the same title ("Contribution to the Study of Larval Competition in *Drosophila*," 1936) on the same subject of larval competition between *white* and wildtype flies. It shows the extent to which their approach had become more subtle. Females from either the wildtype or the *white* strain were allowed to lay eggs in food vials for 24 hours, and the other strain was then allowed to lay competing eggs for another 24 hours. There was thus a "first arrived" and a "second arrived" experiment. The authors also performed a "simultaneously arrived" experiment. The experiment was repeated many times, all three kinds of competition being run simultaneously to eliminate temporal variation in fecundity. The wildtype strain was always very successful. However, it showed a negative correlation in simultaneous runs between its "first-arrived" and "second-arrived" success. Wildtype flies always did better in the "second-arrived" experiment when they did poorly in the "first-arrived" experiment. The authors concluded that the competitive pressure of wildtype larvae was very high. It helped them win when they arrived first, but it led them to die when they arrived second because their number only amplified the competitive pressure due to the other strain. In other words, the relative density of the two strains changed the conditions of competition so as to disadvantage the most frequent strain. This intuition is probably at the root of L'Héritier and Teissier's rejection of constant fitness parameters and would explain why they readily interpreted their first population genetics results in terms of frequency-dependent selection.

4.7 Genetic Evolution in Experimental Populations

The third paper ("An Experiment of Natural Selection: Elimination Curve of the *Bar* Gene in a *Drosophila* Population at Equilibrium," 1934b) reported the first confrontation of two alleles in the same cage. In a particularly honest introduction, L'Héritier and Teissier admitted that this experiment had been unintended and had begun after an accidental contamination between demometers (in the course of experiments using pure strains, a *bar* population cage was contaminated by a wildtype genotype). They followed the dynamics of the polymorphic population for 5 months until the experiment accidentally stopped. They counted the genotypes of the whole population every month. A weakness of the article is its lack of an explicit presentation of materials, methods, and raw data. Instead, L'Héritier and Teissier presented an interpolated curve between the observed points from which they reconstructed data for 20-day intervals (twice the generation time). Despite these weaknesses, this was the first demonstration that natural selection can be studied in the laboratory.

An important aspect of this brief paper is the authors' claim that competition is lower when the new mutant is at a low frequency. This statement was echoed by a comment found in L'Héritier's small book on Fisher and Wright, which was published the same year:

> The advantage of a mutation can be measured by the expected number c of mutant genes in the offspring, where the value of the same parameter for the normal gene is 1. In practical terms, this advantage is unlikely to remain constant over time. Because of the progress of the mutant gene, the change in the population must modify this

advantage: it is reasonable to admit that, when mutant individuals are still relatively few in number, and therefore struggle with less fit normal individuals, they have more offspring than at the end of the replacement process, when they compete with individuals having the same value as themselves. If we ignore this variation of c, we only have a rough approximation of an otherwise interesting phenomenon. (L'Héritier 1934, p. 22)

There is a striking parallel between this comment in a book exposing the theory, and the final speculation contained in the experimental paper published the same year. It appears that, from the very beginning, constant fitness coefficients were not taken seriously by the two experimentalists. Faced with the elaborate mathematical theory of the founding fathers of population genetics, they could only oppose either verbal arguments or experimental results. It is remarkable that, although skilled mathematicians, they appeared to prefer biological intuition. Or maybe *because* they were skilled mathematicians, they understood constant fitness coefficients as legitimate approximations.

It thus appears that the use of the demometer was never trivial. It was never an "illustration" of the equations of population genetics, using actual organisms. From the beginning, it was intended to challenge a mathematical model of adaptive values that could only be solved by experiments.

The same experiment was replicated, this time intentionally, 3 years later. The demometer was followed over a longer time (600 days instead of 150) and in two replicates. In each case, the result was the virtual elimination of *bar*. However, the mutant did not become extinct. This is the conclusion of the paper:

It can be concluded, without any assumptions about the nature of the factors determining the superiority of the normal gene over *bar* in the struggle for life, that this superiority tends to disappear when *bar* has become rare in the population. If this superiority cancels out before the total disappearence of *bar*, populations should show an indefinite state of stability, and would now show only random variations in their genetic make up. (L'Héritier and Teissier 1937a)

They admitted that they could not prove this hypothesis using population cages: the precarious stability of laboratory populations could lead to the chance elimination of the low-frequency allele in a way that could not be distinguished from selective elimination.

4.8 Fitness and the Stability of Polymorphisms

The same year, L'Héritier and Teissier published another experiment on the "elimination of mutant forms" (1937b), which turned out to be another paradoxical case of nonelimination. A population cage containing a mixture of wildtype and *ebony* flies was observed for nearly 2 years. Over this period, the frequency of the mutant decreased from 0.90 to 0.14. They stopped the experiment after calculating that it would take 7 years for the frequency of *ebony* to reach 1%. They went a step further in their interpretations than they had done for *bar* and proposed that *ebony* was maintained by heterozygote advantage – a mechanism first suggested by Fisher (1922) from purely mathematical considerations.[6] The same year, during a discussion following a talk given by Haldane in Paris, L'Héritier summarized the technique of population cages and discussed the results obtained with *ebony*. The informal context meant he could be less cautious: "such an equilibrium is explicable only by a superiority of heterozygous individuals over both homozygotes. We have been unable, however, to obtain a direct proof of the existence of this superiority" (L'Héritier 1938, p. 523).

Why did L'Héritier and Teissier propose different explanations for the maintenance of *bar* and of *ebony*? The published papers give no indication. At the beginning of the *ebony* paper (1937b), the authors note that *bar* is semidominant, whereas *ebony* is recessive. Semidominance suggests a "genic" pattern involving additive variation. In this case, they would have expected the $q/(1-q)$ ratio to follow a linear dynamic if fitness coefficients were constant. In their laboratory notebooks, Teissier surveys the available models of gametic and zygotic selection with explicit reference to Wright. Although the 1937 papers were elliptic, it is clear that L'Héritier and Teissier regarded the *bar* experiment as an important argument in favor of frequency-dependent selection, whereas the *ebony* experiment was the first evidence of a case of heteryzogote advantage. Beside these interpretations, the two studies showed that natural selection acts to maintain variability in populations, whereas genetic polymorphism was still commonly thought of as "transient polymorphism" that affected alleles on their way to fixation or elimination.

4.9 Species Coexistence in the Demometer

Each of the very short four-page papers published by L'Héritier and Teissier was truly original and showed that a great deal of thought had gone into them. An innovative demometer experiment on competition between two species (*Drosophila melanogaster* and *D. funebris*) was published in 1935. The experiment was run in two replicates starting with one of the species in a majority in each case.[7] After a year, both replicates reached an equilibrium of about one *D. funebris* for five *D. melanogaster*. This experiment revealed the authors' interests in population dynamics. In his *Lessons on the Mathematical Theory of the Struggle for Life*, Vito Volterra (1931) had claimed, on the basis of mathematical considerations, that the competition of two species for the same food would always end up with the elimination of one of them (Volterra 1931, pp. 9–13). L'Héritier and Teissier did not explicitly refer to Volterra (indeed, they almost never cited anyone in their papers). They did, however, conclude with this unambiguous statement:

> Two species living on the same milieu . . . can survive side by side in a state of approximate equilibrium. It is noteworthy that most mathematical theories of the

struggle for life imply, on the contrary, that one species must completely replace the other. Our result however agrees with those recently obtained by Gause in competition experiments between Protozoan species.[8]

4.10 Opening the Population Cage in the Wild

The experiment reported in L'Héritier, Neefs, and Teissier ("Insect Apterism and Natural Selection," 1937) provided its authors with lasting fame in the polemical context of French studies on evolution. The authors took a population cage containing wildtype and *vestigial* flies (a recessive wingless mutant) to the Roscoff marine biological station in Brittany and removed the lid. They left the box exposed to the sea wind for 40 days. This article, which appeared in the *Comptes Rendus de l'Académie des Sciences* contains little methodological detail: for instance, the genotype structure of the population is not given. The reader simply learns that a previous experiment in closed cages had shown that larval competition quickly eliminated *vestigial*, whereas in the open air, larval competition was effectively counteracted by the loss of wildtype flies. Although their competitors were gone with the wind, the *vg/vg* homozygotes fluctuated between 12.5 and 48% for three generations. The original figure showing the dynamics of the *vg/vg* genotypic frequency is reproduced in Wright's (1977) Figure 9.2.

Their laboratory notebook shows that this experiment had been very carefully prepared. The box was covered every night with a sheet and opened again every morning. The genotypic structure and weather conditions (including the average wind speed) were recorded every day. The experimental purpose of the authors was clearly to introduce a very well controlled physical factor in addition to the biotic factors that were acting in the experimental population.

The article begins with a long citation from Darwin's *Origin of Species* on the evolutionary cause of flightlessness in island insects:

> For during many successive generations each individual beetle which flew least, either from its wings having been ever so little less perfectly developed or from indolent habits, will have had the best chance of surviving from not being blown out to sea; and, on the other hand, those beetles which most readily took to flight would oftenest have been blown to sea, and thus destroyed. (Darwin 1859, p. 136)

The concluding paragraph discusses the fact that no mutation is disadvantageous in itself, for it depends on the environmental context, and that natural selection is not only conservative but also innovative. Population geneticists of the time must have been frustrated by the fact that most known mutations were defects. This experiment was an effective reply to adversaries of Darwinism. It as much seemed an artist's performance as a scientific investigation, its purpose probably being to ridicule the numerous French opponents of "transformism." Denying natural selection was like flying in the face of the wind.

4.11 The Rise of a New Mutant

During World War II, Teissier's demometer received a blessing from nature. An advantageous mutant appeared in the population cage. Over eight generations, the new gene increased to the substantial frequency of 22% and stabilized at this level for 6 months between 20 and 25%. It was a new allele of *ebony*, a body color mutant. In his report ("Appearance and Fixation of a Mutant in a Stationary *Drosophila* population," 1943) Teissier again stressed that evolution could be genetically innovative. He put forward two possible explanations. The first was that the mutant and wildtype flies had slightly different "needs" and that the polymorphism was balanced in the cage for ecological reasons. The second explanation was heterosis. This gene was extensively studied by Teissier and his followers, culminating in Anxolabéhère's thesis (1978) on frequency-dependent selection.

4.12 A Closer Look at the Maintenance of Polymorphism

In the early 1940s, while L'Héritier was in Clermont–Ferrand, Teissier had two technicians and two students (Mademoiselle Helman and Madame Pérez)[9] to care for the demometers. Teissier designed competition experiments between genotypes that would not immediately eliminate each other and coexist long enough to be studied. One experiment involved *Curly* and *vestigial*. The *Curly* chromosome carries two inversions that act as recombination inhibitors. This chromosome is homozygous lethal but dominant for a morphological alteration of the wing shape and can thus be detected when in the heterozygous state. It is rapidly eliminated by wildtype chromosomes. However, the $Cy, +/+, vg$ heterozygote is more successful than the vg/vg homozygote and stabilizes at a 65% frequency,[10] thus ensuring the maintenance of the polymorphism as two externally distinguishable competitive phenotypes. This situation lasts until recombination – a rare event because of the inversion – breaks up the arrangement and recreates the wildtype genotype. Teissier was able to follow the equilibrium for 80 days ("Persistence of a Lethal Gene in a *Drosophila* Population," 1942b). In another experiment, he associated *sepia* and *ebony* mutants ("Variation of the Frequency of *Ebony* in a Stationary *Drosophila* Population," 1947b). One of Teissier's important observations was that the equilibrium frequency sometimes varied spontaneously and unpredictably to a new asymptotic value. He concluded that the selective values of the observed genotypes fluctuated in response to modifications in other allelic systems. His last paper on the subject ("Natural Selection and Genic Fluctuation," 1954b) concluded with the conviction that, possibly, "natural selection could ensure the maintenance of the genetic diversity of natural populations and give this polymorphism its essentially varying nature." The point here is not the impossibility of using perfectly "isogenized" experimental stocks in studying individual genes but rather whether or not it makes sense to try and isolate a single locus from its genomic context.

One modern textbook illustrates its chapter on selection by a graph showing the steady decrease in frequency of a disadvantageous mutant in a population cage (Crow 1986). This figure was published in *Genetics* in 1960. It proves the accuracy of the basic equations, which every student learns in population genetics on "determinist models." Population cages are thus reduced to experimental designs for classroom demonstrations. The real story is quite different. Sixty years ago, the first population cages were an impertinent challenge to theory. They showed that the decrease in frequency of deleterious mutants did not systematically lead to extinction. This has yet to be disproved.

4.13 The End of the Heroic Period

Before World War II, L'Héritier and Teissier's work had developed in the privileged surroundings of the ENS, which effectively protected them from the hostility of most French zoologists towards Darwinism. This institution hosted many briliant, mathematically trained biologists, including Gustave Malécot before he got his professorship in Lyons and left biology to study economics. Malécot published his first article in 1937 and defended his doctoral thesis in 1939 (Malécot 1937, 1939). His thesis was supervised by a mathematician, Darmois, who also examined L'Héritier's thesis. These facts, among others, show that L'Héritier and Teissier were not isolated. Their joint work took place in a context of genuine enthusiasm shown by a number of young scholars for genetics and population biology.

After the war, Teissier's work was continued by brilliant students such as Lamotte, Petit, Bocquet, and Boesiger, who worked in his laboratory, although he seems to have been a somewhat absent supervisor. Lamotte's thesis gave a new impetus to Malécot's research. However, the new orientation gave rise to a new school that was more concerned with natural populational aspects. Lamotte had received Wright's book from Teissier with the suggestion that he apply it to natural populations. Lamotte chose to verify Wright's distribution models in the land snail *Cepea nemoralis.* Malécot provided decisive help to Lamotte by extending Wright's model to render it compatible with Lamotte's empirical data. Similarly, Hoestlandt, Bocquet, Levi, and Lejuez decided to test models of isolation by distance on the color polymorphism of an intertidal crustacean, *Sphaeroma serratum.*[11] Shore invertebrates are linearly distributed for the simple reason that it is in the nature of their environment. This situation was expected to increase the possibility of isolation by distance according to theoretical predictions that the French group simply called the "Wright effect." Although Lamotte and Bocquet were concerned with natural populations, Boesiger and Petit took over the *Drosophila* experimental program to test hypotheses on the maintenance of polymorphism. Boesiger worked on the hypothesis that sexual selection favored the most heterozygous males, whereas Petit studied frequency-dependent selection and found evidence of a phenomenon she called the rare-male effect.

Those who learned population genetics under Teissier or L'Héritier all confirm that there was a central theoretical thread. An account of the work of the

"Teissier school of population genetics," as perceived on the basis of the lectures given between the late 1940s and the early 1950s, was later provided by these researchers (see for instance Lamotte 1974, Petit 1995). It is notable that it was Teissier, and only he, who was at the origin of this school.

4.14 Was the L'Héritier–Teissier School Really Isolated?

Even a cursory reading of Teissier and L'Héritier's articles shows that they made little effort to bring their work to the notice of their foreign colleagues. There was no French genetics journal. Their papers were all published in two generalist journals, the *Comptes Rendus de l'Académie des Sciences de Paris* and the *Comptes Rendus de la Société de Biologie*, which only published short notes no more than four pages long and with no division into sections. There were no materials and methods, no results or discussion sections, no references, and even no summaries! They barely included a table or a graph. The conclusions were formulated in a few sentences and looked more like reflections than supported arguments. There is a striking contrast between the long series of numbers, statistics, and equations recorded in L'Héritier and Teissier's bench notebooks and the virtual absence of such data in their papers. At the same time, Teissier was the unchallenged French specialist in biological statistics, which were introduced to France by textbooks written by his students, L'Héritier and Lamotte. The virtual absence of statistics in Teissier and L'Héritier's papers can only be explained by the fact that their publications were intended to be read by French biologists who were generally not familiar with the field.

Their papers were all written in French and nobody seems to have worried about publishing in other journals. On three occasions Teissier published in a foreign language. His paper in *Nature* with Huxley on relative growth was published in French, English, and German (Huxley and Teissier 1936a,b,c) and followed an abundant three months' correspondence during which the two authors (who did not meet) each wrote in his own language. In 1945, Teissier published a paper in defence of Darwinism, "Mécanisme de l'évolution," in a Communist party theoretical journal, *La Pensée*. His curriculum (Teissier 1967) mentions that this paper was translated into English by J. B. S. Haldane for the *Modern Quarterly* and into German by L. von Bertalanffy for *Wort und Tort*. In 1956, Teissier reported to the *International Genetics Symposia* on the existence of a racial differentiation between European and Japanese populations of *Drosophila melanogaster*, and this appeared in English in the proceedings of the congress (1957).

Teissier's English seems to have been very limited,[12] and he made little effort to change the situation. In 1952, the editorial board of the recently founded journal *Evolution* rejected his request to submit manuscripts not written in English. However, there is plenty of evidence that before the war he corresponded with Haldane, Huxley, and Wright, and, after the war, with Mayr. There were some misunderstandings. For instance, when Dobzhansky went to Paris in the 1960s and visited Teissier's laboratory, Teissier did not show up! Witnesses

(Krimbas, Lamotte, Petit) confirm that Teissier hated Dobzhansky. However, Dobzhansky was able to meet Teissier's students Petit and Boesiger. Nevertheless, Teissier was not ignored as a scientist by his foreign colleagues. As shown by several international references of the late 1930s, L'Héritier and Teissier's work on population cages was known and duly discussed.[13]

Why was there so little exchange between the French and American scientists? Communication was clearly established in 1939 with Wright's visit to Europe and the publication of his book in Paris. This event was crucial for providing the French with a full version of Wright's view of the genetics of natural populations. The first paper of the GNP series had appeared only the year before, and Teissier may not have realized the importance of Dobzhansky's program, both in its general implications and in its future development into a new school of population genetics. Reciprocally, Wright may have learned about the population cages on this occasion. During the war, when all communications between France and the United States were blocked, the 1941–43 correspondance between Wright and Dobzhansky extensively refers to trying "L'Héritier's population boxes." Several boxes appear to have been built by Mrs. Diederich (1941) and by Dobzhansky.

Apart from the fact that up until the late 1940s, English was not yet the language of science, one reason why L'Héritier and Teissier were so little interested in publishing their results in English was probably the lack of pressure from the academic system of the time. At the end of the war, Teissier had a great deal of power in the French university system. He had been a national commander of the FTP[14] headquarters, the best organized resistance group during the Nazi occupation. This, as much as scientific reasons, weighed heavily in the creation of the first chair of genetics at the Sorbonne after the war. Teissier was the director of the CNRS (*Centre National de la Recherche Scientifique,* the main French research institute) until he was fired because of his communist beliefs at the beginning of the Cold War. However, he retained a lot of power, including the directorship of several laboratories: Gif-sur-Yvette, the *Laboratoire de Zoologie* of the Paris University and the Roscoff marine biology station in Britanny. He was also nominated to the *Académie des Sciences.* This meant influence in journals like the *Cahiers de Biologie Marine* and the *Archives de Biologie,* which published most of Teissier's students' work. However, this system did not encourage people to communicate with the rest of the world.

4.15 The "French Context"

To close this discussion of the L'Héritier–Teissier collaboration, a few words on the context are necessary. Many people – including L'Héritier, Teissier, and Boesiger – have said that in the 1920s both genetics and Darwinism were in a terrible state in France (see, for instance, L'Héritier 1981; Teissier 1945, 1952, 1961; Boesiger 1980). However, we should not take this for granted. First, one should not confound genetics and Darwinism. Darwinism was certainly unpopular among French biologists in the 1930s. The situation was more complicated for genetics, however. As shown by Burian, Gayon, and Zallen (1988), teaching and research were subject to different pressures. There is no doubt that genetics was poorly taught in France in the interwar years. But it was precisely in the mid-1930s that a number of young and brilliant French biologists adopted the genetic paradigm and played a major international role in its renewal. Lwoff and Wollman began the study of bacterial genetics while Ephrussi developed his crucial studies of *Drosophila* eye pigmentation in collaboration with Beadle. It was also in this period that the young Monod went to the United States with Ephrussi and decided to work under the direction of Teissier when he returned to France. Another contextual element was the quality and quantity of publications. In the 1930s, Ephrussi, like Teissier, was an associate editor of the Hermann publishing company and helped publish a series of monographs in genetics.[15]

4.16 The Idea Was in the Cage

The systematic study of genetic changes in *Drosophila* populations began almost simultaneously in five different countries. Working in Berlin, Timofeef-Ressovsky published his first paper in 1932 (Timofeef-Ressovsky 1932). In France, L'Héritier and Teissier's work began in 1932, and the first paper was published in 1933 (L'Héritier and Teissier 1933). In England, Gordon published his first papers in 1935 and 1936 (Gordon 1935, 1936). Dubinin and collaborators' first work on Russian populations of *D. melanogaster* was published in 1936. In America, Dobzhansky's work on *D. pseudoobscura* began in 1936, and the first paper appeared in 1938 (Dobzhansky and Queal 1938). From the very beginning, the French program included an emphasis on the selective maintenance of polymorphism and a critical view of the constant fitness coefficients that were postulated by mathematical models. However, their prime concern was not evolution but the coexistence of individuals (from the same species or from different species) in the ecological milieu. Their conceptual reference, not to say their everlasting target, was Volterra's exclusion principle. They believed that genetically different individuals had different "needs" and so could not completely displace each other. They also believed that individuals from the same class were more in competition between themselves than they were with others. For this reason, the "struggle for life" was frequency dependent.

With these few principles in mind, they challenged the theory and, in fact, succeeded perfectly. However, their small cages were not able to go much beyond the simple observation that fitness coefficients were not constant and that equilibrium frequencies shifted from time to time with changing linkage between alleles at different loci.

When Wright and Dobzhansky came to the conclusion that balancing selection also explained inversion polymorphism in *D. pseudoobscura*, they used the cages that had already proved effective in this respect.

The main difference between the two groups was that the experimental device merely helped the two French biologists to a better understanding of a process that they had accepted a priori. For the American researchers, the story was different. Dobzhansky adopted the view that polymorphisms could be adaptive after falsifying his original idea that inversion polymorphism was neutral.

We initially addressed the question of how and why two schools of population genetics, in France and in America, reached similar conclusions, and we needed to know to what extent they were really isolated from each other.

The starting point of the two schools was virtually the same. Their purpose was to test the validity of theoretical population genetics. They both chose *Drosophila* as a research material because they wanted to record allele frequency changes directly. This was a substantial departure from traditional experimental zoology, which would have preferred more adaptively meaningful characters like morphological or physiological phenotypes.

They also ended up in much the same place, as shown by the culmination of their work in the 1950s and 1960s. Both schools used population cages. Claudine Petit in France, and Lee Ehrman in the United States, were both studying frequency-dependent selection through the rare-male effect. Theodosius Dobzhansky and Ernest Boesiger wrote a book together, *Essai sur l'Evolution*, which was published in Paris in 1968. It attempted to present the evolutionary synthesis by referring quasi-exclusively to Teissier's, L'Héritier's, and Dobzhansky's results. The population cages of each group were shown photographed side by side. It shows that the two groups admitted that they were fundamentally identical.

Between the starting point and the final conclusion, however, the two groups did not feel their work was convergent. This cannot be explained solely by their different publication traditions or by isolation due to the war because we have evidence that they knew of each other's research. The main isolating factor was the difference in their approaches. The American school was interested in the stability of natural polymorphisms. Their initial purpose was to record segregating polymorphism in populations on the assumption that it was selectively neutral. When they recognized that it was not, the issue became one of providing selective explanations for the maintenance of segregating polymorphism. Up until the discovery of allozyme polymorphism in the 1960s (Lewontin 1974), inversion and recessive lethals were the most suitable material for a study of natural polymorphism. To summarize, Dobzhansky's school was interested in the responses of the genome to selection originating from the environment. On the contrary, the French group was interested in biotic factors involved in selection. It focused on the competitive pressure resulting from the population itself. This group was less interested in equilibrium frequencies than in the way they changed and used laboratory experiments where such factors could be controlled. This concern could accommodate any amount of segregating variation. It was also compatible with the study of laboratory mutations, although these were generally gross abnormalities that did not occur in the wild. In fact, one of this school's favorite materials was the *sepia* mutation, which could be studied in competition with the wildtype stock from which it originated. The only time this school introduced an external factor was in the Roscoff experiment using wingless flies. It did it very carefully in what could be briefly described as a laboratory experiment using external wind just as some experiments use room temperature or the natural photoperiod. In a way, each group was trying to exclude the factors that the other valued the most, and they were isolated for intellectual reasons. They genuinely represented two independent routes, each leading to the main conclusion of *Drosophila* population genetics in the 1950s.

4.17 Acknowledgments

We thank Maxime Lamotte and Claudine Petit for being interviewed as part of the preparatory work for this study, Françoise Benhamou and Lily Joly for help and advice, and Matthew Cobb for his comments, which greatly improved the clarity of the manuscript.

Notes

1. See, for example, Mayr and Provine 1980, Lewontin et al. 1981, and Provine 1986. The only available synthesis on the origins of the French school of population genetics is Petit 1995. Partial information can also be found in Dorst 1974; Boesiger 1980; Mayr 1980; Burian, Gayon, and Zallen 1988; Burian and Gayon 1989–90; Gayon 1992; Gillois 1996; Krimbas 1995.
2. Interviews with Claudine Petit, 12 December 1996 and Maxime Lamotte, 23 December 1996.
3. Source: correspondence of L'Héritier with Teissier, Teissier archives, Université Paris 6 – Ecole Pratique des Hautes Etudes.
4. For a detailed bibliography on these two subjects, see Teissier 1958b.
5. This does not mean that L'Héritier and Teissier ignored population genetic theory. On the contrary, their notebooks show that this formula was developed after a long discussion of the respective merits of Haldane, Fisher, and Wright. At this time, they were more strongly influenced by Haldane for reasons that we will discuss in more detail elsewhere.
6. For contextual reasons, it is interesting to note that Malécot, then at the ENS, published a theoretical paper on this subject the same year in which he tried to generalize Fisher's demonstration (Malécot 1937).
7. In fact, in the second case, they put some *D. funebris* in a population of *white D. melanogaster.*
8. This comment refers to the publication of Gause's *Experimental Verifications of the Mathematical Theory of the Struggle for Life* (1935, in French) in Teissier's Hermann collection. This essential and rare book was very different from *The Struggle for Existence* (1934) published the year before in Pearl's collection.
9. Source: interview with Maxime Lamotte, 23 December 1996.
10. See Figure 9.2 in Wright 1977.
11. For detailed bibliography, see the reference section under Bocquet, Goudeau, Hoestlandt, Levi, and Lejuez.
12. In 1936, Teissier wrote to Huxley: Il me serait très agréable de vous rencontrer, mais je ne sais si j'aurai bientôt l'occasion d'aller à Londres. Je dois vous avouer au surplus que je ne parle pas un mot d'Anglais et que par conséquent nous ne

pourrions examiner les questions qui nous intéressent qu'en français ou en faisant appel à un interprête. (I would be most delighted to meet you, but I don't know if I will be going to London in the near future. Besides, I must confess that I do not speak a word of English and that consequently we could only discuss the questions that interest us in French, or through an interpreter.)

13. See especially Timofeeff-Ressovsky's comment at the 1939 Edinburgh Conference (Timofeeff-Ressovsky 1940). L'Héritier and Teissier also gave separate talks on their work at the 1937 *Réunion Internationale de Physique-Chimie-Biologie*, where Haldane, Muller, and Timofeeff-Ressovsky were present (Haldane 1938).
14. FTP, *Franc-Tireurs et Partisans*, was the Communist party resistance organization.
15. Ephrussi's collection ("Genetics") included books by Dobzhansky, Sturtevant, himself, and others, whereas Teissier's collection ("Biometrics and Biological Statistics") included books by Gause, Wright, Volterra, himself, and others.

REFERENCES

Primary Sources

Anxolabéhère, D. (1978). Analyse expérimentale et théorique du rôle des valeurs sélectives dans le maintien du polymorphisme au locus *sepia* chez *Drosophila melanogaster*. Ph.D. dissertation, Paris.

Bocquet, C. (1967). Remarques sur le polychromatisme géographique de *Sphaeroma serratum*, F. et description du phenotype nouveau *Bimaculatum*, *L'année biologique*, 4e série, T. VI, fasc. 9–10:435–44.

Bocquet, C., Lejuez, R., and Teissier, G. (1960). Génétique des populations de *Sphaeroma serratum*, F.. III. – Comparaison des populations mères et des populations filles pour les Sphéromes du Contentin, *Cahiers de biologie marine*, 1:279–94.

Bocquet, C., Lejuez, R., and Teissier, G. (1964). Génétique des populations de *Sphaeroma serratum*, F.. V. – Etude des populations entre Barfleur et l'embouchure de la Seine, *Cahiers de biologie marine*, 5:1–16.

Bocquet, C., Lejuez, R., and Teissier, G. (1965). Génétique des populations de *Sphaeroma serratum*, F.. VI. – Mise en évidence de la panmixie chez *Sphaeroma serratum*, *Cahiers de biologie marine*, 6:195–200.

Bocquet, C., Lejuez, R., and Teissier, G. (1966). Génétique des populations de *Sphaeroma serratum*, F.. VII. – Données complémentaires sur la panmixie, *Cahiers de biologie marine*, 7:23–30.

Bocquet, C., Lévi, C., and Teissier, G. (1950a). Déterminisme génétique des types de coloration chez *Sphaeroma serratum*, Isopode flabellifère, *C.R. Ac. Sc.*, 230:871–3.

Bocquet, C., Lévi, C., and Teissier, G. (1950b). Distribution des types de coloration dans quelques populations de *Sphaeroma serratum* des côtes de Bretagne, *C.R. Ac. Sc.*, 230:1004–6.

Bocquet, C., Lévi, C., and Teissier, G. (1951). Recherches sur le polychromatisme de *Sphaeroma serratum*, F.. Description, étude génétique et distribution sur les côtes de Bretagne des divers types de coloration, *Archives de zoologie expérimentale et générale*, 87:245–98.

Bocquet, C., and Teissier, G. (1960a). Génétique des populations de *Sphaeroma serratum*, F.. I. – Stabilité du polychromatisme local, *Cahiers de biologie marine*, 1:103–11.

Bocquet, C., and Teissier, G. (1960b). Génétique des populations de *Sphaeroma serratum*, F.. II. – Calcul des fréquences géniques, *Cahiers de biologie marine*, 1:221–30.

Boesiger, E. (1967). Signification évolutive du polygénotypisme des populations naturelles, *L'année biologique*, 4e série, T. VI, fasc. 9–10:445–64.

Bregliano, J.-C., and Kidwell, M. G. (1983). Hybrid dysgenesis determinants. In *Mobile Genetic Elements.* J. A. Shapiro, ed. Academic Press: 363–410.

Crow, J. F. (1986). *Basic Concepts in Population, Quantitative, and Evolutionary Genetics.* W. H. Freeman and Co., New York.

Darwin, C. (1859). *On the Origin of Species.* London: Murray.

Diederich, G. W. (1941). Non-random mating between yellow-white and wild Type *Drosophila melanogaster, Genetics,* 26:148.

Dobzhansky, Th. (1946). Experimental reproduction of some of the changes caused by natural selection in certain populations of *Drosophila pseudoobscura, Genetics,* 31:125–56.

Dobzhansky, Th., and Queal, M. L. (1938). Chromosome variation in populations of *Drosophila pseudoobscura* inhabiting isolated mountain ranges. *Genetics,* 23:239–51.

Dobzhansky, Th. (1965). "Wild" and "domestic" species of *Drosophila,* in H. G. Baker and G. Ledyard Stebbins, eds., *The Genetics of Colonizing Species,* Academic Press, New York and London.

Dobzhansky, Th., Conférences données à la Faculté des Sciences d'Orsay. Cited in Boesiger 1967.

Dobzhansky, Th., and Boesiger, E. (1968). *Essais sur l'évolution,* Paris Masson.

Fisher, R. A. (1922). On the dominance ratio. *Proc. Roy. Soc.* Edinburgh. 42:321–41.

Fisher, R. A. (1930). *The Genetical Theory of Natural Selection,* Oxford, UK, Clarendon Press.

Gause, G. F. (1934). *The Struggle for Existence,* Baltimore, Williams and Wilkins.

Gause, G. F. (1935). *Vérifications Expérimentales de la Théorie Mathématique de la Lutte pour la Vie,* Paris, Hermann.

Gordon, C. (1935). An analysis of two wild *Drosophila* populations. *American Naturalist,* 69:381.

Gordon, C. (1936). The frequency of heterozygosis in free-living populations. *Journal of Genetics,* 33:25–60.

Goudeau, M. (1966). Génétique des populations de *Sphaeroma serratum,* F.. VIII. Nouvelles observations sur la stabilité du polychromatisme local, *Cahiers de biologie marine,* 7:23–30.

Haldane, J. B. S. (1938). L'analyse génétique des populations naturelles, in *Réunion Internationale de Physique-Chimie-Biologie,* Congrès du Palais de la Découverte, Paris, 1937, Paris, Hermann: 105–12.

Helman, B. (1949). Etude dela vitalité relative du génotype sauvage *Oregon* et de génotypes comportant le gène *Stubble* chez *Drosophila melanogaster. C.R. Acad. Sci. Paris,* 228:2057–8.

Hoestlandt, H. (1954). Recherches complémentaires sur le polychromatisme de populations de l'Isopode marin *Sphaeroma serratum,* le long des côtes d'Irlande, *C.R. Ac. Sc.,* 238:2360–2.

Hoestlandt, H. (1955). Limite nordique de l'extension d'un Crustacé marin de la faune lusitanienne, *Sphaeroma serratum, C.R. Ac. Sc.,* 240:683–5.

Hoestlandt, H. (1955). Etude de populations de *Sphaeroma serratum* le long du littoral de la Grande Bretagne, *C.R. Ac. Sc.,* 240:916–19.

Hoestlandt, H. (1956). Examen de populations de *Sphaeroma serratum,* sur le côtes dela péninsule ibérique, *C.R. Ac. Sc.,* 243:1561–3.

Hoestlandt, H. (1956). Etude de populations de *Sphaeroma serratum* sur les côtes de l'Archipel des Açores, *C.R. Ac. Sc.*, 243:1680–3.

Hoestlandt, H. (1957). Aspect phénotypique de populations de *Sphaeroma serratum* sur les côtes de Madère, des Canaries et du Maroc Atlantique, *C.R. Ac. Sc.*, 245:2410–13.

Hoestlandt, H. (1962). Examens de populations de l'Isopode marin *Sphaeroma serratum* aux limites méridionales de son extension, *C.R. Ac. Sc.*, 254:3584–6.

Hoestlandt, H., and Teissier, G. (1952). Sur le polychromatisme des *Sphaeroma serratum* du littoral Boulonnais, *C.R. Ac. Sc.*, 234:667–9.

Hoestlandt, H., and Teissier, G. (1952). Sur le polychromatisme des *Sphaeroma serratum* le long des côtes d'Irlande, *C.R. Ac. Sc.*, 235:1052–4.

Huxley, J. S., and Teissier, G. (1936a). Terminology of relative growth. *Nature*, 137: 780–1.

Huxley, J. S., and Teissier, G. (1936b). Terminologie et notations dans la description de la croissance relative. *Comptes Rendus des Séances de la Société de Biologie*, 121:934–7.

Huxley, J. S., and Teissier, G. (1936c). Zur Terminologie des relativen Grössenwachstums. *Biol. Zentralbl.*, 56:381.

Kidwell, M. G. (1979). Hybrid dysgenesis in *Drosophila melanogaster*. The relationship between the P–M and I–R interaction systems. *Genetical Research* 33:205–17.

Lamotte, M. (1948). *Introduction à la Biologie quantitative. Présentation et interprétation statistiques des données numériques*, Paris, Masson.

Lamotte, M. (1951). Recherches sur la structure génétique des populations naturelles de *Cepæa nemoralis* L., *Bull. Biol. de France et de Belgique*, suppl. 35:1–239.

Lamotte, M. (1959). Polymorphism of Natural Populations of *Cepæa nemoralis, Cold Spring Harbor Symposium on Quantitative Biology*, 24:65–86.

Lamotte, M., éd. (1972). *Le Polymorphisme dans le règne animal, hommage à Georges Teissier, Mémoires de la Société Zoologique de France*, No. 37.

Lamotte, M. (1974). Introduction, in *Le polymorphisme dans le règne animal. – Volume publié par la Société Zoologique de France sous la direction de Maxime Lamotte à la mémoire de Georges Teissier, Mémoires de la Société Zoologique de France*, No. 37:11–13.

Lejuez, R. (1958). Sur le polychromatisme de *Sphaeroma serratum*, F. le long du littoral occidental du Cotentin, *C.R. Ac. Sc.*, 247:659–61.

Lejuez, R. (1959). Distribution des types de coloration de *Sphaeroma serratum* sur la côte occidentale du Cotentin, *Bulletin de la Société Linéenne de Normandie*, 9e série, 10e vol.: 39–57.

Lejuez, R. (1960). Sur le poychromatisme de *Sphaeroma serratum*, F. le long du littoral septentrional du Cotentin, *C.R. Ac. Sc.*, 251:1244–6.

Lejuez, R. (1961). Génétique des populations de *Sphaeroma serratum*, F.. IV. – Etude des populations de la côte septentrionale du Cotentin, *Cahiers de biologie maritime*, 2:327–42.

Lewontin, R. C., Moore, J. A., Provine, W. B., and Wallace, B., eds. (1981). *Dobzhansky's Genetics of Natural Populations I–XLIII*, New York, Columbia University Press.

L'Héritier, P. (1932). Comparaison de cinq lignées de Drosophile, type sauvage, au point de vue de leur survivance en présence d'une nourriture toxique, *C.R. Soc. Biol.*, 111:982–4.

L'Héritier, P. (1934). *Génétique et évolution: Analyse de quelques études mathématiques sur la sélection naturelle*, Paris, Hermann, Actualités Scientifiques et Industrielles.

L'Héritier, P. (1937). *Étude de variations quantitatives au sein d'une espèce: Drosophila melanogaster* Meig., Paris, *Éditions des archives de zoologie expérimentale*: 255–356.

L'Héritier, P. (1938). Discussion, following Haldane, 1938, in *Réunion Internationale de Physique-Chimie-Biologie*, Congrès du Palais de la Découverte, Paris, 1937, Paris, Hermann: 522–4., L'Héritier summarizes the results of the various experiments made with population cages in collaboration with Teissier.

L'Héritier, P. (1949). *Les méthodes statistiques dans l'expérimentation biologique.* Publications du CNRS, Paris.

L'Héritier, P. (1954). *Traité de Génétique*, t. 2, *La Génétique des populations*, Paris, Presses Universitaires de France.

L'Héritier, P. (1981). Souvenirs d'un généticien, *Revue de Synthèse*, 102:331–50.

L'Héritier, P., Neefs, Y., and Teissier, G. (1937). Aptérisme des insectes et sélection naturelle, *C.R. Ac. Sc.*, 204:907–9.

L'Héritier, P., and Teissier, G. (1933). Etude d'une population de Drosophiles en équilibre, *C.R. Acad. Sc.*, 197:1765–7.

L'Héritier, P., and Teissier, G. (1934a). Sur quelques facteurs du succès dans la concurrence larvaire chez *Drosophila melanogaster*, *C.R. Soc. Biol.*, 116:306.

L'Héritier, P., and Teissier, G. (1934b). Une expérience de sélection naturelle, Courbe d'élimination du gène *Bar* dans une population de *Drosophila melanogaster*, *C.R. Soc. Biol.*, 117:1049.

L'Héritier, P., and Teissier, G. (1935). Recherches sur la concurrence vitale. Etude de populations mixtes de *Drosophila melanogaster* et de *Drosophila funebris*, *C.R. Soc. Biol.*, 118:1396–8.

L'Héritier, P., and Teissier, G. (1936). Contribution à l'étude de la concurrence larvaire chez les Drosophiles, *C.R. Soc. Biol.*, 122:264–7.

L'Héritier, P., and Teissier, G. (1937a). Elimination des formes mutantes dans les populations de Drosophiles. Cas des Drosophiles bar, *C.R. Soc. Biol.*, 124:880–2.

L'Héritier, P., and Teissier, G. (1937b). Elimination des formes mutantes dans lès populations de Drosophiles Cas des Drosophiles Ebony, *C.R. Soc. Biol.*, 124:882–4.

L'Héritier, P., and Teissier, G. (1937c). L'élimination des formes mutantes dans les populations de Drosophiles, *Soixante-dixième Congrès des Sociétés Savantes*: 297–302.

L'Héritier, P., and Teissier, G. (1937d). Une anomalie physiologique héréditaire chez la Drosophile, *C.R. Acad. Sc.*, 205:1099–1101.

L'Héritier, P., and Teissier, G. (1938a). Transmission héréditaire de la sensibilité au gaz carbonique chez la Drosophile, *C.R. Ac. Sc.*, 206:1683–5.

L'Héritier, P., and Teissier, G. (1938b). Un mécanisme héréditaire aberrant chez la Drosophile, *C.R. Ac. Sc.*, 206:1193–4.

L'Héritier, P., and Teissier, G. (1940). Une anomalie physiologique héréditaire chez la Drosophile, *Proceedings of the VIIth International Congress of Genetics*, Edinburgh.

L'Héritier, P., and Teissier, G. (1944). Transmission héréditaire de la sensibilité au gaz carbonique chez *Drosophila melanogaster*, Travaux des Laboratoires de l'E.N.S., Biologie, fasc. 1:35–74.

Malécot, G. (1937). Quelques conséquences de l'hérédité mendélienne, *C.R. Acad. Sc.*, 204:619–22.

Malécot, G. (1939). *Théorie mathématique mendélienne généralisée*, Thèse de doctorat ès sciences, Paris, Imp. Guilhot, 1939.

Malécot, G. (1948). *Les mathématiques de l'hérédité*, Paris, Masson.

Pearl, R. (1925). *The Biology of Population Growth*, New York, Knopf.

Pearl, R. (1927). The growth of populations, *Quarterly Review of Biology*, 2:532–48.

Pearl, R. (1928). *The Rate of Living.* New York: A. A. Knopf.

Petit, C. (1951). Le rôle de l'isolement sexuel dans l'évolution des populations de *Drosophila melanogaster*, *Bull. Biol. de la France et de la Belgique*, 85:391–418.

Petit C. (1958). Le déterminisme génétique et psycho-physiologique de la compétition sexuelle chez *Drosophila melanogaster*, *Bull. Biol. de la France et de la Belgique*, 92:248–329.

Picard, G., and L'Héritier, P. (1971). A maternally inherited factor inducing sterility in *Drosophila melanogaster*. *Dros. Inf. Serv.* 46:54.

Teissier, G. (1934). Recherches sur le vieillissement et sur les lois de la mortalité. I. – Introduction historique, *Annales de physiologie et de physicochimie biologique*, 10:236–84.

Teissier, G. (1938). Discussion, following Haldane, 1938, in *Réunion Internationale de Physique-Chimie-Biologie*, Congrès du Palais de la Découverte, Paris, 1937, Paris, Hermann: 524–5.

Teissier, G. (1942a). Vitalité et fécondité relative de diverses combinaisons génétiques comportant un gène léthal chez la Drosophile, *C.R. Ac. Sc.*, 214:241–4.

Teissier, G. (1942b). Persistance d'un gène léthal dans une population de Drosophiles, *C.R. Ac. Sc.*, 214:327.

Teissier, G. (1943). Apparition et fixation d'un gène mutant dans une population stationnaire de Drosophiles, *C.R. Acad. Sc.*, 216:88–90.

Teissier, G. (1944). Equilibre des gènes léthaux dans les populations stationnaires panmictiques, *Revue Scientifique*, 82:145–59.

Teissier, G. (1945). Mécanisme de l'évolution, *La Pensée*, 2:3–19; 3:5–31.

Teissier, G. (1947a). Variations de la fréquence du gène *sepia* dans une population stationnaire de Drosophiles, *C.R. Acad. Sc.*, 224:676–7.

Teissier, G. (1947b). Variations de la fréquence du gène *ebony* dans une population stationnaire de Drosophiles, *C.R. Acad. Sc.*, 224:1788–9.

Teissier, G. (1952). Dynamique des populations et taxinomie, *Annales de la Société Royale Zoologique de Belgique*, Fasc. 1, Tome LXXXIII: 23–44.

Teissier, G. (1953). Variations de l'équilibre des gènes dans les populations expérimentales de Drosophiles, *Proceedings of the XIVth International Congress of Zoology*: 141–4.

Teissier, G. (1954a). Conditions d'équilibre d'un couple d'allèles et supériorité des hétérozygotes, *C.R. Ac. Sc.*, 238:621–3.

Teissier, G. (1954b). Sélection naturelle et fluctuation génétique, *C.R. Ac. Sc.*, 238: 1929–31.

Teissier, G. (1957). Discriminative Biometrical Characters in French and Japanese *Drosophila melanogaster*, *Proceedings of the International Genetics Symposia*, 1956:502–5.

Teissier, G. (1958a). Distinction biométrique des *Drosophila melanogaster* françaises et japonaises, *Annales de génétique*, 1:2–10.

Teissier, G. (1958b). *Titres et travaux scientifiques*, Paris, Prieur et Robin.

Teissier, G. (1961). Transformisme d'aujourd'hui, *Editions de la Station biologique de Roscoff*. Repris dans *L'Année biologique*, 4è série, 1, 1962:359–75.

Teissier, G. (1962). *Supplément aux titres et travaux scientifiques*, Paris, Robin et Mareuge.

Teissier, G. (1967). *Titres et travaux scientifiques. Supplément*, Paris, Robin et Mareuge.

Teissier, G. (1969). Génétique des populations de *Sphaeroma serratum*. Isopode littoral, *Publ. Staz. Zool. Napoli*, 37, suppl.: 135–45.

Timofeeff-Ressovsky, N. (1932). Verschiedenheit der normallen Allele der white-Serie aus geographisch getrennten Populationen von *Drosophila melanogaster*. *Biol. Zentralbl.* 52:468.

Timofeeff-Ressovsky, N. (1934). Über die Vitalität einiger genmutation und ihrer Kombinationen bei *Drosophila funebris* und ihre Abhängigkeit vom genotypischen und vom äusseren Milieu, *Z. indukti. Abstamm. – und Vererbungs-Lehre*, 66.

Timofeeff-Ressovsky, N. (1940). Mutations and geographical variation, in J. Huxley, ed., *The New Systematics*: 73–135.

Volterra, V. (1931). *Leçons sur la théorie mathématique de la lutte pour la vie*, rédigées par Marcel Brelot, Paris, Editions Gauthier-Villars. Fac-simile: Paris, Editions Jacques Gabay, 1990. The lessons were actually given in Paris between 1928 and 1929.

Wright, S. (1931). Evolution in Mendelian Populations, *Genetics*, 16:97–159.

Wright, S. (1932). The Roles of Mutation, Inbreeding, Crossbreeding, and Selection in Evolution, *Proceedings of the 6th International Congress of Genetics*, 1:352–66.

Wright, S. (1939). *Statistical Genetics in Relation to Evolution*, Paris, Hermann.

Wright, S. (1977). *Evolution and the Genetics of Natural Populations*, vol. 3, *Experimental Results and Evolutionary Deductions*, Chicago and London, The Chicago University Press.

Historical Studies

Boesiger, E. (1980). Evolutionary biology in France at the time of the evolutionary synthesis, in E. Mayr and W. B. Provine, 1980:309–21.

Burian, R. M., Gayon, J., et Zallen, D. (1988). The Singular Fate of Genetics in the History of French Biology, 1900–1940, *Journal of the History of Biology*, 21:357–402.

Burian, R. M., and Gayon, J. (1989–90). Genetics after World War II: The Laboratories at Gif, in *Cahiers pour l'histoire du CNRS*, 6, 1989:108–10; 7, 1990:25–48.

Dorst, J. (1974). Notice sur la vie et l'oeuvre de Georges Teissier (1900–1972), membre de la section de zoologie, Paris, Palais de l'Institut.

Gayon, J. (1992). *Darwin et l'après-Darwin: une histoire de l'hypothèse de sélection dans la théorie de l'évolution*, Paris, Kimé.

Gillois, M. (1996). Malécot, in P. Tort, ed., *Dictionnaire du darwinisme et de l'évolution*, Paris, Presses Universitaires de France: 1768–1785.

Krimbas, C. (1995). Resistance and acceptance: Tracing Dobzhansky's influence in Europe. *Genetics of Natural Populations: The Continuing Importance of Theodosius Dobzhansky*, L. Levine, ed., New York, Columbia University Press: 23–36.

Lewontin, R. C. (1974). *The Genetic Basis of Evolutionary Change.* New York, Columbia University Press.

Lewontin, R. C. (1981). Introduction: The Scientific Work of Th. Dobzhansky, in *Dobzhansky's Genetics of Natural Populations I–XLIII*, R. C. Lewontin, J. A. Moore, W. B. Provine, and B. Wallace eds., New York, Columbia University Press: 93–115.

L'Héritier, P. (1981). Souvenirs d'un généticien-Revue de Synthèse, 102:331–50.

Mayr, E. (1980). The Arrival of Neo-Darwinism in France, in E. Mayr and W. B. Provine, 1980:321–2.

Mayr, E., and Provine, W. B., eds. (1980). *The Evolutionary Synthesis: Perspectives on the Unification of Biology*, Cambridge, MA, Harvard University Press.

Petit, C. (1995). L'implantation de la génétique des populations en France, in *Nature, histoire, société; Essais en hommage à Jacques Roger*, C. Blanckaert, J. L. Fischer, R. Rey, eds., Paris, Klinsieck, pp. 147–60.

Provine, W. B. (1986). *Sewall Wright and Evolutionary Biology*, Chicago, The University of Chicago Press.

CHAPTER FIVE

Did Eugenics Rest on an Elementary Mistake?

DIANE B. PAUL AND HAMISH G. SPENCER

On the evidence of many genetics texts, of books on biology and society, and of histories of science, eugenicists were guilty of an astoundingly simple mistake. According to conventional accounts, which vary only in details, eugenics enthusiasts thought they could eliminate mental deficiency by segregating or sterilizing affected individuals. But a basic understanding of the Hardy–Weinberg principle suffices to destroy that illusion.

Eugenicists in the 1910s and 1920s attributed most mental defect to a recessive Mendelian factor (or in today's parlance, allele). But it is evident from the simple equation $p^2 + 2pq + q^2 = 1$ that if a trait is rare, the vast majority of deleterious genes will be hidden in apparently normal carriers. Selection against the affected themselves will thus be ineffectual. For example, even if all the affected were prevented from breeding, in a single generation the incidence of a trait at an initial frequency of 0.000100 would be reduced to just 0.000098 (and the allele frequency from 0.0100 to 0.0099). To reduce the incidence to half its original value (i.e., 0.000050) would require some 41 generations, or about 1000 years. Tables in numerous genetics textbooks serve to make the point that hundreds of generations are required before a rare deleterious trait would disappear. Because a human generation lasts about 25 years, eugenical selection would be futile over any meaningful period. P. B. and J. S. Medawar express a common view: the eugenicists were ignorant and muddled (as well as foolish and inhumane). "Only a minority of the offending genes are locked up in the mentally deficient themselves," they explain, "so sterilizing them would not be effective" (Medawar and Medawar 1977, p. 60).

That selection is slow when genes are rare is not a new insight. Indeed, Roll-Hansen (1980) has noted that, as early as 1914, the Norwegian psychiatrist Ragnar Vogt had emphasized the difference between "positive" (dominant) and "negative" (recessive) hereditary diseases, noting that the former could be eradicated by preventing those who have the disease from reproducing, whereas the latter "are transmitted through one or more healthy intermediate links, and only a small part of the diseased individuals have diseased parents. The socially most important hereditary diseases, such as certain kinds of deaf-muteness, feeblemindedness, and mental illness appear to behave like negative (recessive) hereditary diseases, and one can therefore not achieve any appreciable eugenic effect by preventing the diseased individuals from having children through prohibition of marriage or similar means" (Vogt 1914, pp. 4–5). Vogt's point seems not to have been appreciated at the time, but in 1917 it was independently apprehended by Harvard geneticist Edward Murray East. East's argument was then refined by R. C. Punnett of Cambridge University, and Punnett's version was popularized by J. B. S. Haldane in Britain and H. S. Jennings in the United States. "To merely cancel the deficient individuals themselves – those actually feebleminded – makes almost no progress toward getting rid of feeblemindedness for later generations," wrote the latter (Jennings 1927, p. 273; Haldane 1928 [pub. in 1932], p. 105). If the futility of sterilization and segregation were exposed so early and often, it might seem that the numerous geneticists who endorsed these policies were a remarkably dim-witted lot.

Whatever their personal and political failings, this explanation is implausible. R. A. Fisher was a social reactionary, as well as ardent eugenicist. But his worst enemies did not think him stupid. He unquestionably understood the implications of Hardy–Weinberg. Moreover, when Punnett first articulated these implications, he did so in an effort to expand eugenics' scope, not demonstrate its futility. Indeed, in the 1920s and 1930s, nearly all geneticists, including those traditionally characterized as opponents of eugenics, took it for granted that "mental defectives" should be prevented from breeding. To see why few geneticists of that period drew the conclusions that seem so obvious to their present-day successors, let us review the original arguments about the threat represented by carriers.

5.1 The "Real Menace" of the Feebleminded

In his 1917 essay "Hidden Feeblemindedness," East argued that neither the character nor scope of the problem of mental defect had been fully appreciated. Although lauding efforts to cut off the stream of "defective germplasm" through segregation or sterilization of the affected, East thought the primary danger lay elsewhere in the vast mass of invisible heterozygotes.

He had been influenced in this view by the American psychologist Henry H. Goddard, author of *The Kallikak Family: A Study in the Heredity of Feeblemindedness* (1912), an impressionistic study of a "degenerate" rural clan, and *Feeble-Mindedness: Its Causes and Consequences* (1914), a much longer work that discussed the meaning of the data for theories of inheritance. In the latter book, Goddard had argued that "normal-mindedness is dominant and is transmitted in accordance with the Mendelian law of inheritance" (Goddard 1914, p. 556). His views were widely accepted. Thus Punnett could write in 1925 that no one "who has studied the numerous pedigrees collected by Goddard and others [could] fail to draw the conclusion that this mental state behaves as a simple recessive to the normal" (Punnett 1925, p. 704). William E. Castle also praised Goddard's research and uncritically reported his results. "Goddard's evidence," he wrote in an influential textbook, "indicates that feeble-mindedness is a recessive unit-character" (Castle 1927, p. 355). As late as 1930, Jennings was able to assert that feeblemindedness was "the clearest case" of a recessive single gene defect (Jennings 1930, p. 238). Although Paul Popenoe and R. H. Johnson did criticize Goddard's assumption that feeblemindedness resulted from a single gene, they accepted his claims that at least two-thirds of those affected owed their condition directly to heredity and that they numbered at least 300,000 (Popenoe and Johnson 1918, pp. 105, 175).

Biometricians such as David Heron of the Galton Laboratory in London disparaged both the methods and logic used to reach this conclusion. In a passionate response to the stream of publications coming out of Charles H. Davenport's Eugenics Record Office in Cold Spring Harbor, Heron attacked almost every aspect of the Americans' work. Although his essay predated publication of *Feeblemindedness,* Heron's critique was as applicable to Goddard as Davenport. He concluded "that the material has been collected in an unsatisfactory manner, that the data have been tabled in a most slipshod fashion, and that the Mendelian conclusions drawn have no justification whatsoever" (Heron 1913, p. 61). Heron and the other biometricians were themselves ardent eugenicists, with "the highest hopes for the new science" (p. 4). But they feared that eugenics would be crippled at birth by the American Mendelians' crude errors. Perhaps because of their unremitting anti-Mendelian rhetoric and personal style of attack, the biometricians' critiques were largely ignored by Mendelian geneticists on both sides of the Atlantic (Spencer and Paul 1998).

Davenport was one of the few Mendelian geneticists to criticize the category of feeblemindedness, which he characterized as a "lumber room" of different (and separately inherited) mental deficiences (Davenport 1912, 1915). He also noted the illogic of expecting a socially defined trait – a feebleminded person was considered "incapable of performing his duties as a member of society in the position of life to which he is born" (Popenoe 1915, p. 32) – to be inherited as a simple Mendelian recessive (Davenport 1912, p. 286; see also Holmes 1923, pp. 121–33; Wiggam 1924, pp. 56–8). But these were minor quarrels. Until the mid-1930s, Thomas Hunt Morgan was the only Mendelian geneticist consistently to repudiate Goddard's claim that social deviance was largely due to bad heredity (Barker 1989). In the 1925 edition of *Evolution and Genetics,* Morgan argued that much of the behavior tagged with that label was probably due to "demoralizing social conditions" rather than to heredity (Morgan 1925, p. 201). But Morgan's critique, like Heron's, had little impact.

East was thus one of many geneticists to conclude that feeblemindedness was genetic and transmitted as a Mendelian recessive. But he was the first to see the implications for eugenics. Even without benefit of Hardy–Weinberg, East understood that the number of apparently normal carriers must be much larger than those affected. In 1912 Davenport could offer the following advice:

> Prevent the feebleminded, drunkards, paupers, sex-offenders, and criminalistic from marrying their like or cousins or any person belonging to a neuropathic strain. Practically it might be well to segregate such persons during the reproductive period for one generation. Then the crop of defectives will be reduced to practically nothing (Davenport 1912, p. 286).

Two years later, a committee of the American Breeders Association concluded almost as optimistically that two generations of segregation and sterilization would largely "eliminate from the race the source of supply of the great anti-social human varieties" (Laughlin 1914, p. 60). East realized that these predictions were wrong. The "real menace" of the feebleminded lay in the huge heterozygotic reserve, which constituted about 7% of the American population, or one in every fourteen individuals. He warned: "Our modern Red Cross Knights have glimpsed but the face of the dragon" (East 1917, p. 215).

East's point was echoed by Punnett, who earlier had suggested that feeblemindedness could be brought under immediate control. Like many other geneticists, he felt "there is every reason to expect that a policy of strict segregation would rapidly bring about the elimination of this character" (Punnett 1912, p. 137). But as a consequence of work for his influential 1915 book, *Mimicry in Butterflies*, he changed his mind.

For his mimicry work, Punnett needed to know how fast a Mendelian factor would spread through a population (Provine 1971, p. 137; Bennett 1983, pp. 8–10). He turned to his Cambridge mathematics colleague H. T. J. Norton for help. Norton prepared a table (which appears as an appendix to Punnett's book) displaying the number of generations required to change the frequency of completely dominant or recessive factors at different selection intensities (Punnett 1915). From the table, Punnett learned both that selection could act with surprising speed and that, when the recessive factor was rare, extreme slowness. Two years after *Mimicry in Butterflies* appeared, Punnett called attention to the implications of the latter point for eugenics. Policies aimed at the affected, he argued, would take a distressingly long time to work.

He employed a relatively well-understood condition to illustrate the nature of the problem:

> Albinism, for example, behaves on whole as a recessive. Nevertheless, albinos appear among the offspring in an appreciable proportion of matings where either one or both parents are normal, and where no consanguinity can be detected. The same is true of feeblemindedness. This becomes less difficult to understand when we realize that the heterozygotes are bound greatly to outnumber the recessives whenever these form a small proportion of a stable population (Punnett 1917, p. 465).[1]

Although that argument had already been made by East, Punnett was able to work out its implications with much greater precision. Applying the Hardy–Weinberg formula, he concluded that over 10% of the population carried the gene for feeblemindedness. With G. H. Hardy's help, he also estimated the rate at which a population could be freed from mental defect by a policy of segregating or sterilizing the affected. He found the results depressing. Even under the unrealistic assumption that all the feebleminded could be prevented from breeding, their proportion in the population would only decline from

1 in 100 to 1 in 1000 in 22 generations
1 in 1000 to 1 in 10,000 in 68 generations

1 in 10,000 to 1 in 100,000 in 216 generations
1 in 100,000 to 1 in 1,000,000 in 684 generations.

In other words, given Goddard's (unchallenged) estimate that about three in every thousand Americans were feebleminded as a result of genetic defect, it would take over 8000 years before their numbers were reduced to 1 in 100,000. Punnett thus concluded that eugenic segregation did not, contrary to his initial belief, offer a hopeful prospect.

Punnett, who served with Fisher on the Council of the Cambridge University Eugenics Society, did not intend to provide an argument against eugenics (Bennett 1983, p. 12). Like East, he concluded that if "that most desirable goal of a world rid of the feeble-minded is to be reached in a reasonable time, some method other than that of the elimination of the feeble-minded themselves must eventually be found" (Punnett 1917, p. 464). That method would take advantage of the phenomenon of partial dominance. East had noted that complete dominance was rare among the characters studied by plant and animal breeders. He speculated that intelligence tests (which Goddard had introduced to America in 1908) could be used to identify heterozygotes, who would likely exhibit a lower mentality than the "pure normals." Punnett took up the suggestion, concluding his paper with a call for research to focus on carriers of defective genes.

Whatever his intention, Punnett's claim about the inefficacy of selection was seized on by critics of eugenic segregation and sterilization. For example, in 1923 the Central Association for Mental Welfare issued a pamphlet opposing sterilization, which it asserted "would have only a very limited effect in preventing mental deficiency" (Central Association for Mental Welfare, 1923, p. 12). In the same year, the Section of Medical Sociology of the British Medical Association sponsored a discussion on the issue of sterilizing mental defectives. The opponents of such a policy were clearly familiar with Punnett's argument. Thus, Dr. Joseph Prideaux, the mental and neurological inspector of the Ministry of Pensions, argued that if the proportion of mental defectives in the population were 3 or 4 per 1000, it would be necessary to sterilize "some 10 percent of the population, who were carriers of mental defect" (a policy he thought absurd) and that, moreover, "no really good result would be forthcoming until a very long period had elapsed" (Prideaux 1923, p. 231). Dr. H. B. Brackenbury, the Section's president, ended his summary of the discussion by remarking that the more the hereditary impact of rigorous segregation "was looked into the more certain aspects of it appeared to be disappointing," and noting that it had formerly been hoped that complete segregation or sterilization would rapidly eliminate the mentally defective population, "but this was not so" (Brackenbury 1923, pp. 233–4).

R. A. Fisher (Fisher 1924) realized that Punnett's calculations were misleading and easily employed to subvert the eugenic goals that he and Punnett shared. If the goal were to rid the world of the last few mental defectives, Fisher noted, the fact that thousands of generations are required to reduce their number to one in a billion would be meaningful. But if the calculations were extended to this point, "the reader would perhaps see the catch, and recognize that it would not matter if it took ten thousand generations to rid the world of its last lone feebleminded individual!" (p. 114). Even on Punnett's unrealistic assumptions of a single gene for mental defect and random mating, Fisher argued, substantial progress could be achieved in the first few generations if affected individuals were prevented from breeding. Expressing the frequency of the defectives as so many per 10,000 easily demonstrates the point:

From 100 to 82.6 in 1 generation
From 82.6 to 69.4 in 1 generation
From 69.4 to 59.2 in 1 generation.

Hence, in the first generation alone, selection could remove more than 17% of the affected persons.

Fisher's estimate is derived from Hardy's table, which represented an abstract calculation of the effects of selection, given assumptions about the initial number of affected. But the starting figures were chosen for ease of presentation rather than their assumed fit with reality. A standard estimate – and the one used by Punnett – was that three in a thousand individuals were feebleminded. Punnett's table could have been even more dramatic had he the skill to recalculate Hardy's numbers based on the lower initial frequency. But he understood little math. (In his 1916 referee's report on Fisher's classic paper, "The Correlation between Relatives on the Supposition of Mendelian Inheritance," Punnett wrote that it was of little interest to biologists but added; "frankly I do not follow it owing to my ignorance of mathematics" (Bennett 1983, p. 116, n. 12). If Fisher had used Punnett's estimate of the frequency of mental defect, the reduction in the first generation would have been about 10%.[2]

Fisher also examined the effects of relaxing Punnett's assumption of random mating. This time, however, the result was more favorable to eugenics. Fisher assumed that the feebleminded constituted a larger proportion (one sixteenth) of a smaller subsection (5%) of the population whose members mated only with others in that subsection. Hence, he incorporated a form of assortative mating into the model. Although it seems reasonable to assume that the feebleminded would tend to mate among themselves, the 5% figure dramatically decreases the frequency of carriers, thus increasing the efficacy of selection. Even starting from the standard frequency of 30 affecteds per 10,000 people, Fisher calculated that mental defect could be reduced by 36% in one generation (p. 115). Nevertheless, Fisher had shown for the first time that any form of assortative mating could help the eugenics cause.

Fisher's argument is often treated dismissively (see Kevles 1985, p. 165; Barker 1989). But Fisher diverged from Punnett and Jennings only in claiming that the affected tended to mate with each other (which would increase the frequency of homozygotes and thus speed selection) and that the trait was multifactorial. Both claims were eminently reasonable and at least as defensible as those of Punnett or Jennings – the conventional heroes of this historical fable – who were also more alarmist than Fisher. It was the progressive Jennings who asserted that "a defective gene – such a thing as produces diabetes, cretinism, feeblemindedness – is a frightful thing; it is the embodiment, the material realization of a demon of evil; a living self-perpetuating creature, invisible, impalpable, that blasts the human being in bud or leaf. Such a thing must be stopped wherever it is recognized" (Jennings 1927, p. 274).

Fisher's primary criticism was leveled at the use of Hardy's table to demonstrate the inefficacy of selection. He was surely right in claiming that it was deceptive. What mattered to most eugenicists was the potential progress of selection in the next few generations. Here, Fisher demonstrated that eugenical policies could make a substantial difference. Even on Punnett's assumption of random mating, a substantial reduction in a single generation was possible.

In fact, all the geneticists agreed that the incidence of mental defect could be reduced by about 10% in the first generation (and on the same reasoning, 19 and 26% by the second and third generations, respectively). Even Haldane, who regarded compulsory sterilization "as a piece of crude Americanism" thought it "would probably cut down the supply of mental defectives in the next generation by something of the order of 10 percent." (Haldane 1938, pp. 80, 88). If some degree of assortative mating is assumed, the estimates would of course be higher. According to Jennings, the ostensible critic: "A reduction in the number of feebleminded by eleven percent [on the assumption of random mating], or still more, a reduction by thirty or forty percent [if mating is assortative], would be a very great achievement. And it could be brought about in no other way than by stopping propagation of the feeble-minded persons" (Jennings 1930, p. 242).

Why are the estimates so high? It is often said that eugenics was based on a mistake about the efficacy of selection against rare genes. But this was not the eugenicists' error. The crucial point is that feeblemindedness was not considered rare, at least in comparison with a trait like albinism. Thus, Davenport wrote that eugenics was prompted by recognition of the "great proportional increase in feeble-mindedness in its protean forms – a great spread of animalistic traits – and of insanity" (Davenport 1912, p. 308). Indeed, the raison d'etre of the eugenics movement was the perceived threat of swamping by a large class of mental defectives. Numerous British and American studies and an increase in the institutionalized population seemed to indicate that the problem was rapidly worsening. In America it was commonly believed that from 300,000 to 1,000,000 persons were feebleminded as a result of genetic defect (Popenoe and Johnson 1918); those figures tended to increase as mental tests came into wider use to evaluate students, prisoners, inmates of poorhouses and training schools, immigrants at Ellis Island, and army draftees. In 1912, Goddard tested New York City schoolchildren and estimated that 2% were probably feebleminded (Goddard 1912). The results of tests administered to army recruits during World War I were even more alarming, for they indicated that nearly half (47.3%) of the white draft – and 89% of the black – was feebleminded

(Yerkes 1921). Moreover, it seemed that the feebleminded were particularly prolific. For example, the British Royal Commission on the Care and Control of the Feeble-Minded reported in 1908 that defectives averaged seven children, whereas normal couples averaged only four; many other studies came to similar conclusions (see Paul 1995, pp. 78, 62).

Contemporary advocates of the futility of eugenics often mention Tay–Sachs disease, phenylketonuria (PKU), or albinism. Selection against such diseases is certainly futile. But these textbook examples are almost invariably rare conditions whose effects are either lethal or minor. Both their frequency and consequences ensure that they would be of little interest to a eugenicist. Individuals with Tay–Sachs and (with a few exceptions) untreated PKU do not leave offspring. Albinos and treated phenylketonurics do reproduce, but these conditions are not disabling. The frequent employment of albinism in texts is probably an unconscious inheritance from Punnett's original article. In 1917, Punnett had few examples to choose among.

Applied to the historical eugenics movement, the argument about the futility of selection against rare genes is simply irrelevant. Given widely shared assumptions about the causes and incidence of mental defect, eugenic policies could be expected to substantially reduce the number of affected. In any case, geneticists in the 1920s would generally have favored such policies whatever their exact effect. In *Heredity and Eugenics,* Ruggles Gates summarized Punnett's argument, concluding that even if all mental defectives were prevented from reproducing, "the most difficult part of the process of eliminating feeblemindedness from the germ plasm of the population would scarcely have begun" (1923, p. 159). But he ends the same chapter with a call for "the prevention of reproduction on the part of undesirables, such as the feebleminded," reasoning that, "such measures are necessary, not so much for the improvement of the race, as for arresting its rapid deterioration through the multiplication of the unfit" (p. 251). Indeed, most geneticists would have assented to Jennings' claim that "to stop the propagation of the feebleminded, by thoroughly effective measures, is a procedure for the welfare of future generations that should be supported by all enlightened persons. Even though it may get rid of but a small proportion of the defective genes, every case saved is a gain, is worth while in itself" (Jennings 1930, p. 238).

Like Jennings, Lancelot Hogben is often portrayed as an opponent of eugenics. He did criticize some advocates of sterilization for exaggerating the urgency of the problem and the results they could achieve – fearing that overstatement would harm the cause. He also invoked Fisher to argue that there is no need to overstate potential results. That we cannot do everything "is not a valid reason for neglecting to do what little can be done" (Hogben 1931, p. 207). That point was echoed by Edwin G. Conklin, who like Jennings and Hogben, criticized some eugenic proposals. Conklin once remarked that sterilizing all the inmates of public institutions was "like burning down a house to get rid of the rats" (Conklin 1916, p. 438). But he did not oppose sterilization of the feebleminded. On the contrary, he asserted that "all modern geneticists approve the segregation or sterilization of those who are known to have serious hereditary defects, such as hereditary feeblemindedness, insanity, etc." Conklin asked of the American Eugenics Society's proposed sterilization policy; "Can any serious objection be urged to such a law?" (Conklin 1930, pp. 577–8). In 1930, this question was unambiguously rhetorical.

Nearly all geneticists of the 1920s and 1930s – including those traditionally characterized as opponents of eugenics – took for granted that the "feebleminded" should be prevented from breeding. Moreover, nearly everyone agreed on the scientific facts. Punnett, East, Fisher, Jennings, and even Haldane made roughly the same estimates as to the speed and scope of eugenical selection. But in respect to social policy, the facts did not speak for themselves. They required interpretation in light of other assumptions and goals. Thus, Haldane opposed sterilization, arguing that "with mental defects as with physical defects, if you once deem it desirable to sterilize I think it is a little difficult to know where you are to stop" (Haldane 1938, p. 89). That is a powerful argument. But it is a social, not a scientific one. Lionel Penrose was an even more vehement and consistent critic of eugenics. An expert in the genetics of mental deficiency, he stressed the heterogeneity of its causes and the modest influence of eugenic measures in reducing its incidence. But his main argument was ethical. Penrose maintained that the best index of a society's health is its willingness to provide adequate care for those unable to care for themselves (Penrose 1949; see also Kevles 1985, esp. 151–63).

The Hardy–Weinberg theorem meant different things to different people. To those already disposed against eugenics, it proved that policies to prevent the feebleminded from breeding were not worth the effort. There is no reason that those disposed in favor of eugenics should draw the same conclusion. Whether a 10% reduction in incidence is large or small is not a question science can answer. Indeed, one may concede that the percent reduction is small yet still think it worthwhile. Thus, at the close of a long discussion of the implications of Hardy–Weinberg, Curt Stern remarked: "To state that reproductive selection against severe physical and mental abnormalities will reduce the number of affected from one generation to the next by only a few percent does not alter the fact that these few percent may mean tens of thousands of unfortunate individuals who, if never born, will be saved untold sorrow" (Stern 1949, p. 538). A similar point was made by the Swedish Commission on Population in its 1936 report on sterilization. After acknowledging the falsity of the earlier belief that sterilization would result in a rapid improvement of the population, the authors note that it would still result in gradual improvement while preventing possible deterioration and that "whenever an eugenic sterilization is carried out... in the specific case the operation will prevent the birth of sick or inferior children or descendants. Owing to this, sterilization of hereditarily sick or inferior human beings is still a justified measure, beneficial to the individual as well as to society" (quoted in Broberg and Tydén 1996, p. 106). Thus, it may not matter if the reduction in absolute numbers is miniscule. Indeed, if one assumes with Jennings that "the prevention of propagation of even one congenitally defective individual puts a period to at least one line of operation of this devil" and that "to fail to do at least so much would be a crime," the rate of selection is simply beside the point (Jennings 1927, p. 274).

We began by asking whether eugenics was based on an elementary mistake. To the extent that support for eugenical segregation and sterilization was based on the assumption that "it would be possible at one fell stroke [to] cut off practically all of the cacogenic varieties of the race," (Laughlin 1914, p. 47) a loose definition of "feeblemindedness," as well as acceptance of Goddard's shoddy data and defective logic, the answer is yes. But it was possible to recognize these flaws and still remain a eugenicist, as the example of David Heron demonstrates. Moreover, what is usually characterized as the eugenicists' most obvious error – a failure to understand the implications of the Hardy–Weinberg theorem – was a mistake few geneticists made after 1917. By the 1920s, they well understood that the bulk of genes for mental defects would be hidden in apparently normal carriers. For most geneticists, this appeared a better reason to widen eugenic efforts than to abandon them.

It is often said that support for eugenics declined in the 1930s as its scientific errors were exposed. But the eugenics movement grew stronger during the Depression (see Paul 1995, pp. 72–90). In the United States, the number of sterilizations climbed. The procedure was legalized in Germany (1933), the Canadian Province of British Columbia (1933), Norway (1934), Sweden (1934), Finland (1935), Estonia (1936), Iceland (1938) and Japan (1940). Denmark, which in 1929 had legalized "voluntary" sterilization, permitted its coercive use on mental defectives in 1934. These laws were generally applauded by geneticists.

In 1918, Popenoe and Johnson wrote that "so few people would now contend that two feeble-minded or epileptic persons have any 'right' to marry and perpetuate their kind, that it is hardly worth while to argue the point" (Popenoe and Johnson 1918, p. 170). Assumptions we now take for granted they thought too absurd even to require challenging. The inversion of these assumptions in recent decades is best explained by political developments. Revelations of Nazi atrocities, the trend toward respect for patients' rights in medicine, and the rise of feminism have converged to make reproductive autonomy a dominant value in our culture. In 1914, a committee of the American Breeders Association asserted that "society must look upon germ-plasm as belonging to society and not solely to the individual who carries it" (Laughlin 1914, p. 16). Few today would profess such a view. A change in values, and not the progress of science, explains why contemporary Swedes would be unlikely to concur with the 1936 commission that criticized as "extremely individualistic" the concept that individuals have a right to control their own bodies (Broberg and Tydén 1996).

It is not our superior quantitative skills that explain why we today draw very different implications from the Hardy–Weinberg theorem. There was nothing wrong with most eugenicists' math. Our concept of rights, however, is much more expansive than theirs. That is why the same equation holds different lessons for them than it does for us.

5.2 Acknowledgments

Having met as research associate (Diane B. Paul) and graduate student (Hamish G. Spencer) in Richard Lewontin's laboratory, the authors are especially grateful to him and for his having outlined for us some of the arguments in the appendix. We also thank Evelyn Fox Keller, J. Maynard Smith, Michael Ruse, Edna Seaman, and Graham Wallis for reading the manuscript; Leila Zenderland for comments on Goddard; Nils Roll-Hansen for bringing Ragnar Vogt's argument to our attention; the Division of Sciences, University of Otago, for financial support; and the staff of the Science Library, University of Otago, for help with references.

APPENDIX

In this appendix we compare the efficacy of selection under Goddard's single-locus model with that under a simple quantitative model as suggested by Fisher. The argument we use in examining the latter follows suggestions made to us by R. C. Lewontin.

Let us first examine the dynamics of the Mendelian model in which the feebleminded are homozygous for a recessive allele. Let F be the proportion of feebleminded in the population, and thus the frequency q of the dysgenic allele is the square root of F. On the assumption that none of the feebleminded reproduce (i.e., that eugenic selection is complete), the allele frequency after one generation is given by the equation

$$q' = \frac{q}{1+q}.$$

Hence, the percentage drop in the incidence of feeblemindedness is given by

$$\begin{aligned} 100(F' - F)/F &= 100(q'^2 - q^2)/q^2 \\ &= -100q(2+q)/(1+q)^2 \\ &= -100\sqrt{F}(2+\sqrt{F})/(1+\sqrt{F})^2 \end{aligned}$$

This function is shown in Figure 5.A1. We can also use this model to see how long eugenic selection takes to reduce the proportion of feebleminded by a certain fraction, say 50%. The allele frequency after n generations of selection is given by

$$q_n = (n + q^{-1})^{-1},$$

which can be rearranged to give

$$n = q_n^{-1} - q^{-1}.$$

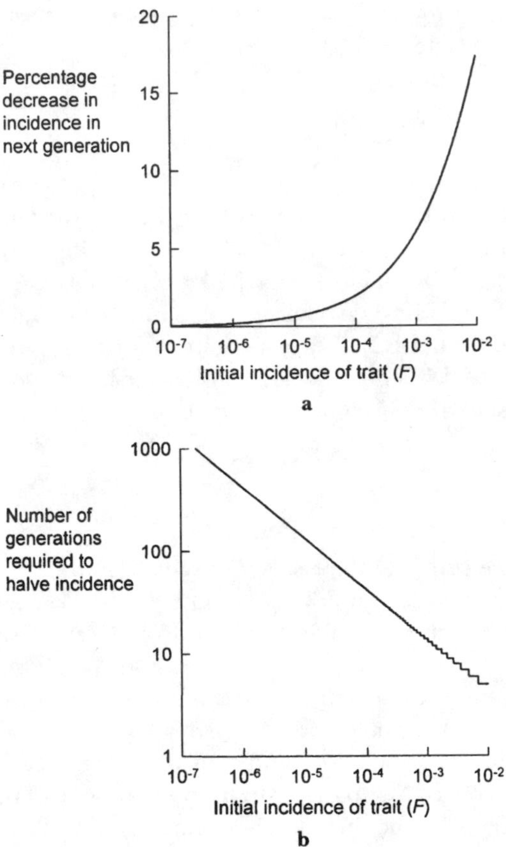

Figure 5.A1. (**a**) The effect of eugenic selection against a recessive Mendelian trait in one generation. (**b**) The time required to at least halve the frequency of a dysgenic recessive Mendelian trait. (The stepped nature of the graph in **b** is because there is no integer solution to the equation; see text.) Note the logarithmic *x*-axis in both figures and the logarithmic *y*-axis in **b**. We have assumed that the eugenic selection is complete (i.e., no affected individuals have children). The frequency q of the dysgenic allele is the square root of the incidence F.

Because halving the incidence of a trait reduces the allele frequency by a factor of $\sqrt{2}$, the number of generations required to halve the incidence is given by

$$(q/\sqrt{2})^{-1} - q^{-1} = (\sqrt{2} - 1)/\sqrt{F}$$

Because n will not in general be a whole number (especially for high incidences), the number of generations required to at least halve the incidence is given by rounding n upwards, that is $[n+1]$. This function is also shown in Figure 5.A1.

Let us now examine the quantitative model in which individuals with a mental ability below a threshold t are considered to be the feebleminded. Suppose that

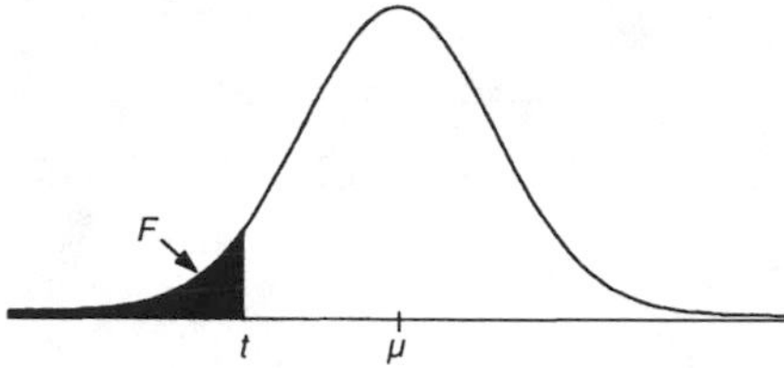

Figure 5.A2. The distribution of mental ability modeled as a quantitative trait with mean μ. The feebleminded are those with mental ability below the threshold t and make up a proportion F of the population equal to the shaded area under the curve.

mental ability has a heritability of 1.0 and is determined by a large number of additive loci, which are two assumptions most favorable to Fisher's argument. The trait will then be normally distributed, with a mean μ and a standard deviation σ. The proportion of feebleminded in the population, F, is the area under the normal curve to the left of t, as shown in Figure 5.A2. This value is easily found from tables of the standard normal distribution. In symbols,

$$F = \int_{-\infty}^{t} \phi(x)\, dx,$$

in which $\phi(x)$ is the normal density function. If all the feebleminded are prevented from breeding, the mean in the subsequent generation, μ', will be larger:

$$\mu' = \mu + h^2(\mu_s - \mu),$$

in which h^2 is the heritability of the trait, and μ_S is the mean of the selected or breeding population. The difference in parentheses is easily found from the properties of the normal distribution, is well known to quantitative geneticists (Falconer 1989, p. 191), and is usually denoted as S:

$$S = \mu_S - \mu = \sigma^2\phi(t)/(1 - F)$$

Now, because the heritability is 1.0, the distribution of the subsequent generation will be normal with a mean μ_S. That is, the distribution of mental ability is moved a distance S to the right. The proportion of feebleminded in the subsequent generation, F', is thus given by

$$F' = \int_{-\infty}^{t} \phi(x - S)\, dx,$$

with the proportional reduction in the proportion of feebleminded given by

$$(F - F')/F.$$

Using the values relevant to Fisher's argument, we find that if $F = 0.01$, then $\phi(t) = 0.0266$ and $S = 0.0269\sigma^2$, giving $F' = 0.0093$. The quantitative model predicts a proportional decrease in feeblemindedness in one generation of eugenic selection, therefore, of about 7% compared with the 17% reduction predicted from the single-locus Mendelian model. The proportional changes for other values of F are shown in Table 5.A1.

Table 5.A1. *Percentage Decrease in Feeblemindedness under*

F	Mendelian Model	Quantitative Model
0.010	17.36	6.95
0.003	10.12	2.76
0.001	6.04	1.13

Notes

1. Indeed most texts continue to treat albinism – as Punnett did – as a single-locus defect. Falconer (1989) is one of the few exceptions. But it has long been known that albinism arises from the actions of recessive genes from at least two loci (McKusick 1992). Consequently, the incidence of homozygous recessives for a particular locus is lower than most texts suggest, and eugenic selection against albinism would be even less efficacious.
2. In emphasizing that his presentation was based on Punnett's assumptions, Fisher traded on this weakness. A reader could easily presume that Fisher employed assumptions favorable to Punnett's case. He did not. "In a single generation," Fisher wrote, "the load of public expenditure and personal misery caused by feeblemindedness would be reduced by over 17 percent." If based on the figures in Punnett's table, this estimate is correct but also misleading. Fisher did add that if the starting point had been thirty instead of a hundred (per ten thousand), the reduction in one generation would be "over 11 per cent" (p. 114). But he failed to note that this is the relevant figure. In fact, Fisher's 11% figure is still inflated. The true reduction is approximately 10.1%. Haldane (1938, p. 88) discreetly gives the correct value – "something of the order of 10 percent" – but less mathematically inclined writers, such as Jennings (1930, p. 242), appear not to have noticed Fisher's error.

REFERENCES

Barker, D. (1989). The biology of stupidity: Genetics, eugenics and mental deficiency in the inter-war years. *Brit. J. Hist. Sci.* 22:347–75.

Bennett, J. H. (1983). *Introduction to Natural Selection, Heredity, and Eugenics.* Oxford, UK: Clarendon Press.

Brackenbury, H. B. (1923). The President's Summary of the Discussion. *British Medical Journal* 2:233–4.

Broberg, G., and Tydén, M. (1996). "Eugenics in Sweden: Efficient Care." In *Eugenics and the Welfare State: Sterilization Policy in Denmark, Sweden, Norway, and Finland,* eds. G. Broberg and N. Roll-Hansen. East Lansing: Michigan State University Press.

Castle, W. E. (1927). *Genetics and Eugenics.* Cambridge, MA: Harvard University Press.

Central Association for Mental Welfare (1923). Sterilization and mental deficiency. *Studies in Mental Inefficiency* 4:10.

Conklin, E. G. (1916). *Heredity and Environment in the Development of Man,* 2d edition. Princeton, NJ: Princeton University Press.

Conklin, E. G. (1930). The purposive improvement of the human race. In *Human Biology and Population Improvement,* ed. E. V. Cowdry, pp. 566–88. New York: Hoeber.

Cowan, R. S. (1977). Nature and nurture: The inter-play of biology and politics in the work of Francis Galton. In *Studies in the History of Biology*, eds. W. Coleman and C. Limoges, pp. 133–208. Baltimore: Johns Hopkins University Press.

Darwin, L. (1928). *What is Eugenics?* London: Watts.

Davenport, C. B. (1912). The inheritance of physical and mental traits of man and their application to eugenics. In *Heredity and Eugenics*, eds. W. E. Castle, J. M. Coulter, C. B. Davenport, E. M. East, W. L. Porter, pp. 269–88. Chicago: University of Chicago Press.

Davenport, C. B. (1915). Review of H. H. Goddard, Feeblemindedness: Its Causes and Consequences. *Science* 42:837–8.

East, E. M. (1917). Hidden feeblemindedness. *Journal of Heredity* 8:215–17.

Falconer, D. S. (1989). *Introduction to Quantitative Genetics.* 3d ed. Harlow, UK: Longman.

Fisher, R. A. (1924). The elimination of mental defect. *Eugenics Review* 26:114–16. (Reprinted in *Collected Papers of R. A. Fisher 1*, ed. J. H. Bennett, 1971 edition. Adelaide, Australia: University of Adelaide.)

Gates, R. R. (1923). *Heredity and Eugenics.* London: Constable.

Goddard, H. H. (1912). *The Kallikak Family: A Study in the Heredity of Feeblemindedness.* New York: MacMillan.

Goddard, H. H. (1914). *Feeblemindedness: Its Causes and Consequences.* New York: MacMillan.

Haldane, J. B. S. (1932). *The Inequality of Man and Other Essays.* London: Chatto and Windus.

Haldane, J. B. S. (1938). *Heredity and Politics.* London: Unwin.

Heron, D. (1913). Mendelism and the problem of mental defect. I. A criticism of recent American work. In *Questions of the Day and of the Fray No. VII.* London: Cambridge University Press.

Hogben, L. (1931). *Genetic Principles in Medicine and Social Science.* London: Williams and Norgate.

Holmes, S. J. (1923). *Studies in Evolution and Genetics.* New York: Harcourt, Brace.

Jennings, H. S. (1927). Health progress and race progress: Are they incompatible? *Journal of Heredity* 18:271–6.

Jennings, H. S. (1930). *The Biological Basis of Human Nature.* London: Faber and Faber.

Kevles, D. J. (1985). *In the Name of Eugenics: Genetics and the Uses of Human Heredity.* New York: Knopf.

Laughlin, H. H. (1914). Report of the Committee to Study and to Report on the Best Practical Means of Cutting Off the Defective Germ-Plasm in the American Population. I. The Scope of the Committee's Work. Eugenics Record Office, Bulletin No. 10A. Cold Spring Harbor, New York.

McKusick, V. A. (1992). *Mendelian Inheritance in Man: Catalogue of Autosomal Dominant, Autosomal Recessive, and X-Linted Phenotypes.* 10th ed. Baltimore: Johns Hopkins University Press.

Medawar, P. B., and Medawar, J. S. (1977). *The Life Sciences: Current Ideas in Biology.* New York: Harper and Row.

Morgan, T. H. (1925). *Evolution and Genetics*, 2d ed. Princeton, NJ: Princeton University Press.

Paul, D. B. (1995). *Controlling Human Heredity: 1865 to the Present.* Amherst, NY: Humanity Press.

Penrose, L. S. (1949). *Biology of Mental Defect.* New York: Grune and Stratton.

Popenoe, P. (1915). Feeblemindedness. *Journal of Heredity* 6:32–6.

Popenoe, P., and Johnson, R. H. (1918). *Applied Eugenics.* New York: MacMillan.

Prideaux, J. F. E. (1923). Comment. *British Medical Journal* 2:231–2.

Provine, W. B. (1971). *The Origins of Theoretical Population Genetics.* Chicago: University of Chicago Press.

Punnett, R. C. (1912). Genetics and eugenics. In *Problems in Eugenics.* First International Eugenics Congress, 1912, pp. 137–8. New York and London: Garland Publishing, 1984.

Punnett, R. C. (1915). *Mimicry in Butterflies.* Cambridge, UK: Cambridge University Press.

Punnett, R. C. (1917). Eliminating feeblemindedness. *Journal of Heredity* 8:464–5.

Punnett, R. C. (1925). As a biologist sees it. *The Nineteenth Century* 97:697–707.

Roll-Hansen, N. (1980). Eugenics before World War II. The case of Norway. *History and Philosophy of the Life Sciences* 2:269–98.

Spencer, H. G., and Paul, D. B. (1998). The failure of a scientific critique: David Heron, Karl Pearson and Mendelian eugenics. *British Journal for the History of Science* 31:441–52.

Stern, C. (1949). *Principles of Human Genetics.* San Francisco: Freeman.

Vogt, R. (1914). Det Norske Videnskabsakademi i Kristiania, Forhandlinger (13 Feb.):4–5.

Wiggam, A. (1924). *The Fruit of the Family Tree.* Indianapolis, IN: Bobbs–Merrill.

Yerkes, R. M. (1921). *Psychological Examining in the United States Army.* vol. 15. Memoirs of the National Academy of Sciences. Washington, D.C.

CHAPTER SIX

Can the Norm of Reaction Save the Gene Concept?

RAPHAEL FALK

The term *norm-of-reaction* was born in the same year as that of the gene. In June 1909, Richard Woltereck (1877–1944) presented his research *Weitere experimentelle Untersuchungen über Artveränderung, speziell über das Wesen quanitativer Artunterschiede bei Daphniden* (Further investigations on change of species, specifically on the nature of quantitative species-differences in Daphnides) to the German Zoological Society. In it Woltereck declared his objective to provide a rejoinder to what he called the Mendelian teaching in the footsteps of Weismann and de Vries, or the de Vries–Johannsen conception of the origin of specific types (specific-kinds, biotypes)[1] by abrupt hereditary changes, that is, by mutations rather than by continuous small changes (Woltereck 1909). Woltereck's work was specifically intended to be a contribution to the reestablishment of the Darwinian conception of gradual and continuous evolution of species as opposed to the growing sentiment for evolution by saltations.

In his *Die Mutationstheorie* de Vries resurrected a typological conception of taxonomy, reaching down to the level of the individual organism. He claimed that until the time of Linnaeus it was actually the genus (*Gattung*) that was the unit of taxonomy: To avoid the lengthy and tiresome lists of variations that accompanied the descriptions of such genera, Linnaeus "raised" species (*Arten*) to the basic entities of taxonomy (de Vries 1902–3, I. pp. 12–13). If so, there was no compelling need to accept the species as the fundamental entity of taxonomy, and the distinct segregating Mendelian unit-characters might be the ones that denote the fundamental discontinuous *type* of taxonomy. It was, however, Wilhelm Johannsen who threatened to consolidate a new typological taxonomy on the foundation offered by genotypic Mendelian discontinuity versus apparent phenotypic continuity (Johannsen 1909). Johannsen's studies in pure lines were closely linked to his concept of grouping and classification of natural entities. His notion was that the genotype, the "constant form-type" of the pure lines, being the purest realization of the basic unit of classification, was actually a direct implementation of Aristotelian typology (Roll-Hansen 1978, pp. 221 ff.). The discontinuity between types (genotypes) is biologically true and meaningful. It reflects the *discontinuous* or saltational character, as opposed to the *continuous* character, of the Darwinian intra- as

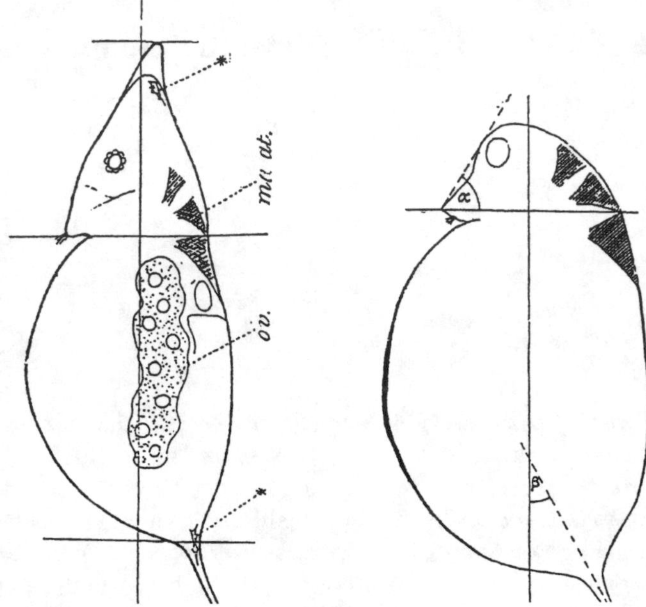

Figure 6.1. Left: Side view of Daphnia (*Hyalodaphnia cucullata* female). The relative height of the helmet is measured along the depicted axis. (*mu. at.* = antennal muscles, *ov.* = ovary.) Right: *Daphnia longispina.* (α and β are another pair of characteristic traits) [after Woltereck 1909].

well as interspecific evolution. It was these ideas that Woltereck set out to counter.

To oppose the claims of stepwise changes between *genetic types*, Woltereck studied local strains, the elementary kinds of *Daphnia* species from different lakes in Germany, which formed morphologically distinct *biotypes* (Woltereck 1909, p. 113). These elementary kinds were pure lines, maintained by many generations of uninterrupted parthenogenetic reproduction. He followed the variation of quantitative characteristics, such as the height of the "helmet" – a protrusion at the tip of the carapace (the chitinous cover) of the animals (Figure 6.1). Woltereck stressed that, for establishing whether the underlying evolutionary changes were continuous or discontinuous, the study of qualitative characteristics was not sensitive enough. A qualitative (present or absent) characteristic could be either completely insensitive to quantitative changes in factors that exert an observable effect on other characteristics, or, what seemed more plausible, it would respond only to the cumulative input of other variables (i.e., there might be a threshold amount of the input factor below which no effect would be detectable in the characteristic). A qualitative characteristic would be an ideal marker for the determination of the presence or absence of a specific input variable exactly because it was indifferent to changes in other variables. However, in the context of the formation of distinct elementary kinds, Woltereck noted, qualitative traits might give a distorted picture of the normal interaction of other variables in the formation of the specific characteristic investigated.

6.1 Norm of Reaction Versus Genotype

Woltereck showed that environmental conditions such as the time of the year, temperature of the water, nutritional conditions, and so forth significantly affect quantitative traits of biotypes (pure lines), whether these are morphological characteristics such as those of the outer shell (carapace) of the *Daphnia* or physiological properties. Thus, for example, the height of the "helmet," which was considered diagnostic for the taxonomists, showed cyclical changes over the year that turned out to dependent on nutrient supplies. "Internal" factors, such as the number of parthenogenetic generations that the biotype went through, also made for a difference in helmet height.

As expected by Woltereck, the response curves of a morphological phenotype of a biotype or an elementary kind varied continuously with changing the relevant environmental variables. However, the unexpected result was that each biotype had its *specific* phenotype response curve. There was no way of predicting the response curve of one biotype from that of another. Woltereck found the histological as well as the physiological changes to be biotype specific,

By having become acquainted with the helmet-building factors, we are now in the position to determine, for each of the tested elementary-kinds, *the full-scale of possibilities of head-heights as dependent on each of the causes that condition* (release, regulate) *the phenotypes.* (Woltereck 1909, p. 130)

Furthermore, Woltereck emphasized that these response curves (for the mean values of a given variable) change for each biotype with every additional variable that is considered. Such additional variables could be the number of generations of parthenogenetic reproduction, the sex of the *Daphnia* (when sexuality had been induced), or the gas and salt content of the water in which the animals were maintained. Thus, knowing the full range of the specific characteristics of a quantitative property such as that of head length would be possible only when its numerical relationship to variation in a very large number of factors had been obtained. The totality of these relationships can be reproduced only through a "nearly infinite" number of tables or schemes of "phenotype curves" (*Phänotypen-kurven*) (Woltereck 1909, p. 135). These inclusive relations may be denoted as the specific and relative *norm of reaction* (*Reaktionsnorm*) of the analyzed traits. As a rule, it would be difficult or even impossible to denote the whole range of the norm of reaction (NOR) of a characteristic, and we must be content with a partial NOR. Significantly, however, the NOR of one biotype cannot be directly deduced from that of another biotype even if these biotypes differ only in alleles of a single gene. For different biotypes the response curves

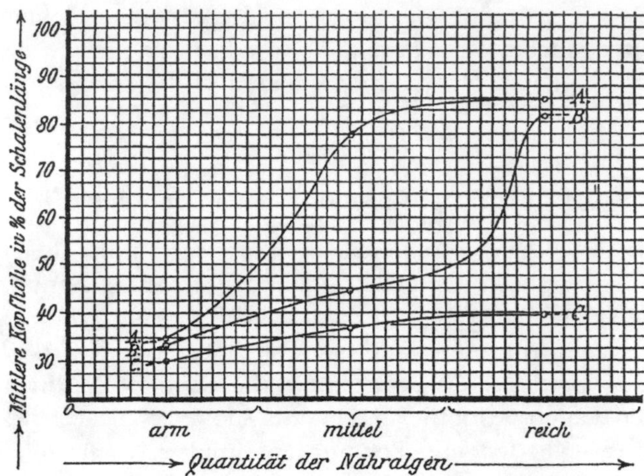

Figure 6.2. Norm of reaction curves of helmet height for three biotypes of *Hyalodaphnia cucullata* females in poor, intermediate, and rich quantities of nutrients (after Woltereck 1909).

of the characteristic, as a function of the environments (and of the genotype), may be absolutely different. Of three biotypes examined for helmet height, one was nearly indifferent to changes in nutrition, another responded to a change from intermediate to rich nutrition, whereas the third responded to a change from poor to intermediate nutrition (Figure 6.2).

As noted, Woltereck wished to examine whether the conception "that changes, even of the quantitative characteristics, are determined exclusively by mutations and without environmental involvement – is satisfactory" (Woltereck 1909, p. 130). Thus, the experiments he carried out were actually only preparatory work for resolving that bigger issue: the role of nurture in shaping the nature of species. He had to be sure that once he controlled the dependence of phenotypic variation on environmental variation, any *residual* variation was *heritable.* What he discovered, however, was that such a segregation of environmental and heritable contributions to phenotypic variability was impossible. There is no typical unit character. *Biotype and environmental variation interact.* It was meaningless to describe any biotype independently of the environmental context. The NOR concept acknowledged the need to describe the phenotypic response "dynamically," in a variety of environments, rather than as a one-to-one relationship between genotype and phenotype. Woltereck's ideas threatened not only to deprive genetics of its major tool for targeting genetic entities but could also be conceived as support for the notion of inheritance of acquired characters. This doomed the concept of NOR for many years.[2]

The idea of the norm of reaction contradicts notions of a deterministic linkage between the genotype and the phenotype. Instead of genes or genotypes that direct the phenotype while environmental influences constitute only disturbing fluctuations, the norm-of-reaction concept stresses that as much of the phenotypic variability is directed by the conditions under which the genotype (the gene) develops or functions as that directed by the genotype itself. Although Johannsen endeavored to overcome the de Vriesian notion of the inheritance of unit characters by calling attention to a "deeper" strata, that of the *genotype* rather than that of the *phenotype*, as the essence of the systematic organization of living forms, Woltereck saw his task as one of refuting the essentialist view of genetic determinism altogether and emphasizing the role of noninherited factors in the shaping of forms and properties. More specifically, the characteristics considered diagnostic of varieties and species were not necessarily hereditary, for they could be modified or induced quite easily by appropriately changing environmental circumstances. Furthermore, Woltereck could provide exactly the relief from the narrow, reductionist sense of the *gene* of which Johannsen warned:

> The segregation of one sort of "gene" may have influence upon the whole organization. Hence the talk of "genes for any particular character" ought to be omitted, even in cases where no danger of confusion seems to exist. . . . It should be a principle of Mendelian workers to minimize the number of different genes as much as possible. (Johannsen 1911, p. 147)

Woltereck's notion of the norm of reaction uncoupled the Gordian knot between the "unit character" and the Mendelian unit of inheritance that had been imposed by de Vries (Falk 1995). Whereas Morgan objected to the narrow view of one-gene–one-phenotype because of the inputs of the rest of the genotype (see Falk and Schwartz 1993), Woltereck called attention to the futility of such a one-to-one relationship because of variation in external as much as in internal inputs to the phenotypes of organisms.

In the final analysis, although Johannsen and Woltereck agreed on rejecting the notion of the unit character, they differed in the role allocated to the genotype in the shaping of the phenotype – especially in the process of evolution and speciation. For Johannsen the genotype was the determinant of the phenotype, and the environment was an unavoidable nuisance that must be reckoned with as such; for Woltereck the environmental variation was a constructive constituent of the individual phenotype. Whereas Johannsen strove toward an essentialist notion of speciation, for Woltereck the diagnostic characteristics of varieties and species were not necessarily hereditary because they could be modified by changing environmental circumstances. Woltereck did not doubt that mutations had a role to play; he wondered, however, whether they should be given an exclusive role in the change of kinds, and his conclusion was an unequivocal no. Because it was fundamental that anything that caused differentiation of species must be transmitted through the germ line and that new species or elementary kinds were initiated only when specific properties were hereditarily changed, it followed that it was the *inclusive* norm of reaction that was inherited as the *Anlagen* of traits.

For Woltereck the NOR was at the root of the speciation controversy: "Is it determined solely through mutations, without environmental-determinations? Or is it continuously hereditarily modified through the impacts of the changing milieu?" (Woltereck 1909, p. 136). He concluded that if mutations are discrete qualitative changes of type that occur at low frequency, as the "selectionists" claimed, then selection could not possibly be the factor that drives the process of evolution. Once a proper mutation occurred, he reasoned (erroneously, as it turned out), it could at most elicit a provisional, temporal selection spree. In other words, Woltereck argued that selection is too inefficient a process to drive an evolution dependent exclusively on discontinuous mutational shifts.[3]

Johannsen agreed with Woltereck as far as the *phenotypic* variability was concerned; he even presented some illuminative examples of his own to genotype–environment interactions in the determination of the phenotype (Johannsen 1911, p. 145).

> But when Woltereck thinks that these facts are inconsistent with the existence of constant differences between the genotypes, he shows himself to have totally misunderstood the question! Of course the *phenotypes* of the special characters, i.e., the *reactions of the genotypical constituents*, may under different conditions exhibit all possible forms of transition or transgression – this has nothing at all to do with constancy or inconstancy of genotypical differences. (Johannsen 1911, p. 145)

For Johannsen the particular organism's "multiple varying reactions are *determined by its 'genotype' interfering with the totality of all incident factors*, may it be external or internal" (Johannsen 1911, p. 133, emphasis added).

6.2 Entrenchment of Genocentricity

An essential point for Johannsen was "that a special genotypical constitution always reacts in the same manner under identical conditions" (Johannsen 1911, p. 146). The significance of this point for the future of genetic research cannot be overestimated. Johannsen provided the young science with the tools to study hereditary transmission experimentally without having to commit itself to its chemicophysical nature. At the same time, he provided the framework for the conceptual isolation of the genotype as the blueprint in the vault that determines development, function, and behavior of creatures and yet is protected from any (adaptive) modification by its carriers. It is this genocentristic notion that was built up until the 1940s and has been dominating genetic thought even after the main thrust of genetic research was directed to functions at the cellular and molecular levels. It is this notion that became extremely difficult to crack even in the 1980s (see, e.g., Buss 1987; Cairns, Overbaugh, and Miller 1988; Jablonka and Lamb 1995).

The quest of geneticists for determinants that may be identified and manipulated, even if we know them only indirectly through their phenotypic effects, made most investigators overlook Woltereck's norm-of-reaction curves in the years to come. At the same time, the quest of animal and plant taxonomists for the "true" entities behind the phenotypic manifestations repeatedly led to promotion of genetically determined typological models that purported to explain away phenotypic continuity and certainly any notion of an integrated genotype–environment variability, as were inherent in the NOR concept.

In the 1920s, the German neurobiologist Oskar Vogt elaborated a typological concept of animal and plant taxonomy very much in the sense of Johannsen's: The real taxonomic entities should be identified not by their similarity or even blood-relatedness but by their common etiology (Vogt 1926). Genera, species, subspecies, and sibships were not continuous with each other but rather genuine types once their etiology was followed in enough detail. Organisms varying from each other by *single gene differences* were fittingly the ultimate, most elementary, real taxonomic entities. Vogt claimed that the mistake of taxonomists in the past was that they attempted to find clues to *whether* organisms form "real" groupings rather than admitting that distinct types exist and accordingly directing their efforts toward the best means of *identifying* them.

According to Vogt, each discontinuous taxonomic entity has a typical (genotypic) characteristic and some amount of fluctuations around this, which genetic and statistical analyses should be able to resolve. Not all genes were proper markers for type-differentiation. Only gene-variants that were manifested in the phenotype under the most varied conditions would be the markers that display the type's saltation or discontinuity. Vogt offered various examples from morphological variability (primarily in pigmentation) of insects and claimed that the continuous monotonous variation of forms and patterns (for which he coined the term *eunomy*) was either interrupted by real discontinuities or showed multipeak frequency distributions. This indicated to him the "reality" of the etiological "types" within species as much as between species.

To overcome the "irrelevant" environmental variations about the hereditary essence, Vogt introduced the terms *penetrance* (the proportion of individuals of a given genotype that show the expected phenotype) and *expressivity* (the degree of variability of the typical phenotype). These phenomenological terms, which explained nothing, allowed, however, the norm-of-reaction interactive conception to be ignored in favor of a flagrant genocentric view. The terms provided a conceptual framework needed to maintain the rigorosity of the one-to-one relationship between genes and phenotypes of the old "unit character." Geneticists used to denote the penetrance and expressivity of each mutation next to its description as an indication of the "reliability" of that gene mutation in crossing experiments: Mutants with low penetrance or variable expressivity were too much subject to disturbing environmental fluctuations in their phenotype to be useful experimental material (see, e.g., Lindsley and Grell 1968).

The asymmetry in the role allocated to the genotype and to the environment in shaping of the phenotype is revealed even more in the notion of *phenocopy* introduced in 1935 by Richard Goldschmidt. Goldschmidt called attention to the *range* of environments that a given genotype might respond to in producing the phenotypic reaction and appropriately pointed out that, theoretically, any change in the phenotype caused by a genotypic change (i.e., any mutation) might be reproduced as a purely phenotypic modification if suitable experimental conditions were found (Goldschmidt 1938). However, the definition of phenocopy as "a nonhereditary, phenotypic modification (caused by special environmental conditions) that mimics a similar phenotype caused by a gene mutation" (Rieger, Michaelis, and Green 1976) explicitly diverted attention from the reciprocal two-dimensional genotype–environmental interaction of phenotypic variation to the primacy of the genotype. Significantly, the complementary concept "genocopy," the production of the same phenotype by different genes ("mimic genes"), suggested by Nachtsheim, never gained popularity.[4]

Thus, in spite of Johannsen's caution about the gene as merely *Etwas* that was present in the gametes and the zygote (Johannsen 1909, p. 124), his typological notion of the genotype actually sustained the conservation of the gene as the determinant of a phenotypic unit character. The explicit protestations of Johannsen, Morgan, and others that a unique trait may be used only as a "marker" for the gene notwithstanding (see Schwartz 2000), the notion of the "unit character" was not eliminated. As late as 1941, in a textbook on *Biology and Human Affairs*, the following reference to genes was made:

> The chromosomes of a fertilized egg cannot contain the characters that will be shown by the adult organism that grows from the egg. . . . All that the egg can contain is the determiners of these things – something in the granules of the chromosomes which causes these characters to develop in the embryo.
>
> In discussion of heredity these determiners are commonly called **genes**. . . . *A gene is the determiner of a hereditary unit character.* (Ritchie 1941, p. 661, italics and bold in original)

The only relevant question that remained was, What were the primary unit characters of the genes and what were secondary phenotypes? Evidence was explored for the spuriousness of "genuine" *pleiotropy,* that is, for the production by one particular mutant gene of apparently multiple effects at the phenotypic level (Grüneberg 1938). Obviously, the remoter consequences of gene action are "more liable to influences by chance environmental factors than the 'original' disturbance," but geneticists went further to conclude that "[a]s unity of gene action is the simpler conception, and as it has been shown to explain fully analysed cases, it may be adopted as a general principle" (Grüneberg 1938, p. 142). The culmination of this approach was Beadle and Tatum's inference based on their "faith in the unproved assumption that a given gene has a single action" (Beadle and Tatum 1941a, p. 115). Once you get to the primary product of the gene, "[i]t is entirely tenable to suppose these genes which are themselves a part of the system, control or regulate specific reactions in the system either by acting directly as enzymes or by determining the specificities of enzymes" (Beadle and Tatum 1941b, p. 499). A one-to-one relationship exists between a gene and its primary phenotype. At the level of the immediate products of the genes there was no room for a norm of reaction; genotype and phenotype were completely coupled. The norm of reaction was deferred to the "remoter consequences" of gene action.

This reductionist conception was further entrenched in the central dogma of molecular genetics: A sequence of DNA is transcribed uniquely into a sequence of RNA, and this messenger-RNA is uniquely translated into a sequence of amino acids or a polypeptide. The discovery of nontranscribed sequences that regulate the very transcription of RNA messages as well as that of transcribed but not translated sequences (introns), even of alternative transcriptions of the same DNA sequence, hardly affected this notion of the unique primary product of the gene.

To the extent that the notion of the norm of reaction was referred to it was relegated to the more distant effects of gene action: It was discussed mainly in relation to the organism as a coadapted system, whether as a result of natural selection or as a consequence of developmental and structural constraints. Thus, when Wright (1931) described the evolution of Mendelian populations, he referred to the environment as acting to *constrain* the phenotypic dependence on the genotype via the very phenotype that emerged rather than merely acting as participant in the linear *construction* of the phenotype. He alluded to "the selection pressures of varied environments within the range of the species," and noted that "individual adaptability is, in fact, distinctly a factor of evolutionary poise. . . . The evolution of complex organisms rests on the attainment of gene combinations which determine a varied repertoire of adaptive cell responses in relation to external conditions" (Wright 1931, p. 147). It seems that the only one who outright rejected the genocentric notion, although he may not have been aware of Woltereck's paper, was Lancelot Hogben. Hogben (1933, pp. 93–8) explicitly stressed that once we accept the concept of the integrated nature and nurture in forming traits such as behavior or intelligence, we are not allowed a direct, linear extrapolation from one set of genotype–environment setting of the phenotype to the outcome of another such interaction. His phenotypic response curves to environmental changes are very similar to Woltereck's *Phänotypen-kurven.*

6.3 Development as an Adaptive Norm

Whereas Woltereck referred to the norms of reaction as the *raw material* of organismic evolution, by the 1940s these were viewed as the *product* of evolution. Waddington (1905–1975) in Britain and Schmalhausen (1884–1963) in the Soviet Union focused attention on the constancy of embryonic development and its regulation in variable environments. Natural selection worked within the range of the "habitual environments" of a taxonomic group for an orderly sequence of developmental processes *in spite* of variation in environmental conditions. Within that range of environmental variations, natural selection buffered development.

In his book *The Strategy of the Genes* Waddington (1957) called attention to the developmental constraints of evolution. He put forward the strategic question,

How does development produce entities which have Form, in the sense of integration or wholeness; how does evolution bring into being organisms which have Ends, in the sense of goal-seeking or directiveness? (Waddington 1957, p. 9)

Genes would do nothing except when a potential for expressing them was also present. The fundamental mechanism of embryonic development must be one by which the different cytoplasms that characterize the "various regions of the egg, act differentially on the nuclei so as to encourage the activity of certain genes in one region, of other genes in other places" (Waddington 1957, p. 14). Waddington emphasized the absolute reciprocality in the roles of genes and environmental circumstances in embryonic development: Differences in cytoplasmic contents would differentially activate genes as much as genes in different environments would bring about different (cytoplasmic) products. Embryonic development is a highly regulated process that evolved by natural selection of genes for their action within a range of specific environmental opportunities. "[T]he most characteristic feature of development is the occurrence of continuous and more or less gradual change.... Embryonic cells appear to be inevitably undergoing processes of alteration: and the factors which control differentiation do so by steering the changing system into particular channels" (Waddington 1957, p. 14). Differentiation is a process of narrowing the range of possible phenotypes as a function of preceding reactions. Canalization is the property of developmental pathways of achieving a standard phenotype in spite of genetic or environmental disturbances.

The whole course of development from the initial stage in the egg up to the final adult condition is a "most favoured path"; that is to say if a mass of material is developing along one such path and is at some time during the course of development forced out of it by some experimental means, it will exhibit "regulative behaviour" and tend to return to the normal path. (Waddington 1957, p. 19)

Waddington described the process of development as a procession along an "epigenetic landscape" with valleys and branching points. The further one proceeds, the more interactions of earlier stages constrain the possibilities of later interactions, and thus the valleys along which development moves deepen and diverge. In the formation of this epigenetic landscape, genes were just one contributor along with many other "environmental" inputs. Deformed phenotypes were brought about by extreme environments, beyond the range in which selection stabilized the norm of reaction, or by mutant alleles whose range of stable norm of reaction was so narrowed or displaced that they interacted "abnormally" in instances at which the NOR of the "normal" allele was well buffered:

> Under the influence of natural selection, development tends to become canalised so that more or less normal organs and tissues are produced even in the face of slight abnormalities of the genotype or of the external environment. (Waddington 1953, p. 118)

On the other hand, phenotypic plasticity of individuals could be a means to extend the range of environments to which the species was adapted. When organisms are exposed to selection pressures that they had not encountered, the interaction of *genotype-and-new-environment* may respond with adaptive "acquired characters." These create the opportunity for the phenotypically acquired character to be tested by natural selection, and – if proved "useful" – to be "assimilated" through the replacement of environmentally induced processes by genotypically guided ones, producing an interaction of *new-genotype-and-environment.*

> if an animal is subjected to unusual circumstances to which it can react in an adaptive manner, the development of the adaptive character might itself become so far canalised that it continued to appear even when the conditions return to the previous norm. (Waddington 1953, p. 118)

Whereas Waddington was primarily interested in properly buffered embryonic development and considered the role of selection in the evolution of this property, Dobzhansky (1900–1975) saw selection of properly adapted phenotypes in the "habitual environments" of a population as the central issue of population genetics within the framework of the "New Synthesis." Already in 1941, in his classical *Genetics and the Origin of Species,* Dobzhansky emphasized the constant mutual interaction of nature and nurture in the shaping of the phenotypes of natural populations.

Dobzhansky first introduced NOR as a prescriptive term in the sense of Woltereck's comprehensive concept of an epiphenomenon of genotype–environment interactions. Although "the norm of reaction of a genotype is at best only incompletely known, . . . [t]he existing variety of environments is immense, and new environments are constantly produced. Invention of a new drug, a new diet, a new type of housing, a new educational system, a new political regime introduces new environments" (Dobzhansky 1955, p. 75; see also Dobzhansky 1951, pp. 21–2). By adopting such a view, genocentricity is sublimed: hereditary defects and diseases become at most socially meaningful terms, not biologically determined ones. They are nothing but "genotypic variants which react to environment as usual for the species or race by production of ill-adapted phenotypes. There is, consequently no hard and fast distinction between 'hereditary' and 'non-hereditary' diseases" (Dobzhansky 1955, p. 76), and it is only a question of time and effort to find the environmental constellation in which a genotype will respond with a phenotype similar to the "adaptive" one.

However, Dobzhansky's attention was directed primarily to the evolution of *natural* populations, and he mainly stressed the role of the norm of reaction within "habitual environments" of individual development. The norm of reaction gradually became the "adaptive norm," the phenotypic response of genotypes in the relatively narrow range of environments that the genotypes encountered during their evolution. Although the two terms are frequently confounded, an important modification accompanied the introduction of the concept of adaptive norms: the prescriptive norm became a descriptive norm. The adaptive norm referred to the response of a *range of genotypes* to a confined array of environments contrary to the notion of the "classic" NOR, which referred to the responses of given genotypes to an open range of environments. The difference becomes especially significant when exploration of new genetic and environmental combinations are considered.

In 1950, Dobzhansky joined E. W. Sinnott and L. C. Dunn, as a coauthor of the fourth edition of *Principles of Genetics.* He introduced the concept of the NOR, or rather the *range of reaction*, as he called it then, as one that provided for a *different* and *specific* response of each genotype to different as well as identical environments,

Every genotype reacts with its environment in its own special manner; but if the same genotype has somewhat different materials to work on, the phenotypes may be appreciably different. *What a genotype determines is the reactions, the responses, of the organism to the environment.* . . . These potentially possible or actually realized phenotypes constitute the *range of reaction* of a given genotype. In practice, one is never certain that the entire range of reaction of any genotype is known, . . . It is, however, evident that, the more thorough becomes our knowledge of the reaction ranges of human genotypes and those of agricultural plants and animals, the greater will be our ability to control them according to our will. (Sinnott, Dunn, and Dobzhansky 1950, p. 22)

Although his wording is still in prescriptive terms, the transition to the descriptive concept of genotype–environment interaction is evident from his examples. In the now classical study of Clausen, Keck, and Hiesey (1940, 1948), the authors took plants of several species from three elevations in California: some from "races" that grew at the Pacific coast, some from "races" of the forest belt of the Sierra Nevada mountains, and some from the "races" of the alpine zone of the Sierra Nevada. Plants from all three locations were cut, each into three parts that constituted a clone of the same genotype. Each clone was planted in the three gardens that were located in the three elevations, respectively. This way a 3×3 matrix was obtained of {altitude-genotype} × {altitude–environments} for different "races" of the same species as well as for that of different species (Dobzhansky 1955, 1962; Sinnott et al. 1950). "The plants native in the alpine zone grew taller in the mid-altitude and sea-level stations than at home, and yet nowhere near as tall as the plants native to these habitats. . . . Each plant has its own norm of reaction, and this determines how it will grow and what it will look like in a given environment" (Dobzhansky 1962, p. 81). These experiments were repeatedly presented by Dobzhansky as illustration of the individual genotypes' *norm of reaction* in all possible environments. However, what was actually shown, and what was of interest to Dobzhansky, was the range of the NOR that was adaptive, that is, the "adaptive norm."

On Dobzhansky's initiative, Schmalhausen's Russian text of *Factors of Evolution: Theory of Stabilizing Selection* was translated into English in 1949 (Schmalhausen 1986). Schmalhausen maintained the genocentric genotype–environment asymmetry. The genotype was the main agent that provided for the regular and orderly development of organisms, whereas "environmental factors act only as agents releasing form building processes and providing conditions necessary for their realization" (Schmalhausen 1986, p. 2). Schmalhausen viewed ontogeny as being in the service of phylogeny (see Gilbert 1994). It was a fact that the expression of a genotype of both the normal and mutant varied in diverse environments, but "every organization, the typical as well as the variant (including mutants), does not pre-exist but develops on a specific hereditary foundation (genotype)" (Schmalhausen 1986, p. 4). Of the wealth of factors that may modify an organism's phenotype, Schmalhausen's attention was directed at those that may bring about the *adaptive modifications.*

> [I]n the process of evolution there have arisen definite optimum norms of growth which are determined by the ecologic position of the organism, especially by its relationship with other organisms. Modifications are possible thus only within the relative narrow limits of this norm. Therefore, it is not the *modification itself* but its confinement within definite limits that should be regarded as an *adaptation.* (Schmalhausen 1986, p. 184–5)

Although Waddington regarded the norm of reaction to extend also beyond the adaptive range as the arena for further genetic assimilation, Schmalhausen, like Dobzhansky, focused on the range of *adaptive responses.* Schmalhausen denoted nonadaptive modifications as *morphoses,* and these were "an entirely different character. They arise as new reactions which have not yet attained a historical basis" (Schmalhausen 1986, p. 8).

The notion of the adaptive norm provided a fertile framework for the development of theoretical models of the evolution of polymorphic populations in manifold environments in space and in time (Bradshaw 1965; Gomulkiewicz and Kirkpatrick 1992; Levins 1968; Roughgarden 1972; Via 1994). It was, however, mainly the examination of the norm of reaction of specific genotypes under varying experimental conditions (Clausen et al. 1940, 1948; Dobzhansky and Spassky 1944, 1963; Gupta and Lewontin 1982) that led Lewontin (1974, 1992) to return to Woltereck's notion of the *unpredictability* of individual phenotypic responses once the genotype–environment interaction over a wide range was considered. In this he deviated from the Schmalhausenian stress on the *predictability* of the phenotypic response in spite, or rather because of, genotype–environment interaction within a range of evolutionary selected contexts (see, e.g., van Noordwijk 1989).

Lewontin points out the fallacy inherent in the hopes of analyzing causes through linear models embodied in the analysis of variance, covariance, and path analysis. The analysis of *interacting causes* has been confounded with the

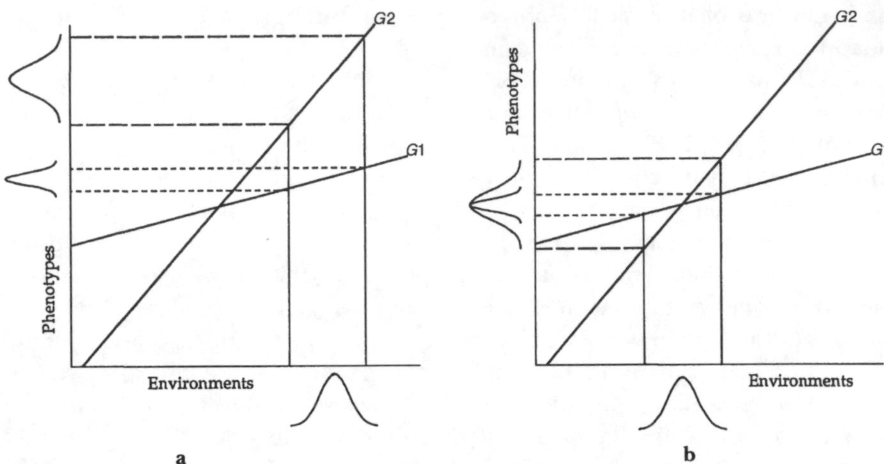

Figure 6.3. Changing the window of observed environments changes the "genetic variance" of the analysis of variance within the same population. (A computerized modification of Figure 8 of Gupta and Lewontin 1982.)

concept of discriminating *alternative causes.* "If an event is the result of the joint operation of a number of causative chains and if these causes 'interact' in any generally accepted meaning of the word, it becomes conceptually impossible to assign quantitative values to the causes of that *individual event*" (Lewontin 1974, p. 402). What we do in the analysis of variation is only to assess how much perturbation of *phenotype* has been the outcome of differences in environment (averaged over all genotypes) and how much the outcome of differences in genotype (averaged over all environments). Such an analysis of deviations from the mean gives a result that depends upon the actual distribution of genotypes and environments in a particular population. As to the methodological inadequacy of the phenotypic analysis of the NOR, Lewontin uses his and Gupta's data on bristle numbers in two lines of *Drosophila pseudoobscura* with nonparallel norm of reaction to temperature shift (Gupta and Lewontin 1982) to demonstrate how the analysis of variance may actually confound environmental and genetic information (Figure 6.3). For given phenotypic response curves, if one range of environmental distribution is chosen on the environmental axis, the genetic component of the phenotypic response of the population composed of the two genotypes may be considerable (Figure 6.3a). Now, a mere shift of the mean of the *environmental* distribution along the environmental axis, may, with the same genetic composition of the population, cause the genetic component of the phenotypic variance to be drastically reduced or disappear (Figure 6.3b). "Thus, *genetic variance* has been destroyed by changing the *environment.*" Conversely, "a change in *genotypic* proportion changes the environmental variance" (Gupta and Lewontin 1982, pp. 944–5).

The analysis of variation is "too specific" in that it completely overlooks the "phenotypic plasticity" exhibited during the life history of each of its individuals;

on the other hand, it is "too general in that it confounds different causative schemes in the same outcome" (Lewontin 1974, p. 403). Lewontin (1992) instead spells out the route leading from mapping the phenotype space of an individual into a genotype space, to the genotype space of its progeny, and further on, into the progenys' phenotype space, and so on. The only stage in this history for which we have a reasonably good theory is the Mendelian theory of parental genotype to progeny genotype transmission. Because the relations between phenotype and genotype and between phenotype and environment are many-to-many relations, in any usual sense of the word, both genotype and environment are always *causes* of phenotypic differences as is a component of genuine random nondeterminacy. Figure 6.4 gives a sample of the possible NORs of two genotypes as a function of environmental variation (with a single, well-ordered environmental variable assumed).

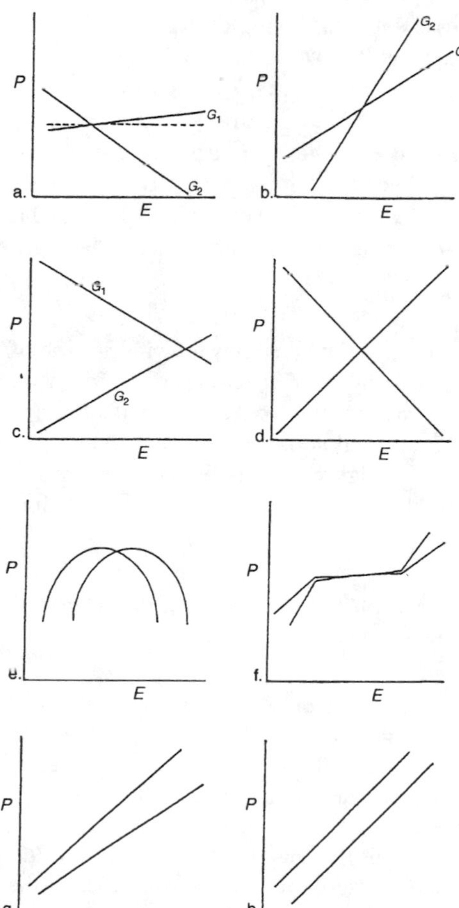

Figure 6.4. Examples of different possible norms of a pair of genotypes in a range of environments. In each case, the phenotype (P) is plotted as a function of environment (E) for the pair of genotypes (G_1, G_2) (after Lewontin 1974).

There is no doubt that the concept of the phenotype that provides a "first approximation" of the genotype was essential in early genetic research. However, as any detailed analysis of genotype–phenotype relationship reveals, such a relation seems to be the exception of which transmission geneticists skillfully took advantage. In many cases the phenotype provides hardly any information at all about the genotype. The concept of the NOR, which stresses the interaction of genotype–environment, suggests the formulation of different approaches to variation that could be highly heritable between individuals, whereas the traits being considered may still be highly malleable. Lewontin points out the open-ended nonpredictability, the epistemology of *possibilities,* that is provided by the NOR notion, whereas most students of the notion of the adaptive norm follow an epistemology of *constraints.* Via's (1994) presentation of the two major types of phenotypic response to the environment, that is, two types of plasticity, is telltaling: "Central to the concept of the reaction norm is the idea that the environments involved are repeatable and predictable aspects of the organism's habitat"; the first kind of plastic responses is to these varying-though-predictable circumstances. The second type of plasticity is "the response to a largely unpredictable variability within environments" (Via 1994, p. 37). Via refers to the first plasticity as the "adaptive reaction norm," or the norm-of-reaction *sensu stricto,* whereas she refers to the second one as "noisy plasticity." From the vantage point of the adaptive norm, the second, "noisy plasticity" is "essentially shot-gun variability that results from developmental instability" (Via 1994, p. 37). For Lewontin, the open-ended "noisy plasticity" is the essence of human nature:

> That the confusion between heritability and lack of malleability is a vital source of misunderstanding and of false claims about social policy becomes clear when we consider the rhetorical question that served as the title for A. R. Jensen's famous article (1969): "How much can we boost IQ and scholastic achievement?" Jensen's answer was "not much," because IQ is said to be highly inheritable! Had Jensen not made that basic error, his entire article could never have come into being. (Schiff and Lewontin 1986, p. 174)

Lewontin's onslaught on the term "development" as a biasing metaphor of ontogeny summarizes his position in a nutshell:

> To describe the life history of an organism as "development" is to prejudice the entire problematic of the investigation and to guarantee that certain explanations will dominate. "Development" means literally an unrolling or unfolding. . . . It means the making of an already predetermined pattern immanent in the fertilized egg. . . . All that is required is the appropriate triggering of the process and the provision of a milieu that allows it to unfold. This . . . reveals the shape of investigation in the field. Genes are everything. The environment is irrelevant except insofar as it allows development. (Lewontin 1995, p. 261)

Two routes are open to break the circularity of the inferences that genotypes depend entirely on observations of phenotypes and deductions of phenotypes depend upon the nature of the underlying genotypes. "One is to acquire detailed information on complex ontogenetic pathways, . . . For the moment this seems utopian. The alternative is to solve part of the problem by *inferring genotypes from phenotypic manifestations that do not have intervening developmental processes.* That is the approach of molecular genetics, which reads the state of the genes from their molecular configuration." (Lewontin 1992, p. 140, my emphasis)[5]

6.4 Norm of Reaction at the Molecular Level of the Gene

Lewontin leads us back to examine the norm of reaction at the level of the gene:

The law of transformation of genotype into phenotype is a law that must contain information not only about the genotype but also about the historical sequence of environments. . . . Two genotypes are then recognized as different only if their norms of reaction differ for some aspect of the environment and some aspect of the phenotype. . . . The environmental contingency of the mapping of genotype onto phenotype depends on the developmental complexity that intervenes between genome and phenome, which is, in part, a function of the molecular details of the control of turning on and off the products of primary protein synthesis. (Lewontin 1992, pp. 140–1)

Of course, even at the most elementary level of molecular biosynthesis there is a many-to-one relation because of the degeneracy of the genetic code. "Nevertheless, the qualitative structure of proteins is essentially independent of environmental contingency, although the amount of their production may be very strongly influenced by environment" (Lewontin 1992, p. 143). It would seem that finally, at the molecular level, a gene may be defined unequivocally, though functionally, as a sequence of DNA that is uniquely transcribed from a start signal to the end signal and, as a rule (though not in such sequences that code for rRNA, tRNAs, etc.), is also translated into a unique polypeptide product. The norm of reaction, the many-to-many relationships of gene–gene as well as gene–environment interaction, seems to have been deferred to a secondary level.

Has the essentialist–reductionist conception succeeded after all? Can the "atoms of genetics" be identified by their structure independently of their specific function? Are not the stretches of the DNA that compose ORFs (open reading frames) – whether interrupted by nonORF sections or not – with initiation signal as well as termination signals, context-independent ontological entities? Indeed, more than half the sequences of putative genes in the completed sequencing of yeast's DNA have been identified as such with neither a recognizable function upon disruption nor a transcript identifiable by its cDNA (see Johnston 1996; Oliver 1996). On second thought, however, such an instrumental definition of the genes would not do. A conditioning of the gene concept on the cellular context (i.e., on the molecular norm of reaction) is unavoidable. We may exclude the untranscribed regulatory sequences from the gene definition but are still left with untranslated processed intron sequences or with the translated but clipped-as-peptides sequences (such as the leader sequences, and the "inteins"). Even more significant, it is not possible to overlook the "overlapping genes," that is, sequences that belong to two partly or completely overlapping transcription units, and especially not the "alternatively spliced genes." From the same transcribed sequences several alternative messenger-RNAs may be processed, each translated into a different polypeptide or into the same polypeptide but at a different level of expression, depending on the sites of splicing. The cues for the alternative transcripts are interactions of specific sites in the DNA with complex nuclear proteins.

The more we learn of the molecular level of DNA organization and function, the more it becomes obvious that there is no unique primary context-independent product of a DNA sequence – even at this level. If we wish to identify a gene with a coding sequence (not exclusively), it is still bound by its norm of reaction under different intracellular or intranuclear "environments." Such a gene may have meaning only in context or within a norm of reaction.

> The emergence of a phenotype is due to a developmental interaction of a genotype with an environment. The term *interaction* does not refer to one or more components in a statistical analysis susceptible to elimination through some transformation of scale applied to group scores and thus to be dismissed as a mere measurement problem. *Interaction*... is the intricate and unique interplay of the numerous causal processes of development from the microlevel of gene interactions and genetic–cellular chemistry to the macrolevel of individual-environment feedback loops. (Platt and Sanislow 1988, p. 254)

6.5 Acknowledgments

I wish to thank Prof. Rishard C. Lewontin and his crew for the pleasant and stimulating sabbatical leave at the Department of Organismic Biology, Museum of Comparative Zoology, Harvard University, Cambridge, Massachusetts. R. C. Lewontin's comments were invaluable. I also enjoyed the helpful comments of Jonathan Harwood.

Notes

1. The German word *Art* means "kind," but it refers here more to the sense of type or species. To be on the safe side I will refer to it as "kind." Thus, *Elementarart* will be translated as "elementary kind."
2. "Woltereck's work was primarily to show whether or not acquired characters are inherited. It was a secondary object to find out whether small variation or distinct

sports occurred in the species" (East 1912, p. 652). And, according to Castle (1913, p. 117), Woltereck "has shown that, among the offspring developed from the unfertilized eggs of the same mother Daphnia, variations do occur which are heritable, so that if one selects extreme variants he obtains a modified race." Castle pointed out that the experiment with Daphnia "is not open to the objection that may be offered to [his own] guinea pig experiment, that it is possibly a result of gametic segregation and recombination, for in Daphnia the reproduction was exclusively by unreduced and unfertilized eggs" (Castle 1913, p. 118–19). In spite of this, he noticed, Woltereck also observed "what the others have failed to observe, that along with the non-inherited variations [of pure lines] occur other similar but less numerous ones which are inherited" (Castle 1913, p. 120). It could not be denied that "in Woltereck's cultures a head form, different from the one with which the culture started, was present at the end of two years, and this did not revert on return to normal conditions"; still, it was possible to claim that "the failure to carry adequately controlled parallel normal lines does not permit a decision as to whether the change is a real one, or due to the progressive selection of a biotype present but obscured in the original population" (Tower, in Castle et al. 1912, p. 211).

3. In some experiments in which lines had been kept for many parthenogenetic generations under specific sets of conditions, later generations showed the pattern characteristic for these conditions even after removing these conditions. It was, however, enough with one sexual cycle to restore the biotype's characteristic norm of reactions. This phenomenon of long-term, quasi-inheritance of acquired characters has been observed also in other invertebrate and protozoa species and was termed *Dauermodifikation.* It was dismissed by many as "artifacts," being the effect of some "irrelevant" cytoplasmic factor that took many cell generations to dilute.
4. New mutants, however, have been subjected to allelism tests with other mutants of similar phenotypes. Such tests determined whether they affected alleles of the same gene, so that any phenotypic *differences* between them were superficial, of whether they affected different genes and the *similarity* between phenotypes was merely superficial. In the 1940s these tests of superficially similar nutrient-deficient mutant of *Neurospora* allowed Beadle and Tatum to construct the metabolic pathways by assuming the "one gene–one enzyme" relationship.
5. Recent developments in sequencing genomes and the "annotation" of the resolved sequences and in proteomics make also the first route less utopian.

Submitted: September 1996 and in revised form: October 1997. Since the revision of this article in October 1997, several publications directly relevant to it have appeared. Most significant are Harwood, J. (1996), Weimar culture and biological theory: A study of Richard Woltereck (1877–1944), *History of Science,* 34:347–77, and Sarkar, S. (1999). From the *Reaktionsnorm* to the adaptive norm: The norm of reaction, 1909–1960, *Biology and Philosophy,* 14(2):235–52. Sarkar refers also to the present article. Piglliuchi and Schlichting have been publishing extensively on the subject in recent years. See their book, Schlichting, C. D., and Pigliucci, M. (1998), *Phenotypic Evolution: A Reaction Norm Perspective,* Sunderland, MA: Sinauer.

REFERENCES

Beadle, G. W., and Tatum, E. L. (1941a). Genetic control of developmental reactions. *The American Naturalist* 75:107–16.

Beadle, G. W., and Tatum, E. L. (1941b). Genetic control of biochemical reactions in Neurospora. *Proceedings of the National Academy of Science, Washington* 27:499–506.

Bradshaw, A. D. (1965). Evolutionary significance of phenotypic plasticity in plants. *Advances in Genetics* 13:115–55.

Buss, L. W. (1987). *The Evolution of Individuality.* Princeton, NJ: Princeton University Press.

Cairns, J., Overbaugh, J., and Miller, S. (1988). The origin of mutants. *Nature* 335: 142–5.

Castle, W. E. (1913). *Heredity.* New York: D. Appleton and Comp.

Castle, W. E., Coulter, J. M., Davenport, C. B., East, E. M., and Tower, W. L. (1912). *Heredity and Eugenics.* Chicago: The University of Chicago Press.

Clausen, J., Keck, D. D., and Hiesey, W. M. (1940). *Experimental Studies on the Nature of Species. I.* Washington, DC: Carnegie Institute of Washington Publication No. 520.

Clausen, J., Keck, D. D., and Hiesey, W. M. (1948). *Experimental Studies on the Nature of Species. III.* Washington, DC: Carnegie Institute of Washington Publication No. 581.

de Vries, H. (1902–3). *Die Mutationstheorie: Versuche und Beobachtungen über die Entstehung von Arten im Pflanzenreich.* Leipzig: Von Veit.

Dobzhansky, T. (1951). *Genetics and the Origin of Species* (3d ed.). New York: Columbia University Press.

Dobzhansky, T. (1955). *Evolution, Genetics, and Man.* New York: John Wiley & Sons.

Dobzhansky, T. (1962). *Mankind Evolving: The Evolution of the Human Species.* New Haven: Yale University Press.

Dobzhansky, T., and Spassky, B. (1944). Genetics of natural populations. XI. Manifestation of genetic variants in *Drosophila pseudoobscura* in different environments. *Genetics* 29:270–90.

Dobzhansky, T., and Spassky, B. (1963). Genetics of natural populations. XXXIV. Adaptive norm, genetic load, and genetic elite in *Drosophila pseudoobscura. Genetics* 48:1467–85.

East, E. M. (1912). The Mendelian notation as a description of physiological facts. *The American Naturalist* 46:633–55.

Falk, R. (1995). The struggle of genetics for independence. *Journal of the History of Biology* 28:219–46.

Falk, R., and Schwartz, S. (1993). Morgan's hypothesis of the genetic control of development. *Genetics* 134:671–4.

Gilbert, S. F. (1994). Dobzhansky, Waddington and Schmalhausen: Embryology and the Modern Synthesis. In M. B. Adams (ed.), *The Evolution of Theodosius Dobzhansky* (pp. 143–54). Princeton, NJ: Princeton University Press.

Goldschmidt, R. B. (1938). *Physiological Genetics.* New York: McGraw–Hill.

Gomulkiewicz, R., and Kirkpatrick, M. (1992). Quantitative genetics and the evolution of reaction norms. *Evolution* 46:390–411.

Grüneberg, H. (1938). An analysis of the "pleiotropic" effects of a new lethal mutation in the rat (*Mus norvegicus*). *Proceedings of the Royal Society, London, series B*, 125:123–44.

Gupta, A. P., and Lewontin, R. C. (1982). A study of reaction norms in natural populations of *Drosophila pseudoobscura*. *Evolution* 36:934–48.

Hogben, L. (1933). *Nature and Nurture*. New York: W. W. Norton.

Jablonka, E., and Lamb, M. J. (1995). *Epigenetic Inheritance and Evolution*. Oxford: Oxford University Press.

Johannsen, W. (1909). *Elemente der exakten Erblichkeitslehre*. Jena: Gustav Fischer.

Johannsen, W. (1911). The genotype conception of heredity. *The American Naturalist* 45:129–59.

Johnston, M. (1996). The complete code for a eukaryotic cell. *Current Biology* 6(5):500–3.

Levins, R. (1968). *Evolution in Changing Environments: Some Theoretical Explorations*. Princeton, NJ: Princeton University Press.

Lewontin, R. C. (1974). The analysis of variance and the analysis of causes. *American Journal of Human Genetics* 26:400–11.

Lewontin, R. C. (1992). Genotype and phenotype. In *Keywords in Evolutionary Biology*, E. F. Keller and E. A. Lloyd (eds.), pp. 137–44. Cambridge, MA: Harvard University Press.

Lewontin, R. C. (1995). À la recherche du temps perdu. *Configuration* 2:257–65.

Lindsley, D. L., and Grell, E. H. (1968). *Genetic Variations of Drosophila melanogaster*. Washington, DC: Carnegie Institute of Washington Publication No. 627.

Oliver, S. G. (1996). From DNA sequence to biological function. *Nature* 379(6566):597–600.

Platt, S. A., and Sanislow III, C. A. (1988). Norm-of-reaction: Definition and misinterpretation of animal research. *Journal of Comparative Psychology* 102:254–61.

Rieger, R., Michaelis, A., and Green, M. M. (1976). *Glossary of Genetics and Cytogenetics* (4th ed.). Berlin: Springer–Verlag.

Ritchie, J. W. (1941). *Biology and Human Affairs*. Yonkers-on-Hudson, NY: World Book Company.

Roll-Hansen, N. (1978). The genotype theory of Wilhelm Johannsen and its relation to plant breeding and the study of evolution. *Centaurus* 22:201–35.

Roughgarden, J. (1972). Evolution of niche width. *The American Naturalist* 106:683–718.

Schiff, M., and Lewontin, R. (1986). *Education and Class. The Irrelevance of IQ Genetic Studies*. Oxford, UK: Clarendon Press.

Schmalhausen, I. I. (1986 [1949]). *Factors of Evolution. The Theory of Stabilizing Selection*. Chicago and London: University of Chicago Press.

Schwartz, S. (2000). The differential concept of the gene: Past and present. In P. J. Beurton, R. Falk and H.-J. Rheinberger (eds.), *The Concept of the Gene in Development and Evolution*. (pp. 26–39) Cambridge: Cambridge University Press.

Sinnott, E. W., Dunn, L. C., and Dobzhansky, T. (1950). *Principles of Genetics* (3d ed.). New York: McGraw–Hill.

van Noordwijk, A. J. (1989). Reaction norms in general ecology. *BioScience* 39:453–8.

Via, S. (1994). The evolution of phenotypic plasticity: What do we really know? In L. A. Real (ed.), *Ecological Genetics* (pp. 35–57). Princeton, NJ: Princeton University Press.

Vogt, O. (1926). *Psychiatrisch wichtige Tatsachen der zoologisch–botanischen Systematik. Zeitschrift für gesamte Neurologie und Psychiatrie* 101:805–32.

Waddington, C. H. (1953). Genetic assimilation of an acquired character. *Evolution* 7:118–26.

Waddington, C. H. (1957). The *Strategy of the Genes. A Discussion of Some Aspects of Theoretical Biology*. London: George Allen and Unwin.

Woltereck, R. (1909). Weitere experimentelle Untersuchungen über Artveränderung, speziell über des Wesen quantitativer Artunterschiede bei Daphnien. *Verhandlungen der Deutschen Zoologischen Gesellschaft* 19:110–73.

Wright, S. (1931). Evolution in Mendelian populations. *Genetics* 16:97–159.

CHAPTER SEVEN

"The Apportionment of Human Diversity" 25 Years Later

MARYELLEN RUVOLO AND MARK SEIELSTAD

Perhaps nothing written by R. C. Lewontin speaks more directly to the human condition than his 1972 article, "The Apportionment of Human Diversity." It may lack the beautiful writing and turns of phrase of his later works, but for sheer scientific impact it is unrivalled. Lewontin's distinctive style of reasoning is brought to focus on the question of human races with all of its characteristic intensity. His findings surprised those who read the paper. Although typological notions of race had been on the decline in anthropology, many scientists and laypeople continued (and a few still continue) to expect substantial genetic differences between the groups they seemed able to recognize visually. By studying allele frequency variation between populations for a collection of polymorphic "classical" (protein-coding) genes, Lewontin observed that most genetic variation was found within populations (85.4%) or between populations within a race (8.3%). Only 6.3% of all human variation was distributed across races. This observation was based on genetic variation in 17 genes encoding blood groups, serum proteins, and red-cell enzymes.

The most important aspect of Lewontin's article is that it asked the right question about human genetic diversity: How great is within-group variation relative to between-group variation? In doing so, it framed the study of human genetic diversity in population genetic terms and brought it in line with studies of other organisms. As Lewontin's paper was being written, Muller's (1950) idea of a finely tuned, little-varying genome was increasingly being disproved, whereas Dobzhansky's (1955) conception of substantial intraspecific heterozygosity was increasingly supported as more and more organisms were surveyed using protein electrophoresis. *Homo sapiens* proved to be similar to most other species in exhibiting considerable heterozygosity (Harris 1967, Lewontin 1967). Lewontin (1972) extended this approach and was the first to estimate the amounts of variation contained in each hierarchic population level.

By its very nature, his article draws attention away from itself and from the problem of human racial "types" by demonstrating that the concept of race has little biological grounding. When asked about it today, Lewontin views the genetics of race as completely uninteresting owing in some small part to his article. Of course, this should not lead us to forget its lesson.

A few years ago, one of us (MR) gave a reading course to a graduate student in the History of Science Department at Harvard, a young man of color whose research focused on pre-Darwinian concepts of variation. He also wanted to learn about Darwin, Mendel, and the current thinking on human genetic variation. After we had developed a certain rapport, he asked a question that had been bothering him since it was raised in one of his undergraduate anthropology courses at another Ivy League institution: Was it true, as his former professor claimed, that a mating between a European and a South African Kung or Hottentot would be infertile? The question was staggering, especially because that particular bit of "information" was probably passed along to an entire class of students no more than 15 years ago. It certainly implied that the professor was ignorant not only of Lewontin's article but also much other work in anthropology and biology!

Every generation of students in anthropology and biology reformulates such questions, and thus returning to this essay and teaching it to undergraduates and graduate students on a regular basis is very important. Indeed, ignoring it and not addressing the biological bases of human races is dangerous because it leaves a vacuum of knowledge that can become filled by unsupported "facts" and imagined "truths."

For a variety of reasons, many anthropologists and evolutionary biologists today are ignorant of the basic biological facts about human genetic differences. Surprisingly, Lewontin's article received relatively little attention when it first appeared and was cited very few times (Figure 7.1). This might have been because the scientific community had already come to believe that human races were effectively a scientific nonissue and that Lewontin's findings were "obvious." Alternatively, the race question might have seemed too politically charged for further investigation. This is a jump from when it first came out to "now." Lewontin's result has permeated popular consciousness to such an extent that at least some supposedly competent commentators can dismiss it as "politically correct" claptrap. Recently, one of us discussed Lewontin's result in an article for *Annual Review of Anthropology* on hominoid genetic diversity (Ruvolo 1997b). An anonymous reviewer, presumably an anthropologist with expertise in genetics, offered the following comments as part of a review:

> There is the standard statement that races don't exist because most variation is within rather than between human populations. We were all trained in this dogma. But is it true? I think that the recent data on DNA sequence variation would suggest not, and instead would suggest that many, if not even most, DNA sequence variations are local and not globally distributed. Even if the *amount* of variation within a group is about the same as that in our whole species, that does not imply that the same variation is found everywhere, and it really is the latter that makes for the politically correct recitation given here. The subject of "race" is given rather superficial treatment here, possibly wrong . . .

Commentary like this, besides engendering a deep sense of sadness because it reflects a basic misunderstanding of Lewontin's and other more recent studies,

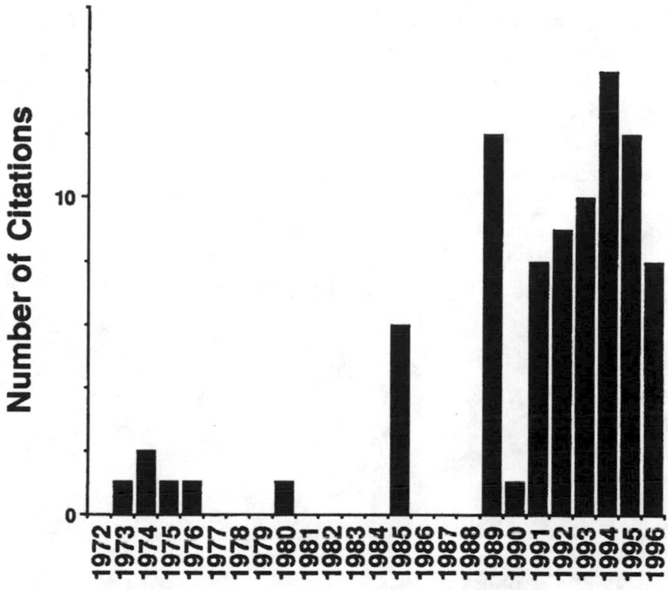

Figure 7.1. Number of citations of Lewontin's 1972 article "The Apportionment of Human Diversity" from 1972–1996 as recorded in the *Science Citation Index.* The sudden peak of interest in 1985 came from journals representing a variety of disciplines: anthropology, theoretical genetics, botany, environmental biology, and statistics. Beginning in 1989, citations were primarily from human genetics and anthropology journals with a smaller number in botanical journals.

emphasizes how important it is to repeat the findings of Lewontin's essay and to go over the logic of its analysis.

The profound implications of the apportionment of human diversity have been difficult to digest for some people. One teaching fellow for an introductory anthropology course on human genetics at Harvard told students (unbeknownst to the professor [MR]) that Lewontin's result was undoubtedly biased because it was based on protein-coding regions of the genome and that natural selection had created this artifactually high degree of genetic overlap among human groups. (This critical interpretation by the teaching fellow was not discovered until grading of the final exams, when many student essays echoed this objection – too late for redress.) Of course, questioning any genetic result which is based on a potentially nonrepresentative subsample of the genome is not unreasonable; but suggesting the result *had to be wrong* (presumably because of some perception of "overwhelming" phenotypic differences among racial types) revealed a startling bias on the part of the teaching fellow.

Until recently it was possible of course to argue that some unique feature of protein polymorphisms had produced this result, and that other genes – perhaps more rapidly evolving noncoding stretches of DNA – would demonstrate racial distinctiveness. Perhaps a relatively lower protein mutation rate or

Table 7.1. *Apportionment of Human Diversity (as Percentage of Total Species Diversity)*

Genetic System (Number of "Races" and Continental Regions)	Within Populations	Within Races Between Populations	Between Races
17 classical polymorphisms (Lewontin 1972; 7 races)	85.4	8.3	6.3
18 classical polymorphisms (Latter 1980; 6 regions)	83.8–87.0	5.5–6.6	7.5–10.4
25 classical polymorphisms (Ryman et al. 1983; 3 races)	86.0	2.8	11.2
79 autosomal RFLPs (Barbujani et al. 1997; 5 regions)	84.5	3.9	11.7
30 autosomal microsatellites (Barbujani et al. 1997; 5 regions)	84.5	5.5	10.0
34 mtDNA restriction sites (Excoffier et al. 1992; 5 groups)	74.5–80.7	3.3–3.6	22.0–15.7
10 Y microsatellites (Seielstad 1998; 5 regions)	67.1	19.2	13.7

some form of selection was to blame, although most readily imagined forms of selection would seem to maximize the genetic differences between races. Selection mediated by climate is the typical explanation for many of the outward phenotypic differences we use in social constructions of race, for example.

However, using the most recent data at the DNA level, Barbujani et al. (1997) have reexamined the apportionment of human genetic diversity. They looked at two classes of DNA polymorphism: microsatellites, with a very high mutation rate (in the range of 10^{-3} to 6×10^{-4} per locus per generation), and nuclear restriction fragment length polymorphisms (RFLPs), whose mutation rate is much lower and more similar to rates for classical genes. Results from these two datasets are remarkably similar and almost identical to all previous studies, including other protein surveys (Table 7.1). Not only are the averages for within-population diversity indistinguishable (85.4% for classical markers in Lewontin's study versus 84.5% for DNA markers), but the values for the individual loci span a similar range: 63.6–99.7% for protein markers versus 54.4–98.3% for DNA markers.

Both of these studies reveal a few loci with low values of within-population diversity, meaning that these loci tend to differentiate populations more than loci with higher values. Sometimes the term "population-specific" is used to refer to the alleles at such loci that show considerably different frequencies between populations. Yet the collection of DNA markers studied to date are globally distributed, not locally restricted (contra the above quotation). One can of course deliberately search for and find loci with low within-population variability in order to differentiate populations maximally (Shriver et al. 1997), an explicitly nonrandom subset of the genome. However, that does not invalidate the clear result that most human variants are present in every population, and populations vary mainly in having different frequencies of those variants. It should also be noted that the utility of any population-specific alleles is limited to those populations for which adequate allele frequency data have been compiled. In other words, alleles that differentiate U. S. whites, blacks, and Hispanics, cannot assign a blood sample of unknown provenance to an Asian or Native American. (This is reminiscent of the problem recognized by Lewontin and Hartl [1991] in which limited database information is used to estimate match probabilities in forensic science.)

The autosomal genes, which include all those mentioned so far, are biparentally inherited. Two parts of our genome, however, are transmitted solely by one sex. Mitochondrial DNA (mtDNA) is almost solely maternally transmitted, whereas the Y chromosome is paternally inherited. Comparing the apportionment of diversity for these two elements with the pattern observed for the autosomes is interesting for at least two reasons: the effective population size (N_e) of each is reduced relative to the autosomes because they are haploid and uniparentally inherited (each has one-quarter the effective population size relative to the autosomes), and they might reveal demographic characteristics that differ between the sexes because each represents just one of the two parental lineages.

The fraction of the genetic variance within populations is lower for both mtDNA (75–81%) and the Y chromosome (67%) (Table 7.1) than for the autosomes, reflecting the relative reduction in N_e. A smaller population size increases the effect of genetic drift, which tends to lower the amount of variation contained in any one population and to differentiate populations that are not exchanging migrants. But the within-population variation of the Y chromosome is reduced even beyond the mtDNA level. A possible explanation for this reduction, which is supported by a study of isolation by distance for the Y chromosome, mtDNA, and the autosomes (Seielstad 1998; Seielstad et al. 1998), is a higher female migration rate relative to that of males (see also Salem et al. 1996; Cavalli-Sforza and Minch 1997; contra the prediction of Hammer and Zegura 1996). Although this is a perhaps counterintuitive result (especially for those holding the image of human males as long-distance marauders, leaving a trail of their genes in their wakes), several predominant cultural practices, such as patrilocality (the tendency for wives to move into the husband's natal household) and hypergamy (rules allowing only women to rise in social status through marriage) would elevate the apparent migration rate of women.

It is traditional for festschrift articles as a genre to praise the distinguished professor while emphasizing how the old man (or woman) did not quite get it right. The author can thereby draw attention back to himself, demonstrating how he or she is just that little bit smarter than the professor and deserves honors equal to those of the honoree (or, at the very least, a job or tenure).

But no hidden agenda of this sort is implied in the following discussion. (Let's face it – any game of intellectual one-upsmanship is bound to be futile when Lewontin is a participant.)

One aspect of Lewontin's 1972 article requiring further attention is a minor discrepancy between his results and some other subsequent studies. Specifically, the amount of variation due to "races" is the smallest component of diversity in Lewontin's study but is somewhat higher (becoming the second greatest source of variation) in subsequent studies (Table 7.1). Why is this?

It may be due to differing ways in which races and groups are defined. After a beautiful discussion of the problems of categorizing human populations into "races," Lewontin settled upon a conservative classification that would not artifactually reduce the between-race component of diversity. He retained the racial categories classically recognized by most anthropologists but with "a few switches based on obvious total genetic divergence," defining a total of seven groups: Caucasians, black Africans, Mongoloids, South Asian aborigines, Amerinds, Oceanians, and Australian aborigines. Later studies (see Table 7.1) used a surrogate for "racial" groups, namely continental classifications, and thus North African Caucasoids, for example, are included with Sub-Saharan Africans. Is this classificatory difference significant enough to produce the differences between the studies? No, because compared with Lewontin's classification, a continental classification should inflate the amount of within-continent or "race" variation.

The difference between the diversity components may also be due solely to the use of different collections of genes and populations. We looked at the effects of alternative classifications on one dataset of the same genes and populations. Results presented in Table 7.1 based on Y chromosome microsatellite data (Seielstad 1998) classified African, European, Pakistani, Asian, and Oceanian populations separately, but grouping Pakistanis with Europeans (as in Lewontin's classification) or, alternatively, with Asians produced almost no effect on the within-continent–race versus the between-continent–race components. Therefore, differences in how races are defined are unlikely to explain the differing amounts of diversity between races and between ethnic groups within races in Lewontin's versus other studies. More likely, because there are no error bars attached to the estimates of 6.3% as the between-race component and 8.3% as the within-race–between-populations component in Lewontin's study, these values may not be significantly different, and they should be interpreted only as being roughly ten times less than the within-population component.

For those of us interested in human evolution, extracting as much information as possible about our species' past from the human genetic data is important. An estimate of the apportionment of human genetic diversity is merely a description of the current static state, a divvying up of the standing crop of genetic variation. Such an estimate provides no history of how that apportioning came about and thus may not tell us much about human evolution. The empirical result for humans, that each population contains most of our species' diversity, can accommodate two very different scenarios for how that existing variation came to be. First, it may be that human racial types are only very recently diverged from each other, and extensive genetic overlap reflects descent from a common ancestral population that lived not long ago. Alternatively, it may be that human racial types emerged long ago, developed, and retained their genetic distinctiveness for some time period and then, as population sizes increased, came back into genetic contact with each other, erasing most of what used to be the outstanding differences between groups. (A variant of this latter scenario assumes that some degree of gene flow occurred throughout the period since population divergence.)

These scenarios correspond to the two major theories of modern human origins. The multiregional model of modern human origins holds that our last common ancestor lived long ago and belonged to the species *Homo erectus*. This species originated in Africa and it was the first hominid to leave Africa roughly 1.8 million years ago. Some populations spread to different corners of the Old World and began to differentiate genetically (although with some unspecified continuing amount of gene flow). Under this model, the transition from *Homo erectus* to *Homo sapiens* was made in several regions of the Old World with gene flow maintaining species' integrity, and human populational and racial differences are thought to be relatively ancient. In this model, among other things, a direct genetic continuity between Chinese *H. erectus* fossils and people living in China today is posited. The rapid replacement model or "out of Africa" model recognizes that *H. erectus* spread out of Africa (as documented by the fossil record) but differs in hypothesizing a transition from one regional population of *erectus* to *sapiens* within Africa relatively recently. Some *Homo sapiens* populations are thought to have migrated subsequently out of Africa into other parts of the Old World, replacing older *H. erectus* populations upon contact. (It is this second hominid migration out of Africa that gives the model its name.)

These theories existed before the introduction of molecular data in human evolutionary studies. When mitochondrial DNA variation among humans was first surveyed (Cann, Stoneking, and Wilson 1987), it became possible to choose between these alternative scenarios by applying molecular clock calculations to infer when all human groups last had a common ancestor. The mitochondrial DNA evidence and the bulk of other molecular data now available from nuclear DNA strongly support a relatively recent origin for modern humans, rather than an ancient one, at roughly 100–300,000 years (although a few loci could support a different picture; see review in Ruvolo 1996). (For the record, the idea of a recent genetic bottleneck in human evolution had been hypothesized much earlier by Haigh and Maynard Smith (1972) on the basis of amino acid differences in globins.)

By itself, the empirical observation on the apportionment of human genetic diversity is consistent with either hypothesis and does not help us choose between them (and indeed, Lewontin's study was not designed to address this question). Yet the apportionment result is valuable to the study of modern human origins because it suggests that finding some as-yet-uncharacterized human population tucked away in a remote corner of the world that is very genetically divergent from all other humans (the existence of which might provide support for the multiregional hypothesis, or at least a relatively ancient common ancestor for modern humans) is unlikely. The overlap among existing human populations in genetic composition is simply too great (Ruvolo et al. 1993).

There is a known ascertainment bias associated with classical polymorphisms, and yet this bias seems not to have affected the diversity estimates in humans. Because most of the classical polymorphisms were originally characterized in laboratory workers of European origin, these loci tend to be more variable in Europeans than in other human groups. The same is true for RFLP markers chosen to be maximally informative in linkage studies, most of which were used on European pedigrees. Microsatellite markers, on the other hand, do not share this bias, and this is evident in measures of within-race (or within-continental) genetic diversity. Classical markers and nuclear RFLPs exhibit greater diversity within Europeans (Latter 1980; Mountain and Cavalli-Sforza 1994) as expected, whereas microsatellites show a different pattern and find Africans to be most diverse (Bowcock et al. 1994; Jorde et al. 1997). Interestingly, although this ascertainment bias could potentially affect within-race diversity estimates, it does not; there is no consistent difference between biased and unbiased studies in their estimates of within-race diversity (Table 7.1). Additionally, all of the genetic studies of autosomal genes, whether based on a biased collection of loci or not, yield the same answer for within-population diversity.

The framework for studying human genetic diversity that Lewontin provided in his landmark 1972 article is just now starting to be applied to studies of genetic diversity in humans' closest relatives. Phylogenetically, chimpanzees are most closely related to humans (Caccone and Powell 1989, Ruvolo 1997). Therefore chimpanzees are especially interesting to examine for their apportionment of diversity so that we can gain a comparative perspective on human evolution. The common chimpanzee (*Pan troglodytes*) has three morphologically and geographically defined subspecies, whereas the pygmy chimpanzee or bonobo (*Pan paniscus*) is monotypic. The one chimpanzee subspecies to have been studied genetically in this way, the eastern common chimpanzee *Pan troglodytes schweinfurthii* resembles *Homo sapiens* in that 80–90% of the subspecies' genetic diversity is contained within populations (Goldberg and Ruvolo 1997). However, moving up to the species level, the within-population component is only 28% (Ruvolo 1997). Gorillas (*Gorilla gorilla*) and orangutans (*Pongo pygmaeus*) are, like common chimpanzees, polytypic hominoid species, each progressively one step more removed phylogenetically from humans and chimpanzees. To the extent that they have been surveyed, they show considerable within-population diversity as a fraction of *subspecies* diversity (50–75% for gorillas; 60–69% for orangutans) but much less within-population diversity as a fraction of species diversity (10% for gorillas; 35–50% for orangutans) (Ruvolo 1997). These have to be viewed as only very preliminary estimates because these species have not been extensively surveyed. Yet the emerging data support the idea that *Homo sapiens* is more like an ape *subspecies* than an ape *species* in its apportionment of genetic diversity. Presumably this pattern has been created because the hominid lineage, unlike those of the apes, has been recently "pruned" of its many side branches during a genetic bottleneck. Evidence of this is elegantly provided by two Neandertal DNA sequences (Krings et al. 1997; Ovchinnikov et al. 2000), which are very different from those of living humans. Had those Neandertal DNA sequences and their molecular descendants survived to the present day, they, together with the sequences of living humans, would resemble those of a typical great ape species in extent of DNA sequence diversity.

The ways in which people respond to Lewontin's article are more than a Rohrschach test of racial biases. They are a reaction to the "surprising" empirical result that to many students seems at odds with their intuitive notions of how different people can be. Stephen J. Gould has pointed out this disparity between human variation observed at the phenotypic versus genetic levels. He likens human morphological variation to the surface of a deformed sphere that has several flattened planes; each plane corresponds to a human race or continental grouping. Each edge between planes represents those populations that seem to fall in between groups, but overall, there are few populations on the edges. Gould's deformed sphere analogy is compelling, but even with this, caution is necessary. Would all anthropologists and biologists agree on the composition of the planes, that is, which populations belong to which races and on the boundaries? No, and this in itself is another reason to question the concept of human races (Lewontin 1982, Gould 1985).

However one carves up human diversity at the morphological level and assigns populations to races, the apparent phenotypic discontinuity between groups contrasts sharply with the genetic picture of extensive overlap among all human populations. It tells us that no matter how different any two humans appear, under the skin we are all very similar. Indeed, the two most genetically divergent humans on the face of the earth are more similar to each other than pairs of gorillas from the same western African lowland forest (Ruvolo et al. 1994) who appear morphologically alike and who do not exhibit the comparable ranges of variation in eye color, hair color, skin color, and body form observed in humans.

Even though Lewontin's study was conducted 25 years ago, and even though it was based on classical human polymorphisms, he got it right – the latest studies using DNA markers confirm his essential result. It was a landmark article because it framed the question of human diversity in the right way – in population genetics terms. It is important to continue teaching this paper, not just for its historical value, but for its revelations about human genetic diversity. Teaching students that human equality is grounded in biology and is therefore "a contingent fact of history" (Gould 1985) and proclaiming this message to the world at large will continue to be important for many years to come. We should all thank R. C. Lewontin for first pointing this out to us.

REFERENCES

Barbujani, G., Magagni, A., Minch, E., and Cavalli-Sforza, L. (1997). An apportionment of human DNA diversity. *Proc. Natl. Acad. Sci. U.S.A.* 94:4516–19.

Bowcock, A. M., Ruiz-Linares, A., Tomfohrde, J., Minch, E., Kidd, J. R., and Cavalli-Sforza, L. L. (1994). High resolution of human evolutionary trees with polymorphic microsatellites. *Nature* 368:455–7.

Caccone, A., and Powell, J. R. (1989). DNA divergence among hominoids. *Evolution* 43:926–42.

Cann, R. L., Stoneking, M., and Wilson, A. C. (1987). Mitochondrial DNA and human evolution. *Nature* 325:31–6.

Cavalli-Sforza, L. L., and Minch, E. (1997). Paleolithic and Neolithic lineages in the European mitochondrial gene pool. *Am. J. Hum. Gen.* 61:247–51.

Dobzhansky, T. (1955). A review of some fundamental concepts and problems of population genetics. Symposium. *Cold Spring Harbor Symp. Quant. Biol.* 20:1–15.

Excoffier, L., Smouse, P. E., and Quattro, J. M. (1992). Analysis of molecular variance inferred from metric distances among DNA haplotypes: Application to human mitochondrial DNA. *Genetics* 131:479–91.

Goldberg, T. L., and Ruvolo, M. (1997). The geographic apportionment of mitochondrial genetic diversity in East African chimpanzees, *Pan troglodytes schweinfurthii*. *Mol. Biol. Evol.* 14:976–84.

Gould, S. J. (1985). Human equality is a contingent fact of history. In *The Flamingo's Smile: Reflections in Natural History*, pp. 185–98. New York: W. W. Norton.

Haigh, J., and Maynard Smith, J. (1972). Population size and protein variation in man. *Genet. Res., Camb.* 19:73–89.

Hammer, M. F., and Zegura, S. L. (1996). The role of the Y chromosome in human evolutionary studies. *Evol. Anthrop.* 5:116–34.

Harris, H. (1967). Enzyme polymorphisms in man. *Proc. Roy. Soc. Lond. B.* 164:298–310.

Jorde, L. B., Rogers, A. R., Bamshad, M., Watkins, W. S., Krakowiak, P., Sung, S., Kere, J., and Harpending, H. C. (1997). Microsatellite diversity and the demographic history of modern humans. *Proc. Natl. Acad. Sci. U.S.A.* 94:3100–3.

Krings, M., Stone, A., Schmitz, R. W., Krainitzki, H., Stoneking, M., and Pääbo, S. (1997). Neandertal DNA sequences and the origin of modern humans. *Cell* 90:19–30.

Latter, B. D. H. (1980). Genetic differences within and between populations of the major human subgroups. *Am. Natu.* 116:220–37.

Lewontin, R. C. (1967). An estimate of average heterozygosity in man. *Am J. Hum. Gen.* 19:681–5.

Lewontin, R. C. (1972). The apportionment of human diversity. *Evol. Biol.* 6:381–98.

Lewontin, R. C. (1982). *Human Diversity*. New York: Scientific American Books, W. H. Freeman.

Lewontin, R. C., and Hartl, D. L. (1991). Population genetics in forensic DNA typing. *Science* 254:1745–50.

Mountain, J. L., and Cavalli-Sforza, L. L. (1994). Inference of human evolution through cladistic analysis of nuclear DNA restriction polymorphisms. *Proc. Natl. Acad. Sci. U.S.A.* 91:6515–19.

Muller, H. J. (1950). Our load of mutations. *Am. J. Hum. Gen.* 2:111–76.

Ovchinnikov, I. V., Götherström, A., Romanova, G. P., Kharitonov, V. M., Lidén, K., and Goodwin, W. (2000). Molecular analysis of Neanderthal DNA from the northern Caucasus. *Nature* 404:490–94.

Ruvolo, M. (1996). A new approach to studying modern human origins: Hypothesis testing with coalescence time distributions. *Mol. Phyl. Evol.* 5:202–19.

Ruvolo, M. (1997). Molecular phylogeny of the hominoids: Inferences from multiple independent DNA sequence data sets. *Mol. Biol. Evol.* 14:248–65.

Ruvolo, M. (1997). Genetic diversity in hominoid primates. *Ann. Rev. Anthropol.* 26:515–40.

Ruvolo, M., Pan, D., Zehr, S., Goldberg, T., Disotell, T. R., and von Dornum, M. (1994). Gene trees and hominoid phylogeny. *Proc. Natl. Acad. Sci. U.S.A.* 91:8900–4.

Ruvolo, M., Zehr, S., von Dornum, M., Pan, D., Chang, B., and Lin, J. (1993). Mitochondrial COII sequences and modern human origins. *Mol. Biol. Evol.* 10:1115–35.

Ryman, N., Chakraborty, R., and Nei, M. (1983). Differences in the relative distribution of human gene diversity between electrophoretic and red and white cell antigen loci. *Hum. Hered.* 33:93–102.

Salem, A.-H., Badr, F. M., Gaballah, M. F., and Paabo, S. (1996). The genetics of traditional living: Y-chromosomal and mitochondrial lineages in the Sinai Peninsula. *Am. J. Hum. Gen.* 59:741–43.

Seielstad, M. T. (1998). Population genetics of the human Y chromosome. Ph.D. dissertation. Cambridge, MA.: Harvard University.

Seielstad, M. T., Minch, E., and Cavalli-Sforza, L. L. (1998). Genetic evidence for a higher female migration rate in humans. *Nature Genetics* 20:278–80.

Shriver, M. D., Smith, M. W., Jin, L., Marcini, A., Akey, J. M., Deka, R., and Ferrell, R. E. (1997). Ethnic-affiliation estimation by use of population-specific DNA markers. *Am. J. Hum. Genet.* 60:957–64.

CHAPTER EIGHT

The Indian Caste System, Human Diversity, and Genetic Determinism

RAMA S. SINGH

> The Indian caste system is the grandest, though perhaps not deliberate, apparently unsuccessful genetic experiment ever performed on human populations . . . a nature–nurture problem of highest complexity . . . an experiment on a grand scale that attempted to breed varieties of men genetically specialised in the performance of different functions.
>
> (Dobzhansky, 1962, pp. 234–5)

8.1 Introduction

Variation is an essential theme in the makeup of our world, and it provides the basis for attempting to understand the world around us. Much of science, particularly the life sciences, makes use of natural variation in the nature of constituent matters, structures, and natural processes to pursue knowledge. When natural variation is absent, we try to create it artificially, as is the case in genetics and molecular biology. If variation is our guide to truth, then human variation must be our most cherished heritage. Genetic variation has occupied a central place in our scientific, social, and political deliberations throughout our history, and in this century it has become a hotly debated topic in matters relating to welfare of individuals and social organizations: fascism and communism, aristocracy and democracy, and the welfare of individuals and groups (Beatty 1994). The concepts of mutations, fitness, and genetic load on the one hand and theories of eugenics, sociobiology, and genetic determinism on the other have made the relation between human diversity and human welfare an important topic of an ongoing debate. Theodosius Dobzhansky played an important role in unravelling genetic variation and relating it to human welfare. He contributed much work and thought to providing a biological basis for such traits as human freedom, civil liberty, and social welfare. Dick Lewontin inherited the problem of genetic variation from Dobzhansky – both as a practitioner of science as well as an educator in the science of variation. He has been the leader in providing a critique of the science of human variation as it relates to the well-being of human institutions. The Indian caste system, by any standard, is the largest social system ever designed along the lines of genetic variation.

I decided to take this opportunity to honor Richard Lewontin and write on the origin and evolution of the Indian caste system and its impact on human diversity and social differentiation of status and power.

The caste system has been the single most dominant factor affecting human relations, from birth to death, of individuals and groups in India. It affects all aspects of life – religious, social, economic, and political. The caste system originated in a racially mixed society segregated by skin color, religion, and cultural practices. With time, as populations grew, the caste system became more intricate and elaborate by being rigidly defined and perpetuated by division of labor and specialization of profession on the one hand and religious and commensal practices and marriage rules (endogamy) on the other. The caste system (or what was then called *varna*, meaning color) changed from a fluid system that allowed freedom of social movement to individuals within (the color) groups to a hierarchically stratified and fixed ladderlike system giving rise to a society that became socially stagnant and economically and politically unequal and unjust. In this society, the individual's social status and power (or lack thereof) became inherited as surely as the color of his or her skin. Yet despite inequality in rank and reward, Indian culture survived unbroken for over three millennia! Not only did it survive but, as some claim, it was the caste-based barrier against outbreeding and the concept of purity and pollution (who eats with whom) that was responsible for protecting Indian society from being overrun by foreign invaders. Foreign invaders came and stayed but were forced to form their own endogamous groups patterned on the local caste system. New rulers only meant new "revenue collectors," but the day-to-day arrangements of social life remained more or less unaffected. An exception was the brief period of the Mughal emperor Aurangjeb (1659–1707) when forced religious conversion was rampant. So powerful has been the caste system that until very recently it was possible to enter into a village only a few miles away from an urban town and feel as if people were living in the middle ages. The caste system behaved like a sticky magical substance that transformed whosoever came into contact with it. Hundreds of Indian social reformers and half a dozen new religious groups have attempted social reform and removal of the caste system. But now they all practice caste! Why is this so?

The purpose of this essay is to provide sufficient background on the religious and social structure of the Indian caste system and compare it with the British class system and American race theory. First, because the Indian caste system cannot be understood without having some knowledge of Hindu religion, I will briefly describe the basic principles of Hindu religion as it relates to the caste system. The emphasis will be on the origin of the caste system, the rules of marriage within and between castes, and on the theory of karma (responsibility for one's actions) which, contrary to the general perception, was *not* invented for the sole purpose of perpetuating the caste system. Second, I will briefly describe the history of the Indian people, summarize the results of genetic variation studies within and between castes, and show that, contrary to the general impression, there is relatively little genetic differentiation between the major castes. The major portion of the genetic differences between castes, if present, is the result of differences in the founding populations. I will also show that the degree of genetic differentiation between North Indians (Indo–Aryans) and South Indians (Dravidians) is also small. Third, I will compare the caste system with the *class* and *race* divisions and discuss the proposition that, although in terms of socioeconomic impact all three systems are rather similar, the Indian caste system differs from the others in that it was based on a nonmaterial (nongenetic) divine paradigm that did not call for "inherent inequality" of human beings. Fourth, I will argue that the Hindu belief in a "God-given" system of social responsibility and socioreligious hierarchy (i.e., *varnashram dharma*) and not in the inherent biological (or genetic) superiority or inferiority of man is the reason why, in India, the eugenic movement (and its later incarnations sociobiology and genetic determinism) did not catch on. Fifth, I will argue that the Indian caste experiment, owing to its grand size, provides us with strong evidence that it is the social and not biological inheritance, i.e., familial inheritance of power and status enforced with differential *samskaras* (religious rites) that has been the major determinant of social status and power differences between the major classes.

8.2 Basic Principles of the Indian Caste System

The following short description of the principles of the caste system are compiled from the Hindu epics *The Laws of Manu*, *The Bhagavad Gita*, and *The Mahabharata* and several secondary sources such as *A Sourcebook in Indian Philosophy* (Radhakrishnan and Moore 1957), *The Wonder that Was India* (Basham 1967), *The Hindu View of Life* (Radhakrishnan 1980), and *Our Heritage* (Radhakrishnan 1973, 1976). The original sources are the Hindu sacred books, the *Vedas* and the *Upanishads*, the latter being more accessible than the former (Box 1).

8.2.1 Four Great Classes

The Hindu religion (*dharma*) as practiced by most Indians today is solidly based on the caste system. It is called the *varnashram dharma; varna* means color (implying class), and *ashrama* refers to the four stages in life (see Section 8.2.2). This immediately means that the *dharma* is not the same for all. As Basham puts it, "there is indeed a common *Dharma*, a general norm of conduct which all must follow equally, but there is also a *dharma* appropriate to each class and to each stage in the life of the individual. The *dharma* of men of high birth is not that of humbler folk, and the *dharma* of the student is not that of an old man" (Basham 1967, p. 138).

The religious scriptures (*Rg Veda* X.90 and *The Laws of Manu* I.31) describe the origin of the caste system as follows: "for the sake of the prosperity of the worlds, he [the Lord] caused the *brahmin*, the *ksatriya*, the *vaisya*, and *sudra* to proceed from his mouth, his arms, his thighs, and his feet" (tranl. Radhakrishnan and Moore 1957). It is interesting to see the similarity between

Box 1. Glossary

ahimsa	noninjury to humans and animals
anuloma marriage	hypergamous marriage – women moving upward
apad–dharma	rules governing the legitimate occupations and activities of Aryans unable to live in the normal manner of their class
asrama	the four stages of life
Brahman	the World Spirit
brahman/brahmin	the priestly class
brahmacarin	a student of the *Vedas*, the first *asrama*
cakravartin	universal emperor
candala	an untouchable
candrayana	penance
dasyu/dasa	an aboriginal; later a slave or serf
dharma	sacred law
dvija	twice-born, the upper three classes
gotra	an exogamous sect
grhstha	householder, the second *asrama*
karma	accumulated effects of good and bad deeds
ksatriya	the warrior class
Law of causation	the Buddhist and Jain theory that the universe in eternal
Law of Samsara	perpetual wandering of the souls as well as of the universe
Law of Karma	determines place of humans and other beings in this cosmic change
mleccha	a barbarian
nirvana	the state of final bliss
pratiloma marriage	hypogamous marriage, women moving downward
reincarnation	the cycle of birth and death
samsara	the cycle of transmigration
samskara	personal ceremonies or sacraments
sannyasin	a wandering religious beggar, the fourth *asrama*
sudra	lowest of the four classes
vanaprastha	a forest hermit, the third *asrama*
varnas	the four classes of Hindu society
varna–shankara	hybrids – progenies from mixed marriages
Varnashram dharma	human journey through four stages of life

this mythology and Plato's fable of the use of different metals: "The God who has created you has put different metals into your composition – gold into those who are fit to be rulers, silver into those who are to act as their executives, and a mixture of iron and brass into those whose task it will be to cultivate the soil or manufacture goods." The first three castes are called twice born (*dvija*), and

Box 2. The Four Major Classes and Their Prescribed Duties

Brahmin

Six prescribed acts for a brahmin are studying, teaching, performing ritual sacrifices for himself and others, making gifts, and receiving gifts. He must know the prescribed means of subsistence and instruct others and himself to live by the rule. He must practice austerity. Under distress from the want of a means of subsistence, he can take up the means of subsistence prescribed for ksatriya but must not do anything that would involve loss of life such as agriculture and selling flesh.

Ksatriya

Of the six acts prescribed for brahmins, three acts are forbidden to kstriya: teaching, sacrificing (rituals) for others, and acceptance of gifts. The ksatriya can carry arms as a means of subsistence; he can study and sacrifice for himself but must protect the people. Under distress from want of a means of subsistence, he can take up the means of subsistence prescribed for the vaisya.

Vaisya

Teaching, sacrificing for others, and acceptance of gifts are also forbidden to a vaisya. The vaisya can study and sacrifice for himself, but his chief function is to breed cattle, till the earth, pursue trade, and lend money. Under distress from the want of a means of subsistence, he can take up the means of subsistence prescribed for sudra.

Sudra

A sudra's duty is to serve the three higher classes, especially learned brahmins. Only under extraordinary situations can a sudra take up something other than service such as handicrafts. According to *Manu*, to serve is innate in the sudra; he cannot be released from servitude.

the second birth refers to a religious initiation. The discharge of prescribed duty (one's *dharma*) is the highest calling. The twice-born men are virtuous and devoted to the law, and they are to establish laws if they are not opposed to the custom of countries, families, and castes. The three twice-born castes can study *vedas*, but the brahmin alone is allowed to teach it. The prescribed duties of the four classes are briefly shown in Box 2.

8.2.2 Four Orders (or *asramas*)

The four orders or stages of life are: *brahmcarin* (the student), *grhstha* (householder), *vanaprastha* (the forest dweller), and *sannyasin* (the wandering ascetic).

These four stages are of course only for the twice born (*dvija*) (i.e., the three upper classes), for only they are allowed to study the *vedas.* The word *brahmcarin* is generally taken to mean abstinence from sex, but it also means abstinence from involvement in worldly affairs. The emphasis is on learning scriptures and one's duties (*dharma*). The householder supports the whole society and is considered superior to all. The emphasis is on means of living, raising families, and participation in social affairs. The forest dweller is one who has fulfilled his duty as a householder and is free from family responsibility but not necessarily free from family attachment. In this stage the emphasis is on personal salvation: meditation, yoga, and knowledge. This stage of life is meant for asking questions and seeking knowledge about the world and about the self. This practice of seeking knowledge, rather than simply believing and following, was responsible for movement away from *Vedic* rituals and for producing the *Upanishads.* The *life-affirming* Vedic religion, based on sacrifice and rituals, slowly gave way to the *life-renunciating* Hindu religion, which puts emphasis on nonpossession of material wealth and movement toward spirituality. This led to the incorporation of a prominent role for the last stage of life: the wandering ascetic, one who is free from possession and free to roam the world in search of knowledge. Such freedom requires freedom not only from material possession but also from family attachment. The emphasis is on treating the whole world as one's family (*Vasudhaiva kutumbakam*) and on seeking salvation through knowledge as Buddha did.

Life-affirmation and life-renunciation are two interwoven threads of the Hindu *dharma,* and the *dharma* practiced today shows both these features. The *Varnashram dharma* (human journey through the four stages of life) promotes belief in the perpetual existence of the world and the possibility that salvation can be achieved by living lawfully as prescribed by the *dharma.* As is the case with the doctrine of *karma,* the *asrama dharma* is meant for all individuals, but the development of the caste system barred it to the *sudras.*

8.2.3 Great Laws of Causation, Karma, and Rebirth

The fourfold order, giving rise to the four great classes, is a fixed ladder of social organization and cannot be changed. As Krishna tells Arjuna in the *Bhagavad Gita,* "the fourfold order was created by Me according to the division of the quality and work. Though I am its creator, know Me to be incapable of action or change" (*Bhagvadgita,* chapter 4, verse 13). In the fourfold order, the emphasis is on *guna* (aptitude) and *karma* (function), and not *jati* (birth). The *varna,* or the class to which we belong, is independent of sex, birth, and breeding (Radhakrishnan and Moore 1957, p. 117). In other words, you take up a caste based on your aptitude.

The permanence of the social order or caste system comes from the *Law of Karma,* which presupposes the *Law of Samsara* and the *Law of Causation.* The *Law of Samsara* holds "perpetual wandering" of souls as well as of the universe. The *Law of Causation* holds that, because "nothing comes from nothing," the present universe has its origin in an earlier universe and goes through the cycles of creation, destruction, and recreation. Humans and all other beings also go through cycles of repeated birth and death (*reincarnation*), and the *Law of Karma* determines the place of humans and all other beings in this cosmic change. One's place in society depends on the totality of one's *karma*, both good and bad, performed in the present as well as the past lives. The theory of caste holds that the *Law of Karma* together with the *Law of Causation* (*reincarnation*) determines whether one will be born as human in the chain of beings and, if human, in which caste. Note that under the Karma theory, as applied to the caste system, caste becomes fixed at the time of birth. Although for the Buddhists salvation depends on the removal of all *karma*, for the Hindus both one's place in society (i.e., caste system) and ultimate salvation depend on good *karma*. It is important to remember that the caste system did not invent the doctrine of *karma*; it merely made use of it. The doctrine of *karma* would apply to people even in a casteless society. No one can escape from his or her past actions, but one can change one's life by performing good *karma*.

8.3 Major Migrations and the History of Indian Populations

The Indian subcontinent has a great deal of human diversity as reflected by variation in stature, facial features, and skin color. The diversity is the result of a great variety of early founding groups and many later migrations that enriched the populations of this subcontinent.

The first evidence of hominid habitation in India has been found in Middle Pleistocene deposits dating from approximately 500,000 years before the present. Objects dated to this era are usually primitive stone tools (Ghosh 1989), but one partial hominid skull from this geological period has been found near Hathnora on the Narmada River floodplain in the state of Madhya Pradesh. One group of investigators identified this skull as an "evolved form" of *Homo erectus*, and another group as an early form of *Homo sapiens* (Lumley and Sonakia 1985; Kennedy et al. 1991). A more recent hominid fossil find from the same area suggests that hominids of that period and location may have been pigmy-sized (Sankhyan 1997). If so, then comparisons with larger European or African hominids of that era could be misleading. It is therefore uncertain when modern *Homo sapiens* first entered the subcontinent, but the earliest remains that have been found date to the late Pleistocene (approximately 16,000 years ago).These were found in a cave in Sabaragamuva province, Sri Lanka and are clearly modern *Homo sapiens* if slightly more robust than modern populations (Kennedy et al. 1987). Unfortunately, although the late Pleistocene and Mesolithic remains have been studied morphologically, they have not been studied using molecular markers to determine their relationship with extant populations (Ghosh 1989).

The major human diversity of India has originated from three major groups: the Indo–Aryans, the Dravidians, and the tribal people. The first major group, which is thought to have migrated into the subcontinent, is the Dravidians;

this group is today mainly identified by the use of languages of the Dravidian family and occasionally by morphology. Although the location of their original homeland is uncertain, this group is believed to have entered the subcontinent through the northwest passes (Cutler 1988). Because the Dravidians were the dominant group in India at the time, the Bronze Age Harappan culture of the Indus Valley (ca. 2500–1700 B.C.) is usually attributed to them. This culture was based on agriculture and trade (Schmidt 1995) and featured some of the largest cities in the world at the time (Harappa, Mohan-jo Daro, and Ganwariwala) that had running water, sewers, and baths. The civilization extended over 1600 km of the Indus Valley, making it the largest culture in the world at the time. A written language was used but unfortunately has not been translated (Shaffer 1998), limiting the amount we can learn about this culture. Archeologists have postulated many theories to explain the end of the Harappan culture, but only two are still under serious consideration: that (relatively gradual) ecological changes of either human or natural origin disrupted Harappan food production and trade or that Aryans migrated into India with higher (Iron Age) technology, eventually destroying the civilization (Cavalli-Sforza, Menozzi, and Piazza 1994; Schmidt 1995). However, it should be noted that these theories are not mutually exclusive. Today, the Dravidian languages are spoken mainly in the four southern provinces of India with pockets of Dravidian speakers elsewhere (Cutler 1988).

The Aryans are generally agreed to have contributed significantly to the modern gene pool of India. An earlier consensus that the Aryans originated to the west or northwest of India has been questioned by scholars, who have noted that there is little archaeological evidence for an external origin (Shaffer 1984). However, linguistic evidence supports the researchers who favor an external origin in that the descendants of the Aryans (Indo–Aryans) speak languages related to European languages of the Indo–European language family (Cavalli-Sforza et al. 1994).

Although it is not clear when the Aryans began migrating into the Indian subcontinent, by 1000 B.C. they controlled much of the territory that had been Harappan. Increased use of iron tools after 800 B.C. (Schmidt 1995) and horse-mounted military units (Cavalli-Sforza et al. 1994) allowed the Indo–Aryans eventually to become the cultural elite throughout most of the Indian subcontinent. Evidence of this includes the fact that elements of Indo–Aryan culture are now found throughout India (the caste system) and that Indo–European languages are now spoken by the majority of the population in all parts of India.

The tribal groups found in many of the provinces of India are usually those outside the typical social system (castes) of the majority of the population but are not part of any otherwise recognizable group. Some of these tribal populations are thought to be the descendants of the earliest groups in the subcontinent, whereas other tribes are recognizable as more recent immigrants (Cavalli-Sforza et al. 1994). Genetic studies of Indian tribes have shown that they are usually most closely related to other tribes, followed by Indian nontribal populations, and lastly world populations – particularly African and Australian groups with morphological similarities (Roychoudhury and Nei 1985). Owing to similarity in haplotypes of the hemoglobin Hb-gene cluster containing the hemoglobin^S (sickle cell) allele, it has been proposed that some Indian tribal populations share a common origin notwithstanding their present large geographical separation (Labie et al. 1989). This could be consistent with the dispersion of a preexisting population by more recently immigrating groups. The tribal people who presently occupy the central and the northeastern provinces of India are supposed to be the original inhabitants of the Indian subcontinent, and in terms of affinity they are referred to as Australoid–Asians. To this group can be also added the numerous tribes that show affinity to, and may have come from, the Dravidians in the south or to the Mongols in the north. It is said that when the Aryans invaded the northwest, India was widely occupied by the Dravidians, who were pushed to the south as the successful Aryans began spreading eastward to the fertile Indo–Gangetic belt of north India.

Besides the three major groups (the Indo–Aryans, the Dravidians, and the Tribals) many other groups have contributed to the genetic diversity of India. The warm climate and the fertility of the land gave rise to prosperous, stable societies and attracted a series of new invaders from the north (326 B.C. to A.D. 1526): the Greeks, the Sakas, the Kusanas, the Hunas, the Arabs, the Turks, and the Mongols. These groups are thought to have had less effect on the gene pool of the continent than previous invasions, for many did not advance beyond northwest India (Schmidt 1995). Most of these groups did not assimilate into the local populations and made their own castes. Very little is known about migration and gene flow into the Indian subcontinent through the northeast; however, the presence of populations speaking Australoasiatic and Sino–Tibetan languages suggests that such migrations have taken place (Cavalli-Sforza et al. 1994).

8.4 Origin and Evolution of the Indian Caste System: How It Works

8.4.1 Origin of the Caste System

The Aryans entered India with a class division in their tribal structure. The earliest hymns make mention of the *ksatra*, the nobility, and the *vis*, the ordinary tribesmen. Thus, tribal aristocracy was a feature of Indo–European society even before the tribes migrated from their home. When they came into contact with the dark-skinned Dravidians (who built the Harappan culture and whom they called *dasas* or *dasyus*, meaning slave or bondman), emphasis on purity of blood and endogamy made the class structure more rigid. Those Aryans who intermarried with the dasas may have been excluded from Aryan society and pushed to the bottom of the class structure (the *sudras*) to which were added Tribals and Dravidians. The sudras may even have been a separate tribe that was conquered and subjugated by the Aryans.

The *brahmins*, the priestly class, gained importance from their professional command of sacrificial rituals, and the *ksatriyas* gained importance for providing protection and law and order. The Aryans' early success cannot be explained except by being better warriors or by having a professional fighting group with fighting machines such as chariots and horses. The Aryan society did not mix with Dravidians and became segregated on the color line, and color division between the Aryans on one hand and the Dravidians and Tribals on the other was the most important building block of the Aryans' caste system. The four classes were crystallizing throughout the period of the *Rigveda*, ca. 1500–2000 B.C. By the end of the vedic period the caste system was in place and considered *fundamental, primeval,* and *divinely* ordained. It is important to note that, although the Hindu scriptures talk of the brahmin and the ksatriya with their dharma-prescribed profession of priesthood and protection respectively, in reality both of these classes were anything but homogeneous. Many ksatriyas were landlords as were many brahmins. Both of them cultivated large estates but of course with hired labor!

The Dravidians in the south had a tribal system of equal status, and the caste system was introduced into their society much later. The south had very few ksatriyas, and nearly the whole of the population consisted of brahmins, vaisya, sudra, and untouchables. The sudras were divided into two groups known as the *right,* representing most of the cultivating and labor castes, and the *left,* representing various castes of craftsmen such as weavers and leather workers, cowherders, and some cultivating castes. There were great rivalries and animosity between left and right. Such was the power of the caste doctrine that no society in India, not even non-Hindus, has escaped the caste system fully. Hindu caste feeling and social hierarchy has found its way in the Muslim, Sikh, and Jain societies despite their denial of the caste system. Even recent converts to Roman Catholicism in south India have carried their caste prejudices with them, and high and low caste feelings prevail.

8.4.2 Marriage Rules and the Great Family Clan (*gotra*)

The Greek Megasthenes who came to India about the year 305 B.C. wrote the following:

> It is not permitted to contract marriage with a person of another caste, nor to change from one profession or trade to another, nor for the same person to undertake more than one, except he is of course the caste of philosophers, when permission is given on account of his dignity [quoted in Ghurye 1957, p. 2].

If social division along the color line was the most important criterion for the origin of the Aryan class system, endogamy, that is, marriage within tribes or subcaste, was the most important mechanism for its preservation. Only within-caste marriages were considered sacred. *The Manu's Laws* talk about "*apad-dharma*," that is, *dharma* during the time of distress. Manu provides a way out of difficult situations but, as we will see, not for all. Thus, when brides cannot be found within caste, brahmin can marry ksatriya, ksatriya marry vaisya, and vaisya marry sudra, but there is no option for the Sudra, for he can only marry within his caste! Even the intercaste marriages between brahmins, ksatriyas, and vaisyas are not free for all: only those marriages are allowed in which the girls move upward in the caste system (*anuloma*, hypergamous marriage). For example, a brahmin can marry his son, but not his daughter, into the ksatriyas. Intercaste marriages in which girls move down the caste ladder (*pratiloma*, hypogamous marriage) were not allowed.

Because all brahmins cannot marry among themselves, nor can all ksatriyas, without incurring the ill effects of close marriage (or so the theory goes), brahmins devised one of the most detailed and intricate marriage systems of outbreeding in which the brahmin and the ksatriya class is divided into family clans or *gotras*. The gotras are exclusive marriage groups; that is, marriages are not allowed between members of the same gotra. There are 49 main gotras, and in varying number they take the name of seven *rsis* (seers, sages) from whom all brahmins are said to have arisen. The gotra system or clans probably predate the caste system, for some ancient Indo–European groups such as the Romans had exogamous clans as well as generally endogamous tribes. Probably this was a feature of all primitive cultures. The gotra system was harmonized in the caste system, and because gotras carried special significance in marriage, it went on splitting into *subgotras* and *sub-subgotras*, taking on the names of many other ancient sages. The gotra system was also adopted by the ksatriyas and the vaisyas with somewhat less emphasis by the latter group. Under extraordinary situations marriage can be allowed within a gotra provided the man and woman have not shared any ancestors for the past seven generations. Occasional marriage within the gotra carries a penance (*candrayana*), a severe fast of a month's duration. However, no stigma attaches to the child of such a marriage – at least in theory.

Although a gotra is not the same as a subcaste, some castelike differentiation has occurred among brahmins and ksatriyas. Although the relevance of a gotra is only at the time of deciding marriage, a significant amount of reverence is attached to the names of the sages after whom the gotras are named. Another basis for higher–lower status among brahmins is whether they are priest or peasant or how many of the four Vedas they have read and learnt. Even the exclusiveness of the "marriage group" has become a status symbol among some brahmins. Some of these factors are responsible for the differentiation of surnames. There are no subcastes in the formal sense among brahmins and ksatriyas.

8.4.3 Occupational Castes and Subcastes

The four great classes continued splitting further and further into castes and subcastes along the professional lines. Subcaste is the genetically most significant unit of social identity and closest to a Mendelian population. India has 3000 or so modern subcastes, and the majority of them are in the lower two great classes, vaisya and sudra. As people took up new professions, they became identified with a new subcaste, and quite often the subcaste is named after the profession. As an example, Figure 8.1 shows a caste census of Bengal in 1931.

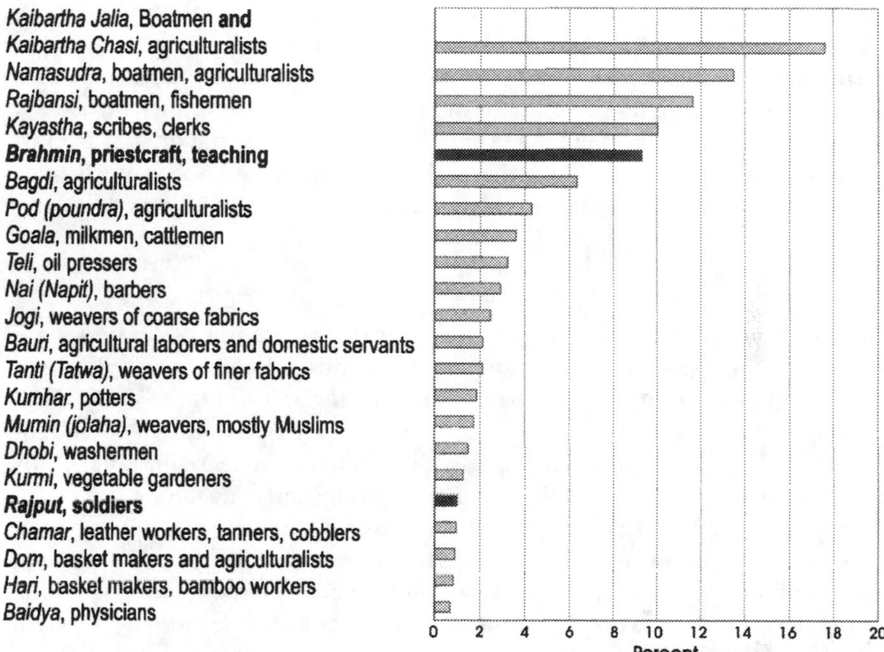

Figure 8.1. Caste composition of West Bengal based on the 1931 census of India. (Modified from Dobzhansky 1962.)

Four castes – brahmins, kayastha, boatmen, and fishermen – each account for 10–20% of the population, and the rest account for less than 5% each. Vaisyas and sudras constitute the bulk of the population and also harbor the most subcastes. The number of subcastes and the number of people belonging to a subcaste very much depend on the makeup of the land and the availability of professions. So it is not surprising to find a great number of agriculturalists in the northwest, boatmen in Bengal, and fishermen in coastal areas. On the other hand, the servile castes such as washermen, barbers, physicians, potters, vegetable gardeners, and oil pressers are generally found in all parts of India. In each village their number was usually small. The relationship between caste and profession is characterized by Bose (1951) as follows:

> The careful ways in which the tradition of close correspondence between caste and occupation was built up is clear indication of what the leaders of Hindu society had in mind. They believed in the hereditary transmissibility of character, and thought it best to fix a man's occupation, as well as his status in life, by means of the family in which he had been born.

It is interesting to note that members of different non-Hindu religious groups were seen more often as different castes than different faiths!

It is also important to note that the Aryan's *varna* (class) system had nothing to do with the caste proper (*jati*), which is a complex social structure arising from tribal affiliation and professional association. The word caste is of Portuguese origin (*casta*, meaning tribes, clan, and families), and the name stuck after the Portuguese invasion of India in the sixteenth century. Most caste proliferation has occurred in the lower two classes, vaisya and sudra. Thus, craft exclusiveness, endogamy, and commensality (the concept of purity or who eats with whom) became the three most-defining characteristics of the caste system.

8.4.4 Intercaste Marriage and the Confusion of Class

The caste system was a very fragile thing. In the beginning the caste system was stable, but in later years the system needed constant enforcement through injunctions against intercaste marriages by making it a sin. Only within-caste marriages were considered sacred; as pointed out earlier, only under special circumstances were intercaste marriages allowed and in such circumstances only between castes next to each other in the hierarchy and only in one direction! However, before the Middle Ages, when the social system was tightened, intercaste (i.e., between the four classes) marriages must have been frequent, giving rise to numerous hybrid, mixed-class, and intermediate castes (*varna–samkara*), although probably it was not as dominant a factor as was once thought (Basham 1967).

In general, the mixed classes were not seen as unclean and enjoyed a position intermediate between that of the two parents. Of all the mixed castes from the hypergamous marriage, only the *nisada*, in theory offspring of a brahmin and a sudra woman, were considered impure. Of the hypogamous marriages, the *candalas*, offspring of a sudra and a brahmin woman, were considered the most impure and became the epitome of the untouchable class!

8.4.5 Untouchables

The untouchables represented a heterogeneous class of people. Its early members included representatives of the early aboriginals who were conquered by the Aryans. There is evidence to show that several centuries before Christ groups of people who served Aryans in the menial and dirty tasks and who were considered outside the *varna* system already existed. This class grew as more and more aboriginal tribes fell to Aryans. Another group of untouchables was the candalas mentioned previously. They were considered the lowest among the untouchables and were not allowed to live in Aryan towns or villages and dwelled outside their boundaries; if living in the same village they were of course segregated and settled at the back of the village. After the Middle Ages, when the caste system became rigid, in some areas candalas became so strictly untouchable that even the touch of their shadow was considered impure, and upon entering towns or villages they were to announce their presence to Aryans by striking a wooden clapper. It is said that in South India they were forced to wear bells around their necks.

Another group of untouchables were those castes whose occupation involved killing or related tasks such as the hunters (*nisada*), the fishermen (*kaivarta*), and leather workers (*chamars* or *karavara*). The untouchables perpetuated their class, and their number expanded as menial jobs – washermen, sweepers, scavengers, executioners, and basket-makers – grew in the society. There was a caste system even among the untouchables! In later India most untouchable groups imagined that some other group was lower than themselves! *Antyavasayins*, offspring of a candala and a nisada, were even despised by the candalas themselves. Some of the sudras who were supposed to serve the society by their labor also fell to the untouchable class. It is important to note that, in the beginning, when the caste system originated, untouchables were outside the caste system. But as more and more of the sudra, on the basis of their menial task, began to fall into the fold of the untouchable class, they all became part of the caste system.

Contrary to the general perception, untouchability had nothing to do with the blood and everything to do with the occupation, conduct, and cleanliness. The *mleccha* (outside barbarians), regardless of their color or race, were also considered untouchables, but their fate was not sealed, for they were allowed to redeem themselves, not individually but as a class collectively, as time passed by adopting orthodox practices and following the rules of the scriptures.

In summary, we can say that the caste system was not created as a comprehensive body of sociopolitical theory and put in practice by some clever people for their own benefit and then perpetuated by the skillful and extensive use of religious and social doctrines. It was also not the result of the *karma* theory, for converted Indian Muslims and noncaste-practicing religions such as Sikhs and Jains continue to have castelike systems within their societies. The growth of the caste system can be explained by the combined effect of Aryans' class and color (or racial) differentiation, occupational proliferation, and the continued arrivals of new aboriginals in the fold. The elasticity or longevity of the caste system can be attributed to three things. The first factor, of course, was the subjugation of the lowest classes by the upper classes – imbalance of power. Second was the rules of marriage (endogamy) because, if one rose against the caste system and married outside his or her caste, his social status would be lost: *when a man is out of one caste he is out of all castes – an outcast!* The third factor was the sense of belonging and security one had in one's caste. The system was not perfect, but as Bose (1951) says:

> It gave economic security in spite of obvious inequality, and this security was guaranteed both by law and by custom. Sudras knew that they cannot rise to a higher status in their present life, but as long the differences were not sky-high, and justice was meted out to all according to his established dues, caste also endured.

8.5 Population Affinities and Genetic Differentiation

During the last quarter of a century a great deal of genetical and anthropological work has been done on the people of India. The most comprehensive and successful studies have been those carried out by the Anthropological

Survey of India and recently published as a multivolume series, *People of India*, by K. S. Singh and his colleagues (Singh, Bhalla, and Kaul 1994). In spite of the opportunity provided by this fascinating material, the amount of systematic genetical work on the Indian populations is relatively sparse but still significant (for recent reviews, see Majumdar and Mukherjee 1993; Singh et al. 1994). Physical anthropological studies in the past have mostly employed anthropometric traits to investigate the problem of the origin and ethnic differentiation of human populations. This choice of the traits created serious limitations because, as elsewhere, the emphasis was on morphometric traits (such as stature and cephalic and nasal index), which are continuous and have limited discriminatory power. Genetic traits (e.g., blood group genes) have been used more recently, but they are suspected to be under natural selection and hence less suitable for studying historical relationships. A few markers (such as the sickle cell gene, color blind gene, and G–6PD) have been widely used, but they too are likely to be under natural selection and do not provide full historical and geneological information. Given the hierarchical complexity of the Indian caste system and its differentiation over space and time, it will be necessary to use molecular markers that can be reasonably assumed to be neutral and to use a sampling scheme that allows the effect of natural selection to be distinguished from that of history. Mitochondrial, microsatellite, and Y chromosome DNA are the most appropriate markers for this kind of work, and only recently have they begun to be used on the Indian population (see Sections 8.5.1 and 8.5.2).

8.5.1 World Human Diversity, Genetic Signatures, and the History of Indian Populations

India has a total of 72 major languages (Figure 8.2) and even a larger number of minor languages and dialects. These languages are distributed into two major families: the Indo-European in the north and the Dravidian in the south. Some languages such as Hindi have a very wide distribution, and some are limited only to a province, but the majority of languages are only spoken locally. The major castes transcend language boundaries, and the major languages transcend caste boundaries. In other words, a widely distributed caste is likely to speak many languages, and a given major language is likely to contain speakers from many castes. On the other hand, some languages are restricted to a region or a group and become a defining characteristic of that group. Both caste and language differentiation have played a major role in the diversification of Indians because they strengthened the barrier against gene flow between castes.

Phylogenetic analysis of genetic markers on a worldwide scale usually places Indian populations relatively close to Europeans. Gene frequency analyses with "classical" markers have had limitations for phylogenetic analysis because these markers were not chosen randomly but have had a European bias in their initial selection. Moreover, most of them are blood-group proteins that are likely to be affected by natural selection. On the other hand, mitochondrial DNA (mtDNA) comparisons with other world populations have shown a clear

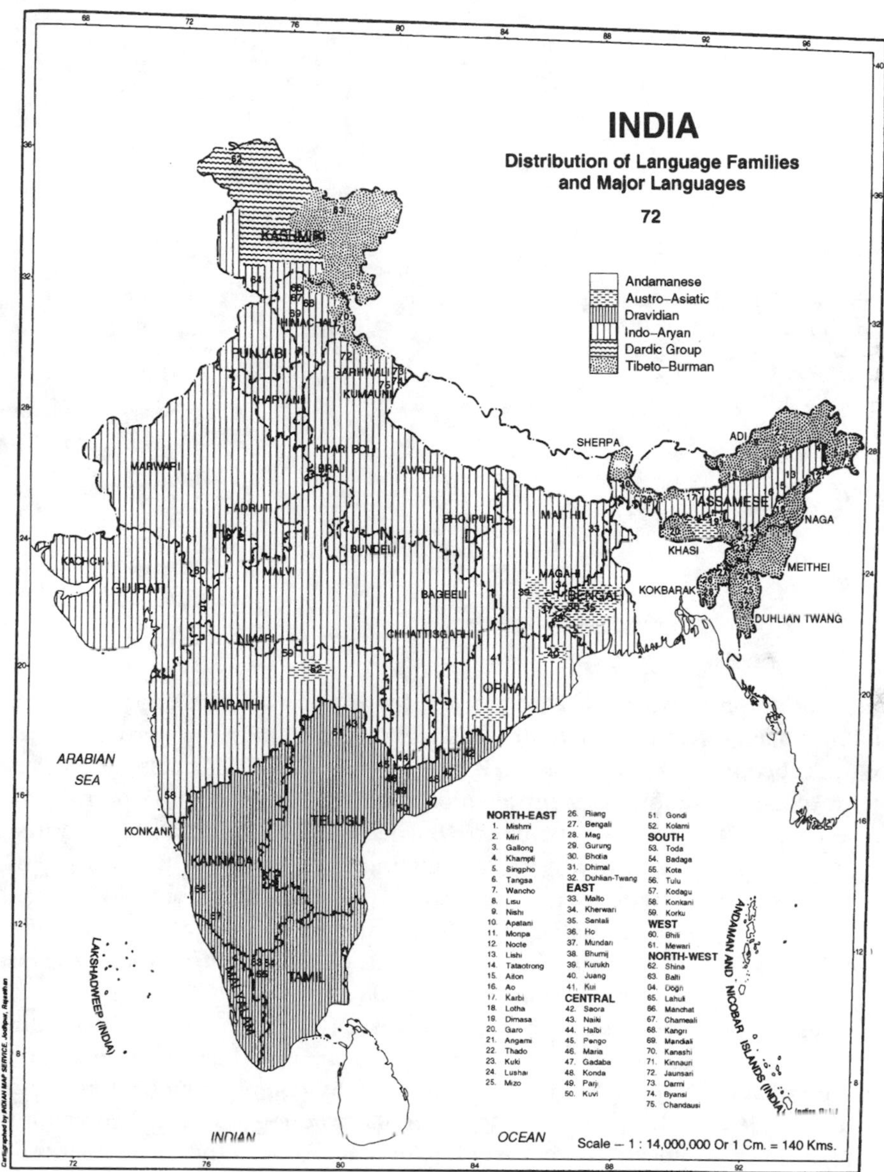

Figure 8.2. Distribution of language families and major languages of India. (From K. S. Singh, (1993). Reprinted by permission of Oxford University Press, New Delhi.)

relationship between Indians and Caucasians with some Asian admixture (Mountain et al. 1995; Barnabas, Apte, and Suresh 1996). Mitochondrial DNA studies, using high-resolution restriction analysis, have shown lower frequencies of certain Caucasian markers in the south than in the north (Passarino et al. 1996). In contrast, low-resolution mtDNA restriction analysis suggested

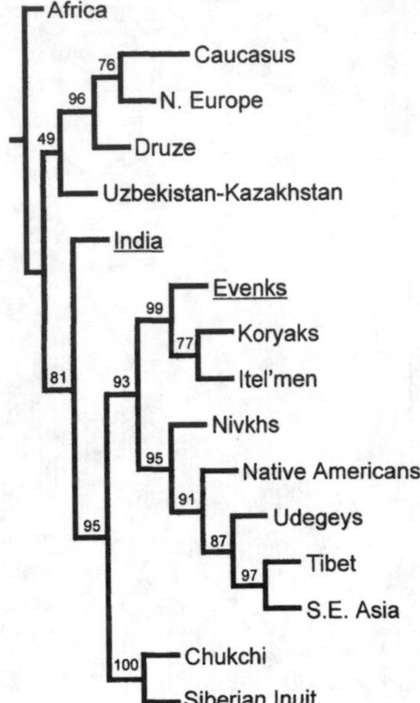

Figure 8.3. Fitch–Margoliash distance tree of India and world populations.

the continuing presence of an older Caucasian migration in southern India that has been infused with a newer Caucasian migration in the northern parts of the subcontinent (Barnabas et al. 1996). Higher levels of nucleotide diversity in northwest and northcentral India than in central and southern India were consistent with this conclusion (Barnabas et al. 1996), although variation in population growth is a factor affecting diversity. A worldwide study of Y-chromosome variation including the Kota (tribal), Madras, and Sri Lankan populations suggested a high affinity between south Indian and north Asian populations such as West Siberians and Kets or possibly Central Asians such as the Altai and Mongolians (Hammer et al. 1998).

A restriction fragment-length polymorphism (RFLP) analysis of mtDNA in our own laboratory has clearly shown *two major sources* for the Indian gene pool: Europe and Central Asia (Figure 8.3). The two most common haplotypes found in India, #30 and #15, are shared (with similar frequencies) with populations from Europe and Siberia, respectively. Excluding northwest Indian populations, which show evidence of recent migration, India is closer to populations from Siberia and Central Asia than Europe (Figure 8.4). The Indian–Central Asian–Siberian connection, also supported by the analysis of Y-chromosome DNA variation (Hammer et al. 1998), clearly shows that the major component of the Indian gene pool has come from Central Asia.

Neither mtDNA–RFLP haplotype data (Thomson 1999) nor microsatellite size variation data (Thampi 1999) showed significant differentiation between populations from north and south India. The tribal populations of India, depending upon their locations, show similarity either to Dravidians in the south or to the East Asian populations.

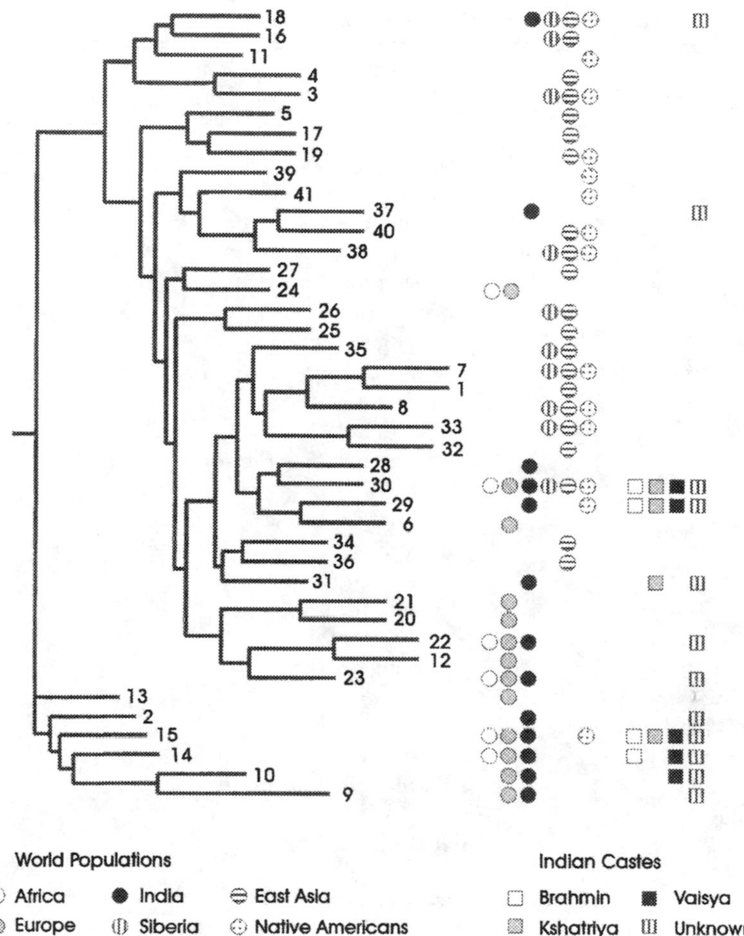

Figure 8.4. Maximum parsimony consensus tree with world haplotype and Indian caste distributions.

8.5.2 Gene Flow and Clinal Variation within India

The repeated invasions of the Indian subcontinent have had visible effects on the present genetic structure of India, as shown by studies with classical blood groups, serum proteins, and red cell enzymes. These effects include similarities between northeast populations (Austo–Asiatic and Sino–Tibetan speakers), tribal populations, and Southeast Asia and similarities between northwestern populations Middle Eastern, and European populations and gene-frequency clines from the northwestern states to the southern and eastern states (Papiha 1996, Thomson 1999).

Physical similarities between populations in northeastern India and those in Southeast Asia had been noted by anthropologists before the advent of classical markers (Papiha 1996). Frequency comparisons of blood group markers

Table 8.1. *Classical Marker Clines Observed within India*

Locus*Allele/Haplotype	Cline Direction	Part of a Larger Eurasian Pattern
*ABO*B*	Decreasing north to south	no
*ABO*O*	Increasing north to south	no
*ABO*A*	Increasing northwest to northeast	no
*Rh*Cde*	Increasing northwest to southeast	no
*Rh*CDe*	Increasing west to east	yes
*Rh*cde*	Decreasing west to east	yes
*MNS*MS*	Decreasing northwest to northeast	yes
*FY*A*[1]	Increasing northwest to south	yes
Haptoglobin		
HP 1*	Decreasing north west to southeast	yes
Adenosine deaminase		yes
*ADA*1*	Increasing northwest to east	
Acid phosphatase *1*		
*ACP1*A*	Decreasing northwest to southeast	yes
Group-Specific Component (Vitamin D binding protein)		
*GC*1F*	Increasing west to east	yes
*GC*1*	Increasing north to south	yes
Adenylate kinase		
AK1 1*	Increasing northwest to northeast	yes

[1] There are small areas of low *FY*A* frequency in the northeast corresponding to Indo–Aryan and Austro–Asiatic speakers (Papiha 1996).
Sources: Cavalli-Sforza et al. 1994, Papiha 1996.

MNS*M, MNS*MS, RH*D, Rh*Cde, and P1*1 support this relationship along with the 6-phosphogluconate dehydrogenase (PGD) allele C. However, it has been noted that Duffy (FY) allele A frequency reaches up to 50% in Indian-Austro-Asiatic-speaking populations, but it is not found in most southeastern or central Asian populations (Papiha 1996, Cavalli-Sforza et al. 1994). Some tribal populations have also been shown to have similarities to Southeast and central Asians with respect to some blood group markers. Acid phosphatase *allele* (ACP1) *allele C* is also missing from the tribal populations that have been sampled thus far, but this perhaps simply shows that there has been little European contribution to these gene pools because this allele is not very common (4–6%) in Europe (Cavalli-Sforza et al. 1994).

A large number of classical markers show gene frequency clines (possibly implying gene flow) within India itself (Papiha 1996). Table 8.1 shows a summary of cline directions within India and whether they are part of a larger pattern. In summary, it appears that classical markers give a picture of allele frequency clines running either east–west or northwest–southeast. With the exception of small pockets of gene frequency in the northeast, no cline appears to run from the northeast to southwest, as would be expected if a major proportion of the Indian gene pool had entered at the northeast of the subcontinent. Most within-India trends (with the exception of HP*1, ABO*O, ABO*B, and Rh*Cde) appear to be part of overall Eurasian gene-frequency trends with (for some alleles) Middle East-characteristic frequencies extending into India farther east than they reach elsewhere (Cavalli-Sforza et al. 1994).

8.5.3 Genetic Differentiation of Castes

Within-India phylogenetic studies using classical markers have had differing results depending upon the sampling locations. Populations in the state of Assam have been shown to cluster based on geographical proximity, implying that the sociocultural hierarchy (of castes and other groups) in this area has not had a large effect on these groups (Majumdar and Mukherjee 1993). However, for populations from northwest India, genetic affinity is somewhat correlated with close sociocultural grouping (Papiha 1996). An analysis of populations from Uttar Pradesh, Andhra, and Gujrat also showed social clustering: the (geographically widely distributed) Muslim groups were closest together followed by Hindu caste groups (sampled from Uttar Pradesh) and then tribal groups from Andhra Pradesh. However, in a six-state Hindu–Muslim comparison, Muslims and Hindus mainly clustered geographically, implying that temporary Islamic dominance did not alter the gene pool of the subcontinent substantially (Papiha 1996). The "broad generalizations" that have emerged with respect to the caste system appear to be that caste groups in some regions differ, particularly those belonging to different *varna* (i.e., to the top and bottom classes) and that geographically clustered castes are more similar irrespective of social hierarchy (Bamshad et al. 1996).

Studies of caste using molecular markers so far have produced inconsistent results, although only southern populations have been sampled. A study of mtDNA sequence (hypervariable segments, HVS I and II) from brahmin, harijan, and tribal groups (Mountain et al. 1995) showed some clustering according to caste but no clear separations into distinct groups. A faster growth rate for the Brahmin population was inferred from differences in the pairwise nucleotide difference distributions, which corresponded to simulation studies of expanding populations. The neighbor-joining tree generated from the Mountain et al. (1995) data supported the interpretation that the common ancestor of the Indian lineages sampled predated the divergence of Eurasian populations and that little gene flow between Indian and Eurasian populations occurred after their separation. Isolated studies of neighboring but socially isolated groups confound the effects of several factors such as varna or color (i.e., the differences in the two major founding populations), population bottlenecks, population growth, and gene flow.

A large study of mtDNA and Y-chromosomal variation in south Indian populations (Bamshad et al. 1998a) showed unimodal mismatch distributions, as in Mountain et al. (1995), but for each caste instead of just the brahmin, suggesting large expansions (but before the last 2,000 years). The south Indian population in general was found to have more affinity to Asian populations than Caucasians (Bamshad et al. 1998b), whereas some individuals in the upper castes were found to have more similarities with Caucasians (Gibbons 1998). A subsequent paper (Bamshad, Watkins, Dixon et al. 1998b) found a positive correlation between caste rank and mtDNA diversity (such that mtDNA diversity was largest in the upper castes) and that mtDNA genetic distance was largest between the highest and lowest castes. Y-chromosomal genetic distance was not correlated with caste, leading the authors to conclude that the large distance and diversity of mtDNA could be best explained by female-specific gene flow between castes (Bamshad et al. 1998b). There is an alternative explanation. The high diversity in the upper castes may have less to do with upward movement of women through intercaste marriage, as the authors have suggested (Bamshad et al. 1998b), and may simply be due to large upper-caste populations and highly structured (owing to repeated splitting) lower-caste populations.

Another study of 36 Hindu males from each of the 4 castes showed 25 mtDNA (HVS II) haplotypes, only 3 of which were shared between castes (Bamshad et al. 1996). Reflecting this, the reported G_{ST} value for the castes was 0.17 ($p < 0.002$) – meaning that 17% of the variance of the mtDNA region sampled occurred between caste groups. This G_{ST} value was not in agreement with previously reported values derived from blood and protein markers, but higher G_{ST} values are often noted for mitochondria owing to differences in population size.

From the preceding results it is clear that because of the repeated splitting and hierarchical structuring of caste populations it is not possible to sort out the effects of varnas, geographical isolation, and marriage restrictions. Molecular studies using samples from wide geographic regions are needed. In our own laboratory, a sequence analysis of mtDNA from the three upper caste populations from all over India did not show significant clustering of same-caste individuals (Figure 8.5, Behara 1995). Individuals belonging to different castes are completely interspersed with each other. Significant differentiation was also not found in an mtDNA restriction analysis of selected polymorphic sites in the preceding sample (Thomson 1999).

8.6 Why Did the Caste Experiment Fail?

As shown in the opening lines of this essay, the caste system has been seen as an experiment that attempted to breed varieties of men and women specialized in the performance of different functions. But the caste system was never an experiment, and even if it were it would have failed. Many have often defended the caste system as a less-than-perfect but nevertheless benevolent social system in which, though people were not equal in ranks and rewards, they all had a

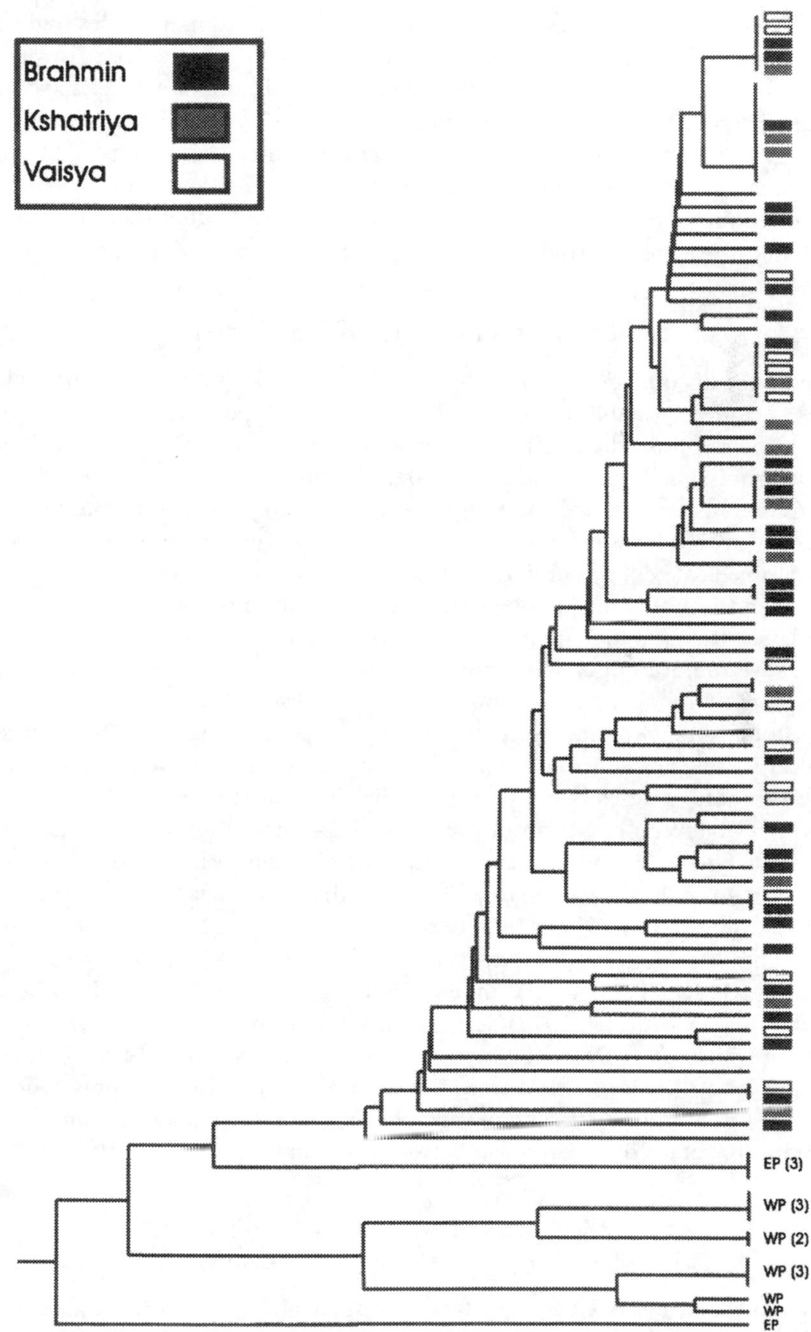

Figure 8.5. UPGMA tree of 84 mitochondrial D-loop sequences employing distances computed according to Kimura's two-parameter model (Kimura 1980). West and East Pygmies (WP, EP) were used as outgroup.

special place in the functioning of the society, and the upper classes had the sacred responsibility to look after the needs of lower classes. The system involved differentiation of economic function, which in itself in not bad, but it gave rise to differentiation of ranks and rewards, which is bad but perhaps not part of the original varna system. As a genetic experiment the caste system had to fail because the value of individual aptitude and achievement ceased to exist, social and economic rewards became fixed in class and family rather than individual merit, and the system barred movements of "genetic elites" between castes.

8.6.1 Fluidity of the Indian Caste System

The absence of consistent and significant genetic differences between the major castes is understandable in view its historical development. Contrary to the general perception, the Indian caste system was not always as rigid as it has been during the last few hundred years. Conquest and subjugation of new tribes, intercaste marriages, occupational caste (*jati*) proliferation, and steady influx of foreign invaders, merchants, and religious groups driven from their homelands have kept the Indian populations intermixing and the caste system fluid. Migration of many people, particularly of brahmins from north to south introduced the caste system into south India.

Because the present caste system is a combination of the original *varna* system and the later occupational caste (*jati*) proliferation, it is important to distinguish between genetic differences arising from these two different sources. In other words, one must determine if the caste populations selected for population genetic studies belong to same or different varnas. Finding genetic differences between castes is not the problem – sometimes we do; sometimes we do not. The real problem is to see if genetic differences have arisen *as a result of the caste system.* All data taken together show that at any level of caste structure there is more variation within than between caste groups. There simply has not been enough time to accumulate caste-bred genetic differences as a result of aptitude-based selection to overcome genetic differences present in the original populations. Lewontin's (1972) demonstration that in human populations the preponderance (over 80%) of genetic variation occurs within and not between populations and races holds true in the case of the Indian caste populations. These results are not surprising in view of the power of even a small amount of gene flow to retard differentiation between populations or castes.

8.6.2 Theories of Caste, Class, and Race and the Nature–Nurture Debate

Human diversity has been a major factor in determining social relations within and between groups of peoples throughout history. Claims of evolutionary association between human diversity and levels of "intelligence," as inferred from variations in social status within societies, or the levels of civilizations between societies (or races) has been the subject of major debate. Preoccupation with human diversity and its social implications began with Columbus' discovery of the New World and the contact between the Europeans and the Native Americans. A serious chapter to this debate was added later by the practice of slavery in America and still later by the development of science and technology, competitive trade practices, and social Darwinism. The treatment of Native Americans and Black Africans almost as separate species and the justification of their exploitation based on their supposed intellectual and civil "inferiority" is a dark chapter in the history of civil society (Gould 1981, 1996).

Among incessant attempts to partition humanity between "superior" and "inferior" groups (see, e.g., Ludmerer 1972, Kelves 1985) are three major ideological movements that have affected human relations and human welfare in unmeasurable ways and have changed the course of history. Two of these ideologies are based on presumed inherent differences in superiority and inferiority of people on the basis of their position in society or their racial and ethnic origins. The British *class system* was based on vertical social and economic stratification, and it preached that pauperism had a biological basis through the inheritance of feeblemindedness (Kelves 1985, Majumdar 1992). The American race policy was based on the presumed biological inferiority of the immigrants from eastern and southern Europe in general and of the Black Americans in particular. Both of these biological ideologies were products of the eugenic movement, and they were later assimilated into one common theory claiming that *class differences in social status and power result from class differences in intelligence.* Educational psychologists have been both the originators and the major supporters of this ideology, and Jensen (1969), Herrnstein (1973), Herrnstein and Murray (1994), and Rushton (1995) are recent examples. The eugenic movement and its goal of "betterment of human stocks" went through a period of intense but defined battle that focused on differences in IQ and its social implications (Kamin 1974, Kelves 1985, Gould 1981, 1996). The eugenic movement is gone, but its goals have become, if anything, more widened and entrenched in the thinking of sociobiology, evolutionary psychology, and the new paradigms of molecular biology. The power of the gene and its explanatory domain has expanded from a few "intelligence" genes affecting IQ and explaining differences between rich and poor to being the root cause of *all* human behavior, human health, and indeed of all human social conditions. A critical history of this new expanded version of the "gene-culture-power" theory can be gained from Rose, Lewontin, and Kamin (1984).

The third major ideological movement has been of course the Indian caste system. Although in India there has never been a sustained school of genetic determinism, there is nevertheless a general tendency to extrapolate from general impressions about the inheritance of physical traits to inheritance of human behavior and social traits. Strong breeding barriers between castes and continuation (inheritance) of family occupations within castes leads to the expectation that genetically controlled occupational differences in ability would accumulate over time – a straightforward prediction from farm breeding. Krimbas (1994, p. 184) points out that Dobzhansky was fascinated with the complexity of the

Indian caste system and performed selection experiments with *Drosophila* "where two populations already selected in contrary directions continued being selected while they underwent a limited exchange of individuals: to a population selected in one direction were added the highest performing individuals in the same direction from the contrary selected population." Dobzhansky named these populations "aristos" and "plebs" (aristocrats and plebeians) and was interested to find out whether a class barrier, permitting a limited exchange of individuals, could sustain a genetic differentiation while different selective pressures were applied to different classes. Even Indian intellectual leaders who fought against the social and economical injustices of the caste system had no trouble in believing in the power of caste-bred differences in occupational abilities. Provided that one did not take advantage of one's position in the caste system (i.e., did not believe in inherent superiority or inferiority of castes), caste-based differences in ability did not pose a problem to the social reformer. On the contrary, aptitude-based differences in professional ability were seen as a plus for a well-functioning society. Furthermore, the more able you were, the more social responsibility you were supposed to shoulder. This was also the position taken by Dobzhansky (1962), who believed in genetic differences in aptitude for professional differentiation but did not believe in unequal status (Paul 1994). Dobzhansky (1973, p. 42) endorsed Marx's dictum "from each according to his ability to each according to his needs." In the Indian case this may have been true in the beginning, but later caste-based exploitation of position and power did occur, and it became a trademark of the caste system. Then it would appear paradoxical that, in spite of the ruthlessness with which the caste system came to be practiced in India, there is neither now nor has there ever been a predominant school of thought that believed in genetic superiority or inferiority of the various castes. Where are the genetic determinists of India?

The answer, of course, lies in the nature of religious beliefs about the origin of the caste system. The caste system is obviously a remnant of the feudal system, which received its legitimacy from beliefs in the Divine plan. Plato's *Republic* comes to mind. With a God-given system in which doing one's sacred (caste-prescribed) duty was considered to be the very essence of *dharma*, there was no reason to erect a materialistic theory to justify the caste system. The *karma theory* was of course utilized to justify the caste system, and this is no coincidence because karma theory is an essential defining ingredient of all Indian religions.

There is a remarkable parallel between how the Indian caste system treated individuals and groups (castes) and what Herrnstein and Murray (1994) say about the treatment of Blacks as individuals and groups in North America. In the caste system, groups, but not individuals, could improve their lot by following scriptures and using pious means. Herrnstein and Murray (1994, p. 117) point out that despite the small association between IQ and measures of social achievements at the individual level, "large differences in social behavior separate groups of people when the groups differ intellectually on the average." As for individuals, they state that "measures of intelligence have reliable statistical relationships with important social phenomena, but they are a limited tool for predicting what to make of any given individual.... IQ score tells little about whether the human being next to you is some one whom you will admire or cherish" (p. 21). Thus, although the caste system allows upward improvement of groups but not individuals, in Herrnstein and Murray's racial doctrine there is no possibility for improvement either for groups or for individuals because both individual differences and group inferiority are genetically determined.

8.6.3 Social Determinism and the Power of Transformative Rituals (*Samskaras* or Rites)

Genetic determinism and dialectical materialism are seen as rival theories in matters of sociobiology, and in both pure environmentalism is more or less abandoned. There is no better example of the role of environment in shaping individual development, position, and power in society than the Indian caste system. To appreciate this one has to observe the importance of *Samskaras* (sacred rites) in the Indian caste system. Although these rites (e.g., birthday ceremonies, name giving, religious initiation, marriage, etc.) were seen by most Hindus as important landmarks that guide proper development of personality, it is their exclusive applications to the upper castes that made them so powerful a tool for instilling self-esteem and pride in oneself. The Indian caste system is an outcome of pure social determinism. This is not to deny genetic differences between castes but only that, regardless of the amount of genetic difference, it was irrelevant to the development of the caste system. The only genetic difference that did matter was, of course, the skin color (varna), which was the primary basis of Indo-Aryans' social stratification and served as a mark of identification between the major classes, especially the upper and the lower.

The power of religious rites or transformative rituals is often unappreciated. The 16 Hindu religious rites, from conception (and birth) to death, are meant to celebrate, to develop self-confidence, and to empower individuals for their dharmic duties in society. The denial of religious rites to lower castes was the denial of their person. Familial inheritance of social position and power, when reinforced with religious rites, becomes the truth that in turn is used to justify the system. As Leo Kamin (see Eysenck and Kamin 1981, p. 182) points out, Plato's fable of the metals mentioned above was created by Socrates, who proposed it as a useful and convenient lie to keep the various social classes in their proper places. When Socrates asked Glaucon, "Do you think there is any way of making them believe it?," the reply was, "Not in the first generation, but you might succeed with the second and later generations." Generations after generations of perpetuation of the doctrine of racial inferiority and the resulting lack of self-confidence among the Blacks and Native Americans have acted as a serious social illness from which these groups are still recovering. The Indian caste system had similar effects, and it endured because each generation's people were socially and religiously conditioned to believe in the naturalness of "rank and reward" difference between castes.

8.7 Conclusions

On the surface, the Indian caste system appears to be a grand social experiment designed to produce groups of people adapted to do specific tasks. Nothing could be further from the truth. It only seems that way, and the idea that it is a designed experiment is simply a post hoc explanation in line with Darwinian selection theory and evolutionary biology. For selective differentiation to occur, there has to be movement of "elites" between classes based on aptitude. It is true that the caste system was initially open-ended and that people could change their castes, but in reality this was true only for the upper classes: the brahmins and ksatriyas. Genetic selection based on aptitudes and achievements, if carried out to its completion, in principle could produce caste differences in ability. But genetic selection, if it ever did operate, would become inefficient when the castes became closed and family profession became a permanent fixed feature. To most people, the caste system represents social hierarchy, economic disparity, bonded labor, and exploitation of social status and power. But these ills have become a feature of all human societies. Then what is new in the caste system? Why does the caste system cause such a strong reaction?

A caste system was not an inherent feature of either of the two main founder societies of India, the Aryans and the Dravidians, although there is some evidence for existence of nobility and labor classes among Aryans before they came to India. What was new to India is that the caste system originated out of the meeting of these two societies, both of which had their own religious and social practices. The peaceful coexistence of Aryans and Dravidians, and later of many other new tribes, would have only been possible with mutual respect (or with the conquering group's imposed will for segregation) with familial (breeding) separation but social and economic interactions among groups. The other options would have been total annihilation or assimilation of the weaker group. Because Aryans are thought of as superior warriors, with their fast-moving chariots and Iron Age technology, they obviously must have dictated the terms of "peaceful" coexistence. To allow a conquered group to live their lives and practice their religion without fear of losing their group identity would appear the best deal one could ask for from a modern standpoint!

Although religious doctrines were employed to justify the caste system and create an efficient society, those at the bottom of the ladder did not fare well – socially, economically, or spiritually. However, in spite of the exploitation of social status and power for personal gains, and in spite of the growing orthodoxy of religious practices, liberalism in thinking has remained as one of the hallmarks of Hinduism. Indians believe in the caste system but not in an *inherent* superiority and inferiority of people. To a Karmic theorist, genetic differences between castes are irrelevant, and to a liberal thinker genetic differences do not translate into unequal status. Thus, although the doctrine of genetic determinism has been gathering momentum in Western scientific thought, social environment, that is, religious rites and family environment (*samskaras*) are held as the most important factors affecting individual development and achievements in India.

The caste system developed in a society in which the welfare of society came before the welfare of the individual. But some social groups did not fare well. Both the failure and the persistence of the caste system can be attributed to the practice of what many think was not part of the *varnashram dharma*: inequality of "ranks and rewards" by forced social and political means. The evolution of the Indian caste system is the perfect example of social determinism.

8.8 Acknowledgments

I was a postdoctoral fellow in Dick Lewontin's lab from 1972–73 (University of Chicago) and from 1973–75 (Harvard). The electrophoresis revolution had made Lewontin's lab a place of pilgrimage to young aspiring population geneticists. At the end of my graduate work at UC Davis, just before leaving for Chicago, I remember putting all my plants and agriculture-related reprints in the hallway with the note "Please help yourself – I am going to Chicago!" Little did I realize that the science and politics of agricultural research would become one of Dick Lewontin's later passions. My recollection of my first meeting with Dick is that I was dumfounded at the depth and breadth of his intellectual interests, and flabbergasted from seeing how he interacted with his students! For most graduate students, and certainly for someone from India, professors' treatment of their students as equals, respect for their interests in the choice of research problems (in Dick's lab many will float for a year or two before finding their niche!), and engaging discussions of both the intellectual excellence and the sociology of science are not the average experience in most labs in north America or elsewhere. Dick Lewontin practices what he preaches. If life's journey is marked with decisive turns, I am glad I took that turn to Chicago! I have learned a lot from Dick.

I wish to thank Costas Krimbas, Subodh Jain, Dick Morton, and Rob Kulathinal for their critical comments on an earlier version of this eassy, and Aaron Thomson for preparing the tables and figures.

REFERENCES

Bamshad, M. J., Fraley, A. E., Crawford, M. H., Cann, R. L., Busi, B. R., Naidu, J. M., and Jorde, L. B. (1996). MtDNA variation in caste populations of Andhra Pradesh, India. *Human Biology* 68:1–28

Bamshad, M. J., Watkins, W. S., Dixon, M. E., Jorde, L. B., Rao, B. B., Naidu, J. M., Prasad, B. V. R., Rasanayagam, A., and Hammer, M. F. (1998b). Female gene flow stratifies Hindu castes. *Nature* 395:651–2.

Bamshad, M., Watkins, W. S., Moore, M. E., Rao, B. B., Naidu, J. M., Prasad, B. V. R., Reddy, P. G., Watkins, C., Rasanayagam, A., Hammer, M. F., and Jorde, L. B. (1998a). MtDNA and Y chromosome variation in South Indian populations. *American Journal of Physical Anthropology* 26(S):27, 67.

Barnabas, S., Apte, R. V., and Suresh, C. G. (1996). Ancestry and interrelationships of the Indians and their relationship with other world populations: A study based on mitochondrial DNA polymorphisms. *Annals of Human Genetics* 60:409–22.

Basham, A. L. (1967, 1981). *The Wonder That Was India.* New Delhi: Rupa & Co.

Beatty, J. (1994). Dobzhansky and the biology of democracy: The moral and political significance of genetic variation, pp. 195–218. In *The Evolution of Theodisius Dobzhansky.* Mark B. Adams (ed.). Princeton, NJ: Princeton University Press.

Behara, A. (1999). Mitochondrial DNA sequence variation and human population structure in the Indian subcontinent. MSc Thesis, McMaster University. Hamilton, Canada.

Bose, N. K. (1951). Caste in India. *Man in India* 31:107–23.

Cavalli-Sforza, L. L., Menozzi, P., and Piazza, A. (1994). *The History and Geography of Human Genes.* pp. 212, 239–42, 290–301. Princeton, NJ: Princeton University Press.

Cavalli-Sforza, L. L., Piazza, A., Menozzi, P., and Mountain, J. (1988). Reconstruction of human evolution: Bringing together genetic, archaeological, and liguistic data. *Proceedings of the National Academy of Sciences of the United States of America* 85:6002–6.

Cutler, N. (1988). Dravidian languages and literatures. In Embree, A. T. (ed.). *Encyclopedia of Asian History.* Scribner, New York; Collier MacMillan, London. 1, 399–402.

Dobzhansky, Th. (1962). *Mankind Evolving.* New Haven, CT: Yale University Press, p. 236.

Dobzhansky, Th. (1973). *Genetic Discovery and Human Equalty.* New York: Basic Books.

Doniger, W., and Smith, B. (1991). *The Laws of Manu.* New Delhi: Penguin.

Eysenck, H. J., and Kamin, L. (1981). *Intelligence: The Battle for the Mind.* London: MacMillan.

Ghosh, A. (1989). *An Encyclopaedia of Indian Archaeology,* 19:316–17. New Delhi: Munshiram Manoharlal Publishers Pvt. Ltd.

Ghurye, G. S. (1957). *Caste and Class in India, 3rd* Ed. Bombay: Popular Book Depot.

Ghurye, G. S. (1969). *Caste and Race in India.* Bombay: Popular Prakashan.

Gibbons, A. (1998). Anthropologists probe genes, brains at annual meeting. *Science* 280:380–381.

Gould, S. J. (1981, 1996). *The Mismeasure of Man.* New York: W. W. Norton & Company.

Hammer, M. F., Karafet, T., Rasanayagam, A., Wood, E. T., Altheide, T. K., Jenkins, T., Griffiths, R. C., Templeton, A. R., and Zegura, S. L. (1998). Out of Africa and back again: Nested cladistic analysis of human Y chromosome variation. *Molecular Biology and Evolution* 15(4):427–41.

Hernstein, R. J. (1973). *I.Q. in the Meritocracy.* Boston: Atlantic–Little, Brown.

Hernstein, R. J., and Murray, C. (1994). *The Bell Curve.* New York: The Free Press.

Jensen, A. R. (1969). How much can we boost I.Q. and scholastic achievement? *Harvard Educational Review* 39:1–23.

Kamin, L. (1974). *The Science and Politics of I.Q.* New York: John Wiley & Sons.

Kelves, D. J. (1985). *In the Name of Eugenics.* Berkeley: The University of California Press.

Kennedy, K. A. R., Deraniyagala, S. U., Roertgen, W. J., Chiment, J., and Sisotell, T. (1987). Upper Pleistocene fossil hominids from Sri Lanka. *American Journal of Physical Anthropology* 72:441–61.

Kennedy, K. A. R., Sonakia, A., Chiment, J., and Verma, K. K. (1991). Is the Narmada Hominid an Indian *Homo erectus? Ameriacn Journal of Physical Anthropology* 86:475–96.

Kimura, M. (1980). A simple method for estimating evolutionary rates of base substitutions through comparative studies of nucleotide sequences. *J. Mol. Evol.* 16:111–120.

Krimbas, C. (1994). The evolutionary worldview of Theodosius Dobzhansky, pp. 179–94. In *The Evolution of Theodosius Dobzhansky*. Mark B. Adams (ed.). Princeton, N.J.: Princeton University Press.

Labie, D., Srinivas, R., Dunda, O., Dode, C., Lapoumeroulie, C., Devi, V., Devi, S., Ramasami, K., Elion, J., Ducrocq, R., Krishnamoorthy, R., and Nagel, R. L. (1989). Haplotypes in tribal Indians bearing the sickle gene: Evidence for the unicentric origin of the B^S mutation and the unicentric origin of the tribal populations of India. *Human Biology* 61(4):479–91.

Lewontin, R. C. (1972). The apportionment of human diversity. *Evolutionary Biology* 6:381–98.

Lewontin, R. C., Rose, S., and Kamin, L. (1984). *Not in Our Genes*. New York: Pantheon.

Ludmerer, K. M. (1972). Genetics and American society: A historical appraisal. Baltimore: The John Hopkins University Press.

Lumley, H. de, and Sonakia, A. (1985). Contexte stratigraphique et archéologique de l'homme de la Narmada, Hathnora, Madhya Pradesh, Inde. *L'Anthropologie* 89(1):3–12.

Majumdar, P. M. H. (1992). *Eugenics, Human Genetics and Human Failings*. London: Routledge.

Majumdar, P. P., and Mukherjee, B. N. (1993). Genetic diversity and affinities among Indian populations: An overview, pp. 255–75, In: Majumdar, P. P. (ed.). *Human Population Genetics*. New York: Plenum.

Mountain, J. L., Hebert, J. M., Bhattacharyya, S., Underhill, P. A., Ottolenghi, C., Gadgil, M., and Cavalli-Sforza, L. L. (1995). Demographic history of India and mtDNA-sequence diversity. *American Journal of Human Genetics* 56:979–92.

Papiha, S. S. (1996). Genetic variation in India. *Human Biology* 68(5):607–28.

Passarino, G., Semino, O., Bernini, L. F., and Santachiara-Benerecetti, A. S. (1996). Pre-Caucasoid and Caucasoid genetic features of the Indian population, Revealed by mtDNA polymophisms. *American Journal of Human Genetics* 59:927–34.

Paul, D. (1994). Dobzhansky in the "nature–nurture" debate, pp. 219–22. In *The Evolution of Theodosius Dobzhansky*, Mark B. Adams (ed.). Princeton, N.J.: Princeton University Press.

Radhakrishnan, S. (1973, 1976). *Our Heritage*. Delhi: Orient Paperbacks.

Radhakrishnan, S. (1980). *The Hindu View of Life*. London: Unwin Paperbacks.

Radhakrishnan, S., and Moore, C. A. (eds.) (1957). *A Sourcebook in Indian Philosophy*. Princeton, N.J.: Princeton University Press.

Rose, S., Lewontin, R. C., and Kamin, L. J. (1984). *Not in Our Genes*. England: Penguin Books.

Roychoudhury, A. K., and Nei, M. (1985). Genetic relationships between Indians and their neighboring populations. *Human Heredity* 35:201–6.

Rushton, J. P. (1995). *Race, Evolution, and Behaviour*. New Brunswick, N.J.: Transaction Publishers.

Sankhyan, A. R. (1997). Fossil clavicle of a Middle Pleistocene hominid from the Central Narmada Valley, India. *Journal of Human Evolution* 32:3–16.

Schmidt, K. J. (1995). *An Atlas and Survey of South Asian History*. New York: M. E. Sharpe, Inc.

Shaffer, J. G. (1984). The Indo-Aryan invasions: Cultural myth and archaeological reality. In Lukacs, J. R. (ed.). *The People of South Asia: The Biological Anthropology of India, Pakistan, and Nepal*, pp. 77–90. New York: Plenum.

Shaffer, J. G. (1998). Indus valley civilization. In *Microsoft Encarta 98 Encyclopedia.* Redmond, WA: Microsoft.

Singh, K. S. (1993). *An Anthropological Atlas. People of India National Series, Vol. XI.* New Delhi: Oxford University Press.

Singh, K. S., Bhalla, V., and Kaul, V. (1994). People of India. Vol. X. *The Biological Variation in Indian Populations.* Delhi: Oxford University Press.

Spurdle, A., Mitchell, J., and Jenkins, T. (1997). Letters to the editor. *Human Biology* 69(3):431–5.

Thampi, S. (1999). Microsatellite DNA variation and genetic structure in human populations from India. MSc Thesis. McMaster University, Hamilton, Canada.

Thomson, A. (1999). Ancestry of modern Indian population. MSc Thesis. McMaster University, Hamilton, Canada.

Wilson, E. O. (1975). *Sociobiology: The New Synthesis.* Cambridge, MA: Harvard University Press.

SECTION B

PHILOSOPHY OF EVOLUTIONARY BIOLOGY

CHAPTER NINE

Selfish Genes or Developmental Systems?

RUSSELL D. GRAY

The price of metaphor is eternal vigilance.

(Lewontin 1983a, p. 36)

Few metaphors have raised more controversy in evolutionary biology than Richard Dawkins' metaphor of the selfish gene (Dawkins 1976). In *The Selfish Gene* Dawkins built on the earlier work of Hamilton (1964) and Williams (1966) to offer a radical revision of the orthodox Darwinian view of nature. Dawkins claimed that genes, not individuals, were the fundamental unit of selection. Central to Dawkins' gene-centered view of evolution was a distinction between two types of entities: replicators and vehicles. Replicators were "any entity in the universe of which copies are made" (Dawkins 1982, p. 293). According to Dawkins, the gene was an obvious replicator, although other entities such as units of culture (memes) may also be replicators. In evolution, these immortal replicators competed through the construction of ephemeral phenotypic vehicles. The better the vehicle, the more replicator copies were produced. Complex adaptations arose through the action of cumulative selection on these replicators. Replicators were thus both the owners of adaptation and the beneficiaries of the selection process. The color of Dawkins' vision is captured in the following often-cited passage:

> Replicators began not merely to exist, but to construct for themselves containers, vehicles for their continued existence.... But do not look for them floating loose in the sea; they gave up that cavalier freedom long ago. Now they swarm in huge colonies, safe inside gigantic lumbering robots, sealed off from the outside world, communicating with it by tortuous indirect routes, manipulating it by remote control. They are in you and in me; they created us, body and mind; and their preservation is the ultimate rationale for our existence. They have come a long way, those replicators. Now they go by the name of genes, and we are their survival machines. (Dawkins 1976, p. 21)

Dawkins' critique of the individual as the unit of selection did not stop with the reduction of individual phenotypes to replicator power. In *The Extended Phenotype* he expanded the power of genes out beyond the skin so that the behavior of other organisms and even the physical environment could all be seen as expressions of replicator power. Thus, "the whole world is potentially part of the phenotypic expression of a gene" (Dawkins 1984, p. 138).

Dawkins' replicator–vehicle distinction continues to play a major role in contemporary debates on evolutionary theory, albeit in the slightly modified replicator–interactor guise. Most critics of Dawkins' views have followed one of two paths – they have either attacked the apparent genetic determinism in Dawkins' writing and emphasized the complexity of genotype–phenotype relations (Bateson 1978, Gould 1980, Wimsatt 1980, Brandon 1988), or they have focused on whether levels above the individual can count as vehicles or interactors (e.g., Wilson and Sober 1994). The critics, however, have generally accepted the replicator–interactor distinction. In contrast, in this essay I will examine the validity of genic selectionism from a developmental systems perspective. I will argue that, although genic selectionism does not necessarily require a commitment to genetic determinism, it is based on mistaken views about the roles of genes in development and evolution. Specifically, the following claims are all either seriously misleading or false:[1]

1. Genes code for phenotypic traits;
2. Genes replicate themselves;
3. Genes are all that is inherited;
4. Genic selectionism is the most general way of representing evolution.

I will argue that replacing all talk of replicators and vehicles with a focus on developmental systems and their contingent reproduction would not only be more accurate, but it would also facilitate the integration of development into evolutionary theory.

9.1 What Is Developmental Systems Theory?

The term *Developmental Systems Theory* (DST) is a label used by Griffiths and Gray (1994) to refer to a range of challenges to gene-centered accounts of development.[2] Developmental Systems Theory is not a theory in the sense of a specific model that produces predictions to be tested against rival models. Instead, DST is a general theoretical perspective on development and evolution at the same level of generality as genic seletionism. This perspective is not attributable to one person or group. Rather, it draws on insights from researchers in a wide range of areas (animal behavior, psychology, population genetics, developmental biology, and molecular biology) who were dissatisfied with genetic determinism and crude, dichotomous accounts of development and who thus attempted to produce a more satisfactory alternative. What follows is a quick sketch of these contributions.[3]

One element of DST is the rejection of all accounts that assume that there are two fundamental types of development – one driven by the genes and the other by the environment. This rejection of developmental dichotomies has its origins in the vigorous debate between comparative psychologists (Schneirla 1956, Lehrman 1970, Gottlieb 1976, Johnston 1987, Johnston and Gottlieb

1990) and ethologists such as Lorenz over the division of behavior into innate and acquired components. The comparative psychologists argued that the concept of innate behavior had many confusing, nonequivalent meanings (e.g., genetic, present at birth, not learned, unchangeable; see Bateson 1991). They emphasized that it was impossible to raise an organism without environmental stimulation, and thus the logic of the isolation experiment that ethologists used to demonstrate innate behavior was fatally flawed. They argued that even Lorenz's modified position (behavior was neither innate nor acquired; instead there were phylogenetic and ontogenetic sources of information for behavior) was untenable. According to Lorenz's modified position, the logic of what were now termed deprivation experiments depended critically on being able to distinguish between environmental conditions that merely provide material support for developmental processes and those that provide specific information. The deprivation experiment allowed the former but aimed to eliminate the latter. However, as Johnston (1987) argued, a precise definition of information does not justify Lorenz's dichotomy between sources of phylogenetic information and background environmental support, and thus even Lorenz's modified position was flawed.

As part of their critique of the logic of isolation experiments, comparative psychologists pointed out that there are many nonobvious experiential contributions to behavioral development that do not fit comfortably into either the innate or acquired category. Gilbert Gottlieb's (1976) classic series of experiments illustrates this very well. Soon after hatching, young ducklings show a clear preference for the maternal call of their own species. This response would satisfy many of the criteria often used to classify a behavior as innate. It is species specific, adaptive, present at birth, and appears to develop without the possibility of prior learning. However, what Gottlieb discovered was that if the ducklings were devocalized *in the egg*, they would no longer show the same clear preference for the call of their own species. Devocalizing the ducklings in the egg deprived them of the opportunity to make calls and thus to stimulate the development of their auditory system. This *prenatal* self-stimulation was therefore necessary for the development of the *postnatal* preferences. However, in hearing its own embryonic call, the duckling is not learning the character of the maternal call (the calls are quite different). Instead, this nonobvious, internally generated experience was stimulating the development of the auditory system that would later be used to detect the species-specific maternal call. As this example shows, development is not the inevitable unfolding of some innate program; instead it involves the contingent coaction of a system of interconnected resources and events. Events at one point in time create the conditions for the next. Stent (1981) likens the contingent self-organizing nature of development to an idealized process of ecological succession.

The preceding arguments convinced most current researchers that the division of behavior into innate and acquired components is too simplistic. However, despite frequent incantations of the word *interaction* and the familiar homily "of course all phenotypes are the joint product of genes and experience," dichotomous views of development refuse to die. In *The Ontogeny of Information*, Oyama (1985) carefully analyzed the ways in which allegedly new "interactionist" conceptions of development fell back into the old nature–nurture traps. She argued that the only way to escape from the seemingly endless replays of the nature–nurture debate is to reject all accounts of development postulating two distinct kinds of development, each guided by preexisting information. Oyama claimed that, rather than developmental information being preformed in the genes or in the environment, it is instead itself constructed by developmental processes. As the title of her book suggests, information has an ontogeny.

Few people have done more to expose the errors of genetic determinism than Richard Lewontin (Lewontin 1974, 1982, 1991; Lewontin, Rose, and Kamin 1984). Lewontin has also been critical of some standard versions of interactionism. A common interactionist view runs something like this. Genes and environment interact to produce the phenotype. It is wrong, therefore, to say that a phenotype (say IQ) is caused either by the genes or the environment. Instead, what we need to determine is their relative causal contributions. How much is IQ due to the genes and how much to the environment? Heritability estimates enable us to infer the relative causal contributions of these two factors. In his brilliant analysis of the errors of this view, Lewontin (1974) made the following points: Heritability estimates calculate the proportion of variation in a population that is attributable to genetic differences. These estimates are specific to a particular population in a particular environment. They do not enable us to infer global causal importance, nor do they indicate how much individual phenotypes can be changed. Lewontin emphasized that heritability estimates assume that genes and environment have simple additive effects. In contrast, evidence from experiments on "norms of reaction" consistently demonstrate the nonadditivity of genetic and environmental factors.

Lewontin's critique of heritability estimates and the associated additive vector model of development led on to his more radical "constructionist" or "dialectical" views of development (Lewontin 1983b, 1991). In this view, not only are the effects of genes and environment contextually and temporally contingent, they are codefining and coconstructing. Although it may still be convenient to talk of "internal" and "external" factors, in reality internal and external factors are not independent variables and do not exist, in any meaningful way, in isolation from one another. An organism's environment plays an essential role in the construction of its phenotype and, reciprocally, the organism selects and modifies (constructs) this environment out of the resources that are available. This "niche construction" (Odling-Smee, Laland, and Feldman 1996) can range from the building of nests, hives, and coral reefs to more subtle forms of construction such as the shedding of leaves and bark and the release of allelochemicals. Earthworms modify the soil, beavers build dams, and we make computers and warm the planet.

The final tenet of DST is an expanded view of inheritance. In contrast to the orthodox view of evolution as just a change in gene frequency, Oyama (1985), Gray (1988, 1992), and Johnston and Gottlieb (1990) all noted that organisms inherit much more than just DNA. Jablonka and colleagues (Jablonka and Lamb 1995, Jablonka and Szathmary 1995) have recently discussed the evolutionary significance of epigenetic inheritance systems such as the inheritance of functional cell states (steady-state inheritance), structural changes in cells that serve as templates for these changes to be replicated in descendant daughter cells, and chromatin marking systems. To this list DST proponents would add the inheritance of ecological factors such as symbiotic microorganisms, social traditions, and aspects of the habitat (e.g., Gray 1992). Developmental processes use all these inherited resources to reproduce the life cycle. It is this lifecycle that is the unit of differential replication in evolution (Griffiths and Gray 1994). In summary, DST is a nondichotomous view of development that emphasizes the themes of contingency, construction, self-organization, and expanded inheritance.

9.2 Genetic Determinism, Vehicles, and Interactors

Armed with this alternative view of development, let us return to the pages of *The Selfish Gene.* Scratch and sniff. Dawkins' talk of "lumbering robots" created "body and mind" by selfish genes smells like crass genetic determinism. Gould (1980), for example, accused Dawkins of the three deadly sins: atomism, reductionism, and determinism. Not guilty as charged responded Dawkins (1982). Although he may have gotten a bit carried away with some of the language and metaphors in *The Selfish Gene,* nothing in his gene-centered view of evolution required a commitment to a gene-centered view of development.

So who is correct here? Certainly much of the language used by Dawkins does facilitate a genetic determinist reading. As Hull (1988) has pointed out, the term *vehicle* does make the phenotype seem causally passive. Vehicles require agents to construct and drive them. By themselves vehicles go nowhere. Viewing vehicles as products of replicator power places genes in the driving seat and thus takes us down the rocky road to genetic determinism. However, Hull's (1980) terminological modification of the replicator–vehicle distinction (to replicators and interactors) maintains Dawkins' central theoretical framework without the implicit genetic determinism.[4] Thus, at least in this respect, nothing in the doctrine of genic selectionism necessitates genetic determinism.

9.3 "Gene for" Locution

Defenders of the gene's-eye view of the world frequently talk of "genes for traits." They are not alone in this practice. Television, newspapers, and respected scientific publications all bombard us with reports of genes for everything from "fruitlessness" (no offspring) in fruit flies to homosexuality in humans. All this talk of genes for traits appears to betray a commitment to genetic determinism, or at least a simple one-to-one mapping between genotype and phenotype. Again, however, I must side with Dawkins and his philosopher defenders (Sterelny and Kitcher 1988) and argue that it does not. Here is why. When biologists speak of "a gene for character X," they often do not mean that there is "a gene for character X." That really would be genetic determinism and preformationism at its worst. Instead, they are just using this phrase as locution or shorthand for the developmentally respectable idea that a genetic difference will cause a difference in an organism's phenotype in a given genetic and extragenetic environment. Saying that a genetic difference leads to a phenotypic difference does not, of course, imply that the trait concerned develops without environmental input. Nor does it imply that the genetic difference has either a unitary molecular basis or only one phenotypic effect (Dawkins 1982, chapter 2). Thus, properly understood, the "gene for" locution does not imply genetic determinism, nor does it require a one-to-one mapping between genotype and phenotype.

Does this mean that all is now sweetness and light? Are the worries about the "gene for" terminology just semantic confusions between friends rather than deep divisions between conceptual rivals? Not quite. As developmental systems theorists have pointed out (Oyama 1985, Gray 1992, Griffiths and Gray 1994) if "a difference that makes a difference" is all that is really meant by the "gene for" locution, then we can equally talk of methylation patterns, cytoplasmic factors, and cultural traditions for traits. After all, a change in any of these extragenetic factors can cause a change in phenotype in a given genetic and environmental context. The "difference that makes a difference" expansion of the "gene for" locution provides no grounds for privileging the role of genetic differences over other developmental factors.

9.4 "Gene for" Information

Although the "gene for" locution is routine in biological and popular discourse, I have yet to encounter a single "extragenetic factor for" locution. (Skeptical readers are invited to deluge me with counterexamples.) I suspect that the reason for this is that most practitioners of the "gene for" expression really believe that there is a sense in which genes can be said to be "for" phenotypes that does not apply to other developmental factors. Thus, in practice the "gene for" expression is not an agnostic shorthand but rather a developmental claim for the primacy of the genes. A justification of this stronger "gene for" position might go something like this. Genes and a whole host of extragenetic factors all have inputs into development. However, there is a fundamental asymmetry between the developmental roles of these inputs. Although genes do not determine development, they contain information, and it is this information that controls or programs development. All the extragenetic factors do is provide material or physical support for development; they do not play an informational role.

The DST response to arguments of this type (see Oyama 1985, Johnston 1987, Gray 1992, Griffiths and Gray 1994) goes something like this. The central issue here is whether the genome has a privileged informational role in development that generates some asymmetry between itself and all other developmental resources. We do not deny that there are distinctions among developmental processes (cf. Schaffner 1998). For example, Gottlieb (1976) suggests that different kinds of interaction may either facilitate, induce, or maintain developmental differences. What we do deny is that there are fundamentally two kinds of developmental processes – one guided by the genes and the other by the environment. (Bateson [1983] suggests that Gottleib's distinctions may apply equally to genetic and nongenetic factors.) We argue that assigning a gene an informational role and relegating the environment to secondary background support cannot be justified using either a mathematical conception of information derived from communication theory (Shannon and Weaver 1949) or a semantic view of information.

Let us start with the mathematical conception of information first. According to this conception of information an event carries information about another event to the extent that it is casually related to it in a systematic fashion. Information is thus said to be conveyed over a "channel" connecting the "sender" with the "receiver" when a change in the "receiver" is causally related to a change in the "sender." Clearly genetic differences are causally related systematically with phenotypic differences in this way. Genetic differences can thus be regarded as the source of information and the environment as background or "channel" conditions. However, the roles can easily be reversed. Environmental differences can be regarded as the source of information relative to a standard genetic background.

Consider the so-called "inborn" metabolic disorder PKU. Infants who are homozygous for the PKU gene have a deficiency in the enzyme phenylalanine hydroxylase. When they are raised on a standard diet this deficiency results in an accumulation of phenylalanine and its metabolites. These substances are toxic to the developing nervous system. The infants thus become profoundly mentally retarded and suffer from vomiting, seizures, hyperactivity, and hyperirritability. If we use the mathematical conception of information, the PKU gene could be said to carry information about the PKU phenotype. Diet is merely a background or channel condition for the information. However, if infants who are homozygous for the PKU gene are raised on a diet that is low in phenylalanine, they do not become profoundly mentally retarded.[5] Differences in diet are thus casually related systematically with phenotypic differences relative to a standard genetic background, and diet can thus equally be said to convey information with the genotype providing the necessary channel conditions. There is a causal symmetry between developmental resources. Indeed, there is something akin to a PKU "phenocopy."[6] Infants of normal genotype whose mothers have PKU and who do not maintain good dietary control during pregnancy can become mentally retarded because of fetal exposure to high levels of phenylalanine.

This kind of informational critique of stronger uses of the "gene for" terminology can be extended all the way down to the molecular level. Genic selectionists are strangely ambivalent about the way in which they define a gene. Dawkins (1982, p. 287) follows Williams (1966) and defines a gene as "that which segregates and recombines with appreciable frequency," or "any hereditary information for which there is a favorable or unfavorable selection bias equal to or many times its rate of endogenous change." The first definition focuses on the gene as a structural unit that is not broken up by recombination, whereas the second defines a gene in functional or informational terms. More recently, Williams appears to have moved to a solely informational view of a gene. "A gene is neither an object nor a property but a weightless package of information that plays an instructional role in development" (Williams 1986, p. 121), and "a gene is not a DNA molecule; it is the transcribable information coded by the molecule" (Williams 1992, p. 11). If genes really were just like the "beads on a string" view of the 1950s, then the structural and functional roles would map on to each other in a simple one-to-one fashion. Sadly, molecular biology has not been kind to genic selectionists. Specifying a DNA sequence does not allow an amino acid sequence to be directly read off. To specify the amino acid sequence one must also specify whether "any nonstandard coding is being used, what reading frame is to be used, that all gene–nongene and intron–exon boundaries are known, and what kinds of RNA editing will take place" (Sarkar 1996, p. 863). Even this is not enough. The state of the cell can influence processes like mRNA editing (Neumann-Held 1999). (For much more detailed discussions of these processes and their implications, see Fogle 1990, Sarkar 1996, Griffiths and Neumann-Held 1999, and Neumann-Held 1999). Thus, even at the molecular level it is inaccurate to view DNA as the sole source of information about even the sequence of amino acids let alone the tertiary structure of the resulting protein. The cell provides more than just passive channel conditions for the expression of a DNA sequence.

9.5 "Gene for" Semantic Information

At this point, genic selectionists might protest that the mathematical conception of information does not capture what they really mean when they use the "gene for" expression. When they talk about a gene containing information for trait X, they are using the term information in an intentional or semantic rather than a causal sense; a gene has an intended meaning rather than just causal consequences. They might then assert that this intentional sense of information reveals the true asymmetry between genes and other developmental resources. Sterelny and Griffiths (1999) explain the distinction between causal and intentional information in the following way:

> A distinctive test of intentional or semantic information is that talk of error or misrepresentation make sense. A map of Sydney carries semantic information about the layout of Sydney. Hence it makes sense to say of any putative map that it is wrong or that it has been misread . . . . Any talk of the genes being misinterpreted, or of the information in the genes being ignored or unused, is a shift from the purely causal notion of information towards something like the intentional notion. (p. 104)

Sterelny and Griffiths point out that one way in which philosophers of mind such as Millikan (1984) have attempted to explain intentional information in material terms is via evolutionary explanations. In "teleosemantic" accounts, intentional content is the product of evolution – specifically, natural selection for a particular function. A gene could thus be said to contain semantic information about a phenotype that it has been selected to produce. If genes were the only developmental resource inherited across generations, then this argument *might* indeed enable genic selectionists to talk of "genes for traits" in a sense that could not be applied to other developmental resources. (For an interesting and detailed discussion of all the complexities and difficulties that would be involved in fleshing out a teleosemantic account of genetic information, see Godfrey-Smith 1999.) However, as we shall see in a later section, the claim that genes are all that is inherited is simply not accurate.

9.6 Genes Are Not Self-Replicating

Although critics often present Richard Dawkins as the type specimen of genetic determinists, this is not entirely fair. In *The Extended Phenotype* he writes as follows:

> But any suggestion that the child's mathematical deficiency might have a genetic origin is likely to be greeted with something approaching despair: if it is in the genes "it is written," it is "determined" and nothing can be done about it; you might as well give up attempting to teach the child mathematics. This is pernicious rubbish on an almost astrological scale. *Genetic causes and environmental causes are in principle no different from each other* (my emphasis). Some influences of both types may be hard to reverse; others may be easy to reverse. Some may be usually hard to reverse but easy if the right agent is applied. The important point is that there is no general reason for expecting genetic influences to be any more irreversible than environmental ones. (1982, p. 13)

Gulp! What happened to the lumbering robots created body and mind by their genes? Has Dawkins lost faith? No, he is merely distinguishing arguments about development from arguments about evolution. When it comes to development, Dawkins adopts a fairly standard "interactionist" approach. However, when it comes to evolutionary questions, Dawkins insists that an exclusive focus on genes is called for. This distinction between developmental and evolutionary concerns helps illuminate an early debate about the validity of genic selectionism. In a review of *The Selfish Gene* Bateson (1978) argued that Dawkins overemphasized the role genes play in organizing development. Paraphrasing Samuel Butler, Dawkins had quipped that an animal was just a gene's way of making more genes. In response Bateson retorted, Could we not equally say a bird was just a nest's way of making more nests? If there can be selfish genes, why not selfish nests? Dawkins (1982) countered by noting that, although nests contribute vitally to avian development, nests cannot function as replicators. They do not pass the critical test for replicators. A mutation in the germ-line DNA will be passed on to the next generation, whereas variation in nest construction is not. Thus,

> ...when we are talking about development it is appropriate to emphasize non-genetic as well as genetic factors. But when we are talking about units of selection a different emphasis is called for, an emphasis on the properties of replicators. The special status of genetic factors rather than non-genetic factors is deserved for one reason only: genetic factors replicate themselves, blemishes and all, but non-genetic factors do not (Dawkins 1982, pp. 98–9).

This statement expresses a fundamental tenet of the gene's-eye view: genes are replicators, but organisms are not. Genetic changes are inherited, whereas organismal changes are not. Most critics of genic selectionism accept this position. However, it contains two flaws – one possibly minor and the other fundamental. The apparently minor error is the claim that genes "replicate themselves." The major error is the claim that genes are all that is inherited.

Let me start with the claim that genes "replicate themselves." As Lewontin (1992, p. 33) notes, "DNA is a dead molecule, among the most nonreactive, chemically inert molecules in the living world." A segment of DNA isolated from the cytoplasmic machinery of ribosomes and proteins has no magic power of self-reproduction. Critics might claim that, despite Dawkins' protestations to the contrary, once again DNA is given a special developmental role. So to what extent is the claim that genes replicate themselves just an empty rhetorical flourish of little substantive consequence? Well, nothing in Dawkins definition of a replicator as "anything in the universe of which copies are made" (Dawkins 1982, p. 83), requires that replicators be *self*-replicating. Indeed, immediately after the preceding definition Dawkins gives two examples of replicators: a DNA molecule and a sheet of photocopied paper. Now not even genic selectionists suggest that pieces of paper photocopy themselves.[7]

According to Dawkins (1982, p. 83) pieces of paper are active replicators in the sense that what is written on them affects their likelihood of being copied; they differ in fitness. They are active replicators in this evolutionary sense rather than in the developmental sense of being *self*-replicating. From a DST perspective, however, the Xerox analogy does reveal a commitment to a privileged developmental role for genes. In the photocopying process, the sheet of paper contains all the information. The Xerox machine merely contributes the materials and channel conditions. Variation that arises in the copying process is noise, not new information.

At this point Dawkins and Williams might concede that both the language of self-replication and the Xerox analogy are a little misleading but claim that this is of no real significance. Perhaps all they were trying to emphasize was that DNA plays a template role in its own reproduction. But then cell membranes act as templates for the production of other cell membranes (Moss 1992). Perhaps they were just trying to highlight the directness and fidelity ("blemishes and all") of genetic replication. However, as Brandon (1988) and Sterelny and Griffiths (1999) point out, proofreading mechanisms edit out many genetic blemishes. In fact, the fidelity of genetic replication depends on these kinds of indirect processes.[8] Perhaps, given that both Dawkins and Williams are primarily interested in the evolution of adaptation, their bottom-line position would go something like this. Cumulative selection requires lineages of copies. Genes are the only entities that are copied across generations, and thus only genes can be the beneficiaries of selection and the owners of adaptation. This is the replicator concept stripped of dubious analogies and loaded language. This might be a good argument if the claim that only genes are inherited were true.[9]

9.7 Genes Are Not All That Is Inherited: The Return of the Selfish Nest

Phenotypes do not persist, they recur. The only biological entities that persisted in the fly culture and the beetle culture are fly genes and beetle genes (Williams 1986, pp. 116–17).

Such is the dominance of genocentric thinking in biology that the claim that only genes are inherited seems like a simple truism – a statement of accepted fact rather than a contestable theoretical position. Indeed, this claim has rarely been challenged in debates on genic selectionism. It forms the bedrock of the gene's-eye view of evolution. Developmental systems theorists reject this bedrock claim and emphasize the importance of extragenetic inheritance. Of course, genic selectionists like Dawkins and Williams have always acknowledged cultural inheritance, but the developmental systems approach goes far beyond that. We argue that the replication of genes is simply one aspect of the reproduction of life cycles. Many other elements such as cytoplasmic gradients, membrane templates, centrioles, methylation patterns (Moss 1992, Jablonka and Lamb 1995), gut microorganisms (Troyer 1984), chemical traces from the maternal diet via maternal milk (Galef and Henderson 1972) or fetal olfaction (Hepper 1988), and different habitats (Gray 1992, Sterelny, Smith, and Dickison 1996) are all inherited down lineages. Genes have many close companions on their evolutionary voyage.

Let me flesh out this argument a little by returning to the debate between Bateson and Dawkins about the evolutionary status of nests. Dawkins (1984, p. 133) asserted that "the accidental incorporation of a pine needle instead of the usual grass is not perpetuated in future 'generations of nests'." However, different nest types often accurately reflect avian phylogenetic relationships. The nest types used by swallows map perfectly on to a phylogeny based on DNA-hybridization data (Winkler and Sheldon 1993). Could variation in the mode of nest construction be passed on by extragenetic as well as genetic means? Sargent (1965) demonstrated that the color of nesting material that fledgling and adult zebra finches are exposed to affects the type of nest they subsequently construct. So variation in nest design can be transmitted extragenetically. Variations in nest design could also be perpetuated as a consequence of habitat shifts. If a population shifted into a new environment (e.g., a native New Zealand bird colonized an exotic pine forest) and tended to remain in that habitat as a consequence of habitat imprinting, then the incorporation of new nest material (e.g., pine needles) from the habitat would be reliably perpetuated. This is not implausible. Klopfer (1963) found that the preference for pine foliage in chipping sparrows may be altered by hand-rearing young birds in oak foliage. Immelmann (1975) discusses a study by Peitzmeier of European mistle thrushes in which the expansion of the species' range from forest to parkland in France and Germany was shown to proceed, not by the spread of several populations, but by the spread of a single population that had become habitat-imprinted on parkland rather than forest. At this point the skeptical genic selectionist might concede the odd selfish nest or two but contend that such cases of extragenetic inheritance are rare and unimportant in evolution. Dawkins (1976) claimed that genes have three important evolutionary characteristics: longevity, fecundity, and fidelity. By longevity, he meant that lineages of gene copies could persist for long periods of evolutionary time; by fecundity, he meant that some lineages would be more productive (fitter or more copies) than others; and by fidelity, he meant that the copying process was very accurate. How does extragenetic inheritance measure up on these three criteria?

Consider the case of certain aphid species and their endosymbiotic *Buchnera* bacteria discussed by Moran and Baumann (1994). These aphids reliably pass on their endosymbionts from the maternal symbiont mass to either the eggs or developing embryo. These bacteria confer a fitness advantage on their hosts. They enable them to utilize what would otherwise be nutritionally unsuitable host plants. Aphids that have been treated with antibiotics to eliminate the bacteria are stunted in growth, reproductively sterile, and die prematurely. A lineage that inherits these bacteria is clearly at an advantage over one that does not. Phylogenetic trees of the bacteria and their aphid hosts are perfectly congruent. This suggests that speciation of the aphids has led to speciation of the bacteria; they have cospeciated. Molecular and fossil evidence suggests that this association is remarkably ancient – between 160 and 280 million years. The *Buchnera* bacteria are thus a vital and persistent extragenetic developmental resource. Their inheritance cannot be reduced to just genetic inheritance as whole organisms are passed on across generations. Their association with the aphids meets Dawkins' three criteria of longevity, fecundity, and fidelity.

The evolutionary significance of the reliable inheritance of symbiotic microorganisms does not stop with insects. Many herbivorous mammals and some reptiles depend on the fermentative activity of gut microorganisms to digest plant fiber (Troyer 1984). Generally, these fermentative microflora cannot survive for long periods outside of the host's body and therefore must be transferred among herbivores by close contact or coprophagy. Troyer (1984) noted that there are thus behavioral adaptations to ensure contact between generations and speculated that the evolution of social systems in organisms as diverse as termites and dinosaurs may have its origin in the need to pass on symbiotic gut microorganisms.

This is perhaps a good place to emphasize that DST is not committed to explanations at the level of individual organisms. Heritable differences between groups can be transmitted by extragenetic means. An interesting example of this occurs in the fire ant *Solenopsis invicta*. In this species there are two distinct forms of social organization: monogynous colonies (single reproductive queen) and polygynous colonies (up to several hundred queens). There are substantial morphological, physiological, and behavioral differences between these colony types. Adult winged queens from monogynous colonies are also heavier and have greater fat reserves than polygynous queens. The difference in fat reserves appears to influence the way colonies are founded and the type of social organization that develops. Keller and Ross (1993) used cross-fostering experiments to demonstrate that the difference between the queens is induced by the type of colony they are raised in. They suggested that exposure to different levels of queen-produced pheromones was the likely inducing factor. The cycle from colony type → pheromone level → queen type → colony type is thus reliably reproduced down colony lineages.

9.8 All This Is True But . . .

There are several strategies that orthodox biologists might follow in an effort to minimize the significance of extragenetic inheritance. Genic selectionists might be tempted to reduce cases of extragenetic inheritance to a dual inheritance model: genes and their cultural equivalent (memes). That could lead us back to a view that there are two sources of development information and thus back to a dichotomous view of development. Most of the examples of extragenetic inheritance I have discussed do not fit this dichotomy (e.g., chemical traces from the maternal diet, epigenetic inheritance, the inheritance of gut symbionts, and the inheritance of fire ant colony type). Another strategy might be to concede that extragenetic inheritance occurs but argue that it only applies in a small number of cases. For example, it could be claimed that changes in chromatin marking in germ-line cells are likely to be reset to their original state during gametogenesis, these changes are unlikely to be of great evolutionary significance. Jablonka and Lamb (1995) disagree and argue that some new epigenetic marks are transmitted through sexual reproduction. Even if this is not the case, sexual reproduction is the exception and not the rule on the tree of life. Asexual reproduction is the dominant reproductive mode in two and half of the three biological kingdoms.

Another way of minimizing the significance of extragenetic inheritance might be to argue that this type of inheritance can only mutate between a limited number of states. In contrast, there are a vast number of possible genetic combinations. Maynard Smith and Szathmary (1995) draw a distinction between limited and unlimited inheritance. In limited inheritance the number of heritable states is less than or equal to the number of organisms. The modular and recursive nature of genetic and linguistic inheritance means that there is a potentially unlimited set of possible states for genomes and utterances. Szathmary and Maynard Smith (1997, p. 559) claim that only organisms with unlimited inheritance can undergo microevolution (small piecemeal steps in sequence space). Although they do not explicitly state it, the implicit claim appears to be that this kind of microevolution is an essential prerequisite for cumulative natural selection and thus the evolution of complex adaptations.

Does this mean that extragenetic inheritance must be consigned to a side show of evolutionary oddities that are interesting but irrelevant to the main event? From a DST perspective I think that there are at least three problem with this argument. First, the significance of the modular structures of DNA and human languages should not be oversold. The fact that different combinations of nucleotide bases or sounds mean different things causally depends on the rest of the developmental system, not just its genetic or linguistic components. DNA by itself is as meaningless as a stream of utterances in a language no one understands. Griffiths (1996) noted that if the rest of the system were organized in such a way that many base-pair combinations collapsed into only a few developmental outcomes, then there would not be an unlimited heredity system. He pointed out that it is not hard to imagine cellular machinery with this result; the existing "genetic code" is already redundant in this way. My second objection to this argument is that it treats genetic and extragenetic inheritance separately. From a systems perspective these sources of heritable variation should be viewed as acting together. Extragenetic inheritance thus *expands* the set of possible heritable combinations rather than merely offering a limited repertoire. Together these combinations facilitate adaptive evolution. For example, as I mentioned earlier, the extragenetic inheritance of gut microorganisms enables organisms to digest cellulose and thus to be herbivores. Given this extragenetic inheritance, selection would favor genetic changes that alter the gut physiology to suit the development of the endosymbionts that function more efficiently on the herbivore's typical diet. This brings me to my third objection. Extragenetic changes can also be piecemeal and incremental. Just as natural selection can favor combinations of genes at different loci, so selection may favor combinations of endosymbionts. Quantitative variations in cytoplasmic factors, nest design, and habitat preferences could also all be passed on extragenetically. Thus, although combinations of these factors are not unlimited, they can be quite large enough to allow a fine-grained response to selection.

9.9 Representing Evolution: Selfish Nucleotides or Developmental Systems?

If the claim that only genes are inherited is the bedrock of the gene's-eye view, then perhaps the central pillar is the claim that genic selection is the most general way of representing evolutionary change. Group selection, individual selection, and the replication of junk DNA can all be *represented* in genic terms. Selection coefficients can be assigned to alleles, and evolutionary change can be calculated as if selection were acting independently at each locus. Individual selection cannot represent changes in junk DNA in this way, nor can group fitness values be calculated in a way to represent individual selection.

Defenders of the selfish gene from Williams (1966) and Dawkins (1976) to Sterelny and Kitcher (1988) have all argued that this generality is one of the main advantages of adopting the gene's-eye view. Critics of the gene's-eye view (e.g., Sober and Lewontin 1982, Sober 1984, Sober and Wilson 1994) concede that all evolutionary change can be represented in genic terms but deny that such "bookkeeping" reflects the real causal nature of selection processes. The genic changes are often just shadows of processes occurring at higher levels.

Although I agree with much of the Sober–Lewontin–Wilson analysis, I think that they have conceded too much. Even the representational argument for genic selectionism is problematic. In a rare moment of equivocation, Dawkins himself provides an excellent argument against representing evolution in genic terms. Dawkins (1982, p. 83) defines an active replicator as "any replicator whose nature has some influence over its probability of being copied." Single nucleotide changes can cause dramatic phenotypic differences. Sickle cell anemia results from a single nucleotide change. Similarly, qualitatively different enzymes can be created by changing just one base and thus altering one amino acid at an active site (Newcomb et al. 1997). Should *The Selfish Gene* thus have really been titled *The Selfish Nucleotide*? In a discussion of exactly what length of DNA counts as a replicator, Dawkins (1982, p. 91) comments that, "while it may not be strictly wrong to say that an adaptation is for the good of the nucleotide, . . . it is not helpful to do so." Further down the page Dawkins attempts to explain why the selfish nucleotide view is not helpful.

> We reject the whole sexual genome as a candidate replicator, because of its high risk of being fragmented at meiosis. The single nucleotide does not suffer from this problem but . . . it raises another problem. It cannot be said to have a phenotypic effect except in the context of the other nucleotides that surround it in the cistron. It is meaningless to speak of the phenotypic effect of adenine. But it is entirely sensible to speak of the phenotypic effect of substituting adenine for cytosine at a named locus within a named cistron. The case of a cistron within a genome is not analogous. Unlike a nucleotide, a cistron is large enough to have a consistent phenotypic effect, relatively . . . independently of where it lies on the chromosome. (Dawkins 1982, pp. 91–2)

This strikes me as a sensible argument. The problem for Dawkins is, as Griffiths and Neumann-Held (1999) note, that it is precisely this argument that he and other defenders of genic selectionism have repeatedly rejected for units of selection above the gene. Critics such as Sober (1984) have emphasized that the effects of alleles depend on the presence and frequency of other alleles. According to Sober, the context sensitivity of individual alleles means that in these cases higher-level units, such as genotypes, are the real units of selection.

The classic case of heterozygote advantage, sickle-cell anemia, provides an excellent example of this. Organisms that are homozygous for the sickling allele are at a substantial disadvantage because they cannot produce normal hemoglobin. However, in malarial environments heterozygotes are resistant to malaria and are thus at an advantage over homozygotes for the normal allele. The sickling and normal alleles do not have constant effects. Their effects vary with the allelic context. Therefore, according to Sober the alleles are not the targets of selection. In his terminology there is selection for genotypes (being heterozygous in malarial environments), not for the individual alleles. Defenders of the selfish gene take a different moral from this story. Sterelny and Kitcher (1988) argue that this example can easily be translated into genic selection by allowing that the fitness of the sickling allele is frequency dependent. In malarial environments the sickling allele is advantageous when it is rare and disadvantageous when it is common.

Thus, there is a dilemma here for Dawkins and his followers. If they maintain that units of selection must have consistent effects, then levels above the gene will often be units of selection. If, on the other hand, the effects of a replicator can be context sensitive, then individual nucleotides must count as replicators. They could attempt to steer a middle course between these positions. In the preceding statement, Dawkins appears to be heading in that direction. The problem for genic selectionists would then be specifying just how much context sensitivity is acceptable. What is the cut-off point beyond which higher level replicators must be invoked? Neither Dawkins nor his defenders have provided a principled justification for a specific cut-off. Without such a cut-off the logical conclusion of the representational argument for genic selection must be that there are only four replicators – the selfish nucleotides A, T, G, and C. From a DST perspective there is another obvious problem with the representational argument for genic selection. Gene frequency changes cannot capture changes in extragenetic inheritance. Therefore, the claim that genic selectionism provides the most general representation of evolution cannot be true. Instead, DST provides the most general account of evolution. In DST, developmental processes and lifecycles are the units of differential replication. Viewing evolution in this way expands the scope of evolutionary explanations. Not only can all that is explained in genic, individual, and group selection be recast in this way, but the replication of relationships, such as the mistle thrush association with parkland habitats discussed above, can also be given specific adaptive and historical explanations (see Griffiths and Gray 1994).

9.10 Does Developmental Systems Theory Go Too Far?

Several recent critical responses have been made to the DST challenge to genocentric views of development and evolution. The general theme of these responses has been that, although some of the points made by proponents of DST are valid, these points are taken to unnecessarily extreme conclusions. Sterelny, Dickison, and Smith (1996), for example, acknowledge the importance of extragenetic inheritance but argue that DST runs the risk of collapsing into unmanageable holism. They suggest that the challenge from DST can be accommodated by substantially expanding the pool of evolutionary replicators.

In their view, genes, nests, methylation patterns, membrane templates, and burrows are all selfish replicators. Griffiths and Gray (1997) have responded to this extended replicator proposal in some detail. In brief, we argued that in practice the extended replicator approach would be at least as unmanageable as they claim DST would be. However, we rejected the claim that DST must collapse into holism. Developmental systems can be defined and individuated in evolutionary terns (Griffiths and Gray 1994). Using Sterelny et al.'s (1996) definition of a replicator, we demonstrated that ultimately Sterelny et al.'s extended replicator vision would converge with DST; lifecycles meet their definition of a replicator.

Schaffner (1998) has also acknowledged some aspects of the DST critique (nonpreformationism and contingency) but has argued that DST goes too far in claiming that other developmental resources are on a par with genes, that developmental causes are indivisible, and that development is unpredictable owing to developmental noise. In response, Griffiths and Knight (1998) argue that most of Schaffner's objections arise in response to mischaracterizations of DST. Perhaps some of the rhetorical excesses of DST proponents (e.g., Gray 1992) are to be blamed for this mischaracterization.

Perhaps the most puzzling response to DST and to Lewontin's constructionist or "dialectical" views of development has come from the philosopher Philip Kitcher (this volume). Kitcher argues that Lewontin's perceptive and accurate criticisms of genetic determinism are at risk of being undermined by his efforts to reconceptualize developmental causation. Kitcher claims that, "in my judgment, no such reconceptualization is needed, and Lewontin's positive proposals are in constant danger of relapsing into the obscurity that he rightly sees as affecting traditional forms of biological holism." According to Kitcher, DST is equally misguided. What I find puzzling about Kitcher's analysis is the picture of the Lewontin–DST challenge he presents. This is a child no parent would recognize. Kitcher claims we argue that

1. the "gene for" locution is incoherent;
2. the notion of a "norm of reaction" is incoherent; and
3. singling out causal factors (e.g., genes) is an unwarranted abstraction from a causally complex situation.

Let us consider each claim in turn. Neither Lewontin nor DST proponents have argued that the "gene for" locution is incoherent. What I have argued (Gray 1992, Griffiths and Gray 1994) is that when it is unpacked in a reasonable way (e.g., Sterelny and Kitcher 1988), it provides no basis for privileging the role of genes in development. *Kitcher concedes this argument.* Neither Lewontin nor DST proponents have argued that the notion of "norms of reaction" is incoherent. Lewontin's writings are peppered with its positive use (Lewontin 1974, 1982, 1983b). I have also used reaction norms in a positive way to illustrate the mutual contingency of genes and environment (Gray 1992, 1997). I have argued that, as a general view of development, reaction norms have some limitations (life-history trajectories across a range of environments would be better). This is an argument for their extension, not their burial.

Neither Lewontin nor DST proponents have claimed that biologists should abandon causal analysis and "perform the impossible feat of considering everything at once." Kitcher claims that there are no conceptual problems with the standard interactionist view of causation. However, one of the reasons why people so consistently misuse heritability statistics is not because of insufficient evidence, as Kitcher suggests, but because of an adherence to an additive model of the interaction between discrete causal factors (Lewontin 1974). We are not abandoning causal analysis but rather changing the underlying model of causation to one in which causes are situated in systems and are thus likely to have nonadditive effects. The practice of changing one variable at a time while holding others constant is not so much wrong as it is incomplete. Additional investigation is required into how a causal factor is coupled in a system of causes and the ways in which these links change over time. This can be done in a piecemeal fashion. It does not require "considering everything at once." Daniel Lehrman's series of experiments on reproduction in the ring dove (e.g., Lehrman 1965) is an excellent example of such an integrative approach. Lehrman and his colleagues investigated the ways in which hormonal changes, behavioral changes, and changes in external stimuli are coupled and change together as a ring dove pair move from courtship to nest building, egg laying, incubation, and chick feeding. They unpacked the interrelationships between physiology, behavior, and the environment by varying one factor at a time. For example, they varied the amount of male courtship a female ring dove was exposed to and measured the resulting changes in endocrine hormones. They then manipulated hormone levels and measured changes in nest-building behavior. They then varied the amount of nest-building behavior and measured the hormonal changes, and so on. There is no unmanageable holism here.

Kitcher claims that there are no conceptual differences between traditional interactionism and the Lewontin–DST versions. He suggests that the nature–nurture debate can be resolved in a quantitative manner. Perhaps genetic determinism is true in 77.1% of all cases. If nothing else, I hope this essay demonstrates that real conceptual issues are involved in these debates – issues that will not be resolved either by partitioning phenotypic variance or by measuring how flat reaction norms are.

9.11 Questions of Practice

In his critique of the Lewontin–DST challenge, Kitcher (this volume) wonders what difference these views might make to our research practices. Quite properly, biologists often want to know the research implications of adopting a particular theoretical perspective. Our metaphors are not neutral with respect to the type of research they encourage. Kitcher has already noted the practical implications of adopting the gene's-eye view. Sterelny and Kitcher (1988) closed their defense of genic selectionism with a revealing concession:

> Genic selectionism can easily slide into naive adaptationism as one comes to credit the individual alleles with powers that enable them to operate independently of one another. The move from the "genes for P" locution to the claims that selection can fashion P independently of other traits of the organism is perennially tempting. But, in our version, genic representations must be constructed in full recognition of the possibilities for constraints in gene–environment coevolution. The dangers of genic selectionism, illustrated in some of Dawkins's own writings, are that the commitment to the complexity of the allelic environment is forgotten in practice.

The message is clear: Dawkinspeak can lead to Dawkinspractice. The price of the selfish gene metaphor is eternal vigilance against the dangers of genetic determinism, atomism, and adaptationism. If the environment is regarded as merely a standard background upon which genes do their causal work, then of course the "complexity of the allelic environment is forgotten in practice." Researchers who adopt a development systems perspective do not need an elaborate system of indexing effects relative to standard environments, and they do not need to be eternally vigilant against the dangers of genetic determinism, atomism, and adaptationism. From a development systems perspective, interaction, the developmental integration of networks of mutually contingent causes, and the mutual construction of organism and environment are the primary objects of study rather than secondary features that we must remember to tack on to make the story work. Here are a few suggestions for DST research tactics:

1. Treat all pseudoexplanatory claims about instincts, genetic programs, and other black boxes as potential research questions for developmental analysis. That is, How does this trait actually develop, what developmental resources does it depend upon, and over what range of parameters is this developmental outcome stable?
2. Study extragenetic inheritance. Conduct studies to investigate the longevity and fidelity of extragenetic inheritance. Experimentally investigate its effects and develop mathematical models of the impact of different types of extragenetic inheritance and their coevolution with genetic change.
3. Investigate whether there are adaptive mechanisms for passing on extragenetic inheritance. If extragenetic inheritance is an adaptive developmental resource, then organisms that reliably and efficiently pass on that resource would be at an advantage. Test these adaptive hypotheses using comparative methods.
4. Study "niche construction." Conduct field experiments to assess the fitness impacts of niche construction activities. Develop models to investigate the evolutionary consequences of the ways in which organisms select and modify their environments. Test predictions from these models using comparative methods.
5. Investigate how the effects of specific causal factors depend on context and study the way changes in these factors are linked together in development. Study the dynamics of such systems.

6. Investigate the extent of this developmental integration. Are some pathways tightly coupled, or is the developmental organization more modular? Does the integration only hold over a restricted range of parameters? Can these parameters be changed by selection? The greater the degree of modularity or quasi-independence, the more selection will be able to act independently on aspects of the phenotype. Are functionally linked traits also linked together in development? Are extragenetic forms of inheritance more or less modular than genetic inheritance in their developmental consequences?

9.12 Conclusion

In this essay I have attempted to evaluate some common justifications for a genocentric focus on development and evolution. I have argued that the standard "gene for" locution provides no basis for assigning genes a primary developmental role and that the idea that genes are self-replicating is false. The two main reasons for adopting the gene's-eye view (the inheritance and representational claims) are also fundamentally flawed. Genes are not all that is inherited, and their effects are not relatively context independent. Taken to their logical conclusion Dawkins' arguments lead not to the selfish gene but to the selfish nucleotide. Given the dangers of the gene's-eye view in practice, I suggest we abandon the selfish replicator and shift our attention to the evolutionary dynamics of developmental systems.

9.13 Acknowledgments

I am grateful to Paul Griffiths and Kim Sterelny for our dialogue on DST and genic selectionism and to Nicola Gavey for her patience while I was writing this chapter. Kendall Clements, Nicola Gavey, Susan Oyama, and Diane Paul made useful comments on earlier drafts. Special thanks are extended to Richard Lewontin, whose insight, passion, and wit are an inspiration even when you live in another hemisphere and seven time zones away.

Notes

1. The inheritance and the representational arguments strike me as the most fundamental objections to the gene's eye view of evolution.
2. As far as I can discover, Johnston (1982) and Oyama (1982) were the first people to use the term "developmental system" in a similar sense.
3. This is obviously a vastly oversimplified account, but I hope it suffices to give readers a brief background to the development of DST. Johnston (1987) gives a more detailed account of the debate between ethologists and comparative psychologists.
4. Hull also places much more emphasis on the evolutionary role of interactors than Dawkins places on vehicles.
5. Compliance with the diet is, however, extremely difficult (Paul 1997), and children attempting to comply with the diet may still show some lesser cognitive deficits (Herzberg and Diamond 1993).

6. The term "phenocopy" is a bit misleading. Phenotypes do not copy genotypes. Phenocopies mimic another phenotype (see Oyama 1981).
7. Williams (1992) also uses the analogy between genes and sheets of paper. However, his language reveals rather different developmental roles for the two "replicators." "One DNA molecule *making another* just like itself would be replication. So would a one-sheet photocopy *used to make* another copy in the same machine" (p. 12, my emphasis).
8. For an interesting critique of the replicator concept that emphasizes the indirectness of genetic replication, see Greisemer (in press). Greisemer argues that we should abandon the replicator concept in favor of *reproducers* with a concomitant emphasis on developmental processes.
9. I say this *might* be a good argument because its validity would depend on the gene concept used. "Genes" as smallish pieces of DNA are replicated with great longevity down lineages. However, "genes" as units of information may or may not be. Most people would agree the developmental information depends not just on the DNA sequence of a specific gene but also on the allelic and extragenetic context. As Sterelny et al. (1996) point out, the reliability of genetic replication is exaggerated by focusing on only the sequence of nucleotides rather than the relational properties of genes. Conversely, given that developmental processes are often buffered against variation in their components, some developmental information might be reliably replicated despite genetic changes. Advocates of the developmental systems perspectives claim that these developmental processes are the real unit of replication (Gray 1992, Griffiths and Gray 1994) or perhaps, more accurately, reproduction (see Greisemer in press).

REFERENCES

Bateson, P. (1978). Book Review: *The Selfish Gene. Animal Behaviour* 26:316–18.

Bateson, P. P. G. (1983). Genes, environment and the development of behaviour. In *Genes, Development and Learning*, eds., P. Slater and T. Halliday, pp. 52–81. Oxford, UK: Blackwell.

Bateson, P. (1991). Are there principles of behavioural development? In *The Development and Integration of Behaviour*, ed., P. Bateson, pp. 19–31. Cambridge, UK: Cambridge University Press.

Brandon, R. (1988). The levels of selection: A hierarchy of interactors. In *The Role of Behavior in Evolution*, ed., H. Plotkin, pp. 51–71. Cambridge, MA, MIT Press.

Dawkins, R. (1976). *The Selfish Gene.* Oxford, UK: Oxford University Press.

Dawkins, R. (1982). *The Extended Phenotype.* Oxford, UK: Oxford University Press.

Dawkins, R. (1984). Replicator selection and the extended phenotype. In *Conceptual Issues in Evolutionary Biology*, ed., E. Sober, pp. 125–41. Cambridge, MA: MIT Press.

Fogle, T. (1990). Are genes units of inheritance? *Biology and Philosophy* 5:349–72.

Galef, B. G., and Henderson, P. W. (1972). Mother's milk: A determinant of the feeding preferences of weaning rat pups. *Journal of Comparative and Physiological Psychology* 78:213–19.

Godfrey-Smith, P. (1999). Genes and codes: Lessons from the philosophy of mind? In *Where Biology Meets Psychology: Philosophical Essays*, ed., V. Hardcastle, pp. 305–31. Cambridge, MA: MIT Press.

Gottlieb, G. (1976). Conceptions of prenatal development: Behavioral embryology. *Psychological Review* 83:215–34.

Gould, S. J. (1980). *The Panda's Thumb.* New York: Norton.

Gray, R. D. (1988). Metaphors and methods: Behavioural ecology, panbiogeography and the evolving synthesis. In *Evolutionary Processes and Metaphors,* eds., M.-W. Ho and S. W. Fox, pp. 209–42. Chichester, UK: Wiley.

Gray, R. (1992). Death of the gene: Developmental systems strike back. In *Trees of Life: Essays of the Philosophy of Biology,* ed., P. Griffiths, pp. 165–209. Dordrecht, The Netherlands: Kluwer.

Gray, R. D. (1997). In the belly of the monster: Feminism, developmental systems and evolutionary explanations. In *Evolutionary Biology and Feminism,* ed., P. A. Gowaty, pp. 385–413. New York: Chapman and Hall.

Griesemer, J. (in press). The informational gene and the substantial body: On the generalization of evolutionary theory by abstraction. In *Varieties of Idealisation,* eds., N. Cartwright and M. Jones. Amsterdam: Editions Rodopi.

Griffiths, P. E. (1996). Developmental system cycles as evolutionary individuals. Talk to The International Congress of Systematic and Evolutionary Biologists, Budapest, Hungary.

Griffiths, P. E., and Gray, R. (1994). Developmental systems and evolutionary explanation. *Journal of Philosophy* XCI:277–304.

Griffiths, P. E., and Gray, R. D. (1997). Replicator II: Judgment day. *Biology and Philosophy* 12:471–92.

Griffiths, P. E., and Knight, R. D. (1998). What is the developmentalist challenge? *Philosophy of Science* 65:253–58.

Griffiths, P. E., and Neumann-Held, E. M. (1999). The many faces of the gene. *Bioscience* 49:656–62.

Hamilton, W. D. (1964). The genetic theory of social behavior, I & II. *Journal of Theoretical Biology* 7:1–52.

Hepper, P. G. (1988). Adaptive fetal learning: Prenatal exposure to garlic affects postnatal preferences. *Animal Behaviour* 36:935–6.

Herzberg, C., and Diamond, A. (1993). Impaired contrast sensitivity in children with treated PKU, presumably due to dopaminergic deficiency. *Society for Neuroscience Abstracts* 19:772.

Hull, D. L. (1980). Individuality and selection. *Annual Review of Ecology and Systematics* 11:311–32.

Hull, D. L. (1988). *Science as a Process.* Chicago: University of Chicago Press.

Immelmann, K. (1975). Ecological significance of imprinting and early learning. *Annual Review of Ecology and Systematics* 6:15–37.

Jablonka, E., and Lamb, M. J. (1995). *Epigenetic Inheritance and Evolution: The Lamarkian Dimension.* Oxford, UK: Oxford, University Press.

Jablonka, E., and Szathmary, E. (1995). The evolution of information storage and heredity. *Trends in Ecology and Evolution* 10:206–11.

Johnston, T. (1982). Learning and the evolution of developmental systems. In *Learning, Development and Culture,* ed., H. C. Plotkin, pp. 411–42. New York: Wiley.

Johnston, T. (1987). The persistence of dichotomies in the study of behavioural development. *Developmental Review* 7:149–82.

Johnston, T. D., and Gottlieb, G. (1990). Neophenogenesis: A developmental theory of phenotypic evolution. *Journal of Theoretical Biology* 147:471–95.

Keller, L., and Ross, K. G. (1993). Phenotypic plasticity and cultural transmission in the fire ant, *Solenopsis invicta. Behavioural Ecology and Sociobiology* 33:121–9.

Klopfer, P. (1963). Behavioral aspects of habitat selection. *The Wilson Bulletin* 75:15–22.

Lehrman, D. S. (1965). Interaction between internal and external environments in the regulation of the reproductive cycle of the ring dove. In *Sex and Behaviour*, ed., F. A. Beach, pp. 335–69, 378–80. New York: Wiley.

Lehrman, D. S. (1970). Scientific and conceptual issues in the nature-nurture problem. In *Development and Evolution of Behavior: Essays in Memory of T. C. Schneirla*, eds., L. R. Aronson, E. Tobach, D. S. Lehrman, and J. S. Rosenblatt, pp. 17–52. San Francisco: Freeman.

Lewontin, R. C. (1974). The analysis of variance and the analysis of causes. *American Journal of Human Genetics* 26:400–11.

Lewontin, R. C. (1982). *Human Diversity*. New York: Scientific American Library.

Lewontin, R. C. (1983a). The corpse in the elevation. *The New York Review* 20, January:34–7.

Lewontin, R. C. (1983b). The organism as the subject and object of evolution. *Scientia* 118:65–82.

Lewontin, R. C. (1991). *Biology as Ideology: The Doctrine of DNA*. New York: Harper.

Lewontin, R. C. (1992). The dream of the human genome. *New York Review of Books* May 28:31–40.

Lewontin, R. C., Rose, S., and Kamin, L. J. (1984). *Not in Our Genes: Biology, Ideology and Human Nature.* New York: Pantheon Books.

Maynard Smith, J., and Szathmary, E. (1995). *The Major Transitions in Evolution.* New York: Freeman.

Millikan, R. G. (1984). *Language, Thought and Other Biological Categories.* Cambridge, MA: MIT Press.

Moran, N., and Baumann, P. (1994). Phylogenetics of cytoplasmically inherited microorganisms of arthropods. *Trends in Ecology and Evolution* 9:15–20.

Moss, L. (1992). A kernel of truth? On the reality of the genetic program. In *Proceedings of the Philosophy of Science Association, Michigan, Philosophy of Science Association. Volume 1,* eds., D. Hull, M. Forbes, and K. Okruhlik, pp. 335–48.

Newcomb, R. D., Campbell, P. M., Ollis, D. L., Cheah, E., Russell, R. J., and Oakeshott, J. G. (1997). A single amino acid substitution converts a carboxylesterase to an organophosphorous hydrolase and confers insecticide resistance on a blowfly. *Proceedings of the National Academy of Science* 94:7464–8.

Neumann-Held, E. M. (1999). The gene is dead – long live the gene: Conceptualizing genes the constructionist way. In *Sociobiology and Bioeconomics: The Theory of Evolution in Biological and Economic Thinking,* ed., P. Koslowski, pp. 105–37. Berlin: Springer.

Odling-Smee, F. J., Laland, K. N., and Feldman, M. W. (1996). Niche construction. *American Naturalist* 147:641–8.

Oyama, S. (1981). What does the phenocopy copy? *Psychological Reports* 48:571–81.

Oyama, S. (1982). A reformulation of the concept of maturation. In *Perspectives in Ethology,* Vol. 5., eds., P. P. G. Bateson and P. H. Klopfer, pp. 101–31. New York: Plenum Press.

Oyama, S. (1985). *The Ontogeny of Information. Developmental Systems and Evolution.* Cambridge, UK: Cambridge University Press.

Paul, D. (1997). Appendix 5. The history of newborn phenylkentonuria screening in the U.S. In *Promoting Safe and Effective Genetic Testing in the United States. Final Report of the Task Force on Genetic Testing,* eds., N. A. Holtzman and M. S. Watson, pp. 137–60.

Sargent, T. D. (1965). The role of experience in the nest building of the zebra finch. *Auk* 82:48–61.

Sarkar, S. (1996). Decoding "coding": Information and DNA. *Bioscience* 46:857–64.

Schaffner, K. (1998). Genes, behavior and developmental emergentism: One process, indivisible? *Philosophy of Science* 49.

Schneirla, T. C. (1956). Interrelationships of the "innate" and the "acquired" in instinctive behavior. In *L'Instinct dans le Compostement des Animaux et de L'Homme*, ed., P.-P. Grasse, pp. 387–452. Masson.

Shannon, C. E., and Weaver, W. (1949). *The Mathematical Theory of Communication.* Urbana, IL: University of Illinois Press.

Sober, E. (1984). *The Nature of Selection: Evolutionary Theory in Philosophical Focus.* Cambridge, MA: MIT Press.

Sober, E., and Lewontin, R. (1982). Artifact, cause and genic selection. *Philosophy of Science* 49:157–80.

Sober, E., and Wilson, D. S. (1994). A critical review of philosophical work on the units of selection problem. *Philosophy of Science* 61:534–55.

Stent, G. (1981). Strength and weakness of the genetic approach to the development of the nervous system. In *Studies in Developmental Neurobiology*, ed., W. M. Cowan, pp. 288–320. Oxford, UK: Oxford University Press.

Sterelny, K., and Griffiths, P. E. (1999). *Sex and Death: An Introduction to Philosophy of Biology.* Chicago: University of Chicago Press.

Sterelny, K., and Kitcher, P. (1988). The return of the gene. *Journal of Philosophy* 85:339–60.

Sterelny, K., Smith, K. C., and Dickison, M. (1996). The extended replicator. *Biology and Philosophy* 11:377–403.

Szathmary, E., and Maynord Smith, J. (1997). From replicators to reproducers: The first major transition leading to life. *Journal of Theoretical Biology* 187:555–71.

Troyer, K. (1984). Microbes, herbivory and the evolution of social behavior. *Journal of Theoretical Biology* 106:157–69.

Williams, G. C. (1966). *Adaptation and Natural Selection.* Princeton, NJ: Princeton University Press.

Williams, G. C. (1986). Comments by George C. Williams on Sober's "The Nature of Selection." *Biology and Philosophy* 1:114–22.

Williams, G. C. (1992). *Natural Selection: Domains, Levels and Challenges.* Oxford: Oxford University Press.

Wilson, D. S., and Sober, E. (1994). Reintroducing group selection to the human behavioral sciences. *Behavioral and Brain Sciences* 17:585–654.

Wimsatt, W. C. (1980). Reductionist research strategies and their biases in the units of selection controversy. In *Scientific Discovery: Case Studies ii*, ed., T. Nickles, pp. 213–59. Dordrecht: Reidel.

Winkler, D. W., and Sheldon, F. H. (1993). The evolution of nest construction in swallows (*Hirundinidae*): A molecular phylogenetic perspective. *Proceedings of the National Academy of Sciences* 90:5705–7.

CHAPTER TEN

The Evolutionary Definition of Selective Agency, Validation of the Theory of Hierarchical Selection, and Fallacy of the Selfish Gene

STEPHEN JAY GOULD

Science thrives upon the continuous correction of error. Most errors either arise from inadequate knowledge of the empirical world or at least persist because we have no means (conceptual or technological) to secure their empirical refutation. For example, we once lacked the technology to discover that buried organic matter might petrify and that wood made of stone might therefore represent the remains of an ancient organism and not the power of rocks to mimic plants and animals by a process analogous to crystallization.

Only rarely, however, do professions get sidetracked by persuing an extensive and longlasting program of research initiated by a pure error in reasoning rather than an inadequacy of empirical knowledge. Yet I think that the gene-centered approach to natural selection – based on the central contention that genes, as persistent and faithful replicators, must be fundamental (or even exclusive) units of selection – represents a pure error of this unusual kind. In beginning with Williams' manifesto (1966), inspired by the remarkable work of Hamilton (1964), and proceeding through the well-written but ill-reasoned codification of Dawkins (1976) to numerous works both popular (especially Cronin 1991) and technical (Dennett 1995), this gene-based approach to selective agency has inspired both fervent following of a quasi-religious nature (see Wright 1994) and strong opposition from many evolutionists, who tend to regard the uncompromising version as a form of Darwinian fundamentalism resurgent (see Gould 1997) variously designated as ultradarwinism (Eldredge 1995) or hyperdarwinism.

I will show in this article that, although genes may be properly identified as fundamental replicators (under one defendable but nonexclusive strategy of research), replicators are not units of selection or, for that matter, causal agents at all under our usual notions of mechanism in science. The misidentification of replicators as causal agents of selection – the foundation of the entire gene-centered approach – rests upon a logical error best characterized as a confusion of bookkeeping with causality.

We commit another form of error when we accept the common conceptual taxonomy that relegates error itself to a purely negative category. Some errors lead only to blind alleys and waste of time. But others, as thoughtful scientists have always recognized, serve as essential prods for progress through correction. Darwin's famous words, distinguishing harmful from salutary error, have frequently been cited in this context: "False facts are highly injurious to the progress of science, for they often endure long; but false views, if supported by some evidence, do little harm, for every one takes a salutary pleasure in proving their falseness" (from the *Descent of Man*). I prefer the stronger statement of the great Italian economist, Vilfredo Pareto: "Give me a fruitful error any time, full of seeds, bursting with its own corrections. You can keep your sterile truth for yourself."

During my career in evolutionary science, I can think of no error more fruitful in Pareto's sense than the gene-centered approach to selection. The central claim, so clearly expressed, forced us to reconceptualize the entire domain of evolutionary causality. The outrageous character of such an ultimate reduction compelled us to rethink our subject by explicitly rejecting the oldest, most traditional, and entirely commonsense notions about our own bodies as agents. (Yet the reductionistic cast of the theory fit so well with conventional preachings about the goals of science that many biologists "caught the spirit" and "followed the program" despite its assault upon ordinary intuition.) Nevertheless, the theory could not work. However stubborn and heroic the attempt, explanation inevitably faltered on the central logical error – especially when selection had clearly worked on emergent properties of higher-level individuals, and no verbal legerdemain could therefore recast the story in terms of genes as causal agents. If "Pareto errors" contain seeds that burst their own structures, then errors of fallacious reason (rather than absent fact) must qualify for this status. The more conventional form of empirical correction usually requires a period of waiting for new technologies or new discoveries (as sources of resolution do not lie within the logical structure of the argument itself), while logical errors always carry the seeds of correction within their own structure.

The fallacy of gene selectionism, and the consequent validity of the alternative (and opposite) hierarchical model of selection, can best be expressed in a series of seven arguments and vignettes, of markedly different length, but all of the same import and to the same purpose. I therefore organize this article in seven subsequent sections.

10.1 Distinction of Replicators and Interactors as a Framework for Discussion

Leading founders of modern gene selectionism as a general view of evolution (Williams 1966 and Dawkins 1976) drew a crucial distinction between reproductive units of heredity, and entities that interact with the environment to bias the transmission of reproductive units into the next generation. Williams viewed nearly all evolution as occurring via genes as reproductive units, with adaptation of organisms (the interacting entities) as a result – a duality that he usually labelled (1966, p. 124 for example) as "genic selection and organic adaptation." Dawkins (1976) agreed entirely and made a more colorful and explicit distinction between "replicators," considered as units of selection and identified as genes – and "vehicles," considered as merely passive repositories built by replicators for their own purposes, and identified as bodies of organisms. In other words, both Williams and Dawkins invoked a criterion of replication to identify genes as the active and fundamental agents of natural selection.

In his 1980 review on "Individuality and Selection," David Hull formalized this distinction in a manner that has – quite usefully and properly in my view – focused the discussion on units of selection ever since. Hull (1980, p. 318) defined a replicator as "an entity that passes on its structure directly in replication"; and an interactor as "an entity that directly interacts as a cohesive whole with its environment in such a way that replication is differential." Hull then defined selection with reference to both attributes as "a process in which the differential extinction and proliferation of interactors cause the differential perpetuation of the replicators that produced them."

Hull then insisted (and continues to maintain in 1994, pp. 627–628) that a causal account of selection must include both concepts (1980, pp. 319–320).

> Evolution of sorts could result from replication alone, but evolution through natural selection requires an interplay between replication and interaction. Both processes are necessary. Neither process by itself is sufficient. Omitting reference to replication leaves out the mechanism by which structure is passed from one generation to the next. Omitting reference to the causal mechanisms that bias the distribution of replicators reduces the evolutionary process to the 'gavotte of the chromosomes,' to use Hamilton's propitious phrase.

10.2 Faithful Replication as the Central Criterion for the Gene-Centered View of Evolution

Williams and Dawkins rejected Hull's dual approach to causality, but chose to search for the unit of selection among replicators rather than interactors. I shall explain under argument three why they followed the correct approach, but made the wrong choice – thus committing the logical but highly fruitful "Pareto error" discussed at the outset of this section. Having thus decided, and understanding that Darwinian selection works on "individuals," what replicating individuals would Williams and Dawkins then designate as units?

We all know that they chose genes as fundamental – and effectively exclusive – replicators, and therefore as *the* unit of selection in Darwinian theory. I will discuss the stated reasons for their choice, but I cannot fully probe the deeper motivations of their philosophical and psychological preferences. I strongly suspect that they, and all defenders of strict gene selectionism, feel special fascination for the traditional reductionism of science. They understood that Darwin himself went as far as he could in this direction, by breaking down the Paleyan edifice of highest-level intentionality (God himself) to the lowest level then practical – to organisms struggling for reproductive success. They also recognized that this breakdown had produced revolutionary consequences for

Western thought, particularly in reconceptualizing all perceived natural "benevolence" (especially the good design of organisms and the harmony of ecosystems) as side-consequences of struggle for personal success among lowest-level individuals, rather than an explicit intention of a loving and omnipotent deity. I imagine that the more thoughtful gene selectionists then made an analogy and reasoned that if they could break causality down even further, below the level of the organism, similarly interesting, and perhaps revolutionary, consequences might ensue. I cannot gainsay either the intuition or the ambition, but I do fault the resulting argument for an erroneous choice of category and level.

If a search for ultimate reduction below the Darwinian body set the deeper motivation for choosing genes as units of selection, what particular rationales did proponents of this theory state? Williams and Dawkins began by arguing that the conventional unit of Darwinian theory – bodies of organisms – cannot properly occupy this role because such bodies lack a key feature that genes possess. The bodies of sexual organisms disaggregate in reproduction, making (so to speak) only half an appearance in the genetic constitution of offspring. How can something so ephemeral be a unit of selection? But genes pass faithful copies of themselves into future generations and therefore maintain the integrity required for an agent of natural selection.

Williams and Dawkins advance the same argument in three steps: (1) the unit of selection must be a replicator; (2) replicators must transmit faithful, or minimally altered, copies of themselves across generations; and (3) sexual organisms disaggregate across generations and therefore cannot be units of selection, whereas genes qualify by faithful replication. Williams developed this argument in his first book (1966) and continues his verbal defense to this day despite remarkable movement, as we shall see (p. 232), towards the position advocated in this chapter. But Williams still hews to the language of gene-selectionism, particularly to the identification of genes as "units of selection" by virtue of faithful replication (so different from Hull's pluralistic view that the definition of a unit must include both replicators and interactors):

> These complications are best handled by regarding individual [i.e., organismic] selection, not as a level of selection in addition to that of the gene, but as the primary mechanism of selection at the genic level. Because genotypes do not replicate themselves in sexual reproduction (cannot be modeled by dendrograms), they cannot be units of selection. (Williams 1992, p. 16)

Dawkins (1978) advances the same argument with the same designation of genes as units of selection:

> However complex and intricate the organism may be, however much we may agree that the organism is a unit of *function*, I still think it misleading to call it a unit of *selection*. Genes may interact, even "blend" in their effects on embryonic development, as much as you please. But they do not blend when it comes to being passed on to future generations.

In buttressing his primary argument based on faithful replication, Dawkins commits a classic error of historical reasoning by arguing that because genes preceded organisms in time and then aggregated to form cells and organisms, genes must therefore control organisms – a striking example of confusing historical priority with current domination (see Gould and Vrba 1982 for a full discussion of this common fallacy). Dawkins' argument collapses for many reasons – most notably on the issue of emergence. A higher unit may form historically by aggregation of lower units. But so long as the higher unit develops emergent properties by nonadditive interaction among parts (lower units), the higher unit becomes, by definition, an independent agent in its own right and not the passive "slave" of controlling constituents. In advancing this false argument, Dawkins closes with a statement that must vie with some choice Haeckelian lines for the dubious distinction of the purplest passage in the history of evolutionary prose:

> Replicators began not merely to exist, but to construct for themselves containers, vehicles for their continued existence. The replicators which survived were the ones which built survival machines for themselves to live in.... Survival machines got bigger and more elaborate, and the process was cumulative and progressive.... Four thousand million years on, what was to be the fate of the ancient replicators? They did not die out, for they are past masters of the survival arts. But do not look for them floating loose in the sea; they gave up that cavalier freedom long ago. Now they swarm in huge colonies, safe inside gigantic lumbering robots, sealed off from the outside world, communicating with it by tortuous indirect routes, manipulating it by remote control. They are in you and me; they created us, body and mind; and their preservation is the ultimate rationale for our existence. They have come a long way, those replicators. Now they go by the name of genes, and we are their survival machines (1976, p. 21).

Dawkins' metaphors, however overblown, do accurately express his false theory of selective agency, for if genes are units of selection, then they are the causal agents of evolution; and if bodies are Darwinian ciphers both for their transiency and by their lethargy relative to "lean and mean" genes living within, then bodies might as well be described as inert and manipulated repositories ("lumbering robots").

10.3 The Rejection of Replication as a Criterion of Agency

The linkage of selective agency to faithful replication has been urged with such force and frequency that the argument has become something of a mantra for many evolutionary biologists. But when we consider the character of natural selection as a causal process, we can only wonder why so many people confused a need for measuring the results of natural selection by counting the differential increase of some hereditary attribute (bookkeeping) with the mechanism that establishes relative reproductive success (causality). Replicators are not causal agents (unless they also happen to be interactors, for only interactors can be agents). Units of selection must be actors within the guts of the mechanism, not items in the calculus of results.

Genes struck many people as promising units for a dual reason that does record something of vital evolutionary significance but bears little relationship to our issue of selective agency. Persistence and replication both count as necessary (but not sufficient) criteria for calling any biological entity an evolutionary individual. Because evolution requires hereditary passage and genes both transmit faithful copies of themselves and represent the smallest functional unit of physical movement between generations of sexual organisms (the kind of individuals we know best for obvious parochial reasons), many biologists assumed that genes must therefore be the basic (or even the only) units of selection.

This interesting error arises from two common fallacies in human reasoning:

1. *The confusion of necessary with sufficient conditions.* We all agree that units of selection must be evolutionary individuals in Darwinian theory, and evolutionary individuality requires adherence to a set of criteria, including hereditary passage and sufficient persistence – properties surely exhibited by genes. But evolutionary individuals, to act as units of selection, must also display other properties that genes generally do not possess. In particular, a unit of selection, to quote Hull's definition once again (1980, p. 318), must interact "directly . . . as a cohesive whole with its environment in such a way that replication is differential."

But in sexual organisms, and in other higher-level individuals, genes generally do not interact *directly* with the environment. Rather, they must operate via the organisms that function as true agents in the "struggle for existence." Organisms live, die, compete, and reproduce; as a result, genes pass differentially to the next generation.

Of course genes influence organisms; one might even say, metaphorically in large part, that genes build organisms. But such statements do not validate the critically necessary claim that, therefore, genes interact directly with the environment when organisms struggle for existence. The issue before us – the venerable problem of "emergence" – is largely philosophical and logical, and only partly empirical. Genes would interact directly only if organisms developed no emergent properties, that is, if genes built organisms entirely in an additive fashion with no nonlinear interaction among genes at all. In such a situation, organisms would be passive repositories, and genes could be generalized and exclusive units of selection, for anything done by organisms could then be causally reduced to properties of individual genes.

This aspect of the question is empirical but also entirely settled (and never really controversial): organisms are replete with emergent properties; our sense of organismic functionality and intentionality arises from these emergent features. Thus, because genes interact with the environment only indirectly through selection upon organisms and selection on organisms operates largely upon emergent characters, genes cannot be units of selection when they function in their customary manner as faithful and differential replicators during the process of ordinary natural selection among organisms. Dawkins' metaphors of selfish genes and manipulated organisms may be colorful, but they could not be more misleading because he has reversed nature's causality: organisms are active units of selection; genes, although lending a helping hand as architects, remain stuck within.

2. *The theory-bound nature of concepts and definitions.* We are drawn to the faithfulness of gene replication and to the contrasting transiency of sexual organisms that must disaggregate to reach the next generation. We might therefore assume that genes become primary candidates for units of selection in their potential immortality, whereas organisms fall from further consideration by the brevity of their coherent lives.

"Sufficient stability" surely ranks as an important criterion for the "evolutionary individuality" required of a "unit of selection." But, in Darwinian theory and the search for units of selection, "sufficient" stability can only be defined as enough coherence to participate as an unchanged individual in the causal process of struggle for differential reproductive success. To be causal units under this requirement, organisms must persist for the single generation of their lifetimes – as they do. This duration may not strike us as a long time in some favored psychological sense, or relative to the persistence of faithful gene replicates, or considered in comparison with geological scales, but these temporal frameworks are irrelevant. Organisms last long enough to be units of selection in the Darwinian process; they therefore possess the "sufficient stability" required of evolutionary individuals.

Of course, evolutionary individuals must be able to pass, differentially and by heredity, their favorable properties into future generations. But nothing about this requirement implies that units of selection must pass copies of *themselves*, bodily and in their entirety, into the next generation. The criterion of heredity only demands that units of selection be able to bias the genetic makeup of the next generation toward features that secured the differential reproductive success of parental individuals. Units of selection only need to magnify their own relative representation in the next generation; they do not have to copy themselves. Sexual organisms happen to magnify by disaggregation and subsequent differential passage of genes and chromosomes. Other kinds of individuals – including genes, asexual organisms, and species – magnify more coherently. This common confusion of magnification with faithful replication has erected a serious stumbling block to proper understanding of the hierarchical theory of selection.

We can best grasp this crucial issue of the relationship between selective agency and the alternative criteria of faithful replication versus differential magnification if we drop, for a moment, the conventional framework of replication versus interaction and return instead to a different metaphor commonly invoked during nineteenth century debates about the nature of Darwinism and natural selection, namely, sieves.

We may use the classical metaphor of sieving to illustrate the inappropriateness of faithful replication as a criterion for defining units of selection. The "goal" of a unit of selection is not unitary persistence (faithful replication), and I do not know why so many people ever tried to formulate the concept in such a manner. The "goal" of a unit of selection is concentration by relative magnification, that is, the differential passage of more of "youness" into the next generation, an increase in relative representation of the heritable part of whatever you are (whether you pass yourself on as a whole, or in disaggregated form, into the future of your lineage).

In the favored metaphor of Darwin's day, selection works as a sieve holding all the individuals of one generation. The sieve is shaken, and particles of a certain size become concentrated while others pass through the webbing (lost by selection). Sieving represents the causal act of selection – the interaction of the environment (shaking the sieve) with varying individuals of a population (particles on the sieve). As a result of this interaction, some individuals live (remain on the sieve), whereas others die (pass through the sieve). Survival depends causally upon variation in emergent properties of the particles (in this simplest case, large particles remain and small particles pass through to oblivion).

In genealogical systems of evolutionary individuals, the surviving particles must reproduce. They may do so by fissioning (faithful passage) or by disaggregation and reconstitution of new individuals as mixtures of hereditary parts of previous individuals. The individuals of the old generation eventually die and evaporate. The individuals of the new generation now live on the sieve and wait for the next shake.

But specifying these various modes for constituting new individuals does not capture what we mean by selection. To rank as an evolutionary individual, you must be able to reproduce, but you don't have to replicate faithful copies of yourself. Rather, you need to be able to magnify – that is, to increase, relative to other individuals – the representation of your heredity in the next generation. Integral "you" may be disaggregated in this process, but so long as the next generation contains a relative increase in your particles, and so long as you qualified as an active causal agent of the Darwinian struggle while you lived, then you are a unit of selection (and a winning unit in this case).

An interesting episode in the history of Darwinism clarifies this concept in a striking way. We all know that Darwin believed in "blending inheritance," or the averaging of parental characteristics in the offspring of sexual reproduction. Now blending inheritance represents an ultimate denial of faithful replication, for the hereditary basis of any selected character gets degraded by half in breeding with an average individual. A historical paradox therefore arises. If units of selection must be faithful replicators, and if Darwin both understood natural selection and believed in blending inheritance, then how could he ever even imagine that selection might work as a mechanism?

The answer must be that faithful replication is not – and never was – the defining characteristic (or even a necessary property) of a unit of selection. Darwin, even given his belief in blending inheritance, could view organisms as units of selection because he understood agency in another way that remains valid today: units of selection are evolutionary individuals that interact with the environment and vary in reproductive success as a causal result.

We may return to the metaphor of sieving: Natural selection can work under blending inheritance because shaking the sieve aids the possessors of favored traits in each generation, for any individual with a phenotype at all biased in the favored direction has a better chance of remaining on the sieve. The offspring of the most favored individuals always blend substantially back to the mean, but this style of inheritance only slows the process of selection – for, as a result of differential reproductive success in each generation, the mean itself still gradually moves in the favored direction.

10.4 Interaction as the Proper Criterion for Identifying Units of Selection

We can only understand the causal nature of selection when we recognize that units of selection must be defined as *interactors, not replicators.* Hull's distinction had great merit, but he fell into overgenerous pluralism in arguing that the specification of causal agency must include statements about both replicators and interactors. Individuals need not replicate themselves to be units of selection. Rather, they must contribute to the next generation by hereditary passage, and they must magnify their contributions relative to those of other individuals. But the contributions themselves can be wholes or parts – faithful replicates or disaggregated bits of functional heredity.

But such differential magnification can only rank as a necessary condition, not a cause. We define selection as occurring when this magnification results *from the causal interaction of an evolutionary individual* (a unit of selection) *with the environment* in a manner that enhances the differential reproductive success of the individual. Thus, and finally, *units of selection must, above all, be interactors.* Selection is a causal process, not a calculus of results, and the causality of selection resides in interaction between evolutionary individuals and surrounding environments.

The study and documentation of group and higher-level selection have been stymied and thrown into disfavor by confusion over these issues – especially in the blind alley of a logically false argument that identified replicators rather than interactors as units of selection and then constructed a fallacious theory, exactly opposite in structure to the hierarchical model, by specifying genes (because they replicate faithfully) as ultimate or exclusive units of selection. In this context, I note with delight that group selection has risen from the ashes to receive a vigorous rehearing (see Lewin 1996 for a popular account) and that this cogent defense rests upon two proposals that can serve as centerpieces for a general theory of macroevolution: (1), the identification of evolutionary individuals as interactors, causal agents, and units of selection; and (2), the validation of a hierarchical theory of natural selection based upon the recognition that evolutionary individuals exist at several levels of organization, including genes, cell lineages, organisms, demes, species, and clades.

D. S. Wilson has vigorously championed this revival (Wilson 1980, 1983), and his collaboration with philosopher E. Sober has produced a particularly important paper on the subject (Wilson and Sober 1994, with 33 accompanying commentaries and the authors's response). Wilson and Sober root their argument by insisting that units of selection must be defined as interactors, not replicators.

I have only one mild quarrel with Wilson and Sober. I agree entirely that units of selection must be interactors, but I prefer a "looser" or "broader" concept of interaction that fosters the proper identification of highest-level individuals in species and clade selection. Wilson and Sober stress the "organism-like" properties of interactors. I would emphasize the potential for a rich panoply of emergent fitnesses and consequent potential for magnification.

A stress on "organism-like" properties leads Wilson and Sober to emphasize direct modes of interaction based on actual contact of sympatric individuals – the old vision of two gladiators duking it out to the finish. But interaction does not require physical contact. Interaction occurs between individuals and environments, not necessarily between individual and individual. The interaction must lead to magnification for causal reasons based on properties that enhance differential reproductive success, but, again, competing individuals need not interact directly with each other. Rather, to speak of selection, competing individuals need only have differential reproductive success based on similar causal interactions with environments. But the environments may be spatially separate and broadly defined. This issue does not often arise at the traditional level of Darwinian individuals or organisms. But higher-level individuals, particularly species and clades, do often compete without contact, and our notion of units of selection must include this important mode of interaction.

Several thoughtful biologists have stressed this point. Consider, for example, the forceful argument of Williams (1992, p. 25), who has come along a long way since formulating the theory of gene selectionism in 1966:

> There are many further questions on the meaning and limits of clade selection. One issue is whether the populations that bear the gene pools need be in ecological competition with each other. I believe that this is not required, any more than individuals within a population need interact ecologically to be subject to individual selection. The reproductive success or failure of a soil arthropod, with an expected lifetime dispersal of a few meters, will hardly influence prospects for a conspecific a hundred meters away. But the descendants of these two individuals might compete, and genes passed on by one may ultimately prevail over those passed on by the other. Selective elimination of one and survival of the other a hundred meters away is individual selection as long as the two arthropods can be assigned to the same population and their genes to the same gene pool.... In the same way, two gene pools in allopatry can be subject to natural selection if, as must always be true, their descendants might be alternatives for representation in the biota.... The ultimate prize for which all clades are in competition is representation in the biota.

10.5 The Internal Incoherence of Gene Selectionism

I regard the heyday of gene selectionism as an odd episode in the history of science, for I am convinced that the theory's central argument is logically incoherent, whatever the attraction (and partial validity) of several tenets, and despite the value of a mental exercise that tries to reconceptualize all nature from a gene's point of view. Close textual analysis of this theory's leading documents reveals consistent internal problems explicitly recognized by authors and invariably met by arguments so flawed in construction that even the perpetrators seem embarrassed or at least well aware of a niggling insufficiency.

I am not alone in noting this peculiar situation and in calling for some serious consideration by historians. Wilson and Sober (1994, p. 590) write:

> The situation is so extraordinary that historians of science should study it in detail: a giant edifice is built on the foundation of genes as replicators, and therefore as the "fundamental" unit of selection, which seems to obliterate the concept of groups as organisms. In truth, however, the replicator concept cannot even account for the organismic properties of individuals. Almost as an afterthought, the vehicle concept is tacked onto the edifice to reflect the harmonious organization of individuals, but it is not extended to the level of groups.

The central problem lies as deep as our definition for the key concept of "cause" in science. Aristotle used a broad concept of causality divided into four aspects, which he called material, efficient, formal, and final (or, roughly, stuff, action, plan, and purpose, that is, the bricks, the mason, the blueprint, and the function, in the classical "parable of the house"). As many historians have noted, modern science may virtually be defined by a revision of this broad view and a restriction of "cause," as a concept and definition, to the aspect that Aristotle called "efficient," (The root of "efficient," comes from the Latin *facere*, meaning to make or to do. Efficient causes are actual movers and doers, appliers of forces. Aristotle's sense does not engage the modern English cognate for doing something well as opposed to doing something at all.)

Modern science, beginning with Newton's generation, banned final cause for physical objects (while retaining the concept of purpose for biological adaptations so long as mechanical causes rather than external conscious agencies could be identified – a problem solved by natural selection in nineteenth century). As for Aristotle's material and formal causes, these notions retained their relevance but lost their role as "causes" under a mechanical worldview that restricted causal status to active mechanisms. The material and formal causes of a house continue to matter: brick or sticks fashion different kinds of buildings, but you never advance beyond a pile of bricks without a plan for construction. However, we no longer refer to these aspects of a totality as "causes." Material and formal attributes have become background conditions or operational constraints in the logic of modern science.

I present this apparent digression because the chief error of gene selectionism lies in failed attempts to depict genes as efficient causes in ordinary natural selection, and the chief "textual mark" of failure can be found in tortuous and clearly discomforting (even to the authors!) arguments advanced by all leading gene selectionists in a valiant struggle to "get through" this impediment. The problem can be simply stated: No matter how you wish to honor genes – as basic units, as carriers of heredity to the next generation, as faithful replicators, or whatever – you cannot deny a fundamental fact of nature: in ordinary, garden variety natural selection – Darwin's observational basis and legacy – organisms, and not genes, are the "things out there" that live and die, reproduce or fail to propagate, in the interaction with environments that we call "natural selection." Organisms play the role of efficient causes – the actors and doers – in the standard form of Darwin's great and universal game.

The gene selectionists know this, of course – so they must then struggle to find a way to say that, even though organisms do the explicit work, genes may somehow still be viewed as primary "units of selection," or causal agents in the Darwinian process. This misguided search arises from a legitimate intuition – that genes are vitally important in evolution, and clearly central to the process of natural selection – followed by the false inference that genes must therefore be designated as primary causes. Genes are, indeed, as central and important as this intuition holds, but genes are not *efficient* causes in the ordinary process of organismic selection. Genes play a primary role as *material* causes, as carriers of continuity to the next generation, but we no longer designate material aspects of natural processes as "causes." Organisms "struggle," as agents or efficient causes; their "reward" may be measured by greater representation of their genes, or material causes, in future generations. Genes are the product, not the agent – the stuff of continuity, not the cause of throughput.

The standard gambit of gene selectionists, in the light of this recognized problem, invokes two arguments, both indefensible.

10.5.1 Attempts to Assign Agency to Genes by Denying Emergent Properties to Organisms

Once you admit, as all gene selectionists do, that genes propagate via selection on organisms as interactors, how then can you possibly ascribe direct causal agency to genes rather than to bodies? Only one logical exit from this conundrum exists: you must assert that each gene is an optimal product in its own place and that bodies impose no consequences upon individual genes beyond providing a home for joint action. If such a view could be defended, then bodies would become passive aggregates of genes – mere packaging – and selection on a body could then be read as a convenient shorthand summary of selection for all resident genes considered individually.

But such a reductionistic view can only be valid if genes build bodies without any nonlinear or nonadditive interactions in developmental architecture. Any nonlinearity precludes the causal decomposition of a body into genes considered individually, for bodies then become, in the old adage, "more than the sum of their parts." In technical parlance, nonlinearity leads to "emergent" properties at the organismic level, and as soon as selection works upon such emergent features, then causal reduction to individual genes and their independent summations becomes logically impossible. I trust that the empirical resolution of this issue will not strike anyone as controversial, for we all understand that organisms are stuffed full of emergent properties. What else is developmental biology but the attempt to elucidate such nonlinearities? The error of gene selectionists does not lie in their stubborn assertion of pure additivity in the face of such knowledge but rather in their conceptual failure to recongnize that this noncontroversial nonlinearity destroys their theory.

Dawkins admits the apparent problem (1976, p. 40):

> But now we seem to have a paradox. If building a baby is such an intricate venture, and if every gene needs several thousands of fellow genes to complete its task, how can we reconcile this with my picture of indivisible genes, springing like immortal chamois from body to body down the ages: the free, untrammeled, and self-seeking agents of life?

Dawkins then tries to resolve his own problem by invoking a parochial Oxbridge metaphor of rowing, with the nine men (eight oarsmen and a cox) as genes, and the boat as a body. Of innumerable candidate rowers, we put together the best boat "by random shuffling of the candidates for each position" and then by running large numbers of trials until the finest combination emerges. Of course the rowers must cooperate in a joint task, but we generate no nonlinearities because localized optimality prevails, and the winning boat ends up with the best possible oarsman in each place. Dawkins then segues back to biology and asserts his view of selection as optimization piece by piece so that each locus (each seat in the boat) eventually houses the best candidate:

> Many a good gene gets into bad company, and finds itself sharing a body with a lethal gene, which kills the body off in childhood. Then the good gene is destroyed along with the rest. But this is only one body, and replicas of the same good gene live on in other bodies which lack the lethal gene.... Many [good genes] perish through other forms of ill luck, say when their body is struck by lightning. But by definition luck, good and bad, strikes at random, and a gene which is *consistently* on the losing side is not unlucky; it is a bad gene (1976, p. 41).

This caricature of selection's power led Dawkins to a later statement that violates what we now know about genetic structure.

> Loci in germ-line chromosomes are hotly contested pieces of property. The contestants are allelomorphic replicators. Most of the replicators in the world have won their places in it by defeating all available alternative alleles. The weapons with which they have won, and the weapons with which their rivals lost, are their respective phenotypic consequences (1982, p. 237).

Such notions of individualized genetic optimality are empirically false, but even if true, these claims still would not support the required argument for nonexistence of emergent organismic properties. Even Dawkins admits (in the quote just above) that selection can only optimize "phenotypic consequences." If phenotypes arise (as they do) by complex nonadditivity among genetic effects, then the genes in your body cannot match the metaphor of optimal goats, hopping happily across the generations.

10.5.2 The Ceteris Paribus Dodge

When the logic of an argument requires a certain construction of the empirical world, but the world answers no, supporters often try to maintain their advocacy by the tactic of conjectural "as if." That is, you admit nature's suboptimal embodiment of your theoretical needs but then claim that you will construe nature "as if" she operated in the required way – all the better to show that your theory works after all. The classical form of "as if" goes by its Latin title of *ceteris paribus* or "all other things being equal." *Ceteris paribus* imposes additivity upon a system made of complexly interacting parts. You isolate one factor and state that you will analyze its independent effects by holding all other factors constant.

Ceteris paribus ranks among the oldest of scholarly devices, an indispensable tactic for any student of complex systems – so I propose no general assault upon this venerable method. Two common circumstances define the legitimate domain of *ceteris paribus*: (1) as a heuristic or exploratory device for approaching systems of such complexity that you do not even know how to think about influences of particular parts unless you can hypothetically assign all others to a theoretical background of constancy, and (2) as a powerful experimental tool when you actually can hold other factors constant and perturb the system by varying your studied factor alone.

But *ceteris paribus* becomes an illegitimate dodge, a false prop to make a potentially false argument unbeatable by definition, in systems dominated by nonadditivity – that is, where the very act of holding all other factors constant will probably make your favored factor work in a manner contrary to its actual operation in a real world of interaction. If A conquers B only when the two entities share a field alone but always loses to B when C also dwells on the field, and if real fields usually house C, then what do we achieve in banning C by *ceteris paribus* and proclaiming A the better letter?

The similar use of *ceteris paribus* to support gene selectionism amounts to denial of a known reality. This tactic represents a fallback position after acknowledging the impossibility of asserting a genuine claim for nonadditivity in the translation of genes to organisms. In other words, you admit that massive nonlinearity actually exists but then state that, for purposes of discussion, you will counterfactually impose *ceteris paribus* so that genes can be named for linear effects. For example, Dawkins explicitly invokes the key phrase (in English rather than Latin) in defending his necessary (but fallacious) notion that genes are "for" particular parts of phenotypes, thus creating the possibility that organisms may be additive aggregates rather than entities defined by nonlinear interaction.

For purposes of argument it will be necessary to speculate about genes "for" doing all sorts of improbable things. If I speak, for example, of a hypothetical gene "for saving companions from drowning," and you find such a concept incredible, . . . recall that we are not talking about the gene as the sole antecedent cause of all the complex muscular contractions, sensory integrations, and even conscious decisions, which are involved in saving somebody from drowning. We are saying nothing about the question of whether learning, experience, or environmental influences enter into the development of the behavior. All you have to concede is that it is possible for a single gene, other things being equal and lots of other essential genes and environmental factors being present, to make a body more likely to save somebody from drowning than its allele would (1976, p. 66).

All major proponents of gene selectionism have provided their own unintended illustrations for why the theory cannot work by trying to "cash out" their system only to fail at the crucial point of assigning causal agency in natural selection. No matter how much they may advocate genes as primary agents or units of selection, gene selectionists cannot deny that nature's Darwinian action generally occurs between whole organisms and environments. They therefore acknowledge this basic fact and either torpedo their system or lapse into a style of verbal obscurity now conventional within the genre.

Consider, as a prime example, Williams' claim that organismic selection should not be viewed "as a level of selection in addition to that of the gene, but as the primary mechanism of selection at the genic level." What can such a statement mean? Williams recognizes organismic selection as the "primary mechanism" by which genes pass differentially to future generations. And primary mechanisms are efficient causes in our usual understanding of science. Williams (1992, p. 38) presents an accurate epitome of selection in the following statement: selection must always operate on interactors (and he knows, as the last quotation emphasizes, that organisms are usually the relevant interactors in cases that he wishes to describe as genic selection); he also recognizes that information must pass to future generations by faithful heredity, and he seems to acknowledge that such biased passage defines the result, not the cause, of selection. Yet he does not take the final logical step in acknowledging that these statements debar gene selectionism as the mechanism of evolution.

Natural selection must always act on physical entities (interactors) that vary in aptitude for reproduction, either because they differ in the machinery of reproduction or in that of survival and resource capture on which reproduction depends. It is also necessary that there be what Darwin called "the strong principle of inheritance," so that events in the material domain can influence the codical record. Offspring must tend to resemble their own parents more than those of other offspring. Whenever these conditions are found there will be natural selection.

Over the years, Dawkins has developed a virtual litany of admissions. Of course organisms are foci of selection, but since biased gene passage emerges as a result, we may identify genes as agents of selection. (But results are not causes, although foci of action surely are):

Just as whole boats win or lose races, it is indeed individuals [organisms] who live or die, and the immediate manifestation of natural selection is nearly always the individual level. But the long-term consequences of nonrandom individual death and reproductive success are manifested in the form of changing gene frequencies in the gene pool (1976, p. 48).

Dawkins then apologizes for framing his descriptions in terms of organisms as causal actors, excusing himself for succumbing to temptations of convenience. (But perhaps we find this mode "convenient" because it provides the best description of a causal reality, whereas the genic mode remains tortuous and uncomfortable because we sense the central error):

In practice it is usually convenient, as an approximation, to regard the individual body as an agent "trying" to increase the number of all its genes in future generations. I shall use the language of convenience. Unless otherwise stated, "altruistic behavior" and "selfish behavior" will mean behavior directed by one animal body towards another (1976, p. 50).... We shall continue to treat the individual as a selfish machine, programmed to do whatever is best for his genes as a whole. This is the language of convenience (1976, p. 71).

I share Wilson and Sober's rising frustration as I read through these catalogues of admission from gene selectionists. Organisms operate as primary Darwinian agents. Organisms (or interactors at other levels, but not replicators) are the causal units of natural selection in any legitimate vernacular or technical sense of mechanism in science. And everyone knows this, whatever obscuring effect the dark glasses of genic reductionism may impose upon language and thought.

10.6 Bookkeeping and Causality: The Fundamental Error of Gene Selectionism

The basic error of gene selectionism, as documented in the preceding sections, can be summarized in a single statement that illustrates the fruitful, "Pareto-like" character of the central fallacy: proponents of gene selectionism have *confused bookkeeping with causality*. The error has been fruitful because changes recorded at the genetic level do play a fundamental role in characterizing evolution, and records of these changes (bookkeeping) do have a legitimately preferred status. The ensuing discussion has therefore indentified and clarified this important aspect of evolution. But the error remains: bookkeeping is not causality; natural selection is a causal process, and units or agents of selection must be overt actors in the mechanism, not merely chosen items for tabulating results.

No one has ever stated the issue more accurately or succinctly than Williams himself (1992, p. 13), thus increasing my puzzlement at his failure to recognize that his own formulation invalidates the gene selectionism still commanding his lip service:

For natural selection to occur and be a factor in evolution, replicators must manifest themselves in interactors, the concrete realities that confront a biologist. The truth and usefulness of a biological theory must be evaluated on the basis of its success in explaining and predicting material phenomena. It is equally true that replicators (codices) are a concept of great interest and usefulness and must be considered with great care for any formal theory of evolution, either cultural or biological.

When we dissect this statement, we note total agreement with the position maintained in this chapter – an attitude that, by general agreement, leads logically and directly to the hierarchical model of selection and the invalidity of one-level, gene-based views. Williams allows that interactors are the "concrete realities" confronting biologists (and chapter 4 of his book eloquently defends the concept of legitimate interactors at several hierarchical levels of increasing genealogical inclusion). He admits that the "truth and usefulness" of a biological theory, natural selection in this case, depends on explaining material phenomena – that is, interactors operating as agents. He does not include replicators – the basis of gene selectionism – in this category, for his last sentence grants them a separate but equal status in evolutionary theory: "It is equally true that replicators (codices) are a concept of great interest" needed "for any formal theory of evolution." Well, if replicators are not causal agents but are vital for any full account of evolution, what are they? I suggest that we view gene-level replicators as basic units for keeping the books of evolutionary change – as "atoms" in the tables of recorded results.

Whereas Williams makes valid separations and defines proper roles but then seems unwilling to own the theoretical consequences, Dawkins seems merely confused. In discussing group selection (1982, p. 115), for example, Dawkins writes:

The end result of the selection discussed is a change in gene frequencies, for example, an increase of "altruistic genes" at the expense of "selfish genes." It is still the genes that are regarded as the replicators which actually survive (or fail to survive) as a consequence of the (vehicle) selection process.

By putting "vehicle" in parentheses as a reminder of selection's intrinsic nature rather than a mere modifying adjective, Dawkins admits that interactors (vehicles in his terminology), not replicators, are the agents of selection. He describes differential survival of replicators as a result of this process – therefore as units for bookkeeping rather than agents of causality – but then fails to disentangle these two different aspects of evolution and continues to grant favored status to genes.

We may indeed, and legitimately as a practical measure, choose to keep track of an organism's success in selection by counting the relative representation of its genes in future generations. (In large part, we choose the genic level in such bookkeeping for the reason always emphasized by Williams and Dawkins: because sexual organisms do not replicate faithfully and therefore cannot be traced as discrete entities across generations). But this practical decision for counting does not deprive the organism of status as a causal agent, nor can such a decision grant causality to the objects counted. The listing of accounts is bookkeeping – a vitally important subject in evolutionary biology but not a form of causality.

If I am correct in stating that gene selectionism represents a rare case in science of an influential theory felled by a logical error – in this case the confusion of bookkeeping with causality – rather than a fallacious proposal about the empirical world, then we must ask why such an error was committed so readily and initially discerned so poorly. I suspect that three major reasons buttressed the error of gene selectionism and also abetted the odd willingness and fervor of so many evolutionists to embrace the concept. The first two reasons are socially based in traditions of scientific inquiry; the third arises for an interesting reason rooted in the logical structure of hierarchies and therefore echoing the framework that should replace gene selectionism.

The two reasons rooted in traditions of scientific procedure include, first, the most general of statements and, second, a preference peculiar to traditions of Anglophonic evolutionary biology. For the generality, I state nothing profound or original in pointing out that a decision to privilege the level of genes plays into the strongest of all preferences in Western science: our traditions of reductionism, or the desire to explain all larger-scale phenomena by properties of the smallest constituent particles.

The allure of reductionism easily engenders the following kind of error: we correctly note that genes play a fundamental role in evolution (as preferred items for a calculus of change – the bookkeeping function); we also recognize that genes lie at the base of a causal cascade in building organisms; finally, and most generally, we view genes as the closest biological approach to an "atom" of basic structure and therefore as a cardinal entity in a reductionistic program. From these statements, we easily slip into the unwarranted inference that genes must be fundamental units or agents in natural selection – the primary causal theory of our profession – all the while forgetting the criteria of individuality and interaction that define units or agents within the theory's logic.

The second, and more particular, reason flows from explicit traditions of the Modern Synthesis, especially in the approach favored by Fisherian population genetics. The heuristics of this field prospered greatly with models that kept track of gene frequencies without worrying much about locus of selective action. A common fallacy in science then conflates the practical basis for success with the causal structure of nature. James Crow (1994, p. 616), one of the world's most thoughtful geneticists, expressed this point particularly well but then also failed to distinguish bookkeeping from causality. Writing in his article "In Praise of Replicators" – and well should we praise them as excellent agents for accounting – Crow explained why our traditions have favored the genetic level of analysis (1994, p. 616):

The reason, I think, is that these pioneers [Fisher and other founders of the Synthesis] and their intellectual heirs have been concerned, not with selection as an end in itself, but with selection as a way of changing gene frequencies. Selection acts in many ways: it can be stabilizing; it can be diversifying; it can be directional; it can be between organelles; it can be between individuals; it can be between groups.... But the bottom line has always been how much selection changes allele frequencies and through these, how much it changes phenotypes. This suggests that we should judge the effectiveness of selection at different levels by its effects on gene frequencies.

I could not ask for a better statement of (unconscious) support for the position here maintained. Again, as noted in several other quotations from gene selectionists, Crow allows that selection, as a causal force ("selection as an end in itself," in his words), operates on interactors at several hierarchical levels of individuality, including groups. He also admits that change in gene frequencies arises as a result of selection and cannot be construed as a causal agent or unit of selection. He then states – and again I do not object – that changing allelic frequencies should be viewed as a "bottom line" in judgments about selection's effect. Nicely said, but a bottom line for what? Crow then gives his crucial answer – for keeping the books of evolutionary change: "we should judge the effectiveness of selection at different levels by its effects on gene frequencies." Thus, changing gene frequencies are results (for bookkeeping), whereas selection (the cause of the changes) operates upon interactors "at different levels" of individuality.

This concept of a "bottom line" also provides an entrée to the third reason for choosing genes as units of bookkeeping: the intrinsically asymmetrical nature of causal flow in hierarchies of inclusion. Thus, the hierarchical view serves as a replacement for gene selectionism and provides the rationale for why many biologists chose – albeit for wrong reasons – to focus on genes in the first place.

We do need to keep the books of evolutionary change, and bookkeeping does require a basic unit of accounting. Candidates for this status must pass the primary criterion always stressed by gene selectionists: faithful replication. But genes are not the only faithful replicators in nature. Asexual organisms and species are also sufficiently faithful. Reductionistic preferences in general, and the relatively greater faithfulness of genetic versus higher-level replication, may set a preference for genes, but another crucial argument, however generally unrecognized or unmentioned, seals the case.

Because bookkeeping cannot be equated with causality and we are not, in establishing a best method for keeping accounts, trying to ascertain the causes for differential success, we want to make sure, above all, that we choose the unit best suited to record *all* evolutionary changes, whatever their causal basis. No single unit of bookkeeping can monitor every conceivable change, but the gene must be our unit of choice because the nature of hierarchies logically implies that genes will provide the most comprehensive record of changes at all levels. (Even so, gene records will miss certain kinds of changes that we generally call evolutionary. For example, as Wilson and Sober [1994] point out, assortative mating of organisms within a population may greatly increase the ratio of homozygotes to heterozygotes at many loci but need not change gene frequencies in the population.)

Hierarchies are generally not fractal, and various levels translate a common set of causes to strikingly different results and frequencies. Moreover, hierarchies are often directional and therefore not indifferent to the nature of the flow of influence. As the most important of all such asymmetries, change at a low level may or may not produce an effect at higher levels – "upward causation" in the standard terminology (see Campbell 1974, Vrba 1989). But change at a higher level must always sort the included units of all lower levels – "downward causation."

If a gene increases in copy number within a genome by duplication and lateral spread (positive gene selection in the genuine sense), phenotypes of organisms may or may not be affected. But selection on higher-level individuals *always* sorts the lower-level individuals included within. If ugly creatures outcompete beautiful conspecifics in organismal selection, then genes for ugliness increase in the population. If stenotypic species prevail over eurytopes in species selection, then genes associated with stenotypy increase within the lineage. If species of polychaetes eliminate species of priapulids in competition over geological time, then polychaete genes increase in the marine biota.

Given this intrinsic asymmetry, what would a good bookkeeper choose for a basic unit? Obviously not the organism, or any high-level individual, because the books will then fail to record many changes at lower levels, and a good bookkeeper wants, as the chief desideratum of his profession, to note all changes. As stated in the previous paragraph, low-level selection need not impose any effect upon higher levels at all. Equally obviously, our optimal bookkeeper will choose genes – not because genes are intrinsically more basic (the reductionist fallacy), not because genes are primary causal agents or causal agents at all (the gene selectionist fallacy), and not because genes replicate faithfully (for other kinds of individuals do so as well) – but, rather, because genes, as the lowest-level individuals in a hierarchy, record the maximum number of evolutionary changes. Thus, the intrinsic nature of hierarchies sets our most important preference for genes as units of bookkeeping, for only genes show near ubiquity in recording changes.

Finally, we must note one other property that, although strongly favoring genes as units of bookkeeping, shows even more clearly why genes cannot be exclusive units of selection, or causal agents. Bookkeepers are objectivists, not sleuths or storytellers. A good bookkeeper wants an unimpeachable record, not a causal hypothesis (that can always be wrong). Books kept in terms of gene frequencies provide the best objective records of "descent with modification" because they do *not* make causal attributions but only count changes ("just the facts, ma'am"). The hierarchical nature of evolutionary mechanics, and the simultaneous action of selection on individuals at several levels, dictates our inability to know the causal basis of change from records of altered gene frequencies alone.

10.7 Gambits of Reform and Retreat by Gene Selectionists

As I have emphasized throughout this discussion, gene selectionism cannot be made coherent as a general philosophy. Yet the allure of the gene remains powerful largely for reasons of general preference in our culture and not for any observed power or intrinsic biological status possessed by evolutionary individuals of this lowest level. When a noncoherent argument remains intriguing and supporters cannot bear the wrench of total abandonment, the argument must then be plastered with compromises and "howevers" or so changed in form that only lip service remains to cover a truly altered substance. Often, given human tendencies to paint a bright face on adversity, gene selectionists have made their necessary retreats but presented them as refinements or elaborations of the original theory. In this closing section, I will show that the two most prominent "revisions" of gene selectionism – Dawkins' extended phenotype (1982) and Williams' codical selection (1992) – represent defeats rather than improvements as advertised.

10.7.1 Dawkins on the "Extended Phenotype"

I have always admired the clarity of Senator Aikens' brilliant solution to the morass of our involvement in the Vietnamese War. At the height of our reverses and misfortunes, he advised that we should simply declare victory and get out. Richard Dawkins got in with his 1976 book *The Selfish Gene.* He declared victory with *The Extended Phenotype* in 1982, but really, at least with respect to the requirement and logic of his original argument, got out.

With admirable clarity, and no ambivalence, Dawkins proclaimed the doctrine of exclusive gene selectionism in 1976:

> I must argue for my belief that the best way to look at evolution is in terms of selection occurring at the lowest level of all . . . . I shall argue that the fundamental unit of selection, and therefore of self-interest, is not the species, nor the group, nor even, strictly, the individual. It is the gene, the unit of heredity (1976, p. 12).

Dawkins presented his later work, *The Extended Phenotype,* as an extension and elaboration of gene selectionism: "This book," he wrote, "is in some ways the sequel to my previous book, *The Selfish Gene*" (1982, p. v). Dawkins had admitted, in his 1976 book, that genes work through phenotypes of the "lumbering robots" (organisms) serving as their passive homes. But if genes are nature's real actors and phenotypes only their means of expression, then why limit phenotypes to bodies? Any consequence of a gene should be equally capable of carrying the gene's interest in a process of selection. Dawkins admitted, of course, that most aspects of this extended phenotype – the footprint of a shorebird in the sand, for example – will be too trivial or ephemeral to be effective in the gene's interest. But other parts of the extended phenotype (with the beaver's dam as Dawkins' favorite example) do contribute to the success of beaver genes and should be included within "the extended phenotype" that the gene – the only and ultimate unit of selection (at least in 1976) – manipulates in its struggle for replicative success.

Dawkins (1982, pp. iv–vii) therefore insists that the viewpoint of *The Extended Phenotype* evolved gradually and progressively from *The Selfish Gene*:

This belief – that if adaptations are to be treated as "for the good of" something, that something is the gene – was the fundamental assumption of my previous book. The present book goes further. To dramatize it a bit, it attempts to free the selfish gene from the individual organism which has been its conceptional prison. The phenotypic effects of a gene are the tools by which it levers itself into the next generation, and these tools may "extend" far outside the body in which the gene sits, even reaching deep into the nervous system of other organisms.

Thus, genes have become even more fundamental and bodies even less consequential: "Fundamentally, what it going on is that replicating molecule ensure their survival by means of phenotypic effects on the world. It is only incidentally true that these phenotypic effects happen to be packaged up into units called individual organisms" (pp. 4–5).

But Dawkins' argument soon begins to unravel. Just when the gene seems poised to swallow the organism entirely as just one incidental aspect of the gene's armamentarium (the fully extended phenotype), Dawkins turns around and tells us that we may treat organisms as focal entitites after all, for we may describe evolution from the organism's point of view just as well:

I am not saying that the selfish organism view is necessarily wrong, but my argument, in its strong form, is that it is looking at the matter the wrong way up. . . . I am pretty confident that to look at life in terms of genetic replicators preserving themselves by means of their extended phenotypes is at least as satisfactory as to look at it in terms of selfish organisms maximizing their inclusive fitness (1982, pp. 6–7).

In other words, Dawkins claims to prefer genes and to find greater insight in this formulation. But he allows that you or I might prefer to stick with organisms – and it really does not matter. Genes and organisms may be viewed, metaphorically, as the two possible resolutions (differing orientations) of the famous optical illusion known as the Necker Cube:

After a few more seconds the mental image flips back and it continues to alternate as long as we look at the picture. The point is that neither of the two perceptions of the cube is the correct or "true" one. They are equally correct. Similarly the vision of life that I advocate, and label with the name of the extended phenotype, is not probably more correct than the orthodox view. It is a different view and I suspect that, at least in some respects, it provides a deeper understanding. But I doubt that there is any experiment that could be done to prove my claim (1982, p. 1).

This perspective of legitimate alternatives pervades the entire book, as in this late passage (1982, p. 232): "The whole story could have been told in . . . the language of individual manipulation. The language of extended genetics is not demonstrably more correct. It is a different way of saying the same thing. The

Necker Cube has flipped. Readers must decide for themselves whether they like the new view better than the old."

This Necker Cube approach has a name in philosophy: *conventionalism*, or the idea that frameworks of explanation cannot be judged as true or false or even as more or less empirically adequate, but only as equally valid in principle and more or less preferable on such nonfactual criteria as depth of insight acquired or satisfaction in understanding gained. Conventionalism often provides an interesting perspective, particularly for some scientific debates that seem especially refractory to empirical resolution and for teaching people that ideas and attitudes influence science and that "naive realism," with theory arising only from observation, represents a silly and bankrupt philosophy of science.

But conventionalism cannot apply to this case. Dawkins has misconstrued his categories in judging gene-based and organism-based viewpoints as alternative versions of the same kind of explanation. The gene-based view works for bookkeeping, whereas the organism-based view represents the standard Darwinian argument for a favored level of causality. In this sense, both are valid, but they are not comparable, and they are not alternatives on an identical playing field of common explanatory intent.

Moreover, Dawkins' shift from the selfish gene to the extended phenotype cannot be defended as a simple extension or elaboration of a consistent and developing viewpoint. Dawkins may try to save face with such a portrayal, but this strategy cannot work. The conventionalism of *The Extended Phenotype* logically entails the negation or denial of gene selectionism as an empirical reality, as presented in *The Selfish Gene*. Dawkins' first book argues in no uncertain terms (see quotation on p. 228) that genes should be viewed as exclusive units of selection (or causal agents) and that bodies, as passive lumbering robots, cannot play such a role. The second book argues that we can view evolution equally well from either the gene's or the organism's point of view, that Dawkins still prefers genes, but that others remain free to favor bodies with just as much claim to empirical adequacy. These two formulations cannot be reconciled as alternative versions of the same attitude, and one view is not a subtler extension of the other. The two formulations represent logically contrasting and mutually exclusive accounts of causality in evolution. I do not happen to regard either as correct, but, in any case, Dawkins' later concept of the extended phenotype controverts his earlier defense of gene selectionism as nature's way.

10.7.2 Williams's Codical Hierarchy

Williams's epochal book of 1966 set the intellectual basis for gene selectionism and may justly be viewed as the founding document for this ultimate version of Darwinian reductionism. But by 1992, Williams had realized that interactors, and not replicators, must be construed as units of selection or causal agents in the usual sense of the term and that hierarchy must hold because no single level of interaction can be designated as exclusive or even fundamental.

Williams, however, did not wish to abandon his former apparatus for viewing genes as fundamental and preferred units of selection. But *que faire*? Genes are replicators in their universal role, and replicators are not units of selection.

Williams therefore pursued an interesting gambit. He admitted that interactors form a hierarchy of evolutionary individuals at several levels (gene, organism, deme, species, clade) and that interactors are units of selection in our usual sense of material entities participating in a causal process. These interactors form a material hierarchy, and gene selectionism cannot apply to this legitimate domain. So Williams then established a separate and parallel hierarchy for abstract units of information (as opposed to material entities), and he construed genes as basic "units of selection" in this alternative and parallel domain, which he called codical (the adjectival form of codices, the plural of codex, his term for a single unit of information). If genes cannot claim exclusivity (or even causal status at all) as units of selection in the usual domain of material objects, then Williams would establish a new and separate hierarchy for nonmaterial units of information – and here the gene could continue to reign.

Williams therefore proposed a fundamental distinction between entities and information, speaking of "two mutually exclusive domains of selection, one that deals with material entities and another that deals with information and might be termed the codical domain" (1992, p. 10). But this codical domain has neither meaning nor existence as a locus for causal units of selection for two reasons.

1. Odd mapping upon legitimate intuitions. Williams continues his allegiance to the logical stumbling block of gene selectionism, the false criterion that has always doomed the theory: faithful replication as the defining property of a "unit of selection" – though now only a unit in the newly formulated codical domain, for Williams has finally admitted that replicators are not causal agents in the usual realm of material entities. Williams now promotes faithful replication as a primary criterion for "unithood" in his codical domain, thus leading to the following peculiar position: *genes* are units of selection (as a replicating consequence in the codical domain of selection upon organisms in the material domain); *gene pools* are also units of selection (as replicating consequences of higher-level selection upon higher-leval entities from demes to clades); but *genotypes*, in an intermediate category, are not units of selection (except in asexual organisms, where replication can be faithful). Thus the codical domain skips a space in the hierarchy and includes no organismic level of selection (except for asexual creatures) because the corresponding codex is impersistent.

2. The old error of confusing bookkeeping with causality. Williams' complex move in devising a separate hierarchy for nonmaterial units of information – and then juxtaposing this new sequence against the conventional hierarchy of evidently material and admittedly causal units – represents a last-ditch effort to rescue the inviable theory of gene selectionism by granting both primacy

and causal status (but only linguistically) to genes as replicators. But nothing new has been added beyond a complex terminology. The old error remains in full force, perhaps even heightened by the counterintuitive complexities and mental manipulations required by the scheme of dual hierarchies.

A parallel hierarchy for nonmaterial entities of information? What can such a concept mean? Pull the codical clothing off the supposed emperor, and whom do we find lurking underneath?: Our old friend, the bookkeeper. Why does he continually try to enter the field of material objects engaged in nature's grand game of causality? Bookkeeping is equally necessary and entirely honorable. The results of causal processes must be tabulated, and we rightly treasure the lists. We continue to stand in awe before "60" in Babe Ruth's home run column for 1927 (but less so after McGwire's 70 in 1998). But 60 is a record, not a cause – a summary of a great achievement, not the bat itself, or the muscles in a pair of strong arms. As nonmaterial objects available for recording, codices are units of bookkeeping.

The run of gene selectionism has been a grand intellectual adventure – from inception as a manifesto (Williams 1966), through numerous excursions into pop culture, to valiant attempts to work through the logical barriers and develop a consistent and workable theory (Dawkins 1982, Williams 1992). "Pareto-errors" always instigate a good race. No one really loses – though false theories like gene selectionism must eventually yield – because the resulting clarifications can only strengthen a field, and interestingly fallacious ideas inspire important insights. Without this debate, we might never have properly clarified the differing roles of replicators and interactors, items for bookkeeping, and units of selection. And we might not have developed a coherent (and probably valid) theory of hierarchical selection without the stimulus of an opposite claim that genes could function as exclusive causal agents.

Some evolutionary biologists, largely (I suspect) in fealty to their own pasts, continue to use the language of gene selectionism, even while their revised accounts elucidate the hierarchical view (see, in particular, Williams' excellent fourth chapter in his 1992 book on selection upon multiple interactors at several levels). Williams, to use a locution of our times, may still be talking the talk of gene selectionism, but he is no longer walking the walk.

Nearly all major participants in this discussion met at Ohio State University in the summer of 1988. There I witnessed a telling vignette that may serve as an epitome for this article. George Williams presented his new views (the substance of his 1992 book) and surprised many people with his conceptual move toward hierarchy (within his unaltered terminology). I could not imagine two more different personalities in the brief interchange that followed. Marjorie Greene – a great student of Aristotle, *grande dame* of philosophy, one of the feistiest and toughest people I have ever known, and a supporter of the hierarchical view – looked at Williams and said: "You've changed a lot." George Williams, one of the calmest and most laconic of men, replied: "It's been a long time."

10.8 Acknowledgment

I first encountered Dick Lewontin by reputation only as the added third author to the revised edition of Simpson and Roe's *Quantitative Zoology*. As a parochial paleontologist, I knew nothing about him beyond this authorship of a book that I greatly admired (and used on an almost daily basis). I first met Dick in an airport lounge as many participants changed planes en route to a conference in California. As a baby first-year assistant professor, I approached him timidly (for he seemed so wise and senior, being more than a full decade older than me). We chatted for a moment, and he then looked at me and said, "Do you think I can get away with talking about Aristotle's distinction between efficient and final causality to a roomful of professional biologists?" Somehow he knew that I had familiarity with humanistic traditions and philosophical literature, and he was bringing me, the most timid of neophytes, into his orbit of unusual competence. The incident seems so tiny, and I am sure that Dick will not remember it. But little affirmations make a world of difference during the first tentative footsteps of a career.

Dick then gave the most articulate and brilliant talk that I had ever heard, and he did include an analysis of Aristotle's views on causality! I have been in awe of his intellect ever since. Among people I have met, I might put a very few others into the same category – Noam Chomsky and Isaiah Berlin come to mind – but I have never met a more brilliant person. Now, 30 years later – with long and close collegiality, including a joint course on evolution and those utterly notorious spandrels of San Marco – I continue to benefit from his generosity and his uncompromising intellectual integrity. I am sometimes sad when I see him downhearted and discouraged about the social realities of science as an institutional enterprise in a commercial world, but I love to see his unfailing brightening whenever an issue of true intellectual substance arises. He remains, as ever, an inspiration to me and to our entire profession.

This chapter, which I dedicate to him (for so much of its content arises from our joint discussions over many years), is an abridged and edited version of part of a chapter from my forthcoming book, *The Structure of Evolutionary Theory*, to be published by Harvard University Press in the auspicious year of 2001.

REFERENCES

Buss, L. W. (1987). *The Evolution of Individuality*. Princeton, NJ: Princeton University Press.

Campbell, D. T. (1974). "Downward causation" in hierarchically organized biological systems. In *Studies in the Philosophy of Biology*, eds. F. J. Ayala and T. Dobzhansky. Macmillan.

Cronin, H. (1991). *The Ant and the Peacock: Altruism and Sexual Selection from Darwin to Today*. Cambridge, UK: Cambridge University Press.

Crow, J. F. (1994). In praise of replicators. *Behavioral and Brain Sciences* 17:616.

Dawkins, R. (1976). *The Selfish Gene*. Oxford, UK: Oxford University Press.

Dawkins, R. (1978). Replicator selection and the extended phenotype. *Zeitschrift für Tierpsychologie* 47:61–76.

Dawkins, R. (1982). *The Extended Phenotype: The Gene as the Unit of Selection.* Oxford, UK: Oxford University Press.

Dennett, D. C. (1995). *Darwin's Dangerous Idea.* New York: Simon and Schuster.

Eldredge, N. (1995). *Reinventing Darwin.* New York: John Wiley & Sons.

Gould, S. J. (1981). *The Mismeasure of Man.* New York: W.W. Norton.

Gould, S. J. (1987). *Time's Arrow, Time's Cycle.* Cambridge, MA: Harvard University Press.

Gould, S. J. (1994). Tempo and mode in the macroevolutionary reconstruction of Darwinism. *Proceedings of the National Academy of Sciences USA* 91:6764–71.

Gould, S. J. (1996). *Full House: The Spread of Excellence from Plato to Darwin.* New York: Harmony Books.

Gould, S. J. (1997). Darwinian fundamentalism. *New York Review of Books* 40(10):34–7.

Gould, S. J., and Vrba, E. S. (1982). Exaptation – a missing term in the science of form. *Paleobiology* 8:4–15.

Hamilton, W. D. (1964). The genetical evolution of social behavior, I and II. *Journal of Theoretical Biology* 7:1–52.

Hull, D. L. (1980). Individuality and selection. *Annual Review of Ecology and Systematics* 11:311–32.

Hull, D. L. (1994). Taking vehicles seriously. *Behavioral and Brain Sciences* 17:627–8.

Lewin, R. (1996). Evolution's new heretics. *Natural History* 105(5):12–17.

Vrba, E. S. (1989). Levels of selection and sorting. *Oxford Surveys in Evolutionary Biology* 6:111–168.

Williams, G. C. (1966). *Adaptation and Natural Selection: A Critique of Some Current Evolutionary Thought.* Princeton: Princeton University Press.

Williams, G. C. (1992). *Natural Selection: Domains, Levels and Challenges.* Oxford, UK: Oxford University Press.

Wilson, D. S. (1980). *The Natural Selection of Populations and Communities.* Benjamin Cummings.

Wilson, D. S. (1983). The group selection controversy: History and current status. *Annual Review of Ecology and Systematics* 14:159–87.

Wilson, D. S., and Sober, E. (1994). Reintroducing group selection to the human behavioral sciences. *Behavioral and Brain Sciences* 17:585–654.

Wright, R. (1994). *The Moral Animal.* New York: Vintage Books.

CHAPTER ELEVEN

Reductionism in Genetics and the Human Genome Project

SAHOTRA SARKAR

11.1 Introduction

In 1886, Ludwig Boltzmann, one of the nineteenth century's most distinguished physicists, somewhat surprisingly announced, "if you ask me for my innermost conviction whether [this century] will one day be called the century of iron, or steam, or electricity, I answer without qualms that it will be named the century of the mechanical view of nature, of Darwin" (1974, p. 15). In a similar vein, if one wonders about biology, if not all of science, it is at least arguable that the twentieth century belongs to Mendel. Though Mendel died in 1884 and his famous paper on hybridization and inheritance had been published in 1866, his insights entered science, medicine, and the public realm only after the recovery of his work around 1900. Since then, they have largely dominated biology.

Mendel posited definite heritable "factors" that were responsible for the inherited "traits" of his experimental organisms. From his work there emerged two rules about the transmission of these factors. These rules provided the basis for the new discipline of genetics: (1) each parent organism has a pair of factors for a trait, exactly one member of a parental pair is transmitted to each offspring, and each such member has equal probability of transmission; and (2) factors responsible for different traits are transmitted independently of each other. Both rules are statistical. The first rule came to be called the "law of segregation of alleles" ("alleles" being Mendel's "factors"). By 1910 it became clear that the second rule, dubbed the "law of independent assortment," was routinely violated. Some factors tended to get inherited together. This phenomenon was eventually called "linkage."

Between 1900 and 1932 it became clear that Mendel's rules, as modified by linkage, provided the missing component, a theory of inheritance, in the framework of the theory of evolution.[1] The origins of many traits could be explained from the assumption that they were the result of specific genes. From about 1900, it was also known that the inheritance of traits is mediated by chromosomes. Morgan and his students discovered that, for a great variety of cases, inheritance and phenogenesis could be decoupled to such an extent that the latter could be relegated to a "black box" during the pursuit of genetics. This permitted the effective use of linkage analysis not only to associate alleles with individual loci on chromosomes but, eventually, to place the loci on definite (physical) positions on those chromosomes. The Morgan school worked with *Drosophila.* By the 1930s, thanks to Bernstein, Fisher, Haldane, Hogben, and others, statistical techniques for human linkage analysis were constructed (Sarkar 1996c). In 1936 Haldane (1936) proposed the first "provisional map of a human chromosome." By the 1940s genetics had become the dominant subdiscipline of biology. It has retained that status ever since.

From the very beginning, the way in which genes are physically specified was a question of obvious interest. Chromosomes consist of proteins and DNA. Until the 1940s proteins were generally believed to be the genetic material. It then became clear that DNA played that role. In 1953 Watson and Crick proposed a double-helix model for DNA. Meanwhile, Pauling initiated a remarkable transformation of biological reasoning whereby the detailed structure of biological macromolecules, especially proteins, was used to explain and predict their behavior. This pattern of reasoning became central to the emerging discipline of "molecular biology" in the late 1950s. In the early 1960s, such explanations were produced with unprecedented success for the regulation of gene expression (in prokaryotes), protein interactions, antibody specificity, and many other phenomena.

Meanwhile, for DNA, it became clear that the sequence of nucleotide bases specified the linear chain of amino acid residues in a protein. This specification came to be viewed as a "coding" relation. That code was deciphered in the 1960s. Genes were identified with segments of DNA. There were two kinds of genes: "structural" genes specifying other macromolecules and "regulatory" genes involved in gene regulation. In the 1960s it was believed that all DNA had one of these two roles. But, that far, prokaryotes had dominated the experimental systems of molecular biology. As eukaryotes began to be systematically studied in the 1970s, this simple picture disappeared. Most eukaryotic DNA had neither of these two roles: it appeared to have no function at all. Even the DNA specifying a single polypeptide chain was not necessarily contiguous. Some DNA specified more than one protein. Ribonucleic acid (RNA), the intermediate molecules between DNA and protein, was routinely modified before protein formation.[2] By the 1980s, it also became clear that the extension of the early successes of molecular explanations to the cellular and higher levels was nontrivial. In particular, it had proved impossible to extend models for the regulation of bacterial genomes to eukaryotes. Partly as a consequence of this, molecular biology failed to provide accounts of pattern formation during development, that is, phenogenesis. The direction in which molecular biology should go was no longer obvious.

In this context, several molecular biologists independently proposed to sequence the entire human genome *blindly*, that is, with no explicit concern for the role of the sequence, and to do so within 15 years as a coordinated project (the Human Genome Project, HGP).[3] This, as they explicitly noted, was a shift away from the traditional way of doing molecular biology because of both the blind sequencing and the plan for coordination. The rationale was that once the sequence was obtained, a new theoretical biology would emerge (Gilbert 1992). Moreover, hereditary (presumably genetic) diseases would also begin to be brought under control. The proposal was met with considerable opposition from both within the biological community and without. The critics did not deny that genomic sequences were of some biological interest or that, eventually, human sequences would be obtained and whatever worries one might have about their interpretation or use would eventually have to be faced. What they questioned was (1) the scientific value of *blind* sequencing and (2) the rationale for sequencing so *rapidly* – perhaps at the expense of pursuing other research programs in biology.[4] Nevertheless, proponents of the HGP were able to convince funding agencies in the United States, Europe, and Japan to sponsor the project.

Three types of objection to the HGP have been raised: (1) scientific and methodological doubts about its value, (2) concerns about how it would siphon off scarce funds from other areas of biology and otherwise affect the ethos of biological research,[5] and (3) concerns about the ethical, legal, political, and social issues raised by the project. This chapter will develop just one methodological objection that has gone surprisingly unnoticed: the undeniable explanatory success of molecular biology is not necessarily a success of genetics. The former vindicates physical reductionism: the thesis that biological processes will succumb to physical explanation. The latter would be a claim of genetic reductionism: that genes best explain biological processes in general. The record of genetic reductionism is relatively dismal. The HGP assumes genetic reductionism. Moreover, to initiate the HGP, its proponents implicitly used the success of physical reductionism to suggest the viability of genetic reductionism. In Section 11.2, the ways in which concepts such as reduction and reductionism are used here will be explained. In Section 11.3, three types of reductionism in genetics, including the two just mentioned, will be distinguished. Then, Section 11.4 will take up the HGP.

11.2 Reduction and Reductionism

The terms *reduction* and *reductionism* are construed in so many ways, especially by philosophers, that it is critically important to specify exactly how they are used here.[6] Reduction is a type of explanation. Reductionism is the thesis that reductionist explanations are the ones to be pursued in a given context; that is, it is a methodological claim that these explanations will answer all pertinent questions. Whether reductionism is tenable in a given research domain is an *empirical* question: its answer is determined by the success of attempts at reduction. Nothing is assumed here about the formal structure of explanation that has been debated ad nauseam by analytic philosophers of science with little resultant insight. The question of reduction is thus separated from that of general explanation in order to focus on the former and to side-step the disputes about the latter. At stake here are the additional criteria an explanation must satisfy to qualify as a reduction.

Nevertheless, four minimal substantive (rather than formal) assumptions about explanation are necessary. These are as follows:

1. Explanation begins with a *representation* of a system. The distinction between a system and a representation is important: what, informally, is the "same system" may have radically different representations for which different rules hold. A cell can be represented as a chemical system or as a cybernetic one. A chromosome can be represented as a group of linked loci (as, for example, it is during linkage analysis) or as a physical object.
2. What is explained is some *feature* of a system as *represented*, some law that it (strictly, the representation) obeys, or some property it displays, and so forth.
3. Given a representation, an explanation involves a process of scientific reasoning that constitutes a *derivation*. No assumption is being made about the structure of derivations.
4. Any explanation invokes a set of factors that bear the *explanatory* weight in the sense that they provide whatever force or insight the explanation has. This set of factors is context-dependent; that is, the context determines which set of factors are the relevant ones and, therefore, bear the explanatory weight. (It follows from this that explanations are themselves context dependent.)

Given these assumptions, three criteria: (1) fundamentalism, (2) abstract hierarchy, and (3) spatial hierarchy are used to characterize different kinds of reduction. These criteria are fully presented in Table 11.1. The discussion here will be restricted to a few comments.

Criterion (1) requires the existence of a different realm from that of the system as initially represented. Unless there is some such difference between where the explanans and the explananda come from, there is little sense in calling an explanation a reduction. Because the explanatory factors come from the new realm, that realm is epistemologically more fundamental (hence, **F**-realm). Its rules are also, in that sense, more fundamental (hence, **F**-rules). This criterion must always be (at least approximately) satisfied in a reduction. All explanations invoke something more epistemologically fundamental. The significance of this criterion is that the **F**-rules come from a different realm. Some reductions only satisfy this criterion and neither of the others. It will be shown in the next section that attempts at genetic explanation based on heritability analysis are reductions of this sort.

However, most types of reduction, including such well-known examples as those in the kinetic theory of matter, also involve an assumption of hierarchical organization in which the explanatory weight is borne successively by entities at progressively lower levels of the hierarchy. Now this hierarchy may or may not be the usual hierarchy in physical space. Criterion (2) requires the existence of a hierarchical organization without any commitment to what rule is used to construct such a hierarchy. Some very interesting reductions only satisfy this criterion (i.e., without satisfying criterion (3)) along with criterion (1). It will be

Table 11.1. *Criteria for Reduction*

Criteria	Explanation
(i) Fundamentalism	The explanation of a feature of a system invokes *only* factors from a different and more fundamental realm (from that of the system as represented). These factors are the ones that bear the explanatory weight. This new realm is the "**F**-realm." The feature to be explained can be derived entirely from the rules of the **F**-realm (the "**F**-rules") and from customary mathematical and statistical procedures (including those of approximation).
(ii) Abstract Hierarchy	The representation of the system has an explicit hierarchical organization with the hierarchy constructed according to some independent criterion (that is, independent of the particular putative explanation). The explanatory factors for some property of an entity at a given level of the hierarchy *only* refer to properties of entities at lower levels. The relevant lower levels constitute the **F**-realm, and these properties define the **F**-rules.
(iii) Spatial Hierarchy	The hierarchical structure is a hierarchy in physical space, that is, entities at lower levels are spatial parts of entities at higher levels. Thus, the independent criterion for hierarchy construction is simply spatial containment.

shown in the next section that the explanations in classical genetics constitute reductions of this sort.

Finally, criterion (3) requires that the hierarchy invoked in criterion (2) be a hierarchy in physical space. (Therefore, criterion (3) cannot be satisfied without the satisfaction of criterion (2).) The type of reduction that has historically received the most attention, which includes those offered by the kinetic theory of matter, requires the satisfaction of all three criteria. In the next section it will be pointed out that this is the type of reduction that has had remarkable success in molecular biology. The **F**-rules are those of macromolecular physics. Thus, the three criteria for reduction introduced above allow distinguishing between at least three types of reduction: those that only satisfy (1); those that satisfy both (1) and (2) but not (3); and those that satisfy all three. Therefore, there can also be three corresponding types of reductionism, each demanding that its type of reduction be pursued.

11.3 Reductionism in Genetics

The criteria introduced in the last section allow distinguishing between three different types of reductionism in genetics. Consider, first, attempts to explain phenotypic variation such as IQ differences on the basis of their "broad heritability," that is, the ratio of the trait's variance that can be attributed to all genetic factors to its total phenotypic variance in a population exposed to a definite range of environmental variation.[7] In this type of explanation a phenotypic feature is considered to be explained by genes if its broad heritability is high and is not so explainable if it is low. This is the only **F**-rule, and the **F**-realm is that of the genes. No assumption is made about what genes are or how they are organized; the representation of the genome has no structure whatsoever. Thus, criterion (1) of Table 11.1 is satisfied, but neither of the other two is invoked. Thus, these putative explanations are reductions of the first of the three types distinguished at the end of the last section.

The trouble with any such explanation, as has been pointed out many times (see Moran 1973, Layzer 1974, Lewontin 1974, Feldman and Lewontin 1975, and Kempthorne 1978), is that broad heritability simply does not perform the task assigned to it. The basic reason is that broad heritability is a property of a population in a given range of environments and can change with the genetic composition of the population and with the introduction of environments beyond the original range. It is not, in *any* sense, a property of an individual. To claim that a high broad heritability can be used to explain from a genetic basis such individual "traits" as intelligence, religiosity, vocational interests, openness, agreeableness, conscientiousness, neuroticism, and extroversion, as some have done (see Bouchard et al. 1990), is epistemologically and ipso facto scientifically irresponsible. That is reason enough to exclude allegedly genetic explanations using heritability analysis from any further consideration. Moreover, the focus of this chapter is on how reductionism in genetics impinges on the HGP and, to their credit, few of the major proponents of the latter have advocated heritability analysis.[8] The rest of this chapter will ignore heritability analysis and this type of reduction.

If heritability analysis exhausted the explanatory repertoire available to it, genetics would be in a sorry state. However, starting with Mendel, the transmission rules for genes (alleles and loci) have generated powerful techniques that allow the explanation of phenogenesis on the basis of inherited alleles. Classical genetics used two such techniques: (1) segregation and (2) linkage analysis. In recent years, three more have been introduced: (3) the allele sharing method, (4) allelic association studies, and (5) quantitative trait loci (QTL) mapping (Lander and Schork 1994). All of these methods are briefly described in Table 11.2. Their use and their role in genetic reductions will be explained here.

Segregation analysis is based on Mendel's law of segregation of alleles. One assumes that a trait is controlled at (usually) one or two loci, each with a definite number of alleles, and a specified pattern of dominance.[9] From this model, the distribution of the trait across a pedigree is predicted. By means of increasingly sophisticated statistical techniques, it is then judged whether the predicted distribution fits observations in actual pedigrees. If the fit is good, then the phenogenesis of the trait is attributed to the alleles at the loci in question. *Linkage analysis* is based on the failure of Mendel's law of independent assortment

Table 11.2. *Methods of Genetic Analysis and Reduction*

Method	Description	**F**-Rules
(i) Segregation Analysis	Inspect a pedigree to test whether observed distribution of a trait agrees with prediction from Mendel's first law.	Law of segregation of alleles; genes produce phenotypes.
(ii) Linkage Analysis	Choose a known locus. Test whether an inherited trait assorts independently.	Violation of law of independent assortment; genes produce phenotypes.
(iii) Allele-Sharing Method	Choose sets of two relatives with the same trait. Test whether they have the same alleles with greater frequency than expected from Mendel's first law.	Violation of law of segregation; genes produce phenotypes.
(iv) Allelic Association Studies	Test for a statistically significant correlation between a trait and an allele.	Genes produce phenotypes.
(v) Quantitative Trait Locus Mapping	Cross two pure lines **A** and **B**. Then backcross offspring with one of them, say **B**. The effects of new alleles introduced by recombination will be seen in the next generation by comparison with the pure **B** line.	Law of independent assortment; genes produce phenotypes.

due to linkage. Starting with a trait attributable to a given locus, one tests whether the inheritance of another is independent of it. If not – and this again is a statistical question – the second trait is presumably attributable to a locus linked to the first. Usually one also estimates the degree of linkage: the higher it is, the more reliable the inference to a genetic explanation.

Both of these methods have led to the genetic explanation of many traits. In humans this includes biochemical conditions such as PKU and morphological variations such as polydactyly. Such explanations assume that the genome has a hierarchical organization: it consists of a group of linkage sets, each linkage set consists of (linked) loci, and there are two alleles at each locus. This is the "genetic representation" of the genome. The **F**-realm is that of genetics. The **F**-rules are Mendel's two laws and a more nebulous assumption that genes (specific alleles at definite loci) are responsible for phenogenesis. This last rule makes the genetic realm more fundamental than the phenotypic, and the first two criteria of Table 11.1 are satisfied. More important, the third criterion is not satisfied. The genetic hierarchy, as a matter of fact, corresponds *approximately* to a physical hierarchy in the chromosome set, but this fact is irrelevant in genetic explanations. Thus, the genetic representation is generally consistent with the "physical representation," that is, consistent in most situations. However, all that is required in these explanations is the genetic representation.

This point is often missed because geneticists have always been interested in the physical nature of the gene. That was part of what led to the explosive growth of molecular biology in the 1950s. Nevertheless, in genetic explanations of the sort just indicated, there is no commitment to what the gene is. This had to be so. When Mendel first invoked his explanatory factors, even the chromosome theory of heredity was yet to be formulated. Much of classical genetics was successfully developed long before it was even known that DNA is the genetic material. Moreover, classical geneticists were well aware that their results could be interpreted using only a formal hierarchy rather than a physical one. The Morgan school said so explicitly even though it was committed to a program of physically locating genes on chromosomes.[10] As late as the 1950s, Lederberg et al. (1951) even interpreted their linkage data to propose a branched structure for a linkage set which, they explicitly noted, could not be interpreted as the physical arrangement of loci on the chromosome. In genetic explanations, the explanatory weight comes from the rules obeyed by alleles and loci as abstract entities and not from their physical instantiations. These genetic explanations will be called "genetic reductions," and the faith that they will be forthcoming for phenogenesis will be called "genetic reductionism."

Segregation and linkage analysis become formidably difficult if more than two loci are involved. However, even the most ardent genetic reductionist knows that few traits can be explained from such a meager repertoire. In recent years, many new methods have been devised for the genetic dissections of such traits. The *allele sharing method* also involves the analysis of pedigrees and looks for a failure of Mendel's first law. The basic strategy is to show that two relatives with the same trait inherit identical alleles at some locus with a frequency greater than what would be expected from that law if that allele had no role in the genesis of that trait. Obviously, explanations inferred from this strategy are genetic reductions similar to those inferred using traditional segregation and linkage analyses.

Allelic association studies try to find statistically significant correlations between specific alleles and traits. Finally, *QTL mapping* uses recombination to generate specific allelic differences to find the effects of alleles at various loci. First, two different pure lines are created. Then they are crossed for one generation. The hybrid offspring are then back-crossed to one of the two parent lines. Though the hybrids initially have one chromosome (of each pair) from each of the pure lines, crossing over induces recombination of alleles from the two different pure lines. Comparing the offspring of the back-cross with the pure line allows an identification of the effects of individual alleles by which the original pure lines differed.

Neither of the last two strategies looks quite like traditional segregation or linkage analysis: one does not begin with a phenotypic feature and then attempt to explain its origin from a genetic basis. The difference is important: as far as research strategy goes, the direction is the reverse of the classical one. These new techniques are indeed important innovations in genetical research. Nevertheless, the explanations that emerge from all three of these techniques, when they are used by themselves, are still instances of genetic reduction. Inferences to explanations using these techniques assume that genes are the appropriate entities to invoke in explanations of phenogenesis and, beyond identifying allelic differences through DNA sequence differences, they do not evoke any of the physical properties of DNA. Allelic association studies do not even explicitly assume a particular mode of inheritance, whereas QTL mapping has similarities to linkage analysis.

Why question genetic reductionism? Segregation and linkage analysis provide (relatively) unequivocal explanations only when one or two loci are involved. These genetic reductions are unproblematic. They were also not only historically important in establishing genetics as a biological subdiscipline of central importance but provided a sort of "null model," deviations from which would indicate etiological complexity in phenogenesis. Nevertheless, such reductions are rare. But, the genetic reductionist claims to be able to explain from a genetic basis the phenogenesis of most traits, including those that are called "complex" because, at the very least, many loci are involved. There are two reasons to be skeptical of this claim:

1. The record so far has generally been one of spectacular failure, especially in the case of complex human behavioral traits which, understandably, have attracted the most publicity. Over the last 10 years, genetic explanations have been offered for schizophrenia, alcoholism, manic depression (bipolar affective disorder), among many other traits, only to turn out to be demonstrably spurious.
2. Even when genetic explanations are unproblematic (i.e., only one or two loci are involved), nongenetic factors must also contribute. What is interesting, and accounts for the power of genetics, is that it makes sense to say that the genes still bear all the explanatory weight. Once more loci begin to be relevant, it is almost certain that the contribution of nongenetic factors also increases. At some point genes *alone* can no longer bear the entire explanatory weight. That is precisely when a genetic reduction fails – the first criterion from Table 11.1 is no longer satisfied. Geneticists are at least implicitly aware of this. For instance, when the last three methods for the genetic dissection of a trait are used with any success, the usual locution is no longer the traditional one, that "the trait is genetic," but rather that "the trait has a genetic component." The last locution is not particularly controversial. But it is also uninteresting. Certainly, it does not suggest that genes rather than other factors should be first pursued in the explanation of phenogenesis. Unfortunately, even when all that genetics can do is provide only *some* of the relevant factors in this sense, genes, alleged to be the

"genes for those traits," have routinely been univocally pursued by biologists in recent years and presented to the general public as being the only factors that bear explanatory weight. When the organism being investigated is *Homo sapiens*, the potential social costs of this move are enormous. That is the most significant problem with genetic reductionism today.

Finally, in sharp contrast to genetic reductionism, the standard explanations of molecular biology are structural. The structures are trivially hierarchical; the hierarchy is the compositional hierarchy of physical space. Cells consist of systems of macromolecules, each consisting of even smaller parts. The physical representation of the genome is the chromosome as a complex macromolecular structure with proteins, DNA, and several chemical moieties. The critical assumptions are that structure determines behavior and what mediates interactions between macromolecules is a lock-and-key fit between the interacting regions. The reason that these explanations are regarded as genuinely successful is that, from these rather simple assumptions, much can be explained. Antibody specificity, why a particular antibody interacts only with its own antigen, is perhaps the simplest of these explanations: the antigen fits exactly into the active site of the antibody. More important, in the late 1950s and early 1960s, explanations of this sort were successful in two domains which were traditionally part of the repertoire of antireductionists: (1) the operon model of bacterial gene regulation fully accounted for regulation through feedback, and (2) the Monod–Wyman–Changeux model of allostery similarly accounted for the behavior of proteins such as hemoglobin whose constituent parts behaved cooperatively.

Because the hierarchies used in these explanations are trivially hierarchies in physical space, explanations of this sort satisfy all three criteria of reduction from Table 11.1. However, there are two interesting points to be made about these reductions: (1) it should be emphasized that the **F**-rules are not the ones used in genetic reductions, and (2) the **F**-realm from which the explanatory factors come is the **F**-realm of macromolecular physics and chemistry rather than physics at any lower level of organization. The **F**-rules, such as the structural rule about lock-and-key fits, are from this realm. In fact, the consistency of these rules with those at lower levels, such as the quantum mechanics of atoms, is an open, nontrivial scientific question. Reductions of this sort will be called *physical reductions* here, and associated with them is the thesis of *physical reductionism*.

11.4 Human Genome Project

The major contention of this chapter is that the HGP, as it was initially conceived in the mid-1980s, was driven by genetic reductionism. If this contention is justified, it has the following consequence: to the extent that genetic reductionism is untenable, the HGP itself is also suspect. A second, *tentative* and less important contention – and one which, because of lack of space, will only be perfunctorily defended – is that, to initiate the HGP, its proponents conflated genetic reductionism with physical reductionism and used the latter's successes to suggest the tenability of the former.[11] This is a major reason why they succeeded, in spite of extended initial debate, in promoting the project to the biological community and, equally important, to the general public as represented by the political institutions that determine how public money is spent.

The discussion here ignores the fact that the HGP has been *significantly* modified from its initial conception. In particular, it has progressively moved away from blind sequencing, though that still remains a major component of the project (see, for example, Collins and Galas 1993, Roberts 1993). These modifications underscore the correctness of the various scientific objections to the original HGP that are being discussed here. Nevertheless, because of these modifications, the critical comments that will be made in the following paragraphs do not always apply to the current version of the HGP. They also do not carry over to the other sequencing projects that have emerged and that may well ultimately be the greatest scientific contribution of the HGP.

The role played by genetic reductionism in the HGP is best explicated by reflecting on what, *specifically*, the HGP was supposed to be: a crash program to sequence the DNA of the entire human genome *blindly*.[12] There was no concern (within the *explicit* original plan of the HGP) to demonstrate the utility of these sequences. There was, instead, an expectation and, often, an explicit promise (e.g., to federal funding agencies in the United States) that knowing the final sequences will result in scientific and medical miracles.[13] The scientific claim was that, once these sequences are available, genes (i.e., particular alleles at definite loci) will be identified and, consequently, explanations of currently recalcitrant phenotypic features will begin to become available. The medical claim was that this will routinely have therapeutic value. The implausibility of the latter claim is strongly suggested by the case of sickle cell disease. For about 50 years the molecular basis of that disease has been known. Yet, to date, no adequate therapeutic intervention exists. There is no case yet in which a therapeutic approach that started with a gene has either led to an effective treatment or has even fared better than traditional empirical approaches. By now, the advocates of the HGP have become guarded about promising therapeutic translations of scientific advances, though, when the political debate on initiating HGP was at its critical stage, they expressed no such reservation – at least in their public pronouncements (see Cook-Deegan 1994).

The suggestion that DNA sequences alone would suffice to provide interesting scientific explanations is an expression of faith in genetic reductionism. This point needs careful elaboration. A major objection to blind sequencing is that as much as 90% of the human genome has no known function; that is, it neither codes for polypeptide chains nor plays any regulatory role. Knowledge of the sequences that mark boundaries between functional and nonfunctional segments remains incomplete. But suppose that this problem is solved and that genes (i.e., functional sequences) can be identified. Then the major use of these sequences will be to begin to look for the phenotypic expression of these alleles (as sequences) through methods such as allele sharing, allelic association studies, and QTL mapping. This constitutes what has come to be called "reverse genetics." It was pointed out in the last section that inferences from explanations of these new techniques, in spite of their novelty, were attempts at genetic reduction. In this sense, the HGP has been driven by genetic reductionism. A faith in the explanatory power of DNA sequences *alone* is not a faith in the explanatory power of physics; rather, it is the classical geneticists' faith in the power of genes. Therefore, the failures of genetic reductionism that were noted in the last section constitute objections to the HGP itself.

In advocating blind sequencing, the proponents of the HGP have, at least implicitly, expressed a faith in genetic reductionism. The reasons given here suggest that this faith is unjustified. However, there is another point that needs to be addressed. Actual research strategies do not use assumptions from classical genetics and molecular biology separately. Almost all classical geneticists were fascinated with the physical basis of heredity, and both they and molecular biologists routinely intertwined formal genetical reasoning with what can be gleaned from physics or chemistry in order to use all available resources for further research. Perhaps, then, a mixture of explanatory strategies from the two domains will be successful and will rescue the HGP. Gilbert (1992), for example, has forcefully argued for such a position. The general idea is that once a DNA sequence is identified as a gene, if it is a structural gene, the polypeptide chain that it specifies can be obtained. From a knowledge of the shape of such a chain, the standard structural rules of molecular biology are used to infer what it does. Ultimately, what is discovered in this way is explained from the **F**-rules of macroscopic physics, though the research process is the reverse of the usual one: one begins with the constituent parts rather than the composite structure. It is easy to envision a similar use of regulatory sequences.

The trouble is that there is no reason to expect success. One elementary source of difficulty is the unsolved protein-folding problem (i.e., that of predicting the three-dimensional conformation of a protein molecule from its amino acid residue sequence). Without even the elementary protein structures in hand, there is little that can be done. Moreover, as the case of sickle cell disease shows, even having these structures might not be enough to allow therapeutic interventions: there are more problems at higher levels of organization. There is no reason to believe that these will be fully solved by 2005, when the HGP was supposed to be completed. *That* is the critical point: it remains possible that, *eventually*, a combination of sequence information and constructive physical reasoning will yield new and significant insights. However, the scientific debate about the HGP was one about research strategy and science policy: whether it made sense to pursue the sequence blindly in a crash program, perhaps at the expense of other types of projects. Reversed physical reductionism is not at present in a position to rescue the HGP.

The last point is underscored by the recent experience with yeast (*Saccharomyces cerevisiae*).[14] In 1992, the publication of a 315-kb sequence for chromosome III of yeast was an important scientific event: it marked the first successful complete sequencing of eukaryotic chromosome. By 1996, the entire nuclear genome of yeast had been sequenced by a consortium of European laboratories.

It was a remarkably technical achievement and was all the more important because the European strategy largely involved cooperation between existing relatively small laboratories and no investment in large-scale sequencing technologies or the creation of special centers (as in the United States). Sequencing was thus introduced without much effect on the ethos of research, whatever it was.

The sequence revealed several surprises, and the most striking one was that, for almost half of all identified protein-specifying regions, no function could be attributed to the encoded proteins. There were no homologs in other organisms or in yeast itself. The newly discovered genes were "orphans." Reverse genetics, with present knowledge, offers little more than hope. As Dujon (1996, p. 266) points out: "it was the discovery of the extent of our ignorance, rather than the discovery of many new genes, that was the most conspicuous finding." Dujon perceptively notes that the orphans deserve the most priority for further investigation. Unless strategies for their exploration are devised, only a rediscovery of ignorance will recur when the genomes of more organisms are sequenced. For Dujon, and many other geneticists, this ignorance correctly signifies scientific opportunity. Nevertheless, the nagging question remains: What use is blind sequencing without much reason to believe that the sequences will reveal new insights in the immediate future? One could have started with orphan sequences that emerged during the course of customary research and designed experimental strategies to explicate their functions. Eventually, these strategies would probably have been successful. Meanwhile, as more sequences, including orphan genes, became available, methods would already have been devised to understand their significance. Why indulge in programs to produce uninterpretable data?

Finally, problems with reductionism do not exhaust the scientific problems with the HGP. Although these are beyond the scope of this chapter, one of them is of such importance that it cannot go without mention. Because of the tremendous polymorphism at the DNA and protein levels, there simply does not exist any single entity that can be called *the* human DNA sequence (or *the* sequence for any other eukaryote) (Sarkar and Tauber 1991). Even a very conservative estimate gives 10^8–10^9 different human DNA sequences that would be fully functional (Sarkar, unpublished calculations). In this sense, the entire goal of the HGP is less than coherent. Some proponents of the HGP have attempted to address this problem by resorting to the locution that what they are after is a "consensus sequence." Leaving aside the dubiety of this move – it simply masks the problem of an incoherent goal by introducing most frequent sequences as "normal" ones by "consensus" – the existence of such variation also makes the systematic use of an individual's sequence even for diagnostic purposes implausible in the near future. Given the current state of computer technology, to screen an individual's sequence against 10^8–10^9 different fully adequate sequences, each 3–4 × 10^9 bases long, and to do it routinely, is little more than a technological chimera.

The discussion here begs the question of why the HGP was pursued. Leaving aside the probably critical role played by sociopolitical ideology and the economic interests of various constituencies, the basic argument has been that the HGP is the "natural development" of molecular biology, possibly of all biology (Gilbert 1992, p. 83), and the revolutionary success of molecular biology cannot be doubted. But, as was pointed out in the last section, that success was one of physical and not genetic reductionism. Thus, this argument involves an implicit conflation of the two. Sometimes, the transition from one to the other is almost imperceptible. Here is one revealing instance from Watson (1992, p. 164): "I have spent my career trying to get a chemical explanation of life, the explanation of why we are human beings and not monkeys. The reason, of course, is our DNA. If you can study life from the level of DNA, you have a real explanation of its processes." The strangeness of these claims is that not only is DNA supposed to be the *only* appropriate locus of "chemical explanation," as if other macromolecules were not epistemologically relevant, but also that one need not, for instance, go any deeper, that is to units smaller than DNA, for "chemical explanation." The only explanation seems to be that genes, encoded by DNA, have replaced the wide variety of molecules that were originally pursued during the so-called "molecularization" of biology. Watson is by no means exceptional – reasoning of this sort was routine during the controversial initiation of the HGP (see Cook-Deegan 1994).

Two further points should be emphasized about the criticisms of the HGP that have been made here:

1. They are based on scientific and philosophical concerns about genetic reductionism, including question of scientific policy. They are, therefore, epistemological rather than sociopolitical.
2. They are limited to being criticisms of the HGP and should not be construed as being directed against all of molecular biology.

As was noted in the last section, the physical reductionist program in molecular biology has been remarkably successful. Whether it has any limit remains an open empirical question.

However, the decision to focus here on epistemological concerns does not deny the critical importance of the social, political, and ethical problems raised by the HGP. For brevity of exposition, these will be lumped together here as the "normative" problems of the HGP.[15] Lewontin (1992) and many others (e.g., Nelkin and Tancredi 1989, Hubbard and Wald 1993) have correctly emphasized these problems, and only two points will be noted here:

1. Genetic reductionism, as exemplified in the HGP, is only the latest manifestation of a long tradition of hereditarian ideology in Western social history (Haller 1984). The HGP, therefore, comes accompanied by all the problems that past instances of that ideology have raised, including, though not limited to, eugenic policies, legal discrimination, and social stigmatization. As the HGP proceeds, and an increasing number of genetically reductionist claims are made for different traits, these attendant normative problems will have to be addressed.[16] It may well be that these problems are sufficient for a compelling case to have been made against the HGP without even considering the epistemological issues

raised in this chapter. In any case, both sets of issues question the wisdom of the HGP at the level of scientific policy.

2. The argument against the initiation of the HGP that was given here is an argument about research strategy. It concerns the questions of whether the (genetic) reductionist claims on which the HGP is based are justified and, if not, whether it makes sense to pursue such a project. It is not suggested here that human DNA sequences are of no interest whatsoever or that they would never be available. It follows that, eventually, these normative problems would have to be faced. So, why criticize the HGP? The scientific criticism of the HGP is an objection to blind sequencing at rapid pace with little concern for the current biological interest of those sequences. Without the HGP, human sequences would have been emerging at a much slower rate. This would provide a more adequate period of time to prepare for the normative problems than is available now.[17] The HGP has tried to address the normative problems by allocating about 5% of its resources to their study. This will no doubt help and is a welcome and unusual acknowledgment by scientists of their normative duties, but this effort is unlikely to have much impact on the normative problems that have already begun to be felt.

11.5 Acknowledgments

It is a pleasure to dedicate this chapter to Dick Lewontin and to acknowledge his help and his ongoing influence on my work on the foundations of biology.

Thanks are due to Ken Schaffner, Abner Shimony, and Bill Wimsatt for discussions about reductionism for over a decade; to Denis Thieffry for discussions about sequencing projects; and to Hans-Jörg Rheinberger for a very helpful critical reading of an earlier version of this manuscript. This work was supported by the Wissenschaftskolleg zu Berlin and the Max-Planck-Insitut für Wissenschaftsgeschichte (Berlin).

Notes

1. This assessment ignores the fact that some organisms are haploid and many, especially among plants, are polyploid. Mendel's rules (even after modification with "linkage") are only strictly correct for diploids. However, the basic pattern of reasoning initiated by Mendel suffices to state the rules of heredity for haploidy (which is trivial) and polyploidy (see Haldane 1930).
2. For more detail on these issues, as well as for a skeptical reading of the genetic "code," see Sarkar (1996a,b).
3. This assessment of the *scientific* origins of the HGP is mainly based on the writings and other public pronouncements of Walter Gilbert. For accounts of other origins such as political, sociological, and technocratic ones, see Cook-Deegan (1994) and Kevles (1997). Unfortunately, both these works mingle detailed historical reconstruction with such uncritical advocacy of the HGP that the arguments of its critics do not get fair play.
4. As Joshua Lederberg put it: "Is it worth the cost? Undoubtedly! Is it the wisest use of this level of expenditure? I have ... doubts (Shapiro 1991, p. 206)."

5. The effect of the HGP on the funding of other biological research remains debatable. Cook-Deegan (1994) suggests a *slight* negative effect. Kevles (1997) suggests no significant effect and also claims that the change in ethos was insignificant. (It should be remembered that both Cook-Deegan and Kevles are rather uncritical partisans of the HGP.)
6. The account of reduction and reductionism given here follows that in Sarkar (1998). It contains discussions of other accounts of these concepts and how they relate to the one presented here.
7. Using IQ differences as an example is not to suggest any commitment to a position that IQ measures anything more important than the ability to perform at IQ tests. It is used here only because of its notoriety.
8. Watson (1992) is a notable exception when he draws on twin studies to argue that certain traits can be explained by genes. Twin studies do not permit genetic analysis beyond heritability analysis.
9. "Dominance" is shown when one value for the trait is exhibited by the heterozygote and one homozygote but not by the other. (It is, thus, a relational property and is usually considered to be a relation between the alleles.)
10. See, for example, Morgan et al. (1915, p. viii) and Morgan, Bridges, and Sturtevant (1925, p. 88).
11. For a fuller – but nevertheless incomplete – discussion of this issue, see Tauber and Sarkar (1992, 1993). A systematic exploration of the soundness of this claim is being left for the future.
12. Of course, it is illegitimate to speak of *the* human genome, and that remains one of the most important scientific arguments against the HGP. This point will be taken up later.
13. Commentators who accept this claim without serious scrutiny of the scientific problems that remain to be solved *even after the sequence is obtained* include Cook-Deegan (1994), Kitcher (1996), and Kevles (1997). These commentaries, even more so than the pronouncements of the HGP's scientific proponents, are the targets at which the criticisms embodied in this paper are directed.
14. The account of yeast sequencing given here is based on the recent review by Dujon (1996).
15. For a detailed evaluation of these problems, see Holtzman (1989).
16. If the claims made in this chapter are correct, then these claims are generally going to be false. Nevertheless, the general perception that these claims are true, because of the increasingly frequent use of genetic terminology in public discourse, is sufficient to create the normative problems.
17. To cite just one example: the demand for genetic counselors will far outgrow the number that will be available by 2005. Were human DNA sequences being obtained at a much slower rate, so that most of the sequences would only be available by say 2030, there would probably have been adequate time to prepare enough genetic counselors.

REFERENCES

Boltzmann, L. (1974). *Theoretical Physics and Philosophical Problems.* Dordrecht, The Netherlands: Reidel.

Bouchard, T. J., Lykken, D. T., McGue, M., Segal, N. L., and Tellegen, A. (1990). Sources of human psychological differences: The Minnesota Study of twins reared apart. *Science* 250:223–8.

Collins, F., and Galas, D. (1993). A new five-year plan for the U.S. Human Genome Project. *Science* 262:43–6.

Cook-Deegan, R. (1994). *The Gene Wars: Science, Politics, and the Human Genome.* New York: W. W. Norton.

Dujon, B. (1996). The yeast genome project: What did we learn? *Trends in Genetics* 12: 263–70.

Feldman, M. W., and Lewontin, R. C. (1975). The heritability hang-up. *Science* 190: 1163–8.

Gilbert, W. (1992). A vision of the grail. In *The Code of Codes: Scientific and Social Issues in the Human Genome Project,* Kevles, D. J. and Hood, L. (eds.), pp. 83–97. Cambridge, MA: Harvard University Press.

Haldane, J. B. S. (1930). Theoretical genetics of autopolyploids. *Journal of Genetics* 22: 359–72.

Haldane, J. B. S. (1936). A provisional map of a human chromosome. *Nature* 137: 398–400.

Haller, M. (1984). *Eugenics: Hereditarian Attitudes in American Thought.* New Brunswick, NJ: Rutgers University Press.

Holtzman, N. (1989). *Proceed with Caution.* Baltimore: Johns Hopkins Press.

Hubbard, R., and Wald, E. (1993). *Exploding the Gene Myth: How Genetic Information Is Produced and Manipulated by Scientists, Physicians, Employers, Insurance Companies, Educators, and Law Enforcers.* Boston: Beacon.

Kempthorne, O. (1978). Logical, epistemological and statistical aspects of nature–nurture data interpretation. *Biometrics* 34:1–23.

Kevles, D. (1997). Big science and big politics in the United States: Reflections on the death of the SSC and the life of the Human Genome Project. *Historical Studies in the Physical and Biological Sciences* 27:269–97.

Kitcher, P. (1996). *The Lives to Come: The Genetic Revolution and Human Possibilities.* New York: Simon and Schuster.

Lander, E. S., and Schork, N. J. (1994). Genetic dissection of complex traits. *Science* 265:2037–48.

Layzer, D. (1974). Heritability analyses of IQ scores: Science or numerology? *Science* 183:1259–66.

Lederberg, J., Lederberg, E. M., Zinder, N. D., and Lively, E. R. (1951). Recombination analysis of bacterial heredity. *Cold Spring Harbor Symposia on Quantitative Biology* 16:413–43.

Lewontin, R. C. (1974). The analysis of variance and the analysis of causes. *American Journal of Human Genetics* 26:400–11.

Lewontin, R. C. (1992). *Biology as Ideology: The Doctrine of DNA.* New York: Harper.

Moran, P. A. P. (1973). A note on heritability and the correlation of relatives. *Annals of Human Genetics* 37:217.

Morgan, T. H., Bridges, C., and Sturtevant, A. H. (1925). The genetics of *Drosophila. Bibliographica Genetica* 2:1–262.

Morgan, T. H., Sturtevant, A. H., Muller, H. J., and Bridges, C. B. (1915). *The Mechanics of Mendelian Heredity.* New York: Henry Holt.

Nelkin, D., and Tancredi, L. (1989). *Dangerous Diagnostics: The Social Power of Biological Information.* New York: Basic Books.

Roberts, L. (1993). Taking stock of the Genome Project. *Science* 262:20–2.

Sarkar, S. (1996a). Biological information: A skeptical look at some central dogmas of molecular biology. *Boston Studies in the Philosophy of Science* 183:187–231.

Sarkar, S. (1996b). Decoding 'coding': Information and DNA. *BioScience* 46:857–63.

Sarkar, S. (1996c). Lancelot Hogben, 1895–1975. *Genetics* 142:655–60.

Sarkar, S. (1998). *Genetics and Reductionism.* New York: Cambridge University Press.

Sarkar, S., and Tauber, A. I. (1991). Fallacious claims for HGP. *Nature* 353:691.

Shapiro, R. (1991). *The Human Blueprint.* New York: St. Martin's.

Tauber, A. I., and Sarkar, S. (1992). The Human Genome Project: Has blind reductionism gone too far? *Perspectives in Biology and Medicine* 35:220–35.

Tauber, A. I., and Sarkar, S. (1993). The ideology of the Human Genome Project. *Journal of the Royal Society of Medicine* 86:537–40.

Watson, J. D. (1992). A personal view of the project. In *The Code of Codes: Scientific and Social Issues in the Human Genome Project*, Kevles, D. J., and Hood, L. (eds.), pp. 164–73. Cambridge, MA: Harvard University Press.

CHAPTER TWELVE

Organism, Environment, and Dialectics

PETER GODFREY-SMITH

12.1 Introduction

Lewontin has argued for many years that the concept of *adaptation*, despite its prevalence in English-speaking biology, is a bad guiding concept for understanding the relations between organisms and environments. In its place, Lewontin urges that we think in terms of organic *construction* of environments, or in terms of an *interpenetration* between the two.

Lewontin also sees his position on this question as a "dialectical" one. As a non-Marxist admirer of Lewontin's work, I have usually tended to evaluate his views in isolation from any connection to dialectics and in isolation from any larger Marxist framework. In this form Lewontin's ideas have had a substantial impact upon my thinking ("impact" being just the right word), and I am one of many within the philosophy of biology whose work has been shaped by his views and concerns. But although I have resisted such an embedding, Lewontin's own view is that it is dialectic that ties many of his themes together. This seems a good occasion to try to work out exactly what role the idea of dialectic plays in Lewontin's work on organisms, environments, adaptation, and construction. My aim here is not to link those topics to broader political concerns but to look at the specific contribution that may be made to the biological issues by dialectical ways of thinking.

Although I will approach these topics by way of Lewontin's ideas, I will also outline elements of a different view about organisms and environments. This view will have some similarities to Lewontin's position but substantial divergences also.

12.2 Lewontin on the Mainstream View: Atomism and Externalism

In Lewontin's view, a mistaken picture of organism–environment relations has become prevalent in evolutionary theory. This reflects a larger mistake that the scientific tradition has made – the adoption of a "Cartesian reductionist" worldview. Lewontin has presented various characterizations of Cartesian reductionism (see especially Levins and Lewontin 1985, p. 269),[1] but perhaps the core of the view is the claim that the best way to understand a system is to break it down into its constituent parts and then work out how the individual attributes of the parts combine to generate the attributes of the whole. When the reductionist does this, it is assumed that the attributes of the parts that are important in that context exist in isolation of the other parts. Then a "breaking down" of the whole, whether literal or analytical, will not destroy the lower-level properties that matter.

A second feature of Cartesian reductionism, according to Lewontin, is the claim that in understanding causal processes we should allocate the roles of "cause" and "effect" in an exclusive way; each factor must play one role or the other. Thus, when something switches between these roles (as many things obviously do), we are supposed to handle each causal interaction independently and in turn.

For Lewontin, mainstream contemporary Darwinism – the synthetic theory as it is currently used and taught, especially in the English-speaking world – is a paradigm of a Cartesian reductionist view.

Mainstream Darwinism is firstly *atomistic*. It assumes, according to Lewontin, that it is possible to break an organism down into individual traits and investigate the evolution of each trait one at a time. When traits interact in their evolution (as of course they do), this problem is solved with an appeal to a "trade-off" between selective demands, various kinds of "constraints," or some other conceptual tool to reintegrate the organism partially.

A second major error is the *externalist* structure of mainstream Darwinism. I call "externalist" any explanation of properties of an organic system in terms of the properties of its environment. For Lewontin mainstream Darwinism treats organisms, in the context of evolution, as passively responding to environmental demands.

This externalist pattern concerns evolutionary processes, not individual development. Development is explained within the orthodox view in a primarily internalist way, in terms of the expression of coded instructions in DNA. Lewontin sees this as another sense in which the organism is cast as an object "alienated" from the forces governing its existence. However, as I understand the structure of Lewontin's analysis, the genes themselves in this picture are ultimately subordinate to the environment; they are a channel through which the environment speaks. The peculiar properties of the genetic system are downplayed.

Under the heading of "externalism" there are two different issues. Lewontin objects first to the idea that the trajectory of evolution can be almost entirely predicted from environmental demands. Second, there is the *asymmetric* nature of the standard picture. In this picture, organisms respond to the environment, but the environment is largely autonomous with respect to the organisms. The environment is seen as either stable (as far as the time scale of the evolutionary process in question is concerned) or else as changing according to its own intrinsic dynamic.

In a more detailed discussion of externalist explanatory patterns (Godfrey–Smith 1996), I use the term "asymmetric externalism" to describe views that both explain organic properties in terms of environmental properties *and* deny, implicitly or explicitly, that these environmental properties are to be explained in terms of other properties of the organic system in question. For Lewontin, a second mistake in mainstream Darwinism is its asymmetric externalist explanatory structure.

Thus, the view of organism–environment relations that Lewontin identifies here has two main features. First, the dominant pattern of explanation is outside-in. Second, the targets of explanation are individual traits, and each is explained in terms of a specific set of environmental demands.[2] This picture is summed up, for Lewontin, in the bad idea of *adaptation* as the basic evolutionary relationship between organism and environment.

Although Lewontin often discusses these two features of the mainstream view together and presents them as a result of Cartesian reductionism, I think that in some respects they are distinct ideas with different sources. This is because I do not see externalism as necessarily associated with the basic ideas of Cartesian reductionism. Even once we have decided to take systems apart and distinguish sharply between causes and effects, it is a further question whether we see organisms as straightforward products of the environment. So I think there is another source for the externalist commitment.

In fact Lewontin does often cite another source for this aspect of adaptationism, and this source is *theological* (Levins and Lewontin 1985, p. 67). In some ways the idea of adaptation is a holdover from an older worldview – a view in which every living thing has a natural place toward which it tends and where it properly belongs. This view is in many respects at tension with the mechanical, reductionist perspective of the scientific revolution. Thus, for Lewontin, when a biologist atomizes an organism into traits and explains each trait in terms of adaptive response to a specific environmental demand, the biologist is acceding to an unholy marriage of scientistic reductionism and prescientific theology.

12.3 Dialectics and Biology

In the place of Cartesian reductionism, Lewontin advocates a "dialectical" approach.

The concept of dialectic has a long and spectacular history. In this context the best place to start may be with Engels, to whom *The Dialectical Biologist* is dedicated. In his unfinished *Dialectics of Nature*, Engels (1940, p. 26) gave a brief three-part formulation of the basic dialectical laws, attributing the discovery of these laws to Hegel, as follows:

1. The law of the transformation of quantity into quality and vice versa.
2. The law of the interpenetration of opposites.
3. The law of the negation of the negation.

The first law asserts that changes in degree are not sharply distinct from changes in kind. The second, as I understand it, has mainly to do with categorization and description. It asserts that objects tend to have pairs of properties that are apparently incompatible but for which there is no real impossibility about the relationship. According to a dialectic view, understanding how such properties can coexist in a single thing is important in understanding how the thing behaves. The third law describes a specific pattern of change. In this pattern one condition gives rise to an incompatible condition that negates or overcomes it only to be overcome by a new condition that it generates in turn. In its most specific and controversial form, this is a *progressive* pattern in which the sequence produces a "resolution" or "synthesis" of the contradiction between the first two conditions (the "thesis" and "antithesis").

Levins and Lewontin give a systematic account of their own understanding of dialectic in the last chapter of *The Dialectical Biologist.* They do not aim for "laws" but do outline a set of basic dialectical principles (1985, pp. 273–5).

1. A whole is a relation of heterogeneous parts that have no prior independent existence as parts.
2. In general, the properties of parts have no prior "alienated" (roughly, independent or autonomous) existence but are acquired by being parts of a particular whole.
3. The interpenetration of parts and wholes is a consequence of the interchangeability of subject and object, of cause and effect.
4. Change is a characteristic of all systems and all aspects of systems.

Various consequences of these principles are also discussed (pp. 278–83). These include the claims that (1) there is no fundamental "basement" level in nature; (2) contradiction is a feature of the world, not just of thought; (3) change can be explained in terms of the interaction of opposing processes, and processes of self-negation; (4) apparently exclusive categories (such as the categories of deterministic and random) coexist and interpenetrate, and (4) complex systems tend to show spontaneous activity. Levins and Lewontin also give a range of methodological prescriptions (pp. 286–88).

It is notable that the first three of Levins and Lewontin's principles concern part–whole relations, and Engels did not have a principle of this type in his three laws. As I understand it, the claims about part–whole relations are part of the content of Engels' second law concerning the interpenetration of opposites. "Part" and "whole" are opposites, and they exist only in relation to each other like "up" and "down" or "positive" and "negative."

Levins and Lewontin do not give a central place to the specific pattern of change associated with the idea of dialectic (Engels' third law). They do assert, more generally, that change is ubiquitous, and they discuss some examples with the classical dialectical pattern. For Lewontin, this does seem to be one important dynamic pattern, but he does not assert that biological change *always,* or "in the important cases," has a specific dialectical form. This is consistent with, and perhaps a consequence of, Lewontin's generally skeptical view of attempts to summarize the behavior of complex biological systems with very simple principles of change. I take this view to be independent of (and perhaps at tension with) his Marxism.

So we should distinguish between (1) dialectic as *a specific pattern of change,* and (2) dialectic as a view about *how to describe the composition and behavior of natural systems.* The latter is more important than the former in Lewontin's biological views.

There is another idea associated with dialectics I have not mentioned yet. This is a historicist view of the assessment of ideas, including philosophical and biological ones. In some versions of this view, we should not assess theories with concepts of truth, accuracy, and justification that are history independent. I understand Lewontin as holding a more moderate version of this idea. According to this view, we can only understand why a theory was seen as justified or unjustified by attending to the historical and social circumstances; however, there is a real difference between accurate and inaccurate representation of the world.

In particular, mainstream Darwinism was a definite advance over earlier views of evolution (development-based or "transformationist" views, as Lewontin calls them). It was a stage through which our understanding maybe even *had* to pass (Levins and Lewontin 1985, p. 106). But it is not, according to Lewontin, an accurate representation of how evolution works, and to attain a more accurate view we must go beyond Cartesian reductionism and externalist explanatory patterns.

12.4 Lewontin's Alternative

The view Lewontin seeks to put in place of the mainstream picture is one in which (1) the integration of parts of organisms is recognized not as dogma but as a common biological state and as the object of study, (2) the organism is viewed as the subject and object of both its development and evolution, and (3) the relation between organism and environment is seen as one of interpenetration and co-construction. Given the topic of this chapter, I will concentrate on the third of these.

Lewontin has summarized his view of organism–environment relations as follows:

> [T]he organism and environment are not actually separately determined. The environment is not a structure imposed on living beings from the outside but is in fact a creation of those beings. The environment is not an autonomous process but a reflection of the biology of the species. Just as there is no organism without an environment, so there is no environment without an organism. (Levins and Lewontin 1985, p. 99)

At various places he has listed a range of aspects of, and arguments for, this position. My list below combines several of these with modifications (see Levins and Lewontin 1985, pp. 53–8, 99–102).

1. Organisms *select* their environments; they determine which specific regions of their surroundings will interact with them. They do this by means of their size and their behavior; for example, an insect can occupy tiny spaces with different properties from most of its surroundings.
2. Organisms determine which properties of external things are *relevant* to them and determine the statistical patterns found in those conditions. An environmental chemical might be toxic to one organism but irrelevant to another.
3. Organisms physically *alter* external things as they interact with them. They crop grass, fertilize and pollute, dam streams, and so on.
4. Organisms transform the *signals* that reach them from their environments. A signal that begins as air vibrations becomes an electrical and then a chemical signal to the rest of the body.

Most will agree that all these things occur. The harder questions concern their significance.

For example, is it correct to see these ideas as diverse arguments pointing towards a single conclusion? Consider arguments (2) and (3). Argument (3) concerns physical action on the environment by organisms. Argument (2), however, concerns the relevance of environmental factors *to* organisms. Lewontin uses the example of gravity; by being very small, an animal can make gravity irrelevant to its life. But that is not altering the operation of gravity in the world in general. Thus, (2) and (3) apparently concern different things: argument (3) concerns the role of the organism in changing the *environment*, and (2) concerns the role of the organism in determining the effect of the environment on *itself*.

From my point of view, Lewontin's arguments are heterogeneous in their content and in the conclusions they support. I will return to this later. But first it will be helpful to consider these issues in a dialectical context. For this is the way to understand why Lewontin sees these claims as constituting a single argument.

Lewontin says that the biologically important properties of living systems are properties that exist only in the context of part–whole relationships. Lewontin is also opposed to rigid separations between organism and environment. So one important type of whole system here is the organism–environment pair, the organism-in-its-context. If we recognize this whole as an important unit (following not just Lewontin but a long and heterogeneous tradition including John Dewey [1938], Susan Oyama [1985], and many others), then from a dialectical viewpoint we should expect its parts – the organism on the one hand and the environment on the other – to exist only in relation to each other. Consequently, changes in one part of this whole will imply (in a logical, not a causal sense) changes in other parts. When the organism changes how it lives or relates to external conditions, this constitutes a change in its environment as well. From this point of view the processes in Lewontin's (2) have a similar status to the processes described in (3) – in both cases, the organism is acting to construct or determine its environment. And in this view a great many things an organism does that have no immediate effects on things outside its skin will constitute "changing its environment." The dialectical perspective is, in this respect, a holistic one – a view in which changes in one part of a system are held to imply widespread changes elsewhere.

Thus, we can discern the common thread in Lewontin's arguments if we accept a dialectical viewpoint and also see the organism–environment pair as constituting a whole about which dialectical principles apply.

12.5 Assessing Lewontin's View

Lewontin's critique is concerned with the overall structure of mainstream Darwinism, with prevailing tendencies in how theory is elaborated, explanations are given, and research is conducted. The existence of exceptions and counterexamples (which can certainly be found) does not settle the question of whether his criticism is fair. In my view, Lewontin has succeeded in describing tendencies that play a powerful role in mainstream Darwinism. He has isolated and made vivid a tendency to mold biological thinking around a particular causal and dynamic pattern. This is the pattern in which organisms and environments are treated in a strongly asymmetrical way, with environmental pressures driving a process of adjustment or accommodation on the part of organisms. That is not the *only* causal pattern that mainstream Darwinists discuss. Indeed, there is one other pattern that is just as important: the repeated splitting of lineages of organisms over time. But these two patterns – the accommodation of environmental demands through cycles of mutation and selection and the splitting of lineages – occupy a unique and central place in the culture of Darwinism. This role is visible in the structure of models, in habits of thought, and in exemplars like the perennial and externalist example of industrial melanism, an example in which moths adjust over evolutionary time to environmental changes they play no role in shaping.

It is harder to say whether the tendency Lewontin describes is really a *problem*, as he claims. Some biologists may reply that a fairly asymmetric treatment of organism and environment is just the right approach, given the real causal differences between any given organism and the totality of factors that constitute its environment. If we are to make some single relationship between organism and environment a "central metaphor" in evolutionary thinking, the best candidate is an externalist one.

Lewontin's work suggests two different lines of reply to this. One reply is that there should not be *any* single pattern placed at the heart of the theory in this way. But another theme in his writing is that a *different* pattern, one featuring organism–environment symmetry, should be made central.

There is tension between these two themes. I view the first as the appropriate one for him. The two themes might be reconciled if the second – the idea of construction as central – were taken less literally than I take it. Lewontin could be seen as overstressing the role of constructive relationships to persuade people to appreciate their importance. A view in which neither relation is seen as central might then emerge (as a "synthesis"!). This is a possible reading, but it is not mine. I think Lewontin is attracted to the pluralist, "no-central-metaphor" view and to some very general arguments that suggest adaptation is a bad concept for biology, whereas construction is a good concept. So the tension is real.

I will return to these topics in the final section. Before that there are some other features of Lewontin's views to discuss.

Next we should look at some apparent concessions Lewontin makes toward the mainstream view – concessions about the "quasi-independence" of biological traits and "continuity" in organism–environment relations (1985, pp. 63–4, 79–80).

Lewontin introduces these concepts after asking what sense his own view can make of the most striking phenomena for an adaptationist: cases of evolutionary convergence, such as the development of fins and finlike appendages in many different aquatic animals. Lewontin accepts that this could only happen if organisms have two abstract properties.

> First, there must be *continuity*, so that very small changes in morphology, physiology and behavior usually have only a small effect on the ecological relations of the organism. Continuous deformations of phenotype should map frequently onto continuous deformations of ecological relation. Second, characters must be *quasi-independent*. That is, there must exist a large number of possible phenotypic correlations between a character change and other aspects of the phenotype. If character correlations are unbreakable, or nearly so, then no single aspect of the phenotype, like fins, could ever develop without totally altering the rest of the organism in generally nonadaptive ways. (Levins and Lewontin 1985, pp. 63–4)

To adaptationist eyes, these are big concessions. Lewontin has criticized mainstream Darwinism for holding an atomistic and externalist view of evolution. But here he seems to concede that both of these views must be approximately correct.

The point is clearest in the case of "quasi-independence." The characters of an organism are quasi-independent if, when one character changes its state, there are many ways that this change can coexist with the states of other characters. To some extent, each character can evolve independently.

"Continuity" is more complicated, but part of what it involves is that organisms' actions on the environment must be limited. If each change made to one characteristic massively altered an organism's relations to its ecological context, nothing resembling Darwinisn evolution would be possible. If any slight change to the shape of a fin changed the nature of the surrounding water, or made it impossible for the organism to lay eggs, the development and fine-tuning of the fin would be impossible. A slight change in the fin must, in most cases, only imply slight changes in the organism's total relationship to its environment.

This requirement of continuity has two parts. First, a slight change to one character must not alter the internal workings of other parts of the organism too much. Second, a slight change to one character must not bring about massive environmental changes. Thus, an organism's "constructive" relationships to its environment must be limited in extent.

If these arguments are good, then the features of mainstream Darwinism that Lewontin opposes are not optional features; the basic structure of the theory requires that atomism and an asymmetric externalist view contain an element of truth. *To the extent* that the relations between parts of an organism, and the relations between organism and environment, show strong integration and interpenetration, Darwinian evolution cannot occur. So the adaptationist can apparently say to Lewontin: Accept an element of truth in atomist and externalist views, or else leave Darwinism altogether. There cannot be a biology that is at once dialectical and that accepts the most basic Darwinian ideas.

Lewontin does not want dialectical biologists to reject Darwinism altogether. As I understand his position, his best reply to this is as follows: The dialectical view does not claim that every system displays the maximum level of integration and interdependence possible. It does not claim that a change in one part of a whole always implies large-scale change to other parts. Rather, the dialectical view treats this degree of integration as a variable that can have high and low values. Organisms can, over evolutionary time, become more tightly and more loosely coupled with respect to their parts. The Cartesian view takes an atomistic state to be the natural, normal condition; the dialectical view takes the degree of independence of parts in a system to be an historically changing property.

If this is indeed Lewontin's reply, we reach a stage at which each side has distanced itself from extreme views, and it is hard to resolve the issue. The mainstream view does not claim that simpleminded reductionism is true; it claims that any Darwinist view requires that organisms *approximate* systems whose parts can change independently and whose effects on the environment are limited. Lewontin's reply, as I envisage it, claims that the dialectical view treats the degree of integration in organisms as a variable whose value is contingent and changes over time. This is a moderate version of the dialectical view, for it admits the importance of situations that its rival sees as central. It is hard to judge which of these two is right.

At least, this is how things stand on the question about integration of parts *of an organism* (a question I will not pursue further here). On the issue of how organisms affect their *environments* there is more to be said.

In the previous section I listed various arguments Lewontin gives for his constructivist view of organism–environment relationships. I claimed that although Lewontin presents these as arguments for a single position, there are large apparent differences between the phenomena he cites. In particular, the arguments I listed as (2) and (3) concern different issues.

2. Organisms determine which properties of external things are *relevant* to them.
3. Organisms physically *alter* external things as they interact with them.

I claimed that (3) concerns the role of the organism in changing or constructing the *environment*, whereas (2) concerns the organism's role in determining the environment's effect on *itself*. From a dialectical point of view, however, the two arguments point in the same direction. We can see the organism–environment pair as a single whole in which organism and environment are parts that codetermine each others' properties. Then a change to the organism will, by changing the relations the organism has to its environment, change the environment itself.

This dialectical perspective makes good sense of Lewontin's position, but we should ask now if this is really the best way to think about organisms and environments. I claim it is not, because if this perspective is taken, too many different sorts of things are lumped together as cases of "organisms constructing their environments." It is better to understand "construction of the environment" as a specific, restricted kind of process – something that occurs only some of the time.

A simple example will make this clear. Consider two types of organism that each encounter a chemical in their environment that is toxic to them. They respond differently. Species A changes its body chemistry in such a way that the toxin no longer has any effect on it. Species B develops a way of destroying the toxin altogether while it is still external to it. Individuals of species B excrete something that interacts with the toxin, or expose it to sunlight, or bury it. The routes taken by species A and B, I claim, are two different ways that organisms respond to environmental conditions. One way is for organisms to change only themselves. The other way is for organisms also to change the intrinsic properties of things external to them – to reconstruct or transform environmental conditions.

The problem with Lewontin's dialectical approach, as I understand it, is that what species A does and what species B does here would *both* qualify as organic reconstruction or transformation of the environment. Species A has done something that falls under category (2) in Lewontin's arguments, and species B has done something that falls under (3). For me, only what species B has done should be put in this category. The relationship whereby the organism "selects out" the environmental features that matter to it is not a relationship in which the organism exerts the kind of power over the environment involved in a constructive relationship.

There is a connection between this argument and the questions concerning "continuity" and "quasi-independence" discussed earlier. In the discussion of quasi-independence I said that Lewontin should stress that the degree of integration of parts in an organism is a variable whose value can change – neither total independence of traits nor complete holistic interdependence are basic "states of nature." The organism's role in constructing its environment should be treated the same way. Some organisms construct and transform more than others; beavers do it more than deer, earthworms do it more than leeches. We should not assume in advance *either* that an organism is passively accommodating environmental conditions it played no part in creating *or* that every aspect of an organism's environment is its own construction.

12.6 Adaptation and Construction

Lewontin has claimed that we must change our way of thinking about organisms and environments in a wholesale way: "The metaphor of adaptation must therefore be replaced by one of construction" (Levins and Lewontin 1985, p. 104). This quote exemplifies what I discussed in the previous section as the second, mistaken theme in his work on these topics. Rather than a replacement, there should be a supplementation. Both adaptation and construction are real relationships that organisms have, in particular instances, to environmental conditions. In this final section I will sketch a framework based on this idea.

The environment of an organism is simply its surroundings, including all the contents of that specific part of the world. By moving about, an organism can help to determine which parts of the world constitute its surroundings. And by changing its own structure, an organism can make things in that part of the world relevant or irrelevant to its life. However, neither of these activities – changing where one lives and changing what is relevant – should be conflated with construction or transformation of environments in the strict sense. To construct, reconstruct, or transform its environment, an organism must *cause changes to the intrinsic properties of external objects.*

When the hypothetical species B, in the example of the previous section, exposed a toxic chemical to sunlight and thereby denatured it, the action is a genuine example of construction. But if the organism changes itself in such a way that the chemical is no longer harmful to it, then no external thing has had its intrinsic properties altered. "Intrinsic" properties here are contrasted with "relational" properties (this is a standard distinction in philosophy). An intrinsic property is a property that something has *independently* of other objects and their characteristics, chemical composition is an intrinsic property, for example.

It is necessary to make this restriction about intrinsic properties to prevent the trivialization of my concept of construction. For instance, *being poisonous to species A* is a relational property of samples of the toxic chemical before any evolutionary changes have occurred. When species A changes its internal chemistry, it can change this relational property of samples of the chemical. We need to restrict environmental construction to changes made to intrinsic properties, or else we find that almost everything counts as construction.

It may be objected at this point that one important thing organisms do is to alter various relational properties of external things. Spatial location is a relational property, and one thing organisms do is to alter the layout of the environment, bringing some things closer together, separating others, and so on. That is certainly correct. Making changes to relational properties is something organisms do, but the cases in which this constitutes construction of the environment are cases in which, in addition to relational properties being changed, *some* external things have their intrinsic properties changed. When an animal digs a burrow, this involves changes to both types of characteristics. Grains of sand have their location changed while remaining intrinsically the same as before. But an area of ground has its intrinsic structure altered.

Although I stress the distinction between changing the organism and changing the environment, it should not be thought that any organism must do one or the other of these exclusively at any time. Species B, in the toxic chemical scenario, made changes both to itself *and* to external things. It is species A that makes the distinction clear by doing one and not the other. Further, the strategies of species A and B could be undertaken simultaneously by some species C; it could reduce its sensitivity and change its environment as well. But species C is then doing two different types of things; it is dealing with the environment in two importantly different ways at once.

An organism adapts *to* some specific environmental condition when it changes itself and leaves that environmental condition unchanged. So to some extent, "adapt" and "construct" should be understood as complementary concepts. But a single evolutionary process can instantiate both adaptation and construction; an organism can adapt to some aspects of its environment by acquiring an ability to reconstruct other aspects. Species B, for example, makes real changes to its local environment, but the impetus behind its acquisition of this capacity did not spring preformed from inside those organisms. The acquisition, over evolutionary time, of the capacity to make environmental changes was an adaptive response to the general problem posed by the biochemical capacities of a particular type of substance.

The view I am defending puts a lot of weight on the distinction between internal and external changes. People sometimes claim, as B. F. Skinner once did, that the skin is not so important a boundary. I claim that it is often a very important boundary in both theoretical and practical matters. But it is important for some purposes and not others. It can be biologically important, for example, without being metaphysically important; the skin does not mark any boundary between real and unreal.

Even within biology, the importance of the skin as a boundary need not be universally important or universally unimportant. In a radical criticism of the concept of inheritance and the role of the genetic, Oyama and other "developmental systems theorists" have argued that what is passed across generations is not just a packet of genetic information but a total set of "developmental resources" (Oyama 1985, Griffiths and Gray 1994). These resources include things internal to organisms and external things as well. External resources can include a habitat on which there is reliable imprinting by each new generation, a particular type of nest structure, and so on. Developmental systems theorists claim that in understanding the reliable reproduction of life cycles across generations, we should not distinguish sharply between resources found inside the skin and resources outside it.

For the most part, this issue is independent of the one I am concerned with here. We could have one view of the importance of the skin as a boundary in investigating development and inheritance and another view in understanding evolutionary aspects of organism–environment relations.

However, there are complex connections between some versions of the developmental systems perspective and the view I defend here. A strong version of the developmental systems perspective claims that the unit in evolution is not a skin-bound organism but a larger system composed of a bundle of internal and external things linked together by their role in life cycles. That is, some versions of this view claim that the traditional organism is the wrong unit of analysis *in general* – the wrong unit for understanding development, inheritance, evolution, and the rest. If this position were to be accepted, then in distinguishing between mere adaptation to the environment (species A) and adaptive reconstruction of it (species B), we would have to use a different boundary from the usual one. We would distinguish changes to factors internal to the developmental system (however this is defined) and changes to things external to it. Within a theory of this kind it would be possible and advisable to relocate the same *type* of distinction as before. If *no* distinction of this type is made, then, as I argued earlier, we lose sight of an important difference between the various relationships living things have to their surroundings. We lose sight of the difference between moths changing their color, over evolutionary time, to fit the color of local trees, and moths developing the capacity to spray-paint the trees a preferred color.

I have argued here for a narrow and restricted sense of "construction of environments." This is a view that (1) takes the commonsense position that an organism's environment is just its surroundings, including all the contents of these surroundings, and (2) distinguishes between changes made to the organism and changes made to the world outside it. I am convinced that this distinction can and should be made in a sharp way. As I said earlier, this is not to claim that organisms will tend to do either one *or* the other of these at a given time. Neither is it to claim that these two types of relationships exhaust the options. But it is a distinction that we should hang onto.

Parts of this final section will have the appearence of philosophical legislation about the scientific use of language – an activity that risks raised eyebrows and worse. I think of this conceptual corralling and reorganizing as having two distinct roles and audiences. First, when the eyebrows have settled there is the possibility that scientists will find that their thinking about the topic is clearer. But second, in addition to the conceptual demands of science there are the conceptual demands of what I would call *philosophy of nature*. This is a loaded term and one associated in some people's minds with failed philosophical attempts to dictate science to the scientists. I do not mean anything like that; I appropriate the term here for a different activity.

When undertaking philosophy of nature in my sense, a writer comments on the overall picture of the natural world that science, and perhaps other types of inquiry, seem to be giving us. But this commentary does not have to use language in the same way that scientists find convenient for their own work. It can use its own categories and concepts. These concepts should be ones suited to the task of describing the world as accurately and precisely as possible when a diverse range of scientific descriptions are to be taken into account and when a philosophical concern with the underlying structure of theories is appropriate. Richard Lewontin, for example, is someone who has the distinction of having contributed in a large-scale and enduring way to both a particular part of science and to this project of philosophy of nature.

12.7 Acknowledgments

I thank Richard Francis, Richard Lewontin, Susan Oyama, Kim Sterelny, and Rasmus Winther for discussions and correspondence on these issues. Comments by members of an audience at Northwestern University also led to several key improvements.

Notes

1. *The Dialectical Biologist* (1985) includes much work that Levins and Lewontin had published independently. They say that the essay "Dialectics" was cowritten especially for the collection, however.
2. When the adaptationist view is combined with atomism in this way, one of Lewontin's arguments against adaptationism can be fended off. Lewontin says that the adaptationist has a problem because the concept of a "niche" is predictively empty, for niches do not exist until organisms fill them (1985, p. 98). But a niche is a *total* way of life involving a whole package of traits. Thus, an atomist, who treats each trait of an organism separately, does not have to predict or explain which total lifestyle possibilities will be found in nature – which niches will be filled.

REFERENCES

Dewey, J. (1938). *Logic: The Theory of Inquiry.* Reprinted in *John Dewey: The Later Works, 1925–1953.* Volume 12:1938, ed. J. A. Boydston. Carbondale, IL: Southern Illinois University Press, 1991.

Engels, F. (1940). *Dialectics of Nature.* Translated and edited by C. Dutt. New York: International Publishers.

Godfrey-Smith, P. (1996). *Complexity and the Function of Mind in Nature.* Cambridge, UK: Cambridge University Press.

Griffiths, P. E., and Gray, R. D. (1994). Developmental systems and evolutionary explanation. *The Journal of Philosophy* 91:277–304.

Levins, R., and Lewontin, R. C. (1985). *The Dialectical Biologist.* Cambridge, MA: Harvard University Press.

Oyama, S. (1985). *The Ontogeny of Information.* Cambridge, UK: Cambridge University Press.

CHAPTER THIRTEEN

Units and Levels of Selection

An Anatomy of the Units of Selection Debates

ELISABETH A. LLOYD

13.1 Introduction

Richard Lewontin was the first to investigate systematically the set of problems raised by a hierarchical expansion of selection theory in his landmark 1970 article. His classic abstraction and analysis of the three principles of evolution by natural selection – phenotypic variation, differential fitness, and heritability of fitness – have served as the launching point for many biologists and philosophers who have wrestled with units of selection problems. Lewontin's critical discussion of the empirical evidence for selection at various biological levels has served as both touchstone and target for later work. But Lewontin's essay was addressed to the efficacy of different units of selection as causes of evolutionary change (1970, p. 7). In the analysis I offer in this chapter, this is but one of four distinct questions involved in the contemporary units of selection debates.

For at least two decades, some participants in the "units of selection" debates have argued that more than one question is at stake. Richard Dawkins, for instance, introduced the terms *replicator* and *vehicle* to stand for different roles in the evolutionary process (1978; 1982a,b). He proceeded to argue that the units of selection debate should not be about vehicles, as it had formerly been, but about replicators. David Hull, in his influential article, "Individuality and Selection" (1980), suggested that Dawkins' "replicator" subsumes two distinct functional roles, and the separate categories of "replicator," "interactor," and "evolver" were born. Brandon, arguing that the force of Hull's distinction had not been appreciated, analyzed the units of selection controversies further, claiming that the question about interactors should more accurately be called the "levels of selection" debate to distinguish it by name from the dispute about replicators, which he allowed to keep the "units of selection" title (1982).[1]

This analysis was first presented in July 1989 at the Plenary session of the International Society for the History, Philosophy, and Social Studies of Biology. In November 1989, it was delivered to the Genetics Colloquium at Harvard and distributed to the NSF Workshop on Development and Evolution, at the Santa Fe Institute.

My purpose in this chapter is to delineate further the various questions pursued by Robert Brandon, Richard Dawkins, James Griesemer, David Hull, Richard Lewontin, John Maynard Smith, Sandra Mitchell, Elliott Sober, Michael Wade, George C. Williams, David S. Wilson, William Wimsatt, Sewall Wright, and many others under the rubric of "units of selection."[2] I will isolate four quite distinct questions that have, in fact, been asked in the context of considering, What is a unit of selection? In Section 13.2, I describe each of these distinct questions. In Section 13.3, I return to the sites of several very confusing, occasionally heated debates about "the" units of selection. I analyze many leading positions on the issues using my taxonomy of questions.

This analysis does not, of course, make differences vanish, but I hope to clarify the terms of the debates. My analysis also does not resolve any of the conflicts about which research questions are most worth pursuing; moreover, I do not attempt to decide which of the questions or combinations of questions I discuss ought to be considered *the* units of selection question. Although I have elsewhere argued that the interactor question (see Section 13.2.1) is a primary question for evolutionary genetics, that claim is intended as historical and descriptive; most evolutionary genetics models that address any version of the units of selection question have focused on which level of interaction must be represented in the model to make it dynamically and empirically adequate (Lloyd 1988, especially Chapters 5 and 6).[3] Furthermore, the mere persistence of the three other questions to be discussed attests to their importance and general interest.

13.2 Four Basic Questions

Four basic questions can be delineated as distinct and separable. As we shall see in Section 13.3, these questions are often used in combination to represent *the* units of selection problem. But before we continue, we need to clarify some terms.

The term *replicator*, originally introduced by Dawkins but since modified by Hull, is used to refer to any entity of which copies are made. Dawkins classifies replicators using two orthogonal distinctions. A "germ-line" replicator, as distinct from a "dead-end" replicator, is "the potential ancestor of an indefinitely long line of descendant replicators" (1982a, p. 46). For instance, DNA in a chicken's egg is a germ-line replicator, whereas that in a chicken's liver is a dead-end replicator. An "active" replicator is "a replicator that has some causal influence on its own probability of being propagated," whereas a "passive" replicator is never transcribed and has no phenotypic expression whatsoever (1982a, p. 47). Dawkins is especially interested in *active germ-line replicators*, "since adaptations 'for' their preservation are expected to fill the world and to characterize living organisms" (1982a, p. 47).

Dawkins also introduced the term *vehicle*, which he defines as "any relatively discrete entity . . . which houses replicators, and which can be regarded as a machine programmed to preserve and propagate the replicators that ride inside it" (1982b, p. 295). According to Dawkins, most replicators' phenotypic effects are represented in vehicles, which are themselves the proximate targets of natural selection (1982a, p. 62).

Hull, in his introduction of the term *interactor*, observes that Dawkins' theory has replicators interacting with their environments in two distinct ways: they produce copies of themselves, *and* they influence their own survival and the survival of their copies through the production of secondary products that ultimately have phenotypic expression. Hull suggests the term *interactor* for entities that function in this second process. An *interactor* denotes that entity which interacts, as a cohesive whole, directly with its environment in such a way that replication is differential – in other words, an entity on which selection acts directly (Hull 1980, p. 318). The process of evolution by natural selection is "a process in which the differential extinction and proliferation of interactors cause the differential perpetuation of the replicators that produced them" (Hull 1980, p. 318; cf. Brandon 1982, pp. 317–8).

Hull also introduced the concept of "evolvers," which are the entities that evolve as a result of selection on interactors; these are usually what Hull calls "lineages" (Hull 1980). So far, no one has directly claimed that evolvers are units of selection. They can be seen, however, to be playing a role in considering the questions of who owns an adaptation and who benefits from evolution by selection, which we will consider in Section 13.3.1 and 13.3.3.

13.2.1 The Interactor Question

In its traditional guise, the interactor question is, What units are being actively selected in a process of natural selection? As such, this question is involved in the oldest forms of the units of selection debates (Darwin 1859 (1964), Haldane 1932, Wright 1945). In his classic review article, Lewontin's purpose was "to contrast the levels of selection, especially as regards their efficiency as causes of evolutionary change" (1970, p. 7). Similarly, Slobodkin and Rapaport assumed that a unit of selection is something that "responds to selective forces as a unit – whether or not this corresponds to a spatially localized deme, family, or population" (1974, p. 184).

Questions about interactors focus on the description of the selection process itself, that is, on the interaction between entity and environment and on how this interaction produces evolution; they do not focus on the outcome of this process (see Wade 1977, Vrba and Gould 1986). The interaction between some interactor and its environment is assumed to be mediated by "traits" that affect the interactor's expected survival and reproductive success. (N.B. – An interactor may be at any level of biological organization, including a group, an organism, a chromosome, a kin group, or a gene.) In other words, some portion of the expected fitness of the interactor is directly correlated with the "value" of the trait in question. The expected fitness of the interactor is commonly expressed in terms of genotypic fitness parameters, that is, in terms of the fitness of combinations of replicators; hence, interactor success is most often reflected in, and counted through, replicator success. Several methods are available for expressing such a correlation between trait and (genotypic or organismic) fitness, including regression and variances, and covariances. Several models are also available for representing interactors; in all of these, the interactor's trait is correlated with replicator fitness values, and the component of the replicator fitnesses attributed to the interactor is not available or reproducible from a lower level of interactor.[4]

In fact, much of the interactor debate has been played out through the construction of mathematical genetical models. The point of building such models is to determine what kinds of selection, operating on which levels, may be effective in producing evolutionary change.

It is widely held, for instance, that the conditions under which group selection can effect evolutionary change are quite stringent and rare. Typically, group selection is seen to require small group size, low migration rate, and extinction of entire demes.[5] Some modelers, however, disagree that these stringent conditions are necessary. Matessi and Jayakar, for example, show that in the evolution of altruism by group selection, very small groups may not be necessary (1976, p. 384; contra Maynard Smith 1964); Wade and McCauley also argue that small effective deme size is not a necessary prerequisite to the operation of group selection (1980, p. 811). Similarly, Boorman shows that strong extinction pressure on demes is not necessary (1978a, p. 1909). And finally, Uyenoyama develops a group selection model that violates all three of the "necessary" conditions usually cited (1979).

That different researchers reach such disparate conclusions about the efficacy of group selection is partly because they are using different models with different parameter values. Wade highlighted several assumptions, routinely used in group selection models, that biased the results of these models against the efficacy of group selection (1978). For example, he noted that many group selection models use a specific mechanism of migration; it is assumed that the migrating individuals mix completely, forming a "migrant pool" from which migrants are assigned to populations randomly. All populations are assumed to contribute migrants to a common pool from which colonists are drawn at random. Under this approach, which is used in all models of group selection prior to 1978, small sample size is needed to get a large genetic variance between populations (Wade 1978, p. 110).

If, in contrast, migration occurs by means of large propagules, higher heritability of traits and a more representative sampling of the parent population will result. Each propagule is made up of individuals derived from a single population, and there is no mixing of colonists from different populations during propagule formation. On the basis of Slatkin and Wade's analysis, much more between-population genetic variance can be maintained with the propagule model (1978, p. 3531). They conclude that, by using propagule pools as the assumption about colonization, one can greatly expand the set of parameter values for which group selection can be effective (Slatkin and Wade 1978, p. 3531; cf. Craig 1982).

My point here is not, however, to survey the various models[6] but rather to illustrate the *level of disagreements* within the interactor question itself. It should also be emphasized that not all discussion regarding which levels of selection are causally efficacious has been quantitative. Many authors have attempted to determine which levels of selection must or should be taken into account through qualitative descriptions of interactors.[7]

Note that what I am calling the "interactor question" does *not involve attributing adaptations or benefits to the interactors.* Interaction at a particular level involves only the presence of a trait at that level with a special relation to genic or genotypic expected success that is not decomposable into fitness components at another level.[8] A claim about interaction indicates only that there is an evolutionarily significant interaction occurring at the level in question; it says nothing about the existence of adaptations at that level. As we will see, the most common error made in interpreting many of the genetical models is that the presence of an interactor at a level is taken to imply that the interactor is also a manifestor of an adaptation at that level.

13.2.2 Replicators

Starting from Dawkins' view, Hull refined and restricted the meaning of "replicator," which he defined as "an entity that passes on its structure directly in replication" (1980, p. 318). I will use the terms *replicator* and *interactor* in Hull's sense throughout the rest of this chapter.

Hull's definition of replicator corresponds more closely than Dawkins' to a long-standing debate in genetics about how large or small a fragment of a genome ought to count as the replicating unit – something that is copied, and which can be treated separately (see especially Lewontin 1970). This debate revolves critically around the issue of linkage disequilibrium and led Lewontin, most prominently, to advocate the usage of parameters referring to the entire genome rather than to allele and genotypic frequencies in genetical models.[9]

The basic point is that with much linkage disequilibrium, individual genes cannot be considered as replicators because they do not behave as separable units during reproduction. Although this debate remains pertinent to the choice of state space for genetical models, it has been eclipsed by concerns about interactors in evolutionary genetics.

13.2.3 Beneficiary

Who benefits from a process of evolution by selection? There are two predominant interpretations of this question: Who benefits ultimately, in the long term, from the evolution by selection process? and Who gets the benefit of possessing adaptations as a result of a selection process?

Take the first of these, the issue of the ultimate beneficiary. There are two obvious answers to this question – two different ways of characterizing the long-term *survivors and beneficiaries* of the evolutionary process. One might say that the species or lineages (Hull's "evolvers") are the ultimate beneficiaries of the evolutionary process. Alternatively, one might say that the lineages characterized on the genic level, that is, the surviving alleles, are the relevant long-term beneficiaries. I have not located any authors holding the first view, but, for Dawkins, the latter interpretation is *the primary fact* about evolution. To arrive at this conclusion, Dawkins adds the requirement of *agency* (cf. Hampe and Morgan, 1988). For Dawkins, a *beneficiary*, by definition, does not simply passively accrue credit in the long term; it must function as the initiator or causal source of a biochemical causal pathway. Under this definition, the *replicator* is causally responsible for all of the various effects that arise further down the biochemical pathway, irrespective of which entities might reap the long-term rewards.[10]

A second and quite distinct version of the "benefit" question involves the notion of adaptation. The evolution by selection process may be said to "benefit" a particular level of entity under selection, though producing *adaptations* at that level (Williams 1966, Maynard Smith 1976, Eldredge 1985, Vrba 1984). On this approach, the level of entity actively selected (the interactor) *benefits* from evolution by selection at that level through its acquisition of adaptations.

I think it is crucial to distinguish the question concerning the level at which adaptations evolve from the question about the identity of the ultimate beneficiaries of that selection process. One can think – and Dawkins does – that organisms have adaptations without thinking that organisms are the "ultimate beneficiaries" of the selection process.[11] I will therefore treat this sense of "beneficiary" that concerns adaptations as a separate issue, discussed in the next section, under the topic of the manifestor of adaptations.

13.2.4 Manifestor of Adaptations

At what level do adaptations occur? Or, as Sober puts this question, "When a population evolves by natural selection, what, if anything, is the entity that does the adapting?" (1984, p. 204).

As mentioned previously, the presence of adaptations at a given level of entity is sometimes taken to be a *requirement* for something to be a unit of selection.[12] Wright, in an absolutely crucial observation, distinguished group selection for "group advantage" from group selection per se (1980); in my terms, he claimed that the combination of the interactor question with the question of what entity had adaptations had created a problem in the group selection debates. Following Wright, I submit that the identification of a unit of selection with the manifestor of an adaptation at that level has caused a great deal of confusion in the units of selection debates in general.

Some, if not most, of this confusion is a result of a very important but neglected duality in the meaning of "adaptation" (in spite of useful discussions in Brandon 1978b, Burian 1983, Krimbas 1984, Sober 1984). Sometimes "adaptation" is taken to signify *any trait at all* that is a direct result of a selection process at that level. In this view, any trait that arises directly from a selection process is claimed to be, *by definition*, an adaptation (e.g., Sober 1984; Brandon 1985, 1990; Arnold and Fristrup 1982).[13] Sometimes, on the other hand, the term "adaptation" is reserved for traits that are "good for" their owners, that is, those that provide a "better fit" with the environment, and that intuitively satisfy some notion of "good engineering."[14] These two meanings, which I call the *selection-product* and *engineering* definitions, respectively, are distinct, and in some cases, incompatible.

Williams, in his extremely influential book, *Adaptation and Natural Selection*, advocated an engineering definition of adaptation (1966). He believed that it was possible to have evolutionary change result from direct selection favoring a trait *without* having to consider that changed trait as an *adaptation*. Consider, for example, his discussion of Waddington's (1956) genetic assimilation experiments. Williams interprets the results of Waddington's experiments in which latent genetic variability was made to express itself phenotypically because of an environmental pressure (1966, pp. 70–81; see the lucid discussion in Sober 1984, pp. 199–201). Williams considers the question of whether the bithorax condition (resulting from direct artificial selection on that trait) should be seen as an adaptive trait, and his answer is that it should not. Williams instead sees the bithorax condition as "a disruption . . . of development," a failure of the organism to respond (1966, pp. 75–8). Hence, Williams draws a wedge between the notion of a trait that is a direct *product* of a selection process and a trait that fits his stronger *engineering* definition of an adaptation (see Gould and Lewontin 1979; Sober 1984, p. 201; cf. Dobzhansky 1956).[15]

This essential distinction between the *selection-product* and *engineering* views of adaptation is far from widely recognized. My claim here is that greater awareness of this distinction and its consequences will contribute to the understanding of several very heated debates in evolutionary theory.

For example, the engineering notion of adaptation is at work in the long dispute over the relationship between natural and sexual selection. Many evolutionists, starting with Darwin, rejected the idea that the products of a sexual selection process should be considered *adaptations*. In fact, analysis of the process of sexual selection is sometimes motivated by the drive to find an explanation for the presence of "maladaptive" traits; hence, the distinction between the selection product and engineering notion of adaptation plays an important role. Kirkpatrick (1987), for instance, uses a notion of adaptedness based on mean survival values in his argument that sexual selection does not always produce adaptations.

Consider for a moment the two schools of sexual selection theory. The "good genes" school claims that mate choice evolves under selection for females to mate with ecologically adaptive genotypes. The assumption here is that even though it appears that the females are basing their mate choice on a nonadaptive character, the character is actually an indication of the male's adaptedness (see, e.g., Vehrencamp and Bradbury 1984, Hamilton et al. 1990). The "nonadaptive" school claims that "preferences frequently cause male traits to evolve in ways that are *not adaptive* with respect to their ecological environment" (Kirkpatrick 1987, p. 44; emphasis added). In other words, the kinds of males preferred by females do not correspond with the kinds of males favored by natural selection. The result is a compromise between natural and sexual selection, the final state being one "that is maladaptive with respect to what natural selection acting alone would produce" (Kirkpatrick 1987, p. 45). Sir Ronald Fisher developed mathematical models showing how preferences for maladaptive males could evolve (1958 [1930]; see discussion in Lande 1980; Spencer and Masters 1992; Cronin 1991).

But an alternate concept of adaptation is available: the sexually selected traits that are advantageous to mating can still be seen as adaptations once the meaning of "adaptation" is adjusted. In this school of thought, the notion of "adaptation" should be broadened to include traits that contribute exclusively to reproductive success even though the more traditional definition is in terms of engineering for survival (e.g., Cronin 1991 versus Bock 1980; see Kirkpatrick 1987).

As these perennial debates about the relation between sexual selection and natural selection show, when asking whether a given level of entity possesses adaptations, it is necessary to state not only the level of selection in question but also which notion of adaptation – either *selection-product* or *engineering* – is being used. This distinction between the two meanings of adaptation also turns out to be pivotal in the debates about the efficacy of higher levels of selection, as we will see in Sections 13.3.1 and 13.3.3.

13.2.5 Summary

In this section, I have described four distinct questions that appear under the rubric of "the units of selection" problem: What is the interactor? What is the replicator? What is the beneficiary? and What entity manifests any adaptations resulting from evolution by selection? I have also discussed the existence of a very serious ambiguity in the meaning of "adaptation." I have no intention of defending one meaning or the other, but I will show that *which* meaning is in play has had deep consequences for both the group selection debates and the species selection debates. In Section 13.3, I will use my taxonomy of questions to sort out the most influential positions in three debates: group selection (Section 13.3.1), genic selection (Section 13.3.2), and species selection (Section 13.3.3).

13.3 An Anatomy of the Debates

13.3.1 Group Selection

Williams' famous near-deathblow to group panselectionism was, oddly enough, about benefit. He was interested in cases in which there was selection among groups and the group as a whole *benefited from* organism-level traits (including behaviors) that seemed disadvantageous to the organism.[16] Williams argued that the presence of a benefit to the group was *not sufficient* to establish the presence of group selection. He did this by showing that a group benefit was not necessarily a group adaptation.[17] His assumption was that a genuine group selection process results in the evolution of a group-level trait – a real adaptation – that serves a design purpose for the group. The mere existence, however, of traits that benefit the group is not enough to show that they are adaptations; in order to be an adaptation, under Williams' view, the trait must be an *engineering* adaptation that evolved by natural selection. Williams argued that group benefits do not, in general, exist *because* they benefit the group; that is, they do not have the appropriate causal history (see Brandon 1981, 1985, p. 81; Sober 1984, p. 262 ff.).

Implicit in Williams' discussion is the assumption that being a unit of selection at the group level requires two things: (1) having the group as an interactor, and (2) having a group-level engineering-type adaptation. That is, Williams combines two different questions, the interactor question and the manifestor-of-adaptation question, and calls this combined set *the* unit of selection question. These requirements for "group selection" make perfect sense given that Williams' prime target was Wynne-Edwards, who promoted a view of group selection that incorporated this same two-pronged definition of a unit of selection.

This combined requirement of "strong" (engineering) group-level adaptations in addition to the existence of an interactor at the group level is a very popular version of the necessary conditions for being a unit of selection within the group selection debates. David Hull claims that the group selection issue hinges on "whether entities more inclusive than organisms exhibit adaptations" (1980, p. 325). John Cassidy states that the unit of selection is determined by "Who or what is best understood as the possessor and beneficiary of the trait" (1978, p. 582). Similarly, Eldredge requires adaptations for an entity to count as a unit of selection, as does Vrba (Eldredge 1985, p. 108; Vrba 1983, 1984).

Maynard Smith (1976) also ties the engineering notion of adaptation into the version of the units of selection question he would like to consider. In an argument separating group and kin selection, Maynard Smith concludes that group selection is favored by small group size, low migration rates, and rapid extinction of groups infected with a selfish allele and that "the ultimate test of the group selection hypothesis will be whether populations having these characteristics tend to show 'self-sacrificing' or 'prudent' behavior more commonly than those which do not" (1976, p. 282). This means that the presence of group selection or the effectiveness of group selection is to be measured by the existence of nonadaptive behavior on the part of individual organisms along with the presence of a corresponding group-level adaptation. Therefore, Maynard Smith does require a group-level adaptation for groups to count as units of selection. As with Williams, it is significant that he assumes the *engineering* notion of adaptation rather than the weaker *selection-product* notion. As Maynard Smith puts it, "an explanation in terms of group advantage should always be explicit, and always calls for some justification in terms of the frequency of group extinction" (1976, p. 278; cf. Wade 1978; Wright 1980).

In contrast to the preceding authors, Sewall Wright separated the interactor and manifestor-of-adaptation questions in his group selection models (cf. Lewontin 1978; Gould and Lewontin 1979). Wright distinguishes between what he calls "intergroup selection," that is, interdemic selection in his shifting balance process, and "group selection for group advantage" (1980, p. 840; cf. Wright 1929, 1931).[18] He cites Haldane (1932) as the originator of the term "altruist" to denote a phenotype "that contributes to group advantage at the expense of disadvantage to itself" (1980, p. 840). Wright connects this debate to Wynne-Edwards, whom he characterizes as asserting the evolutionary importance of "group selection for group advantage." He argues that Hamilton's kin selection model is "very different" from "group selection for the uniform advantage of a group" (1980, p. 841; see Arnold and Fristrup 1982; Damuth and Heisler 1987, 1988).

Wright takes Maynard Smith, Williams, and Dawkins to task for mistakenly thinking that because they have successfully criticized group selection for group advantage, they can conclude that "natural selection is practically wholly genic." Wright argues, "none of them discussed group selection for organismic advantage to individuals, the dynamic factor in the shifting balance process, although this process, based on irreversible local peak-shifts is not fragile at all, in contrast with the fairly obvious fragility of group selection for group advantage, which they considered worthy of extensive discussion before rejection" (1980, p. 841).

This is a fair criticism of Maynard Smith, Williams, and Dawkins. My diagnosis of this problem is that these authors failed to distinguish between two questions: the interactor question and the manifestor-of-adaptation question. Wright's interdemic group selection model involves groups only as interactors, not as manifestors of group-level adaptations. Further, he is interested only in the effect the groups have on organismic adaptedness and expected reproductive success. More recently, modelers following Sewall Wright's interest in structured populations have created a new set of genetical models that are also called "group selection" models and in which the questions of group adaptations and group benefit play little or no role.[19]

For a period spanning two decades, however, Maynard Smith, Williams, and Dawkins did not acknowledge that the position they attacked, namely, Wynne-Edwards', is significantly different from other available approaches to group selection, such as Wright's, Wade's, Wilson's, Uyenoyama's, or Lewontin's. Ultimately, however, both G. C. Williams and Maynard Smith recognized the significance of the distinction between the interactor question and the manifestor-of-adaptation question. In 1985 Williams wrote, "If some populations of a species are doing better than others at persistence and reproduction, and if such differences are caused in part by genetic differences, this selection at the population level must play a role in the evolution of a species," while concluding that group selection "is unimportant for the origin and maintenance of adaptation" (Williams 1985, pp. 7–8).

And in 1987, Maynard Smith made an extraordinary concession:

There has been some semantic confusion about the phrase "group selection," for which I may be partly responsible. For me, the debate about levels of selection was initiated by Wynne-Edwards' book. He argued that there are group-level adaptations . . . which inform individuals of the size of the population so that they can adjust their breeding for the good of the population. He saw clearly that such adaptations could evolve *only* if populations were units of evolution. . . . Perhaps unfortunately, he referred to the process as "group selection." As a consequence, for me and for many others who engaged in this debate, the phrase came to imply that groups were sufficiently isolated from one another reproductively to act as units of evolution, and not merely that selection acted on groups.

The importance of this debate lay in the fact that group-adaptationist thinking was at that time widespread among biologists. It was therefore important to establish that there is no reason to expect groups to evolve traits ensuring their own survival unless they are sufficiently isolated for like to beget like. . . . When Wilson (1975) introduced his trait-group model, I was for a long time bewildered by his wish to treat it as a case of group selection, and doubly so by the fact that his original model . . . had interesting results only when the members of the groups were genetically related, a process I had been calling kin selection for ten years. I think that these semantic difficulties are now largely over. (1987, p. 123)

Dawkins also seems to have rediscovered the evolutionary efficacy of higher-level selection processes in an article on artificial life. In this article, he is primarily concerned with modeling the course of selection processes, and he offers a species-level selection interpretation for an aggregate species-level trait (Dawkins 1989a). Still, he seems not to have recognized the connection between this evolutionary dynamic and the controversies surrounding group selection because in his second edition of *The Selfish Gene* (Dawkins 1989b) he had yet to accept the distinction made so clearly by Wright in 1980. This was in spite of the fact that, by 1987, the importance of distinguishing between evolution by selection processes and any strong adaptations produced by those processes had been acknowledged by the workers Dawkins claimed to be following most closely, Williams and Maynard Smith.

13.3.2 Genic Selection

One may understandably think that Dawkins is interested in the replicator question because he claims that the unit of selection ought to be the replicator. This would be a mistake. Dawkins is interested primarily in a specific ontological issue about benefit. He is asking a special version of the beneficiary question, and his answer to that question dictates his answers to the other three questions under consideration in this chapter.

Briefly, Dawkins argues that because replicators are the only entities that "survive" the evolutionary process, they must be the beneficiaries. What happens in the process of evolution by natural selection happens *for their sake*, for their benefit. Hence, interactors interact for the replicators' benefit, and adaptations belong to the replicators. Replicators are the only entities with real agency as initiators of biochemical causal chains; hence, they accrue the credit and are the real units of selection.

Dawkins' version of the units of selection question amounts to a combination of the *beneficiary* question plus the *manifestor-of-adaptation* question. He does not somehow mistakenly think that he is answering the more predominant interactor question; rather, he argues that people who focus on interactors are laboring under a misunderstanding of evolutionary theory. One reason he thinks this, I submit, is that he takes as his opponents those who hold a combination of the interactor plus manifestor-of-adaptations definition of a unit of selection (e.g., Wynne-Edwards). Unfortunately, Dawkins thereby ignores those who are pursuing the interactor question alone; these researchers are not vulnerable to the criticisms he poses against the combined interactor–adaptation view.

I will discuss two aspects of Dawkins' own version of the units of selection issue. The first is his own preferred interpretation of the "real" units of selection problem. Here, I will attempt to clarify the key issues of interest to Dawkins and to relate these to the issues of interest to others. The second significant aspect of Dawkins' treatment of the units question is his characterization of the alternative views. The radical position Dawkins takes on units of selection makes more sense once his characterization of his opponents' questions becomes clear. In sum, he attributes to his opponents – sometimes incorrectly – a rich definition of a unit of selection involving not just the interactor question but also the beneficiary and manifestor-of-adaptation questions.

13.3.2.1 Dawkins' Preferred Question: Beneficiary

Dawkins believes that interactors, which he calls "vehicles," are not relevant to *the* units of selection problem. The *real* unit of selection, he argues, should be replicators, "the units that actually survive or fail to survive" (1982b, pp. 113–16). Organisms or groups as "vehicles" may be seen as the unit of *function* in the selection process, but they should not, he argues, be seen as the units of *selection* because the characteristics they acquire are not passed on (1982b, p. 99). Here he is following Williams' line. Genotypes have limited lives and fail to reproduce themselves because they are destroyed in every generation by meiosis and recombination in sexually reproducing species; they are only temporary (Williams 1966, p. 109). Hence, genes are the only units that survive in the selection process. The gene (replicator) is the real unit because it is an "indivisible fragment," it is "potentially immortal" (Williams 1966, pp. 23–4; Dawkins 1982b, p. 97).

The issue, for Dawkins, is "Whether, when we talk about a unit of selection, we ought to mean a vehicle at all, or a replicator" (1982b, p. 82). He clearly distinguishes the dispute he would like to generate from the group-versus-organismic selection controversy, which he characterizes as a disagreement "about the rival claims of two suggested kinds of vehicles" (1982b, p. 82). In his view, replicator selection should be seen as an alternative framework for both organismic and group selection models.

There are two mistakes that Dawkins is not making. First, he does not deny that interactors are involved in the evolutionary process. He emphasizes that it is not necessary, under his view, to believe that replicators are directly "visible" to selection forces (1982b, p. 176). Once again, his "vehicles" are conceived as the units of function in the selection process (like interactors) but not as the units of selection.[20] Dawkins has recognized from the beginning that *his* question is completely distinct from the interactor question. He remarks, in fact, that the debate about group versus individual selection is "a factual dispute about the level at which selection is most effective in nature," whereas his own point is "about what we ought to *mean* when we talk about a unit of selection" (1982a, p. 46). He also realizes that genes or other replicators do not "literally face the cutting edge of natural selection. It is their phenotypic effects that are the proximal subjects of selection" (1982a, p. 47). He suggests changing his own terminology to "replicator survival" to avoid this confusion but does not seem to have followed up.

Second, Dawkins does not specify how large a chunk of the genome he will allow as a replicator; there is no commitment to the notion that single genes are the only possible replicators. He argues that if Lewontin, Franklin, Slatkin, and others are right, his view will not be affected (see Section 13.2.2). If linkage disequilibrium is very strong, then the "effective replicator will be a very large chunk of DNA" (Dawkins 1982b, p. 89). We can conclude from this that Dawkins is not interested in the replicator question at all; his claim here is that his framework can accommodate any of its possible answers.

On what basis, then, does Dawkins reject the questions about interactors? I think the answer lies in the particular question in which he is interested, namely, What is "the nature of the entity *for whose benefit* adaptations may be said to exist"?[21]

On the face of it, it is certainly conceivable that one might identify the beneficiary of the adaptations as – in some cases, anyway – the individual organism or group that exhibits the phenotypic trait taken to be the adaptation. In fact, Williams seems to have done just that in his discussion of group selection.[22] But Dawkins rejects this move, introducing an *additional* qualification to be fulfilled by a unit of selection; it must be "the unit that actually survives or fails to survive" (1982a, p. 60). Because organisms, groups, and even genomes are destroyed during selection and reproduction, the answer to the survival question must be the replicator. (Strictly speaking, this is false; it is *copies* of the replicators that survive. He therefore must mean replicators in some sense of information and not as biological entities [see Hampe and Morgan 1988; cf. Griesemer in press]).

But there is still a problem. Although Dawkins concludes, "there should be no controversy over replicators versus vehicles. Replicator survival and vehicle selection are two aspects of the same process" (1982a, p. 60), he does not just leave the vehicle selection debate alone. Instead, he argues that we do not need the concept of discrete vehicles at all. I have shown elsewhere that if vehicles are understood strictly as interactors, Dawkins (and everyone else) *cannot* do without them (Lloyd 1988, Ch. 7).

In Dawkins' analysis, the fact that replicators are the only "survivors" of the evolution-by-selection process automatically answers the question of who owns the adaptations. He claims that adaptations *must* be seen as being designed for the good of the active germ-line replicator for the simple reason that replicators are the only entities around long enough to enjoy them over the course of natural selection. He acknowledges that the phenotype is "the all important instrument of replicator preservation," and that genes' phenotypic effects are organized into organisms (that thereby may benefit from them during their own lifetimes) (1982b, p. 114). But because only the active germ-line replicators survive, they are the true *locus of adaptations* (1982b, p. 113). The other things that *benefit* over the short term (e.g., organisms with adaptive traits) are merely the tools of the real survivors, the real owners. Hence, Dawkins rejects the vehicle approach partly because he identifies it with the manifestor-of-adaptation question, which he has answered by definition, in terms of long-term beneficiary.[23]

13.3.2.2 Dawkins' Characterization of Other Approaches

As discussed earlier, Dawkins is aware that the vehicle concept is "fundamental to the predominant orthodox approach to natural selection" (1982b, p. 116). He rejects this approach in *The Extended Phenotype*, claiming, "the main purpose of this book is to draw attention to the weaknesses of the whole vehicle concept" (1982b, p. 115). I will argue in the following paragraphs that his "vehicle approach" is *not* equivalent to what I have called the "interactor question" but encompasses a much more restricted approach.

In particular, when Dawkins argues against "the vehicle concept," he is arguing against the desirability of seeing the individual organism as the one and only possible vehicle. His target is explicitly those who hold what he calls the "Central Theorem," which says that *individual organisms should be seen as maximizing their own inclusive fitness* (1982b, pp. 5, 55). Dawkins' arguments are indeed damaging to the Central Theorem, but they are ineffective against other approaches that define units of selection more generally, that is, as interactors.

One way to interpret the Central Theorem is that it implies that the individual organism is always the beneficiary of any selection process; Dawkins seems to mean by "beneficiary" *both* the manifestor of an adaptation and that which survives to reap the rewards of the evolutionary process. He argues, rightly and persuasively, I think, that it does not make sense *always* to consider the individual organism to be the beneficiary of a selection process.

Note, however, that Dawkins is not arguing against the importance of the interactor question in general but rather against a particular definition of a unit of selection. The view he is criticizing assumes that the individual organism is the interactor, *and* the beneficiary, *and* the manifestor-of-adaptations. Consider his main argument against the utility of considering vehicles; the primary reason to abandon thinking about vehicle selection is that it confuses people (1982b, p. 189). But look at his examples; their point is that it is inappropriate always to ask how an organism's behavior benefits that organism's inclusive fitness. We should instead ask, says Dawkins, "whose inclusive fitness the behavior is benefiting" (1982b, p. 80). He states that his purpose in the book is to show that "theoretical dangers attend the assumption that adaptations are for the good of . . . the individual organism" (1982b, p. 91).

So, Dawkins is quite clear about what he means by the "vehicle selection approach"; he advances powerful arguments against the assumption that the individual is always the interactor cum beneficiary cum manifestor-of-adaptations. This approach is clearly *not* equivalent to the approach to units of selection I have characterized as the interactor question. Unfortunately, Dawkins extends his conclusions to these other approaches, which he has, in fact, not addressed. Dawkins' lack of consideration of the interactor definition of a unit of selection leads to two grave problems with his views.

One problem is that he has a tendency to interpret all group selectionist claims as being about beneficiaries and manifestors-of-adaptations as well as interactors; this is a serious misreading of authors who are pursuing the interactor question alone. Consider, for example, Dawkins' argument that groups should not be considered units of selection:

> To the extent that active germ-line replicators benefit from the survival of the group of individuals in which they sit, over and above the [effects of individual traits and altruism], we may expect to see adaptations for the preservation of the group. But all these adaptations will exist, fundamentally, through differential replicator survival. The basic beneficiary of any adaptation is the active germ-line replicator (1982b, p. 85).

Notice that Dawkins begins by admitting that groups can function as interactors and even that group selection may effectively produce group-level adaptations. The argument that groups should not be considered real units of selection *amounts to the claim* that the groups are not the ultimate beneficiaries. To counteract the intuition that the groups do, of course, benefit, in some sense, from the adaptations, Dawkins uses the terms "fundamentally" and "basic," thus signifying what he considers the most important level. Even if a group-level trait is affecting a change in gene frequencies, "it is still genes that are regarded as the replicators which actually survive (or fail to survive) as a consequence of the (vehicle) selection process" (1982b, p. 115). Dawkins argues, "a population . . . is not stable and unitary enough to be 'selected' in preference to another population" (1982b, p. 100).

Saying all this does not, however, address the fact that other researchers investigating group selection are asking the interactor question and sometimes also the manifestors-of-adaptations question rather than Dawkins' special version of the (ultimate) beneficiary question. Dawkins gives no additional reason to reject these other questions as legitimate; he simply reasserts the superiority of his own preferred units-of-selection question.

This is fair enough, provided that Dawkins keeps the different questions clear. But he seems instead to misinterpret the claims of group selectionists. For instance, Dawkins believes that group selectionists hold an expanded version of the Central Theorem, that is, that group-inclusive fitness is "that property of a group which will appear to be maximized when what is really being maximized is gene survival" (1982b, p. 187). Although Wynne-Edwards might be characterized as holding this view, Wade and Wilson certainly cannot. I think that Dawkins rejects their projects because he does not distinguish them from Wynne-Edwards's program (1978, pp. 73–4; 1982b, p. 115).

The other, more serious problem is that Dawkins fails to address, in his own theory, the interactor question itself: Which entities can and should be delineated as having traits or properties by means of which they interact with the environment in ways that affect the process of evolution by natural selection? In his desire to eliminate empirically and theoretically unjustified claims about beneficiaries of the selection process, Dawkins omits consideration of relevant questions about phenotypes that are addressed by other theoreticians. His attempt to circumvent the problems of a very restricted approach to selection that focuses on the organismic phenotype leads him, unfortunately, to gloss over a gap in his own view – specifically, how to delineate the "extended phenotype."

In discussing the extended phenotype, Dawkins is interested exclusively in traits that "might conceivably influence, positively or negatively, the replication success of the gene or genes concerned" (1982b, p. 234). Other incidental traits are "of no interest to the student of natural selection"; therefore, they are not included in the extended phenotype (1982b, p. 207). This makes perfect sense; he focuses on traits on which selection operates. But does he offer a principle for identifying these traits? Sterelny and Kitcher (1988) managed to distill a method of determining significant phenotype out of Dawkins. It turns out, not surprisingly, that this method is a simple version of the same principle widely used to delineate *interactors* by those theorists working on the (pure) interactor question (Lloyd 1988, Chapter 7). Hence, Dawkins can accommodate a viable method for delineating the extended phenotype but only through landing himself squarely in the middle of the interactor debate.

In conclusion, Dawkins' objections are effective against a particular view of the units of selection in which the unit is an interactor plus the beneficiary plus the manifestor of an adaptation. These objections are ineffective, however, against the more sophisticated views about units of selection widely held among the geneticists developing hierarchical selection models based on the interactor question.

The tension between Dawkins' picture of his opponents and the actual range of views they hold has led to an unfortunate and severe case of arguing at cross-purposes. The genic selectionists never actually make contact with some of their supposed opponents, and the hierarchal modelers similarly fail to realize that the appropriately limited version of the genic selectionist claims is no threat to them. Dawkins' concern to establish the ontological priority of genes over the perceived theoretical hegemony of individual organisms falls on deaf ears among the geneticists, who do, after all, provide the inspiration for Dawkins' view in the first place.

13.3.3 Species Selection

Ambiguities about the definition of a unit of selection have also snarled the debate about selection processes at the species level. I argue in this section that a combination of the interactor question and the manifestor-of-adaptation question (in the engineering sense) led to the rejection of research aimed at considering the role of species as interactors, *simpliciter*, in evolution. Once it is understood that species-level interactors may or may not possess design-type adaptations, it becomes possible to distinguish two research questions: Do species function as interactors, playing an active and significant role in evolution by selection? and Does the evolution of species-level interactors produce species-level engineering adaptations and, if so, how often?

For most of the history of the species selection debate, these questions have been lumped together; asking whether species could be units of selection meant asking whether they fulfilled *both* the interactor and manifestor-of-adaptation roles. For example, Vrba used Maynard Smith's treatment of the evolution of altruism as a touchstone in her definition of species selection (1984). Maynard Smith argued that kin selection can cause the spread of altruistic genes but that it should not be called group selection (1976).[24] Vrba agreed that the spread of altruism should not be considered a case of group selection because "there is no group aptation involved; altruism is not emergent at the group level" (1984, p. 319; Maynard Smith gives different reasons for his rejection). This amounts to assuming that there must be group benefit in the sense of a design-type group-level adaptation. Vrba's view was that evolution by selection is not happening at a given level unless there is a benefit or adaptation at that level. That is, her definition of a unit of selection is a combination of an interactor and a manifestor-of-adaptations. She explicitly equates units of selection with the existence of an adaptation at that level (1983, p. 388); furthermore, it seems that she has adopted the stronger *engineering* definition of adaptation.

Eldredge also argues that species selection does not happen unless there are species level adaptations (1985, pp. 196, 134). Eldredge rejects certain cases as higher-level *selection processes* because "frequencies of the properties of lower-level individuals which are part of a higher-level individual simply do not make convincing higher-level adaptations" (1985, p. 133).

Vrba, Eldredge, and Gould all defined a unit of selection as requiring an emergent, adaptive property (Vrba 1983, 1984; Vrba and Eldredge 1984; Vrba and Gould 1986). In my analysis, this amounts to asking a combination of the interactor and manifestor-of-adaptations question.

But consider the lineage or species-wide trait of variability. Although this may come as a surprise to some, treating species as interactors has a long tradition (Dobzhansky 1956, Thoday 1953, Lewontin 1958). If species are conceived as interactors (and not necessarily manifestors-of-adaptations), then the notion of species selection is not vulnerable to Williams's original antigroup-selection objections.[25] The old idea was that lineages with certain properties of being able to respond to environmental stresses would be selected for, that the trait of variability itself would be selected for, and that it would spread in the population of populations. In other words, lineages were treated as interactors. The earlier researchers spoke loosely of adaptations, where adaptations were defined in the weak sense as equivalent simply to the outcome of selection processes (at any level). They were explicitly *not* concerned with the effect of species selection on organismic level traits but with the effect on species-level characters such as speciation rates, lineage-level survival, and extinction rates of species. I have argued, with Gould, that this sort of case represents a perfectly good form of species selection even though some balk at the thought that variability would then be considered, under a weak definition, a species-level adaptation (Lloyd and Gould 1993, Lloyd 1988).

Vrba has also recognized the advantages of keeping the interactor question separate from a requirement for an engineering-type adaptation. In her more recent review article, she has dropped her former requirement that, in order for species to be units of selection, they must possess species-level adaptations (1989). Ultimately, her current definition of species selection is in conformity with a simple interactor interpretation of a unit of selection (cf. Damuth and Heisler 1988, Lloyd 1988).

It is easy to understand how the two-pronged definition of a unit of selection – as interactor and manifestor-of-adaptation – held sway for so long in the species selection debates. After all, it dominated much of group selection debates until just recently. And I have argued that some of the confusion and conflict over higher-level units of selection arose because of an historical contingency – Wynne-Edwards implicit definition of a unit of selection and the responses it provoked.

13.4 Conclusion

It makes no sense to treat different answers as competitors if they are answering distinct questions. I have offered a framework of four questions with which the debates appearing under the rubric of "units of selection" can be classified and clarified. I follow Dawkins, Hull, and Brandon in separating the classic question about the level of selection or interaction (*the interactor question*) from the issue of how large a chunk of the genome functions as a replicating unit (*the replicator question*). I also separate the interactor question from the question of which entity should be seen as acquiring adaptations as a result of the selection process (*the manifestor-of-adaptations question*). In addition, I insist that there is a crucial ambiguity in the meaning of *adaptation* that is routinely ignored in these debates: adaptation as a selection product and adaptation as an engineering design. Finally, I suggest distinguishing the issue of the entity that ultimately benefits from the selection process (*the beneficiary question*) from the other three questions.

I have used this set of distinctions to analyze leading points of view about the units of selection and to clarify precisely the question or combination of questions with which each of the protagonists is concerned. I conclude that there are many points in the debates in which misunderstandings may be avoided by a precise characterization of *which* of *the* units of selection questions is being addressed.

13.5 Acknowledgments

I am grateful to the members of those audiences, and to Richard Dawkins, Steve Gould, Marjorie Grene, Jim Griesemer, Bill Hamilton, David Hull, Dick Lewontin, John Maynard Smith, Hamish Spencer, Mike Wade, and George Williams for helpful suggestions on previous drafts. A draft of this article was assigned to a Philosophy of Biology class at UC Davis (Winter 1991) by Michael Dietrich, and I thank him for his comments and suggestions. Some of this material was also presented informally in an interview in December 1989 with Werner Callebaut, parts of which were published in his *Taking the Naturalistic Turn: How Real Philosophy of Science Is Done* (1993). An abbreviated presentation of the four basic questions (Section 2 of this chapter) is contained in my "Unit of Selection" essay for *Keywords in Evolutionary Biology* (1992), and I would like to thank Evelyn Fox Keller for insightful suggestions for improving the presentation of these ideas.

Notes

1. Mitchell also argues for the importance of keeping the interactor issue separate from issues involving replicators (1987).
2. For some of the pivotal arguments in the debates, see Brandon 1982, 1985, 1990; Dawkins 1978, 1982a,b, 1986; Hull 1980; Lewontin 1970; Maynard Smith 1964, 1976; Sober 1984; Sober and Lewontin 1982; Wade 1978, 1985; Wilson 1975, 1980; Wright 1980.
3. I have found nearly two hundred references to papers and books by biologists and philosophers that treat what is called here "the interactor question" (Lloyd 1988/1994); this represents just a fraction of the literature on the topic. The second most prominent interpretation of "the units of selection" question is "the replicator question" discussed in Section 2.2.
4. For example, Arnold and Fristrup 1982; Colwell 1981; Craig 1982; Crow and Aoki 1982; Crow and Kimura 1970; Damuth and Heisler 1988; Hamilton 1975; Heisler and Damuth 1987; Lande and Arnold 1983; Li 1967; Ohta 1983; Price 1972; Uyenoyama 1979; Uyenoyama and Feldman 1980; Wade 1978, 1980, 1985; Wade and Breden 1981; Wade and McCauley 1980; Wilson 1983; Wilson and Colwell 1981; Wimsatt 1980, 1981. See discussion in Lloyd 1988.
5. See Aoki 1982; Boorman and Levitt 1973; Fisher 1930; Ghiselin 1974; Leigh 1977; Levin and Kilmer 1974; Maynard Smith 1964, 1976; Uyenoyama 1979; Williams 1966.
6. See Lloyd 1988, Chapter 4.
7. For example, Arnold and Fristrup 1982; Brandon 1982; Colwell 1981; Damuth and Heisler 1988; Eldredge 1985; Griesemer and Wade 1988; Heisler and

Damuth 1987; Lewontin 1970; Lloyd 1986, 1988, 1989; Sober 1981, 1984; Sober and Lewontin 1982; Vrba and Gould 1986; Wade 1985; Wilson 1980; Wimsatt 1980, 1981.

8. See the models cited in n.5 above for various technical approaches to expressing this special relation between fitness and trait.
9. Lewontin 1970, 1974; Franklin and Lewontin 1970; Slatkin 1972; see discussion in Wimsatt 1980; Brandon 1982.
10. The sense of agency assumed by Dawkins is worth investigating in detail. I will not, however, address this issue directly here. See Griesemer and Wade 1988. Related issues are discussed in Section 3.2.
11. Brandon (1985) argues that such a view, which separates the level of adaptation from that of beneficiary, cannot be explanatory. Although I sympathize with Brandon's conclusions, they follow only under his set of definitions, which Dawkins and other genic selectionists would certainly reject.
12. For explicit assumptions that being a unit of selection involves having an adaptation at that level, see Brandon 1982, 1985; Burian 1983; Mitchell 1987; Maynard Smith 1976; Vrba 1984.
13. Oddly, Williams writes, "natural selection would produce or maintain adaptation as a matter of definition" (1966, p. 25; cf. Mayr 1976). However, Williams is committed to an engineering definition of adaptation (personal communication 1989).
14. For example, Williams 1966; Bock 1980; Dunbar 1982; Ghiselin 1974; Gould and Lewontin 1979; Hull 1980; Lewontin 1978; Mayr 1978.
15. Note that Williams says that "natural selection would produce or maintain adaptation as a matter of definition" (1966, p. 25; cf. Mayr 1976). This comment conflicts with the conclusions Williams draws in this discussion of Waddington; however, Williams later retracts this bithorax analysis (1985).
16. Similarly for Maynard Smith (1964).
17. Hence, Williams is here using the term *benefit* to signify the manifestation of an adaptation at the group level.
18. It is worth remembering at this point that under Wright's view, interdemic group selection provided the means for attaining greater organismic adaptation; groups that are favored are "those local populations that happen to have acquired superior coadaptive systems of genes" (1980, p. 841).
19. For example, Damuth and Heisler 1987; Heisler and Damuth 1988; Slatkin and Wade 1978; Uyenoyama 1979; Uyenoyama and Feldman 1980; Wade 1978, 1985; Wilson 1983.
20. Hull has also argued that his own "interactors" and Dawkins' "vehicles" are not the same things (1980).
21. (1982b, p. 81, my emphasis; cf. pp. 4, 5, 52, 84, 91, 113, 114). Compare an alternative formulation of Dawkins' central question: "When we say that an adaptation is 'for the good of' something, what is that something? . . . I am suggesting that the appropriate 'something,' the 'unit of selection' in that sense, is the active germ-line replicator" (1982a, p. 47). This particular formulation, I think, asks two questions, one about who the beneficiary of the selection process is and one about who possesses adaptations. Griesemer and Wimsatt's studies (1989) on Weismannism are of great help here.
22. Note that Williams, even though he "keeps his books" in terms of genes, argued against the notion that particular group traits were group adaptations *because* these group traits are not properly understood as *benefiting the group* in the proper historical selection scenario (Williams 1966).

23. Mitchell arrives at a similar conclusion through different arguments. She highlights the impact of the notion of adaptation that seems to depend on the roles that Dawkins assigns to replicators (1987, pp. 359–62).
24. Again, this was because the groups were not interpreted as possessing design-type adaptations *themselves.*
25. As Williams himself has acknowledged, in a discussion on species selection: "The answer to all these difficulties must be Lloyd's . . . idea that higher levels of selection depend, not on emergent characters, but on any and all emergent fitnesses" (1992, p. 27).

REFERENCES

Aoki, K. (1982). A condition for group selection to prevail over counteracting individual selection. *Evolution* 36:832–42.

Arnold, A. J., and Fristrup, K. (1982). The theory of evolution by natural selection: A hierarchical expansion. *Paleobiology* 8:113–29.

Bock, W. (1980). The definition and recognition of biological adaptation. *American Zoologist* 20:217–27.

Boorman, S. A., and Levitt, P. R. (1973). Group selection on the boundary of a stable population. *Theoretical Population Biology* 4:85–128.

Boorman, S. A. (1978a). Mathematical theory of group selection: Structure of group selection in founder populations determined from. . . . *Proceedings of the National Academy of Sciences, USA* 69:1909–13.

Brandon, R. N. (1978b). Adaptation and evolutionary theory. *Studies in History and Philosophy of Science* 9 (3):181–206.

Brandon, R. N. (1981). Biological teleology: Questions and explanations. *Studies in History and Philosophy of Science* 12 (2):91–105.

Brandon, R. N. (1982). The levels of selection. *Proceedings of the Philosophy of Science Association* 1982 1:315–23.

Brandon, R. N. (1985). Adaptation explanations: Are adaptations for the good of replicators or interactors? In *Evolution at a Crossroads: The New Biology and the New Philosophy of Science,* D. Depew and B. Weber (eds.), pp. 81–96. Cambridge, MA: MIT Press/Bradford.

Brandon, R. N. (1988). Levels of selection: A hierarchy of interactors. In *The Role of Behavior in Evolution,* H. C. Plotkin (ed.), pp. 51–71. Cambridge, MA: MIT Press.

Brandon, R. N. (1990). *Adaptation and Environment.* Princeton, NJ: Princeton University Press.

Burian, R. M. (1983). Adaptation. In *Dimensions of Darwinism,* M. Greene (ed.), pp. 287–314. Cambridge, UK: Cambridge University Press.

Callebaut, W. (ed.) (1993). *The Naturalistic Turn: How Real Philosophy of Science Is Done.* Chicago, IL: University of Chicago Press.

Cassidy, J. (1978). Philosophical aspects of the group selection controversy. *Philosophy of Science* 45:575–94.

Colwell, R. K. (1981). Evolution of female-based sex ratios: The essential role of group selection. *Nature* 290:401–4.

Craig, D. M. (1982). Group selection versus individual selection: An experimental analysis. *Evolution* 36:271–82.

Cronin, H. (1991). *The Peacock's Tail.* Oxford, UK: Oxford University Press.

Crow, J. F., and Kimura, M. (1970). *An Introduction to Population Genetics.* New York: Harper and Row.

Crow, J. F., and Aoki, K. (1982). Group selection for a polygenetic behavioral trait: A differential proliferation model. *Proceedings of the National Academy of Sciences, USA* 79:2628–31.

Damuth, J., and Heisler, I. L. (1988). Alternative formulations of multilevel selection. *Biology and Philosophy* 3:407–30.

Darwin, C. (1964). *On the Origin of Species* (Facsimile of 1st ed., 1869, edited by E. Mayr). Cambridge, MA: Harvard University Press.

Dawkins, R. (1978). Replicator selection and the extended phenotype. *Zeitschrift für Tierpsychologie* 47:61–76.

Dawkins, R. (1982a). Replicators and vehicles. In King's College Sociobiology Group, Cambridge, *Current Problems in Sociobiology*. Cambridge, UK: Cambridge University Press, pp. 45–64.

Dawkins, R. (1982b). *The Extended Phenotype*. New York: Oxford University Press.

Dawkins, R. (1986). *The Blind Watchmaker*. New York: Norton.

Dawkins, R. (1989a). The evolution of evolvability. In *Artificial Life, Santa Fe Institute Studies in the Sciences of Complexity*, C. Langdon (ed.), pp. 201–20. Reading, MA: Addison–Wesley.

Dawkins, R. (1989b). *The Selfish Gene*, Revised edition. New York: Oxford.

Dobzhansky, T. (1956). What is an adaptive trait? *American Naturalist* 40 (855):337–47.

Dobzhansky, T. (1968). Adaptedness and Fitness. In *Population Biology and Evolution*, R. C. Lewontin (ed.), pp. 109–21. Syracuse, NY: Syracuse University Press.

Dunbar, R. I. M. (1982). Adaptation, fitness and the evolutionary tautology. In King's College Sociobiology Group, Cambridge, *Current Problems in Sociobiology*. Cambridge, UK: Cambridge University Press, pp. 9–28.

Eldredge, N. (1985). *Unfinished Synthesis: Biological Hierarchies and Modern Evolutionary Thought*. New York: Oxford University Press.

Fisher, R. A. (1930). *The Genetical Theory of Natural Selection*. London: Oxford University Press. (Revised ed., 1958).

Franklin, I., and Lewontin, R. C. (1970). Is the gene the unit of selection? *Genetics* 65:707–34.

Ghiselin, M. T. (1974). *The Economy of Nature and the Evolution of Sex*. Berkeley, CA: University of California Press.

Gould, S. J., and Lewontin, R. C. (1979). The spandrels of San Marco and the Panglossian paradigm: A critique of the Adaptationist Programme. *Proceedings of the Royal Society of London* B205:581–98.

Gould, S. J., and Vrba, E. S. (1982). Exaptation – A missing term in the science of form. *Paleobiology* 8:4–15.

Griesemer, J. R. (in press). The informational gene and the substantial body: On the generalization of evolutionary theory by abstraction. In N. Cartwright and M. Jones (eds.), *Varieties of Idealization*. Poznan Studies. Amsterdam: Rodopi Publishers.

Griesemer, J. R., and Wade, M. J. (1988). Laboratory models, causal explanation, and group selection. *Biology and Philosophy* 3:67–96.

Griesemer, J. R., and Wimsatt, W. (1989). Picturing Weismannism: A case study of conceptual evolution. In M. Ruse (ed.), *What the Philosophy of Biology Is, Essays for David Hull*. Dordrecht, The Netherlands: Kluwer, pp. 75–137.

Haldane, J. B. S. (1932). *The Causes of Evolution*. London: Longmans, Green.

Hamilton, W. D. (1975). Innate social aptitudes in man: An approach from evolutionary genetics. In R. Fox (ed.), *Biosocial Anthropology*. New York: Wiley, pp. 133–55.

Hamilton, W. D., Axelrod, R., and Tanese, R. (1990). Sexual reproduction as an adaptation to resist parasites (A review). *Proceedings of the National Academy of Sciences, U.S.A.* 87:3566–73.

Hampe, M., and Morgan, S. R. (1988). Two consequences of Richard Dawkins' View of Genes and organisms. *Studies in History and Philosophy of Science* 19:119–38.

Heisler, I. L., and Damuth, J. (1988). A method for analyzing selection in hierarchically structured populations. *American Naturalist* 130:582–602.

Hull, D. L. (1980). Individuality and selection. *Annual Review of Ecology and Systematics* 11:311–32.

Keller, E. F., and Lloyd, E. A. (1992). *Keywords in Evolutionary Biology.* Cambridge, MA: Harvard University Press.

Kirkpatrick, M. (1987). Sexual selection by female choice in polygynous animals. *Annual Review of Ecology and Systematics* 187:43–70.

Krimbas, C. B. (1984). On Adaptation, Neo-Darwinian Tautology, and Population Fitness. In *Evolutionary Biology*, M. K. Hecht, B. Wallace, and G. T. Prance (eds.), 17:1–57.

Lande, R. (1980). Sexual dimorphism, sexual selection, and adaptation in polygenic characters. *Evolution* 34 (2):292–305.

Lande, R., and Arnold, S. J. (1983). The measurement of selection on correlated characters. *Evolution* 37:1210–27.

Leigh, E. G. (1977). How does selection reconcile individual advantage with the good of the group? *Proceedings of the National Academy of Sciences, USA* 74:4542–6.

Levin, B. R., and Kilmer, W. L. (1974). Interdemic selection and the evolution of altruism: A computer simulation study. *Evolution* 28:527–45.

Lewontin, R. C. (1958). A general method for investigating the equilibrium of gene frequency in a population. *Genetics* 43:421–33.

Lewontin, R. C. (1970). The units of selection. *Annual Review of Ecology and Systematics* 1:1–18.

Lewontin, R. C. (1974). *The Genetic Basis of Evolutionary Change.* New York: Columbia University Press.

Lewontin, R. C. (1978). Adaptation. *Scientific American* 239:156–69.

Li, C. C. (1967). Fundamental theorem of natural selection. *Nature* 214:505–6.

Lloyd, E. A. (1986). Evaluation of evidence in group selection debates. *Proceedings of the Philosophy of Science Association* 1986 1:483–93.

Lloyd, E. A. (1988). *The Structure and Confirmation of Evolutionary Theory.* Westport, CT: Greenwood Press. Paperback edition with new preface, Princeton University Press, 1994.

Lloyd, E. A. (1989). A structural approach to defining units of selection. *Philosophy of Science* 56:395–418.

Lloyd, E. A. (1992). Unit of Selection. In *Keywords in Evolutionary Biology.* E. F. Keller and E. A. Lloyd (eds.), pp. 334–40. Cambridge, MA: Harvard University Press.

Lloyd, E. A., and Gould, S. J. (1993). Species selection on variability. In *Proceedings of the National Academy of Sciences, USA* 90:595–9.

Matessi, C., and Jayakar, S. D. (1976). Conditions for the evolution of altruism under Darwinian selection. *Theoretical Population Biology* 9:360–87.

Maynard Smith, J. (1964). Group selection and kin selection: A rejoinder. *Nature* 201:1145–7.

Maynard Smith, J. (1976). Group selection. *Quarterly Review of Biology* 51:277–83.

Maynard Smith, J. (1981). The evolution of social behavior and classification of models.

In *Current Problems in Sociobiology*, Kings College Sociobiology Group (eds.), pp. 29–44. Cambridge, UK: Cambridge University Press.

Maynard Smith, J. (1984). The population as a unit of selection. In *Evolutionary Ecology*, 23rd British Ecological Society Symposium, B. Shorrocks, ed., pp. 195–202. Oxford, UK: Blackwell.

Maynard Smith, J. (1987). Evolutionary progress and levels of selection. In *The Latest on the Best*: J. Dupre (ed.). Cambridge, MA: MIT Press.

Mayr, E. (1976). *Evolution and the Diversity of Life*. Cambridge, MA: Harvard University Press.

Mayr, E. (1978). Evolution. *Scientific American* 239:49–55.

Mayr, E. (1982). Adaptation and selection. *Biologisches Zentralblatt* 101:161–74.

Mitchell, S. (1987). Competing units of selection?: A case of symbiosis. *Philosophy of Science* 54:351–67.

Munson, R. (1971). Biological adaptation. *Philosophy of Science* 38:200–15.

Ohta, K. (1983). Hierarchical theory of selection: The covariance formula of selection and its application. *Bulletin of the Biometrical Society of Japan* 4:25–33.

Price, G. R. (1972). Extension of covariance selection mathematics. *Annals of Human Genetics* 35:485–90.

Slatkin, M. (1972). On treating the chromosome as the unit of selection. *Genetics* 72:157–68.

Slatkin, M. (1981). A diffusion model of species selection. *Paleobiology* 7 (4):421–5.

Slatkin, M., and Wade, M. (1978). Group selection on a quantitative character. *Proceedings of the National Academy of Sciences, USA* 75:3531–4.

Slobodkin, L. B., and Rapoport, A. (1974). An optimal strategy of evolution. *Quarterly Review of Biology* 49:181–200.

Sober, E. (1981). Holism, individualism, and the units of selection. *Proceedings of the Philosophy of Science Association 1980* 2:93–121.

Sober, E. (1984). *The Nature of Selection*. Cambridge, MA: MIT Press.

Sober, E., and Lewontin, R. C. (1982). Artifact, cause and genic selection. *Philosophy of Science* 49:157–80.

Spencer, H., and Masters, J. (1992). Sexual selection: Contemporary debates. In E. F. Keller and E. A. Lloyd (eds.). *Keywords in Evolutionary Biology*.

Sterelny, K., and Kitcher, P. (1988). The return of the gene. *Journal of Philosophy* 85:339–61.

Thoday, J. M. (1953). Components of fitness. *Symp. Soc. Exp. Biol.* 7:96–113.

Uyenoyama, M. K. (1979). Evolution of altruism under group selection in large and small populations in fluctuating environments. *Theoretical Population Biology* 15:58–85.

Uyenoyama, M. K., and Feldman, M. W. (1980). Evolution of altruism under group selection in large and small populations in fluctuating environments. *Theoretical Population Biology* 17:380–414.

Vehrencamp, S. L., and Bradbury, J. W (1984). Mating systems and ecology. In J. R. Krebs (ed.). pp. 251–78. *Behavioural Ecology: An Evolutionary Approach*. Sunderland, MA: Sinauer.

Vrba, E. (1983). Macroevolutionary trends: New perspectives on the roles of adaptation and incidental effect. *Science* 221:387–9.

Vrba, E. (1984). What is species selection? *Systematic Zoology* 33:318–28.

Vrba, E. (1989). Levels of selection and sorting with special reference to the species level. *Oxford Surveys in Evolutionary Biology* 6:111–68.

Vrba, E., and Eldredge, N. (1984). Individuals, hierarchies and processes: Towards a more complete evolutionary theory. *Paleobiology* 10:146–71.
Vrba, E., and Gould, S. J. (1986). The hierarchical expansion of sorting and selection: Sorting and selection cannot be equated. *Paleobiology* 12:217–28.
Waddington, C. H. (1956). Genetic assimilation of the bithorax phenotype. *Evolution* 10:1–13.
Wade, M. J. (1977). An experimental study of group selection. *Evolution* 31:134–53.
Wade, M. J. (1978). A critical review of the models of group selection. *Quarterly Review of Biology* 53:101–14.
Wade, M. J. (1980). Kin selection: Its components. *Science* 210:665–7.
Wade, M. J. (1985). Soft selection, hard selection, kin selection, and group selection. *American Naturalist* 125:61–73.
Wade, M. J., and Breden, F. (1981). The effect of inbreeding on the evolution of altruistic behavior by kin selection. *Evolution* 35:844–58.
Wade, M. J., and McCauley, D. E. (1980). Group selection: The phenotypic and genotypic differentiation of small populations. *Evolution* 34:799–812.
Williams, G. C. (1966). *Adaptation and Natural Selection.* Princeton, NJ: Princeton University Press.
Williams, G. C. (1985). A defense of reductionism in evolutionary biology. *Oxford Surveys in Evolutionary Biology* 2:1–27.
Williams, G. C. (1992). *Natural Selection: Domains, Levels, and Challenges.* New York: Oxford University Press.
Wilson, D. S. (1975). A general theory of group selection. *Proceedings of the National Academy of Sciences, USA* 72:143–46.
Wilson, D. S. (1980). *The Natural Selection of Populations and Communities.* Menlo Park, CA: Benjamin Cummings.
Wilson, D. S. (1983). Group selection controversy: History and current status. *Annual Review of Ecology and Systematics* 14:159–87.
Wilson, D. S., and Colwell, R. K. (1981). Evolution of sex ratio in structured demes. *Evolution* 35:882–97.
Wimsatt, W. (1980). Reductionist research strategies and their biases in the units of selection controversy. In T. Nickles (ed.), pp. 213–59. *Scientific Discovery: Case Studies.* Dordrecht, The Netherlands: Reidel.
Wimsatt, W. (1981). Units of selection and the structure of the multi-level genome. *Proceedings of the Philosophy of Science Association 1980* 2:122–83.
Wright, S. (1929). Evolution in a Mendelian population. *Anat. Rec.* 44:287.
Wright, S. (1931). Evolution in Mendelian populations. *Genetics* 10:97–159.
Wright, S. (1945). Tempo and mode in evolution: A critical review. *Ecology* 26:415–9.
Wright, S. (1980). Genic and organismic selection. *Evolution* 34:825–43.
Wynne-Edwards, V. C. (1962). *Animal Dispersion in Relation to Social Behavior.* Edinburgh: Oliver and Boyd.

CHAPTER FOURTEEN

In Defense of Neo-Darwinism

Popper's "Darwinism as a Metaphysical Programme" Revisited

COSTAS B. KRIMBAS

It is a great pleasure for me to participate in this volume in honor of R. C. Lewontin, who has been a friend for nearly 40 years and who has influenced my thoughts and views. Curiously enough, he introduced me also to the work of Karl Popper, a philosopher who exerted a great influence on the active scientists of the second half of the twentieth century. Thus, in 1971, I read *The Poverty of Historicism* and *The Open Society and its Enemies*; later on I became familiar with his other works. Further, in 1974, the proceedings of a symposium on reductionism organised by my late teacher Theodosius Dobzhansky with his student Ayala were published, and they contain a significant contribution by Popper (Ayala and Dobzhansky 1974).

14.1 Neo-Darwinism as a Metaphysical Programme

Popper examined problems of evolutionary biology and, to my knowledge, published his views at least three times (Popper 1972, 1974, and 1978). Although he drastically modified and even retracted his earlier views in his 1978 article, I think that it is of great interest to examine his 1974 seminal text. There he stated that neo-Darwinism is a metaphysical program, "almost tautological," that could be described as applied situational logic. His article triggered several criticisms and protests (e.g., Halstead 1980, Little 1980, Anonymous 1981, Stamos 1996).

A metaphysical research program is not subject to falsification but to criticism and argumentation. According to Popper, Darwinism (neo-Darwinism to be precise) is applied situational logic; that is, once the situation or the conditions are provided, the outcome may be inferred by application of formal logic. Therefore, "neo-Darwinism is not a testable scientific theory but a possible framework for testable scientific theories." (Popper 1974, vol. 1, p. 134)

We could condense the neo-Darwinian mechanism of the evolutionary process into a three-assertion argument (according to Maynard Smith 1972; Hull 1974; Lewontin 1977, 1978; Krimbas 1984; see also Popper 1974 for a slightly different formulation). These assertions, or principles, according to Lewontin, are "necessary and sufficient conditions for evolutionary change by natural selection," and they explain what Popper termed "applied situational logic" (Popper 1974, vol. 1, p. 134). They refer to the evolutionary mechanism, not to the fact that evolution took place. This last assertion, that is, that evolution took place, was formulated in Popper's "first conjecture" as follows:

> The great variety of the forms of life on earth originate from very few forms, perhaps even from a single organism: there is an evolutionary tree, an evolutionary history. (Popper 1974, Vol. 1, p. 135)

14.2 Testing Popper's First Conjecture

Popper's first conjecture is subject to test and could eventually be falsified in different ways. One of them is the comparison of phylogenetic (evolutionary) trees of the same taxonomic units (e.g., the same species) derived from independent sets of valid data and constructed by the use of an adequate and efficient algorithm. Several such algorithms were proposed.[1] They permit the construction of binary trees. For their relative efficiencies, see Nei (1991). If evolution really took place, these trees should display the same topology; an unexpected dissimilarity should be considered a disproof of evolution. Similarity among trees comprising taxa from all five kingdoms could be considered a corroboration of the monophyletic origin of life; a disparity, on the contrary, should be taken as a proof against it.

By valid data, I mean accurate data deriving from uncorrelated variables (traits) that are numerous enough to lead to statistical inferences and providing unbiased estimations of genetic distances between the species or the taxa examined, when distances are used, or adequate clustering, when discrete character data are dealt with; that is, data expected to provide robust topologies. If the speciation events are very old, the relevant information may be lost; all traits – for example, all DNA sequences – differ among themselves in the same measure, and the topology cannot be resolved. Further, the traits used should not show homoplasy, that is similarity due to evolutionary convergence often an outcome of similar selective processes, seldom of chance effects (convergences, parallelisms, reversals). Similar selective pressures on different taxa produce similar phenotypes, and their similarity is not a measure of phylogenetic relatedness but of an ecological similarity. Similarities due to homoplasy do not necessarily change monotonically (increase) with the time elapsed since the splitting of the two taxa under consideration. Morphological, anatomical, physiological, and behavioral traits or characteristics can in principle show homoplasy. There are methods to clean the data and exclude those showing convergence. However, this operation may be justly considered circular when our task is to test the hypothesis of evolution, since the aberrant data those falsifying the evolutionary hypothesis, are the ones excluded. There are, on the other hand, some data considered more "safe," data that a priori seem less prone to produce homoplasies: the DNA base sequence outside the genes or eventually inside the introns that is in locations considered selectively neutral (although even this is actually disputed; see, for example, Shields et al. 1988). Care should also be taken to avoid repetitive DNA sequences, which owe their origin to mobile elements, for these elements do not necessarily follow the evolutionary history of the species or taxa but become incorporated in the DNA by some kind of horizontal transfer. Other good quality data are the banding sequences of giant chromosomes encountered in some Diptera. Wallace (1966) argued that these data may support the case for evolution as strongly as evidence provided in court such as identifications of fingerprints or of impressions on bullets. Here, too, some constraints have to be respected as discussed in Krimbas and Powell (1992). In the case of DNA sequences, similar constraints exist not yet mentioned here. In both cases a polymorphism conserved transpecifically with alternative variants fixed in some but not all clades in the reverse order of their appearance may lead to erroneous tree constructions (see also Miyamoto and Cracraft 1991). Some sources of data may not be rich enough in information to resolve nodes in the tree. In the case of gene arrangements, we may thus obtain collapsed trees; additional information may permit a more detailed resolution (see Brehm and Krimbas 1993).

I fully realize that in trying to exorcise the devil I have indicated too many constraints; however, more should be added: a complete examination of all possible trees to select the best one, in some of the methods already mentioned, is not a simple procedure; some shortcuts (branch and bound algorithms) facilitate this task, but the difficulties arising from an increased number of taxa are far from being resolved. When DNA base sequences are used, the alignment procedure for establishing the homologous positions (when small deletions and additions are considered) is a process that may induce errors and lead to the construction of incorrect trees. Furthermore, parsimony methods have proven inconsistent in some cases [e.g., for four taxa when evolutionary rates are unequal, for five taxa with equal rates when the topology is such that that root is on the central pendant edge adjacent to a short and long edges, for six taxa when rates of change are low, for a large number of taxa when all branch lengths (internal and pendant) are both small and equal: see Penny, Hendy, and Steel (1991)]. When a long-short-long edge pattern is encountered, parsimony methods may face a "long edge attract" problem affecting and distorting the topology. Although there is a belief that, as data accumulate, we may be reaching the correct topology (and for this there are indications: the longer the sequences the greater is the similarity among trees), parsimony methods may, however, lead to a wrong tree, as explained above.

Not only parsimony methods present problems of accuracy. The algorithms provide topologies for dichotomous trees; when the ancestral species gives rise to more than two species almost simultaneously, the resolution is poor and polytomous trees should be expected (Hoelzer and Melnick 1994). Problems are numerous, and the precautions to be taken also numerous. One may hope that when all necessary precautions are taken (if this is possible) and the data are adequate, we may rightly argue that a statistically unexpected dissimilarity among the topologies of evolutionary trees produced by independent data is a good indication against evolution. In the cases in which the topologies are identical, the degree of corroboration increases with the number of independently derived trees and the number of species or taxonomic units involved (OTUs).

This procedure was already proposed and used by Penny, Foulds and Hendy (1982). They presented trees comprising eleven species from the amino acid sequence of five proteins and thus from five independent sets of data (later on they extended them to twelve species and six proteins). Taking into consideration the statistical errors originating from the restricted sampling sizes, they concluded that the trees produced are not more dissimilar than expected from the evolutionary hypothesis (and this is a weak formulation of the argument). These trees actually show similarities unexpected when random data are used (but even this is not the strong formulation, that of the identity of topologies). The authors concluded that their results were consistent with the theory of evolution. At the end of their article they referred to Lakatos' programme and remarked:

> An interesting philosophical question would arise if the results of this work had falsified the prediction that the trees would be similar. Would this disprove the theory of evolution, or could it just mean that the sequences had changed so rapidly that they had lost all information about their earlier history [...]. However, it is probably true that specific predictions form hypotheses, rather that hypotheses themselves, are falsifiable.

The acceptance of this conjecture will render hypotheses similar to metaphysical statements (for a recent discussion of the subject mostly oriented to DNA sequences, see also Penny et al. 1991). This may well be the case, and it is not a specificity of the theory of organic evolution.

Regarding evolutionary theory in several studies dealing with the congruence of phylogenetic tree topologies we witness indeed the existence of a dogmatic belief in the theory. An overwhelming accumulation of supportive data of the evolutionary theory resulted in a belief that hardened with time and this rendered it inaccessible to test. The theory is accepted in advance untested.

14.3 The Case of *Drosophila melanogaster* Group Phylogeny

To illustrate the last statement, I will mention only one example, a well-studied case: that of the species of the genus *Drosophila* belonging to the *melanogaster* group. This cluster includes eight closely related species generally subdivided in three species complexes (the erecta complex: *D. erecta, D. orena*; the yakuba complex: *D. teissieri, D. yakuba*; and the *melanogaster* one: *D. melanogaster, D. mauritiana, D. simulans, D. sechellia*). We have at our disposal several phylogenies based on a variety of different data[2] (see Figure 14.1). Furthermore, we have data on the importance of the pre- and postzygotic genetic barrier among these species.

All the trees produced did not show the same topology, and at least some of the differences were not due to sampling errors. Here I would like to make

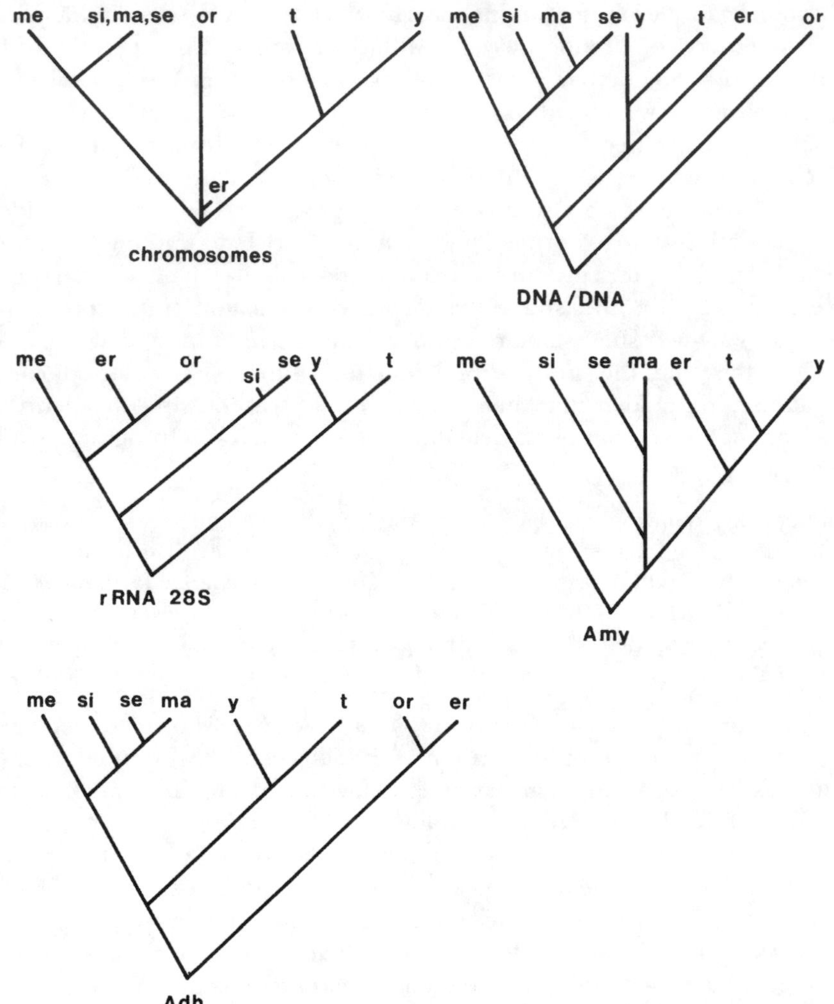

Figure 14.1. Seven phylogenetic trees for the species belonging to the *melanogaster* cluster of *Drosophila* species based on different types of data indicated in each tree (based on literature mentioned in Note 2) and a consensus tree. Abbreviations are as follows: me: *D. melanogaster*, si: *D. simulans*, ma: *D. mauritiana*, se: *D. sechellia*, or: *D. orena*, er: *D. erecta*, t: *D. teissieri*, y: *D. yakuba*.

a digression. Cladogenesis is generally considered to occur by a series of dichotomous branching events. This assumption is violated when introgressive hybridization occurs between incipient species. Then the phylogenetic history should resemble more a netlike pattern than a tree branching in a dichotomous way. One of the astonishing differences among trees for the *Drosophila melanogaster* group of species is encountered in the topology of the tree originating from the mitochondrial DNA data. Mitochondria are cell organelles transmitted exclusively by the mother (however, exceptions to this rule have been

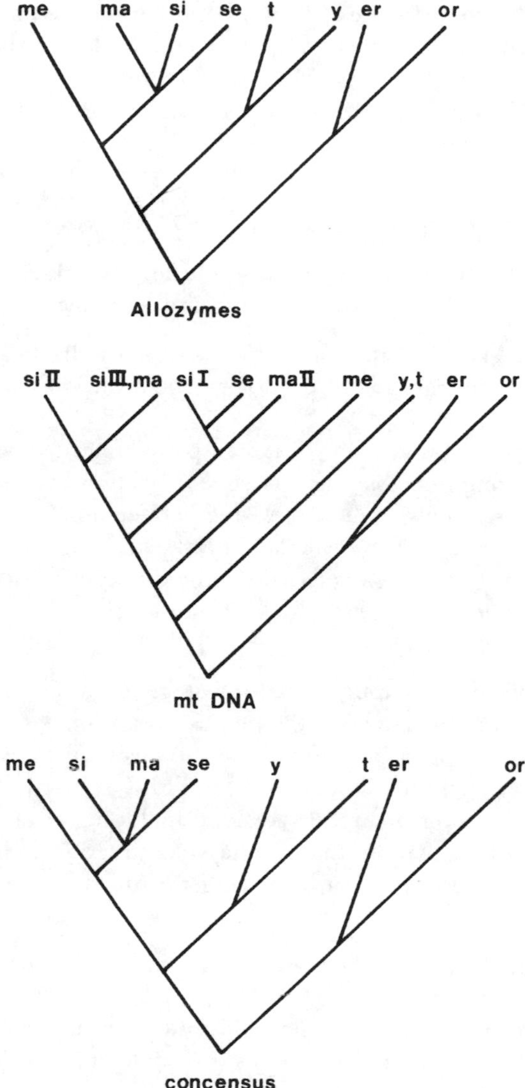

Figure 14.1. *(continued)*

reported). In this tree, there was no satisfactory grouping of strains belonging to the same species: different strains of *Drosophila mauritiana* and *D. simulans* grouped in a mixture in such a way that strains of the same species were not clustered together. These findings were interpreted to reveal the occurrence of previous introgressive hybridization(s) – events remaining undetected when nuclear data were used. This way of proceeding indicates that scientists are not really concerned to test the validity of the hypothesis of evolution; they actually do not question it anymore. They use the data only to unravel the historical details of the phylogenetic process. They are more interested in constructing a consensus tree from all available data – a tree encapsulating all relevant information. Therefore, even though the evolutionary hypothesis may in principle be tested and subject to falsification, in practice it is not. In this sense we are obliged to agree with Popper's remarks.

14.4 Evolutionary Change: The Three Assertions

We now turn to the mechanism of the evolutionary process and its three-assertion formulation. These three assertions are as follows:

The first assertion. There is (significant) variation among individuals from a population of a species in morphology, physiology, and behavior. This is the principle of variation.

The second assertion. This variation is at least partly inherited, and thus, on the average, offspring resemble their parents or relatives more than they do to randomly chosen individuals of the same population. The existence of this phenotypic correlation between relatives is the principle of heredity.

The third assertion. Different variants leave different numbers of offspring either immediately or in remote generations. This is the principle of natural selection or of differential fitness.

These assertions or assumptions lead inevitably to an evolutionary change except for the case in which the population has already reached an equilibrium point. In this sense, Popper's argument that the mechanism proposed by neo-Darwinism is an applied situational logic seems fully justified.

A comment is in order: I carefully avoided including what certain authors call the principle of adaptation; that is, I have refrained from formulating the argument as a four-step assertion, replacing the third one by two others as follows:

3. Some variants are better adapted than others to their environment (principle of adaptation).
4. The better-adapted variants leave more offspring and thus contribute more to the genetic makeup of future generations (principle of fitness or of selection).

The reason for avoiding the four-step argument is the impossibility of defining adaptation independently of selection. In a lengthy essay (Krimbas 1984), I demonstrated that it is logically impossible to arrive at an independent definition of adaptation; I have also shown that, in practice, all evolutionists have explicitly or implicitly defined adaptation in terms of selection (or of fitness, which is equivalent). Therefore, I will not pursue this matter further.

14.5 Falsifiability of Neo-Darwinism

It seems that there are different ways to falsify the neo-Darwinian mechanism once the reality of the evolutionary process is accepted. The first is to report evolutionary products incompatible with neo-Darwinism. We can conceive of the existence of such cases. This is shown by an imaginary example introduced by J. Maynard Smith (1972, p. 87):

> If one discovers a deep-sea fish with varying numbers of luminous dots on its tail, the number at any one time having the property of being always a prime number, I should regard this as rather strong evidence against neo-Darwinism. And if the dots took up in turn the exact configuration of various heavenly constellations, I should regard it as an adequate disproof.

That this example is so elaborate in order to be, according to Maynard Smith, an adequate counterexample, reveals a real problem. It is very difficult to assert that a given organ, structure, or trait has not been shaped by natural selection. The environmental factors and their combinations are so numerous and mostly unknown that it is impossible to make a strong assertion regarding the selective neutrality of a trait. Furthermore, we usually ignore the conditions prevailing in the past. Adaptations, that is the end products of selective processes, are not perfect, and even Darwin knew it. Thus, we cannot use an engineering approach, which permits to design in advance the optimal solution to an environmental challenge and then compare it with the one encountered in nature. Adaptations may not be the optimal solutions to environmental problems. Furthermore, we generally ignore the kind of problem to which this supposed solution corresponds, that is the way an environmental factor is articulated to the trait selected.

Many scientists use the argument of an optimal response from the part of the organism. When an engineering or another type of approach fails, scientists tend to modify the nature of the problem the organism faces, making it different or more complex until they reach an accord between the structure and the environmental challenge. This approach may have a heuristic value but at the same time transforms neo-Darwinism into a metaphysical enterprise. For this reason, Gould and Lewontin (1979) severely criticized this modus operandi in their well-known essay "The Spandrels of San Marco and the Panglossian Paradigm: A Critique of the Adaptionist Program."

There is of course a second way to falsify neo-Darwinism, which is to show that one or more of its three assertions is invalid. Then the situational part is modified, the logical application of the three assertions is invalid, and a neo-Darwinian mechanism does not necessarily predict the outcome.

14.6 Counterinstances to the Three Assertions

The three assertions are falsifiable existential claims that are expected to have a general application. Each one is testable and nontrivial. Regarding the first assertion, that of the presence of variation, we should notice that long ago the existence of significant variation in natural populations was not perceived. By *significant* we mean any genetic variation that could eventually lead or contribute to the modification of a species into a new one or more generally that could lead to an evolutionary change. Living organisms are complex systems and therefore, as some authors have argued, unique by their own nature and thus subject to variation. But this assertion is neither true nor constitutes a logical necessity. Different populations harbor different amounts of variation, lines or strains derived from a strong inbreeding may become completely monomorphic, and clonal reproduction multiplies exactly the same genotype. Monozygotic twins are genetically completely similar.

The neo-Darwinian mechanism considers the first and second assertions (of variation and of heredity) in the light of the genetics of Mendel, Morgan, Muller, and Timoféeff-Ressovsky. According to these assertions, mutations generate new variation, and genetic recombination through sexual reproduction, produces new combinations of genetic variants. Mutation and recombination are random processes at the experimental scale of the population geneticist and evolutionist. Any departure from these assumptions should be regarded as contrary to neo-Darwinism. According to this view, mutation is random in the sense that the mutation rates of specific genes producing alleles needed for a genetic improvement are independent of the selective forces acting on them. There is no direct genetic imprinting or direct genetic modification from the environment to render the individual fitter. Neo-Darwinism does not accept the existence of a Lamarckian mechanism for inheritance of acquired characters.

14.7 The Long Tradition for a Heritability of Acquired Characters

The noninheritance of acquired characteristics is not an obvious truth. On the contrary, several authors have supported Lamarckian or pseudo-Lamarckian mechanisms of inheritance, which should be considered as counterexamples to the neo-Darwinian process. Long after the publication of Lamarck's *Philosophie zoologique*, Brown-Sequard (1860) supported the inheritance of epilepsy produced by nervous system lesions in guinea pigs. The original results could not be reproduced 15 years later by Romanes. Standfuss (1900–1901) asserted the existence of a hereditary effect of the heat-induced modification in coloration of butterflies. Stockard in 1923 found inherited effects produced by fumes of alcohol in rats. We could continue this list with the well-known and disreputable experiments of Kammerer (1924) on salamanders, olms, and midwife toads; with d'Herelle's (1926) explanation of bacterial resistance to phages; with McDougall's experiments (1927, 1930, 1938; Rhine and McDougall 1933) on the inheritance of maze learning by mice, which was shown to be subject to a different interpretation by Agar and his collaborators (Agar et al. 1954). The extraordinary devices of squids reported as *bouton pression*, which permit the perfect fit and tight closure of their mantle to seawater, were interpreted as products of an inherited acquired character. Cuénot (1932) could not conceive of another possible mechanism responsible for their origin. Of course, neo-Darwinists actually have a different view. Frazzetta (1975) and Dawkins (1986) are to be consulted for a modern argument supporting the creation of complex adaptations by a neo-Darwinian process involving the accumulation of successive small improvements. Let us continue this list with the modification of the lacustral mollusks *Limnaea* interpreted by Jean Piaget, the well-known psychologist, in his 1929 thesis published in *Revue suisse de Zoologie* as the inheritance of an acquired character. Later Piaget modified his position, adapting it to Waddington's mechanism of incorporating acquired characters. Even Waddington's (1953a) experiments on decanalization and selection (genetic assimilation of an acquired character) have been considered by some authors as supportive of a Lamarckian mechanism. Of course, this is not the case (in this context, see also Waddington's 1953b paper on the Baldwin effect).

The recent claim of an environmental modification of immune traits and their subsequent inheritance (another case of a Lamarckian inheritance of acquired characters) has been thoroughly discredited. Burnet (1959), in his clonal selection theory, explained in neo-Darwinian terms the phenomena of acquired immunity, that is, of the specific immune responses and the immunological memory. A specific immune response is achieved by a selection of the cells producing the adequate antibody from a multitude of different lymphocyte cells, each one producing its own type of antibody, and the subsequent proliferation of the cells of the selected line. "Random mutation" is the mechanism responsible for the initial production of this immense genetic diversity among the lymphocytes, or rather their ancestor cells. Steele (1979) had proposed a Lamarckian mechanism, which would render this acquired immune response inherited: a vector could extract the appropriate genes from selected and thus numerous lymphocytes and would transport, introduce, and incorporate them into the genetic material of the gametes, the cells for sexual reproduction. The vector was envisaged to be a retrovirus, an RNA virus, which, however, is endowed with the enzyme reverse transcriptase, permitting the virus to transcribe its RNA to DNA. Guttmann and Aust (1963) and Gorczynski and Steele (1980, 1981) claimed the existence of cases of inherited transmission of acquired immunological tolerance. However, in a series of articles, several other authors (Brent, Rayfield et al. 1981; McLaren et al. 1981; Brent, Chandler et al. 1981; Nisbet-Brown and Wegmann 1981) could not repeat their assertion (see also the comments of Howard 1981). Another recent claim regarding epigenetic inheritance systems (EIS) by Jablonka and Lamb (1995, 1998) has been heavily criticized (see several critics in *Journal of Evolutionary Biology* vol. 1, no. 2).

14.8 Another Challenge to Neo-Darwinism: Directed Mutations

Contrary to the previously mentioned cases, a discussion is ongoing for the so-called *directed mutations* in bacteria and in yeast. When faced with an unusual environment in starvation conditions but in presence of a new source of energy, a new sugar, bacteria tend to adapt and take advantage of the new resource by acquiring the adequate enzymatic machinery for this purpose – a machinery they originally lacked. This may be explained by the combination of random mutations responsible for the appearance of the adequate gene, permitting the metabolism of the sugar, and of selective pressures responsible for their amplification; that is, the proliferation of these variants accompanied by the death of the alternative ancestor types. This typically neo-Darwinian mechanism of random mutations followed by selection explained all kinds of changes in bacteria such as the acquisition of resistance to antibiotics and to phages as well as the acquisition of an enzymatic machinery (Luria and Delbruck 1943, Demerec and Fano 1945, Ledeberg and Ledeberg 1952). However, recently some authors have observed an increased mutation rate in those alleles, enabling the bacteria to use the new resources under starvation conditions; this was interpreted as a sort of directed mutation permitting the bacteria to become adapted. A molecular mechanism to account for this was proposed: the cell would produce a highly variable set of messenger RNA molecules (mRNA) and reverse transcribe the one that made the best protein. For this, a reverse transcripts is needed (provided by some retrovirus) and a way to monitor the protein product of each mRNA variant and decide that that variant should be transcribed (Cairns, Overbaugh, and Miller 1988). An alternative explanation was put forward: in the stress conditions of a new environment containing mainly a novel source of energy, which bacteria cannot yet use, and in starvation conditions, the accuracy of DNA copying is reduced and repair mechanisms of DNA do not function properly; thus, all mutation rates are increased and not only those that specifically give origin to the favorable alleles. Therefore, there is no directed mutation.

It is not clear yet which of the two mechanisms is correct, but it seems most probable that regions of the DNA are amplified even under starvation conditions. These amplified regions are inherently unstable; if they acquire a useful mutation, the cell containing it leaves the stationary state, starts dividing, and incorporates this region in its genetic material. This mechanism seems intermediate between the two hypotheses, but, in my opinion, is in better accord with the neo-Darwinian mechanism (Foster and Cairns 1992; see also Boe 1990; Hall 1991; Benson, DeCloux, and Munro 1991; for a not so recent but accurate review see Lenski 1989).

Whatever the correct answer is, the example shows that some of the three neo-Darwinian assertions may be questioned. It shows that there is no logical impossibility for falsifying the widely accepted neo-Darwinian evolutionary mechanism.

14.9 The Capacity of the Theory to Incorporate Alternative Views

Even the generality of the Mendelian mechanism (on which the second assertion rests) may be questioned. Mendel's "law" of segregation is the cornerstone of Mendelism; exceptions to this rule (e.g., the segregation distortion loci) have been reported. This may be taken as a limitation (or even a falsification) of the Mendelian mechanism. Another limitation of the Mendelian–Morganian genetics is the existence of mobile elements, of genetic entities, which do not continuously occupy a well-defined position (locus) on the chromosome but may jump from one to another or enter and leave the chromosome. The presence of mobile elements could also be regarded as an invalidation or a limitation of the classical Mendelian–Morganian mechanism of inheritance. What is remarkable in this respect is not so much the impossibility of falsifying the basic tenets of neo-Darwinism, but its outstanding capacity for transformation by incorporating and engulfing more or less widely differing mechanisms into a general scheme, a general theory of the mechanisms of the evolutionary change.

An example of this tendency of an all-engulfing capacity is the incorporation into the general scheme of neo-Darwinism of its alternative, Motoo's Kimura neutral theory of evolution. This is an alternative to the third assertion: a genetic change of a population may occur by a neutral, that is, a nonselective process. Mutation, generating variability by originating new alleles, produces genic polymorphism. A restricted population size favors random walk changes in allele frequencies from one to the next generation. The reason for this is that the gametes participating to form the individuals of the next generation are few and do not permit the exact mirroring of the original situation. Thus, the departure from the original frequency occurs as a result of the restricted population size. Generation after generation, this random walk leads to fixation or elimination of alleles. Even in the presence of selection (i.e., of systematic differences in reproductive capacity under defined environmental conditions), random walk may lead the path when selection is weak. It is quite possible that several molecular variants are subject to neutral conditions, and thus they permit us to measure those changes that are strictly proportional to the time elapsed. One of the concerns of modern evolutionists is to distinguish between these two alternatives, selection or genetic drift, or rather, to estimate their relative importance in the evolutionary process. The matter is not easily resolved because selection (selective coefficients), randomly changing direction and intensity, may closely mimic stochastic neutral processes. Several methods have been proposed to distinguish the causes of genetic changes – either when electrophoretic data are available or when DNA sequence data are provided.

All these comments reveal another important aspect: the theory of evolution is subject to falsification (both in Popper's first conjecture and in the form of its three assertions), but scientists are not really interested in this. Even worse, when specific observations seem prima facie to falsify or to modify the original theoretical construction, this construction is not rejected. On the contrary, the new facts as well as some alternative mechanisms tend to be incorporated into the preexisting construction; they are engulfed in it, changing it in an even broader scheme able to explain all possible specific instances of the evolutionary processes. In this sense the theory plays a role of a possible framework for testable scientific theories or rather of models, as Lloyd (1988) remarked. The theory no longer resembles a set of universal assertions regarding how evolution occurs but consists of several ever increasing models, each mirroring a particular situation. Each model is a calculating device permitting the prediction of the outcome in the near or even in the remote future. Thus, these models can be tested; that is, in particular circumstances we may be able to arrive at the conclusion that a certain model faithfully represents this particular situation. As Lewontin (1991) noticed, the testing procedure is linked also to heuristic activities. When the model fails to pass the test we may discover what is wrong in this particular case. Even when the model is confirmed, it may, in the process of confirmation, generate results that have been seen in nature but not properly understood. This model-testing activity linked to discovery and to a deeper understanding of nature is what really interests scientists in evolutionary biology. In the light of the preceding discussion, it is, therefore, misleading to take physics as a model science, a paradigmatic science, when trying to study how evolutionary or population biology proceeds and evolves.

14.10 Evolution, History, and Prediction

Of course the deeper reason for this is that evolutionary processes are historical, and each process is unique. This is why it is usually argued that history can be narrated but not predicted. This statement is not absolutely correct. In experimental laboratory situations as well as in nature, some future steps can be foreseen when the proper model of the situation is applied. When the model is robust (i.e., it applies to a vast array of conditions), the predictive power is greater. Instances of historical processes that permit prediction are discussed in Krimbas (1984). One example will be mentioned here. Cave animals manifest similarities in their phenotypic traits, rudimentary eyes, depigmentation, absence of wings, development of chemoreceptors, low metabolism, paucity of eggs laid, and control of population size by the extension of life span. Underground animals also show some of these characteristics. Although all animals may not display all these characteristics to the same extent, the rule seems to be generally obeyed. A similar and stable environment produces the same selective pressures, which result in the appearance of similar phenotypes. Thus, a prediction is possible.

The situation is different when rules or laws of an absolute generality are searched for. These do not exist, or, when some such rules are formulated, they are practically devoid of information, as in Van Valen's Red Queen hypothesis, which states that every species has a priori the same probability (or propensity) of extinction, regardless of how long it has existed.

A similar situation is encountered in human history. Of course short-term demographic or economic predictions seem to be possible and are actually practiced in everyday politics. Popper (1957) argued that prediction in human history is impossible for the following reason: There is no way to anticipate what we will know tomorrow. Because the course of human history is strongly influenced by the growth and the content of human knowledge, it seems impossible to predict the future course of human history. Strictly speaking this is an absolutely correct argument; however, it leaves us with the possibility of a more fuzzy second-order prediction, a statistical one. We might not be able to know in advance the content of the future human knowledge in chemistry, but we may be able to predict more or less accurately the number of research publications in chemistry in the next decade. These kinds of second-order and short-term predictions – not necessarily precise but better than nothing – may be encountered in statistical laws or rules. Of course, in this sense, human history, although showing some similarity with evolutionary history, seems to be even more idiosyncratic.

14.11 Acknowledgments

I would like to thank Dr. Marie-Louise Cariou for providing data on the *melanogaster* species group, Sandra Krimba for drawing the figures, and Georgios Papagounos for comments.

Notes

1. Such algorithms are the Unweighted Pair Group Method with Arithmetic Means (originally Sokal and Michener, then of Sneath and Sokal, UPGMA), the Transformed Distance Method of Farris, the method of Fitch and Margoliash, the Minimum Evolution Method of Cavalli-Sforza and Edwards, the Distance Wagner Method, the Neighborliness Method of Sattah and Tversky, the Neighbour-Joining Method described recently again by Saitou and Nei, the Maximum Parsimony Method (Wagner Parsimony Method), the Evolutionary Parsimony Method of Lake, and the Maximum Likelihood Method Felsenstein.
2. These phylogenies are based on morphological data, on data of gene arrangement structure revealed from the banding pattern of their giant chromosomes (Ashburner, Bodmer, and Lemeunier 1984), allozyme data (Cariou 1987), data on DNA–DNA hybridization (Caccone, Amato, and Powell 1988), data regarding the sequences of mitochondrial DNA (Monneret, Solignac and Wolstenholme 1990, Satta and Takahata 1990), those regarding ribosomal RNA 28S (Pelandakis et al. 1990, Satta and Takahata 1990), base sequence data of the alcohol dehydrogenase (ADH) gene and the pseudogenes (Sullivan, Atkinson, and Starmer 1990; Jeffs and Ashburner 1991), those of the amylase gene (Shibata and Yamazaki 1995), those of the period gene (only for the species of the *melanogaster* complex; Kliman and Hey 1993).

REFERENCES

Agar, W. E., Drummond, F. H., Tiiegs, O. W., and Gunson, M. M. (1954). Fourth (final) report on a test of McDougal's Lamarckian experiment on the training of rats. *Journal of Experimental Biology* 31:307–21.

Anonymous (1981). How true is the theory of evolution? *Nature* 290:75–6.

Ashburner, M., Bodmer, M., and Lemeunier, F. (1984). On the evolutionary relationships of *Drosophila melanogaster. Dev. Gen.* 4:295–312.

Ayala, F. J., and Dobzhansky, Th. (eds.) (1974). *Studies in the Philosophy of Biology* (Reduction and Related Problems). London: MacMillan.

Benson, S. A., DeCloux, A. M., and Munro, J. (1991). Mutant bias in nonlethal selections results from selective recovery of mutants. *Genetics* 129:647–58.

Boe, L. (1990). Mechanism for induction of adaptive mutations in *Eshcheria coli. Mol. Microbiol.* 4:597–601.

Brehm, A., and Krimbas, C. B. (1993). The phylogeny of nine species of the *Drosophila obscura* group inferred by the banding homologies of the chromosomal regions. IV. Element C. *Heredity* 70:214–20.

Brent, L., Chandler, P., Fierz, W., Medawar, P. B., Rayfield, L. S., and Simpson, E. (1981). Further studies on the supposed lamarckian inheritance of immunological tolerance. *Nature* 295:242–4.

Brent, L., Rayfield, L. S., Chandler, R., Fierz, W., Medawar, P. B., and Simpson, E. (1981). Supposed Lamarckian inheritance of immunological tolerance. *Nature* 290:508–12.

Brown-Sequard, C. E. (1860). Hereditary transmission of an epileptiform affection accidentally produced. *Proc. Roy. Soc. Lond.* 10:297–8.

Burnet, F. M. (1959). *The Clonal Selection Theory of Acquired Immunity.* London: Cambridge University Press.

Caccone, A., Amato, G. D., and Powell, J. R. (1988). Rates and patterns of scnDNA and mtDNA divergence within the *Drosophila melanogaster* subgroup. *Genetics* 118:671–83.

Cairns, J., and Foster, P. L. (1991). Adaptive reversion of a frameshift mutation in *Eschcerichia coli. Genetics* 128:695–701.

Cairns, J., Overbaugh, J., and Miller, S. (1988). The origin of mutants. *Nature* 335:142–5.

Cariou, M. L. (1987). Biochemical phylogeny of eight species in the *Drosophila melanogaster* subgroup, including *D. sechellia* and *D. orena. Genet. Res. Cambr.* 50:181–5.

Cuénot, L. (1932). Génétique et adaptation. In *Eugènique et Selection* (Bibliothèque Génèrale des Sciences Sociales) Paris: Felix Alcan, pp. 201–47.

Dawkins, R. (1986). *The Blind Watchmaker.* London: Longman Scientific and Technical.

Demerec, M., and Fano, U. (1945). Bacteriophage resistant mutants in *Escherichia coli. Genetics* 30:119–36.

d'Herelle, F. (1926). *The Bacteriophage and its Behavior.* Baltimore: Williams and Wilkins.

Felsenstein, J. (1984). Phylogenies from molecular sequences: inference and reliability. *Ann. Rev. Genet.* 22:521–65.

Foster, P. L., and Cairns, J. (1992). Mechanisms of directed mutation. *Genetics* 131:783–9.

Frazzetta, T. H. (1975). *Complex Adaptations in Evolving Populations.* Sunderland, MA: Sinauer.

Gorczynski, R. M., and Steele, E. J. (1980). Inheritance of acquired immunological tolerance to foreign histocompatibility antigen in mice. *Proc. Natl. Acad. Sci. USA* 77:2871–5.

Gorczynski, R. M., and Steele, E. J. (1981). Simultaneous yet independent inheritance of somatically acquired tolerance to two distinct H-2 antigenic haplotype determinants in mice. *Nature* 289:678–81.

Gould, S. J., and Lewontin, R. C. (1979). The spandrels of San Marco and the Panglossian paradigm: A critique of the adaptationist programme. *Proc. R. Soc. Lond.* B 205:581–98.

Guttmann, R. D., and Aust, J. B. (1963). A germplasm-transmitted alteration of histocompatibility in the progeny of homograft tolerant mice. *Nature* 197:1220–1.

Hall, E. B. G., (1991). Adaptive evolution that requires multiple spontaneous mutations: mutations involving base substitutions. *Proc. Natl. Acad. Sci. USA* 88:5882–6.

Halstead, B. (1980). Popper: good philosophy, bad science? *New Scientist* 17 July 1980:215–17.

Hoelzer, G. A., and Melnick, D. J. (1994). Patterns of speciation and limits to phylogenetic resolution. *TREE* 9:104–7.

Howard, J. C. (1981). A tropical volute shell and the Icarus syndrome. *Nature* 290:441–2.

Hull, D. L. (1974). *Philosophy of Biological Science.* Englewood Cliffs, NJ: Prentice–Hall.

Jablonka, E., and Lamb, M. J. (1995). *Epigenetic Inheritance and Evoluion: The Lamarckian Dimension.* Oxford: Oxford University Press.

Jablonka, E., and Lamb, M. J. (1998). Epigenetic Inheritance in Evolution. *J. Evol. Biology* 711:159–83.

Jeffs, P., and Ashburner, M. (1991). Processed pseudogenes in *Drosophila*. *Proc. R. Soc. Lond.* B 208:151–9.

Kammerer, P. C. (1924). *The Inheritance of Acquired Characteristics.* New York: Boni and Liveright.

Kliman, R. M., and Hey, J. (1993). DNA sequence variation at the period locus within and among species. *Genetics* 133:375–87.

Krimbas, C. B. (1984). On Adaptation, Neo-Darwinian, tautology, and population fitness. *Evolutionary Biology* 17:1–57.

Krimbas, C. B., and Powell, J. R. (1992). Introduction. In *Drosophila Inversion Polymorphism,* (eds.) C. B. Krimbas and J. R. Powell, pp. 1–52. Boca Raton, FL: CRC Press.

Ledeberg, J., and Ledeberg, E. M. (1952). Mutations of bacteria from virus sensitivity to virus resistance. *J. Bacteriol.* 63:399–406.

Lenski, R. E. (1989). Are some mutations directed? *TREE* 4:148–50.

Lewontin, R. C. (1977). Adattamento. In *Enciclopedia Einaudi Torino* vol. 1, pp. 198–214, Torino: Istituto dell' Enciclopedia Italiana.

Lewontin, R. C. (1978). Adaptation. *Scientific American* 239 (September):156–69.

Lewontin, R. C. (1991). The structure and confirmation of evolution theory. *Biology and Philosophy* 6:461–6.

Little, J. (1980). Evolution: Myth, metaphysics, or science? *New Scientist* 4 September 1980:708–9.

Lloyd, E. A. (1988). *The Structure and Confirmation of Evolutionary Theory.* New York: Greenwood Press.

Luria, S. E., and Delbruck, M. (1943). Mutations of bacteria form virus sensitivity to virus resistance. *Genetics* 28:491–511.

Maynard Smith, J. (1972). *On Evolution.* Edinburgh University Press.

McDougall, W. (1927). An experiment for the testing of the hypothesis of Lamarck. *Brit. J. Psychol.* 17:267–304.

McDougall, W. (1930). Second report on a Lamarckian experiment. *Brit. J. Psychol.* 20:201–18.

McDougall, W. (1938). Fourth report on a Lamarckian experiment. *Brit. J. Psychol.* 28:321–45.

McLaren, A., Chandler, P., Buehr, M., Friez, W., and Simpson, E. (1981). Immune reactivity of progeny of tetraparental male mice. *Nature* 290:513–14.

Miyamoto, M. M., and Cracraft, J. (1991). Phylogenetic inference, DNA sequence analysis, and the future of molecular systematics. In *Phylogenetic Analysis of DNA Sequences,* (eds.) M. M. Miyamoto and J. Cracraft, pp. 3–17. Oxford, UK: Oxford University Press.

Monneret, M., Solignac, M., and Wolstenholme, D. R. (1990). Discrepancy in divergence of the mitochondrial and nuclear genomes of *Drosophila teissieri* and *D. yakuba*. *J. Mol. Evol.* 30:500–8.

Nei, M. (1991). Relative efficiencies of different tree-making methods for molecular data. In *Phylogenetic Analysis of DNA Sequences,* (eds.) M. M. Miyamoto and J. Cracraft. pp. 90–128. Oxford, UK: Oxford University Press.

Nisbet-Brown, E., and Wegmann, Th. G. (1981). Is acquired immunological tolerance genetically transmissible? *Proc. Natl. Acad. Sci. USA* 78:5826–8.

Pelandakis, M., Higgins, D. G., and Solignac, M. (1991). Molecular phylogeny of the subgenus *Sophophora* of *Drosophila* derived from large subunit of ribosomal RNA sequences. *Genetics* 84:87–94.

Penny, D., Foulds, L. H., and Hendy, M. D. (1982). Testing the theory of evolution by comparing phylogenetic trees constructed from five different protein sequences. *Nature* 297:197–200.

Penny, D., Hendy, M. D., and Steel, M. A. (1991). Testing the theory of descent. In *Phylogenetic Analysis of DNA Sequences,* (eds.) M. M. Miyamoto and J. Cracraft, pp. 155–83. Oxford, UK: Oxford University Press.

Piaget, J. (1929). L'adaptation de la *Limnaea stagnalis* aux milieux lacustres de la Suisse romande. *Revue suisse de Zoologie* 36:263–531.

Popper, K. R. (1945). *The Open Society and its Enemies.* London: Routledge and Paul Kegan (1969).

Popper, K. R. (1957). *The Poverty of Historicism.* New York and Evanston: Harper and Row.

Popper, K. R. (1972). *Objective Knowledge: An Evolutionary Approach.* Oxford, UK: Claredon Press.

Popper, K. R. (1974). Autobiography. In P. A. Schilpp (ed.). *The Philosophy of Karl Popper,* La Salle, IL: Open Court. Vol. 1. pp. 1–181 (Darwinism as a metaphysical research programme, pp. 133–143).

Popper, K. R. (1978). Natural selection and the emergence of mind. *Dialectica* 22:339–55, reprinted in G. Radnitzky and W. W. Bartley III (eds.) (1987). *Evolutionary Epistemology, Rationality, and Sociology of Knowledge* (second printing 1988). La Salle, IL: Open Court, pp. 138–55.

Rhine, J. B., and McDougall, W. (1993). Third report on a Lamarckian experiment. *Brit. J. Psychol.* 24:213–35.

Satta, Y., and Takahata, N. (1990). Evolution of *Drosophila* mitochondrial DNA and the history of the *melanogaster* subgroup. *Proc. Natl. Acad. Sci. USA* 87:9558–62.

Shibata, H., and Yamazaki, T. (1995). Molecular evolution of the duplicated *Amy* locus in *Drosophila melanogaster* subgroup: concerted evolution only in coding region, and excess of nonsynonymous substitutions in speciation. *Genetics* 141:223–36.

Shields, D. S., Sharp, P. M., Higgins, D. G., and Wright, F. (1988). "Silent" sites in *Drosophila* genes are not neutral: Evidence of selection among synonymous codons. *Mol. Biol. Evol.* 5:704–16.

Solignac, M., Monnerot, M., and Mounolou, J. C. (1986). Mitochondrial DNA evolution in the *melanogaster* species subgroup of *Drosophila. J. Mol. Evol.* 23:31–40.

Stamos, D. N. (1996). Popper, falsifibility, and evolutionary biology. *Biology and Philosophy* 11:161–91.

Standfuss, M. (1900–1901). Synopsis of the experiments on hybridization and temperature-made modifications with Lepidoptera up to the end of 1898. *Entomologist* 30:161–7, 283–92, 340–8; 31:11–13, 75–84.

Steele, D. F., and Jinks-Robertson, S. (1992). An examination of adaptive reversion in *Saccharomyces cerevisiae. Genetics* 132:9–21.

Steele, E. J. (1979). *Somatic Selection and Adaptive Evolution.* London: Williams-Wallace, Crown Helm.

Stockard, C. R. (1923). Experimental modification of the germ plasm and its bearing on the inheritance of acquired characters. *Proc. Am. Phil. Soc.* 42:311–24.

Sullivan, D. T., Atkinson, P. W., and Starmer (1990). Molecular evolution in the alcohol dehydrogenase genes in the genus *Drosophila. Evol. Biol.* 24:107–47.

Waddington, C. H. (1953a). Genetic assimilation of an acquired character. *Evolution* 7:118–26.

Waddington, C. H. (1953b). The "Baldwin effect," "Genetic Assimilation" and "Homeostasis." *Evolution* 7:386–7.

Wallace, B. (1966). *Choromosomes, Giant Molecules and Evolution.* New York: Norton.

CHAPTER FIFTEEN

The Two Faces of Fitness

ELLIOTT SOBER

The concept of fitness began its career in biology long before evolutionary theory was mathematized. Fitness was used to describe an organism's vigor, or the degree to which organisms "fit" into their environments. An organism's success in avoiding predators and in building a nest obviously contributes to its fitness and to the fitness of its offspring, but the peacock's gaudy tail seemed to be in an entirely different line of work. Fitness, as a term in ordinary language (as in "physical fitness") and in its original biological meaning, applied to the survival of an organism and its offspring, not to sheer reproductive output (Cronin 1991, Paul 1992). Darwin's separation of natural from sexual selection may sound odd from a modern perspective, but it made sense from this earlier point of view.

Biologists came to see that this limit on the concept of fitness is theoretically unjustified. Fitness is relevant to evolution because of the process of natural selection. Selection has an impact on the traits that determine how likely it is for an organism to survive from the egg stage to adulthood, but it equally has an impact on the traits that determine how successful an adult organism is likely to be in having offspring. Success concerns not just the robustness of offspring but their number. As a result, we now regard viability and fertility as two components of fitness. If p is the probability that an organism at the egg stage will reach adulthood, and e is the expected number of offspring that the adult organism will have, then the organism's *overall fitness* is the product pe, which is itself a mathematical expectation. Thus, a trait that enhances an organism's viability but renders it sterile has an overall fitness of zero. And a trait that slightly reduces viability, while dramatically augmenting fertility, may be very fit overall.

The expansion of the concept of fitness to encompass both viability and fertility resulted from the interaction of two roles that the concept of fitness plays in evolutionary theory. It describes the relationship of an organism to its environment. It also has a mathematical representation that allows predictions and explanations to be formulated. *Fitness is both an ecological descriptor and a mathematical predictor.* The descriptive ecological content of the concept was widened to bring it into correspondence with the role that fitness increasingly played as a mathematical parameter in the theory of natural selection.

In this chapter I want to discuss several challenges that have arisen in connection with idea that fitness should be defined as expected number of offspring. Most of them are discussed in an interesting article by Beatty and Finsen (1989). Ten years earlier, they had championed a view they dubbed "the propensity interpretation of fitness" (Mills and Beatty 1979; see also Brandon 1978). In the more recent article, they "turn critics." Should fitness be defined in terms of a one-generation time frame – why focus on expected number of *offspring* rather than *grand*offspring, or more distant descendants still? And is the concept of mathematical expectation the right one to use? The details of my answers to these questions differ in some respects from those suggested by Beatty and Finsen, but my bottom line will be the same – expected number of offspring is not always the right way to define fitness.

In what follows, I will talk about an *organism*'s fitness even though evolutionary theory shows scant interest in individual organisms but prefers to talk about the fitness values of *traits* (Sober 1984). Charlie the Tuna is not a particularly interesting object of study, but tuna dorsal fins are. Still, for the theory of natural selection to apply to the concrete lives of individual organisms, it is essential that the fitness values assigned to traits have implications concerning the reproductive prospects of the individuals that have those traits. How are trait fitnesses and individual fitnesses connected? Because individuals that share one trait may differ with respect to others, it would be unreasonable to demand that individuals that share a trait have identical fitness values. Rather, the customary connection is that the fitness value of a trait is the average of the fitness values of the individuals that have the trait. For this reason, my talk in what follows about the fitness of organisms will be a harmless stylistic convenience.

To begin, let us remind ourselves of what the idea of a mathematical expectation means. An organism's expected number of offspring is not necessarily the number of offspring one expects the organism to have. For example, suppose an organism has the following probabilities of having different numbers of offspring:

number (i) of offspring	0	1	2	3
probability (p_i) of having exactly i offspring	0.5	0.25	0.125	0.125

The expected number of offspring is $\sum ip_i = 0(0.5) + 1(0.25) + 2(0.125) + 3(0.125) = 0.875$, but we do not expect the organism to have precisely 7/8ths of an offspring. Rather, "expectation" means *mathematical* expectation, a technical term; the expected value is, roughly, the (arithmetic) average number that the individual would have if it got to live its life again and again in identical circumstances. This is less weird than it sounds; a fair coin has 3.5 as the expected number of times it will land heads if it is tossed 7 times.

In this example, the expected number of offspring will not exactly predict an individual's reproductive output, but it will probably come pretty close.

However, there are cases in which the expected value provides a very misleading picture as to what one should expect. Lewontin and Cohen (1969) develop this idea in connection with models of population growth. Suppose, to use one of their examples, that each year a population has a probability of 0.9 of having a growth rate of 1.1 and a probability of 0.1 of having a growth rate of 0.3. The expected (arithmetic mean) growth rate per year is $(0.9)(1.1) + (0.1)(0.3) = 1.02$; thus, the expected size of the population increases by 2% per year. At the end of a long stretch of time, the population's expected size will be much larger than its initial size. However, the fact of the matter is that the population is virtually certain to go extinct in the long run. This can be seen by computing the geometric mean growth rate. The geometric mean of n numbers is the nth root of their product; because $[(1.1)^9(0.3)]^{1/10}$ is less than unity, we expect the population to go extinct. To see what is going on here, imagine a very large number of populations that each obey the specified pattern of growth. If we follow this ensemble for, say, 1000 years, what we will find is that almost all of the populations will go extinct, but a very small number will become huge; averaging over these end results, we will obtain the result that, on average, populations grow by 2% a year. Lewontin and Cohen point out that this anomaly is characteristic of multiplicative processes.

A simpler and more extreme example that illustrates the same point is a population that begins with a census size of 10 individuals and each year has a 0.5 chance of tripling in size and a 0.5 chance of going extinct. After 3 years, the probability is 7/8 that the population has gone extinct, but there is a probability of 1/8 that the population has achieved a census size of $(3)(3)(3)10 = 270$. The expected size of the population is $(7/8)(0) + (1/8)(270) = 33.75$. This expected size can be computed by taking the expected yearly growth rate of $(0.5)(3) + (0.5)0 = 1.5$ and raising it to the third power; $(1.5)(1.5)(1.5)10 = 33.75$. In expectation, the population increases by 50% per year, but you should expect the population to go extinct.

Probabilists will see in this phenomenon an analogue of the St. Petersburg paradox (Jeffrey 1983). Suppose you are offered a wager in which you toss a coin repeatedly until tails appears, at which point the game is over. You will receive 2^n dollars, where n is the number of tosses it takes for tails to appear. If the coin is fair, the expected payoff of the wager is

$$(1/2)\$2 + (1/4)\$4 + (1/8)\$8 + \cdots$$

The expected value of this wager is infinite, but very few people would spend more than, say, $10 to buy into it. If rationality means maximizing expected utility, then people seem to be irrational – they allegedly should be prepared to pay a zillion dollars for such a golden opportunity. Regardless of whether this normative point is correct, I suspect that people may be focusing on what will *probably* happen, not on what the average payoff is over all possible outcomes, no matter how improbable. Notice that the probability is only 1/8 that the game will last more than three rounds. What we expect to be paid in this game deviates enormously from the expected payoff.

For both ecologists and gamblers, the same advice is relevant: Caveat emptor! If you want to make predictions about the outcome of a probabilistic process, think carefully before you settle on expected value as the quantity you will compute.

15.1 The Long-Term and the Short-Term

The definition of fitness as expected number of offspring has a one-generation time scale. Why think of fitness in this way rather than as having a longer time horizon? Consider Figure 15.1 adapted from Beatty and Finsen (1989). Trait A produces more offspring than trait B (in expectation) before time t^*; however, after t^*, A produces fewer offspring than B, and in fact A eventually produces zero offspring. The puzzle is that A seems to be fitter than B in the short term, whereas B seems to be fitter than A in the long term. Which of these descriptions is correct?

The issue of whether fitness should be defined as a short-term or a long-term quantity will be familiar to biologists from the work of Thoday (1953, 1958), who argued that fitness should be defined as the probability of leaving descendants in the very long run; he suggests 10^8 years as an appropriate time scale. Thoday (1958, p. 317) says that a long-term measure is needed to obtain a definition of evolutionary progress. This reason for requiring a long-term concept will not appeal to those who think that progress is not a scientific concept at all (see, for example, discussion in Nitecki 1988 and Sober 1994). Thoday's argument also has the drawback that it repeatedly adverts to the good of the species without recognizing that this may conflict with what is good for individual organisms.

Setting aside Thoday's reason for wanting a long-term concept of fitness, does this concept make sense? Brandon (1990, pp. 24–5) criticizes Thoday's approach and the similar approach of Cooper (1984) on the grounds that

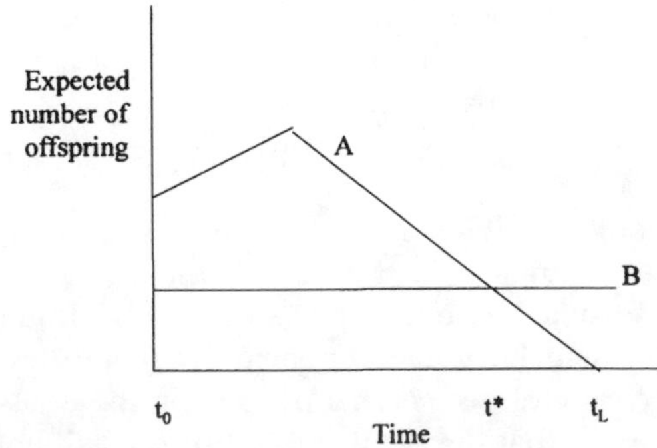

Figure 15.1. Trait A is fitter than trait B initially, but later on the reverse is true. This means that B has a higher long-term fitness than A.

selection "proceeds through generational time" and "has no foresight." I think both these criticisms miss the mark. Long-term probabilities imply foresight no more than short-term probabilities do. And the fact that selection occurs one generation at a time does not mean that it is wrong to define a quantity that describes a trait's long-term expected fate. Brandon also faults Thoday's proposal for failing to be operational. How are we to estimate the probability that a present organism or species will have descendants in the distant future? The point is well taken when the inference is *prospective*; in this case, the short-term is more knowable than the long-term future. However, when we make *retrospective* inferences, the situation reverses. An inferred phylogeny may reveal that a derived character displaced an ancestral character in one or more lineages. This information may provide evidence for the claim that the derived trait had the higher long-term fitness. In contrast, the one-generation fitnesses that obtained 60 million years ago may be quite beyond our ken.

Rather than rejecting a long-term concept of fitness and defending a short-term measure, I suggest that there is frequently no need to choose. In the accompanying figure, the y values for A and B at a given time tell us which trait had the higher short-term fitness at that time. The long-term fitness of a trait – its fitness, say, from t_0 to t^* or from t_0 to t_L – is a statistic that summarizes the relevant short-term values. There is no paradox in the fact that A has the higher short-term fitness whereas B has the higher long-term fitness. The same pattern can be found in two babies. The first has the higher probability of reaching age 20, whereas the second has the higher probability of surviving to age 60. The probability of a baby's reaching age 60 is a product – Pr(surviving to age 20 | you are a baby)Pr(surviving to age 60 | you have survived to age 20) = $(s_1)(s_2)$. The first baby may have a higher value on s_1 than the second, whereas the second has a higher value on s_2 than the first; overall, the first baby's product may be lower than that of the second. Long-term fitness is a coherent concept that may be useful in the context of certain problems; however, its coherence and desirability do not undermine the concept of short-term fitness.

15.2 When a One-Generation Time Frame Is Inadequate

The concept of short-term fitness discussed so far has a one-generation time frame – an organism at the egg stage has a probability p of reaching reproductive age and, once it is an adult, it has e as its expected number of offspring – the product pe is its overall fitness. However, a one-generation time frame will not always be satisfactory for the concept of short-term fitness. Fisher's (1930) model of sex ratio shows why (Sober 1984). If, in expectation, one female has 5 sons and 5 daughters whereas another produces 10 daughters and 0 sons, how can their different sex-ratio strategies make a difference in their fitnesses? Fisher saw that the answer is invisible if we think one generation ahead but falls into place if we consider two. The sex ratio exhibited by a female's progeny influences how many *grand*offspring she will have.

Other examples may be constructed of the same type. Parental care is a familiar biological phenomenon, but let us consider its extension – care of grandoffspring. If A individuals care for their grandoffspring, but B individuals do not, it may turn out that A individuals are fitter. However, the advantage of A over B surfaces only if we consider the expected numbers of grandoffspring that survive to adulthood. This example may be more of a logical possibility than a biological reality; still, it and sex ratio illustrate the same point. In principle, there is no a priori limit on the size of the time frame over which the concept of fitness may have to be stretched. If what an organism does in its lifetime affects the life prospects of organisms in succeeding generations, the concept of fitness may have to encompass those far-reaching effects.

15.3 Stochastic Variation in Offspring Number

Let us leave the question of short-term versus long-term behind and turn now to the question of whether fitness should be defined as a mathematical expectation. This is not an adequate definition when there is stochastic variation in viability or fertility. Dempster (1955), Haldane and Jayakar (1963), and Gillespie (1973, 1974, 1977) consider stochastic variation among generations; Gillespie (1974, 1977) addresses the issue of within-generation variation. These cases turn out to have different mathematical consequences for how fitness should be defined. However, in both of them, selection favors traits that have lower variances. In what follows, I will not attempt to reproduce the arguments these authors give for drawing this conclusion. Rather, I will describe two simple examples that exhibit the relevant qualitative features.

Let us begin with the case of stochastic variation among generations. Suppose a population begins with two A individuals and two B's. A individuals always have two offspring, whereas the B individuals in a given generation all have one offspring or all have three, with equal probability. Notice that the expected (arithmetic average) offspring number is the same for both traits – 2. However, we will see that the expected frequency of B declines in the next generation.

Assume that these individuals reproduce asexually and then die and that offspring always resemble their parents. Given the numbers just described, there will be four A individuals in the next generation and either two B individuals or six, with equal probability. Although the two traits begin with the same population frequency and have the same expected number of offspring, their expected frequencies in the next generation differ:

$$\text{Expected frequency of A} = (1/2)(4/6 + 4/10) = 0.535$$
$$\text{Expected frequency of B} = (1/2)(2/6 + 6/10) = 0.465$$

The trait with the lower variance can be expected to increase in frequency. The appropriate measure for fitness in this case is the geometric mean of offspring number averaged over time; this is the same as the expected log of the number of offspring. Trait B has the lower geometric mean because $[(3)(1)]^{\frac{1}{2}} = 1.7 < [(2)(2)]^{\frac{1}{2}} = 2$. The geometric mean is approximately the arithmetic expected number minus $\sigma^2/2$.

Let us now consider the case of within-generation variance in offspring number. Gillespie (1974) describes the example of a bird whose nest has a probability of escaping predators of about 0.1. Should this bird put all its eggs in one nest or establish separate nests? If the bird lays 10 eggs in just one nest, it has a probability of 0.9 of having 0 offspring and a probability of 0.1 of having 10. Alternatively, if the bird creates 2 nests containing 5 eggs each, it has a probability of $(0.9)^2$ of having 0 offspring, a probability of 2(0.9)(0.1) of having 5, and a probability of $(0.1)^2$ of having 10. The expected value is the same in both cases – 1.0 offspring – but the strategy of putting all eggs in one nest has the higher variance in outcomes. This example illustrates the idea of within-generation variance because two individuals in the same generation who follow the same strategy may have different numbers of offspring.

Does the process of natural selection vindicate the maxim that there is a disadvantage in putting all one's eggs in one basket? The answer is yes. To see why, let us examine a population that begins with two A individuals and two C's. A individuals always have two offspring, whereas each C individual has a 50% chance of having 1 offspring and a 50% chance of having 3. Here C individuals in the same generation may vary in fitness, but the expected value in one generation is the same as in any other. In the next generation, there will be four A individuals. There are four equiprobable arrangements of fitnesses for the two C individuals, and thus there are four equiprobable answers to the question of how many C individuals there will be in the next generation – two, four, four, and six. The expected number of C individuals in the next generation is four, but the expected frequencies of the two traits change:

$$\text{Expected frequency of A} = (1/4)(4/6 + 4/8 + 4/8 + 4/10) = 0.52$$

$$\text{Expected frequency of C} = (1/4)(2/6 + 4/8 + 4/8 + 6/10) = 0.48$$

Once again, the trait with the lower variance can be expected to increase in frequency.

In this example, the population grows from four individuals in the first generation to somewhere between 6 and 10 individuals in the second. Suppose we require that population size remain constant; after the four parents reproduce, random sampling reduces the offspring generation to four individuals. When this occurs, the trait with the higher variance has the higher probability of going extinct.

Gillespie (1974, 1977) constructed a model to describe the effect of within-generation variance. A trait's variance (σ^2) influences what happens only when population size (N) is finite; in the infinite limit, variance plays no role. On the basis of this model, Gillespie says that a trait's fitness is approximately its arithmetic mean number of offspring minus the quantity σ^2/N. Notice that this correction factor will be smaller than the one required for between-generation variance if $N > 2$.

Why, in the case of within-generation variance, does the number of individuals (N) in the whole *population* appear in the expression that describes the fitness of a single *trait*, which may be one of many traits represented in the population? In our example, why does the fitness of C depend on the total number of C and A individuals? And why does the effect of selecting for lower variance decline as population size increases? The reasons can be glimpsed in the simple calculation just described. To figure out the expected frequency of C, we summed over the four possible configurations that the population has in the next generation. There is a considerable difference among these four possibilities – trait C's absolute frequency is either 2/6, 4/8, 4/8, or 6/10. In contrast, if there were 2 C parents but 100 A's, there still would be four fractions to consider, but their values would be 2/202, 4/204, 4/204, and 6/206; these differ among themselves much less than the four that pertain to the case of 2 A's and 2 C's. The same diminution occurs if we increase the number of C parents; there would then be a larger number of possible configurations of the next generation to consider, and these would differ among themselves less than the four described initially. In the limit, if the population were infinitely large, there would be no difference, on average, among the different possible future configurations.

The presence of N in the definition of fitness for the case of within-generation variance suggests that the selection process under discussion is density dependent. Indeed, Gillespie (1974, p. 602) says that the population he is describing is "density-regulated," for a fixed population size is maintained. However, we need to recognize two differences between the case he is describing and the more standard notion of density dependence that is used, for example, to describe the effects of crowding. In the case of crowding, the size of the population has a causal impact on an organism's expected number of offspring. However, the point of Gillespie's analysis of within-generation variance is to show that fitness should not be defined as expected number of offspring. In addition, the case he is describing does not require that the size of the population have any causal influence on the reproductive behavior of individuals. The two A's and two C's in my example might be four cows standing in the four corners of a large pasture; the two A's have two calves each, whereas each of the C's flips a coin to decide whether she will have one calf or three. The cows are causally isolated from each other, but the fitnesses of the two strategies reflect population size.

In the two examples just presented, within-generation variance and between-generation variance have been understood in such a way that the former entails the latter, but not conversely. Because each C individual in each generation tosses a coin to determine whether she will have one offspring or three, it is possible for the mean offspring number produced by C parents in one generation to differ from the mean produced by the C parents in another. However,

B parents in the same generation always have the same number of offspring. What this means is that B is a strategy that produces a purely between-generation variance, whereas C is a strategy that produces both within- and between-generation variance.

In both of the examples I have described, the argument that fitness must reflect variance as well as the (arithmetic) mean number of offspring depends on the assumption that fitnesses should predict *frequencies* of traits. If, instead, one merely demanded that the fitness of a trait should allow one to compute the expected *number* of individuals that will have the trait in the future, given the number of individuals that have the trait initially, the argument would not go through. The expected number of individuals in some future generation is computed by using the arithmetic mean number of offspring. When the population begins with two B individuals or with two C individuals, the expected number of B or C individuals in the next generation is four. The value that generates this next-generation prediction is two – the arithmetic mean of one and three. Note that the variance in offspring number and the size of the whole population (N) are irrelevant to this calculation.

That fitness is influenced by variance may seem paradoxical at first, but it makes sense in the light of a simple mathematical consideration. If traits X and Y are exclusive and exhaustive, then the number of X and Y individuals in a given generation determines the frequencies with which the two types occur at that time; however, it is not true that the *expected* number of X and Y individuals determines their *expected* frequencies. The reason is that frequency is a quotient:

frequency of X individuals =

(number of X individuals)/(total number of individuals).

The important point is that the expected value of a quotient is not identical with the quotient of expected values:

E(frequency of X individuals) ≠

E(number of X individuals)/E(total number of individuals).

This is why a general definition of fitness cannot equate fitness with expected offspring number. The fitness values of traits, along with the number of individuals initially possessing each trait, are supposed to entail the expected frequencies of the traits one or more generations in the future (if selection is the only force influencing evolutionary change). Expected number of offspring determines the value of the quotient on the right, but the expected frequency is left open.

Notice that this point about the definition of fitness differs from the one that Lewontin and Cohen (1969) made concerning population growth. Their point was to warn against using the *expected number* of individuals as a predictor. The present idea is that if one wants to predict the *expected frequencies* of traits, something beyond the expected number of individuals having the different traits must be taken into account.

15.4 Conclusion

Evolutionists are often interested in long-term trends rather than in short-term events. However, this fact about the interests of *theorists* does not mean that the *theory* enshrines an autonomous concept called "long-term fitness." The long-term is a function of what happens in successive short terms. This metaphysical principle is alive and well in evolutionary theory. However, traits like sex ratio show that the short term sometimes has to be longer than a single generation.

The example of sex ratio aside, we may begin thinking about the fitness of a trait by considering a total probability distribution, which specifies an individual's probability of having 0, 1, 2, 3. . . offspring. The expected value is a summary statistic of this distribution. Although this statistic sometimes is sufficient to predict expected frequencies, it is not always a sufficient predictor; when there is stochastic variation in offspring number, the variance is relevant as well.

Are the mean and variance together sufficient to define the concept of fitness? Beatty and Finsen (1989) point out that the skew of the distribution is sometimes relevant. In principle, fitness may depend on all the details of the probability distribution. However, Gillespie's analysis of within-generation variance leads to a more radical conclusion. When there is stochastic variation within generations, Gillespie says that the fitness of a trait is approximately the mean offspring number minus σ^2/N. Notice that the correction factor adverts to N, the population size; this is a piece of information not contained in the probability distribution associated with the trait. It is surprising that population size exerts a general and positive effect on fitness.

The results of Dempster, Haldane and Jayakar, and Gillespie show how the mathematical development of a theoretical concept can lead to a reconceptualization of its empirical meaning. In Newtonian mechanics, an object's mass does not depend on its velocity or on the speed of light; in relativity theory, this classical concept is replaced with relativistic mass, which is the classical mass divided by $(1 - v^2/c^2)^{\frac{1}{2}}$. As an object's velocity approaches zero, its relativistic mass approaches the classical value. In similar fashion, the corrected definition of fitness approaches the "classical" definition as σ^2 approaches zero. People reacted to Einstein's reconceptualization of mass by saying that it is strange and unintuitive, but the enhanced predictive power of relativity theory meant that these intuitions had to be re-educated. A definition of fitness that reflects the expected number of offspring, the variance in offspring number, and the population size yields more accurate predictions of expected population frequencies than the classical concept, and so it is preferable for the same reason.

It is sometimes said that relativity theory would not be needed if all objects moved slowly. After all, the correction factor $(1 - v^2/c^2)^{\frac{1}{2}}$ makes only a trivial difference when $v \ll c$. The claim is correct when the issue is prediction, but science has goals beyond that of making accurate predictions. There is the goal of understanding nature – of grasping what reality is like. Here we want to know which laws are true, and relativity theory has value here, whether or not we need to use that theory to make reasonable predictions. A similar point may apply to the corrected definition of fitness; perhaps evolving traits rarely differ significantly in their values of σ^2; if so, the corrected definitions will not be very useful when the goal is to predict new trait frequencies. This is an empirical question whose answer depends not just on how traits differ with respect to their variances but on the population size; after all, even modest differences in fitness can be important in large populations. But quite apart from the goal of making predictions, there is the goal of understanding nature – we want to understand what fitness is. In this theoretical context, the corrected definition of fitness is interesting.

What is the upshot of this discussion for the "propensity interpretation of fitness?" This interpretation has both a nonmathematical and a mathematical component. The nonmathematical idea is that an organism's fitness is its propensity to survive and be reproductively successful. Propensities are probabilistic dispositions. An organism's fitness is like a coin's probability of landing heads when tossed. Just as a coin's probability of landing heads depends on how it is tossed, so an organism's fitness depends on the environment in which it lives. And just as a coin's probability may fail to coincide exactly with the actual frequency of heads in a run of tosses, so an organism's fitness need not coincide exactly with the actual number of offspring it produces.

These ideas about fitness are not threatened by the foregoing discussion. However, the propensity interpretation also has its mathematical side, and this is standardly expressed by saying that fitness is a mathematical expectation (see, for example, Brandon 1978, Mills and Beatty 1979, Sober 1984). As we have seen, this characterization is not adequate in general, although it is correct in special circumstances. But perhaps all we need do is modify the mathematical characterization of fitness while retaining the idea that fitness is a propensity (Brandon 1990, p. 20).

This modest modification seems unobjectionable when there is between-generation variation in fitness; after all, if an organism's expected (= arithmetic mean) number of offspring reflects a "propensity" that it has, so too does its geometric mean averaged over time. However, when there is within-generation variation, the propensity interpretation is more problematic. The problem is the role of population size (N) in the definition. To say that a coin is fair – that $p = 1/2$, where p is the coin's probability of landing heads when tossed – is to describe a dispositional property that it has. However, suppose I define a new quantity, which is the coin's probability of landing heads minus σ^2/N, where N is the number of coins in some population that happens to contain the coin of interest. This new quantity $(p - \sigma^2/N)$ does not describe a property (just) of the coin. The coin is described by p and by σ^2, but N adverts to a property that is quite extrinsic to the coin.

Is it really tenable to say that p describes a propensity that the coin has but that $(p - \sigma^2/N)$ does not? After all, the coin's value for p reflects a fact about how the coin is tossed just as much as it reflects a fact about the coin's internal composition. Perhaps the propensity is more appropriately attributed to the entire coin-tossing device. However, $(p - \sigma^2/N)$ brings in a feature of the environment – N – that has no causal impact whatever on the coin's behavior when it is tossed. It is for this reason that we should decline to say that $(p - \sigma^2/N)$ represents a propensity of the coin.

I conclude that an organism's fitness is not a propensity that it has – at least not when fitness must reflect the existence of within-generation variance in offspring number. In this context, fitness becomes a more "holistic" quantity; it reflects properties of the organism's relation to its environment that affect how many offspring the organism has; but fitness also reflects a property of the containing population – namely, its census size – that may have no effect on the organism's reproductive behavior. Of course, the old idea that fitness is a mathematical expectation was consistent with the possibility that this expectation might be influenced by various properties of the population; frequency-dependent and density-dependent fitnesses are nothing new. What is new is that the *definition* of fitness, not just the factors that sometimes affect an individual's expected number of offspring, includes reference to census size.

15.5 Acknowledgments

I am very much in Dick Lewontin's debt. I spent my first sabbatical (1980–81) in his laboratory at the Museum of Comparative Zoology at Harvard. I had written one or two pieces on philosophy of biology by then, but I was very much a rookie in the subject. Dick was enormously generous with his time – we talked endlessly – and I came away convinced that evolutionary biology was fertile ground for philosophical reflection. While in his laboratory, I worked on the units of selection problem and on the use of a parsimony criterion in phylogenetic inference. I still have not been able to stop thinking and writing about these topics. Thanks to Dick, 1980–81 was the most intellectually stimulating year of my life.

Dick is a "natural philosopher." I do not mean this in the old-fashioned sense that he is a *scientist* (though of course he is that) but in the sense that he is a *natural at doing philosophy*. It was a striking experience during that year to find that Dick, a scientist, was interested in the philosophical problems I was thinking about and that he was prepared to consider the possibility that they might be relevant to scientific questions. I came to the laboratory with the rather "theoretical" conviction that there should be common ground between science and philosophy, but the experience I had in the laboratory made me see that this could be true, not just in theory, but in practice.

During that year, I attended Dick's courses in biostatistics and population genetics; I gradually started to see how deeply the concept of probability figures in evolutionary biology. The present chapter, I think, is on a subject that is up Dick's alley. It is a pleasure to contribute this chapter to a volume that honors him.

My thanks to Martin Barrett, John Beatty, James Crow, Carter Denniston, Branden Fitelson, John Gillespie, David Lorvick, Steve Orzack, and to Dick as well for useful discussion of earlier drafts of this chapter.

REFERENCES

Beatty, J., and Finsen, S. (1989). Rethinking the propensity interpretation – a peek inside Pandora's box. In *What the Philosophy of Biology Is*, ed. M. Ruse, pp. 17–30. Dordrecht, The Netherlands: Kluwer Publishers,

Brandon, R. (1978). Adaptation and evolutionary theory. *Studies in the History and Philosophy of Science* 9:181–206.

Brandon, R. (1990). *Adaptation and Environment.* Princeton, NJ: Princeton University Press.

Cooper, W. (1984). Expected time to extinction and the concept of fundamental fitness. *Journal of Theoretical Biology* 107:603–29.

Cronin, H. (1991). *The Ant and the Peacock.* Cambridge, UK: Cambridge University Press.

Dempster, E. R. (1955). Maintenance of genetic heterogeneity. *Cold Spring Harbor Symposium on Quantitative Biology* 20:25–32.

Fisher, R. (1930). *The Genetical Theory of Natural Selection.* New York: Dover, 2d edition, 1957.

Gillespie, J. (1973). Natural selection with varying selection coefficients – a haploid model. *Genetical Research* 21:115–20.

Gillespie, J. (1974). Natural selection for within-generation variance in offspring. *Genetics* 76:601–6.

Gillespie, J. (1977). Natural selection for variances in offspring numbers – a new evolutionary principle. *American Naturalist* 111:1010–14.

Haldane, J. B. S., and Jayakar, S. D. (1963). Polymorphism due to selection of varying direction. *Journal of Genetics* 58:237–42.

Jeffrey, R. (1983). *The Logic of Decision.* Chicago: University of Chicago Press. 2d edition.

Lewontin, R., and Cohen, D. (1969). On population growth in a randomly varying environment. *Proceedings of the National Academy of Sciences* 62:1056–60.

Mills, S., and Beatty, J. (1979). The propensity interpretation of fitness. *Philosophy of Science* 46:263–86.

Nitecki, M. (ed.) (1988). *Evolutionary Progress.* Chicago: University of Chicago Press.

Paul, D. (1992). Fitness – historical perspectives. In *Keywords in Evolutionary Biology*, ed. E. Keller and E. Lloyd, pp. 112–14. Cambridge, MA: Harvard University Press.

Sober, E. (1984). *The Nature of Selection.* Cambridge: MIT Press. 2d edition, Chicago: University of Chicago Press, 1994.

Sober, E. (1994). Progress and direction in evolution. In *Progressive Evolution?*, ed. J. Campbell. Boston: Jones and Bartlett.

Thoday, J. (1953). Components of fitness. *Symposium of the Society for Experimental Biology* 7:96–113.

Thoday, J. (1958). Natural selection and biological process. In *A Century of Darwin*, ed. S. Barnett, pp. 313–33. London: Heinemann.

CHAPTER SIXTEEN

Evolvability

Adaptation and Modularity

JEFFREY C. SCHANK AND WILLIAM C. WIMSATT

The eye and other such organs of "extreme perfection" deeply concerned Darwin (1859). They are highly complex and well suited in a multiplicity of ways to the environments in which they are found. To many of his contemporaries they were paradigmatic of intelligent design – touchstones of natural theology and ways of learning about the creator. Darwin viewed such organs with special interest as the most critical and difficult test of his theory (Lewontin 1978): he had to account for them to win its acceptance, and they have remained a focal point for critics of evolution ever since.[1]

Heritable variation in fitness among phenotypic characters, "Darwin's Principles" after Lewontin (1970), are the key elements of Darwin's theory of evolution by natural selection. These principles provide an abstract schema that Lewontin and others following him have exploited to good advantage in delineating a multiplicity of possible selection processes. Campbell's (1974) similar abstractions characterize a diversity of selection processes in psychology and culture.[2] But these abstractions have a cost: to account for any specific adaptation, Darwin's abstract principles must be filled in with specific mechanisms of, and constraints on, heritability, variation, and fitness. They are completely silent on what kinds of morphologies, physiologies, and behaviors *can* evolve. Nor is this answered by the detailed stories we tell in particular cases. This is the problem of *evolvability*, identifying the kinds of mechanisms and constraints by which complex morphologies, physiologies, and behaviors can and do evolve.

Darwin was well aware that specific mechanisms were required for his abstract principles. His proposed mechanism of inheritance, pangenesis, was replaced by Mendelism, but other components of his theory have remained at least superficially the same. Small heritable variations in morphologies, physiologies, and behaviors place their possessors in different *ecological relationships* with their environments, some proportion of which provide slight advantages in survival and reproduction over competitors – causing their increase in frequency in the population. Repeated in successive generations, this produces a cumulative process of adaptation, yielding organisms and character complexes well suited to their environments.

Processes of adaptation allow the abstract theory of natural selection to explain organs of "extreme perfection," given adequate mechanisms of inheritance and variation. But although this notion is fundamental to the evolution of complex and finely tuned adaptations, characters can and do evolve that have no adaptive function (Gould and Lewontin 1978). Some features of gene regulatory systems (Kauffman 1993) or morphological architecture (Newman 1994) may be generic – unavoidable characteristics of complex developmental systems. Self-organization is important at both genotypic and phenotypic levels, generating complexity from simplicity. Certain features of selection regimes (e.g., systematic differences among fitness contributions of genes *within* an organism) may be generic consequences of genotypic–phenotypic organization and necessary for the evolution of complex adaptations (Wimsatt and Schank 1988). These are all pivotal for explaining Darwin's organs of "extreme perfection" but have ambiguous or complex relations to adaptation.

In this chapter, we discuss Lewontin's (1978) two constraints, *continuity* and *quasi-independence*, on the process of adaptation. We argue that *modularity* in the mapping of genotypes into phenotypes is fundamental to the process of adaptation. Our discussion of modularity generates questions and predictions about the evolution of complex adaptations. How modular is the genotype–phenotype mapping? What evolutionary mechanisms modularize it? If variability requires modularity, are there limits on the integrative complexity of adaptations or architectural design principles in their mode of articulation?

16.1 Adaptation: Modularity and Variability

Lewontin (1978, 1984; Gould and Lewontin 1978; Levins and Lewontin 1983) discussed various problems with the concept of adaptation (e.g., morphological and developmental constraints; the identification of niches and their dynamical character, leading to changing ecological relations between phenotype and environment) and with the adaptationist program (e.g., treating all biological characters as adaptations and adaptation as a passive response to the environment). Nevertheless, he took processes of adaptation very seriously.[3] Near the end of his 1978 and 1984 articles, he set out two new and fundamental constraints on the evolution of adaptations:

> The mapping of character states into net reproductive fitness must have two characteristics: *continuity* and *quasi-independence*. By continuity we mean that very small changes in a character result in very small changes in the ecological relations of the organism and therefore very small changes in reproductive fitness. Neighborhoods in character space map into neighborhoods in fitness-space. So a very slight change in the shape of a fin or mammalian appendage to make it finlike cannot cause a dramatic change in the sexual recognition pattern, or make the organism attractive to a completely new set of predators. By quasi-independence we mean that there exist a large variety of developmental paths by which a given character may change, and

although some of these may give rise to countervailing changes in other organs and in other aspects of the ecological relations of the organism, *a non-negligible proportion of these paths will not result in countervailing effects of sufficient magnitude to overcome the increase in fitness from the adaptation* [italics added]. In genetic terms, quasi-independence means that a variety of mutations may occur, all with the same effect on the primary character but with different effects on other characters, some set of which will not be at a net disadvantage. (Lewontin 1984, p. 247)

Continuity is a constraint on the relations of phenotypic characters to their environments and how these relations affect fitness. Continuity requires that small changes in a character produce only small changes in fitness. Quasi-independence consists of two related constraints. One constraint is developmental, requiring a sufficient number of developmental pathways along which variant characters can develop and only weakly interact with the developmental pathways of other phenotypic characters. The other constraint is genetic, requiring that mutations in the genotype causally influence specific phenotypic characters and only weakly interact with others. Put together, quasi-independence requires that there must be epigenetic pathways from genes to characters that have weak interactions with other genes and characters.[4] Finally, although continuity and quasi-independence are not the same, they are also not independent. The failure of character variants to be quasi-independent will decrease the likelihood of continuity because with increasing interdependence comes the decreasing likelihood that the totality of the direct and indirect costs and benefits of a character change will be small.

Quasi-independence is a constraint on variability in adaptive evolution. Genetically it requires that the mapping of genotypes into phenotypes be such that mutations will typically have only relatively local affects on the phenotype, thereby determining in part what kinds of changes are likely to have relatively few countervailing effects. *Modularity* – as we use the term here – is a form of quasi-independence modulating the mappings between genotypes and phenotypes, allowing characters to vary in ways more likely to satisfy continuity and thereby increasing the proportion of phenotypic characters that may be adaptive. Thus, we analyze modularity in terms of the pleiotropic interactions of genes mapping to phenotypes. Not all mutations are equal in their consequences for fitness, and thus, independent of modularity concerns, not all mutations will satisfy the constraint of continuity. But, even if we assume the best case scenario for continuity (all mutations have the same or similar small effect on fitness), pleitotropic interactions can geometrically increase the fitness effects of specific mutations, as we will see later, which violates continuity. Thus, in its genetic sense, modularity is fundamentally important for continuity in adaptive evolution. It is also a fundamental "meta-engineering" principle of genotype–phenotype relationships. In the remainder of this chapter we will examine some of the limits, constraints, and implications of modularity.

16.2 Complex Adaptations and Modularity

Adaptations such as the eye are character complexes generated by many genes, gene products, and changing physical conditions that interact during development to produce a well-adapted eye (Halder, Callaerts, and Gehring 1995). Continuity and quasi-independence constrain the genotype–phenotype and phenotype–fitness mappings of characters within these complexes. If a mutation affecting a given character of the eye affects many other characters of the eye as well (e.g., the ability to detect color patterns in potential mates), the net effect on reproductive fitness is likely to be very disadvantageous. From the discussion in last section, we should conclude that at least some genotype–phenotype mappings within a character complex must be modular in order for variability to fall within the constraint of continuity. Consider a hypothetical character complex with maximal pleiotropy – each gene affects all characters. Mutations in genes with such broad effects (unless they represent coordinated allometric tuning of parameterized relationships) are more likely to scramble the ecological relationships of characters and thus more likely to violate continuity than cases with lower pleiotropy. Thus, just as the genotype–phenotype relations of an organism must be modular with respect to adaptive complexes, there must also be modularity within complex adaptations.

A fundamental problem for the evolution of complex adaptations is how to avoid increasing pleiotropy as a consequence of the potential combinatorial explosion resulting from the interaction of parts (Wagner and Altenberg 1996). This problem suggests that the evolution of modularity must be an ongoing process continually decoupling characters in complex adaptations and allowing them to evolve by small changes (Wimsatt 1981).

The most fundamental mechanism for introducing modularity into complex adaptations is parcellation (Wagner 1996; Wagner and Altenberg 1996; see also Raff's 1996 concept of *dissociation*). Parcellation reduces pleiotropic effects between the genotype–phenotype mapping of character complexes and thereby introduces modularity. But how can this happen? Altenberg and Wagner (Wagner and Altenberg 1996; Wagner 1996) argued convincingly that parcellation could occur by directional selection on a character or some subset of characters with stabilizing selection acting on the other characters. Under these conditions, modifier genes would be favored that decouple pleiotropic effects. But are there limits on parcellation and the evolution of modularity?[5]

16.3 Accretion and Pleiotropic Entrenchment

Continuity and quasi-independence imply that a fundamental mode for the evolution of increasingly complex characters must be via the accumulation and integration of small variations into previously successful adaptations. One mechanism meeting these constraints is *accretion*. By accretion we mean that new characters are accumulated at the "pleiotropic periphery" of the adaptive

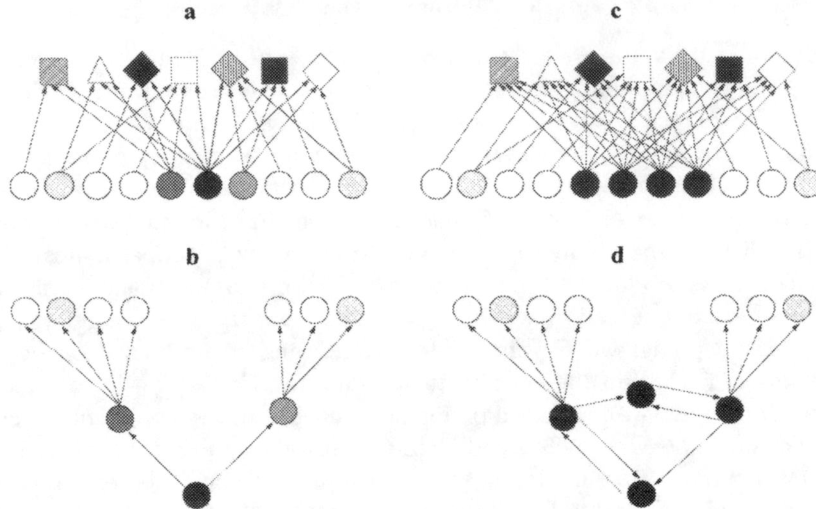

Figure 16.1. The process of accretion in the evolution of complex adaptations can be explained by considering the mapping of regulatory genes into themselves and structural genes. (**a**) An illustration of a hypothetical gene-regulatory mapping and its pleiotropic effects. The shading of genes indicates the scope of pleiotropic effects of regulatory genes. Arrows indicate the effect of a gene on a character, and multiple arrows from a regulatory gene to characters, whether inside or crossing modules, indicates pleiotropy. Genes in the center of the array have the largest number of pleiotropic effects. These differences in (**a**) are products of the regulatory-gene mapping (**b**). Here a hierarchical cascade of three regulatory genes controls the structural genes directly affecting the phenotypic characters. The high pleiotropy of the three center genes in (**a**) is due to their location in the regulatory cascade (**b**). (**c**) An illustration of a more complicated case with even more pleiotropic interactions: four central genes affect all phenotypic characters. The underlying regulatory structure (**d**) reveals why: the many pleiotropic interactions arise because they mutually regulate each other and thus indirectly interact with all phenotypic characters in the complex.

complex (i.e., characters and new genes are integrated into the genotype–phenotype map in ways that tend to minimize the introduction of pleiotropic effects). Such additions, as required by continuity, must produce relatively little change in the overall fitness of phenotypic characters.

Accretion is a piecemeal process of co-opting and adding to the gene-regulatory regulatory structure (Figure 16.1). To the extent that structural genes are the most "downstream" components of this mapping, they will typically have less pleiotropic effects than "upstream" regulatory genes and will be the most modular parts of genotype–phenotype relations among character complexes.[6] Genes deeper in the regulatory structure will have greater *pleiotropic entrenchment* (a special case of generative entrenchment, Wimsatt 1986), and their action will be less modular in the genotype–phenotype mapping.

Accretion as a mechanism of pleiotropic entrenchment leads to two predictions about the evolution of complex adaptations. First, adaptations can grow at the periphery by co-opting new structural genes or adding regulatory genes with low pleiotropic entrenchment. Regulatory genes with relatively high pleiotropic entrenchment should be evolutionarily conserved if the regulatory network continues to serve its regulatory and developmental functions in the generation of a complex adaptation. Second, co-opting structural and regulatory genes requires some mechanism for linking genes in a regulatory manner. For example, the linking of genes for synchronous expression could occur if the co-opted genes acquired the same transcription factors. Thus, the evolution of complex adaptations may require the conservation of specific transcription factors (such as those for *hox* and *pax* genes). From accretion, pleiotropic entrenchment, and the requirement for mechanisms linking genes for synchronous (or asynchronous) gene expression, we predict that one should find both phylogenetically conserved transcription factors and pleiotropically entrenched regulatory structures in homologous complex adaptations.

Indeed, Halder et al. (1995; also see Raff 1996) have argued that *Pax-6* is a highly conserved transcription factor in eye development in mammals, flies, and mollusks, suggesting a common evolutionary origin in very different eye types. Their proposed mechanism for evolution of the eye in each of these phyla is the process we have labeled *accretion.* With a core regulatory structure and *Pax-6* transcription factors, new functional characteristics could be co-opted by genes acquiring *Pax-6* regulatory elements. These authors offer as an example the "evolutionary tinkering" of the lens of crystalline genes, which encode the structural proteins of the lens. These proteins were not evolved specifically for the lens but were co-opted from other enzymes or small heat-shock proteins while still performing their other functions (a phenomenon called gene sharing; Piatigorsky and Wistow 1989). Richardson, Cvekl, and Wistow (1995) have shown that these lens elements were recruited by acquiring *Pax-6* response elements. If this theory of the evolution of the eye is correct, it would be a paradigm for the evolution of complex adaptations from simpler adaptive elements through a process of accretion. Moreover, it suggests how multiple independent inventions of complex adaptations may be less surprising than one might think (Halder et al. 1995; Raff 1996).

Accretion in the evolution of complex adaptation adds dependencies to any upstream portions of gene–regulatory and genotype–phenotype mappings. Genes with broader pleiotropic effects are relatively more *entrenched* than those with more narrow effects. In Figure 16.1 we see that such genes may be regulatory genes controlling many other regulatory or structural genes. Pleiotropic entrenchment also makes "deep" changes in complex adaptations exceedingly unlikely, for deeper changes are more likely to violate continuity and quasi-independence. This explains "frozen accidents" but also suggests that complex adaptations can only be modified at their periphery. This implication of pleiotropic entrenchment, however, has trouble explaining apparently deep developmental modifications like heterochrony, progenesis, and neoteny. Without other mechanisms satisfying continuity and quasi-independence, pleiotropic entrenchment predicts that heterochrony should be exceedingly rare.

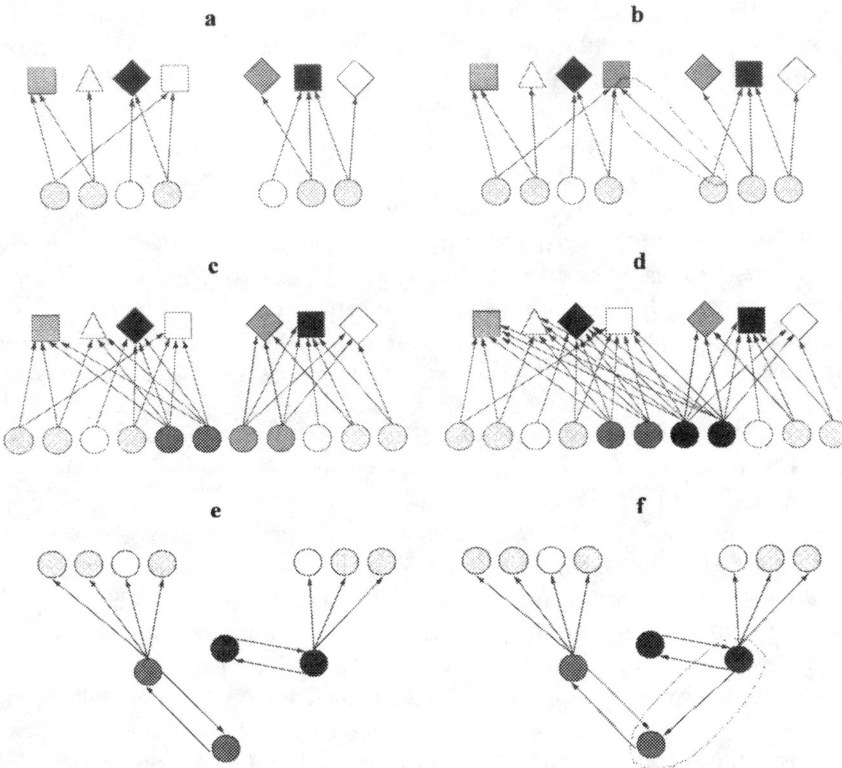

Figure 16.2. (**a**) An illustration of two modules with no pleiotropic interactions between them. (**b**) An illustration of a hypothetical mutation that pleiotropically *integrates* them (indicated by the dotted line around the integrating mutation). Shallow *parcellations* are illustrated by the reverse transition. The mutations in both cases are shallow because only one (or relatively few) pleiotropic interaction is (are) produced. (**c**) An illustration of two modules, this time with considerably more intramodule pleiotropic interactions than in (**a**). (**d**) An illustration of a hypothetical integrating mutation that affects all of the phenotypic characters in the mapping. The combinatorial explosion in pleiotropic interactions can be explained by examining the underlying genotype–regulatory mappings in (**e**) and (**f**), which illustrates two simple regulatory networks underlying the genotype–phenotype mapping in (**c**) and (**d**), respectively. The circled regulatory connection in (**f**) is introduced by mutation. The new regulatory connection is a *deep integration* because it is between two pleiotropically entrenched genes and results in an increase in pleiotropic entrenchment of elements of the regulatory network. Similarly, *deep parcellation* is illustrated by going in the opposite direction (**d**) to (**f**) explained by (**c**) and (**e**), respectively.

Pleiotropic entrenchment can also be explained by its implications for two mechanisms proposed for the evolution of modularity in the genotype–phenotype mapping: integration and parcellation (see Wagner 1996; Wagner and Altenberg 1996). We distinguish between "shallow" and "deep" integration and parcellation (Figure 16.2). With shallow integration, the integrating mutation produces relatively few pleiotropic interactions.[7] Deep integration and parcellation, however, are problematic for adaptive evolution: mutations that produce combinatorial explosions in pleiotropic interactions are far more likely to violate continuity and quasi-independence (Figure 16.2). This suggests that accretion is the only effective mode for the evolution of complex adaptations from simpler characters. But although it may be the dominant mode, it is not universal: through gene duplication the constraints of pleiotropic entrenchment can sometimes be avoided.

16.4 Gene Duplication and Evolution of Complex Adaptations

The difficulty of making deep changes indicates important limits to the evolutionary processes of parcellation and integration. Pleiotropically entrenched genes have numerous consequences, making it more likely that changes in them will violate quasi-independence and continuity. Ruddle et al. (1994; also see Raff 1996) has argued that, as genes become increasingly integrated into developmental pathways, it becomes more difficult for them to mutate or take on new functions without disrupting development. We had already noted this as *generative entrenchment:* because as either genes or characters accumulate developmental dependencies, they become less able to mutate successfully and more likely to be conserved (Wimsatt 1986; Schank and Wimsatt 1987). Pleiotropic entrenchment is a special case of generative entrenchment in which the entrenched element is a gene and the entrenching factor is the multiplicity of adaptive pleiotropic effects.

Gene duplication, however, can permit deep integration and parcellation and the rapid evolution of complex adaptations. Ohno (1970) first suggested it as an important evolutionary mechanism via polyploidy. Gene duplication potentially avoids the obstacles produced by deep pleiotropic entrenchment through redundancy: if one block of genes can continue to serve current biological functions, the duplicate blocks are free to be co-opted, integrated, or parcellated into new developmental pathways and adaptive functions (Ruddle et al. 1994). If large blocks of genes can be duplicated with few phenotypic consequences, then the block of duplicated genes can be modified without violating continuity and quasi-independence. This, however, gives rise to important questions, including,[8] How much of the cluster of genes or regulatory network must be duplicated in order not to violate continuity and quasi-independence?

Wagner (1994) examined this question with a simple but elegant model: gene networks are dynamical systems regulating each other's transcription. The degree of interaction is determined by a matrix of interaction coefficients. The output of the model is an array of transcriptional products corresponding to the genes in the gene network. Using mathematical analysis and computer simulations, Wagner analyzed how different sized duplications would produce deviations in the transcriptional products matrix. Both approaches showed that maximal deviation occurred when only 40% of the gene network was duplicated, and the smallest deviation occurred when either the whole network was duplicated or a single gene. Thus, duplication events that satisfy continuity and quasi-independence will tend to involve either single genes (or very small subsets of genes) or whole or nearly whole gene networks or clusters.

Gene duplications are richly implicated in adaptive evolution. *First*, duplication of whole gene networks releases the duplicate network from pleiotropic entrenchment, permitting deep integrations with other gene networks or deep parcellations. Holland et al. (1994) argued that gene duplications (i.e., Hox gene clusters, other homeobox gene families, Wnt genes, and insulin-related genes) and their integrations into developmental pathways were the critical events in vertebrate evolution. Ruddle et al. (1994) argued that gene deletions – a mechanism for deep parcellation – following duplication events have occurred in Hox clusters in both echinoderms and lower chordates. *Second*, gene duplication provides a mechanism for the rapid evolution of new complex adaptations: redundancy allows the duplicated network to be co-opted into new biological functions, and mutation and selection can tinker with an already functional regulatory network and transcription factors. This prediction of gene duplication is realized in the repetition of Hox gene clusters and other gene families in metazoans (Ruddle et al. 1994). *Third*, single gene duplications can facilitate accretion by providing a mechanism for introducing genes with appropriate transcription factors, which can then be modified by selection or can be deeply integrated into new or old gene networks, increasing developmental and adaptive possibilities geometrically (Ruddle et al. 1994). Finally, gene duplication may be an "adaptation pump" for increasing the complexity of organisms,[9] but with the increasing complexity of the genotype–phenotype mapping comes the problem of how selection can maintain increasing numbers of complex adaptations.

The problem of increasing numbers of complex adaptations can be expressed in two theoretical but robust constraints (Wimsatt and Schank 1988). First, if the core of a regulatory network for a complex adaptation is to be conserved, the relative contributions to fitness of these genes must be greater than that of other noncore genes. Without asymmetries in fitness contributions, complexity catastrophes can ensue (Kauffman 1993). However, even with moderate fitness asymmetries, selection can preserve fitter subsets of genes at or near fixation even when many or most of the slightly less fit genes are lost (Schank and Wimsatt 1987). Meeting this constraint follows naturally, almost by definition, from pleiotropic entrenchment. Deeply pleiotropically (or generatively) entrenched characters will show very large selection coefficients for their preservation. Indeed, their substantial modification will commonly produce lethals.

Complexity catastrophes are a second constraint on the evolution of numerous complex adaptations: they become increasingly likely as the number of core networks increases by the same logic that demonstrated that some can be preserved in an all or none manner. This constraint suggests a limit on the accumulation of complex adaptations in an organism. Elsewhere, we have discussed several mechanisms by which this constraint can be relaxed (Wimsatt and Schank 1988). These include changes in genome architecture that convert genetic loads from larger forms of genetic load to smaller ones; for example, tandem duplication followed by recombination bringing heterotic alleles onto the same chromosome, forming a small and tightly linked supergene (Spofford 1972). Such conversion could produce load reductions of two orders of magnitude or more through the conversion of segregational load into much smaller mutational plus recombinational loads. Two mechanisms involving generative entrenchment and the relative timing of developmental events are capable of converting hard selection into soft selection and reducing the hard selected load. Reduction in the mutation rate can also increase the number of genes that can be maintained by selection.

Finally, asymmetries in fitness contributions of genes may be induced and complexity catastrophes avoided by the integrative process of co-opting genes into different adaptive complexes, forcing them to serve multiple functions (i.e., gene sharing). By serving diverse and multiple functions, asymmetries in fitness contributions among genes are almost certain, and fewer genes may be required to support multiple complex adaptations, which in turn may lessen the possibility of complexity catastrophes. However, by co-opting genes into different adaptive complexes, pleiotropy will likely increase and modularity decrease. Thus, constraints on the number of complex adaptations may require reduction in modularity in direct conflict with continuity and quasi-independence, which require it. Therefore, although continuity and quasi-independence require the evolution of modularity, constraints on the number of complex adaptations that can be maintained in an organism likely limit the degree of modularity in its genotype–phenotype relationship.

16.5 Conclusions

Adaptation is fundamental to a theory of evolvability, providing evolutionary explanations for Darwin's organs of "extreme perfection." Modularity, a form of quasi-independence in genotype–phenotype mappings, is fundamental to continuity in adaptive evolution. We have argued that the evolution of complex adaptations will normally proceed by a process of accretion; this introduces pleiotropic entrenchment, which renders deeply entrenched genes nearly immutable to "deep" modification. This limits modularity within complex adaptations. Gene duplication is a mechanism for avoiding the constraint of pleiotropic entrenchment and allowing deep integration and parcellation. In addition, gene duplication facilitates accretion in the evolution of complex adaptations from simpler characters and may function as a complex adaptation "pump" within the as yet unknown limits of selection to maintain complex adaptations (Wimsatt and Schank 1988).

In our view, modularity is a fundamental constraint on variability in character complexes and thus is a fundamental constraint for the evolvability of complex adaptations. But, although modularity is a fundamental constraint on evolvability, it is not without limits. As we and others have argued, the internal genotype–phenotype mapping of a character complex cannot be so tightly coupled that there are a negligible number of small mutational variations that satisfy these constraints. Modularity, however, cannot be unlimited. Indeed, adaptive variability may actually be enhanced by a conserved-core gene-regulatory structure, either by providing a stable base on which to build complex adaptation through accretion or through the mechanisms of gene duplication.

Several predictions are derivable from the evolutionary processes of accretion, pleiotropic entrenchment, modularity, and gene duplication that are confirmed or consistent with the current state of knowledge about metazoans:

1. Pleiotropic mappings from genes to phenotypic characters within a character complex and gene regulatory networks cannot be so tightly coupled that small variations have a negligible chance to be adaptive. At least some combinations of characters must be able to change with relatively little effect on other characters within the complex.
2. Accretion favors the evolutionary conservation of transcription sequences (e.g., *hox* and *pax* genes) because this facilitates co-opting new genes at the pleiotropic periphery.
3. Accretion in turn should increase pleiotropic entrenchment, and thus core gene–regulatory networks will also tend to be conserved by pleiotropic entrenchment.
4. Without gene duplication, the evolution of complex adaptations would be a slow process of evolutionary tinkering at the pleiotropic periphery of a growing character complex.
5. Gene duplication allows deeper modifications and integrations in complex adaptations.
6. Gene duplication allows complex developmental regulatory structures to be co-opted for new complex adaptations. This is an alternative mode for the evolution of complex adaptations, an adaptation "pump" increasing the frequency of larger adaptive jumps that do not violate continuity and quasi-independence.

Returning to Lewontin's formulation of quasi-independence (above), we noted that he distinguished between its developmental and genetic aspects. Our discussion of modularity focused primarily on its genetic aspects. We took a small step toward development by integrating relatively simple aspects of gene regulation into the genetic notion of modularity. A fuller understanding of quasi-independence and modularity in its developmental aspects is a far more difficult task. Organisms do not merely evolve adaptations for adult environments; they also face a succession of ontogenetic environments with their own specific adaptive requirements. During development, a change in a character can have countervailing effects on the ecological relations of the developing organism in any or all of these successive ontogenetic environments.

16.6 Acknowledgments

This work was supported by the Center for the Integrative Study of Animal Behavior, Indiana University (Schank) and the Chicago Humanities Institute, University of Chicago (Wimsatt). We thank Günter Wagner and Lee Altenberg for comments on a draft of this paper.

Notes

1. Richards (1983) argues that Darwin's inability to explain the sterile castes in colonial insects was pivotal in his infamous delay in publishing his work on evolution, citing his remark in 1843 that "this is the rock upon which my theory could founder." Darwin apparently thought of colony selection as his way out only within the year before publication of the *Origin.* Sterile castes of social insects were a long-standing paradigm of natural theologians. The Scheuchzer *Kupfer-bibel* (1731–35) depicts carpenter ants and termites in one plate, referring the reader to a biblical passage where one is enjoined "Look thou o sluggard to the ant, and learn from the wisdom of her ways." So also for Paley's famous case. The *Kupfer-bibel* depicts dissections of eye and ear in two separate plates – both keyed to a biblical passage that begins, "Think you that he who made the eye cannot see, that he who made the ear cannot hear."
2. Lewontin's opposition to evolutionary theories of culture and simplistic reductionisms are well known. We accept most of his critiques of existing theories of cultural evolution but would argue that better mechanistic (but nonreductionistic) theories are possible.
3. Lewontin's emphasis on the way that organisms interact with their environments rather than passively react to them is, if anything, strengthened by ideas of generative entrenchment (Wimsatt 1986, Schank and Wimsatt 1987). The emphasis on generativity fits naturally with an interactionist view of adaptation. But more crucially, understanding the role of generative entrenchment in adaptive processes involves the recognition that either inner or outer factors and structures and their relationships over different time scales can become entrenched. External entrenchments most strikingly include oxygen, nitrogen fixation, and the cycles for other organically key elements through which organisms have acted as element pumps and compound constructors. These products and others have entered the food web and become constructively entrenched in inner and outer aspects of most other life forms. See Levins and Lewontin (1983) chapters 2 and 3. We would urge a significantly larger role for such processes than Godfrey-Smith recognizes in his contribution to this volume. We are less happy with the metaphor of construction, however, because it seems to suggest a single unified project. It is indeed rather more like the tower of Babel would have been if construction had continued: some populations (like termite colonies and nest-building species) building organized rooms and others temporary shelters on orthogonal plans, intersecting as often through their conflicts as through their common interests. Much of the construction by diverse builders on a variety of different time scales arises through adaptation to the efflux from their diverse factories, or perhaps like the later cities of Troy, constructed on (and using) the middens of the earlier (now ruined) layers. This is not to say that there are no larger couplings or Gaia-like regulations – we do not take a position on this either way, but we want to emphasize that the structure of the environment is likely even more of a bricolage – and in most ways less integrated – than the structure of multilayered internal kluges: the myriad overlapping exaptations turned adaptations that we call adaptive design. Rather than "construction," we suggest *cogeneration,* a term suggesting that organism and environment over time play important (and generative and thus informational) roles. It does not require that they play equivalent or even symmetric roles. But it gets us away from the idea that the genes are the centers of the biological universe and toward looking also at the multidimensional temporal and

spatial structure of the environment and, most centrally, at the relations between the two.

4. We do not assume that the causal relationships from genotypes to phenotypes are either necessary or sufficient for phenotypic characters they affect. Phenotypic characters are codependent upon the interacting developmental trajectories of the organism, its environments, and the pattern of activation of its genes during development. Instead we view the causal relationships from genotype to phenotype to be *complex causal relationships,* indicating the often tortuous and inevitably context-dependent epigenetic pathways from gene products to phenotypic characters. Parallel comments apply to the complex relationship between phenotype and fitness. Thus, the *niche* could be thought of as a complex relationship between phenotype and environment, indicating dimensions of interaction particularly relevant to the determination of fitness.
5. This raises additional questions. How modular can a phenotype become? If modularity is good, is maximal modularity (i.e., no pleiotropy in the genotype–phenotype map) better? Or is it even possible? Too much modularity would seem to conflict with biological integration. Can any character complex be pleiotropically decoupled? What happens when we consider the underlying gene-regulatory structure of development in the genotype–phenotype map?
6. Structural genes can also have many pleiotropic effects, especially if co-opted into different adaptive complexes (e.g., consider the myriad effects for sickle cell anemia and malaria resistance of substituting Hb_S or Hb_C for Hb_A).
7. We should also distinguish "weak" and "strong" pleiotropic interactions in terms of the magnitude of the interactions – with corresponding "weak" and "strong" effects on the mapping of phenotypic characters into fitness. Stronger pleiotropic interactions will be more resistant to parcellation and integration than weak pleiotropic interactions.
8. Gene duplication provides redundancy – a more frequent way of escaping the effects of pleiotropic entrenchment, and thus it emerges as an important mechanism of evolution. But we must also consider constraints on how much of a gene cluster must be duplicated, dosage effects, and other limitations on gene duplication as an adaptation pump for increasing phenotypic complexity.
9. This is only true on average because duplication also facilitates gene deletion and parcellation. But these processes should lag the "adaptation pump" effect in a steady-state equilibration process, and if duplication is massive, as in chromosome duplication or increases in ploidy, they should lag it substantially.

REFERENCES

Campbell, D. T. (1974). Evolutionary epistemology. In *The Philosophy of Karl Popper Vol. 2,* (ed.) P. A. Schilpp, pp. 413–63. Chicago: Open Court.

Darwin, C. (1859). *The Origin of Species: Facsimile Reprint of 1st Edition.* Cambridge, MA: Harvard University Press.

Gould, S. J., and Lewontin, R. C. (1978). The spandrels of San Marco and the panglossian paradigm: A critique of the adaptationist programme. *Proceedings of the Royal Society of London.* 205:281–98.

Halder, G., Callaerts, P., and Gehring, W. J. (1995). New perspectives on eye evolution. *Current Opinion in Genetics & Development.* 5:602–9.

Holland, P. W. H., Garcia-Fernàndez, J., Williams, N. A., and Sidow, A. (1994). Gene duplications and the origins of vertebrate development. *Development 1994 Supplement* 125–33.

Kauffman, S. A. (1993). *The Origins of Order: Self-Organization and Selection in Evolution.* Oxford, UK: Oxford University Press.

Levins, R., and Lewontin, R. C. (1983). *The Dialectical Biologist.* Cambridge, MA: Harvard University Press.

Lewontin, R. C. (1970). The units of selection. *Ann. Rev. Ecol. System.* 1:1–18.

Lewontin, R. C. (1978). Adaptation. *Scientific American.* 239:212–30.

Lewontin, R. C. (1984). Adaptation. In *Conceptual Issues in Evolutionary Biology.* (ed.) E. Sober, pp. 235–51, Cambridge, MA: MIT Press.

Newman, S. A. (1994). Generic physical mechanisms of tissue morphogenesis: A common basis for development and evolution. *J. Evol. Biol.* 7:467–88.

Ohno, S. (1970). *Evolution by Gene Duplication.* Heidelberg: Springer–Verlag.

Piatigorsky, J., and Wistow, G. J. (1989). Enzyme-crystallins: Gene sharing as an evolutionary strategy. *Cell.* 57:197–9.

Raff, R. A. (1996). *The Shape of Life.* Chicago: University of Chicago Press.

Richards, R. C. (1983). Why Darwin delayed, or interesting problems and models in the history of science. *Journal of the History of the Behavioral Sciences.* 19:45–53.

Richardson, J., Cvekl, A., and Wistow, G. (1995). Pax-6 is essential for lens-specific expression of V-crystallin. *Proc. Natl. Acad. Sci. USA.* 92:4676–80.

Ruddle, F. H., Bentley, K. L., Murtha, M. T., and Risch, N. (1994). Gene loss and gain in the evolution of the vertebrates. *Development 1994 Supplement.* 155–61.

Scheuchzer, J. J. (1731–35). *Kupfer-bibel, in Welcha die Physica sacra, oder heiligte Naturwissenschaft der Heil.* Augsburg, Germany.

Schank, J. C., and Wimsatt, W. C. (1987). Generative entrenchment and evolution. In *PSA 1986,* Vol. 2, (eds.) A. Fine and P. K. Machamer, pp. 33–60, Lansing, MI: Philosophy of Science Association.

Spofford, J. (1972). A heterotic model for the evolution of duplications. In *Evolution of Genetic Systems,* (ed.) H. H. Smith, pp. 121–43, New York: Gordon and Breach.

Wagner, A. (1994). Evolution of gene networks by gene duplications: A mathematical model and its implications on genome organization. *Proc. Natl. Acad. Sci. USA.* 91:4387–91.

Wagner, G. P. (1996). Homologues, natural kinds and the evolution of modularity. *American Zoologist.* 36:36–43.

Wagner, G. P., and Altenberg, L. (1996). Complex adaptations and the evolution of evolvability. *Evolution.* 50:967–76.

Wimsatt, W. C. (1981). Units of selection and the structure of the multi-level genome. In *PSA 1980,* Vol. 2, (ed.) P. D. Asquith and R. N. Giere, pp. 122–83, Lansing, MI: Philosophy of Science Association.

Wimsatt, W. C. (1986). Developmental constraints, generative entrenchment, and the innate-acquired distinction. In *Integrating Scientific Disciplines,* (ed.) P. W. Bechtel, pp. 185–208, Dordrecht, The Netherlands: Martinus–Nijhoff.

Wimsatt, W. C., and Schank, J. C. (1988). Two constraints on the evolution of complex adaptations and the means for their avoidance. In *Progress in Evolution,* (ed.) M. Nitecki, pp. 231–73, Chicago: University of Chicago Press.

CHAPTER SEVENTEEN

Organism and Environment Revisited

ROBERT N. BRANDON

17.1 Introduction

In a series of talks and articles dating back over 20 years now, Richard Lewontin (1978, 1982, 1983a,b) has forcefully argued for a radical reconception of the fundamental metaphors of evolutionary biology. The first metaphor, coming to us from Mendel, is that organisms are the ontogenetic results of the *unfolding* of factors, genes, internal to the organisms themselves. The second, from Darwin, is that the extant distribution of organisms results from the phylogenetic process of an autonomous external environment *selecting* the variants that best solve the problems posed by this independently existing factor. Thus, organisms are the *objects* of both the internal forces of development and the external forces of environmental selection. The first metaphor is still widely accepted by most developmental geneticists, though some developmental biologists have produced impressive examples contradicting this view (Nijhout and Emlen 1998, Klingenberg and Nijhout 1998). It is the second metaphor, the central metaphor of the theory of evolution by natural selection, that is the focus of this chapter.

More fully, the second metaphor is this: the external environment, both biotic and abiotic, exists prior to and independently of the evolving organisms of interest. To be evolutionarily successful, they must adapt to the preexisting environment just as a locksmith cuts and files a key to fit a preexisting lock. In other words, ecological *niches* are out there waiting for organisms to evolve to fill them.

But, so Lewontin argues, the idea of unfilled niches is conceptually confused. Niches are defined by the activity of the organisms in question. For example, the eastern five-lined skink eats small insects, burrows in leaf litter or dirt, and escapes predation either by being buried or by its extremely fast movement. The eastern fence lizard coexists in many of the same geographical areas and is insectivorous, but this lizard certainly does not inhabit the same niche. It climbs trees and fences (hence its name) and relies in large part on camouflage to escape predation. Thus

(1) *Organisms determine which parts of their external environment are relevant.* The bark of pine trees is a relevant part of the fence lizard's environment but not of the skink's.

(2) *Organisms modify their environment in ways that are important to their lives.* These modifications can have positive or negative effects. Certainly, beavers do better in the ponds they create than when they are denied the opportunity to create such ponds. Similarly, many pine species lay down heavy layers of pine needles that prevent pine seedlings from taking hold but that also provide fuel for rapidly burning fires which, when they occur regularly, wipe out competing species and create favorable conditions for new pine seedlings. However, when these fires do not occur, hardwoods come up that eventually replace the pines. Thus, in this case the same modification of the environment – the layer of pine needles – either creates conditions favorable to the continued existence of the pine population (when regular fires occur) or leads to the local extinction of the pines. (Ecological succession is based on species modifying their environment in ways that lead to their own local demise.)

(3) *Organisms create a statistical pattern of their environment different from the pattern in the external world.* For instance, skinks, which are exotherms, do not experience the full range of temperature changes in a given day. On a cool morning they may sun themselves and later bury themselves during the heat of the afternoon. Here, active behavior dampens temperature oscillations; in endotherms, such as ourselves, internal mechanisms accomplish much the same thing (of course endotherms also actively escape extremes of temperature). Virtually all plants and animals store vital resources, such as food and water, so that they do not experience the full range of oscillations in the resources in the external world. The time scale over which this averaging of environmental resources takes place varies considerably. We humans cannot manage very long without water – a few days at most – but a cactus can easily go many months without it.

The preceding points do not exhaust Lewontin's arguments against the lock and key metaphor for evolution by natural selection, but they do, I think, capture the gist of them. The alternative metaphor Lewontin suggests is that of *construction.* Organisms, by their choices, activities, physiologies, and so forth, construct their own environments. The physical features of my back yard are there for both the skink and the fence lizard, but they construct importantly different environments, or niches, from them. Likewise, the physical features of the Sonoran desert are there for both cactus and annual desert flowers, but they deal with the rarity and unpredictability of rain in very different ways. The cactus is built for storing large quantities of water for long periods of time, while the annual flower produces seeds that can lay dormant for many years waiting for a soaking rain to cause them to germinate quickly, grow, and reproduce. Thus, there is no preexisting lock to which one must adapt a key. Rather, organisms construct their own environments that selectively favor some variants over others, which can then change the relevant environment, and so on.

The old metaphor suggests that evolution be modeled by two simultaneous differential equation systems. The first describes the way (populations of) organisms O evolve in response to the environment E:

$$dO/dt = f(O, E)$$

The second describes the *autonomous* change of the environment:

$$dE/dt = g(E)$$

(Thus, the lock and key metaphor by no means assumes a static E but just that changes in E are not a function of, nor an effect of, O.)

The constructionist alternative states that we need to model evolution by natural selection using a pair of coupled differential equations:

$$dO/dt = f(O, E) \quad \text{and} \quad dE/dt = g(O, E)$$

In the next section, we will explore this alternative. In particular, we will look at the only two efforts I know of to model constructionism explicitly. In Section 3, I will argue that some of the most interesting work in recent evolutionary biology, work involving frequency-dependent selection, exemplifies Lewontin's constructionism quite well, even though that point has not been made explicit in this work. In the final section, we will examine the consequences of contructionism for adaptationism.

17.2 Recent Efforts to Model Organism–Environment Coevolution

If Lewontin's arguments are cogent, and I think they are, it is reasonable to ask what impact they have had over the last 20 or so years. At first glance the answer to this question appears disheartening, for I know of only two explicit attempts to model the sort of organism–environment coevolution that constructionism describes. In this section, I will briefly review these two explicit explorations of constructionism. In the next section, I will argue that things are not so bleak as they first appear, that much path breaking work in evolutionary biology over the last 20 years does in fact fit the constructionist model even though that remains to be made explicit.

Laland, Odling-Smee, and Feldman (1996) approached the problem of modeling what they call "niche construction" through a genetically sophisticated two-locus population genetic model. We will discuss this shortly, but first I want to turn to work I have been involved with in collaboration with Janis Antonovics (Brandon and Antonovics 1995, 1996). We too developed a population genetic model of what we call "organism–environment coevolution," but we also made conceptual and methodological points about constructionism – points we need to consider briefly before moving on to the models.

Given that there has been considerable controversy historically over the basic definition of the term *niche* (see Colwell 1992 and Griesemer 1992), it is not surprising that Lewontin's claim that there are no empty niches is itself somewhat unclear and controversial. In one conception of niche, the populational conception, a niche is an attribute of the relevant population – its food sources, predators, activities, and so forth. In this conception, it is an analytic truth (i.e., a statement that is true simply by virtue of the meanings of its component words) that no empty niches exist. Without the population, the collection of its attributes, its niche, is also clearly nonexistent. In the other major conception of niche, the environmental conception, a niche is a set of biotic and abiotic factors independent of the organisms of interest. Here, obviously, empty niches are at least a conceptual possibility.

How, then, should we interpret Lewontin's claim? Antonovics and I argue that this question can best be approached by distinguishing among three different conceptions of environment operative in population biology. The first is that of the *external environment*. The external environment consists of the sum total of factors, both biotic and abiotic, external to the population of interest. Significantly, these factors can be measured independently of the organisms of interest. For example, we can measure the nitrogen concentrations along some transect in a field without involving our target organisms – say a population of some species of grass. This concept more or less equates to the environmental conception of niche.

Unfortunately, the external environment is only indirectly related to organismic evolutionary dynamics. The reason for this is that the factor, or factors, we pick out to measure may or may not affect our organisms' survival and reproduction.[1] If we are concerned with that, we need to take the organisms' point of view via "phytometer experiments."[2] Here we use the target organisms as measuring devices. For instance, we might clone up multiple copies of a single genotype of a grass and plant out these copies along a transect to see how the grass experiences that transect with respect to survival and reproduction.[3] This measures what we call the *ecological environment*.

To get to the concept of environment most relevant to evolutionary studies, we must go one step further. Natural selection is essentially comparative – without variation there can be no selection. The *selective environment* is measured in terms of the *relative* reproductive success of different genotypes (or phenotypes) across space or time (what is known in quantitative genetics as genotype X environment interaction). For example, if we were to clone up multiple copies of two different genotypes and plant multiple copies of both along our transect, we would see whether or not the *relative* performance of the two genotypes differed with different positions along our transect. Differing relative performance (relative realized fitness) indicates selective environmental *heterogeneity*. In contrast, areas of space (or time) in which relative fitness remains constant are selectively *homogeneous*.

How are these three types of environments related? Variation in the external environment is neither necessary nor sufficient for variation in the ecological environment. It is not sufficient because, as pointed out already, variation in some external factor may simply not affect the target organisms at all. It is not necessary because variation in the ecological environment may be due to density- or frequency-dependent factors that are *internal* to the population of interest and so not part of the external environment.[4] Variation in the ecological environment is not sufficient for variation in the selective environment because, in those cases, in which all genotypes respond to the ecological environment in the same way, *relative fitness* will remain unchanged. It is, however, necessary because, for the relative fitness of two or more genotypes to vary, the absolute fitness of at least one genotype must vary.[5] Thus, variation in the external environment is neither necessary nor sufficient for variation in the selective environment.

Given that it is the selective environment that is most directly related to evolution by natural selection, and given that the selective environment is defined in terms of the relative performances of different types across space or time, it follows that from an evolutionary point of view environments have no existence independent of the organisms that inhabit them. This is a purely conceptual point. It corresponds to the analytic truth that, in the populational conception of niche, empty niches do not exist. If that were the point of Lewontin's constructionism, then it would hardly be revolutionary. However, Antonovics and I argue that the best way to interpret Lewontin's constructionism is as a contingent claim about external environments and the organisms that occupy them. The contingent claim is, we argue, that organisms and their external environments (which are conceptually independent) are related in a reciprocal causal evolutionary dynamic so that each is the cause of the other as accounted for in Lewontin's coupled set of differential equations above. In other words, organisms and their external environments coevolve.

We illustrate this point with a computer model which, though fairly simple, has two features that are arguably biologically realistic and of considerable generality. The model is genetically quite simple – it is of a haploid population with two genotypes A and a. The fitness functions are negatively frequency dependent. The two interesting features of the model are that (1) we explicitly model the spatial structure of the frequency dependence, and (2) the frequency-dependence has a one-generation time lag. The first point will be discussed further in the next section; suffice it to say here that if in nature selection is frequency dependent, it is highly unlikely that the relevant frequencies will be those in the whole population. Rather, selection will be frequency dependent because of interaction among organisms (trying to mate, fighting off parasites or predators, feeding, etc.), and these interactions are highly unlikely to occur randomly in the population as a whole. One, plausibly quite general case of nonrandom interaction is where there is a spatial structure to interactions (you interact only with closest neighbors). Thus, in our model the relevant relative frequencies are those of the three closest neighbors to the target site but with another twist. We look not at the neighbors in the present generation, but in the last generation – point (2). This is realistic for many cases, but I will mention what is plausibly the most general one. In cases of host–parasite coevolution in which parasites have a shorter generation time than hosts, parasites tend to become well adapted to the locally common host genotype(s). (This will be discussed more fully later.) In such cases there is negative frequency-dependent selection with a one-generation time lag.

It is essential to the goals of our model that we represent the spatial substructuring of the selective environment because we want to show how that can lead to spatial structuring of the external environment, where the external environment at site i in generation t is defined as the number of A's in the triplet neighborhood centered on i in generation $t - 1$. (To think more concretely about this, conceptualize the external environment as the number and type of genotype-specific pathogens in the surrounding soil for a plant located at site i.) Runs of our model do indeed show how a spatially structured external environment can evolve from one with no spatial structure. Among other results, I will mention just two. First, runs of the model show cases in which the ecological environment – measured in terms of how the "average" genotype perceives the site – can be quite uniform but in which there is considerable variability of both the selective and external environment. We take this to illustrate an important point for community ecology – namely, that one cannot take ecological homogeneity as a sign of a lack of evolutionary action; it may hide considerable selective and external environmental heterogeneity that may ultimately feed forward to destroy the ecological homogeneity (as it did in the particular run in question). Thus, if one wants to understand the fate of a particular population within an ecological community, one cannot treat that population as a genetic black box.[6] Second, we observed multiple runs with both genotypes starting at fairly intermediate frequencies but in which one of the two fails to persist. This contradicts the expectation from standard (nonspatially structured) population genetics, where with negative frequency dependence an equilibrium with intermediate frequencies is predicted. (For obvious reasons, when type A gets rare, it becomes more fit and increases in frequency; when it becomes too common, it becomes less fit and decreases in frequency.) Comins, Hassell, and May (1992) have shown how spatial substructuring can greatly affect equilibria in ecological models. It would be interesting to extend the sort of model discussed here to examine the effect of spatial substructuring on equilibria under frequency-dependent selection.

It is one thing to develop a model of organism–environment coevolution; it is quite another to study it empirically. The final point Antonovics and I make with respect to constructionism is that the "phytometer method," the method of taking the organism's point of view in studying the environment, is an empirical method that promises success in studying organism–environment coevolution, where success is attainable. (For further discussion of this, see Brandon and Antonovics 1996, pp. 174–6.)

Concurrently, but independently, Laland et al. (1996) developed an evolutionary model illustrating the consequences of niche construction. Their approach is quite different, as are the results, which are complementary to ours. Whereas our model is genetically simple (one haploid locus with two alleles), theirs is more sophisticated (diploid, two diallelic loci). On the other hand, our model has the ecological complexity of spatial substructuring, whereas theirs takes the global frequencies as the relevant frequencies in determining frequency-dependent fitness.

In their model, the **E** (for environment) locus determines the niche-constructing activities of the population. In particular, a resource **R** is assumed to be a function of the frequency of alleles at the **E** locus. The **A** locus has a fitness contribution determined by the amount of **R** available to the population.

That is, different diploid genotypes at the **A** locus make different contributions to the two-locus fitnesses, depending of the value of **R**. For instance, the *AA* genotype may make the highest contribution to fitness, *Aa* intermediate, and *aa* lowest with high levels of **R**, whereas the reverse fitness relation holds for low levels of **R**. (This is only an example. Other relations, including overdominance, are possible in the model.) These fitness contributions are fixed (for a given value of **R**) and thus not frequency dependent. In contrast, the value of **R** is dependent on the frequency of alleles at the **E** locus. (The form of the functional dependence of **R** on **E** can be varied, reflecting different weightings of past generations.) Hence, that component of fitness of the two-locus genotypes is frequency dependent.

Among the novel results derived from this model are that niche construction can "(1) cause evolutionary inertia and momentum, (2) lead to the fixation of otherwise deleterious alleles, (3) support stable polymorphisms where none are expected, (4) eliminate what would otherwise be stable polymorphisms, and (5) influence disequilibrium." (Laland et al. 1996, p. 293). I will briefly comment on (1) and (2).

As mentioned earlier, the value of **R** is some function of the frequency of alleles at the **E** locus. In some cases (e.g., in which **E** alleles control the quality, quantity, or both of the web a spider spins which in turn determines **R**), the effect of **E** on **R** is immediate (in generational time). In other cases, it is the cumulative effect over many generations of niche construction that determines **R**. A good example of this would be the borrowing behavior of earthworms and the effect this has on soil composition. In such cases there is a time lag between change in niche-constructing activities and the resulting change in selection pressures. Interestingly, this time lag is similar to the sorts of time lags that have been demonstrated in theoretical studies of maternal inheritance (Kirkpatrick and Lande 1989) and gene-culture coevolution (Feldman and Cavalli-Sforza 1976). This suggests (what should be intuitively plausible anyway) that maternal provisioning and cultural transmission are types of niche construction.[7]

With regard to point (2), Laland et al. (1996) show that the niche construction controlled by the **E** locus can overcome external selection pressures on the **A** locus; that is, alleles at the **A** locus that otherwise would be selected against are selected for because of the effect of the **E** locus. This, they argue, is likely to be quite general. Plausible examples include the sorts of buffering of external environmental heterogeneity discussed in the introductory section of this article; for example, the ways endotherms and exotherms buffer themselves from the full range of temperature variations in the external environment. Thus, we can imagine an allele at the **A** locus that would be selected against when the organism experiences the full range of temperature variations in the external environment but that is not selected against (maybe even is selected for) when behaviors, physiological mechanisms, or both buffer the organism from that variation. (It is often remarked that air conditioning – a good example of niche construction – made the American South livable. But my ancestors lived here long before air conditioning.)

If two models and some interesting results are all there is to show 20 years after Lewontin first promoted constructionism, then it seems we must conclude one of two things: either (1) evolutionary biologists have been remiss in largely ignoring Lewontin's important idea, or (2) the idea is not so important after all. As someone who has worked on developing constructionism, it might not be surprising to find out that I do not favor (2). In the next section, I will suggest that (1) is not fully warranted either; that in fact a good deal of cutting-edge research in evolutionary biology does fit the constructionist mold. Making this point explicit is the purpose of the next section.

17.3 Frequency-Dependent Selection: Coevolution and the Emergence of Higher Levels of Selection

Selection is frequency dependent when the fitness of one type is a function of the relative frequency of that and other types. Broadly speaking, these types may or may not belong to the same population. Perhaps the best known evolutionary argument employing frequency-dependent selection is Fisher's (1930) argument that a 1:1 sex ratio is favored by selection.[8] Here the relevant frequencies are those of males and females within a single population. But another well-known case of frequency-dependent selection involves Batesian mimicry (e.g., the viceroy butterfly's mimicry of monarchs). Here, the fitness of the mimic depends on the ratio of mimic to model and hence involves the relative frequencies of types in two populations.[9] Thus, we can distinguish two types of frequency-dependent selection: the first case having a dynamic governed by factors internal to the evolving population, and the second being a coevolutionary dynamic between two populations.[10]

A second distinction, orthogonal to the first, is between negative frequency-dependent selection and positive frequency-dependent selection. Negative frequency-dependent selection occurs when a type is favored when rare and selected against when common. Both examples above, Fisher's sex ratio argument and Batesian mimicry, are examples of negative frequency-dependent selection. Two other examples, both plausibly quite general, are host–parasite coevolution, which we will discuss shortly, and differential resource utilization. The latter case may be either coevolutionary (e.g., when coexisting species divide up environmental resources so as to avoid competition) or may occur within a single population (e.g., when different genotypes have slightly different resource requirements, perhaps leading to a stable polymorphism). The Brandon–Antonovics model discussed in the previous section employed negative frequency-dependent fitnesses.

Positive frequency dependence occurs when a type becomes increasingly fit as it becomes more common. In one way this is theoretically less interesting than negative frequency dependence because, everything else being equal, the initially most common type will soon evolve to fixation. Empirically, we are less likely to observe such cases because, once at fixation, selection no longer operates, and they become unobservable.[11] But the very interesting model of

Laland, Odling-Smee, and Feldman discussed above involves a sort of positive frequency dependence between the alleles at the two loci.[12] They argue that the sort of niche-constructing activity represented by this model is likely to be common in nature (e.g., alleles at a locus that affect dam-building activity in beavers will have a positive feedback on alleles at another loci affecting swimming behavior). Similarly, the example mentioned earlier of pine trees that deposit pine needles, that lead to regular fast-burning fires, that suppress other competing species and lead to the germination of pine seedlings is clearly a case of positive frequency dependence. At the coevolutionary level, symbiotic relationships arguably all involve positive frequency-dependent selection.

Returning to our first distinction – that between within-population examples of frequency dependence and coevolutionary examples – I now want to argue that both are exemplary cases of Lewontin's constructionism.

Clearly, in cases of within-populational frequency-dependent selection, the evolving population is *constructing* the conditions that determine its own selective environment. This, of course, occurs within an external environment that may be exerting other selection pressures on the population. But it is the state of the population that determines the frequency-dependent selection. The state of the population also responds to that selection (if it is assumed the population is not at an equilibrium), leading to a new selective environment. In Lewontin's terms, **O**, the state of the population, is a function of **E**, which is in turn a function of **O**, accounting for the pair of coupled differential equations.

One might at first think that the coevolutionary cases fit the old lock and key metaphor because the population the target population is adapting to is part of the external environment. Although that is correct, the point is that this part of the external environment does not have a dynamic that is autonomous from that of the target population. To take the example of host–parasite coevolution, hosts evolve in response to their parasites, but the parasites are also evolving in response to evolving host defenses. Thus, again the set of coupled differential equations is applicable (where, from the point of view of the host, the parasites constitute **E**, but vice versa from the point of view of the parasites). And so frequency-dependent selection of both the within-populational and coevolutionary varieties are clear cases of Lewontin's constructionism.[13]

Thus far in this section, I have argued for two points: (1) frequency-dependent selection nicely fits Lewontin's model of constructionism, and (2) frequency-dependent selection is common in nature. Although one certainly cannot reasonably claim that studies of frequency-dependent selection have numerically dominated studies in evolutionary genetics during the last 20 years, I will suggest that such studies have produced some of the most exciting work during that time.

I have already referred to host–parasite coevolution several times. As early as 1983, Lewontin pointed out the strong parallel between such systems and his constructionist view (Lewontin 1983b). Since that time there have been important theoretical treatments of the problem (May and Anderson 1983, Seger and Hamilton 1988) as well as several empirical investigations demonstrating this coevolutionary dynamic (see Seger and Hamilton 1988, pp. 187–9 for a review of relevant empirical studies; also see Lively 1989 and Lively, Craddock, and Vrijenhoek 1990). I can think of at least three reasons to be excited by such work. First, studying the coevolutionary dynamic is likely to force us to revise conclusions drawn from much simpler evolutionary models. (Recall above that in the Brandon–Antonovics model we observed alleles with negative frequency-dependent fitness starting at moderately high relative frequencies yet disappearing in a run, contrary to standard population genetic expectations.) Second, parasitism is nearly ubiquitous, and thus host–parasite coevolution is likely to be of considerable importance in understanding evolution.[14] Third, some (e.g., Seger and Hamilton 1988) have argued that host–parasite coevolution is key to understanding one of the central problems of evolutionary biology: the evolution of sex. The basic idea is that the sexual production of a genetically variable array of offspring will be favored over the asexual production of genetically identical offspring when parasites are evolving rapidly to overcome the host's genotype-specific defenses. In the terms introduced earlier, sex will be favored (in the host) when the evolution of parasites produces temporal or spatial variation, or both, in the selective environment of sufficient magnitude to overcome the "cost of meiosis" (Antonovics et al. 1988).

A second broad area of revolutionary work in recent evolutionary biology concerns the evolution of higher levels of selection. There are two closely related questions here. One is, How do higher-level entities *emerge* in the process of evolution? For example, how did complex multicellular organisms evolve from simple single-celled organisms (see, e.g., Buss 1987 and Maynard Smith and Szathmáry 1995)? Second, once evolved, how do we describe the dynamics of these entities' further evolution. For example, given that interacting groups of organisms have emerged, how do we describe the dynamics of evolution by group selection? Or, more generally, how do we describe the dynamics of multilevel selection? Space limitations preclude anything even approaching an adequate overview of this area here. All I want to argue is that the evolution of higher levels of selection essentially involves the types of frequency-dependent selection we have been discussing.

In an early review of the models of group selection, Uyenoyama and Feldman define a group as "the smallest collection of individuals within a population defined such that genotypic fitness calculated within each group is not a (frequency-dependent) function of the composition of any other group" (1980, p. 395). In other words, group selection occurs only when a population is subdivided into local groups for which *individual* fitness is a function of the local, not global, frequencies.[15] (Recall how frequency dependence works in the Brandon–Antonovics model discussed earlier.) This is true whether we are talking about inter- or intrademic group selection (Wade 1978), trait group selection (which is a type of intrademic group selection) (Wilson 1980), or kin group selection (Michod 1982).

Although these mathematical models have clearly been articulated and established in the literature (surely one of the most revolutionary developments in theoretical evolutionary genetics of the last 20 years), empirical investigations into their possible applications in nature have been relatively few (but see McCauley 1994 and Stevens, Goodnight, and Kalisz 1995). The reasons for this are partly sociological (for years group selection was held in about the same esteem as Lamarckism) but also partly due to the difficulty of identifying the sorts of traits "whose fitness is likely to be influenced by local frequency and a population structure likely to generate considerable variation among local subunits in the frequency of that trait and yet allow sufficient dispersal for the trait to spread" (McCauley and Taylor 1997, p. 407; they identify sex expression in gynodioecious plants as such a trait). I can think of no other area in evolutionary biology in which exploring the theory–experiment interface is as challenging and as likely to yield important results.

If one considers the potential levels of selection (for a list of these see Brandon 1988, p. 64), clearly the level of the individual organism has been most thoroughly investigated theoretically and empirically. Theoretically, group selection is the next most thoroughly investigated, while empirically suborganismic levels are the next most completely documented (e.g., meiotic drive). Thus, the general claim I want to make about the evolutionary emergence of higher levels of selection essentially involving frequency-dependent selection is, at present, a speculative claim, but one that, I think, is plausible. What makes some entity emerge as an "interactor" (see Hull 1980) with respect to selection is that its parts are *integrated* in a way that would make selection among the parts frequency-dependent. That is, the evolutionary fate of one part is a frequency-dependent function of the other parts of the interactor.[16]

The final point of this section is (3) that studies of frequency-dependent selection, in particular studies of coevolution and of the emergence of new levels of selection, have been, and promise to continue to be, among the most exciting areas of contemporary evolutionary genetics. (I do not pretend to have established this point conclusively but only to have made it plausible.) If this is so, and if, as argued earlier in this section, frequency-dependent selection is an exemplar of Lewontin's constructionism, then we cannot say that contemporary evolutionary biology has ignored the phenomena of constructionism. Rather, the connection between the areas of study just discussed and constructionism has not generally been made explicit. That is what I have tried to do here.

17.4 Philosophical Implications: Constructionism and Adaptationism

As the coauthor of "The Spandrels of San Marco and the Panglossian Paradigm: A Critique of the Adaptationist Programme" (Gould and Lewontin 1979), Richard Lewontin might be viewed as an antiadaptationist and his constructionist view as serving to reinforce this attack. This, I think would be a serious oversimplification of his views. In this concluding section, I will argue that any form of adaptationism worth having is not only compatible with Lewontin's constructionism but is actually strengthened by it.

The San Marco article is, contrary to its popular perception, a deeply conservative piece. It reminded evolutionary biologists of what they already knew, namely that (1) natural selection is not the only factor in evolutionary change; (2) in consequence, not all traits are adaptations (i.e., solely the products of evolution by natural selection); and, therefore, (3) any methodology that automatically assumes for any trait that it is an adaptation is unjustifiably biased. (Thus, I think the so-called "adaptationist programme" is any position that denies any or all of these three points.) In calling the article conservative I do not intend to devalue it. I do believe that this position needed to be championed in 1979, just as we still occasionally need to be reminded of it.[17] But it is antiadaptationist only if Darwin is antiadaptationist.

In contrast, Lewontin's constructionism is, I believe, highly novel. What are its implications for adaptationism? The answer to that, of course, depends on what one means by "adaptationism." Recently there have been some efforts to state adaptationism as a testable empirical generalization (Orzack and Sober 1994, 1996; Brandon and Rausher 1996). Basically the idea is this: most traits are mainly the product of evolution by natural selection. (Making this idea precise enough to test is difficult and controversial, as evidenced by the articles referenced above.) I think this effort is worth pursuing, but my bets are that any empirical generalization deserving of the name adaptationism is going to be false.

Others, such as Mayr (1983), view adaptationism as a useful methodological heuristic. According to this view, asking adaptationist questions (what-for questions) is a useful first step in approaching biological phenomena. I have no quarrel with this so long as it does not ultimately bias one's results toward the conclusion that a trait is an adaptation (as, I am afraid, sometimes in practice it does).

But there is a third view that might reasonably be termed adaptationism.[18] That is the view that the process of adaptation and its results are evolutionary phenomena of particular interest. In this sense Darwin was clearly an adaptationist. One of the two major problems tackled in the *Origin* is explaining the exquisite fit of organisms to their environments (see, e.g., Darwin 1859, pp. 3 and 186–94).[19] One should not confuse questions of the relative importance or relative interest of phenomena with the sort of relative frequency question Gould and Lewontin discuss. A type of phenomena may be extremely important yet rare (one might think of large asteroid impacts with Earth in this regard; see Brandon 1990, p. 152 for further discussion). Thus, it begs no questions in, for example, the selectionist–neutralist debate to say that evolution by natural selection is a particularly important or interesting phenomenon. Why is it so interesting? Because we have no other scientifically validated theory to explain Darwin's "organs of extreme perfection and complication."

The phenomenon of adaptation becomes no less interesting once we realize, as Lewontin's constructionism shows, that it is considerably more complicated than we once thought. If we previously admired the torpedo shape of penguins, which allows them to move through water with speed and efficiency, we should find it no less admirable when we see that it is the active behavior of penguin ancestors, and not the inevitable force of the external environment, that makes the properties of water a relevant part of the penguin's environment. If organismic adaptation justly excited our interest, should that interest wane once we realize, as multilevel selection theory teaches, that each new level of selection is capable of producing new levels of adaptation? Coevolutionary adaptation, whether with positive frequency dependence (e.g., symbiosis) or negative frequency dependence (e.g., host–parasite coevolution) is both ubiquitous and productive of Darwinian organs of extreme complication and perfection. This is so even if that perfection is in many cases temporary and does not relate directly to any feature of the autonomous external environment.

Adaptationism, as I have just characterized it, is neither an ontological nor a methodological view. Rather it singles out one phenomenon, adaptation, as particularly worthy of study. In this sense of the term I am happy to label myself an adaptationist. Whether Lewontin would care to so label himself I leave for him to say. But I will say that I got my adaptationism from him.[20]

Notes

1. There are two importantly different sorts of cases here. One, some external factors (e.g., the relative position of the planets) do not affect the target organisms at all. Two, other factors (e.g., temperature) may significantly affect the organisms' physiology but not affect them differentially and so not be a factor relevant to (current) evolutionary dynamics.
2. See Antonovics, Ellstrand, and Brandon 1988; Brandon 1990; as well as Brandon and Antonovics 1995, 1996 for further discussion of such experiments. For a recent example see Stratton and Bennington 1998.
3. One would want to use multiple copies of a single genotype, not multiple genotypes, so as not to confound genetic differences with environmental differences. Alternatively, one could clone up multiple copies of the genotypes extant in the field, and plant them out in numbers proportional to their field frequencies. This would measure how the "average" genotype experiences the transect.
4. The interesting *empirical* issue here is how often such density- and frequency-dependent factors are *mediated* by some factor in the external environment versus cases in which there is no such external environmental mediation. For instance, many of the best-studied cases of negative frequency-dependent selection in plants (there are also excellent animal examples that I cite in the next section, but let us think about plants for the moment) involve genotype-specific pathogens evolving in the soil of the host plant. That makes that bit of soil a bad place for other copies of that genotype (and so is negative frequency-dependent selection). Here selective environmental heterogeneity, and perhaps ecological environmental heterogeneity, are mediated by external environmental heterogeneity. See Bever 1994 for a nice example. But in other cases (e.g., minority mating advantage in fruit flies), there is no reason to think that the selective and ecological heterogeneity is underlain by any external environmental heterogeneity. The point made in the text is conceptual – external environmental heterogeneity is not necessary for heterogeneity of the ecological environment.
5. Thus heterogeneity of the ecological environment, when it is measured by a single genotype, is necessary for heterogeneity of the selective environment. This is not

true when the ecological environment is measured by the "average" genotype. See Brandon and Antonovics 1996, p. 166, for further elaboration.

6. A point made by Colwell 1984 when he says, "a rose is not a rose is not a rose." In our terminology, genetically different roses will respond differently to the same external environment.
7. Indeed, the distinction between maternal or, more broadly, phenotypic inheritance and cultural transmission should in many cases be quite difficult to make. In an early article on cultural transmission, Brandon 1985, I gave a very abstract characterization of "phenotypic transmission," which would include both cultural transmission and maternal inheritance.
8. More precisely, what is favored is a 1:1 ratio in the investments put into the two sexes. The basics of Fisher's argument are simple. Because in sexually reproducing species the *total* reproductive success of the two sexes must be equal, the sex in the minority must have the higher *average* reproductive success and hence be favored by selection.
9. Batesian mimicry occurs when one palatable species, in this case viceroys, mimics an unpalatable one, monarchs. If predators, here birds, encounter mainly monarchs, they will learn to associate that phenotype with distastefulness and so will avoid eating that phenotype. If, on the other hand, they encounter mainly viceroys they will not avoid the phenotype. Thus, the adaptedness of the viceroy phenotype depends on its being rare relative to monarchs.
10. Nothing precludes more than two populations being involved, but we will not consider any such cases here.
11. John Maynard Smith in his excellent textbook on evolutionary genetics defines frequency-dependent selection in terms that only fit cases of negative frequency dependence (1989, p. 69).
12. The E allele at the **E** locus produces more **R** than e. With increasing **R**, A is fitter than a at the **A** locus. Without overdominance this, depending on initial frequencies, leads either to the fixation of the *EEAA* genotype or the *eeaa* genotype.
13. Having made the distinction between within-populational and coevolutionary types of frequency dependence, let me remind the reader of footnote 4 above in which I raised the empirical issue of how often apparent cases of within-population, frequency-dependent selection are mediated through the external environment versus instances in which they are not so mediated. Clearly there are at least some examples of both. But in those cases in which there is external environmental mediation (as virtually all imaginable plant examples would be), if that mediation is biotic (as in the buildup of genotype-specific pathogens in the soil surrounding a plant), then we are really dealing with a coevolutionary case. Note also that I have here only dealt with frequency-dependent, and not density-dependent, selection. Similar arguments could be made concerning density-dependent selection, but I will not give them here because I think frequency dependence has delivered a much richer yield in recent evolutionary biology.
14. If one accepts Ray's (1991) argument that parasitism is an expected outcome of evolution in general, not just on this planet, then the generality of this subject is not limited to Earth-bound evolution. I, for one, accept his conclusion, even though I do not find the argument completely compelling.
15. This is a necessary, but not sufficient, condition for group selection. For necessary and sufficient conditions for group selection, see Brandon (1990), chapter 3; also see Lloyd (1988), chapter 5.

16. Which is *not* to say, as Sober (1984, pp. 319, 340, 345) has claimed, that all the parts must have an identical fitness component determined by the higher level whole but only that the higher-level fitness component of the parts be a frequency-dependent function of the parts of the whole. Increased fitness of the whole may well be achieved at the expense of *some* of the parts (see Brandon 1990, pp. 125–7 for such an example and further discussion).
17. In 1979 the prime target was sociobiology. In saying that all evolutionary biologists already knew (1)–(3), I do not mean to say everyone in all the satellite disciplines around evolutionary biology (e.g., human sociobiology then and evolutionary psychology now) knew (1)–(3). For a clear recent example of someone who needs to hear (1)–(3), see Dennett (1995).
18. The threefold distinction I have made more or less mirrors the distinction among three kinds of adaptationism made by Godfrey-Smith (forthcoming).
19. The other major problem he took on was showing how, starting from a single origin of life, descent with modification could account for the extant distributions of life forms, which explains the "origin of species" part of the problematic of his book with that title.
20. I was a graduate student in the Department of Philosophy at Harvard from 1974–9. I took several courses from Dick, and he was extremely generous with his time in advising me on my dissertation. He was then, and remains now, a constant source of stimulation and inspiration.

REFERENCES

Antonovics, J., Ellstrand, N. C., and Brandon, R. N. (1988). Genetic variation and environmental variation: Expectations and experiments. In *Plant Evolutionary Biology*, (eds.) L. D. Gottlieb and S. K. Jain, pp. 275–303. London: Chapman and Hall.

Bever, J. D. (1994). Feedback between plants and their soil communities in an old field community. *Ecology* 75:1965–77.

Brandon, R. N. (1985). Phenotypic plasticity, cultural transmission and human sociobiology. In *Sociobiology and Epistemology*, (ed.) J. Fetzer, pp. 57–73. Dordrecht, The Netherlands: D. Reidel Publishing Co.

Brandon, R. N. (1988). The levels of selection: A hierarchy of interactors. In *The Role of Behavior in Evolution*, (ed.) H. C. Plotkin, pp. 51–71. Cambridge, MA: MIT Press. [Reprinted in Brandon 1996.]

Brandon, R. N. (1990). *Adaptation and Environment*. Princeton, NJ: Princeton University Press.

Brandon, R. N. (1996). *Concepts and Methods in Evolutionary Biology*. Cambridge, UK: Cambridge University Press.

Brandon, R. N., and Antonovics, J. (1995). The coevolution of organism and environment. In *Concepts, Theories, and Rationality in the Biological Sciences*, (eds.) G. Wolters and J. G. Lennox, pp. 211–32. Konstanz, Germany: UVK-Universitätsverlag Konstanz GmbH.

Brandon, R. N., and Antonovics, J. (1996). The coevolution of organism and environment [expanded version of Brandon and Antonovics 1995]. In R. Brandon, *Concepts and Methods in Evolutionary Biology*. Cambridge, UK: Cambridge University Press.

Brandon, R. N., and Rausher, M. D. (1996). Testing adaptationism: A comment on Orzack and Sober. *American Naturalist* 148:189–201.

Buss, L. W. (1987). *The Evolution of Individuality.* Princeton, NJ: Princeton University Press.

Colwell, R. K. (1984). What's new? Community ecology discovers biology. In *A New Ecology: Novel Approaches to Interactive Systems,* (eds.) P. W. Price, C. N. Solbodchikoff, and W. S. Gaud, pp. 387–96. New York: John Wiley & Sons.

Colwell, R. K. (1992). Niche: A bifurcation in the conceptual lineage of the term. In *Keywords in Evolutionary Biology,* (eds.) E. F. Keller and E. A. Lloyd, pp. 241–8. Cambridge, MA: Harvard University Press.

Comins, H. N., Hassell, M. P., and May, R. M. (1992). The spatial dynamics of host parasitoid systems. *Journal of Animal Ecology* 61:735–48.

Darwin, C. (1859). *On the Origin of Species.* London: John Murray.

Dennett, D. C. (1995). *Darwin's Dangerous Idea.* New York: Simon & Schuster.

Feldman, M. W., and Cavalli-Sforza, L. L. (1976). Cultural and biological evolutionary processes, selection for a trait under complex transmission. *Theoretical Population Biology* 9:238–59.

Fisher, R. A. (1930). *The Genetical Theory of Natural Selection.* Oxford, UK: Clarendon Press.

Godfrey-Smith, P. (forthcoming). Three kinds of adaptationism. In *Optimality and Adaptationism,* (eds.) S. Orzack and E. Sober, Cambridge, UK: Cambridge University Press.

Gould, S. J., and Lewontin, R. C. (1979). The spandrels of San Marco and the panglossian paradigm: A critique of the adaptationist programme. *Proceedings of the Royal Society London* B 205:581–98.

Griesemer, J. R. (1992). Niche: Historical perspectives. In *Keywords in Evolutionary Biology,* (eds.) E. F. Keller and E. A. Lloyd, pp. 231–40. Cambridge, MA: Harvard University Press.

Hull, D. (1980). Individuality and selection. *Annual Review of Ecology and Systematics* 11:311–32.

Kirkpatrick, M., and Lande, R. (1989). The evolution of maternal characters. *Evolution* 43:485–503.

Klingenberg, C. P., and Nijhout, H. F. (1998). Competition among growing organs and developmental control of morphological asymmetry. *Proceedings of the Royal Society London B* 265:136.1–136.5.

Laland, K. N., Odling-Smee, F. J., and Feldman, M. W. (1996). The evolutionary consequences of niche construction: A theoretical investigation using two-locus theory. *Journal of Evolutionary Biology* 9:293–316.

Lewontin, R. C. (1978). Adaptation. *Scientific American* 239:156–69.

Lewontin, R. C. (1982). Organism and environment. In *Learning, Development and Culture,* (ed.) H. C. Plotkin, pp. 151–72. New York: Wiley.

Lewontin, R. C. (1983a). Gene, organism and environment. In *Evolution from Molecules to Men,* (ed.) D. S. Bendall, pp. 273–85. Cambridge, UK: Cambridge University Press.

Lewontin, R. C. (1983b). The organism as the subject and object of evolution. *Scientia* 118:63–82.

Lively, C. M. (1989). Adaptation by a parasitic trematode to local populations of its snail host. *Evolution* 43:705–22.

Lively, C. M., Craddock, C., and Vrijenhoek, R. C. (1990). Red queen hypothesis supported by parasitism in sexual and clonal fish. *Nature* 344:864–6.

Lloyd, E. (1988). *The Structure and Confirmation of Evolutionary Theory.* New York: Greenwood Press.

May, R. M., and Anderson, R. M. (1983). Epidemiology and genetics in the coevolution of parasites and their hosts. *Proceedings of the Royal Society London B* 219:281–313.

Maynard Smith, J. (1989). *Evolutionary Genetics.* Oxford, UK: Oxford University Press.

Maynard Smith, J., and Szathmáry, E. (1995). *The Major Transitions in Evolution.* Oxford, UK: Freeman & Co.

Mayr, E. (1983). How to carry out the adaptationist program. *American Naturalist* 121:324–34.

McCauley, D. E. (1994). Interademic group selection imposed by a parasitoid–host interaction. *American Naturalist* 144:1–13.

McCauley, D. E., and Taylor, D. R. (1997). Local population structure and sex ratio: Evolution in gynodioecious plants. *American Naturalist* 150:406–19.

Michod, R. E. (1982). The theory of kin selection. *Annual Review of Ecology and Systematics* 13:23–55.

Nijhout, H. F., and Emlen, D. J. (1998). Competition among body parts in the development and evolution of insect morphology. *Proceedings of the National Academy of Sciences USA* 95:3685–9.

Odling-Smee, F. J. (1988). Niche constructing phenotypes. In *The Role of Behavior in Evolution,* (ed.) H. C. Plotkin, pp. 73–132. Cambridge, MA: The MIT Press.

Orzack, S. H., and Sober, E. (1994). Optimality models and the test of adaptationism. *American Naturalist* 143:361–80.

Orzack, S. H., and Sober, E. (1996). How to formulate and test adaptationism. *American Naturalist* 148:202–10.

Ray, T. S. (1991). An approach to the synthesis of life. In *Artificial Life II, SFI Studies in the Sciences of Complexity, vol. X,* (eds.) C. G. Langton, C. Taylor, J. D. Farmer, and S. Rasmussen. Reading MA: Addison–Wesley.

Seger, J., and Hamilton, W. D. (1988). Parasites and sex. In *The Evolution of Sex,* (eds.) R. E. Michod and B. R. Levin, pp. 176–93. Sunderland, MA: Sinauer.

Sober, E. (1984). *The Nature of Selection.* Cambridge, MA: The MIT Press.

Stevens, L., Goodnight, C. J., and Kalisz, S. (1995). Multilevel selection in natural populations of *Impatiens capensis. American Naturalist* 145:513–26.

Stratton, D. A., and Bennington, C. C. (1998). Fine-grained spatial and temporal variation in selection does not maintain genetic variation in *Erigeron annuus. Evolution* 52:678–91.

Uyenoyama, M., and Feldman, M. W. (1980). Theories of kin and group selection: A population genetics perspective. *Theoretical Population Biology* 19:87–123.

Wade, M. (1978). A critical review of the models of group selection. *Quarterly Review of Biology* 53:101–114.

Wilson, D. S. (1980). *The Natural Selection of Populations and Communities.* Menlo Park, CA: Benjamin/Cummings.

Wilson, D. S. (1983). The group selection controversy: History and current status. *Annual Review of Ecology and Systematics* 14:159–87.

CHAPTER EIGHTEEN

An "Irreducible" Component of Cognition

MASSIMO PIATTELLI-PALMARINI

18.1 Introduction (with a Caveat)

The line of argument that I intend to develop here, basically, could be summarized by the old saying that no "ought" can be derived from "is." A less bare-bones summary could be the following: Truths of reason and abstract principles that appear to us to possess necessary and universal validity are of such a nature that they cannot be explained by, or reduced to, the laws of the natural sciences. The cogency of scientific reasoning is, in fact, itself seen as an instantiation of the more general powers of correct reasoning. The latter cannot, therefore, be explained, or "justified," by the former. I will try to show that there is a constitutive normative component in the study of our cognitive systems and that this component surely cannot be explained in terms of gradualistic adaptation, even if it is supposed that other components of cognition can. This greatly reduces the interest of *any* gradualist adaptationist approach to our cognitive capacities. As a whole, this is an argument against the so-called evolutionary epistemology.

I offer these reflections as an homage to my old friend Dick Lewontin, from whom I have learned so much over the years, and I am happy to participate in a volume dedicated to him. At the outset, however, I think I owe him, and the reader, a caveat: When I have presented similar arguments to various audiences, I have been often accused of defending a Platonistic position. For these allegations, I have two counters. One consists in just reporting what George Boolos once told me about the nature of logical and mathematical truths: "It has never been properly specified what the alternative to Platonism is." The other is an open challenge: Tell me what your reasonable candidate for the origins of (the justification for) *necessary* truths of any sort is, and I will show you that *it* has to be at the very heart of the study of cognition. My central point here is that evolutionary epistemology cannot offer a reasonable candidate. If there is a valuable alternative that avoids the perils of Platonism and the *voie sans issue* of gradualistic, adaptive "hill climbing" (Dawkins 1996), I will be happy to accept it. What is important, and what I intend to show here, is that nothing short of it will do. With these qualifications, if my position may still be dubbed as Platonist, well, let it be. I can promise that acceptance of the theses I defend here does not presuppose, nor imply, the acceptance of Platonism. Any sound philosophy of logic and mathematics will do, provided it is not a variant of evolutionary epistemology.

18.2 On Cognitive Science in General

A peculiar trait of cognitive science is its constant, albeit often tacit, reliance on particular background systems of universal and necessarily valid principles and laws over and beyond what is generically required in every science by the so-called canons of scientific methodology. As we will see, cognitive science has a special need, *in the very statement of its working hypotheses and in the proper characterization of its primary data,* to combine empirical generalizations with truths of reason and with rationally cogent deductions from such combinations. For brevity, I will hereinafter refer to this component as the normative component.

Cognitive science is, on the one hand, part of biology because it investigates species-specific systems of knowledge decomposable into classes of specialized mental representations, computations, algorithms, and inferential heuristics. On the other hand, at least part of it interfaces with various elements of the theory of rationality. Mental processes that would probably appear very peculiar to a Martian scientist (even to one fully informed of our contingent goals, utilities, and beliefs) systematically interact in each individual subject with reflective thought, considered judgment, rational choice, and carefully planned action. The complete picture must, therefore, incorporate both contingent mechanisms and postulates of reason, both species-specific heuristics and universally valid schemes of inference. The experimental data themselves stand out against a background of logical principles (for instance *modus ponens*, syllogistic schemata, etc.) of axioms and theorems in probability theory, in the theory of games, in utility theory, of Bayesian confirmation rules, and so on. (In a moment, some simple standard examples will suffice to make this point clear.)

Interesting epistemological problems are thus generated by the heterogeneity of these two components. Although the specialized psychological mechanisms can safely be attributed to our biological constitution, and are undoubtely the product of *some* evolutionary process, the normative principles are, strictly speaking, uncaused, or, from a more general standpoint, alien to the very notions of being caused or uncaused. Any lawful, projectable, and empirically adequate mixed construal of these two components will inevitably reproduce this fundamental heterogeneity.

This leads us to a central consideration that originated with Descartes and has been rightly insisted upon by Noam Chomsky: Many of our cognitive states and cognitive processes are neither caused by the external stimuli nor independent of them. They are systematically related to these stimuli, and systematically, sometimes even nomologically, interrelated, but the category of strict causation is, in many instances at least, strictly inapplicable to them. We rather have to admit, in those instances, that what we say, think, and do is appropriate to the stimulus, or to the prompting situation, but not *caused* by it. This notion of appropriateness, and the explanations based upon it, may appear not very satisfactory to the "hard" scientist even if we concede the spurious requirements of a truly special science. Yet, this relation of appropriateness, vague as it is, is a (arguably *the*) fundamental epistemological notion of cognitive science. It has proved very problematic (to say the least) to reduce the conceptual role played by this notion to more standard patterns of causality. Not because neuroscience, physics, evolutionary biology, biochemistry, and so forth, are still inadequate or insufficiently sophisticated but because it is the *kind* of relation that we are entitled to doubt could in principle be reduced to strict causality. I intend to try to clarify this special state of affairs.

18.3 On Modules and Central Processors

Cognitive science definitely meets the criteria for being included among the "special sciences." It is, in fact, a natural science among others motivated by proprietary concerns and guided by certain explanatory criteria and justifiable idealizations; it pays special attention to certain classes of data and aims to construct nomological explanations in terms of the relevant regularities in the mental–neuronal makeup of our species. In this respect, it resembles more, say, geology or immunology than the truly basic sciences like cosmology, particle physics, or quantum theory. It is characteristic of experiments in cognitive science to elicit a certain class of reproducible and statistically significant responses from subjects exposed to a certain specific class of "prompting" situations. The response can sometimes be manifest and public, like an utterance, a gesture, a measurable time delay in pushing a button, and so on. But often what really counts is something more abstract, more internal, and more significant than a manifest "response" (for instance, a piece of reasoning, a specific interpretation, a proposition expressing a certain judgment, a bona fide report of certain subjective beliefs, a system of expressed preferences, or even the willingness to bet real money on a certain possible outcome). These can be primary data for cognitive science, in spite of (or, should I rather say, because of) their abstract characterizations. The candidate explanations of these data stand in need of systematic, well-characterized, internal correlates – for instance, specific mental representations and computations, suitably constrained and concatenated, lawfully and predictably accounting for what we observe. These internal correlates are, of course, expected to meet the standard scientific criteria of generality, projectability, systematic coverage, explanatory power, elegance and simplicity.

We witness that, indeed, in certain respects, the task facing the cognitive scientist appears quite similar in kind to the one facing any natural scientist such as the biologist. The task is to discover certain classes of structural properties and internal individual propensities possessed by certain organisms that can best explain a wide variety of relevant experimental facts. Very often, indeed, an adequate explanation in cognitive science amounts, so to speak, to just that, making, up to a point, its characterization as a proper subfield of biology legitimate. We are dealing with certain species-specific capacities at a suitably abstract level (see, for instance, the positions expressed by Chomsky and the ensuing discussions in Piattelli-Palmarini 1980). Much of the study of modular input-systems (in the sense of Fodor 1983) can be thus characterized, and successful examples of this strictly naturalistic approach are to be found in much of (though by no means all) the study of sensory perception, motor control, mental, imagery, and speech processing. This biological perspective also characterizes some subfields of linguistic inquiry (notably psycholinguistics, the study of language acquisition, natural language parsing, etc.).

In many other cases, however, the task facing the cognitive scientist is significantly different and more elaborate because even the most specialized, bullheaded, cognitively impenetrable (in the sense of Pylyshyn 1984), and encapsulated knowledge system must eventually interface with the central processor (again in the sense of Fodor 1983). In less guarded language, we may state that several of our modular cognitive subsystems must eventually interface with judgment, reasoning, and decision making. Many modules, in fact, systematically interact with reasoning and judgment, and their transient outputs may be significantly redirected upon the intervention of the central processor. In understanding and producing sentences, for instance, some operations of the linguistic modules interface with reflective thought, up to and including, in linguistically extreme examples, the use of paper and pencil. (We all think that we unproblematically and correctly understand the sentence *No eye injury is too trivial to ignore,* yet it requires paper and pencil to show that this sentence, if properly analyzed, means the exact opposite of what we think it means.) Barring perverse examples (like this one), our understanding of current linguistic expressions depends on the successful cooperation of automatic bullheaded syntactic routines with truth-sensitive stepwise logical derivations. Careful linguistic analysis typically displays a systematic combination of abstract logical criteria with strictly modular components (for a discussion, see Higginbotham 1987).

In order to witness a clear mobilization of the resources of the central processor, let us consider a very simple and well-reproducible phenomenon concerning syllogistic reasoning.

18.4 Figural Effects and Mental Models

No one, even at a very young age, shows any hesitation in inferring from *All Ruritanians are hunters* and *John is a Ruritanian* that *John is a hunter.*

Slightly less straightforward, but still basically unproblematic, is the task of inferring from *No sailor is a rider* and *All Ruritanians are sailors* that *No Ruritanian is a rider.*

Things change dramatically, however, even for highly intelligent and cultivated subjects, when we consider the following syllogistic scheme, presented as referring to effective states of affairs in a certain real country:

All the bankers are athletes.
None of the stationers is a banker.

What must we conclude??

Some respondents crank out, as if by reflex, *No stationer is an athlete,* or *No athlete is a stationer* but then most of the time spontaneously recant and look for something better. A significant number of respondents, *after long and careful reflection,* conclude that there is *no relevant inference at all* to be drawn (Johnson-Laird 1983, Johnson-Laird and Steedman 1978). This is also a wrong answer because there *is* a correct and mandatory inference, namely: *Some of the athletes are not stationers.*[1]

These wrong answers, taken together, practically exhaust the range of actual responses (Johnson-Laird and Wason 1970; Wason and Johnson-Laird 1972).

The special difficulty we experience with this syllogistic scheme derives (to put it in a nutshell) from the figural effect produced by the heterogeneity of the quantifiers (*all, none, some*), by a sort of cross-over in the order of terms in the premises (*bankers* appears first in the first premise, second in the second, and nowhere in the conclusion), and by the thrust of the deduction, which proceeds *from* two universal quantifiers *to* a narrow existential quantifier followed, moreover, by a negation.

Cognitive scientists, to account for the psychological anomaly of this particular syllogistic scheme, have proposed specific hypotheses concerning our spontaneous inferential procedures and the way we compulsively organize our baseline representational mental models (Johnson-Laird 1983). Our systematic failures with syllogisms of this kind seem to be caused by longer sequences of derivational steps and by the construction of several successive mental models, all of these representing *for us humans* a heavy computational burden. The various solutions that have been proposed differ somewhat, but we do not need to enter into these details.[2] Despite some technical disagreements about the characterization of the task and the solutions (Boolos 1984, Guyote and Sternberg 1981, Kyburg 1983) what really counts here, in my opinion, is that the anomaly of this particular syllogistic scheme is purely cognitive, not logical. An ideal logical mind would fail to see any anomaly in the scheme. The relevant categories, to such a mind, could not possibly be anything like "figural disposition," "quantificational heterogeneity," or "after long and careful reflection" but only those of valid and invalid inferential schemes. Considerations that come so naturally, and so relevantly, to the cognitive scientist (what is the computational burden, how quantificationally heterogeneous is the scheme, how long does it take to eliminate such and such fallacious intermediate inferential step, etc.) would be totally alien to the ideal logical mind (ILM). The ILM, not particularly interested in the peculiarities of our mental setup, would fail to see here what the *facts* are. These are facts *for* the cognitive scientist, not for the pure logician. Nonetheless, the cognitive scientist always has to keep the categories and the standards of the ILM in his or her background. We have to find a specific psychological explanation, in this case, because there is a (normatively) correct conclusion to the syllogism. The fact, the *explanandum*, is that we do not see *it*.

This simple example already allows us to emphasize a quite general and salient epistemological consideration: The primary data are not just constituted by the mere recordable statement that "No inference can be drawn," but by the more complex fact that such is the response, although there is a perfectly valid inference to be drawn, and that the vast majority of intelligent and cultivated human beings, even *after careful reflection*, fail to draw it.

Token response statements (such as I do not see any consequence, or There is no conclusion at all) are classified as instances of a type response that must make reference to a normative logical scheme. Only against this normative background of valid schemes, we develop specific expectations, and the truly relevant facts become *observables* for the cognitive scientist.

18.5 Enter Epistemic Boundedness

Fodor has characterized the central processor as nondomain specific, unencapsulated, and Quinean. This is a fact that, he stresses, is empirically interesting, because it might well have turned out to be otherwise (Fodor 1983). As a matter of psychological fact (not as a result of definitions), the fixation of our central beliefs is relevantly similar to the construction of theories in science, and this potentially applies to any domain whatsoever, from balancing one's checkbook to repairing marine engines. These belief-fixation processes typically rely, at least in principle, on *all* the available evidence. Other theorists (notably Chomsky) have maintained that the central processor may well be modular too; that is, it can potentially come to know only a subset of all the truths that could be known, given all the evidence that is actually (not just potentially) available to the subject. We are epistemically bounded in the special sense that we will forever only access a limited subclass of all the ideally conceivable hypotheses because, by our very nature, and unbeknownst to us, we are confined to mandatory procedures that use *less* than all the available evidence (Chomsky 1986). We may not realize that it is so, but it is. An epistemically bounded creature has great difficulty in realizing that it is epistemically bounded. A caricature that makes this point vivid is the one of the mouse that has to learn its way in a maze governed by the laws of prime numbers. It is not to be expected that the mouse, after experiencing endless frustrations, secretly snaps, "Damn! This bloody contraption must be based on the laws of prime numbers. And I do not possess the concept of a prime number!" The poor mouse would just go on indefinitely, failing to find its way, and trying out (presumably) all sorts of crazy, low-level hypotheses.

As the simple example of our anomalous syllogism already shows, the discovery of a variety of cognitive heuristics and strategies, which appear to constrain our natural reasoning and guide us *even* in certain domains of reflective judgment, lends support to the notion of epistemic boundedness. By simple induction, because all other creatures in the biological realm, and intelligent machines as well, are epistemically bounded in one way or another, it is reasonable to suppose that we are in the same situation, albeit in ways different from theirs. Moreover, cognitive neuropsychologists, examining both normal and pathological cases, have harvested ample and diversified evidence for some modularity in the domains of memory, attention, arithmetical skills, grammar, lexical access, and the semantics of natural languages (for a review, see Shallice 1991 and the ensuing commentaries).

Epistemic boundedness might still be compatible with the thesis that the normative background, with respect to which these psychological mechanisms are singled out for what they are, is nondomain specific and unencapsulated and that it might, therefore, constitute the truly "central" component of the central processor. Our constitutively limited individual psychological *access* to this component may be what accounts for our epistemic boundedness. As I will show, however, even this claim meets serious difficulties. We will eventually have to conclude that, if the central processor really embodies nondomain-oriented, unencapsulated, isotropic and Quinean systems of knowledge, they will have to be found elsewhere. More will be said about this in a moment.

Further evidence for some modularity in the central processor can readily be presented, even if in summary, by considering a few selected cases of decision making in situations of uncertainty. We will subsequently see how the epistemological considerations developed for these cases carry over to many other domains of cognitive science, allowing us to draw some interesting conclusions on the workings of the central processor.

18.6 Some Heuristics and Biases

It has been shown, for instance, that most people make decisions on the basis of a widespread and almost universal cognitive strategy called framing. This psychological mechanism dictates choices and preferences to us that flatly contradict the idealized norms of rationality (for instance, the transitivity of preferences, the maximization of expected utility, the axioms of probability, etc.).

Let us examine a simplified version of certain typical problems and answers reported in a vast literature (for comprehensive coverage, see Arkes and Hammond 1986; Elster 1986; Gärdenfors and Sahlin 1988; Kahneman, Slovic, and Tversky 1982; Osherson and Smith 1996; for nontechnical introductions, see Piattelli-Palmarini 1994; Sutherland 1992).

A certain country in Southeast Asia is faced with a deadly epidemic that threatens the lives of up to 600 people.

Two alternative programs to combat the disease have been proposed. Let us call them A and B, respectively.

- If we adopt program A, 200 people will be saved.
- If we adopt program B, there is a 1/3 probability of saving 600 people and a 2/3 probability of saving no one.

Which program would you recommend?

The median of the choices expressed by a variety of experimental subjects is 72% for A and 28% for B.

The basic phenomenon of the framing of choices becomes fully evident when the problem is reformulated in the following way:

The situation is the same as above, but the choice problem is now rephrased as follows:

There are two programs, C and D.

- If we adopt program C, 400 people are going to die for sure.
- If we adopt program D, there is a 1/3 probability that no one will die and a 2/3 probability that 600 people will die.

Which program would you recommend?

Here the median of choices shows a reversed pattern: 22% for C and 78% for D.

According to the theory of expected value, a rational subject ought to remain perfectly indifferent between A, B, C, and D. (In terms of expected value, $100 for sure is the same as $200 with a probability of 0.5). A more "human" calibration, based on the theory of expected utility (not expected value), may well condone a preference for certainty over uncertainty. Yet, there still is something paradoxical in choosing *both* A over B *and* D over C. In fact, we have the *same* factual situation described differently, and truthfully, in both cases. Truthful and complete, yet differently phrased, descriptions of a given choice should *not* induce different preferences. It is not rational, no matter how you compute your expected utility, to make different decisions solely as an effect of a different presentation of the same facts. Note, in passing, that the facts are characterized as "the same" on normative grounds because psychologically they are not at all "the same." In fact, this paradox can be explained in terms of the formation of mentally represented different baseline situations (persons saved for sure as the baseline in the first frame versus persons dying for sure in the second frame) and by the well-known psychological *asymmetry* between gains (further lives saved) and losses (further lives lost). It still remains the case that such psychological mechanisms lead us into paradoxes. These can be psychologically explained but not dissolved (Piattelli-Palmarini 1994, Tversky and Kahneman 1986).

Another rich repertory of experiments deals with lotteries and choices among pairs of lotteries (called prospects). Many such experiments are presented to the subject as make-believe lotteries, but it has been confirmed that many respondents are consistently willing to maintain their choice even if offered the opportunity to bet for real, and some actually volunteer to bet real money.

You are asked to choose between

a) A sure gain of $75
b) 75% chances to win $100 and 25% chances to win nothing at all

Now, you are asked to choose between

a′) A sure loss of $75
b′) A 75% chance to lose $100 and a 25% chance to lose nothing at all

The expected values for a) and b), and for a′) and b′) are identical, but the majority of respondents choose a) over b), *and* b′) over a′). Even when we substitute expected values with expected utilities (the latter being a concave monetary function of the former) and allow for the intrinsically greater value psychologically of certainty over uncertainty, the *combination* of these two preferences represents a paradox from a normative point of view.

A more refined and even more revealing experiment, a fully numerical one, is the following:

There are two lotteries:

a) 50% chance to win $20 and 50% chance to lose $5
b) 50% chance to win X and 50% chance to lose $15

What is the *minimum* value (in dollars) for X that you would put in a sealed-bid auction for the opportunity to exchange b) for a)?

NB: If you ask for too much, the opportunity to exchange b) for a) will be adjudicated to someone else.

The median response is $X = \$60$. That is, a minimum increment of $40 in the possible gain is required to compensate for an increment of $10 in the possible loss.

But the situation changes dramatically in the following (normatively) strictly analogous choice:

a′) 50% chance to win $15 and 50% chance to win $40
b′) 50% chance to win $5 and 50% chance to win Y

What is the *minimum* value (in dollars) for Y that you would put in a sealed-bid auction for the opportunity to exchange b′) for a′)?

NB: Again, you must not be too greedy.

In this case (all gains, no losses) the median response is $X = \$52$. That is, a difference of $12 between the two best outcomes *suffices* to compensate for a difference of $10 between the two inferior outcomes. The departure from normative criteria of rational choice is striking (Tversky and Kahneman 1992).

The task facing the cognitive scientist here is one of predictably, generally, and counterfactually accounting for the choices made by the respondents. In the current theories, a marked asymmetry is postulated between the subjective expected utility (SEU) for gains versus the SEU of losses. The latter is a steeper function than the former. This experiment shows it very clearly and numerically. Very complex nonlinear functionals, with separate functionals for gains and losses, have been proposed to account for these subjective preferences.

Simplifying drastically, for our present purposes, several cognitively and mathematically highly sophisticated models (Shafir 1988; Shafir, Osherson, and Smith 1989; Kahneman and Tversky 1979; Tversky and Kahneman 1986, 1992), we can intuitively see what happens. Framing effects, that is, the respondent's susceptibility to inconsequential variations in the formulation of available choices, are incompatible with the normative concept of a rational economic subject, yet the rules of framing play an essential role in the explanation of actual choices. People *do* show this sensitivity to superficially different formulations of the available choices – even highly intelligent people when asked to risk real money and even when the stakes are quite high. It is worth stressing that the criteria dictating these choices are (in Tversky's felicitous wording) "neither 'rational,' nor merely capricious." There are certain definite rules, even if they are not easy to formalize, and even if they allow for considerable vagueness.

As Tversky and Kahneman (1992) explain,

> The framing process is governed by two rules of mental economy: acceptance and segregation. The acceptance rule states that, given a reasonable formulation of a choice problem, decision makers are prone to accept the problem as presented to them, and do not spontaneously generate alternative representations. Acceptance explains why different formulations of the same problem often yield different preferences. . . . Numerous demonstrations of framing in the literature attest to the strength of the tendency to accept problem formulations as they are given. . . .
>
> A second rule of mental economy is the segregation of the decision problem at hand from its broader context. In accord with this rule, people frame decision problems by focusing on those acts, outcomes and contingencies that appear most directly relevant to the choice under consideration. . . .
>
> The prime example of segregation of outcomes is the nearly universal practice of thinking about choice problems in terms of gains and losses, rather than in terms of wealth or final asset positions. In this representation, the part of the outcome that depends on the choice is isolated from pre-existing wealth. Decision-dependent outcomes are also segregated from accompanying gifts or penalties.

This is a characteristic example of an interesting and well-corroborated psychological (i.e., "descriptive," not normative) explanation of our choices. Here is a typical confirming instance:

Being first offered a bonus of $300, you then face the following choice:

a) Win $100 for sure
b) 50% chance to win $200 and 50% chance to win nothing

Most subjects choose a) (this is typical of risk aversion for gains).

Being first offered a bonus of $500, you then face the following choice:

a′) A sure loss of $100
b′) 50% chance to lose $200 and 50% chance to lose nothing at all

Most subjects now choose b′) (this is typical of risk seeking for losses).

The two situations, of course, are perfectly equivalent in terms of final assets (the expected value of the overall final net gain is $400 in both cases and for both choices). Yet, the marked and well-reproducible preferences of the majority of respondents exemplify the conjunction of two widespread and well-documented cognitive strategies: segregation and risk aversiveness in situations of *gain*; segregation and risk seeking in situations of *loss*.

Note, again, that the psychological explanations are *defined* with respect to the normative baseline. The very notions of being risk averse or risk seeking obviously presuppose that we know what *would* count as being risk neutral. Probability calculus and the calculations of expected values offer the required baseline. The next step, in order to construct a psychological explanation, is to infer what the subject would, in fact, choose, if it is *assumed* that he or she is risk averse (or risk seeking, or whatever other propensity your psychological theory hypothesizes). A certain rationality in the subject's processes of decision making must be preserved in spite of localized points of deviance. Otherwise, no psychological explanation could be found. Specific heuristics and biases can be inferred only thanks to the otherwise normatively *correct* links, in the subject's head, between his or her propensities and utilities, no matter how deviant, and the decisional consequences in the context of other beliefs and utilities.

Identical considerations apply to other heuristics and biases, which I have reviewed elsewhere (Piattelli-Palmarini 1994) outside the domain of lotteries and monetary choices. The literature describes and analyzes a whole array of well-documented and thoroughly reproducible mental heuristics called "availability," "anchoring," "neglect of base rates," "overconfidence," "typicality," and so on (Kahneman, Slovic, and Tversky 1982). These constitute interesting explanations of how people, as a matter of fact, think because they rest on a composite background of spontaneous human cognitive propensities *and* of normative theories that tell us how people *ought* to think. In other words, the very hope of being able to construct a descriptive (i.e., psychological) theory of decision making presupposes that we understand what a *fully* rational agent *would* choose, *if* he or she entertained these strange beliefs and followed these strange inferential rules.

18.7 More on Descriptive Versus Normative Explanations

The psychological hypotheses that explain the kinds of heuristics, biases, and cognitive strategies we have just encountered are rightly, even if somewhat too modestly, called "descriptive" – a term that marks the contrast with "normative." In fact, an ideal rational subject, in order to avoid inconsistency, instability, and the possibility of being "Dutch-bookable" (i.e., of *willingly* accepting bets that he or she is destined to lose no matter what) should show none of these heuristics and strategies, none of these propensities. The normative theories impose, for each of these situations, radically different strategies of choice, severally derived from set theory, formal logic, expected utility theory, the axioms of preference,

Kolmogoroff's axioms, Bayesian confirmation rules, and so forth. Yet, the rules that actually govern people's choices and judgments are often different from, and sometimes flatly incompatible with, the rules that we can severally derive from these normative theories.

In the descriptive theories, in fact, typically an array of cognitive strategies, representational formats, and computational procedures are postulated that constitute for our species universal (or at least quite general) properties. These are contingent facts about a certain class of organisms. Our Martian scientist would undoubtedly find these cognitive rules very awkward and would probably look down to us (to use an expression from Hilary Putnam) "in great sadness." We happen to instantiate these heuristics and biases in virtue of the fact that we are the sort of organisms that we are. When we are dealing with bona fide modular systems, that is all there is to say. However, when we deal with the central processor, when we eventually connect the output of these heuristics and biases with judgment and reasoning, these psychological mechanisms, though necessary, are no longer sufficient to explain what is going on. As we already remarked, the normative theories offer the necessary background of concepts, theorems, schemes of inference, and optimization principles that allow the cognitive scientist to understand what *kind* of cognitive processes are responsible for what *kinds* of phenomena.

The main epistemological difference between the study of our cognitive modules (i.e., between the branches of cognitive science that properly speaking belong to biology) and these other branches of cognitive science is that in the former we do not have to invoke any idealized background norm of rationality. Our objective is to discover what certain natural systems are and how they work, whatever that turns out to be. In the domains of cognitive science that deal, directly or indirectly, with the workings of the central processor we have to presuppose the general reasonableness of the subject. An important fraction of the ceteris paribus conditions, which we take to be obviously satisfied when we analyze the actual responses of the experimental subjects, constitute instances of the charity principle.

In the light of what we have seen earlier, we should not (at the very least not necessarily) attribute to any real human subject a full-blown rationality (i.e., conscious access to all the normative principles and to their deductive consequences), but we have to attribute a minimum of understanding of the means-to-ends relations and a minimum of internal consistency between actually entertained beliefs. When expressing a certain choice, the respondent genuinely believes that he or she is optimizing *something* and that he or she will, in *some* sense, be better off as a result of the choice. There must always be, even in the descriptive theory, an underlying assumption of minimal consequentiality between subjective utilities and the means chosen to obtain them. For instance, some working knowledge of *some* version of *modus ponens*, and of some probabilistic metric, must be attributed to the subject. We have to assume this much; otherwise, we would not be able to understand *what* the respondent believes to be the case, what he or she *thinks* that he or she is doing, which heuristic we reckon he or she is actually adopting, and what *would* be a better heuristic.

18.8 Central (and the Not-So-Central) Components of the Central Processor

This assumed generic background of reasonableness is mandatory but also highly problematic because I take it to be uncontroversial that we do not even begin to possess a complete and satisfactory theory of human reasonableness (this term I have loosely borrowed from Putnam 1981). We do possess rather satisfactory axiomatic theories for some idealized domains of rationality, but we are entitled to be quite skeptical even about the possibility of ever attaining a finite and completely axiomatized *global* theory of normative rationality. And even this would only be a specialized province of what I have here called reasonableness, which would also have to include pragmatic rules of prudence, a host of contingent ceteris paribus conditions, updated versions of particular beliefs, and transient utilities, not to mention knowledge about facts of the world and other people's beliefs and desires, feelings, moods, attitudes, and so forth. Nonetheless, in cognitive science we are condemned to be charitable in this very loose sense. This appears to be, to all intents and purposes, the really central component of the central processor. Here we have to pretend (at least) that we pay attention to all the available evidence.

The items of propositional knowledge and the particular skills and abilities that form this backround of reasonableness are totally unencapsulated. In fact, it is impossible to circumscribe and protect from revision, or from updating, even a fragment of this background – even narrowing it down to those rules, skills, and beliefs that figure directly and explicitly in the causal account of the relevant piece of reasoning, or the action that we are examining, or both. As Fodor has rightly stressed (Fodor 1990), succeeding in this segregation would amount to having found a solution to the frame problem, which is something we have independent reasons to think is unsolvable (Pylyshyn 1987).

The general background of reasonableness is, therefore, truly not-domain-specific and truly unencapsulated. We are entitled to be moderately pessimistic about ever being able to produce a satisfactory descriptive theory of *its* structure. Not surprisingly, our high degree of scientific pessimism matches the one we entertain about the perspectives of a formal theory of general pragmatics.

The normative background, however, seems to possess a different status. The axioms, rules, and inferential schemes that figure in it are, by necessity, invariant under the actions and thoughts in the etiology of which they factually or counterfactually appear. Although they may be open to piecemeal theoretical revision in the long run, these norms and schemes are certainly encapsulated with respect to their contingent psychological instantiations. Any change in those components is brought about by specialists through painstaking, rigorous, intersubjective processes and counts as a theoretical discovery or sometimes as a triumph of critical rational thinking. Reconsiderations of, say, probability axioms, or utility theory, do not come from experiments in cognitive science just as revisions in physical theories are not induced by psychological data in our naive conceptions of the physical world. They enter the "bridge laws" (Fodor 1975) of cognitive science as internally consistent fragments of a more basic science. At the level of the cognitive theory of the central processor, the normative background is, at any one time, to be taken as fixed, that is, totally encapsulated (in the sense of Pylyshyn 1984). Moreover, each normative subdomain that appears in a given, specific, full-blown psychological explanation is *also* domain oriented (in the sense of Fodor 1983). Kolmogoroff's axioms have nothing to say about the scope of quantifiers in natural languages, and syllogistic schemes have little to say about choosing between specific lotteries.

The structure of the central processor, therefore, appears rather composite. We find in it contingent descriptive rules that are functionally characterized, relatively shallow, largely insensitive to beliefs and utilities, domain specific, and (to a large extent) encapsulated. These heuristics and biases (marginally self-corrigible) arguably constitute a kosher modular component *of the central processor.*

Next, we find normative rules that are propositionally (not functionally) characterized, domain specific, and encapsulated. They also arguably constitute a modular component, though perhaps not so kosher (at least according to Fodor's characterization). In fact, they are (unlike bona fide modules) cognitively accessible at every interlevel, definitely not shallow, and sensitive to all the evidence there is *in their proper domain of application.* We will have more to say about them.

Finally, we also have a truly isotropic (i.e., not-domain-specific), large, and poorly articulated component of totally unencapsulated pragmatic norms. Here, indeed, revision can strike anywhere, at any time, and is also conditional upon the contingent results of actions immediately governed by these norms. Updating can well be induced, moment by moment, by the very actions that they govern.

Seen in the light of this complex and diversified background, it makes sense that the *explanandum* is, at bottom, not just a certain empirically ascertained response but this response qualified for what it is against such background. Our subtheories of the various modules, combined with our theory of the central processor, allow us to derive certain specific expectations based on our presumption of general reasonableness and on some specific, counterfactual-supporting, normative component. What makes the token utterance (or the token reaction time, etc.) a fact *for* cognitive science is this elaborate system of specific expectations. It will repay to examine more closely the causal status of these semimodular normative rules in our explanations. Before we do that, it will be useful to emphasize that an analogous normative component is also to be found in many other domains of cognitive science – for instance, even in the study of sensory perception.

18.9 Principle of Realism

The kinds of phenomena that are evidenced in many experiments on visual and auditory perception and on motor planning and motor control are often perceived for what they are because we judge that something goes wrong with respect to the actual situation, that is, the objective physical characterization of the stimulus. In "apparent motions" we cannot help seeing objects move even when we know full well that nothing is actually moving (Kolers 1972). In the retinex system of color perception (better known as the Land effect), we see colors even when a perfectly black-and-white Mondrian pattern is illuminated by a rigorously monochromatic light (Ottoson and Zeki 1985). In the "phonemic restoration effect" we hear linguistic sounds that are actually not in the stimulus. Cases of such perceptual anomalies are abundant and are to be found in a variety of domains. For instance, there is a subjective feeling of a "backward walk" by subjects actually walking forward inside a cylinder, with black stripes on its walls, and rotating at certain angular velocities different from the angular velocity of the base on which they walk. In these devices, the experimentalist can even generate in the subject the feeling of being "heavy on one's legs."

In general, what we observe in these experiments is either that the subject perceives something that is "not really there," or that he or she misinterprets some information, or mentally fills in some objectively missing element (for a critique of the very notion of "filling-in," see, however, Dennett 1991). The judgment that something goes wrong always depends *on a presupposition of strong realism.* The full *explanandum* is, once again, constituted by what the respondent subjectively perceives to be the case *and* by its *not* being what objectively *is* the case. The cognitive processes invoked to account for the actual response must typically redress another tipped balance: We must be able to explain why the principle of realism is *not* obeyed and then account for these misperceptions.

Epistemically speaking, the status of these various overarching (and rather obvious) principles of realism in the study of perception is equally normative, and the explantory role they play is strictly analogous to the one played by the normative theories in other domains of cognitive science. I do not mean to say that the philosophical status of the laws of physics is the same as that of logical or mathematical truths. I am only saying that the normative background role they play in the explanations offered by different domains of cognitive science is analogous. The pieces of basic science that appear in the characterization of the observable facts and in the bridge laws of these branches of cognitive science are usually pieces of physics just as the pieces of basic science that appear in other domains of cognitive science are pieces of logic, mathematics, or decision theory. What counts, in the present context, is that the assumed baseline norm of adequacy vis-à-vis objective reality is some standard physicalistic postulate that is not, in turn, explicable on psychological or neurological grounds. It is simply taken for granted, and is, once again, cognitively uncaused. What can be genetically, neurophysiologically, psychologically, or computationally caused is some specific *deviation from* the principle of realism, and some ensuing compensation, but surely not the principle itself. The principle of realism, that is, the default assumption that there are things, and properties thereof, to be perceived, is caused (if at all) by "the way the world is," that is, by whatever we severally deem to be the cause of the laws of physics (if anything is). It should be evident that the causes of the *good* correspondence between percepts and objective facts-of-the-matter cannot be attributed to any adaptive evolutionary mechanism. Let us spend another moment on this point.

It hardly requires elaborate argumentation to see that the principles governing the physical world cannot be explained as a consequence of biological adaptation. If anything, adaptationism may attempt to explain how higher organisms end up developing an adequate internal representation of some of these principles. Pace the evolutionary epistemologist, adaptive mechanisms can (sometimes) explain the adaptive value (when indeed there is one) of *specific* cognitive processes but not the adaptive value of a *general* disposition to perceive correctly what is, in fact, the case. Any such explanation, because of its purported generality, will tacitly presuppose some even more general principle of adequacy, of which the adequacy of the mechanisms of evolutionary adaptiveness must itself be an instantiation. No story can count as a success story unless the criteria for success are characterized independently. But then, either we have to admit that this higher-level principle of adequacy is not in turn justifiable on naturalistic grounds, and from the evolutionary–epistemological point of view we are back to square one, or we generate an intrinsic circularity in the explanation. In other words, evolutionary adaptation may (sometimes) account for this or that specific deviation from the baseline norm of realistic adequacy, but the epistemological need for a standard, undeviated principle of realism cannot in turn be explained on evolutionary grounds. It is a truth of reason that no explanation based on a law of nature can rationally justify the necessary validity of an abstract scheme of inference of which that particular law is itself an instance. Nothing in nature can be the cause of this necessary validity. Based as it is on a truth of reason, the cause "overflows" the explanations and the justifications provided by any natural science.

The implicit baseline assumption, to be taken as a default value, is that, unless we positively ascertain it to be otherwise, there *is* a match between reality and the outcomes of perceptual and cognitive processes. The very idea of evolutionary success for perception and cognition is based on the antecedent and constitutive idea of this match. The overarching notion that adequacy of fit counts as success, and greater adequacy as greater success, must be presupposed. It cannot in turn be explained by means of adaptive mechanisms. Local and circumscribed failures of this baseline assumption are always interesting and lead the cognitive scientist to postulate intriguing and systematic mechanisms that explain the intriguing and systematic mismatch. These (and only these) anomalous mechanisms can eventually, in turn, be justified on evolutionary grounds (though not obligatorily in terms of adaptation). But in the ordinary or default case there is nothing to justify on evolutionary grounds. *Adequacy is what happens when nothing special happens.* It will usually repay to understand *how* a certain perceptual system manages to attain adequacy, in actual and in counterfactual situations, but adequacy per se needs no explanation. The presumption of adequacy is what we have to start with.

This quite natural and often tacit principle of realism, for all its vagueness, does play a role in the explanations of cognitive science. The principle is a normative a priori one possessing the epistemic status of a postulate of reason, and it is, accordingly, uncaused.

This digression on physicalistic postulates and on their explanatory role in cognitive science may have been useful in emphasizing the presence of a normative causal, but uncaused, component *even* in the background of many strictly modular input systems. Let us now return to the central processor.

18.10 Downward Causation

Also in the realm of the central processor there are quite natural "default assumptions." Unless we reliably ascertain otherwise, we take for granted that people will in general accept what is true, assent to a sound argument, and abide by a transparently valid scheme of inference. We would be uncomfortably placed when asked to justify such assumptions or explain such propensities from within the realm of natural science. Simply, there would be no firmer *scientific* ground on which to build such explanations and justifications. We enter the domain of metaphysics and leave that of the natural sciences. As we saw earlier, scientific explanations and justifications can be found for empirical observations that refute the default assumptions when in particular circumstances we witness particular propensities to accept falsehoods, to assent to fallacious arguments, or to reject valid schemata. These deviations can be explained, as we saw, through certain psychological mechanisms. But reasonableness, being the default case, cannot. Once again, in analogy with the case of perceptual mechanisms, when things go well, we can only examine, within the realm of natural science, *how* this happens, but not justify *its happening*.[3]

Specific systems of mental representations and specific algorithms can be found responsible for the ways in which we attain this fit but not for the fact that *there* is a fit. If some mapping of contingent psychological states onto the abstract, universal, normative schemata that constitute the default baseline has been made correctly, then it can explain why we are successful in our reasoning. It falls outside the realm of cognitive science, and of natural science in general, to explain *why* being rational leads to success. Properties in virtue of which a valid rational argument preserves truth from the premises to the conclusion are not psychological nor in any way "naturalizable." The principle that genuinely sound reasoning, by preserving truth from premises to conclusions, is apt to lead to successful decisions, is what the entire psychology of reasoning is ultimately based upon. The psychology of reasoning (as opposed to normative theories of reason) takes off precisely when we witness systematic and interesting exceptions to it. From a metaphysical point of view, because the nature of such baseline principles consists in abstract properties of abstract entities, no natural instantiation of them can constitute an explanation of, or a justification for, what they are. The reason why their contingent psychological instantiations have the causal power they have on other mental states, or on actions modifying the outside world, or both is to be found in the ways in which these instantiations (somehow) materialize, reflect, or participate in those abstract properties.[4]

The direction of causation, and therefore of explanation and justification, is *from* the universal and abstract *to* the psychological and contingent, not the other way around. It is the validity of, say, *modus ponens* that explains why its tokening here and now in John's chain of thoughts causes John to arrive at true conclusions from true premises. No property of John's mind–brain can explain the validity of *modus ponens.* This causal "downward" relation is, therefore, strongly asymmetric, and it reproduces one level farther "up" the "downward causation" from mental-computational laws to their contingent physical instantiations, as countenanced in the standard versions of mental functionalism. The notion of multiple instantiability, on which standard functionalism is often based (Dennett 1978, Fodor 1975, Putnam 1975), applies here too (in ways that have been pointed out by Putnam in his more recent writings). To see this more clearly, let us consider the following, very elementary case.

Take the following collection of devices: an old-style gears-and-pegs arithmetical machine (a Pascaline), an ordinary pocket calculator, a Macintosh personal computer, a Cray mainframe computer, and a parallel distributed processor (say, the CM–5 perfected by the Thinking Machines Corporation). Let us imagine that each one of these five devices sits in a different room and that in each room a different human operator attempts to perform the following "crazy" calculation: 100 divided by 0. In each room something untoward will happen. One screen will show E, or INF, another will fill up with strange characters, another machine will start beeping at the operator, another machine may enter into an infinite loop, another may get clogged or give no output at all, and so on. Because these five devices are differently built (i.e., they have different hardwares and, presumably, different emergency escape strategies), no *one* physicalistic explanation will uniformly apply to them all. But the brute juxtaposition of five distinct physicalistic chains of events would miss something important, that is, an important computational generalization of which all these physical anomalies are contingent instantiations: The computational anomaly represented by the instruction to divide 100 (or any other finite number) by 0. We have five physical stories, but just one computational anomaly. It is not a fact for any of the five physicalist stories that *all* five devices produce an anomalous output. This higher-level collective anomaly is invisible to the physicalist stories. So far, I have just restated the standard doctrine of functionalism.

I will now follow the spirit of Putnam's more recent reconsideration of functionalism (Putnam 1987, 1988) and urge us to go a step beyond. Let us in fact consider the individual computational instantiations of the story. Suppose that one machine works in octal, with step-by-step updating of the memory unit, whereas another works in floating-point operations on a virtual disk, and another works through parallel units, and so on. It seems that we can repeat the move we already made with respect to the physicalistic stories: Go one level higher in abstraction and appeal to the general-theory-of-computation part of mathematics. Or we may as well go into arithmetic, where it is a *number-theoretical fact* that a division by zero is illegitimate (there is a divergent asymptote for any succession of divisions of a finite number by ever smaller numbers). Again, in a less guarded language, let us say that every finite number divided by zero "makes" infinity.

This abstract "fact" obviously remains invariant across all the contingent computational architectures, across all the contingent representational formats for numbers and arithmetic operations, and across all the contingent computational algorithms. We are tempted to say that the "real," bedrock explanation of why *every* computing device "experiences" some anomaly when prompted to divide by zero lies in this abstract, number-theoretical "fact." What, therefore, figures prominently in the *explanans* is not a law of physics, or a generalization couched in terms of instructions and programs, but rather a universally, necessarily invalid computation, a "pathological" operation on numbers.

The example of the division by zero is only a dramatization. Any other computation would do as well mutatis mutandis. These machines severally deploy certain physical events possessing abstract common invariants precisely because they have all been built as physical instantiations of abstract operations on abstract entities. The properties and laws of numbers are what "explain," at a suitably general level, what these machines actually do. In a sense, therefore, the ultimate cause of all these physicalistic *and* algorithmic anomalies is not itself physical nor strictly speaking computational. The ultimate bedrock cause of this single overarching fact is an abstract property of an abstract operation. This is the proper cause, *if anything is.*[5]

18.11 Ditto for Many Explanations in Cognitive Science

As we have seen, many explanations routinely offered by cognitive scientists appeal to abstract properties and laws that possess the same irreducible abstract status. It is, for instance, significant that, when philosophers of psychology and cognitive science are forced to offer an instance of a truly general, exceptionless psychological "law," they almost invariably produce something like *modus ponens* (if one believes that "if p, then q" and one also believes that p, then one must believe that q) (Sanford 1989; Stalnaker 1968, 1984). As a matter of brute psychological fact, even in our actual use of this simple inferential scheme something can often go wrong as a vast literature on the psychology of reasoning on conditionals shows (Cosmides and Tooby 1989; Girotto, Blaye, and Farioli 1989; Johnson-Laird and Wason 1970; Tooby and Cosmides 1988; Wason 1983; Wason and Johnson-Laird 1972; for a recent summary, see Newstead and Evans 1995). But it is indeed correct to assume that some rule very close to *modus ponens* always looms large even in the background of people's actual reasoning on conditionals. The descriptive theory (couched, say, in terms of "invited inferences," "cheater detectors," "familiarity with the instance," "pragmatic effects," "deontic rules," and so on) will account for the deviation of the actual rule of inference from the laws of logic, but it will still presuppose that some modified version of *modus ponens* does play a causal role in people's thinking about certain specific instances of conditionals.

The explanatory role of the unmodified *modus ponens* in counterfactually building those accounts will not itself be psychological. The cogency of *modus ponens* is logical, not psychological, and it ultimately explains why even bastardized versions of it still carry some subjectively perceived cogency. As we saw earlier, there can be no functional or computational justification for this fact. We may hope ultimately to find the neurological correlates of framing, risk aversiveness, invited inferences, and cheater detectors, and we may hope to account for the causal role that these heuristics play in our mental life in terms of some neurological causality. (I do not know if even this is a plausible expectation, but let me concede it, for the sake of the present argument.) For all that (as I have heard Fodor correctly stress) we do not expect to find the neurological localization of *modus ponens.* And even if, by momentuous luck, we did find that our brain shows certain patterns of neurological events when, and *only* when, we reason in terms of *modus ponens,* they would have to count as effects, not as causes. Just as in our division by zero, these neurological events surely cannot explain why *modus ponens* is a necessarily valid inferential scheme. No neurological mechanism can *cause* the validity of an abstract deductive scheme. It would be the other way around. A certain specific (and for the time being purely science-fictional) pattern of neuronal firing would have a certain *correct* causal role in our reasoning and on our actions *because* it instantiates a necessarily valid scheme of inference.

More generally, the abstract properties of abstract entities such as sets, numbers, functions, propositions, valid deductive schemes, and so on as well as their role in explaining the way we reason cannot be identified with facts about mental representations or with facts about neuronal firings. On the contrary, they are what the corresponding mental representations are representations *of* (remarks in the same spirit are to be found in Higginbotham 1987, 1988). They are abstract contents, not contingent formats for those contents.

From all this, it follows that there can be no complete functionalist account of the central processor because there is no functionalist account for (at least) one component: the normative background. We have to admit that there is an uncaused component in the etiology of our thoughts and actions and that there is a downward causality proceeding from abstract entities and properties to their contingent psychological and neuronal instantiations. In particular, it follows that it is hopeless in principle (not just hard as a matter of fact) to account for this component by means of an evolutionary explanation.

I am not for a moment suggesting that this kind of causation is unproblematic nor even that we can seriously claim that we understand it. But I *am* suggesting (pace the evolutionary epistemologist) that *it* lies at the very heart of cognitive science. We do not quite understand what kind of causality that one is and how it works, but we have to presuppose it, and we can recognize it when we see it at work.

In sum, although cognitive science shares significant features with biology, it appears as a truth of reason that this kind of downward causality is in principle irreducible to biological causes, to facts about our psychological and biological makeup. This is not a statement about any limitation in the ontology of our natural sciences; it is rather a statement about severe limitations in our present theories of abstraction. We are perhaps found here banging against the wall of our epistemic boundedness.

18.12 Conclusion: Back to Appropriateness, Systematicity, and Free Will

The psychological mechanisms, the representational formats, and the computational constraints typically postulated in the descriptive theories of cognitive science can be seen as strictly caused by the makeup of our organism, but, as we have seen, the normative component that figures in many explanations of cognitive science cannot be caused by our biology. An ideally complete biological explanation could well account for the fact that we manage to gain mental *access* to those abstract, necessarily valid rules and constraints. It might even explain *how* we manage, but it will never explain the causal power that thoughts and actions instantiating them have on other thoughts and actions.

Their status as abstract, uncaused entities is compatible with (though surely not an explanation of) the fact that many of our reflective thoughts and actions are uncaused. They are systematically related to their eliciting prompts and systematically interrelated because there are composite, lawlike constructs out of psychological mechanisms and abstract properties governed by axioms and theorems that constitute a complex background of beliefs, norms, utilities, and pieces of reasoning based upon them. The contents of our psychological states are (at least in part) functionally individuated in terms of their systematic causal effects on other mental states and on actions. Their causal powers, therefore, figure prominently in the etiology of our thoughts and behaviors, accounting for their appropriateness and their counterfactual-supporting systematicity. But, at bottom, at the very source, many of our thoughts and actions are uncaused because at least one component of their etiology (that is, the normative component) is itself uncaused.

Of course, there may be other components in the etiology of our thoughts and behaviors that are equally uncaused on other grounds. I am not pretending here to solve the mystery of free will cheaply. I have only tried to elucidate one reason why many of our thoughts and behaviors are *both* appropriate to the stimulus, systematically interconnected, *and* uncaused. It is perhaps a blessing or perhaps a curse, for a special science to have to cope with this unique state of affairs.

18.13 Acknowledgments

I am grateful to Paul Horwich and Daniel Kahneman for many valuable comments on a previous draft. I have fond memories of my conversations with the late George Boolos on the topics covered in this chapter, and I like to think that I have managed to incorporate his comments and suggestions.

Notes

1. To meet initial criticism voiced by logicians, special care has been taken, in the actual presentation of this problem, to point out explicitly that *there is* at least one member in the class of bankers by using the definite article "the." Moreover, it is specified explicitly to the subjects that the premises refer to a real state of affairs in a certain real country. This piece of additional information, redundant and obvious as it may appear, is thereby phrased explicitly and again implied by the use of the definite article in such a way as not to distract the subjects from the truly interesting task. It is necessary to state explicitly that such groups are not "empty" in order to meet the logician's requirement for existential quantifiers that properly license the conclusion (see the brisk exchange between George Boolos and Philip Johnson-Laird in *Cognition* [Boolos 1984]. This controversy is, in my opinion, a clear example of the pivotal role that logical principles play in determining what the *psychological* basic facts *are*).
2. Some psychologists believe that there is such a thing as a general abstract mental logic, whereas some, like Philip Johnson-Laird, deny that there is anything of the sort.
3. Whether or not the larger project of "naturalizing epistemology" (Quine 1960; Campbell 1974) is workable (and I do not think it is), what I am here denying is that epistemology *can* be naturalized through an evolutionary account of cognitive science and the neurosciences.
4. I realize that I am here, indeed, using Plato's metaphors of *mimesis, metesis,* and *parousia.* However, the point I wish to stress is independent of the metaphors one chooses.
5. Hilary Putnam some years ago had offered a most persuasive argument having the same thrust. He had considered a square peg of side *d*, which obviously cannot pass into a round hole of diameter *d*, while a round peg of diameter *d* can pass into a square hole of side *d.* Though this is a perfectly "physical" fact, it eludes any explanation couched in physical laws. Plainly, Putnam's and mine are both examples of facts instantiated by physical systems but that can *only* be explained in terms of abstract nonphysical concepts and principles.

REFERENCES

Arkes, H. R., and Hammond, K. R. (1986). *Judgment and Decision Making: An Interdisciplinary Reader.* Cambridge, UK: Cambridge University Press.

Boolos, G. (1984). On "syllogistic inference." *Cognition* 17:181–2.

Campbell, D. T. (1974). Evolutionary epistemology. In Schilpp, P. A. (ed.). *The Philosophy of Karl Popper,* pp. 413–63. La Salle, IL: Open Court.

Chomsky, N. (1986). *Knowledge of Language: Its Nature, Origin, and Use.* New York: Praeger Scientific.

Cosmides, L., and Tooby, J. (1989). Evolutionary psychology and the generation of culture, part II: A computational theory of social exchange. *Ethology and Sociobiology* 10:51–97.

Dawkins, R. (1996). *Climbing Mount Improbable.*

Dennett, D. C. (1978). *Brainstorms.* Cambridge, MA: Bradford Books/The MIT Press.

Dennett, D. C. (1991). Real patterns. *The Journal of Philosophy* 91:27–51.

Elster, J. (ed.) (1986). *Rational Choice.* New York: New York University Press.

Fodor, J. (1975). *The Language of Thought.* New York: Thomas Y. Crowell.

Fodor, J. (1983). *The Modularity of Mind.* Cambridge, MA: Bradford Books/The MIT Press.

Fodor, J. (1990). *A Theory of Content and Other Essays.* Cambridge, MA: Bradford Book/The MIT Press.

Gärdenfors, P., and Sahlin, N.-E. (eds.) (1988). *Decision, Probability and Utility: Selected Readings.* Cambridge, UK: Cambridge University Press.

Girotto, V., Blaye, A., and Farioli, F. (1989). A reason to reason: pragmatic basis of children's search for counterexamples. *Cahiers de Psychologie Cognitive/European Bulletin of Cognitive Psychology* 9(3):297–321.

Guyote, M. J., and Sternberg, R. J. (1981). A transitive-chain theory of syllogistic reasoning. *Cognitive Psychology* 13:461–525.

Higginbotham, J. T. (1987). The autonomy of syntax and semantics. In *Modularity in Knowledge Representation and Natural-Language Understanding,* (ed.) J. L. Garfield, pp. 119–31. Cambridge, MA: Bradford Book/The MIT Press.

Higginbotham, J. T. (1988). Knowledge of reference. In *Reflections on Chomsky,* (ed.) A. George. Oxford, UK: Basil Blackwell.

Johnson-Laird, P. N. (1983). *Mental Models. Towards a Cognitive Science of Language, Inference, and Consciousness.* Cambridge, UK: Cambridge University Press.

Johnson-Laird, P. N., and Steedman, M. J. (1978). The psychology of syllogisms. *Cognitive Psychology* 10:64–99.

Johnson-Laird, P. N., and Wason, P. C. (1970). Insight into a logical relation. *Quarterly Journal of Experimental Psychology* 22:49–61.

Kahneman, D., Slovic, P., and Tversky, A. (1982). *Judgment under Uncertainty: Heuristics and Biases.* Cambridge, UK: Cambridge University Press.

Kahneman, D., and Tversky, A. (1979). Prospect theory: An analysis of decision under risk. *Econometrica* 47:263–91.

Kolers, P. A. (1972). *Aspects of Motion Perception.* Oxford, UK: Pergamon Press.

Kyburg, H. E. J. (1983). Author's response: The role of logic in reason, inference and decision. *Behavioral and Brain Science* 6:263–73.

Manketlow, K. I., and Over, D. E. (1993). *Rationality: Psychological and Philosophical Perspectives.* London: Rutledge.

Newstead, S. E., and J. St. B. Evans (eds.) (1995). *Perspectives on Thinking and Reasoning: Essays in Honour of Peter Wason,* Hillsdale, NJ: L. Erlbaum Associates.

Osherson, D. N., and Smith, E. E. (1996). *An Invitation to Cognitive Science. Volume III: "Thinking"* (2d ed.). Cambridge, MA: Bradford Books/The MIT Press.

Ottoson, D., and Zeki, S. (eds.) (1985). *Central and Peripheral Mechanisms of Colour Vision.* Basingstoke: Macmillan.

Piattelli-Palmarini, M. (1980). *Language and Learning: The Debate between Jean Piaget and Noam Chomsky*. Cambridge, MA: Harvard University Press.

Piattelli-Palmarini, M. (1989). Evolution, selection and cognition: from "learning" to parameter setting in biology and in the study of language. *Cognition* 31:1–44.

Piattelli-Palmarini, M. (1994). *Inevitable Illusions: How Mistakes of Reason Rule Our Minds*, New York: J. Wiley & Sons.

Putnam, H. (1975). Philosophy and over mental life. In *Mind, Language, and Reality – Philosophical Papers*, vol. 2. Cambridge, UK: Cambridge University Press.

Putnam, H. (1975). *Mind, Language and Reality (Philosophical Papers, Vol. II)*. Cambridge, UK: Cambridge University Press.

Putnam, H. (1981). *Reason, Truth and History*. Cambridge, UK: Cambridge University Press.

Putnam, H. (1987). *The Many Faces of Realism (The Paul Carus Lectures)*. La Salle, IL: Open Court.

Putnam, H. (1988). *Representation and Reality*. Cambridge, MA: Bradford Books/MIT Press.

Pylyshyn, Z. W. (1984). *Computation and Cognition: Toward a Foundation for Cognitive Science*. Cambridge, MA: The MIT Press.

Pylyshyn, Z. W. (ed.) (1987). *The Robot's Dilemma: The Frame Problem in Artificial Intelligence*. Norwood, NJ: Ablex.

Quine, W. V. O. (1960). *Word and Object*. Cambridge, MA: The MIT Press.

Sanford, D. H. (1989). *If p, then q: Conditionals and the Foundations of Reasoning*. London: Routledge.

Shafir, E. B. (1988). *An Advantage Model of Risky Choice*. Unpublished Ph.D. Thesis, MIT, Department of Brain and Cognitive Sciences.

Shafir, E. B., Osherson, D. N., and Smith, E. E. (1989). An advantage model of choice. *Journal of Behavioral Decision Making* 2:1–23.

Shallice, T. (1991). Précis of "From Neuropsychology to Mental Structure." *Behavioral and Brain Sciences* 14:429–69.

Stalnaker, R. (1968). A theory of conditionals. In *Studies in Logical Theory*. N. Rescher (ed.), Oxford, UK: Basil Blackwell.

Stalnaker, R. (1984). *Inquiry*. Cambridge, MA: The MIT Press.

Sutherland, S. (1992). *Irrationality: Why We Don't Think Straight*. New Brunswick, NJ: Rutgers University Press.

Tooby, J., and Cosmides, L. (1988). *Can Non-Universal Mental Organs Evolve? Constraints from Genetics, Adaptation, and the Evolution of Sex* (88-2). Institute for Evolutionary Studies. Technical Report 88-2.

Tversky, A., and Kahneman, D. (1986). Rational choice and the framing of decisions. *The Journal of Business* 59(4), Part 2:251–58.

Tversky, A., and Kahneman, D. (1992). Advances in prospect theory: Cumulative representation of uncertainty. *Journal of Uncertainty* 5:297–323.

Wason, P. C. (1983). Rationality and the selection task. In *Thinking and Reasoning*, J. S. B. T. Evans (ed.), London: Routledge and Kegan Paul.

Wason, P. C., and Johnson-Laird, P. N. (1972). *Psychology of Reasoning: Structure and Content*. Cambridge, MA: Harvard University Press.

CHAPTER NINETEEN

From Natural Selection to Natural Construction to Disciplining Unruly Complexity

The Challenge of Integrating Ecological Dynamics into Evolutionary Theory

PETER J. TAYLOR

I bring into focus the challenge of making evolutionary theory more ecological, which includes recognizing that ecological dynamics are implicit in any evolutionary theory. Writing in the spirit of Lewontin's essays on organisms as the subject and object of evolution, I do not present a well-formed program of ecological evolutionary theory but point to the existence of problems. My aim is to provoke further, much needed, discussion. First, I examine the problems in identifying which characters of an organism confer fitness, the criteria by which nature "selects." I argue that the problems warrant replacing the metaphor of natural selection. In the center of any historical account we should see a lineage of active organisms – organisms that construct their responses to situations that earlier responses in their lifetime and their lineage's earlier responses have helped construct. "Natural construction," as I use the term, folds construction in the sense of developmental–ecological interactions during an organism's lifetime into a wider evolutionary construal of the term. With this as background, I review approaches to theorizing ecological organization and the ways evolutionary theory fits explicitly and implicitly into those approaches. I conclude that to meet the challenge of integrating ecology into evolutionary theory, natural constructionists would need to recognize that ecological complexity is more "unruly" than it is structured or "system-like."

19.1 Introduction

In the third chapter of *On the Origin of Species*, Darwin introduced his concept of natural selection by noting that, given the struggle for existence, "any variation, however slight and from whatever cause proceeding, if it is in any degree profitable to an individual of any species in *its infinitely complex relations to other organic beings and to external nature*, will tend to the preservation of that individual, and will generally be inherited by its offspring" (Darwin 1859, p. 61, my emphasis). That is, all evolution occurs *in an ecological context*. The structure and dynamics of evolution's ecological context have not, however, been well integrated into evolutionary theory. Population genetic evolutionary theory, most notably, has avoided unraveling ecological complexity by compressing organism–organism and organism–environment relationships into the fitness conferred on an organism by its characters. The center stage in theory could then be occupied by the genetic basis and differential representation of characters within single species. In turn, speciation could become a process of genetic divergence in which the environment mostly took the role of raising and lowering barriers to gene flow.

In this chapter I bring into focus the challenges of making evolutionary theory more ecological. Or, given that ecological dynamics are implicit in any evolutionary theory, I might say, the challenges of making these dynamics explicit.[1] Writing in the spirit of Lewontin's essays on organisms as the subject and object of evolution (Lewontin 1983; see also 1982, 1985), I do not present a well-formed program of ecological evolutionary theory but point to the existence of problems. My aim is to provoke further, much needed, discussion.[2]

There are two strands to my argument, which correspond to two interpretations of the quote from Darwin above. Read one way, Darwin was deflecting attention from a major weakness in the conceptual structure of natural selection as a theory of evolutionary change. "Do not ask me," Darwin in effect is saying, "to identify which characters of an organism confer fitness; there are too many indirect interactions and feedbacks to do this reliably. Just take it as self-evident that there must be such characteristics." Suppose instead that we focus attention on identifying such characters, the criteria by which nature "selects." In Section 19.2 I examine the resulting problems and argue that they warrant replacing the metaphor of natural selection; "natural construction" is my proposal. Of these problems, the one that concerns me most here is the nonintegration of ecological dynamics into evolutionary theory.

A second, more charitable and forward-looking, reading of the passage quoted is that Darwin foreshadowed an integrated ecological–evolutionary theory. In this spirit, Section 19.3 of this chapter reviews approaches to theorizing ecological organization with the goal of identifying (1) ways evolutionary theory fits into them (explicitly and implicitly) and (2) more precisely the shape of the challenge of integrating ecological dynamics into evolutionary theory. I conclude that to meet the challenge of integrating ecology into evolutionary theory, natural constructionists would need to recognize that ecological complexity is more "unruly" than it is structured or "system-like."

19.2 From Natural Selection to Natural Construction

As a form of historical explanation, natural selection is very restrictive; the world and its history must have a particular and atypical shape to fit it (Taylor 1987). To develop this claim, let me start with Darwin himself, who, in the first four chapters of *On the Origin of Species* set out an argument for natural selection as a means of evolution. Part of this argument involves straightforward deductions. *If* there exist (some modern terms are inserted here)

Variation among organisms in characters;
Inheritance of characters; and
Hyperfecundity, which ensures that not all can survive to reproduce and that there will be a *struggle for existence*;
then there will be differential representation in lineages of organisms of variant characters over time, that is, modification by descent or evolution. Now, if there is also
Natural selection, that is, greater survival and reproduction of those organisms with characters that fit them better to their environment (including potential mates), then evolution will result in improvement of adaptation to conditions of existence.[3]

This schema is put to use in four ways, but only the fourth approach can lead to a well-confirmed explanation (Lloyd 1988) of actual observed evolutionary changes:

1. *Forward speculation.* If there existed heritable characters (or genes, genotypes...) that resulted in certain, differential survival and reproductive success rates (the inaptly named "Darwinian fitness" values), then ... (Here the philosopher or mathematician fills in the result.)

2. *Backward fitting.* If we assume there exist heritable characters with certain differential fitness values, then, given a specified model of their genetics (linkage, number of loci, number of alleles, etc.), the fitness values and other parameters would have to be ... in order to match the observed data. (Here statisticians or empirical population geneticists fill in their estimates, at least for the cases in which they can overcome the substantial problems of estimation; Lewontin 1974.)

3. *Hybrid speculation-fitting.* For a given speculative scenario, the biologist finds examples that appear to fit the predicted result. Conversely, given a certain observation, a speculative scenario is fashioned that produces that outcome. A third hybrid mode is to use the estimates from backward fitting for a given model and, in an empirically grounded form of forward speculation, predict the most likely future.

4. *Integrated interpretation*, that is, empirical observation and rational interpretation simultaneously. For this, we need not assume natural selection when we speculate or interpret observations but demonstrate that some actual observed character change was produced by a process of natural selection. Unfortunately, the form of the theory of natural selection makes this very difficult to do properly. Let me explain.

Natural selection holds that organisms enjoy differential survival and reproductive success *because of the effect of some character they possess.*[4] When seeking to demonstrate natural selection, it is not sufficient to point to differential representation of a character. This is, at best, a promissory note for a natural selective account. (There are a multitude of uncashed notes in the literature – most notably, in Endler's (1986) impressive compilation of purported cases of natural selection in the wild.) Now the existence of natural selection in the short term, say, for a generation, is not my issue; I am not positing chance and genetic drift instead of natural selection. The question is whether the generation-by-generation natural selection adds up over time to an outcome that we can, *retrospectively*, assign to natural selection associated with some specific character that increased in frequency in the population.

To demonstrate natural selection in this sense requires at a minimum demonstrating both a functional and a temporal correlation between the specific character and the differential survival and reproductive success. That is, we need an analysis of the character's effect with respect to the environmental circumstances and consistency between the origin in time of those circumstances and the increase in the character's frequency. To appreciate this dual requirement, consider, hypothetically, that we notice that the angle of a flower and of the hovering of its hummingbird pollinator coincide. To establish a functional correlation we could perturb the flower angles experimentally. If the seed set of the perturbed plants is lower, then we might suspect that the match of the normal plants to the hummingbird hovering angle is a result of natural selection. But what if we then find (using cladistic analysis) that the closest relative of the plant has the same angle without having the hummingbird pollinator in its environment? If the ancestor of both has neither the angle or the pollinator, it is probably the case that the angle arose earlier than the environmental circumstance of the hummingbird pollinator.

In recent years, emphasis on the temporal correlation has increased, at least at the macroevolutionary level (Coddington 1988, Donoghue 1989, Brooks and McLennan 1994), but functional correlations still dominate most people's thinking about evolutionary explanation. In implicit recognition of the difficulty of demonstrating both the functional and temporal correlations, researchers often seek out (or speculatively invent) cases whose special conditions enhance the possibility of natural selection's serving as an explanation of the historical change in the frequency of the character in question. These special conditions include the following:

- the character has a single effect (enabling one to discount other effects that might have complicated the functional correlation);
- the character and its effects are effectively independent from those of other characters (again so one can discount other effects in establishing the functional correlation; e.g., the survival advantage of heavy metal tolerance in plants on mine tailings eclipses the costs of the severe physiological effects);
- the character is reproduced over time (so one does not have to deal with changing characters and the changing functional correlations; e.g., single-locus phenotypes are preferred to complex, developmentally–environmentally modulated characters);
- the organism–environment relationship is consistent over time (again so one does not have to deal with changing functional correlations); and

- the time span is limited (reducing the chance of changing functional correlations).

From a knowledge of biology, we should agree that these special conditions are rare or not necessarily generalizable. The consequences for evolutionary theorizing have been several. Biologists (and others) often

- collapse "selection," using the term as a synonym for differential representation of characters[5];
- rely on claims about current functionality without evidence of historical (temporal) change;
- accept milder standards of evidence (e.g., the historical process has been observed in some cases of natural selection, and thus it is plausible that it occurred for the character whose current function has been demonstrated; or one works at a coarse level of resolution of characters, environments, and change so that departures of the detailed mechanisms from the special conditions are not evident[6]);
- invoke repeatedly the same few textbook cases of natural selection; or
- perform laboratory or other experimental work in which selection literally, not metaphorically, takes place.

I concede that applying the strong standards of evidence I have outlined may result in few natural selective accounts qualifying as adequate historical explanations. Nevertheless, it should also be recognized that squeezing evolutionary change so it can fit the special conditions above has the effect of discounting many important aspects of biology:

- characters that are not singled out in living activity of organisms;
- the development or the reconstruction during an organism's lifetime of its characters (over and above the transmission of genetic and other material to the zygote and in contrast to snapshots of characters at some point of time in the lifecycle);
- the broader conditions for "recurrency" of characters, which depend not only on the genetics implicated in the development of characters but also on the persistence of environmental conditions at least insofar as the organisms modulate or "construct" those conditions (Lewontin 1982, 1983, 1985);
- the contribution of developmental and ecological flexibility to the evolutionary origin of characters (Taylor 1987)[7]; and, more generally,
- the structured, yet changing, ecological dynamics to which organisms both respond and contribute.

In short, characters are part of structured processes. For some biologists and philosophers of biology these complexities of biology mitigate against the coherent accumulation of change over time; they believe that only when the special conditions more or less apply does evolution lead to identifiable adaptive outcomes. Others attempt to incorporate some of these aspects of biology by adjusting the theory of natural selection, as in, for example, frequency-dependent selection. This tinkering, however, preserves room for the almost conventional moves back and forward in evolutionary thought among forward speculation and backward fitting.[8] My preference is to free ourselves from the restrictive, and thus widely misused, form of the natural selective explanation.[9] The investigation and interpretation of historical change in biology can be opened up by reenvisioning it as, I suggest, a process of "natural construction."

In proposing construction as a metaphor, I am not, of course, suggesting that there is any superintending, goal-directed Natural Constructor. Moreover, I accept that biological evolution is a population-variational, trial-and-error phenomenon (Lewontin 1982, 1983) and that natural selection can be an adequate explanation in special cases. However, I advocate that we place in the center of any historical account a lineage of active organisms – organisms that construct their responses to situations that earlier responses in their lifetime and their lineage's earlier responses have helped construct. Natural construction, therefore, folds construction in the sense of developmental–ecological interactions during an organism's lifetime (Lewontin 1982, 1983, 1985)[10] into evolutionary construction (Oyama 1988). The common thread is a sense of construction as building, which connotes processes in which many diverse resources are mobilized in turn. In building upon what is already constructed, organisms are enabled and constrained and are engaged in multiple cross-linked or intersecting processes, which ensures that the effects of any aspect of living are polyvalent (Taylor 1995). There are no off-the-shelf models for interpreting such linked construction processes. To resist slipping back into the special cases for which natural selective explanations might suffice, we need, among other things, to

- recognize that any process involving multiple contributing causes is difficult to partition into single kinds of causes ("genetic," "environmental," "developmental constraints");
- trace, using careful phyolgenetic analysis, sequences in time of character–environmental correlations (Donoghue 1989);
- identify regularities in the form of intrinsic structural properties that relate to function (detected as regularities in historical patterns of functional morphological change; Lauder 1990); and
- identify historical–evolutionary regularities in developmental–ecological interactions[11] and "resources" (Griffiths and Gray 1994).

This last item means that we need more work theorizing (1) the construction and reconstruction of characters, as against their genetic determination and transmission; (2) the dynamics of ecological organization, including its ongoing restructuring; and (3) the intersections of (1) and (2). Since the early 1980s, several biologists have sought to bring embryology and development back into evolutionary theorizing (Gilbert, Opitz, and Raff 1996). Less attention, however, has been given to theorizing evolution's ecological context, the subject of Section 19.3, and to theorizing developmental–environmental interactions (though see Gilbert 1997).

19.3 The Unruly Complexity of Evolution's Ecological Context

Integrating the structure and dynamics of evolution's ecological context remains a neglected project within evolutionary theory. Nevertheless, the different approaches to theorizing ecological organization can still be read in terms of the ways that evolutionary theory fits into them, whether or not this is made explicit. Table 19.1 provides a classification of five basic orientations.

Central to the first three orientations is the notion of system, which I use in the strong sense of an entity that has clearly defined boundaries and has coherent internal dynamics – dynamics that govern the system's responses to external influences and determine its structure, stability, and development over time.[12] System in this sense can refer not only to the basic units of systems ecology but also to the guilds and communities of community ecology. These three orientations differ according to the relative time scales of ecological and evolutionary processes. In contrast to viewing ecological organization as system-like, various ecologists have emphasized what I call its "unruly complexity." That is, organisms and processes transgress the boundaries of any unit of ecological structure, spanning levels and scales; natural categories for, and reduction of, the complexity are elusive; ecological structures are subject to restructuring; control and generalization are difficult. The two nonsystem orientations differ according to whether this unruly complexity can be disciplined theoretically. Let me illustrate the distinctions in Table 19.1 through a schematic review of twentieth-century theories of ecological organization. This review will also motivate my formulation of the challenges of integrating ecological dynamics into evolutionary theorizing.

Table 19.1. *Five Orientations to Theorizing Ecological Organization and Evolution*

Focus on system (or community)

C. system evolves as a **C**oherent whole
Fast return to equilibrium; slow change or evolution of system

Focus on individuals in context of system

S. **S**table system
Fast return to equilibrium; intermediate speed evolution of population of individuals; slow change of system

R. system transient, yet **R**egularly recurring
Fast passing of transient context (e.g., succession); intermediate speed evolution of population of individuals; slow change in nature of transient context

Focus on ecological organization as not system-like

Relevant processes not separable into "ecological" and "evolutionary" time scales

AT. **A**nti-**T**heory

D. unruly complexity can be **D**isciplined

19.3.1 Plant Communities, Associations, and Populations

Early this century, Frederic Clements proposed that the community (or "formation") of plants colonizing an uninhabited site is like a developing organism; it passes through a predictable succession of stages, each providing the conditions for the following stage, and results finally in a stable, self-sustaining "climax" (Clements 1916). The interactions among the species (especially competitive interactions) and the changes the species effect on the habitat provide the cause for this development. At the same time the climax is determined by the habitat and climate, suggesting the possibility of a "physiological" analysis of the complex "organism's" response to its external environment. Three of the five orientations, therefore, are evident in Clements' ecological theory: R – succession repeats itself in similar conditions, and thus the context within which successional species find themselves is regular; S – the endpoint of succession, the climax, is a stable situation in which populations can evolve; and C – the climax changes over time as the climate changes.

In the responses of Henry Gleason to Clements' community concept we find almost all the opposing terms that have arisen in debates in American and English ecology since Clements. For the integrated complex organism, Gleason substituted the shifting association of populations. The properties of these associations are very contingent on the physical environment and the patterns of immigration from the surrounds. That is, the association depends on particular conditions that are not controlled by it. The ecologist should expose regularities by analyzing variable responses of individual populations to variable environmental conditions rather than the Clementsian approach of classifying a limited set of types of communities and successional sequences. This follows because "succession is an extraordinarily mobile phenomenon, whose processes are not to be stated as fixed laws, but only as general principles of exceedingly broad nature, and whose results need not, and frequently do not, ensue in any definitely predictable way" (Gleason 1926).

Although Gleason's views invite a nonsystem orientation, we need to consider their actual implementation. Gleason's views were not popular in his own time, but the individual–population thrust now dominates the theories of plant population ecology. The focus has been on demographic strategies of individual plant species – how they colonize, grow, survive, disperse, and so on (Dirzo and Sarukhan 1984) – or sometimes more specifically on the plant "behavior" (form and physiology) that is optimal, in the given conditions, for energy capture and growth (Givnish 1986). For most plant evolutionists the individualistic approach appears to have eliminated the need for theory about ecological organization. For example, in demographic strategy theory the ecological context is evident mostly in the form of "variable environments" for which the strategies exhibit phenotypic plasticity (Sultan 1987). Theory about ecological organization is, nevertheless, hidden behind the appearances. In order for plastic or optimal strategies to have evolved, as it is argued, through natural selection, the ecological context must have had a more or less consistent effect over a period of time. Givnish (1988), for example, recognizes community level regularities but claims that explanations of them can be derived from models of the adaptiveness of observed variants or "morphs." Yet, he admits, the optimal behavior of a morph may depend "strongly on the mixture of competing morphs it faces and can in turn affect the composition of the mixture" (Givnish 1986, p. 8). He then invokes game theory as a means of determining the optimal interactive strategy for a morph, but this assumes, as before, that the rules of the game (i.e., the context of the interaction) have remained consistent. Without this assumption the plant population ecologist still faces the issue of evolution occurring in contexts that may change over ecological time; the need to inquire further into the structure and dynamics of those contexts remains. In the meantime, individualistic plant ecology's implicit view of ecological organization fits, not the AT, but the R, or perhaps even the S, orientation.

19.3.2 Animal and General Ecology

In these fields the central aspect of Clements' community theory, namely, the "systemness" of nature, remained productive of theory for longer than was the case in plant ecology. Two conceptual trajectories can be distinguished: systems or ecosystem ecology and community ecology.

Systems ecology focuses on nutrient and energy flows among compartments, living and nonliving, of entire ecosystems. For H. T. Odum, one of the field's pioneers, the intermeshing of biological and physical aspects of systems opened the way for him in the mid-1950s to begin proposing analogies based on properties of electrical circuits and, using energy as the universal currency, to advance theoretical propositions about living systems based upon thermodynamics, such as, ecosystems will develop until they maximize total energy flow per unit time (maximum power; a C formulation).[13] Systems can be understood without needing the intimate knowledge of the range and flexibility of particular species' behavior; they are simply structures of energy stores and flows. Given this natural reduction of complexity, the evolution not only of ecosystems as a whole, but also of organisms is governed by their efficiency at trapping and processing energy (either in R or S situations).

In Odum, and in subsequent systems ecology, a central role was given to measurement (of the different flows of energy or nutrients in an ecosystem), in contrast to making experiments on well-controlled parts of the system. After collecting data for an entire system and summarizing them in a flow diagram, the systems ecologist could act as if the diagram represented the system's dynamic relations. When translated into computer models, the flow diagram becomes a means to generate predictions about an otherwise unmanageable complexity (Taylor 1988). Systems modeling in this spirit has been pragmatic in its placement of boundaries and internal aggregations (e.g., species into trophic compartments). More generally, theoretical principles have been less important than organizing measurements on a huge variety of systems, an emphasis beginning in the late 1960s with the International Biological Program. Unifying theory such as the maximum power principle of Odum has not been necessary for most systems ecologists. Hierarchy theorists (O'Neill et al. 1986) are an exception. They claim that (1) data analytic methods can expose the natural boundaries and aggregations (measurement again being central here), and (2) at different spatial and temporal scales a system will be stable in its composition or homeorhetic in its processes (S or R).

Although systems ecologists differ in their emphasis on theory versus bookkeeping of measurements, they generally give a lower priority to analysis of mechanisms of the component subsystems than do other ecologists. The behavior of any subsystem measured in isolation (e.g., the throughput of some nutrient in an organism) is considered to be an inaccurate guide to the behavior of the subsystem when constrained by the rest of the system. Alternatively, the task of integrating knowledge of the separate mechanisms is considered to be practically impossible. In any case, evolution of and within populations as subsystems presumably occurs in the R or S forms.

In contrast to systems ecology, community ecology has focused on parts of, rather than entire, ecosystems, namely, on communities of interacting populations. The boundaries of communities can be drawn so that they include simply a host–parasite pair or several species in a "guild" of animals requiring a similar type of food or other resource, or in a web of predators and prey. From the late 1950s on, MacArthur championed the search for general theoretical propositions about regulation of population sizes and distributions (in space and in characters) through interspecific interaction (chiefly, competition for limiting resources); he also popularized the use of models, often mathematical, to formulate and investigate these propositions (Kingsland 1995). Although rarely discussed in these terms, a strong system view is also essential to MacArthurian community ecology. Studying a community of competitors or using a simple model requires the dynamics of the community to be effectively independent of its context of other species, resources, nutrient cycles, and so on. This independence requires either that the community have a separate time scale or location in space or that the external interactions be weak or consistent compared with the interactions within the community. The mathematician represents the system by the variables of a model, whereas the external context enters only through the parameters; without this separation mathematicians cannot apply their tools, especially the analysis of the stability of equilibria, to communities. Furthermore, this system view promotes a particular hierarchical view of complexity, namely, that it is effectively decomposable into loosely linked systems. Although context independence may be achievable in well-controlled laboratory experiments, it is likely to represent a special situation if it occurs in the field.

Microevolution fits readily into this framework. Provided that competition does not lead to extinction of a population, the frequency of characters evolves in response to the same intra- and interspecies interactions as govern the ecology. This is the domain of evolutionary ecology, including coevolutionary studies (Futuyma and Slatkin 1983). An early and popular idea of evolutionary ecology is that evolution can be dichotomized into r and K selection of characters enabling, respectively, rapid colonization of newly available or temporary sites (R) versus efficient use of resources in crowded, well-established communities (S). Studies of demographic strategies, however, have yielded a more complex picture (especially for plants). Furthermore, as was the case with individualistic plant ecology, theory in this field depends on the interacting species' being effectively the only players and the ecological conditions not changing at a rate faster than evolutionary change can occur (R and S; but see Herrera 1986).

A well-developed, nonsystem-view alternative to MacArthurian community ecology has existed since the 1950s in the work of the Australian ecologists Andrewartha and Birch. To explain the distribution and abundance of species, ecologists should, they propose, focus on one species at a time and, using their knowledge of natural history and of the animal's physiology and behavior, trace the direct and indirect influences on the animal's chance to survive and reproduce. This "envirogram" of influences consists of physical factors and other species, often acting independently of the density of the focal species. Although, presumably, the other species have their own envirograms, the resulting loops of influences or feedback are not traced or theorized (AT). Evolutionary theory is not entirely absent, however. Recognizing the variation of environments (among locations and over time) and the risk of local extinction of any local population, evolutionary studies of species in this kind of evolutionary context focus on describing the genetic and phenotypic variations that reduce the risk of the extinction of the entire population (Andrewartha and Birch 1984). The benefits of risk reduction strategies might make us hope that organisms employ them, yet how they evolve is not clear (remember that regularities of ecological–evolutionary context are ruled out). Perhaps, just as in individualistic plant ecology, S or R consistency of context is implicitly required.

Andrewartha and Birch's perspective have not had a strong direct influence on community ecology. However, in the 1980s, both the generality and usefulness of ecological theory expressible in simple models was vigorously questioned (Strong et al. 1984). The skepticism was supported by comparing actual patterns of species' coexistence or morphological differences with those generated by a "null" model, that is, a model lacking the species interactions of the MacArthurian model, and showing that the null model fits the observations just as well as the model based on species interactions. Such results fuel a particularistic view of ecology (Simberloff 1982): There are many factors operating in nature, and in any particular case at least some of these will be significant. A model cannot capture these and still have general application. Instead, ecologists should investigate particular situations, and experimentally test specific hypotheses about these situations guided by and adding to knowledge about similar cases. Under a particularistic view, no theory of ecological organization and evolution within that context is possible (AT).

Testing hypotheses will tell us little, however, about how theory is generated in the first place, whereas models used in an exploratory fashion have served to stimulate new formulations and questions (Taylor 1989). For example, mathematical exploration of how complexity of communities may be related to their persistence or stability has contributed to shifting theory onto new grounds. Originally, the possibility of achieving equilibrium in a community and ecological complexity was held to result from the underlying stability of the ecological system. Mathematical analysis shows, however, that complexity works strongly against stability (unless the complexity is nearly decomposable, i.e., consists of loosely linked subsystems). Subsequently, a "landscape" view has arisen, which holds that a community may persist in a landscape of interconnected patches even though the community is transient in each of the patches (DeAngelis and Waterhouse 1987). Metapopulation theory, an actively explored variant, examines the persistence, not of communities, but of populations (or phoretic associations of communities on carrier species; Wilson 1983) in such a landscape (Hastings and Harrison 1994). From a different angle, models that distinguish among individual organisms (in their characteristics and spatial location) have been shown to generate certain observed ecological patterns, such as patterns of change in size distribution of individuals in a population over time, where large-scale, aggregated models have not (DeAngelis and Gross 1992). And, the effects mediated through the populations not immediately in focus, or, more generally, through "hidden variables," upset the methodology of observing the direct interactions among populations and confound many principles derived on that basis (Wootton 1994). Clearly, the theory generated by exploratory modeling is moving from the S and R of MacArthurian ecology to nonsystem formulations.

The picture of communities as contingent constructions, permeable to invasions and reconstruction, and requiring analysis of the responses of individual species – a Gleasonian view – has been developed not only in modeling but in ecological theory more generally during the last decade. Historical contingency, nonequilibrium formulations, local context, and individual detail have been reasserted.[14] In patch dynamic studies, for example, the scale and frequency of disturbances (that create open "patches") is now emphasized as much as species interactions in the periods between disturbances (Pickett and White 1985). Studies of succession and of the immigration and extinction dynamics for habitat patches pay attention to the particulars of species dispersal and the habitat being colonized and how these determine successful colonization for different species (Gray, Crawley, and Edwards 1987). Similarly, but on a larger scale, such a shift in focus is supported by biogeographic comparisons that show that continental floras and faunas are not necessarily in equilibrium with the extant environmental conditions (Haila and Järvinen 1990).

These changes in theoretical emphasis have subdued the ambitions many ecological theorists had in the 1960s and 1970s for identifying general principles (Kingsland 1995). Ecological theorists have become increasingly aware that situations may vary according to historical trajectories that have led them, that particularities of place and connections among places matter, that time and place are a matter of scale that differs among species; that variation among individuals can qualitatively alter the ecological process; that this variation is a result of ongoing differentiation occurring within populations (which are specifically located and interconnected); and that interactions among the species under study can be artifacts of the indirect effects of other "hidden" species.

Although these perspectives are becoming widely appreciated, the appropriate way to theorize the resulting nonsystem-like complexity remains far from settled. The antitheoretical current persists, but others hope for some way of synthesizing the insights or heuristics from the different angles of investigation and theory (Taylor 1989). In other words the unruly complexity may be disciplined (D). The means, however, for weaving together heuristics has yet to be articulated. For example, how does the community ecological idea that there is a limit to the similarity of coexisting species combine with the ideas that spatial heterogeneity or an intermediate level of disturbance promote diversity? Whether or not this kind of question can be answered, that is, whether D can hold out against both AT and the approaches that treat complexity as system-like, depends, I believe, on theorists of ecological organization finding rules or regularities of historical construction. These rules would have to account for observed patterns in ecological organization in a way that finds the appropriate level of generality given the particularity of different species and physical conditions. If complexity can be disciplined without being suppressed, natural constructionists would have some of the ecological theory they need for explanations of evolution. Theories of developmental–ecological interactions and the other aspects of natural construction mentioned in Section 19.2 could then build upon that. Eventually, accounts might be produced explaining how evolution proceeds in the context of these intersecting processes without relying on the theoretical separation of "genetic (transmission)," "developmental," "ecological," and "evolutionary" time scales.

The reductionist program and engineering efforts of molecular biology and biotechnology are incredibly productive of new knowledge, technical capabilities, and research questions. Given the dominance of genetics within biology, it would at this point in history take more than generalities about processes of ecological organization to shift the focus of evolutionary studies from genetics, differential representation of characters, and individual populations. Nevertheless, for almost four billion years organisms have constructed their living within unruly ecological complexity; the challenge of integrating ecology into evolutionary theory should eventually be addressed.

19.4 Acknowledgments

The comments of Susan Oyama, Deborah Gordon, and an anonymous referee helped me clarify my exposition.

Notes

1. I applaud recent work in comparative biology and in the new field of "historical ecology" that is careful to match phylogenetic information with information about the timing of changes in the environment (Brooks and McLennan 1984;

see discussion later in this essay). Historical ecologists, however, do not emphasize ecological dynamics or examine how these dynamics complicate their framework of speciation through adaptation by natural selection.

2. Implications for social thought of the ideas developed in this essay are discussed in Taylor (1998).
3. I prefer Darwin's original formulation of these conditions to modern revisions – for example, Lewontin (1982, p. 153; 1985, p. 76) and Lloyd (1988) – because Darwin's formulation makes clear that differential representation and evolution are not equivalent to natural selection (see note 4). Further qualifications to Darwin's formulation can be made. For example, modification by descent would occur without hyperfecundity as long as organisms left different numbers of offspring sometimes, and differential representation over time could occur without inheritance. Darwin's basic version suffices, however, to develop my critique of natural selection.
4. This follows Braidie and Gromko (1981), who document the inadequate definitions abounding in biology and evolutionary textbooks. In their definition, the characters must be heritable, but I consider this to be not definitional but rather a condition that contributes to the possibility that natural selection, sensu Darwin, accumulates over time to produce a discernible adaptational change; see the subsequent paragraphs of this section.
5. A variant of this is the idea that a character could be the cause of natural selection for organisms having that character if the character has a positive effect on survival and reproduction averaged over the range of contexts in which the character occurs. This conceptual move underwrites theories about natural selection of suborganismic units and, sometimes, superorganismic units.
6. In the hypothetical example above, if we had not noticed the plant's relative that had the same angle but not the hummingbird pollinator, but we were aware of the ancestor, natural selection would have been both functionally and temporally plausible. At some point, however, we balk at allowing a lessening of resolution to support natural selective explanation. We know, for example, that it is too coarse to correlate bird feathers both functionally and temporally to the bird lineage's move into the air.
7. To fully account for the direction of the observed historical change one should – even in a natural selective explanation – study the origin of the character. This explanatory requirement might account for biologists' reliance on another special condition, namely, the origin of characters by mutation or recombination, which make this origin random with respect to environmental circumstance.
8. As evident, in particular, in debates about "fitness," units of selection, and levels of selection (see relevant entries in Keller and Lloyd 1992).
9. It follows that I also propose abandoning the concept of adaptation in both its senses, that is, the character that has been the causal focus of a natural selection account and the process of its evolutionary development. My thinking along these lines drew at a key stage on Lewontin's critique of the concept (see Lewontin 1983). For a discussion of adaptation engaging more closely with his critique, see Godfrey-Smith, "Organism and environment," this volume.
10. See, in particular, Lewontin's discussion of organisms constructing their environments in the senses of determining what is relevant, transducing physical signals, modulating patterns of variation, and altering their environment (1983, p. 76–77; see also Godfrey-Smith, Chapter 12 in this volume).

11. See note 1.
12. This distinction, along with much of the material in the remainder of this section, has been adapted from Taylor (1992).
13. In a similar spirit of mimicking the determinism of change in some physical systems, information-theoretic (Margalef 1968) or "phenomenological" (Ulanowicz 1986) formulations of ecosystem development have been proposed.
14. Not all the field has embraced the shift; the legacy of MacArthur and the responses to critics of community as a unit of organization are well represented in Diamond and Case (1986) and Gee and Giller (1987).

REFERENCES

Andrewartha, H. G., and Birch, L. C. (1984). *The Ecological Web.* Chicago: University of Chicago Press.

Bradie, M., and Gromko, M. (1981). The status of the principle of natural selection. *Nature and System* 3:3–12.

Brooks, D. R., and McLennan, D. A. (1994). Historical ecology as a research programme: Scope, limitations and the future. In *Phylogenetics and Ecology*, (eds.) P. Eggleton and R. I. Vane-Wright, pp. 1–28. London: Academic Press.

Clements, F. E. (1916). *Plant Succession: An Analysis of the Development of Vegetation.* Washington, DC: Carnegie Institution of Washington.

Coddington, J. A. (1988). Cladistic tests of adaptational hypotheses. *Cladistics* 4:3–22.

Darwin, C. 1859 (1964). *On the Origin of Species.* Cambridge, MA: Harvard University Press.

DeAngelis, D. L., and Gross, L. J. (eds.) (1992). *Populations and Communities: An Individual-Based Perspective.* New York: Chapman and Hall.

DeAngelis, D. L., and Waterhouse, J. C. (1987). Equilibrium and nonequilibrium concepts in ecological models. *Ecological Monographs* 57:1–21.

Diamond, J., and Case, T. J. (eds.) (1986). *Community Ecology.* New York: Harper and Row.

Dirzo, R., and Sarukhan, J. (eds.) (1984). *Perspectives on Plant Population Ecology.* Sunderland, MA: Sinauer.

Donoghue, M. J. (1989). Phylogenies and the analysis of evolutionary sequences with examples from the seed plants. *Evolution* 43:1137–56.

Endler, J. A. (1986). *Natural Selection in the Wild.* Princeton, NJ: Princeton University Press.

Futuyma, D. J., and Slatkin, M. (eds.) (1983). *Coevolution.* Sunderland, MA: Sinauer.

Gee, J. H. R., and P. S. Giller (eds.) (1987). *Organization of Communities: Past and Present.* Oxford, UK: Blackwell.

Gilbert, S. (1997). Environmental regulation of animal development. In *Developmental Biology.* Sunderland, MA: Sinauer.

Gilbert, S., Opitz, J., and Raff, R. (1996). Resynthesizing evolutionary and developmental biology. *Developmental Biology* 173:357–72.

Givnish, T. J. (ed.) (1986). *On the Economy of Plant Form and Function.* Cambridge, UK: Cambridge University Press.

Givnish, T. J. (1988). Theory and observations in plant community ecology: A review of recent approaches. Address to the ISEM Symposium, University of California, Davis, August 15.

Gleason, H. (1926). The individualistic concept of the plant association. *Bulletin of the Torrey Botanical Club* 53:1–20.

Gray, A. J., Crawley, M. J., and Edwards, P. J. (eds.) (1987). *Colonization, Succession and Stability.* Oxford, UK: Blackwell.

Griffiths, P. E., and Gray, R. D. (1994). Developmental systems and evolutionary explanation. *The Journal of Philosophy* 91:277–304.

Haila, Y., and Järvinen, O. (1990). Northern conifer forests and their bird species assemblages. In *Biogeography and Ecology of Forest Bird Communities,* (ed.) A. Keast, pp. 61–85. The Hague: SPB Acad. Publishing.

Hastings, A., and Harrison, S. (1994). Metapopulation dynamics and genetics. *Annual Review of Ecology and Systematics* 25:167–88.

Herrera, C. M. (1986). Vertebrate-dispersed plants: Why they don't behave the way they should. In *Frugivores and Seed Dispersal,* (eds.) A. Estrada and T. H. Fleming, pp. 5–18. Dordrecht, The Netherlands: Dr. W. Junk.

Keller, E. F., and Lloyd, E. A. (eds.) (1992). *Keywords in Evolutionary Biology.* Cambridge, MA: Harvard University Press.

Kingsland, S. (1995). Chapter 8 and Afterword. In *Modeling Nature: Episodes in the History of Population Biology,* 2d ed. Chicago: University of Chicago Press.

Lauder, G. V. (1990). Functional morphology and systematics: Studying functional patterns in an historical context. *Annual Review of Ecology and Systematics* 21:317–40.

Lewontin, R. C. (1974). *The Genetic Basis of Evolutionary Change.* New York: Columbia University Press.

Lewontin, R. C. (1982). Organism and environment. In *Learning, Development, and Culture,* (ed.) H. C. Plotkin, pp. 151–70. New York: John Wiley & Sons.

Lewontin, R. C. (1983). The organism as the subject and object of evolution. *Scientia* 118:63–82. Reprinted in *The Dialectical Biologist,* (ed.) R. Levins and R. C. Lewontin, pp. 85–106. Cambridge, MA: Harvard University Press.

Lewontin, R. C. (1985). Adaptation. In *The Dialectical Biologist,* (eds.) R. Levins and R. C. Lewontin, pp. 65–84. Cambridge, MA: Harvard University Press.

Lloyd, E. A. (1988). *The Structure and Confirmation of Evolutionary Theory.* New York: Greenwood Press.

Margalef, R. (1968). *Perspectives in Ecological Theory.* Chicago: University of Chicago Press.

O'Neill, R. V., DeAngelis, D. L., Waide, J. B., and Allen, T. F. H. (1986). *A Hierarchical Concept of the Ecosystem.* Princeton, NJ: Princeton University Press.

Oyama, S. (1988). Stasis, development and heredity. In *Evolutionary Processes and Metaphors,* (eds.) M.-W. Ho and S. Fox, pp. 255–74. New York: John Wiley & Sons.

Pickett, S. T. A., and White, P. S. (eds.) (1985). *The Ecology of Natural Disturbance and Patch Dynamics.* Orlando, FL: Academic Press.

Simberloff, D. (1982). The status of competition theory in ecology. *Annales Zoologici Fennici* 19:241–53.

Strong, D. R., Simberloff, D., Abele, L. G., and Thistle, A. B. (eds.) (1984). *Ecological Communities: Conceptual Issues and the Evidence.* Princeton, NJ: Princeton University Press.

Sultan, S. E. (1987). Evolutionary implications of phenotypic plasticity in plants. *Evolutionary Biology* 21:127–78.

Taylor, P. J. (1987). Historical versus selectionist explanations in evolutionary biology. *Cladistics* 3:1–13.

Taylor, P. J. (1988). Technocratic optimism, H.T. Odum, and the partial transformation of ecological metaphor after World War II. *Journal of the History of Biology* 21:213–44.

Taylor, P. J. (1989). Revising models and generating theory. *Oikos* 54:121–6.

Taylor, P. J. (1992). Community. In *Keywords in Evolutionary Biology*, (eds.) E. F. Keller and E. A. Lloyd, pp. 52–60. Cambridge, MA: Harvard University Press.

Taylor, P. J. (1995). Building in construction: An exploration of heterogeneous constructionism, using an analogy from psychology and a sketch from socio-economic modelling. *Perspectives on Science* 3:66–98.

Taylor, P. J. (1998). Natural selection: A heavy hand in biological and social thought. *Science as Culture* 7:5–32.

Ulanowicz, R. E. (1986). *Growth and Development.* New York: Springer–Verlag.

Wilson, D. S. (1983). The group selection controversy: History and current status. *Annual Review of Ecology and Systematics* 14:159–87.

Wootton, J. T. (1994). The nature and consequences of indirect effects in ecological communities. *Annual Review of Ecology and Systematics* 25:443–66.

SECTION C

THE POLITICS OF EVOLUTIONARY BIOLOGY

CHAPTER TWENTY

Battling the Undead

How (and How Not) to Resist Genetic Determinism*

PHILIP KITCHER

> "But wait," the exasperated reader cries, "everyone nowadays knows that development is a matter of interaction. You're beating a dead horse."
>
> I reply, "I would like nothing better than to stop beating him, but every time I think I am free of him he kicks me and does rude things to the intellectual and political environment. He seems to be a phantom horse with a thousand incarnations, and he gets more and more subtle each time around.... What we need here, to switch metaphors in midstream, is the stake-in-the-heart move, and the heart is the notion that some influences are more equal than others. (Oyama 1985, pp. 26–7)

Nobody has done more to combat genetic determinism than Richard Lewontin, whose writings, from the original IQ controversy to present debates about the implications of human molecular genetics, diagnose errors that have seduced influential scholars and their readers into believing vulgar slogans about genes and destiny.[1] Lewontin's reward for his decades of effort has often been the irritated response that what he claims is uncontroversial: once the intellectual poverty of a version of genetic determinism has been exposed, there is a rush to denial ("That is not what we meant; that is not what we meant at all"). Yet, within months or years, some new version of the view that human behavior is largely shaped by the genes returns, inspiring a new rash of popular discussions and, in some instances, framing debates about social policy. It is small wonder, then, that people appalled by the sloppy thinking Lewontin has exposed yearn for the "stake-in-the-heart move."

Lewontin's own response to the continued reemergence of genetic determinism has been to deny the correctness of the interactionist credo, the conventional wisdom to which purveyors of determinist claims retreat in the face of criticism (see Levins and Lewontin 1985, Chapter 3 and Conclusion; Lewontin, Rose, and Kamin 1984, Chapter 10; Lewontin 1991, especially pp. 3–37). Although many of his best arguments consist in demonstrating how determinists

*It's an honor and a pleasure to dedicate this essay to Dick Lewontin who has inspired so many people in so many ways.

have ignored interactions among genetic and nongenetic factors, Lewontin seems to believe that acceptance of these arguments is not enough, that we need to free ourselves from the grip of the Cartesian picture of the world as a machine, that we should recognize the interdependence between organism and environment, and that we should formulate a "dialectical biology." He is convinced that there has to be a fundamental error – an error that can be corrected only by reconceptualizing some parts of biology.

In my judgment, no such reconceptualization is needed, and Lewontin's positive proposals are in constant danger of relapsing into the obscurity that he rightly sees as affecting traditional forms of biological holism (Lewontin, Rose, and Kamin 1984, p. 279; Lewontin 1991, pp. 14–15). Genetic determinism persists not because of some subtle error in conventional ideas about the general character of biological causation but because biologists who are studying complicated traits in complex organisms are prone to misapply correct general views. Ironically, the existence of this tendency to error testifies to the social pressures on biological practice – pressures that Lewontin has been at some pains to point out. The search for the stake-in-the-heart rests on a misunderstanding of the problem and may even undermine the effectiveness of the more limited measures that Lewontin and others have crafted.

It is high time to back up assertion with argument. Let us begin more slowly by asking what the thesis of genetic determinism claims.

Here is a first version. To suppose that a particular trait in an organism is genetically determined is to maintain that there is some gene, or group of genes, such that any organism of that species developing from a zygote that possessed a certain form (set of forms) of that gene (or a certain set of forms of those genes) would come to have the trait in question, whatever the other properties of the zygote and whatever the sequence of environments through which the developing organism passed. Although this is a relatively simple way to articulate the idea that genetic causes take priority, it is of little use for reconstructing the debates about genetic determinism. Perhaps, with sufficient ingenuity, one can discover traits that are genetically determined in this sense, but any such traits will be causally "close" to the immediate biochemistry in which DNA is involved – they will not be the characteristics for which we wonder about the rival contributions of nature and nurture. Even if we apply the definition to a relatively uncontroversial exemplar, investigating whether it counts Huntington's disease (HD) as genetically determined, we encounter trouble. True enough, human beings who carry abnormally long CAG repeats at a particular locus near the tip of chromosome 4 undergo neural degeneration, typically between the ages of 30 and 50, and doctors know of no preventative treatment. Does this mean that no way is known of contriving an environment in which the terrible decay does not occur? Not really. Huntington's disease could be forestalled by giving those with the long repeats the opportunity to end their lives before the onset of the disease, and it is overwhelmingly probable that some people with such repeats have suffered accidental death early in life. Hence, strictly speaking, there are environments in which people who have abnormally long CAG repeats at the HD locus do not develop HD, and thus, according to the definition, HD would not count as genetically determined.

Of course, the existence of environments in which the expression of the HD phenotype is forestalled by early death is hardly comforting, and it would be reasonable to suggest that the account of genetic determinism ought to be refined in one of two obvious ways: (a) by replacing the demand that the trait be acquired in *all* environments with something weaker ("almost all") and (b) by restricting attention to complexes of causal factors that enable the organism to develop to the age at which the trait would normally first appear. But it seems more illuminating to make explicit the strategy that underlies the proposed definition. That strategy begins by isolating certain properties of organisms for exploration of their causal impact, regarding the phenotype as the product of contributions from particular kinds of DNA sequences, on the one hand, and from *everything else*, on the other. It goes on to inquire how the phenotype varies as the DNA sequences are held constant and as other factors (the cytoplasmic constitution of the zygote, the molecules passed across cell membranes, etc.) change. The graphical representation of this, the *norm of reaction* of the genotype, is a familiar concept in genetics, and the crudest sort of genetic determinism consists in claiming that the norm of reaction for the trait of interest is flat (see Figure 20.1). Just the kinds of difficulties that appeared in the HD example make doctrines of so simple a form implausible, but the pictorial style of representation suggests plenty of ways in which the genetic factors can be seen as playing important causal roles – perhaps the norm of reaction will be flat almost everywhere, perhaps it will vary only slightly, perhaps the norms of reactions for different genotypes will show a universal relation, perhaps the norm of reaction will be flat if we restrict ourselves to those complexes of other

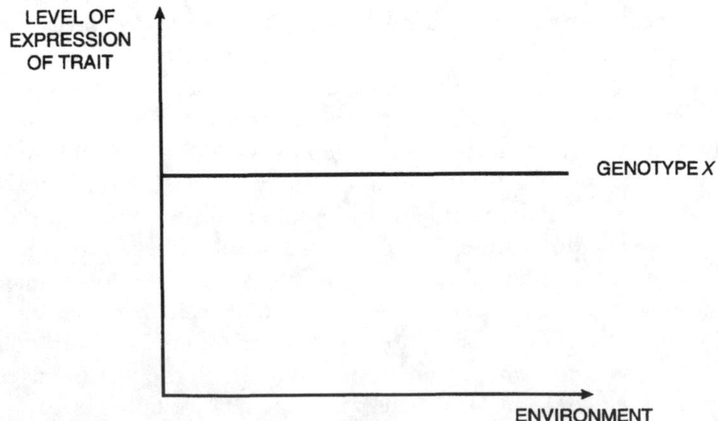

Figure 20.1. A graphical representation of the simplest type of genetic determinism. The level of expression of the phenotypic trait of interest in individuals with the focal genotype ("Genotype *X*") remains constant no matter how the environment varies.

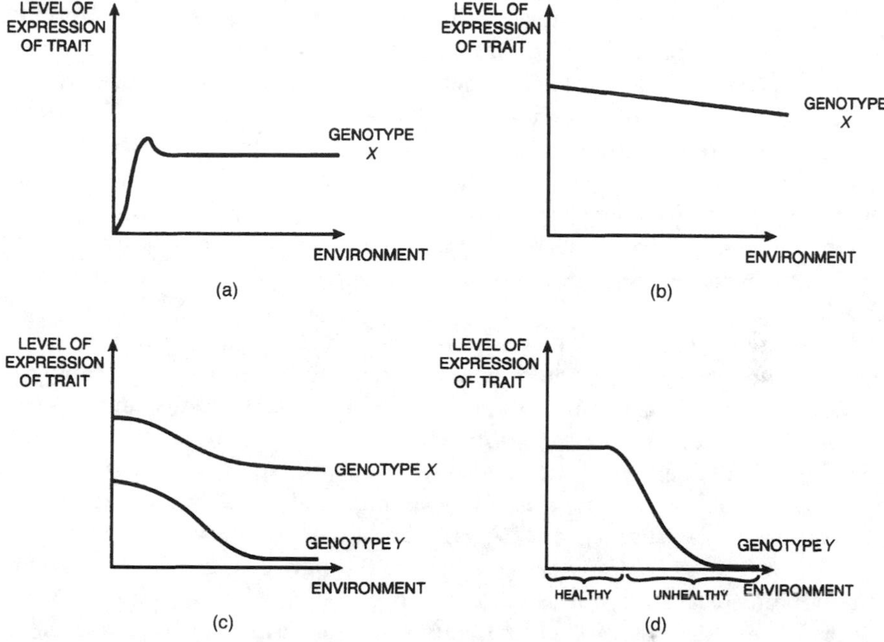

Figure 20.2. Some determinist themes. In (**a**), the level of expression of the trait is constant (for genotype *X*) in almost all environments; in (**b**), the level of expression is almost constant across all environments; in (**c**), despite variation in levels associated with genotypes *X* and *Y*, the level of expression for individuals with *X* is always greater than that for *Y*, no matter what the environment; in (**d**), there is considerable variation in the level of expression but only in environments that are unhealthy. These themes admit further refinements, combinations, and variations.

factors that we think of as healthy for the organism. (See Figure 20.2.) We might thus see genetic determination as a matter of degree, and, instead of quibbling about the proper definition of genetic determinism, investigate the shapes of the norms of reaction in the cases of interest to us.

One of the great insights of Lewontin's early discussions of these questions was his recognition of this as the real issue to which claims of genetic determination were directed (Levins and Lewontin 1985, p. 114). Moreover, Lewontin explained with admirable lucidity why the methods employed to establish those claims could not deliver such conclusions. Estimates of heritability do not reveal the contours of norms of reaction; cross-cultural surveys are only likely to do better if one can be confident that the entire space of nongenetic causal variables is covered (Lewontin, Rose, and Kamin 1984, pp. 245–51). If, as I believe, Lewontin was right in his diagnosis of the errors of popular behavior genetics (most evident in doctrines about the determination of IQ) and popular human sociobiology (manifested in conclusions about the ineradicability of sexual differences in behavior), then the besetting sin was the tendency to draw certain kinds of pictures on the basis of woefully inadequate evidence.

It should now be obvious how a weary critic of hasty generalizations about norms of reaction might go further. Perhaps the tendency to suppose that the relative invariance of a phenotypic trait, given a particular genotype across a manageable range of environments, indicates a flat norm of reaction might be scotched by denying the legitimacy of any such representation altogether. During the 1980s and 1990s, Lewontin and others (most prominently Susan Oyama, Paul Griffiths, and Russell Gray) began to argue that our entire view of genotype–phenotype relations needed to be changed, and that the framework within which I have been posing issues about genetic determination ought to be rebuilt.

Is the notion of a norm of reaction well defined? The writings of those who demand a new conception of nature and nurture – a "dialectical biology" (Lewontin) or "developmental systems theory" (Oyama, Griffiths, and Gray) – suggest several worries about the notion and its relatives (such as the standard genetic idiom of a gene "for" such-and-such a trait). Organism and environment, it is said, are interdependent; there is "developmental noise" in the production of phenotypes; the singling out of genes as causal factors is an unwarranted abstraction from a complex causal situation wrongly giving priority to some determinants of the phenotype; the notion of a gene "for" a trait cannot be coherently reconstructed. These are important concerns, and I will take them up in order.

Lewontin has argued that an organism's environment should not be thought of as identifiable prior to the organism and its distinctive forms of behavior:

Are the stones and the grass in my garden part of the environment of a bird? The grass is certainly part of the environment of a phoebe that gathers dry grass to make a nest. But the stone around which the grass is growing means nothing to the phoebe. On the other hand, the stone is part of the environment of a thrush that may come along with a garden snail and break the shell of the snail against the stone. Neither the grass nor the stone are part of the environment of a woodpecker that is living in a hole in a tree. That is, bits and pieces of the world outside of these organisms are made relevant to them by their own life activities. (Lewontin 1991, pp. 109–10).

The facts reported here are uncontroversial, and the last sentence strikes me as completely correct. What exactly follows?

Lewontin uses these observations to oppose both the idea that we can think of organisms adapting to environments that are independent of them and the idea that we can think of the phenotype as dependent on causal interactions between genotype and environment. The latter conception is the principal concern here, although similar remarks apply to both types of criticism. Lewontin is moved by a principle about causes and causal dependence: *C* cannot be a causal factor in the production of *P* if *C* is dependent on *P*. Applying his conclusions about the dependence of environment on organism, he maintains that we cannot see the environment as a causal factor in the production of the phenotype, and thus the idea of a norm of reaction, with its partitioning of causal variables along different axes, is confused.

There are two related points to be made about this line of reasoning: first, it is not obvious what notion of dependence figures in the causal principle, and, second, it is not clear just one notion of environment is pertinent here. Consider the notion of dependence. In one very obvious sense, the stone in Lewontin's garden is *in*dependent of the presence of phoebe, thrush, and woodpecker – it sits there before the arrival of the birds, before the eruption of fledglings from the nest. So, if we understand "dependence" to mean that the existence of one thing is an effect of the presence of the other, then Lewontin's principle, although plausible, does not apply to the case at hand: there is no reason to think that the contents of the garden cannot play roles in the formation of phenotypes. On the other hand, if we understand "dependence" to mean that the causal relevance of one thing varies with the properties of the other, then the principle does apply to the relations between the birds and the garden. Whether grass, stones, or holes in trees are causally relevant to the development of the birds varies with the properties of the birds, as Lewontin's illustrations show. But now there is no great plausibility in the causal principle itself, for, eleaborated, it says that if the causal relevance of C to P varies with the properties of the bearers of P, then C cannot be a causal factor in the production of P in *any* case, and this claim seems to verge on paradox.

The point can be clarified further by focusing on the other murky term in the argument, "environment." Biologists typically think of environments as those parts of the world outside the organism that are causally pertinent, and in this, the *functional* environment, great tracts of nature are not part of the organism's environment. Lewontin's observations reveal very clearly that an organism's functional environment can depend on what the organism does. However, when we think about the development of an organism, we can pick out some potential causal factors – say the organism's DNA – and take the environment, the *total* environment, to be everything else. In Lewontin's phrase, the total environment is all "the bits and pieces of the world outside the organism" plus some more "bits and pieces" – to wit those inside the organism but not the DNA.[2] The phenotype the organism acquires is determined together by the genotype (the DNA sequences) and the total environment, and, of course, a large part of the total environment will be causally irrelevant. Furthermore, it is quite correct to note that the functional environment, the bits and pieces that are pertinent, depends on (in the sense of varying with) the properties of the developing organism. But this is quite compatible with the causal analysis of phenotypes in terms of genotypes and total environments and with the attempt to draw norms of reactions that identify the causal contributions.

Yet there is an important point behind Lewontin's argument, one that becomes misfocused because of his eagerness to drive a stake into the heart of genetic determinism. To produce a picture indicating the shape of a norm of reaction is to advertise oneself as understanding how to order environments along the axis, and that is typically false advertising. In most instances, we only have the most rudimentary knowledge of how to identify the functional environment, and our ignorance affects the pictures and the conclusions drawn from them.[3] Typically, we can divide the factors outside the DNA into three categories: those we can identify and know to be causally relevant, those we can identify and know not to be causally relevant, and those we either cannot pick out or whose relevance we do not know. (It is, of course, quite possible for us to realize that there is much about which we are likely to be ignorant.) Confronted with a claim about the genetic determination of human propensities to violent behavior (for example), modesty should urge us to think that the last category is quite large, and thus a demonstration that the norm of reaction for a genotype remains flat over a wide range of the nongenetic variables known to be relevant ought not to inspire much confidence that the result would survive a more detailed and fine-grained partitioning. Thus, the right point to make is that we should not leap to premature conclusions about the character of the functional environment, that we should recall the fragility of our representations of the nongenetic causal factors, and that, in consequence, even though the notion of a norm of reaction is perfectly well defined, even though norms of reaction are just what we are trying to discover, knowledge of such norms is very hard to come by. Lewontin has miscast the important methodological point about the difficulty of settling the questions of concern (the shapes of norms of reaction) as an incorrect conceptual point about the incoherence of the notion of a norm of reaction.

The second concern about interactionism focuses on the possibility of "developmental noise." Lewontin argues that even knowledge "of the genes of a developing organism and the complete sequence of its environments" (1991, p. 26) would not allow prediction of the phenotype. In support of this claim, he notes that fruitflies typically have different numbers of bristles at the left and right sides of their thorax, that the difference cannot be explained by a difference in genotype and is not traceable to differences in environment.

> Moreover, the tiny size of a developing fruitfly and the place it develops guarantee that both left and right sides have had the same humidity, the same oxygen, the same temperature. The differences between left and right side are caused neither by genetic nor by environmental differences but by random variation in growth and division of cells during development: *developmental noise.* (1991, p. 27)

Once again, it is important to ask what is being counted as part of the environment and what standards are being used to assess identity of environment.

There are three main types of answer to the question Why do fruitflies have different numbers of bristles at the left and right sides of their thorax? One is to suggest that Lewontin has just counted environments as the same in too coarse a fashion. Perhaps the temperature on the left is the same as that on the right so long as we measure to two or three significant figures, but there are minute differences from side to side, and, at crucial stages of cell division, these differences make a difference. This answer would broadly accept Lewontin's conception of the environment but would eliminate the notion of developmental noise in terms of a more precise understanding of the environmental variables.

The second response would take advantage of the fact that, when interactionists undertake causal analysis of phenotypes in terms of the contributions of DNA and other factors, some of these other factors might be internal to the organism. One of the principal achievements of developmental biology in recent years has been the demonstration of how initial asymmetries in the cytoplasm interact with the DNA in the first stages of ontogeny to produce patterns in early embryos (worked out in greatest detail so far for *Drosophila*). It is quite possible that the differences in rates of cell division do account for the difference in bristle number and that these rate differences are remote effects of the inhomogeneity of the zygote. Although we could reasonably describe them as "random" in the sense that there is no uniform process that determines the distribution of molecules in the cytoplasm of the ovum – so that the initial state of the zygote is the result of contingencies of the formation of a particular egg – they are not *irreducibly* random. A fine-grained specification of the total environment of the DNA would provide a causal explanation of the asymmetry in bristle number. Thus, once again, the form of the phenotype can be viewed as fixed by the genotype and the environment provided that we conceive of the environment in the proper (total) fashion. There is no need to invoke developmental noise or to think that the notion of a norm of reaction breaks down here.

The last possibility is that even the initial distribution of molecules throughout the zygote together with the fine-grained structure of the sequence of environments through which the fly develops does not determine the bristle number. Perhaps the asymmetry is irreducibly random in that no further introduction of causal factors will account for it. I do not know if Lewontin has this possibility in mind, but the existence of fundamental indeterminacies in quantum physics makes it necessary to consider it. There are no well-established instances of quantum events playing a significant role in ontogeny, and many biologists and philosophers seem convinced that subatomic indeterminacies will wash out because of the enormous numbers of molecules that play a role in the development of an organism (the law of large numbers is often thought to be suggestive here). If irreducible randomness does not "percolate up" from the quantum level, then, of course, there is no challenge to the notion of a norm of reaction and no reason to think that subatomic indeterminacies are a source of developmental noise. But, even if some differences in phenotypes ultimately trace to random subatomic events, a simple revision would save the concept of a norm of reaction. Instead of thinking in terms of a single phenotype, fixed by the genotype and (total) environment, we would have to suppose that this congeries of factors determines a probability distribution of phenotypes: pictorial representations would thus illustrate expected values of phenotypes, and, given the elusiveness of quantum effects at the phenotypic level, it would be entirely reasonable to suppose that the spread around the mean was very small.

I turn now to the third worry, the idea that singling out the genotype and considering its effects against background environmental conditions is misguided abstraction from a complex causal situation. No interactionist denies that many causal factors are involved in development (that, after all, is the point of interactionism). However, interactionists defend the legitimacy of a general strategy of causal analysis – the strategy of isolating some of the causal factors, holding them constant, and investigating how the effect varies when other factors are altered. Interactionists ought to support a principle of causal democracy: if the effect E is the product of factors in set S, then, for any $C \in S$, it is legitimate to investigate the dependence of E on C when the other factors in S are allowed to vary. Taking E to be a phenotypic trait, C to be a particular genotype, and S to be a large (probably mostly unknown) set of factors in the total environment (that is factors in the rest of nature outside the genotype), the democracy principle endorses the legitimacy of seeking norms of reaction for phenotypic traits. But it should already be clear that the democracy principle endorses lots of other ways of undertaking causal analysis. For example, we might consider a particular environmental factor and investigate what happens to the phenotype when we vary the genotype and other parts of the environment or we might pick out some mix of genotypic and (total) environmental factors, investigating how the phenotype varies with respect to the rest of the causal factors. The democracy principle accords no special privilege to the representations that foreground the role of genes.

But why, then, do we always end up discussing whether genotypes are all-powerful in development? Why does democracy in principle always translate into elitism in practice? As we shall see, the answers turn out to be complicated, but, for the present, the interactionist's claim is simply that we should not suppose that efforts to investigate the effects of some factors, while others are allowed to vary, are incoherent or illegitimate. Complex causal situations do not demand that we perform the impossible feat of considering everything at once; rather they challenge us to find ways of making these factors manageable.[4] One defense of the prevalence of efforts to chart genotype–phenotype relations against the background of other variables would cite the epistemic benefits of such investigations: this is something we know how to do and that we can expect to prove informative. I will argue below that this cannot be the whole story.

For the moment, we can move on from the blanket charge that any kind of separation out of causal factors does violence to the causal complexities of development and turn to the last line of objection. Russell Gray (both writing on his own and in collaboration with Paul Griffiths) has provided the sharpest version of the charge that thinking in terms of genes "for" traits is a confusion. Alluding to an earlier attempt to suggest that talk of genes "for" traits always presupposes a relativization to "standard" genetic backgrounds and "standard" environments, Griffiths and Gray offer the following counter:

> Consider the DNA in an acorn. If this codes for anything, it is for an oak tree. But the vast majority of acorns simply rot. So "standard environment" cannot be interpreted statistically. The only interpretation of "standard" that will work is "such as to produce evolved developmental outcomes" or "of the sort possessed by successful

ancestors." With this interpretation of "standard environment," however, we can talk with equal legitimacy of cytoplasmic or landscape features coding for traits in standard genetic backgrounds. No basis has been provided for privileging the genes over other developmental resources. (Griffiths and Gray 1994, p. 283).

There is much here with which I agree, although the last sentence contains an ambiguity that enables Griffiths and Gray to arrive at more exciting conclusions than those to which they are entitled.

In (Sterelny and Kitcher 1988), Kim Sterelny and I proposed to reconstruct the everyday talk of genes "for" traits by developing the intuitive idea that "we can speak of genes for *X* if substitutions on a chromosome would lead, in the relevant environments, to a difference in the *X*-ishness of the phenotype" (Sterelny and Kitcher 1988, p. 348). (We avoided the idea of "coding for" that Griffiths and Gray attribute to us.[5]) The notion of environment we appealed to was that of *total* environment conceived as everything outside the locus (or loci) of interest, and we sketched accounts of standardness for the genetic background and for the part of the environment that does not consist of other parts of the DNA. With respect to the extraorganismal environment, we offered three theses: (1) there are alternative ways of explicating the notion of "standard conditions," (2) one of these ways is to count as standard those environments frequently encountered by organisms of the species under study, and (3) another is to count as standard only those environments that do not substantially reduce population mean fitness (Sterelny and Kitcher 1988, p. 350).[6]

Although (1) remains untouched, Griffiths and Gray have shown that (2) and (3) are problematic if the aim is to reconstruct standard genetic discourse. Botanists studying the oak genome want to identify some loci as affecting particular structures in the mature tree, but, for most acorns, genetic substitutions at the pertinent loci do not affect the form of the related structures because those acorns rot. In the accounts of standard environment offered in both (2) and (3), individuals with genetic differences at the loci do not manifest any phenotypic differences in the trait that is supposed to be influenced when they grow in standard environments because the most frequent environments, which also happen to be environments that do not reduce population mean fitness, are environments in which no mature tree grows.

Consider a locus "for" root proliferation. A botanist declares that the allele *A* is "for" root proliferation, meaning thereby that *AB* trees generate more roots than *BB* trees given standard complements of genes at other loci, standard distributions of molecules in the zygotes, and standard sequences of environments. Suppose now that we interpret "standard" in the fashion of (2) or (3). We have to acknowledge that, in most standard environments, the number of roots generated by an organism growing from an *AB* zygote is no greater than that generated by an organism growing from a *BB* zygote (both numbers are zero). However, the botanist could still claim that, for any standard environment, the number of roots generated by the organisms developing from *AB* zygotes is never less than the corresponding number for *BB* zygotes, and, in some standard environments, it is greater. So let the allele *A* be "for" root proliferation just in case in all standard (total) environments the number of roots generated by *AB* individuals is greater than or equal to the number of roots generated by *BB* individuals with the inequalty holding strictly in some cases. Obvious challenge: surely, by luck, the sole oak tree growing in one environment might be *BB* whereas thousands of acorns around (some *BB*, some *AB*) rot; thus, in that environment, the inequality would be reversed. Response: once again, we have to be careful to individuate environments; at the fine-grained level, the environment encountered by the lucky acorn is different from that encountered by the unlucky ones, and, if an *AB* acorn had found itself in precisely that fortunate environment, then it would have generated more roots than its *BB* counterpart.

This strategy for reconstructing the "gene for *X*" locution allows us to retain the interpretation of "standard" as "statistically normal" by weakening the demand that genes "for" *X* promote *X*-ishness in every standard environment. Alternatively, we could decide that a standard environment is one that allows for the development of those features required for the manifestation of the general (determinable) property of which the trait on which we are focusing is a particular (determinate) instance. So, in the case at hand, to talk about genes "for" root proliferation is to suggest that there are differences among individuals with various genotypes – *specifically individuals that have the capacity for producing roots (that is, trees).* Environments that prevent the maturing organisms from manifesting the general property (exhibiting any form of the trait) are thus ruled out as nonstandard, but, in accordance with the pluralistic line offered in our thesis (1), that demarcation will vary with the kind of trait in which we are interested.

I conclude that talk of genes "for" traits can be coherently reconstructed (indeed along the lines that Sterelny and I originally suggested). However, Griffiths and Gray are right to note that a similar form of reconstruction would enable us to speak of "cytoplasmic or landscape features" for traits (here I drop their reference to "coding" since it is a rhetorical flourish irrelevant to the discussion). Indeed, the molecular developmental genetics of *Drosophila* has already begun to emphasize the causal role of proteins deposited by the mother in the cytoplasm of the ovum: to say that the *Bicoid* protein is "for" head–tail polarity is to note that variations in the forms or concentrations of that protein will lead zygotes with standard complements of genes, given environments standard in other respects, to develop variation with respect to the anterior and posterior structures. Moreover, we can speak of some environments as "stunting" the growth of plants of particular taxa, meaning that plants with standard complements of genes, grown in those environments, will be shorter than those grown in different environments. Far from being a *reductio* of the interactionist view, this point simply testifies to the democracy principle introduced above. Interactionists want to allow for various ways of analyzing the complex processes of development, *one* among which is the identification of norms of reaction for genotypes, or the discovery of genes "for" traits. (See Figure 20.3.)

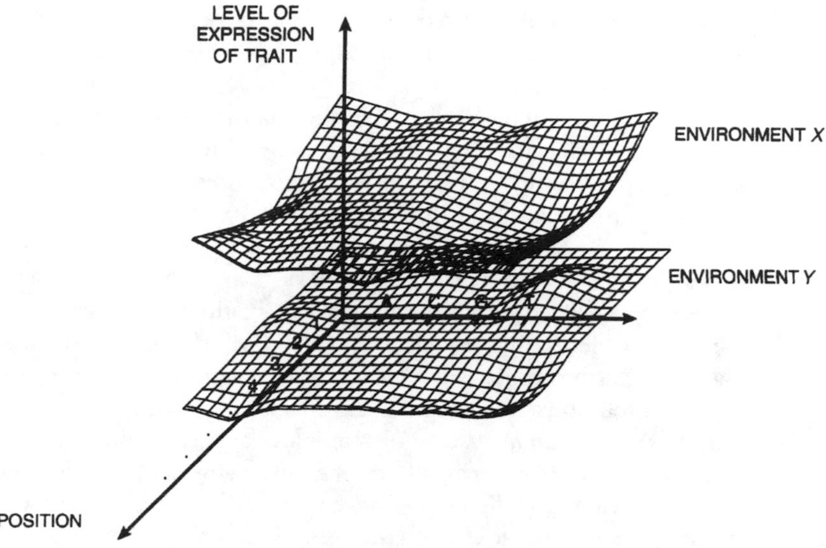

Figure 20.3. Graphical representation of a different style of causal analysis. In the plane of the two horizontal axes, we code genotypes at the locus of interest by specifying the nucleotide at each position. For a fixed environment, the variation of the level of expression of the trait, as the genotype varies, is represented by a surface in the space. (This can be thought of as a dual to the notion of norm of reaction.) For the example shown, the level of the trait for environment *X* is always greater than the level for environment *Y*.

There is a standard temptation to think that all scientific disputes can be readily resolved into differences of principle. Finding that people who advance genetic determinist claims assent to interactionism, critics of genetic determinism want to find some substantive thesis that separates the two camps, and this accounts, I believe, for the repudiation of interactionism. At bottom, however, this dispute, like other significant debates in contemporary biology, is not quite like this.[7] Instead of thinking of two groups of biologists who differ on general principles, we should view biological practice as supplying a toolkit that different people draw from in different ways. Faced with the complexities of ontogeny, biologists have some techniques of causal analysis – of the many forms sanctioned by the democracy principle. For reasons that will be probed shortly, the model of causal analysis that looks at the effects of a single genotype across varying environments is attractive when people are trying to fathom the causes of human behavior, but working out rigorous conclusions about the pertinent norms of reaction proves very difficult, and it is easy to leap to conclusions. Many of Lewontin's most pointed critiques expose the ease with which scholars have leaped to conclusions.

One moral we might draw is that we have a defective instrument, but that, I have been urging, is incorrect. There is nothing the matter with the type of model that has been applied. Rather, the trouble lies in the difficulty of the task and the tendency for the impetuous to bungle. Of course, we might do better if we had different tools – so maybe, after all, there is a case for moving beyond interactionism (not now dismissed as false or incoherent doctrine, but as a source of models too primitive for the important tasks of fathoming human ontogeny) toward "dialectical biology" or "developmental systems theory."

A different set of models for analyzing human development would be welcome, especially if they could be used to achieve insights into the causes of complex capacities and disabilities. Unfortunately, neither Lewontin's "dialectical biology" nor the "developmental systems theory" pioneered by Oyama offer anything that aspiring researchers can put to work.[8] If we want to understand why people become addicted or resist addiction, have the sexual orientations they do, give way to violence or live peacefully (and I will consider, shortly, why we might want insight into these issues), then both versions of the transinteractionist approach to nature and nurture leave us helpless. In effect, they are primarily critiques of the past misuses of old tools and at best blueprints for new tools that we might develop. When problems of analyzing human behavior seem socially urgent, and when investigators believe that new advances in molecular genetics have given new scope to the old models, pleas for "dialectical biology" or "developmental systems theory" are likely to fall on deaf ears.

There is a profound irony here. Nobody has been more sensitive than Lewontin to the social pressures that shape biological research – especially in attempts to evaluate the contributions of nature and nurture. Oyama, too, clearly recognizes these pressures. Unless there are cogent reasons for thinking that past methods of analysis are fundamentally flawed (and I have argued that there are not) rather than simply misapplied in the episodes that Lewontin and Oyama view (rightly) as politically mischievous, then the social pressures to find answers will make fledgling ventures in transinteractionism seem vague and underdeveloped rivals to well-articulated techniques that promise resolution of important questions. Furthermore, the critics of conclusions about the important effects of genotype on phenotype will be seen as taking refuge in nebulous appeals for a new general view of the causation of behavior and as driven to this predicament solely by their sense of outrage at the determinist claims.

Contemporary human genetics, including human behavior genetics, is full of promises largely because of the possibilities of using sequencing techniques to identify shared alleles (combinations of alleles) in different people.[9] Instead of the dubious passages from heritability to conclusions about causation, genetic research can hope to discover norms of reaction more directly by finding large numbers of individuals who share a genotype and tracking the variation in phenotype across environment.[10] Of course, our pervasive ignorance of the causally relevant features of the extraorganismic environment, to which I alluded earlier, should lead us to be tentative in evaluating the results, for we may well be overlooking some crucial environmental variable. Yet this is precisely the point on which the critique of genetic determinism should focus, and it would be unfortunate, perhaps even tragic, if we were to overlook it because the only way of opposing determinist theses was seen as the acceptance of some underdeveloped transinteractionist biology.

The confident behavior geneticist believes that new molecular techniques will enhance our understanding of socially important facets of human behavior. Sometimes the motivation for applying those techniques is impeccable. Researchers into addiction or alcoholism want to understand the causal pathways so that they can prevent human misery: in these areas, many investigations are continuous with attempts to fathom mechanisms behind diseases.[11] They begin with genetic causes not because they are convinced that these are the most important (that the norms of reaction for certain "addictive" genotypes are virtually flat) but because they want to unravel the neurochemistry, and they see the investigation of genotypes as a thread that will lead them into the tangle. For they know how to sequence DNA, and, by finding allelic sequences that correlate with addiction, they may be able to see how abnormal proteins make a difference to certain reactions in the brain and thus understand the molecular details of the interactions between organism and environment that go differently in addicts and in others. There is no question of "privileging" the genes in this kind of inquiry but rather a pragmatic criterion for using a particular type of model and a readily comprehensible, even admirable, medical motivation.[12]

At its best, research in behavior genetics is driven by a morally defensible motivation (that of alleviating human suffering). The investigator tries to understand the plight of the unfortunate by beginning with particular alleles and tracing how the associated phenotypes vary across environments because this is a readily applicable strategy. Yet the goal is to move from singling out certain loci as playing a causal role to identifying differences in the chemical reactions that occur in the formation of healthy and unhealthy phenotypes and from there to discovering what kinds of contributions the environment makes. For, at the end of the day, the goal is to bring relief by adjusting the input from the environment.

However, the reasons for entering on a program of genetic research are not always so easily defensible. Consider the much-disputed example of the genetics of violent behavior. There are good reasons to suspect that the environmental factors causally relevant to eruptions of violence are complex and varied, that there are fine-grained differences in environments that can have large effects, and that, in consequence, our attempts to construct the norms of reaction for "violence" genotypes will be highly fallible. Further, unless we are profoundly deceived, there are some readily identifiable features of the physical and social environment that have major impact: rates of crime are much higher in decaying inner cities, but I doubt that there is a "violence" allele that has the pleiotropic effect of sending its bearers into grim urban environments. Thus, there is an obvious form of causal analysis that could harness the techniques of molecular genetics and that would be sanctioned by the democracy principle enunciated above. Perhaps students of the causes of violent behavior should show how immersion in hostile environments generates greater levels of violence when genotypes vary, compared with sequestration in the leafy suburbs (see Figure 20.4). These students could expect to show something important

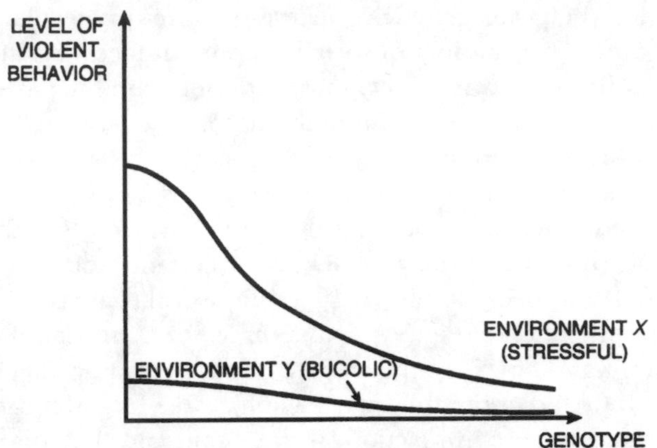

Figure 20.4. Representation of a pronounced environmental effect on tendencies to violence. The two axes for representing genotypes of Figure 20.3 have been condensed into one (surely more plausible than representing environmental variation on a single axis!), and the graphs show the variation in levels of violence for fixed environments as the genotype varies. The representation is purely hypothetical, but it is worth noting that the pronounced role of the environment is compatible with discoveries of "violence alleles"; individuals with genotypes near the origin who encounter stressful environments show much greater levels of violence than others – even those who share similar environments. This does not detract from the obvious fact that there is a very marked effect of environment.

about the causes of violence and to support their conclusions with greater rigor than the hunters of "violence alleles." Yet, there has not been any notable impetus to do the work.

And, of course, we know why. In a society that consistently and callously turns its back on programs that might aid the unfortunate and that sees taxation as a form of robbery rather than a necessary means to social cooperation, the investigation I have outlined has no obvious point. (It might, after all, lead to campaigns for expensive new social programs.) Better, then, to take a different tack, to find out who the people are who are likely to become violent and do something about them in advance. Thus, a politically palatable solution would be to discover genotypes whose norms of reaction show a high propensity for violent behavior virtually invariant across environment. Perhaps there are a few such rare genotypes (the possibility should not be excluded), but the overwhelming likelihood is that we will mistakenly come to believe that they are far more frequent than they are (because of our massive ignorance about how to partition environments) and that these conclusions will reinforce the prevailing sense that social solutions are hopeless.[13]

In fact, the motivations for the study are doubly illicit because they are blind both to the serious dangers of reaching erroneous conclusions and, when articulated, the practical policies are morally disreputable. What precisely is the "something" that is to be done to those who bear the "violent" genotypes? Are they to be branded as criminals, labeled from childhood up, even before they have done anything? Should they be forcibly restrained or treated with tranquilizing chemicals? It is precisely because the motivations for the investigation of the genetics of criminality are economic – after all, we could spend money and invest in jobs for inner city youths, clean up their environments, and make hope possible – that we know in advance that the solution has to be cheap. Hence, we cannot anticipate that great moral niceties (always expensive) are likely to bulk large in the application of "discoveries" about "violence genotypes." Add the difficulty of discovering such genotypes (or, at least, common genotypes), and the potential for injustice is obvious.

Thus, there are two major questions that we ought to ask about proposals to unearth genes "for" complex human traits (including dispositions to forms of behavior that prove either personally or socially disruptive). First, is the investigation informed by the interactionist's commitment to explore the impact of some factors, while others vary, in a way that recognizes our ignorance about environmental causes and that pragmatically deploys the genetic techniques to remedy that ignorance? Second, does the information to be acquired lead to a social policy that is both applicable and morally defensible? As my pair of examples indicates, the answers will be quite different in different instances, and there is no shortcut for considering cases individually.[14]

Some scientists bridle at the thought that my second question should ever figure in the evaluation of a program of scientific research, insisting that the business of science is to uncover the truth, however unpalatable, and that inquiry cannot be subordinated to moral critique. Lewontin has often been criticized for introducing extraneous "political" considerations into discussions of biological investigations,[15] but, in my judgment, his recognition of the wider framework in which science is done is profoundly correct. Researchers cannot hide from themselves the fact that their findings will be applied, often by people who do not grasp the nuances of their positions, nor can they take refuge in the division of labor proposed by Tom Lehrer's brilliant song:

"When the rockets go up, who cares where they come down?
That's not my department," says Werner von Braun.

Many workers in contemporary human genetics, including the genetics of behavioral traits, are convinced that their inquiries will promote human well-being, although critical discussion of the ways in which genetical information can affect people's lives may sometimes undermine their confidence. Unless we have a scientifically informed and ethically sophisticated public discourse about possible programs of genetic research, we are likely either to lose important benefits or, more likely, by accepting the most extravagant promises at face value, mix in significant social harms with the improvements we seek.

Because he sees the latter possibility so clearly, Lewontin has come to advocate a "dialectical biology" that will move beyond interactionism. I have tried to argue that the critiques of interactionism are flawed, that they do not respond to the genuine problems of using biology to promote human good, and that there is no substitute for a detailed examination of the merits of individual cases. It is appropriate to close by noting that, in carrying out the much-needed piecemeal critique, there is no better paradigm than the writings of Richard Lewontin.

20.1 Acknowledgments

I am extremely grateful to Peter Godfrey-Smith for valuable discussion, to Paul Griffiths and Richard Lewontin for illuminating correspondence, and to the editors for their suggestions for improvement. I have sometimes followed my readers' advice but surely not as resolutely or as frequently as they would have wished. E-mail exchanges with Lewontin have convinced me that the position ascribed to him here is not always his own, but he and I agree that discussions of genetic determinism and of the notion of a norm of reaction have been marked by important confusion. So, while I must apologize for my misreading, I hope that this essay clarifies some of the underlying issues.

Notes

1. For prominent examples, see Lewontin (1974; reprinted as Chapter 4 of Levins and Lewontin 1985), Lewontin, Rose, and Kamin (1984, especially Chapters 5 and 9), and Lewontin (1991).
2. The importance of this point is clear from the development studies in *Drosophila* pioneered by Christiane Nusslein-Volhard and her coworkers (see Lawrence 1992); the general point has been made very forcefully by Evelyn Fox Keller in her recent Tanner Lecture.
3. There are a few biological instances in which we might be entitled to some confidence that we have identified important environmental determinants – in studies of the growth of corn plants or of the development of particular structures and behaviors in fruitflies. But it is easy to be misled by these simple cases, concluding that we have a more general ability to map the environmental axis in representing norms of reaction. (Of course, with respect to many of the traits in which we are interested, it is probably grossly inaccurate to think in terms of a single environmental dimension: to distinguish environments adequately would require coding them by some vector in a space of large dimension.)
4. "Dialectical biology," "developmental systems theory," or both might come to do this, providing better ways of abstracting from the mix of causal factors. But until they do so, we are stuck with *different* schemes of abstraction. Our predicament is discussed in more detail below.
5. Paul Griffiths has acknowledged, in correspondence, that the addition of the "coding" idea goes beyond what Sterelny and I actually said. Conversations with Peter Godfrey-Smith have, however, convinced me that there is a rationale for the Griffiths–Gray attribution, for the "standard view" current in *contemporary* biology does seem to make the addition. Hence, according to Godfrey-Smith, the reconstruction offered here does not defend the "standard view" but presents a position intermediate between that view and the approach of the critics – a position to which (perhaps) the "standard view" ought to modulate. I leave it to readers to judge if this is right.

6. As a reader of an earlier version noted, there is a fourth possibility, namely that "standard" is defined by an arbitrary convention. I will not consider this possibility here.
7. In particular, three of the most heated disputes in recent theoretical biology seem to swirl around issues of which kinds of models are to be taken as defaults, which to be seen as exceptional. I have in mind the controversy about punctuated equilibrium, the adaptationist controversy, and the debate over human sociobiology. Stephen Jay Gould has been very clear in seeing that the issue about punctuated equilibirum is a matter of frequency, that it concerns what paleontologists take to be the "standard" situation. Similarly, in Gould and Lewontin (1979), Gould and Lewontin opposed a tendency to think that any trait that strikes the evolutionary biologist's eye should be assumed to be an adaptation until reasons are supplied to the contrary. I argued for a similar approach to sociobiology, urging that problems come not because of the flaws of the theoretical tools for building models but because of the ways in which these tools are used (Kitcher 1985, pp. 117–21).
8. Here I give little detail. But it may help to imagine a sympathetic biologist reading Lewontin, Oyama, or both. The obvious question that would arise for this biologist would be, How do I put these ideas to work in concrete situations?, and to that, neither Lewontin nor Oyama supplies much by way of answer. This does not mean that "dialectical biology" and "developmental systems theory" should be abandoned but that the kinds of work needed to make them viable pieces of biological theory are specific models for tackling interesting problems (rather than philosophical diagnoses of previous errors). It would be *very* interesting, for example, to see a developmental systems analysis of early development in *Drosophila* or a dialectical biology substitute for some part of population genetics.
9. For investigations of traits in which a single locus (or a small number of loci) is assumed to have major effects, this may make monozygotic twins redundant. Perhaps twin studies will only continue to be useful when the trait is assumed to be highly polygenic.
10. Despite the arrival of more direct methods that make the announcement of heritability measures irrelevant, some behavioral geneticists continue to include such measures in reporting their investigations. This seems to be an unfortunate tic from which they cannot free themselves. At a behavioral genetics workshop, organized by the National Academy of Sciences, David Goldman gave a presentation in which he accompanied interesting findings by using molecular approaches with heritability estimates. When asked by Marcus Feldman why he had included the heritability measures, and whether he would be inclined to use heritability estimates as a prelude to molecular investigations of, say, the genetics of religious belief, Goldman replied that he would take this to be suggestive. Feldman clearly viewed this as a *reductio* of the continued deployment of heritability measures. In my opinion, he was quite right to do so.
11. It is certainly wrong to conjure up a sterotypical behavior geneticist – socially insensitive and methodologically crass. Irving Gottesman (to single out one example) has been formulating standards for careful analysis for decades, and his own work has been sensitive to the social uses to which behavior genetics might be put.
12. Of course, Lewontin has expressed skepticism about the role that molecular genetics can be expected to play in future therapies (see 1991, pp. 67–8). I agree

with his assessment that there is no *automatic* translation of molecular insights into practical treatments but resist the idea that molecular genetics is either a universal panacea or else useless. We can expect varying degrees of therapeutic success in different cases, and it is impossible to predict in advance where molecular insights will be fruitful (see Kitcher 1996a, pp. 105–12).

13. One recurrent problem is the propensity for lumping, supposing that the causes of all cases of conditions that share a name are the same. Thus, discovering a "violence" allele in a family with a history of antisocial behavior not only induces the conviction that any environmental factors can be ignored but also inclines people to think that there are lots of other "violence" alleles waiting to be discovered.
14. The case of sexual orientation seems intermediate in character because the human consequences of the research are so uncertain. LeVay (1996) gives a sensitive response to the moral questions surrounding the search for biological causes of sexual orientation, although, as I argue Kitcher (1996b), it is not clear that his defense of research in this area is successful.
15. See the comments by David Botstein quoted in (Burr 1996, p. 274).

REFERENCES

Burr, C. (1996). *A Separate Creation.* New York: Hyperion.

Gould, S. J., and Lewontin, R. C. (1979). The spandrels of San Marco and the panglossian paradigm: A critique of the adaptationist programme. *Proceedings of the Royal Society of London* B 205:581–98.

Griffiths, P. E., and Gray, R. D. (1994). Developmental systems and evolutionary explanation. *Journal of Philosophy* 91:277–304.

Kitcher, P. S. (1985). *Vaulting Ambition: Sociobiology and the Quest for Human Nature.* Cambridge, MA: MIT Press.

Kitcher, P. S. (1996a). *The Lives to Come: The Genetic Revolution and Human Possibilities.* New York: Simon & Schuster.

Kitcher, P. S. (1996b). Review of (LeVay 1996), *The Sciences.* November–December 1996.

Lawrence, P. (1992). *The Making of a Fly.* Oxford, UK: Blackwell.

LeVay, S. (1996). *Queer Science.* Cambridge, MA: MIT Press.

Levins, R., and Lewontin, R. C. (1985). *The Dialectical Biologist.* Cambridge, MA: Harvard University Press.

Lewontin, R. C. (1991). *Biology as Ideology.* New York: Harper.

Lewontin, R. C., Rose, S. E., and Kamin, L. (1984). *Not in Our Genes.* New York: Pantheon.

Oyama, S. (1985). *The Ontogeny of Information.* Cambridge, UK: Cambridge University Press.

Sterelny, K., and Kitcher, P. S. (1988). The return of the gene. *Journal of Philosophy* 85:339–61.

CHAPTER TWENTY-ONE

The Poverty of Reductionism*

STEVEN ROSE

21.1 The Rise of Neurogenetic Determinism

From its Baconian inception, modern science has been about knowledge and power – above all, the power to control and dominate nature, including human nature. Nowhere perhaps has this Faustian pact been made so explicit as in the program that has shaped molecular biology since its origins. Its very name was invented as long ago as the 1930s by Warren Weaver of the Rockefeller Foundation as part of a coherent policy by one of the major fundgivers in the field. That policy, drawing on prevalent eugenic considerations of race betterment, was intended specifically to achieve a "science of man" that was also a science of social control. To this end, Rockefeller concentrated its resources on the sciences of psychobiology and heredity in the firm belief, fostered by Weaver, that such control would come about through the study of the "ultimate littleness of things" (Kay 1993).

This vision has been immensely productive in both scientific knowledge and technologies, the products of the Baconian alliance. Today, we can see its lineage in the mushrooming biotechnology companies in the United States, Japan, and Europe and in the Human Genome Project and the Decade of the Brain. But to naturalize it as if it is the only way of understand the living world and to ignore its explicit goals of social control and implicit eugenic agenda are to fail to grasp the directions in which it is leading us, as if modern science has simply transcended the ideologies that shaped its past. Today's molecular biology is, however unreflectingly, heir to this past and cannot simply shrug it off. Thus, the dramatic advances in knowledge of the past decades have been accompanied by ever more strident claims that the new genetics, molecular biology, and neuroscience are first about to explain, and in due course, to modify the human condition and in doing so usher in a new era of what some years ago one of the enthusiasts for the new biology called a "psychocivilized society."

*This text is an edited and shortened version of a chapter in my book *Lifelines* published by Allen Lane, The Penguin Press in 1997, and Oxford University Press (New York) in 1998.

There should be tattooed on the forehead of every young person a symbol showing possession of the sickle-cell gene or whatever other similar gene . . . It is my opinion that legislation along this line, compulsory testing for defective genes before marriage, and some form of semi-public display of this possession, should be adopted.

The date of this quasi-Nazi proposal? Not the 1930s, but 1968. And its author? Hero of antiwar and alternative health movements, twice Nobel prizewinner, once for chemistry and once for peace, Linus Pauling (Kay 1993, p. 276).

Week after week newspapers report what are seen as major breakthroughs in biological and medical understanding. Here is a random sampling: "Stress, Anxiety, Depression: The New Science of Evolutionary Psychology Finds the Roots of Modern Maladies in the Genes" was the cover story for *Time* magazine for August 28, 1995. "Gene Hunters Pursue Elusive and Complex Traits of Mind," claimed the *New York Times* on October 31, 1995. "Studies Link One Gene to a Specific Personality" offered the *Talahassee Democrat* in January 1996. In July 1993, the London *Daily Mail* offered an "abortion hope after 'gay genes' finding." The London *Independent* described "How Genes Shape the Mind" (November 1, 1995). More circumspectly, the London *Guardian* on February 1, 1996, described Robert Plomin's (newly appointed from the United States to a professorship at London's Institute of Psychiatry) hunt for "intelligence genes" as "the search for the clever stuff" and listed the "losers in life's genetic lottery." Plomin himself, in an article for the German *Bild der Wissenschaft* for September 1996 suggested that childhood accidents are 50% genetic in cause and divorce in midlife 30%.

The emerging synthesis of genetics and the brain sciences – *neurogenetics* – and its philosophical and political offspring, which we may call *neurogenetic determinsm*, offers the prospect of identifying, ascribing causal power to, and eventually modifying genes affecting brain and behavior. Neurogenetics claims to be able to answer the question of where, in a world full of individual pain and social disorder, we should look not merely to explain but even more potently to change our condition. Although only the most extreme reductionist would claim that we should look for the origins of the Bosnian war in deficiencies in neurotransmitter mechanisms in Dr. Karadzic's brain, and its cure by the mass prescription of Prozac, many of the arguments offered by neurogenetic determinism are not far removed from such extremes. Give the social its due, the claim runs, but in the last analysis the determinants are surely biological.

Urban violence, addiction, and psychic distress, to which solutions are required, are desperately serious features of life in Europe and the United States today. Thus, the argument against hunting for neurogenetic explanations is not that it is immoral or unethical to do so; it is simply that despite the seductive power of reductionism, neurogenetics is the wrong level at which to find answers to many of the problems that confront us. It then becomes at best an inappropriate use of scarce human and financial resources and at worst a substitute for social action. I need to reiterate this strongly if only because I find it so persistently, even perversely, misunderstood. I am distressed by the arrogance with which some biologists claim for their – our – discipline explanatory and interventive powers it certainly does not possess and so cavalierly dismiss the counterevidence.

21.2 Reductionism as Ideology

This is not a new debate; it has recurred in each generation at least since Darwin's day and most recently in the form of the polemical disputes over the explanatory powers of sociobiology in the 1970s and 1980s. It is not my intention to go over that old ground again (Rose, Lewontin, and Kamin 1984). What is new today, however, is the way in which the mystique of the new genetics is seen as strengthening the reductionist argument. At its simplest, neurogenetic determinism argues a directly causal relationship between gene and behavior. In a social and political environment conducive to such claims and that has largely despaired of finding social solutions to social problems (although no one to my knowledge is researching the genetic "causes" of homophobia, racism, or financial fraud), apparently scientific assertions become magnified by press and politicians and researchers may argue that their more modest claims are traduced beyond their intentions. Yet this is hard to credit when so much effort is put by researchers themselves into sales talk. The press releases surrounding the publication of LeVay's (1993) book *The Sexual Brain*, which claimed, on the basis of his studies of the postmortem brains of several presumed gay men who had died of AIDS, to have located a specific region of the brain that differed in presumed gay from presumed straight men, or Hamer's claim to have identified a "gay gene" (Hamer et al. 1993), were couched in language that left little need for media magnification.

These are scarcely minor concerns. We all want to know where to look to explain our personal successes and failures, our foibles and vices, to say nothing of the chronic crises of the world around us. For such problems we have the choice of invoking either social or personal explanations. If social, we can seek solutions through social action by improving the economy, changing the law, or working to alter the social structures of power and privilege. If personal, we can explore our own individual life history by way of psychotherapy. Or we can invoke the biological and claim that the roots of the problem we confront lie within individual brain structure, biochemistry, or genetics. If the causes of our pleasures and our pains, our virtues and our vices, lie predominantly within the biological realm, then it is to neurogenetics that we should look for explanation and to pharmacology and molecular engineering that we should turn for solutions. This simplification, with its implication that the world is divided into mutually incommensurable realms of causation, in which explanations are *either* social *or* "biological," and with its cheaply seductive dichotomies of nature *or* nurture, genes *or* environment, is fallacious. The phenomena of life are always and inexorably simultaneously about nature and nurture, and of human existence and experience always and inexorably simultaneously biological *and* social. Adequate explanations must involve both. Yet again and again one finds the reductionist claim, unqualified, making the headlines and setting the research agenda.

Neurogenetic determinism, I argue, is based on a faulty reductive sequence whose steps include reification and arbitrary agglomeration, improper quantification, belief in statistical "normality," spurious localization, misplaced causality, dichotomous partitioning between genetic and environmental causes, and confounding metaphor with homology. It is not necessarily the case that any individual step in this sequence is inevitably in error; it is just that each is slippery and the dangers of tumbling very great. The issue at stake here is the question of the appropriate level of organization of matter at which to seek causally effective determinants of the behavior of individuals and societies. The structure of the argument is similar whether the discussion focuses on intelligence, sexuality, or violence, and I will here base my analysis mainly around these themes.

21.3 Reification and Agglomeration

The first step in the process is *reification.* Reification converts a dynamic process into a static phenomenon, a phenotype. Violence is the term used to describe sequences of interactions between persons, or even between a person and his or her nonhuman environment. That is, it is a process. Reification transforms the process into a fixed thing, *aggression*, which can be abstracted from the dynamically interactive system in which it appears and studied in isolation, as it were, in the test-tube. Yet, if the activity described by the term *violence*, or *altruism*, or *sexuality*, can only be expressed in an interaction between individuals, to reify the process and pretend that it is in any sense an isolable character is to lose its meaning.

Arbitrary agglomeration carries reification a step further, lumping together many different reified interactions as if they were all exemplars of the one character. Thus, *aggression* becomes the term used to describe processes as disparate as a man abusing his lover or child, fights between football fans, strikers resisting police, racist attacks on ethnic minorities, and civil and national wars. Agglomeration proceeds by assuming each of these social processes is merely a reified manifestation of some unitary underlying property of the individuals, and thus that identical biological mechanisms are involved in, or even cause, each. This is well illustrated in the article by Brunner et al. (1993) describing a Dutch pedigree including eight men "living in different parts of the country at different times" across three generations who showed an "abnormal behavioral phenotype." The types of behavior included "aggressive outbursts, arson, attempted rape, and exhibitionism." Can such widely differing types of behavior, described so baldly as to isolate them from social context, appropriately be subsumed under the single heading of aggression? It is unlikely that such an assertion, if made in the context of a study of nonhuman animal behavior, would pass muster (I certainly could not get away with reporting a study involving such varied behavior in eight chicks!). Yet Brunner's article was published in one of the world's most prestigious journals with considerable surrounding publicity.

Much attention was devoted to its report that each of these "violent" individuals also carries a mutation in the gene coding for the enzyme monoamine oxidase (MAOA), which, among other functions, is associated with the metabolism of a particular neurotransmitter and is believed to be site of action of several psychotropic drugs. Could this mutation then be the "cause" of the reported violence? Brunner himself subsequently disclaimed the direct link and indeed dissociated himself from the public claims that his group had identified a "gene for aggression," claiming that this was merely a journalistic distortion (Brunner 1996). Yet the article is now widely cited in the research literature, and what Brunner described as "abnormal" now becomes "aggressive behavior." Thus, an article with this title, describing mice lacking the monoamine oxidase A enzyme, appeared in *Science* 2 years after the original Brunner article (Cases et al. 1995). The authors describe the mouse pups as showing "trembling, difficulty in righting, and fearfulness . . . frantic running and falling over . . . (disturbed) sleep . . . propensity to bite the experimenter . . . hunched posture," and so on. Of all these features of disturbed development, the authors chose only to highlight aggression in their article's title and to conclude their account by claiming that these results "support" the idea that the particularly aggressive behavior of the few known human males lacking MAOA . . . is a more direct consequence of MAO deficiency." When I pointed out in a letter to *Science* that what Cases and colleagues headlined as aggression was a minor and scarcely surprising aspect of this grossly disturbed developmental pattern, one of the authors telephoned me explaining that their article had been highlighted this way because it seemed the best way of drawing attention to their results.

More disturbingly, this type of evidence, slight though it may seem, has at once become part of the arsenal of argument employed, for example, by the U.S. Federal Violence Initiative, and aimed at identifying inner-city children regarded as "at risk" of becoming violent in later life as the result of predisposing biochemical or genetic factors. This program, proposed originally by the then director of the U.S. National Institute of Mental Health, Frederick Goodwin, originally ran into a hostile barrage of publicity over its potentially racist overtones with its repeated coded references to "high-impact inner-city" youth, and not long afterwards Goodwin left his directorship. Proposals to hold a meeting to discuss his proposals were several times abandoned. Nonetheless, aspects of the research program have gone ahead in the United States focused largely on Chicago (Breggin and Breggin 1994).

As with each step in the reductionist cascade I am describing, the problem lies not in the fact that as researchers, within the methodology available to us, we need to classify and group together different types of observation as belonging in some way together. These are not inevitably illegitimate steps. Science seems often to proceed by alternately grouping together different phenomena as aspects of the same thing (lumping) and recognizing differences between them (splitting). Lumping is, however, inappropriate as applied to "violence," as in these examples. Grouping arson and exhibitionism in the same category is not likely to make much sense to either a criminologist or a judge and jury in court.

To get around this difficulty, some researchers have recently relabeled these cases so that they no longer appear as examples of "violence" but of a different category of "antisocial behavior." But far from solving the problem, such relabeling only makes it worse. Just as agglomeration lumps disparate activities, so the identical act may be regarded as socially acceptable or unacceptable depending on the circumstances. Bombing a government building if you are a pilot and your nation is at war with those you are bombing is socially praiseworthy; on the other hand, if you are a national of the society whose buildings you bomb you are guilty of the antisocial behavior called terrorism. Contrast the medals given to U.S. pilots during the Gulf War with the criminal charges against the bombers of the federal office in Oklahoma City. Perhaps the clearest-cut example comes from an episode in Northern Ireland in the early 1990s. A British soldier, Lee Clegg, was on duty at an army checkpoint when a stolen car crashed through the roadblock. Clegg lifted his rifle and shot dead one of the occupants of the car, a teenage girl who had been joyriding. He was charged and convicted of murder, perhaps the ultimate in antisocial behavior. The army, supported vociferously by the English tabloid press, was outraged and waged a vigorous and ultimately successful campaign for his release and reinstatement. He was, they argued, doing his duty; the car might after all have held IRA terrorists, not joyriding kids, in which case he might even have been given a medal. So the identical act can be defined either as socially approved or antisocial behavior, depending now not on the act itself but on the perception of those who observe it. How can this conceivably form the basis for a biological, individually based categorization in which we look for unusual genes for neurotransmitter enzymes in Private Clegg's brain to explain what has happened?

21.4 Improper Quantification

Improper quantification argues that reified and agglomerated characters can be given numerical values. If a person is violent, or intelligent, one can ask how violent, how intelligent, by comparison with other people. This assumption that any phenomenon can be measured and scored is reflective of the belief that to mathematize something is in some way to capture and control it. The best-known example is IQ. The first steps involve parallel reifications and agglomerations to those described for violence. "Intelligent behavior," essentially an interactive process between an individual and others, or with the social, living and inanimate world, becomes fixed as a unitary character. Many different examples of such behavior are then all taken to be manifestations of something called, as if finally to freeze dynamics into statics, "crystallized intelligence," and given a special symbol, *g*, originally introduced by the psychologist Charles Spearman in the 1920s (is it only coincidence that this is also the symbol for one of the most hallowed of physical forces, that of gravity?). Tests are then devised to measure this inferred hidden constant. Of all the assumptions built into this process, for the moment I want to consider only one, the extraordinary belief that the multiple aspects of behavior (even reified and agglomerated behavior) that go to compose what we may recognize as intelligence – speed and accuracy of responding to new information, skill at deriving meanings from ambiguous social situations, capacity to innovate in novel environments, and many others as well – can all be reduced to a single number so that the entire human population can be ranked by it just as they might be if we were to line them all up by height.

Of course, to achieve this type of mathematical reduction it is necessary to discount many of these richly interacting human capacities even though to most people they would seem as among the most salient aspects of what is called intelligence. Instead psychometricians retreat into an private world inhabited only by a group of like-minded devotees to the art of counting. Indeed, they find it difficult to relate to other brain and behavioral scientists, who mostly look askance at psychometry's commitment to arbitrary numerology (which means in practice that the only other discipline to which they can relate, and with which psychometry has historically been linked, is a certain subarea of behavioral genetics, indeed, the two, psychometry and behavioral genetics, are the twin offspring of the eugenic movements of the first part of the twentieth century). To see this cavalier rejection of anything other than their own reduction of intelligence at its most arbitrary, one need go no further that the first chapter of Herrnstein and Murray's *The Bell Curve* (1994), which, faced with the voluminous critiques, from many different perspectives, of such reduction of intelligence to a single score, sweeps all opposition to their dictate aside. Intelligence, they insist, is not to be confounded with talent, insight, creativity, capacity to find or solve problems or resolve difficulties any more than it has anything to do with musical, spatial, mathematical or kinesthetic ability, sensitivity, charm or persuasiveness:

There is such a thing as a general factor of cognitive ability on which human beings differ. All standardized tests of academic aptitude or achievement measure this ability to some degree, but IQ tests expressly designed for that purpose measure it most accurately. IQ scores match, to a first degree, whatever it is that people mean when they use the word intelligent or smart in ordinary language. (p. 22)

Thus, intelligence is what intelligence tests measure, and, if other tests, constructed on different principles, fail to conform by providing a measure compatible with this unitary view of *g*, they are simply dismissed as being beneath consideration.

21.5 Statistics and the Norm

Belief in *statistical normality* assumes that in any given population the distribution of such behavioral scores is Gaussian. Again the best known example is IQ, the tests for which were refined and remolded by successive generations of psychometricians until IQ was made to fit (almost) the approved statistical shape.

That is, the tests that did not result in distributing the population according to the curve were rejected, or test items within them modified, until they fitted the curve – a feat achieved between the wars in the various revisions of what became known as the Stanford–Binet test, which was originally developed in the 1920s. Yet, the assumption that the entire population can be distributed along a single dimension is to confuse a statistical manipulation for a biological phenomenon. There is no biological necessity for such a unidimensional distribution nor for one in which the population shows such a convenient spread.

The power of this reified statistic should not be underestimated. It conveniently conflates two different concepts of "normality." The statistical sense of the term does not have a "value" attached to it, for it merely describes a particular shape of curve that has the property that 95% of its area is to be found within two standard deviations of the mean. But in common parlance the term normal is indeed normative. It describes not merely how things are, but how they ought to be; to lie more than two standard deviations from the mean in a Gaussian distribution is to be abnormal, with all that this implies. When Herrnstein and Murray called their book *The Bell Curve*, they played precisely into these multiple meanings of reified normality.

21.6 Spurious Localization

Once processes have been reified into objects and arbitrarily quantified, the reified object ceases to be a property even of the individual but instead becomes that of a part of the person. This accounts for the penchant for speaking of, for example, schizophrenic brains, genes – or even urine – rather than of brains, genes, or urine derived from a person diagnosed as suffering from schizophrenia. Of course, everyone knows that this is a shorthand, but the resonance of "gay brains" or "selfish genes" does more than merely sell books for their scientific authors; it both reflects and endorses the modes of thought and explanation that constitute neurogenetic determinism, for it disarticulates the complex properties of individuals into isolated and localized lumps of biology.

Thus, recent years have seen an unusually polemical debate, more reminiscent of the early days of nineteenth-century phrenology than of modern research, among different neuroanatomists, each claiming to have found *the* brain seat of homosexuality. Two regions in particular have contended for the honor of conveying male same-sex preference: the corpus callosum and the hypothalamus. I do not want here to enter into a detailed analysis of the empirical evidence, which has fortunately been subject to detailed and stringent empirical criticism by Fausto-Sterling (1992). My concern is once more with the structure of the argument deployed by those seeking to locate homosexuality in a bit of the brain or an aberrant gene, for it shows all the features I have already described for violence and intelligence, and more besides. The expression of same-sex preference is scarcely a stable category either within an individual's lifetime or historically – indeed, that it might be used as a term to describe an individual rather than part of a continuum of sexual activities and preferences available to all, seems to have been a relatively modern development. What the reductionist argument does is to remove the description of sexual activity or preference as part of a relationship between two individuals, reifies it, and turns it into the phenotypic "character" resulting from one or more abnormal, gay genes. As always, it deprives the term of personal, social, or historical meaning, as if to engage in same-sex erotic activity or even to express a same-sex preferred orientation meant the same in Plato's Greece, Victorian England, or San Francisco in the 1960s.

Just as homosexuality is "located" to the hypothalamus, so aggression had been "located" in another set of structures within the brain, the limbic system, and in particular one part thereof, the amygdala. In the 1970s, two psychosurgeons proposed to treat inner-city violence by amygdalectomizing militant ringleaders from America's inner-city ghettos. I used to believe that things were a little more sophisticated today, but the sight, in a television documentary made in 1995, of California-based psychologist Adrian Raine standing in front of two PET scans of human brains and explaining that one, the brain of "a murderer," showed "low activity" in the frontal cortex by comparison with the other, the "normal," led me gloomily to conclude that the days of Lombroso were not that long past (Moir and Jessel 1995). Raine was theorizing that the function of the "more evolved" cortex in humans was to control the "primeval" limbic system and therefore that, when frontal activity is low, the amygdala and other limbic systems are out of control and, left to their own devices, will drive their owners to violence. It is not clear whether a similar finding would apply to scans of the brains of the war heroes who have been responsible for some of the greatest massacres of modern times, Stormin' Norman and the killings of fleeing Iraqi troops on the Basra Road in 1991 or Ratko Mladic and the mass graves of the Muslim men of Srebenica in 1995. What is certain is that a view of the brain as composed of "less-" and "more-" evolved structures is yet another of those evolutionary fantasies. The great mass of the cerebral cortex, in humans and other mammals, shows evolutionary descent from the olfactory bulb still there in present-day reptiles. But that does not mean that we think by smelling.

Raine's claims return to an older tradition in "localizing" reified properties. More frequently these days, that localization takes the form, not of a brain structure, but of an abnormality in some brain chemical – a neurotransmitter or an enzyme – or the gene responsible for its production. The particular substance in question tends to fluctuate with the fashionable molecule of the moment. Thus, a few years ago much attention was paid to one particular neurotransmitter, GABA, as being particularly associated with aggressive behavior. Today, aggression is more likely to be explained as being "caused" by a disorder of serotonin metabolism (specifically, the reuptake of the secreted neurotransmitter). Abnormalities of serotonin reuptake mechanisms are blamed for everything from depression and suicide to "impulsive behavior" and violence, and the universal panacea is Prozac, one of a family of drugs that selectively inhibits serotonin reuptake (Kramer 1994).

21.7 Misplaced Causation

It is at this point that neurogenetic determinism introduces its misplaced sense of causality. It is of course probable – indeed in some contexts certain – that during aggressive encounters people show dramatic changes in, for instance, the levels of circulating steroid hormones and adrenalin in their bloodstream and the release of neurotransmitters in their brains, all of which can be affected by drug treatments. People whose life history includes many such encounters are likely to show lasting differences in a variety of brain and body markers. But to describe such changes as if they were the *causes* of particular behaviors is to mistake correlation or even consequence for cause. When you have a cold, your nose runs. Yet despite the invariable correlation of the two, it would be a mistake to believe that the cold was caused by the nasal mucus; the chain of cause–effect runs in the reverse direction. Nor, even though Prozac both inhibits serotonin reuptake mechanisms and may diminish the likelihood of your committing suicide or murder, does this mean that the level of serotonin release in your brain is the cause of your desire to kill yourself or someone else. After all, when one has a toothache, the pain can be alleviated by taking aspirin, but it does not follow that the cause of the toothache is too little aspirin in the brain. This issue has dogged interpretation of the biochemical and brain correlates of psychiatric disorders for decades, yet it still continues. Thus, similar claims that an abnormality in the dopamine receptor could underlie susceptibility to substance abuse were countered by the argument that the abnormality was the result, not the cause, of drug taking (Holden 1994, Chipkin 1994). Such beliefs are, however an almost inevitable consequence of the processes of reification and agglomeration, for if there is one single thing called, for instance, "alcoholism," then it becomes appropriate to seek a single causative agent.

21.8 Dichotomous Partitioning

If aggression, or antisocial behavior, or homosexuality are "caused" by some "abnormality" in brain structure, biochemistry, or hormonal imbalance, what "causes" these in their turn? They could of course be the consequences of some feature in the environment (and if so are usually argued to result from some aspect of early rearing or poor diet, as when infant "temperament" in early months is claimed to predict later poor performance at school or adult violence). More often, though, attention turns to those well-known first causes, the genes, and the apparatus of heritability studies is wheeled out. For even if there is difficulty in regarding such socially defined attributes as simple phenotypes, if they correlate with a "real" measure such as the level of an enzyme or neurotransmitter, then the heritability of this can surely be determined. A good example of this mode of thinking is the claim that IQ test scores correlate with a more neurophysiological measure referred to as "inspection time." That a heritability measure is rarely applicable to the human situation, is widely misunderstood and in most cases meaningless, has not prevented behavioral geneticists and psychometricians from endeavoring to apply it nor deprived it of its ideological resonance, as when it is reiterated that the heritability of intelligence – or rather of IQ test score – is as high as 80%. Political orientation, neuroticism, and attitudes to military drill, royalty, censorship, and divorce, among many others, are all supposed to show relatively high heritability. Indeed, it becomes hard to find any human attribute or belief, even the most seemingly trivial, to which the heritability statistics fail to yield significance. Newly sophisticated statistical techniques, such as that known as quantitative trait locus analysis (Plomin, Owen, and McGuffin 1994), are employed and purport to show that even those conditions for which major genetic causation cannot be shown (Alzheimer's disease is a good example, where only about 5% of the cases are clearly associated with a specific genetic dysfunction) are in fact the result of the small additive effects of many genes. And although no one claims that heritability equals destiny nor that the statement of probability provides information about any specific individual rather than measures variance within a population, the whole tenor of the approach is to transfer the burden of explanation and, if appropriate, of intervention, from the social or even personal level to that of pharmacological or genetic control.

21.9 Confounding Metaphor with Homology

If first causes are genetic, the adaptationist paradigm within ultra-Darwinism must seek to account for how they may have evolved. It then becomes appropriate to seek for equivalents of the human behavior under consideration in the nonhuman animal world; that is, to find an animal model in which the behavior can be more readily controlled, manipulated, and quantified. Place an unfamiliar mouse into a cage occupied by a rat, and the rat is likely, eventually, to kill the mouse. The time taken for the rat to perform this act is taken as a surrogate for the rat's aggression; some rats will kill quickly and others slowly or even not at all. The rat that kills in 30 seconds is on this scale twice as aggressive as the rat that takes a minute. Such a measure, dignified as *muricidal behavior*, serves as a quantitative index for the study of aggression and ignores the many other aspects of the rat–mouse interaction such as the dimensions, shape, and degree of familiarity of the cage environment to the participants in the muricidal interaction, whether there are opportunities for retreat or escape, and the prior history of interactions between the pair. And it is not that these are merely speculative variables, for many of them have been studied in detail by ethologists and shown to affect the nature of the relationships between the animals profoundly. But the reductive procedure goes further, for it then assumed that, just as time to kill becomes a surrogate for a measure of aggression, so this behavior in the rat is transmogrified into an analogue of the aggression shown by drive-by gangs shooting up a district in Los Angeles, as in the concluding sentences to the Cases et al. (1995) article cited earlier. That is, if one can find physiological or biochemical mechanisms (brain regions, neurotransmitters, or genes) associated with the so-called "aggression" in mouse-killing rats, then there should be equivalent or identical brain regions, neurotransmitters, or genes involved in human alcoholism (Crabbe, Bellknap, and Buck 1994) or "aggression" (Johnson 1996). This type of evolutionary fantasy at best confounds a metaphor or analogue with a homologue. At worst it simply makes a bad pun on different meanings of the word "aggression." But it has become the vital, ultimate link in the chainmail armor of reductive ideology.

21.10 Consequences of Reductionist Fallacies

From the birth of modern science, methodological reductionism has proved a powerful and effective lever with which to move the world. We owe to it many of the most penetrating insights into mechanisms in every field of science, including biology. But especially in biology, complexity and dynamics, open rather than closed systems, are norms rather than exceptions, and the methodology of reductionism, however powerful, has difficulties in dealing with complexity – indeed it may be positively misleading.

Reductionism as an ideology, insisting on trying to replace higher-level descriptions with lower-level properties, hinders biologists from thinking adequately about the phenomena we wish to understand. Two consequences at least lie in the social and political domain rather than the scientific. The ideology serves to relocate social problems to the individual, thus "blaming the victim" rather than exploring the societal roots and determinants of the phenomena that concern us. Violence in modern society is no longer to be explained in terms of inner-city squalor, unemployment, extremes of wealth and poverty and the loss of the hope that by collective effort we might create a better society. Rather, it is a problem resulting from the presence of individual violent persons, themselves violent as a result of disorders in their biochemical or genetic constitution.

But in a strange way, the blame is simultaneously placed upon them and lifted from them. Where once a murderer might have been regarded as morally culpable, or the cause of his (as it almost invariably is) violence sought in an unhappy or abused childhood, now it is argued to be due to lower "frontal activity" or chemical imbalances in his brain, themselves the consequence of faulty genes or birthing difficulties. Thus, in a much-discussed U.S. court case, the lawyer for Stephen Mobley, sentenced to death for the violent slaughter of the manager of a pizza parlor, sought permission to mount a genetic defense against the sentence, claiming that Mobley might have been endowed with the same mutation in his monoamine oxidase gene that Brunner reported, in which case, Mobley would not be "responsible" for the murder he committed. "It was not I, it was my genes." Similarly, if homosexuality is "in the genes," a gay man should not, in a homophobic society, be regarded as morally culpable, still less guilty of criminal behavior, for following his genetic dictates. It is not surprising, therefore, that certain sections of the gay and lesbian community have actively welcomed the determinist claims of LeVay and Hamer or that both the Christian fundamentalist right and the judiciary are worried about just how far the determinist argument can be stretched.

The second immediate social consequence of reductionist ideology is that attention and funding are diverted from the social to the molecular. If the streets of Russia are full of vodka-soaked drunks and rates of alcoholism are catastrophically high among native Americans or Australian aborigines, the ideology demands funding research into the genetics and biochemistry of alcoholism. And it becomes more productive to study the roots of violent "temperament" in babies and young children than to legislate to remove handguns from society. The point is that, as the whole of my argument up till now has stressed, for any phenomenon in the living world in general and the human social world in particular, one can offer multiple forms of explanation, of which the reductionist one, properly formulated, is a legitimate one. But for any such phenomenon there are also *determining levels* of explanation – those that most clearly account for the specificity of the phenomenon and also point to potential sites of intervention into it.

To come back to the violence example again: crimes of violence are more frequently carried out by men than by women (although this picture is changing in both the United States and the United Kingdom). One may argue that this says something about the Y chromosome, but the overwhelming majority of men are not violent criminals, and thus the policy implications of research seeking to explore the Y chromosome in the context of crime – short of selective abortion of all male fetuses – are negligible. Violent crime is much higher in the United States than in Europe – higher, for instance, than in Britain and much higher than in Sweden. Could this be accounted for by some unique feature of the American genotype? Well, possibly, but this is pretty unlikely given that much of the American population originated by migration from Europe. But also, the rates of violent crime change dramatically over quite short time periods. For instance, the death rate from homicide among young U.S. males increased by 54% between 1985 and 1994. No biologically based explanation can account for this increase, and thus it becomes more helpful to ask instead, What has changed in the United States over this period that might account for such an increase? What is different about the organization of U.S. society from that of Europe? Could one important difference between the United States and Europe be the estimated 280 million handguns in personal possession in the United States? Unlike reductionist ones, such hypotheses may give clues for meaningful intervention.

So although in an ontologically unitary universe, *of course* it is axiomatic that there is something different about the biochemical and physiological states of someone who is in the process of committing a murder from those states in the same person when he or she is in a prison cell, and probably between the murdering individual and someone who in similar circumstances does not murder, this difference cannot be relevant to answering questions about the causes and responses to social violence. Nor, therefore, can it represent the appropriate level at which to intervene if we wish to reduce the amount of violence on the streets. A program devoted to the detection of which levels of serotonin may predispose a person to an increased statistical possibility of engaging in one of a number of activities, from suicide through depression to murder, followed by the mass screening of individual children to identify at-risk individuals and their drugging throughout life, or raising them in environments designed to alter their serotonin levels, or both – which is after all the action program that would result from an attempt to define the genetic–biochemical factors as the right level for intervention – only has to be enunciated to demonstrate its fatuity. Good, effective science requires a better recognition of determining explanation and hence the determining level at which to intervene. Failing this, it becomes a waste of human ingenuity and resource, a powerful ideological strategy of victim blaming, and a distraction from the real tasks that both science and society require.

REFERENCES

Breggin, P. R., and Breggin, G. R. (1994). A biomedical program for urban violence control in the U.S.; the dangers of psychiatric social control. *Center for the Study of Psychiatry*, Mineo, together with evidence given by the Breggins to the NIH panel on violence research.

Brunner, H. G. (1996). Discussion in *Genetics of Criminal and Antisocial Behavior*, Ciba Foundation Symposium 194, G. R. Bock and J. A. Goode (eds.), pp. 155–67. London: Wiley.

Brunner, H. G., Nelen, M., Breakfield, X. O., Ropers, H. H., and van Oost, B. A. (1993). Abnormal behavior associated with a point mutation in the structural gene for monoamine oxidase A. *Science* 262:578–80

Cases, O., and 10 others (1995). Aggressive behavior and altered amounts of brain serotonin in mice lacking MAOA. *Science* 269:1763–8.

Chipkin, R. E. (1994). D2 receptor genes – The cause or consequence of substance abuse? *Trends in Neuroscience* 17:50.

Crabbe, J. C., Belknap, J. K., and Buck, K. J. (1994). Genetic animal models of alcohol and drug abuse. *Science* 264:1715–23.

Fausto-Sterling, A. (1992). *Myths of Gender: Biological Theories about Women and Men*, 2d edition. New York: Basic Books.

Hamer, D. H., Hu, S., Magnuson, V. L., Hu, N., and Pattatucci, A. M. L. (1993). A linkage between DNA markers on the X chromosome and male sexual orientation. *Science* 261:321–7.

Herrnstein, R. J., and Murray, C. (1994). *The Bell Curve*. New York: The Free Press.

Holden, C. (1994). A cautionary genetic tale: The sobering story of D2. *Science* 264:1696–7.

Johnson, H. C. (1996). Violence and biology: A review of the literature. *Families in Society: The Journal of Contemporary Human Services* 77:3–17.

Kay, L. E. (1993). *The Molecular Vision of Life: Caltech, the Rockefeller Foundation and the Rise of the New Biology*. Oxford, New York: Oxford University Press.

Kramer, P. D. (1994). *Listening to Prozac*. London: Fourth Estate.

LeVay, S. (1993). *The Sexual Brain*. Cambridge, MA: MIT Press.

Moir, A., and Jessel, D. (1995). *A Mind to Crime*. London: Michael Joseph.

Plomin, R., Owen, M. J., and McGuffin, P. (1994). The genetic basis of complex human behaviors. *Science* 264:1733–6.

Rose, S. P. R., Lewontin, R., and Kamin, L. (1984). *Not in Our Genes*. Harmondsworth: Penguin.

CHAPTER TWENTY-TWO

Behavior Genetics

Galen's Prophecy or Malpighi's Legacy?

EVAN BALABAN

The suggestion that these two temperaments [reactive and nonreactive] are under some genetic control is based on the degree of behavioral similarity of identical twins and the modest but consistent associations with sympathetic reactivity, asymmetry of cerebral activation, facial structure, body build, eye color, and atopic allergies. It is difficult to imagine the environmental experiences that could produce, in infants growing up with well-educated, economically secure, affectionate parents, the combination of a high sleeping heart rate at two weeks, and, at twenty-one months, cries of fear at the entrance of a clown. (Jerome Kagan, *Galen's Prophecy*, Basic Books 1994, pp. 261–2)

To me, clowns aren't funny. In fact, they're kinda scary. I've wondered where this started, and I think it goes back to the time I went to the circus and a clown killed my dad.

(Jack Handey, *Deep Thoughts*, Berkley Publishing Group, 1992, p. 6)

Behavior is all the more difficult to study scientifically because everyone seems convinced that they know what they are studying. There is no other aspect of an organism for which the assumptions and personal biases that underlie quantitative treatment can wield so much power over scientific and public psyches. Just as the Italian anatomist Marcello Malpighi (1628–1694) confirmed the existence of adult body organs preformed on the blastodiscs of unincubated chicken eggs (Malpighi 1686), most observers see in the behavior of other animals or fellow humans what they most expect to find there.

The American animal behaviorist Donald Griffin coined the term *simplicity filters* to describe this mindset (Griffin 1976). Simplicity filters are produced by convolving observations with preconceptions; they can artifactually supply components of the behavioral "signal" that are not really there. More usually, they simply remove those details of the behavioral signal that do not fit into the current theoretical zeitgeist.

Consider a swarm of bees in a glass observation hive. Observation hives have existed in Europe since the time of Pliny (his *Natural History* [XI, 16, 49] mentions a hive with horn windows). In the eighteenth century, hives with glass windows such as those described by the French naturalist de Réaumur (1740) became de rigeur for the budding social pastime of bee-watching, which remained popular in France, Germany, and England into the beginning of the twentieth century. As a result, many careful observers chronicled the most intimate details of bee behavior.

The gyrating movements of worker bees were frequently noted phenomena. The German pastor Spitzner, writing in 1788, noted:

> When a bee has come upon a good supply of honey anywhere, on her return home she makes this known in a peculiar way to the others. Full of joy she twists in circles about those in the hive.... for many of them soon follow when she goes out again. I observed this in the glassed hive when I put some honey not far away on the grass and brought only two bees to it. (Spitzner 1788, p. 102)

And 35 years later, his compatriot Unhoch noted:

> Without warning, an individual bee will force its way suddenly in among 3 or 4 motionless ones, bend its head toward the surface, spread its wings, and shiver its raised abdomen a little while.... The dance mistress often repeats her dance four or five times at different places.... What this dance really means I cannot yet comprehend. (Unhoch 1823, p. 115)

Why, then, did it take so long for someone to divine the structure and significance of these movements? Even after Karl von Frisch published his work on the bee dance (von Frisch 1946), many bee experts refused to believe that such "simple" animals could communicate the location of food sources to their nestmates. The English ethologist William Thorpe was struck by how easy it was to confirm the dance language in spite of his initial skepticism (Thorpe 1949). Griffin (1950) concurred, adding, "Now that he has told us what to look for in the seething turmoil of bees creeping over the honeycomb, now that his insight has made order where there seemed to be utter chaos, anyone with a little patience and a hive of bees can test the principal conclusions for himself" (p. 10).

It may not be surprising that a bit of knowledge changes the perception of bee behavior because bees are such alien creatures to most of us. It is more disturbing to consider how assumptions affect the way we view the behavior of the fellow creatures we all consider ourselves to be experts in. For example, the changing criteria for psychiatric diagnoses during the course of the twentieth century provide an instructive example of simplicity filters operating on the perception of human behavior. Compare Ernst Kretschmer's (1936) physical "types" to the diagnostic categories in the latest edition of the *Diagnostic and Statistical Manual of Mental Disorders* for a demonstration of how jarringly different categories that purport to be dividing up the behavioral world for the same purpose can be.

Although all scientific endeavors are filtered by the reasons a researcher chooses to study a given question in the first place, the training and background researchers bring to the study, and the choice of a measurement system for the study, behavior seems especially vulnerable to these effects. This chapter considers some of the ways simplicity filters operate in measuring behavior for genetic studies and in correlating genotypes and behavioral phenotypes. I conclude by offering some general suggestions on how we can deconvolve them out of future behavioral studies.

22.1 Three Filters: Goals, Training, Measurement

Table 22.1 lists the methods most commonly applied to the study of behavioral genetics broken down according to the entity that the method focuses on (populations or individuals) and whether the method involves correlations of genetic and behavioral variation or experimental manipulation of genetic variables to examine their behavioral consequences. These are simply the tools investigators have at their disposal: how they get wielded and interpreted depends on the scientific goals of the users.

Three basic approaches are used by behavioral geneticists. The first two deal with genes as causal entities with direct effects on behavior. One looks to genetic variation to explain behavioral pathologies, whereas the other tries to deduce something about the causation of normal behavioral variation from examining genes. Both of these frequently rely on correlative methods. The third approach uses the correlation between genes and behavior as a way of uncovering developmental pathways or mechanistic details of brain function – behavior and genetics are used as tools to study something else.

22.1.1 Filter One: What Is the Point of Doing Behavioral Genetics?

Each of the three behavior–genetic approaches articulates its research goals in different ways.

For those seeking genes that contribute to behavioral "diseases," behavioral genetics is about individual or public health – specifically, about identifying individuals or groups who are at "risk" for certain behavioral conditions. Why is such identification important? Identifying "disease" genes may enable the creation of developmental interventions to minimize the probability that a person or a population with a certain genotype will develop negative behavioral attributes (or maximize the probability of developing positive attributes). Yet it is never made clear whose job it is to develop the interventions once the individuals or groups who are "at risk" can be identified nor who will decide when such interventions are or are not practical. Another frequently stated idea is that simply identifying genes may provide some clues for "fixing" the behavioral pathology. Practitioners of this approach also like to dwell on the notion of ascertaining the causal responsibility for a behavioral condition, citing the "damage" done by "environmental determinists" who have fixed the blame for a particular condition such as autism or schizophrenia on parental behavior (for instance "a cold, distant mother") when it is really the fault of the "disease gene." Advocates of this position do not discuss the potential "damage" that

Behavioral similarity correlated with degree of genetic similarity	**Natural allelic variation among individuals with behavioral performance**
Behavioral differences between populations, strains, or species that differ genetically	
Quantitative trait loci studies	
Linkage studies (sib-pair and other techniques)	
Artificial selection and response to selection	**Using individuals with natural mutations o experimentally altered genomes to iden "candidate genes" for a behavioral phe**
Transgenic studies using alleles inserted into different genetic backgrounds to examine how genetic background affects the behavioral phenotype correlated with a particular allele of a gene	**Using behavior to find functional differen animals with genetically engineered mut gene structure or gene expression**
	Mosaics: construction of animals with gen cells or tissues to relate differences in th of particular cell groups to differences i
	Genetically manipulating cell parameters (receptors, signaling pathways, secretions peptides, hormones, or growth factors) cell groups and using behavior as a tool understanding brain function

discovery of disease genes whose effects are not currently remediable creates in the parental blame department if parents carrying these "disease" genes decide to have children.

Individuals who look to allelic differences to explain variation in normal behavior use similar reasoning but frequently dress it up in different language. By identifying alleles at "candidate" genes (or "marker" DNA sequences close to them) that are correlated with behavioral variation, they argue that they are identifying some of the "causes" of that variation. The goal of identifying such genes and their alleles tends to be subject and investigator specific. For instance, in the case of aggression (Hen 1996, Tecott and Barondes 1996), it is hoped that the specific genes uncovered will give hints for pharmacological interventions that can control undesirable violent behavior. In the case of sexual orientation (Hamer et al. 1993, Hu et al. 1995), the goal seems to be to answer questions about the development of different sexual orientations: Are people gay or straight because they are "born" that way? In the case of genetic markers associated with high and low IQ (Plomin, McClearn et al. 1994) the justification given is that

> even a small handhold on the genetic contribution to individual differences will help in the climb toward understanding how genes interact with neural, psychological, and environmental processes in cognitive development.... In a broader perspective, it will help integrate genetic research on human and nonhuman animals at the universal level of DNA. It will also help integrate the increasingly fractionated biological and behavioral sciences. (pp. 116–7)

Oddly enough, no one ever mentions paltry issues like money or prestige in these first two approaches, regardless of the fact that in the United States there are parties such as insurance companies, educational industries, and major employers who would pay a lot for the ability to predict even roughly costly mental illnesses, student's performance on educational tests, and similar attributes. Reasoned discussions of what researchers may personally gain from performing these kinds of studies (both monetarily and in terms of conventional career advancement), and thus where their conflicts of interest lie, are virtually nonexistant (Balaban 1996).

For practitioners of the third approach (using genes and behavior to study neural development), the goals of the enterprise are quite different. Genes are seen as one input into the developmental system and behavior as a principal output. Altering a genetic input may help to identify some of the components of the system and enable one to study their interaction. If altering a gene via mutation causes a major problem in development, yielding either lethality or a gross anatomical or physiological malfunction, then it is not necessary to study an organism's behavior. However, many gene alterations do not produce any obvious changes in externally observable phenotypes. Behavior provides a sensitive, global indicator of neural function because of the complex nature of behavioral phenotypes, which depend on the simultaneous operation of a plethora of intra- and extracellular functions. When one of these functions is slightly compromised, behavior may be the most easily observed aspect of the nervous system that one can use to assay its functional integrity. Using behavioral phenotypes also allows the study of a range of mutations in a single gene, some of which may compromise function so severely as to be lethal and others of which may have extremely subtle effects, allowing a more complete documentation of the range of biological processes that a single gene product directly or indirectly participates in (Hall 1995).

Given their different ideas about why they are doing behavior genetics, members of these different communities may present and interpret the same phenomena in very different ways, as we will see below.

22.1.2 Filter Two: Who Does Behavioral Genetics?

A second source of filtering comes from training in how to study behavior. One of the major tensions in twentieth-century behavioral science is between schools of thought that devise simple "instruments" for characterizing behavioral variation among individuals, frequently under "controlled" conditions (psychology, psychiatry, sociology), versus those that try to examine variation in the performance of spontaneously occurring behavior (ethology, ecology, evolutionary biology). Behavioral genetics work has a disproportionate representation of the first type of approach.

Even though they tend to use similar methods, people studying both behavioral pathology and "normal" variation come from two different scientific backgrounds. The majority of them were trained in the Anglo–American psychological or biological psychiatric tradition with reliance on psychometric tools and simple "tests" to characterize behavior. The second group, which is becoming more numerous, consists of people whose background is predominantly in molecular or cellular biology. Many in this group have neither the training nor the appreciation for the problems inherent in quantifying complicated whole-organism attributes. They use the tests developed by the first group in the sense of a protocol used to separate RNA from DNA: "simply follow the recipe." The people who use genes and behavior to study development tend to come from a background that emphasizes biology or zoology with a smattering of experimental psychology; there are those who study naturalistic forms of behavior such as courtship (Hall 1994) as well as those that rely to a greater extent on simple laboratory tests.

The simple tests used in animal studies, such as mazes, artificial stressful situations, or conditioning or instrumental learning paradigms involving food rewards or aversive stimuli, are typically designed with little regard for the lives the test subjects normally lead. The primary human methods are interviews or psychometric instruments (both are frequently combined with diagnostic categories). In one sense such procedures are good because they provide a concrete embodiment of the simple assumptions that investigators base their behavioral ideas upon. But the mantle of experimental rigor that they provide can conceal the fact that the actual combinations of behavioral factors deployed in the tests are not readily quantifiable, nor are these factors necessarily consistent from subject to subject or even within retests of the same subject. Contextual variables with major impacts on ongoing behavior that occur outside of the narrow confines of the experiment (such as social interactions with other subjects, differential societal pressures, developmental conditions, etc.) are frequently not controlled or even measured at all.

Much behavior–genetic work in humans has relied on methods borrowed from quantitative genetics. This work, which continues today (Bouchard 1994; Plomin, Owen, and McGuffin 1994), relies on correlations measured in populations of subjects to estimate how much phenotypic variation can be statistically assigned in a particular transmission model to the effects of "genes" in aggregate – so-called heritability statistics. Heritability is a population measure that has proven empirically useful for animal breeders in situations in which they can control the environment and matings of their breeding stock and for measuring effects of artificial or natural selection (Falconer 1989; Lande 1984; Arnold 1990, 1994; Boake 1994). It is based on measures of phenotypes, assumptions about environments, and mean genetic similarities of different classes of relatives derived from Mendelian genetics: it does not incorporate any direct measure involving genes. Those who use heritabilities in human behavioral analysis insist that they are studying the causation of "individual differences" in behavior. They claim that heritabilites can prove that genes play a role in the behavioral variation one is analyzing and that heritabilities also estimate the degree to which variation in the behavior relies on genetic variation (Plomin, DeFries, and McClearn 1990; Neale and Cardon 1992; Plomin, Owen et al. 1994).

These reasons are not very compelling. The outdated question of whether genes play some vague role in behavioral variation among individuals may provide ammunition in arguments with social scientists, but it is not informative in genetic work. There is a trivial level at which genes play a role in any behavioral variation: each individual in a natural population (except monozygotic twins, which are relatively rare in humans) is genetically unique, and each individual is also behaviorally unique. What makes these genetic differences interpretable with regard to behavioral variation is the determination of the pathway(s) through which given genes affect behavior. This allows us to understand whether a particular gene plays the same significant and consistent role in a behavioral system across different individuals or simply has a few allelic variants that can screw up brain function. Understanding the pathway may also enable one to predict whether the variation in behavior such genes contribute to will be very consistent from one genotype to the next or one environmental situation to the next. Heritabilities, on the other hand, provide no direct information relevant to finding particular genes, tracing the pathway from genotype to phenotype, nor do they answer the more general question of whether "genetic" effects on a behavior show consistency across different individuals.

Heritability statistics were designed to avoid all of the messy details of genetic systems and developmental biology; that is their strength. But they were designed with a very limited, circumscribed job in mind: to help describe what happens when phenotypic selection is applied to a particular population in a defined environment. Heritabilities remain useful for the tasks for which they were designed: aids for artificially selecting mean-trait values in populations in controlled environments or studying evolutionary questions about natural or artificial selection.

Recently, Plomin, Owen et al. (1994) have claimed that heritabilities (they call them "quantitative genetic research") are "needed to inform molecular research":

> Most fundamentally, quantitative genetic research can steer molecular genetic research toward the most heritable syndromes and combinations of symptoms. Genes are less likely to be identified for complex behaviors that show little genetic influence in the population unless some aspect of the trait can be found that is highly heritable, as in the case of breast cancer. (p. 1735)

I have enough respect for Plomin et al.'s mastery of genetics to assert that they know this is not true. Depending on many variables (especially the effects particular alleles have in heterozygotes), phenotypic factors that exhibit classical Mendelian behavior (and can therefore be easily identified in lineage studies in humans) may not show significant heritabilities in a family study (nor will "maternal effect" genes that act on an organism's phenotype through its mother's phenotype). I would assert that the heretofore identified human genes linked with particular conditions of medical interest owe all of their success to their Mendelian patterns of inheritance and none of their success to a reliance on heritablilities.

Despite increasing statistical sophistication in the models for estimating heritabilities (Neale and Cardon 1992), a significant heritability does not guarantee the existence of identifiable genes affecting variation in a particular phenotype – heritabilities contain no information about how a "genetic effect" is distributed in a population. We do not know if a heritability value that is significantly greater than zero may be due to the same few alleles of strong effect in some lineages but not in others, to a few alleles of strong effect that are different in different lineages, to many alleles of weak effect that are the same in all lineages, to many alleles of weak effect that are different in every lineage, to no alleles at all but rather to a failure to detect environmental covariation with relatedness in some subset of the study population, or to some complex mixture of all of these possibilities (see Lewontin 1974 and Wahlsten 1990 for an extended discussion of many other problems with heritability analyses).

The survival and popularity of heritabilities in behavioral genetics is perhaps best explained by the persistence of academic traditions and the filtering done by these traditions. You do not have to know many "messy" details about biology, development, or the functional side of genetics to employ heritabilities, which makes them ideal for sociologists, psychologists, and clinical practitioners who use heritabilities the most and who do not receive much training in, or exposure to, areas they regard as someone else's job. On the contrary, these researchers receive considerable training in the types of statistical treatments that spawned heritability in the first place.

22.1.3 Filter Three: The Choice of How to Measure Behavior

From the point of view of an observer, behavior is an incredibly "overdetermined" system. Any behavioral attribute has multiple dimensions, most of which are loosely correlated. There are therefore many rival ways of describing variation in a particular behavior that are not necessarily equivalent. For instance, if I wanted to measure the intelligence of everyone reading this essay, I could (to use a deliberately extreme example) either give them an IQ test or release them naked, one at a time, in the middle of the Gobi desert and measure how long they survive. Clearly, these tests differ in what is at stake for the testee, and in other ancillary skills that they call upon that we would not consider to be a direct part of intelligence (motivation, attention, physical condition, stamina, etc.), as do any tests examining any aspect(s) of the behavioral phenotype. It is rarely possible to describe or test any aspect of an organism's behavioral performance in isolation from other aspects, and it is always possible to find measures of single behavioral attributes that agree with each other or that disagree with each other.

Behavioral measurements can be very ad hoc, and people who study behavior tend to be very defensive about this. Some of this defensiveness may have to do with the inferiority complex ("physics envy") that most behavioral scientists exhibit when trying to justify why what they do is science. Behavioral researchers within the disciplines of biology or medicine have spent the better part of this century convincing skeptical colleagues that it is possible to measure and study behavioral attributes "objectively." Psychology has had to battle for its very legitimacy in similar terms. Many of the problems of measurement and quantification were swept under the rug in this fight for legitimization to be argued among the cognoscenti but not to be aired to the outside world.

There are no generally accepted yardsticks in the study of behavior that can be used to decide which scales of measurement and which measures are "good" and "bad"; other external criteria are always brought into service. For instance, many psychiatric classifications base their validity in part upon demonstrations of a correlation of symptoms with a biological variable (response to medication, etc.) or with genetic variation (linkage or association studies, or evidence from family–twin studies). The practice of selectively validating those behavioral measures that best correlate with responses to drugs, particular molecular markers or with variation in relatedness is troubling because of its circularity – especially when different studies have to play with behavioral definitions in different ways to achieve this kind of a fit (Kidd 1993).

One consistent feature of behavioral genetic studies has been the insistence on the part of investigators that their work is objective and that any scientific, social, or political preconceptions on their part are irrelevant to evaluating its validity. Yet, most behavioral scientists with any training in psychology are familiar with Robert Rosenthal's work on experimenter effects in behavioral research (Rosenthal 1966). Rosenthal reported findings like the "Pygmalion effect," in which students perform up or down in response to their teacher's expectations. He also found expectancy effects in experiments done with animals by laboratory assistants and students in which their expectations bias the results of their behavioral measurement (for instance, giving them mice from the same strain for learning experiments but identifying certain mice to be from a "bright" strain and others from a "dull" strain). It is easy to dismiss such accounts because, by and large, they do not study seasoned researchers alerted to the dangers of subjectivity; careful researchers usually adopt "blind" or "double-blind" designs to avoid this type of bias.

Yet subjectivity should neither be denied nor uniformly regarded as negative. It is unavoidable in research on multidimensional systems like behavior at both an intentional and an unintentional level. At the intentional level subjectivity enters into decisions on what aspects of a behavior to measure – which ancillary covariates to quantify and how they are quantified. Unintentional biases usually creep into the measures actually employed on the behavior itself.

22.1.3.1 What to Measure

The decision to measure a particular behavior in a particular way is usually highly subjective – the product of legitimate scientific assumptions that researchers have to make every day in order to carry out research. Such assumptions are driven by a researcher's view of the causation of the behavior, which assumes that some sources of variation are relevant and that others are not. Why should genetic studies of spatial learning in rodents choose the Morris water maze rather than a radial arm maze, foraging, or food-caching paradigm – tests on which the same individual can exhibit very different behavioral profiles? Is aggressive behavior in humans better measured by histories of violent acts or by the kinds of fantasies a person has? Is human intelligence better measured by artificial passive tests or by performance in real-life situations? There are no good scientific criteria upon which to base such decisions; it is therefore important that the subjective evaluations on which these judgments are based be made explicit. All too often, this is not the case.

22.1.3.2 Dealing with Covariates

Subjectivity always comes into play in decisions about how events that are ancillary to the quantified behavior are treated. For instance, which aspects of a subject's personal history (socioeconomic status, cultural background, general stress level, etc.) are relevant covariates and which are not? How do we determine whether the measures employed for such contexts or covariates perform as intended? For instance, is counting the number of books in a home (Bouchard et al. 1990) a decent measure of the "richness" of the human educational environment? Again, this problem has not been given the attention it requires.

22.1.3.3 Unintentional Measurement Bias

Most physical scientists examine the bias in their measuring instruments so that they can design an "inverse filter" that will correct data for the bias.To measure the frequency content of human speech, one needs to know the frequency response of the microphone and tape recorder so that these biases can be removed from the Fourier analysis of the speech. It is vital to know to what extent the "instruments" that measure behavior bias the data they generate.

The only arena in which this kind of bias has received any attention is sampling bias in population studies. Four kinds of bias have been identified that distort the picture given by population studies: (1) the practice of selecting subpopulations for a study based on the perceived "severity" of their behavioral phenotype; (2) cohort effects, which are a problem in many heritability studies of human behavior; (3) the retrospective collection of behavioral data, which is a special problem in studies using pedigrees; and (4) the validity of extrapolating from studies on special classes of relatives such as mono- or dizygotic twins to the general population.

But surely bias is a more fundamental problem in behavioral studies of individuals as well. For instance, bias is automatically incorporated into behavioral measurements by the "scale" of behavioral resolution that they pay attention to. Scale biases are a problem in any system in which events that are heterogeneous in time or space are averaged. At very "coarse" levels of measurement (averaging over a lot of time or space), all members of a population, species, or genus can be very similar behaviorally, whereas at very fine levels, different performances of the same behavior by the same individual can be remarkably different. For any given behavior that is measured, different scales of measurement can give very different profiles of behavioral variation: investigators always choose the level of variation that "best suits their purpose." The ways in which scale choices can affect the pattern of behavioral variation are in theory amenable to modeling if the underlying "fine structure" of the behavioral variation is known: behavior genetics is badly in need of such studies.

A second kind of unintentional bias is a direct outcome of observer's conceptions of what it is that unifies the set of attributes they are measuring as reflected in the labels that are attached to certain behaviors. For instance, the American psychologist Calvin S. Hall (1934) noted that, when he was very fearful, his autonomic functions went haywire, especially his lower visceral functions: he had to go to the bathroom. Hall's legacy is still with us today in a behavioral genetic paradigm that measures "fearful behavior" in rodents via micturation and defecation – the "open field" test (Whitney 1970, Flint et al. 1995). Those conversant in rodent behavior will know that micturating and defecating in rodents and humans are not easily homologized. Few people reading this chapter could, like mice, divine the gender or individual identity of a person on the street from their urine or stools or tell how they had fared in a previous altercation. Furthermore, urinating and defecating in rodents is not necessarily only a sign of fear – males who are not particularly fearful engage in this behavior at great frequency when marking a piece of turf that they will later defend. Many genes that have an impact on urinating and defecating (as older male readers with prostate conditions will appreciate) may have little to do with emotional responsiveness, and yet, according to this test, they all become by definition genes for "emotional behavior."

The label put on a behavior can lead to bias in both the measurement of that behavior and in the perception of the general significance of a piece of work. Because much human and animal research in behavioral genetics uses simple tests applied under very unnatural conditions that actually measure a complex conglomeration of influences, these labels are frequently questionable and mislead both researchers and those who interpret the research.

22.2 The Filters in Action in Behavior–Genetic Research

What are some of the effects of the filtering action of goals, training, and chosen methodology in behavior genetic research? We will look at three broad areas: the assessment of different kinds of effect specificity, the creation of "virtual behaviors," and the study of development.

22.2.1 The Problem of Specificity: What Is Being Measured, What Genes Are Affecting, and How They Are Affecting It

22.2.1.1 Behavioral Specificity

There is no such thing as a "pure" measurement of a single behavioral system; all behavioral measures are aggregates of perceptual, motor, emotional, attentional, decision-making, and other components. When one examines the correlation between genetic variation and behavioral variation, it is thus important to bear in mind that even the simplest of behavioral performances will be impacted by the actions of many genes – a characteristic called polygeny. Most people who do behavioral genetics today have a good appreciation of polygeny, even if they do not deal with it very effectively. They typically fare much worse when they consider the opposite side of the coin, the number of different behavioral processes that are impacted by variation in a single gene, which was formally referred to as pleiotropy. These concepts are done the least justice in studies involving single genes. We will examine problems related to pleiotropy in single-gene studies in this section and come back to problems related to polygeny in the next section on genetic specificity.

Single-gene effects on behavior seem like they should be rather straightforward to characterize and interpret. Yet this is not the case because all behavioral measurements are aggregates of many different brain and bodily processes, and gene effects on these different processes may mimic each other. For instance, genes that affect perceptual processes or motor behavior by altering the time course of ionic conductances in sensory transduction cells or muscles may affect behavioral performance in a "learning test" in the same way as genes that affect the function of neural cells more directly involved in learning or memory.

Said in another way, gene effects on a particular behavioral process may be direct (mediated by the relevant genes affecting brain cells specifically involved in producing the behavior in question) or indirect (affecting some more global parameter in a much larger number of cells that in some way debilitates the organism, or affecting a parameter that is required for the behavioral assay but is not an integral part of the behavior itself, like motor coordination required for certain learning tests). The only way to find out which kind of gene effect one is dealing with is to look at many behaviors (to see not only the variety of what is affected but also what the affected behaviors may have in common) and to look at the identity and function of the gene product in more than one of the diverse developmental and functional processes it participates in. This is precisely what most behavior–genetic studies fail to do.

Hall (1995) cogently discusses the problem of pleiotropy from the *Drosophila* biologist's point of view. He notes that even people who study gene effects on morphological structures generally want the genes they focus on to exhibit "specificity of action":

> They seem afraid, on the one hand, to admit that the factor acts pleiotropically (one must almost phenol-extract the relevant information out of the publications or the investigators in question) and yet, on the other, are sullenly reluctant to follow up a discovery of broad mutational effects or gene-product expression by experimentally addressing questions like: "What is my 'soulneurin' protein doing in the ass end of the fly as well as in the brain, where I first found it and want it to be?" (pp. 15–16).

This seems to apply with even more force to behavioral genetics work. A prime example of the negative effects of filtering by goals, training, and methods in behavioral characterizations comes from recent papers on behavioral phenotypes associated with structural deficiencies in the a form of the enzyme calcium–calmodulin–kinase–2 [α–CAMKII] (produced by engineering a mutation in the gene for this enzyme; see the next section).

Silva et al. (1992) presented the case for this enzyme's being specifically involved in spatial learning and not in nonspatial forms of learning. This demonstration involved comparing the performance of mutant and control mice in the Morris water maze (in which the animal must use spatial cues present in the room to remember the location of a slightly submerged platform) and in a task in which animals were trained to turn right or left (does not require the use of spatial cues). According to the authors, the mutants were significantly impaired on the Morris water maze task and not at all impaired on the right–left task, implying that spatial learning or memory was specifically impaired. However, a closer examination of the behavioral tests discloses the operation of simplicity filters.

The dependent variable in the Morris water maze task was the time it took animals to find the area where the submerged platform was located (Figure 22.1B). Learning is usually defined as an experience-dependent modification in behavior. Lack of learning would mean that animals should take about the same amount of time on subsequent exposures to find the submerged platform. An examination of the published figure from this report reveals that this is not

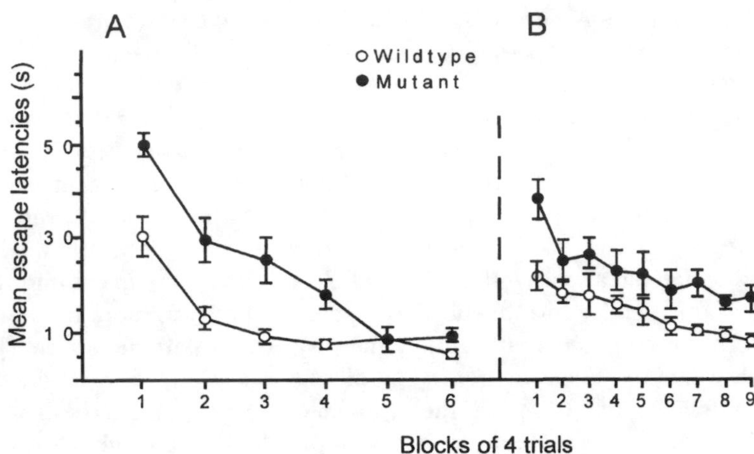

Figure 22.1. Escape latencies for animals in the Morris water maze (from Silva et al. [1992], redrawn from the original). Original text of this figure follows. "(**A**) In the first phase of the experiment, wildtype controls ($n = 14$) and α–CaMKII mutant mice ($n = 11$) were trained to navigate to a randomly located visible platform. The platform was rendered visible by attaching a small white flag to its top. Each animal was first trained to climb on the platform and given a 15-s proactive swim to ensure that all animals could swim. On each trial, a subject was placed in the pool for 60 s. Once a subject found the platform, it was allowed to remain there for 45 s. Animals were given 12 trials a day, in blocks of 4 trials, on 2 consecutive days. The α–CaMKII mutant mice were initially impaired at locating the visible platform but overcame this deficit and learned to locate as rapidly as controls. (**B**) All animals were then trained to find a hidden platform located in a fixed location. The top of the platform was 1-cm below the surface of the water. Wildtype and α–CaMKII mutant mice were given either 3 or 5 days of training as described above (only 3 days are shown). The figure shows that wildtype controls had lower escape latencies than the mutants. The bars indicate the SEM." Figure and caption reprinted with permission from Silva, A. J., Paylor, R., Wehner, J. M., and Tonegawa, S. (1992). Impaired spatial learning in α-calcium-calmodulin kinase II mutant mice. *Science* 257:206–11.

the case: both normal and mutant animals took less time to find the platform on successive exposures to the maze, and their "rate of improvement" (the slope of the time versus successive exposure graph, Figure 22.1B) appears to be roughly equivalent. What is different about them is that the mutant mice are slower than the controls. In the right–left task, time is not the dependent variable, but some conglomeration of the rate and success of learning is: animals are not scored on how long they take but rather on the proportion of correct choices they make after some initial learning period. Here, mutant and normal mice perform no differently because time is no longer the variable used to differentiate their performance.

I would conclude that Silva et al. (1992) actually demonstrate that α–CAMKII is not essential for spatial learning but that it may compromise motor function or sensorimotor communication. The control for impaired motor function is supposed to be shown by Figure 22.1A: given enough practice, α–CaMKII mutant mice can find a visible (not hidden) platform in the same "minimum" time as normal mice. But this seems to be an easier task for the normal mice, too: they reach an asymptote in performance by the second or third block of trials. It takes them eight or nine blocks of trials in Figure 22.1B to reach this same asymptote. Animals with impaired motor function might be expected to perform differently in the two tests: in the task represented in Figure 22.1A, the "impaired" animals can devote all their attention to swimming because they can see their goal at all times. In the task in Figure 22.1B, this is no longer the case.

A second α–CAMKII article from the same laboratory on aggressive behavior in these mutant mice (Chen et al. 1994) talked about a "well-circumscribed syndrome of behavioral abnormalities, consisting primarily of decreased fear response" [mutant mice do not flinch as much in a conditioned fear response test, and if you put them in the center of an open arena, they do not move out of the center as quickly; they also defecate less in both of these tests than "normal" mice], "and an increase in defensive aggression" [if you isolate animals for a month and throw them in another animal's cage, they bite more in response to being attacked than "normal" mice], "in the absence of any measured cognitive deficits" (p. 291). These "well-circumscribed" effects were only seen in animals heterozygous for the mutation: homozygous animals (used exclusively in the first article on learning, which claims a "specific" effect) are now described as "displaying abnormal behavior in all paradigms tested" (p. 291). However, both heterozygotes and homozygotes exhibit decreased mating success relative to "normal" mice, which is a classic sign of a mutation with general deleterious effects.

Heterozygous mice are said to have no cognitive deficits because of their "normal" performance in a Lashley Type III maze (what happened to all of the learning tests used in the Silva et al. (1992) article?), whereas homozygotes did not do well in this task ("homozygotes often lapped repetitively within the same alley" [Chen et al. 1994, note 24, p. 294]). In the Type III maze (Lashley 1929, p. 31), the animal must learn a sequence of five alternating right and left turns (RLRLR) to get from a starting box to a reward box. It is unclear if homozygotes simply took longer to get to the end of the maze but eventually reached it nevertheless. In this test (as in the Morris water maze), animals are scored on the time it takes to reach a goal, the number of "turns" it takes them to reach it rather than on how many times they successfully reach the goal, or both factors.

In neither of these articles is the possibility of a more general, "less-well-circumscribed" effect considered despite the presence of ample clues in both studies suggesting that such effects could explain many of the results. Impaired motor function or impaired sensorimotor communication immediately leap to mind. The reduced "fear" responses (slower and fewer movements, less defecation), for instance, could be interpreted as being produced by lowering metabolism or impairing motor function or sensorimotor communication. The increased "defensive aggression" may be due to the fact that heterozygote animals are not so nimble, and thus they have learned to rely more on biting than on acrobatic wrestling or trying to escape to make up for their lack of agility. That one can find different behavioral tests nominally in the same domain ("learning" in the first case and "aggression" in the second), whereas in some tests mutant individuals are different and in others they are the same as "normal" individuals, does not make behavioral effects "distinct" or "well circumscribed." These effects may simply be due to more general processes that are differentially utilized by the tests; the effects are not very specific at all.

On paper, it looks as if much behavioral work has been done on α–CAMKII mutant mice, but that work is very superficial. This case was deliberately chosen because it was published twice in a prestigious journal, was performed in the laboratory of a Nobel laureate, and has one of the most extensive characterizations of the behavioral phenotype for its genre. Most other studies, such as the recent work on mice with introduced mutations in the neural nitric oxide synthase gene (Nelson et al. 1995) [or "the mice that roared . . . and killed and raped, due to a genetic defect," as *Newsweek* put it (Begley 1995)], simply measure one behavior (in this case aggression) and go on to describe the gene effects as "specific" to that behavior. A recent survey of genetic and neurobiological work on aggressive behavior (Balaban et al. 1996) found that in the majority of cases "behavioral pleiotropy" is either totally neglected or underreported and underemphasized. A prime example is a much-publicized Dutch study of a family in which males had a mutated form of the gene for the enzyme monoamine oxidase A (Brunner, Nelen, Breakefield et al. 1993; Brunner, Nelen, van Zandvoort et al. 1993): no behavioral data were provided beyond the level of anecdotes about memorable violent acts gathered from family members.

One case in which pleiotropy has been examined is the so-called *Drosophila* "learning" mutants. Initially presented as affecting only learning and memory (Aceves-Pina et al. 1983), these mutations in cellular second messenger systems have since been shown to have a variety of effects on the excitable behavior of muscle, on sensory transduction, and on development (Corfas and Dudai 1990, Tully 1991, Zhong and Wu 1991, Delgado et al. 1991). It was only when researchers examined mutant flies in other contexts and the biochemical pathways that these gene products participated in that the more general nature of the deficiencies caused by compromising the efficacy of intracellular signaling functions was recognized. It is to the credit of *Drosophila* researchers that these effects were examined in such detail; such mutations are perhaps more accurately described as "intracellular signaling mutants." It is not surprising that functions such as learning (which involve signal transduction) are affected by mutations that compromise the efficacy of intracellular chemical messenger systems mediating cellular responses to transduced signals. It is also not surprising that such mutations compromise an array of organismal functions at many levels. (I think that what explains the willingness of the *Drosophila* community to examine these issues and the failure of the researchers in the preceding examples to do so, are differences in training and goals. Fewer individuals in the *Drosophila* community have ideological axes to grind about linking genes directly to behavior: they are more interested in developmental processes.)

Behavioral specificity is a problem that has been very much ignored by the community of researchers pursuing gene linkage and association studies in humans (Lander and Schork 1994, Lander and Kruglyak 1995, Risch and Zhang 1995, Risch and Merikangas 1996). There are real limitations to how many behavioral aspects of the phenotype one can hope to measure in single individuals, but the problems inherent in using just one behavioral condition to sift through genotypes should be addressed. As a start one could seriously look for "suites" of behavioral traits that covary in a similar way with the "candidate" genes. Although there have been and continue to be "multivariate" behavioral studies in heritability research (DeFries and Fulker 1986; Neale and Cardon 1992; Plomin, Owen and McGuffin 1994), this tendency does not seem to have taken root in the linkage or association study community.

Heritability studies that look for suites of traits that go together (so-called genetic correlations) do not seriously address the question of pleiotropy because they cannot tell you the number of different "genetic" (and "nongenetic") factors that covary with the phenotypes. The emerging practice of using many different "dimensions" on a personality test to sift through groups that differ in the alleles of a particular gene and looking for any statistical correspondance without correcting probability tests for multiple comparisons (Cloninger, Adolfsson, and Svrakic 1996; Ebstein et al. 1996; Benjamin et al. 1996) is neither methodologically nor theoretically sound. The supposed gene effect in these cases is a small shift in mean test scores that have been statistically "massaged" from the original data (the magnitude of the difference is about 2.5 raw test score points in "novelty seeking" (from 15.5 to 17.9 [Ebstein et al. 1996]); or 4 "corrected" score points on "extraversion" (from 53.4 to 57.3), and about 3 points on "conscientiousness" (from 45.9 to 43.2 [Benjamin et al. 1996]). The measurements that produce these dimensions are the answers to a long set of questions such as "I have sometimes done things just for kicks or thrills" versus "I often try new things just for fun and thrills," or "I think things through before coming to a decision" versus "I like to think about things for a long time before I make a decision." This is a "virtual" rather than a "real-world" behavioral measure – it mixes together an incredible conglomeration of behavioral processes, compounds the mixture by self-reporting, and statistically manipulates the numbers describing an individual's performance before the final behavioral "index" is calculated (we will discuss "virtual" behavioral tests in more detail below).

In human linkage or association studies, information on the "pleiotropy" of real, measurable behaviors could be very valuable in evaluating the specificity and the robustness of the phenotypic effects of allelic or chromosome segment differences, the probable identity of the gene(s) within a chromosome segment that mediate(s) the effect, and something about the developmental pathways the gene or genes are acting through. Perhaps the filtering influence of focusing on the "gene" side of the story in linkage and association studies has rendered their primary users blind to the potential contributions of more sophisticated behavioral characterizations.

22.2.1.2 Genetic Specificity

Specificity problems also extend to the genomic side of the equation. Despite the extensive development of new methods for detecting gene linkages with "complex" phenotypes in both humans and experimental animals (Kidd 1993; Crowe 1993; Crabbe, Belknap, and Buck 1994; Takahashi, Pinto, and Vitaterna 1994; Lander and Schork 1994; Lander and Kruglyak 1995; Risch and Zhang 1995; Risch and Merikangas 1996), a substantial problem remains after such linkages are established. One must still determine the pathway that mediates these effects and their behavioral specificity (pleiotropy, which we have discussed above). Yet it is also vital to determine the extent to which the behavioral effects of these "candidate" genes depend on the identity of alleles at other genetic loci throughout the rest of the genome – the question of so-called modifiers or genetic background (the problem of polygeny). This problem has been highlighted by recent experimental work on a single-gene mutation that affects the brain and behavior of *Drosophila melanogaster* (De Belle and Heisenberg 1996), for behavioral and anatomical phenotypes associated with the same mutant allele at one genetic locus can be markedly affected by allelic differences at other loci in the genome carrying the mutation.

Just as pleiotropy seems invisible to many behavior–genetic investigators, so does polygeny at a concrete level. If your goal ends with identifying single-gene linkages to behavior, then it is easy to see why you would think polygeny is not problematic. One just keeps identifying linkages until all the relevant genes are obtained. But this ignores an important subtlety about the conditional way in which polygeny works. Linkages identify alleles that are the "tip" of a proverbial iceberg. In order for the allele of a gene at a particular site on the chromosome to show the behavioral effect that it has, particular alleles at other genes in the genome must also be present in a certain form, or the phenotype may be very different. This is a possible explanation for why linkages between particular alleles and behavioral pathologies found in one set of lineages may not be found in other lineages. Linkages do not show you the full suite of genetic differences that yield a particular behavioral difference. They only give you a point of entry into a developmental or functional pathway. To understand "single-gene effects" completely, we need to learn how allelles at other genes in the genome contribute to the phenotypic differences shown by alternative alleles of the "focal" gene.

Uncovering the developmental pathway through which a particular allele achieves its behavioral effects can be uniquely informative. Are its effects directly mediated by an absence or excess of gene product, or are they instead mediated indirectly by a regulatory system in the developing organism attempting to correct for the gene product's absence or excess? In this latter case, we may find that the primary behavioral "effect" from an altered gene has nothing to do with the "normal" function of the gene itself. The effect can actually come from a regulatory system altering the expression of a totally different set of gene products. A mutation may also work as a negative "gain-of-function" through which an altered gene product now interferes with biological processes it formerly had nothing to do with.

In animals it is possible to conduct manipulative studies on genes to examine their effects on behavior at this detailed level – so-called reverse genetics. Recently, several genre-specific problems of technique and interpretation with these studies have been described (Gerlai 1996a,b; Crawley 1996; Lathe 1996; Crusio 1996; Morris and Nosten-Bertrand 1996). These problems primarily concern one's ability to ascribe causation of a phenotypic change to a single gene versus multiple genetic factors and the use of appropriate control animals to ascertain behavioral effects. The problems arise because of the details of how one engineers mutations in particular "target genes" like α–CAMKII discussed above.

"Gene targeting" involves genetically manipulating tissue-cultured "stem" cells, which, if placed into early embryos, can combine with normal embryonic cells to form any of the cell types in the body. The problem is that the strain of mice whose stem cells work the best (i.e., are good at "invading" host embryos) are not very viable as breeding animals. In fact, this strain (called "129") is notorious for brain and behavioral anomalies: having impaired performance in spatial learning tasks, being motorically "passive," and having gross abnormalities of the corpus callosum, the main neural pathway between the two brain hemispheres (Gerlai 1996a). In normal practice one performs the genetic manipulation on these stem cells in cell culture and then makes a "chimera" by taking a very early mouse "host" embryo from another, more viable strain, combining the altered stem cells with it, and implanting it into a "host" strain mother. When the chimeras become sexually mature, one looks for animals whose germ cells (those that make sperm or eggs) are derivatives of the introduced "stem" cells. By mating these animals to members of the "host" strain (or sometimes even a third strain) and recovering individuals heterozygous for the introduced mutation, one can generate a line of animals that will produce offspring who are heterozygous or homozygous for the introduced mutation (as well as offspring who lack it altogether) when the mice are mated to each other.

The problem is that the sexual reproduction needed to generate the line of heterozygous animals from the cross between the germline chimera and the host strain animal (and the continuing crosses needed to perpetuate the heterozygous line) will also involve random recombination events, and thus chromosomes from the genome of the "host" embryo strain will recombine with those from the genome of the "stem cell" strain. Thus, the "mutant" mice are hybrids that will differ from both the host and "stem-cell" strains at many different loci. Recombination also occurs on the chromosomes containing the introduced mutation. Without much additional breeding work (which these studies typically do not carry out), it is hard to say that the behavioral effects one observes are due to the altered gene per se rather than other recombinant elements in the genome or to effects that the altered piece of DNA might have on the expression of "normal" genes near the altered gene on the chromosome. Recall that the major phenotypes claimed in the α–CAMKII articles were also features of the "stem-cell" strain of mice: cognitive impairment and motor deficiencies. It is also hard to decide what the appropriate control animals are. Ideally, they should only differ genetically in the identity of the allele at the targeted locus, but such animals do not exist because of the genetic heterogeneity introduced by the recombinations (Gerlai 1996a,b).

It is interesting that gene-targeting studies have gone on for about 10 years with authors routinely claiming to demonstrate cleanly the effects of single-gene manipulations on both behavioral and nonbehavioral phenotypes without anyone's pointing out that this was not the case. This is a classic example of simplicity filters.

22.2.1.3 Behavioral Autonomy and the "Environment"

Behavioral researchers always have to deal with the question of the conditionality of behavioral phenotypes. Does a particular phenotype depend on interactions between the organism and identifiable factors external to it (including other organisms), or is it more or less "autonomous" to the organism?

Biologists have known since the work of Richard Woltereck (1909) that an organism's phenotype is a function of both intrinsic and extrinsic factors (see Falk, this volume). Some of the intrinsic factors are the organism's own genetic makeup and the genetic makeup of the organism's mother (which directly mediates certain features of the organization of the early embryo and characteristics of the milieu that the developing embryo finds itself in). But there are also many nongenetic "intrinsic" factors, which in mammals include the diet of the mother (influenced by events external to the mother), her hormonal states (induced both by events external to the mother and by the mother's nervous system), and stochastic events within both the embryo and the mother that can influence embryonic development (Gilbert 1994). "Nature" and "Nurture" were abandoned as explanatory concepts by developmental biologists because they do not adequately capture the complexity of the influences operating on the development of organisms – "intrinsic" factors in development are not readily separable from "extrinsic" ones.

When we look at development from the point of view of genes, we find the same problems in classifying causation. The biochemical processes mediated by gene products are sensitive to the milieu in which those reactions occur. A biochemical reaction catalyzed by a particular gene product will change its kinetics and even its outcome depending upon the local presence or absence of the products of other genes. The milieu external to the cell in which these reactions occur will also affect the reactions either directly (through diffusion of substances into the cell that affect the molecules participating in the reaction) or indirectly (through the presence or absence of factors that cause the cell to change its internal chemical milieu actively by opening or closing ionic channels in the cell or by changing which genes the cell is expressing). Even in a single-celled organism, the effects of a gene change on the cell's phenotype are as much a statement about "intrinsic" gene effects as they are about the "extrinsic" effects of a cellular milieu or environment.

Jumping back to the level of behavior, we have the additional problem of documenting whether a particular behavioral effect is intrinsic or extrinsic to the whole organism. Social behaviors depend upon interactions with other organisms. Many of the phenotypes used in behavioral genetics are byproducts of such social behaviors. To what extent are differences in these interactive behaviors autonomous to one of the participants? Consider the case of an aggressive interaction. Suppose that we observe certain structured interactions between people (or animals), and our measure of aggression is the number of aggressive displays or even fights that result. Certain behavioral or physical attributes of a target subject that are independent of aggression (such as an odor, a facial expression, a way of holding the body) could alter the behavior of social partners in a way that then elicits increased "aggression" from the subject. The subject would not show such aggression if placed with social partners that did not react to these other behavioral or physical attributes. We would not know this was the case until we actually checked to see if the behavior of the partners played a role in the appearance of aggressive behavior from our subject. Even if an effect appears to be "intrinsic" to a subject at this level of analysis, that does not mean that its genesis is intrinsic. Recall the example of aggression in α–CAMKII mice in which it was suggested that mice whose agility is impaired may develop the strategy of biting more in response to attack (increased aggression) because they do not have the agility to wrestle or escape as effectively if reared together with more "normal" animals. The only way to really understand the question of behavioral autonomy is to characterize both developmental and situational aspects of behaviors.

Behavioral geneticists have not paid much attention to such subtleties. The problem of behavioral autonomy has mainly been addressed by people who measure heritabilities, whose primary concern is with how behavioral autonomy will affect the statistical assignment of proportions to the "sources of variation" in heritability models. The approach taken here ascribes behavioral variation to two sources: "heredity" and "the environment":

It is now generally agreed that both nature and nurture play a role in determining behavior. However, the mistaken notions of the nature–nurture argument have often been replaced with the equally mistaken notion that the effects of heredity and environment cannot be analyzed separately, a view called interactionism.... Obviously, there can be no behavior without both an organism and an environment. The scientifically useful question is: For a particular behavior, what causes differences among individuals?... Various environmental hypotheses leap to mind.... However, genetic hypotheses should also be considered. Research in behavior genetics is directed toward understanding differences in behavior. Methods are employed that consider both genetic and environmental influences, rather than assuming that one or the other is solely important. As a first step, behavioral genetics research studies whether individual differences are influenced by hereditary differences and estimates the relative influences of genetic and environmental factors. (Plomin et al. 1990, p. 5)

As pointed out earlier, "heredity" here has nothing to do with genes that one can measure directly. It is a statistical entity estimated from the degree to which behavioral similarity among classes of relatives correlates with their predicted average (not actual) genetic similarity:

The fundamental tenet of quantitative genetic theory is that genetic differences among individuals can lead to phenotypic differences, even for complex traits (such as behavior) that are highly polygenic and influenced as well by nongenetic factors. Genetic influence does not imply genetic determinism in the sense of a direct or close relationship between genes and their effects on behavior. The pathways from gene expression through cells, tissues, and organs to behavior are likely to be very complex. Genetic influence means only that genetic differences among individuals relate to behavioral differences observed among them; no specific genetic mechanisms or gene-behavior pathways are implied.... That is, quantitative genetic methods assess the total impact of inherited genetic variability of any kind, regardless of its molecular source. (Plomin et al. 1990, pp. 247–8)

The "environment" is whatever residual variance has not been wrung out of the phenotype after removing the variation supposedly due to "heredity":

It should be mentioned that quantitative genetics employs a much broader definition of environment than is usual is psychology. "Environment" in this context literally means "nongenetic," in the sense that it is that portion of variance that cannot be accounted for by heredity. This definition of environment thus includes biological factors, such as anoxia at birth, prenatal effects of drugs, and even environmental influences on DNA itself, as well as traditional environmental factors, such as childrearing, school environments, and peers. (Plomin et al. 1990, pp. 248–9)

This view of the world quickly encounters some troubling ambiguities. For instance, if one finds that identical twins reared separately are behaviorally similar, this could be due to their "family environment" being similar by chance in the separate cases – an "environmental" effect. It could also happen if the behavioral parameter one measures in the twins is not responsive to "environmental" influence – a "genetic" effect. A third possibility is that the twins' genotype influences the "family environments" in a similar way, and thus this ostensibly "environmental" effect is really a "genetic" one (the possibility that the intrauterine environment, where most of neural development occurs, may play some role in behavioral similarities is rarely considered).

A solution to these ambiguous situations was initially proposed by Sandra Scarr and her colleagues, who posited that the "environment" (the left-over variance) may be in large part elicited by the genetically influenced characteristics of individuals themselves, or that genetically influenced characters would cause individuals to actively choose particular "environments." Other researchers have been attempting to define and measure such influences, or, as Plomin and Bergeman (1991) have put it, the "nature of nurture" (Scarr and McCartney 1983; Plomin and Bergemann 1991 [and associated commentaries]; Braungart, Fulker, and Plomin 1992; Chipuer et al. 1993; Kendler et al. 1993).

Kendler et al. (1993) introduced an article titled "A Twin Study of Recent Life Events and Difficulties" with the following statement (a "life event" is a death, illness, injury, or personal crisis in a relative or friend, or personal tragedies such as marital, work or interpersonal difficulties, being robbed or assaulted, financial or legal problems, illness or injury):

We do not wish to imply that genes "code" for life events as they do for eye color or blood group. It is, however, plausible that a number of human traits, like personality, which are influenced by genetic factors, affect the probability of experiencing life events. These questions are interesting because a major focus of psychiatric research has been to understand the relationship between environmental stress . . . and the onset or recurrence of psychopathic conditions. It will be difficult to reach a fuller understanding of this key issue if we do not understand the major influences on the occurrence of adverse life events themselves. (p. 791)

In a penultimate section of the article called "Implications," they conclude:

In particular, our results argue strongly against the validity of a random model of life events, in which having lots of life events is just a reflection of "bad luck." We found that more than 40% of the total variance in life events was due to genetic or familial-environmental factors. . . . The genetic or familial–environmental influences on mental illness may not "directly" increase vulnerability to illness but may instead increase risk for psychiatric illness by predisposing individuals to *create for themselves high-risk environments.* (p. 795, emphasis as in original)

Kendler et al. (1993) believe that their results

argue for a rethinking of our concept of the "environment". . . . Epidemiologists often conceptualize the environment as something "out there," which impinges on the organism in a unidirectional manner. Our results require a more dynamic concept of the environment where the individual and the environment influence one another in a bidirectional fashion. (p. 795)

Plomin, Owen, and McGuffin (1994) feel that this type of research has made two important "discoveries":

First, the way in which the environment influences behavioral development contradicts socialization theories from Freud onward. For example, the fact that psychopathology runs in families has reasonably, but wrongly, been interpreted to indicate that psychopathology is under environmental control. Research shows that genetics generally accounts for this familial resemblence. Environmental influences on most behavioral disorders and dimensions serve to make children growing up in the same family different, not similar. . . . The second discovery . . . is that ostensible measures of the environment appear to assess genetically influenced characteristics of individuals. To some extent, individuals create their own experiences for genetic reasons. In addition, genetic factors contribute to the prediction of developmental outcomes from environmental measures. (p. 1735)

The "environment" to an ecologist or to a population, evolutionary, or developmental biologist does not consist of a garbage can in which the residual terms of a statistical model are thrown, nor has it been regarded as something that an organism's behavior cannot alter. It is a real, physical entity with a countably infinite set of identifiable features of import for organisms like gravity, temperature, pH, concentrations of substances that cross the placenta, and the genotypes of other individuals: it may be messy and multidimensional, but one can take some of these dimensions, independently measure them, and try to understand the way they interact with the developing organism. It is in fact the measurement of the "environment" that most strongly illustrates the difference in worldview between the authors of these heritability studies and most biologists.

To document the "environment" validly, one must measure its variation independently from measurements of the variation in the organism's phenotype (especially if one thinks that the two interact). The "environmental measures" employed in these heritability studies fail to do this. Here, "environments" are normally assessed either retrospectively by "rating scales," which use questions such as, Did you receive more attention from your mother or father? or Did you have stricter discipline as a child or as an adolescent? or by retrospectively rating the amount of general categories of parental behavior, behavior with siblings and friends, things about the home, and so forth, that subjects remember. Similar interviews may also be conducted with parents and family members of the subjects. "Environment" can also be assessed "more directly" by "interview instruments" such as the Home Observation for Measurement of the Environment (HOME, Caldwell and Bradley 1978). It is not surprising that these measures show "heritabilities" similar to the behaviors measured in the same studies because they are the same kind of "virtual" behavioral measures (see the next section), often from the same subjects as the ones whose own "behavior" (responses to questions about oneself rather than about one's parents, etc.) is being studied.

If you cannot (or do not want to) independently measure an "environment" to document whether individuals develop in the same environments or not, you can still attempt to "control" these environments. If you can ensure the equality of "environmental" parameters by deliberately controlling them, then you at least have a rationale for treating the "environment" like a noise term, as heritability models do. The key here is to ensure that there is no covariation in the "environment" (as defined by a biologist) between subjects that differ genetically. If you cannot ensure this, phenotypes are a fundamentally contaminated and therefore unreliable species of data with respect to "genetic" and "environmental" sources of variation. Post hoc attempts to partition causation between these two presumed sources will be without meaning.

Of course, the "goals" filter may be strongly at work here. Empirical biological research trying to specify the complex pathways from genes to behavior may use different rules of evidence than the enterprise of telling social scientists that they have the wrong theoretical orientation. This may be why researchers who use heritabilities have adopted an indirect definition of "heredity" and a definition of "environment" that is at odds with the rest of biology. By treating the deep problem of how to deal with behavioral autonomy as an opportunity for throwing additional "variance" from the bin marked "environment" into the bin marked "genetic," heritability studies have failed to address it at all.

We need to keep the worthy idea that an individual's characteristics will affect the qualities of his or her own experiences through effects on the person's "environment" (which has been kicking around behavioral biology since the ninteenth century) but to focus more on how it is that those characteristics develop in the first place in a system in which individual characteristics and the forces that affect them always interact.

22.2.2 "Virtual Behavior": When the Phenotype Is No Longer the Actual Behavior

Indeed, we bet that watching *Jeopardy* does not show genetic influence because it is too specific a response. (Plomin and Bergemann 1991, p. 418)

A curious feature of many of the behavioral phenotypes studied in behavioral genetics is that they are fictional quantities distilled from actual behavioral measurements. These could be termed "virtual behaviors." We will touch upon three classes of virtual behaviors: making the difference between an organism and the mean of the population it is assigned to into the phenotype being studied; "thresholding" – taking a behavioral characteristic that is continuously distributed and introducing a theoretical variable that makes the distribution discontinuous, the theoretical variable then becoming the "phenotype" under study; and the making of "conglomerate" indices such as IQ or personality tests.

22.2.2.1 Distance from the Population Mean

Human and animal behavioral genetic studies frequently abandon direct behavioral measures in favor of a subtly different approach that still seems like a direct measurement but is not. Behavioral measures are "normalized" so that the actual phenotype is the distance of individuals from the mean of the "population" they belong to. The phenotype is now a population rather than an individual character because it depends on the states of other members of the study population. The same subject will have a different "phenotype" depending on the mixture of other subjects it is pooled with. Frequently, "population" membership is determined by subjective criteria; that is, populations can be fictive creations of the investigators themselves.

In presenting and discussing such measures, authors often conflate population-level and individual-level characteristics. For instance, Neale and Cardon (1992) write as follows in their book *Methodology for Genetic Studies of Twins and Families*:

It is vital to remember that almost every result in this book, and every conclusion that others obtain using these methods, relate to the causes of human differences, and may have almost nothing to do with the processes that account for the development of the mean expression of a trait in a particular population. We are necessarily concerned with what makes people vary around the mean of the population, race, or species from which they are sampled. . . . This point is stressed because, whatever subsequent genetic research on population and species differences might establish, there is no necessary connection between what is true of the typical human and what causes variation around the central tendency. For this reason, it is important to avoid short-hand expressions as "height is genetic" when we really mean "individual differences in height are mainly genetic." (pp. 4–5)

Note that it is misleading to state that such techniques study "individual differences." Explaining the distance of different individuals from the population mean is not the same thing as explaining why individuals differ from each other. Individuals do not have one set of developmental processes "that account for the development of the mean expression of the trait" and another set accounting for variation about this mean. Population quantities like mean and variance are averages of a particular mixture of events occuring within individuals in the population. It is both illogical and unbiological to make up "virtual" phenotypes of trait means and trait variances. Behavioral phenotypes are individual and not population attributes.

Population measures like the distance from the mean are appropriate in the realm of animal or plant breeding, where the characteristics under study really are populational rather than individual (the mean phenotype of the population) and where the desired goal is to select artificially for populations whose mean is some distance away from that of a reference "parental" population. This measure is not suited to most behavioral genetic studies that are concerned primarily with explaining individual (not population) differences. "Distance from the mean" measures are only applied because of the extension of "heritability" techniques into a domain they were never designed for.

22.2.2.2 Thresholding and the "Risk" of Behavioral Phenotypes

Thresholding consists of imposing a discontinuity on behavioral variation that breaks it into different discrete classes. One then hypothesizes that there is a normally distributed liability variable (or variables) that influence(s) the probability that an individual falls into one or another of these behavioral classes depending on where the "threshold" for the hypothetical liability curve falls. An individual's "liability" becomes the phenotype under study. If one assumes that the threshold falls in the same place for all groups of individuals being compared and that the variance in the liability curves of all groups is the same, then one can see if groups differ in their "mean liability."

The thing that makes liability analysis simultaneously appealing and appalling is the ease with which biologically plausible "liability factors" can be generated. All that such a model requires is that "underlying variables that combine to give the liability could be made normal by a scale transformation" (Falconer 1989). However, if "liability" is not unimodally distributed, no amount of transformation can render it so. Falconer discusses two ways in which this could theoretically happen: single genes of large phenotypic effect (which have been historically rare in studies of the genetics of "normal" behavior but not in the genetics of disease), or environmental factors whose net effect is large in relation to gene effects on the phenotype (a case that Falconer subsequently ignores). In human work, when "environmental factors" are not systematically controlled or even measured, this latter situation is highly likely to occur. Falconer does give one example of an environmental factor with an extremely large effect in the context of disease: exposure to a pathogen.

Liability analysis may be largely inapplicable to behavioral characters except in rare cases in which there is an overwhelming linear combination of gene effects in a particular direction. Because of the extreme generality of the model, we frequently have the problem of more degrees of freedom in the "explanation" than in the data it is trying to explain. Liabilities then become "fudge factors" like penetrance and expressivity in Mendelian models of autosomal dominant traits: free parameters one can fiddle with to make the model fit any set of observations.

22.2.2.3 Conglomerate Indices

The previous two examples could simply be the product of simplicity filters of training and methods: people warping behavioral measures into virtual entities to apply their favorite techniques to situations in which their applicability is dubious. This last category has a lot more to do with scaling, labeling, and bias in behavioral measurements as well as with the goals of research.

Let us suppose our goal is to prove to a doubting world that genes have any sort of effect on human behavior (there is no distinction here between trivial and specific effects – any effect will do). There is a way that we could bias our behavioral measurements so that most of the time we would be assured of finding genes that affect behavior. From a cynical reading of recent literature in behavioral genetics, the strategy could be caricatured as follows: make up phony "behaviors" that are a huge conglomeration of an unidentifiable number of very different neural and nonneural physiological processes and label them as if they were measures of one thing that people really care about.

As suggested by Plomin and Bergeman in the quote that opened this section, do not look for a behavior that is too "specific" – one wants to lump as many effects under a single label as possible to maximize the probability of finding genetic variation that impacts on this "behavior." As long as one can derive a performance measure based on the conglomeratation that corresponds to current prejudices about attributes like "intelligence" or "personality," one can play the game of saying that what is being measured is not really important because performance on the test reliably "predicts" individuals with more or less of the quality named by the label. As long as you can maintain the fiction that the virtual behavior is a good stand-in for any deep knowledge about the real entities one is alluding to, you are home safe, because if you spread your virtual behavior widely enough you can certainly find genes that will impact on it. You will win because you will find genes, even if you have to change your own rules of evidence to find them, as Plomin, McClearn et al. (1994) did in their study of markers for high and low IQ. They used two independent groups of subjects: one as a primary group in which to test gene associations with IQ, and the second as a control group for sampling error in which to independently replicate marker associations found in the first group. When none of the associations replicated, they claimed that this justified lumping the groups together and reporting the combined results as significant – long live independent replication!

There is a twisted and shifting logic applied to these situations whereby watching *Jeopardy* is too specific a response to show "genetic influence," but "willingness to drive when drunk" (Martin and Boomsma 1989) is not (and "religious affiliation" [Eaves, Martin, and Heath 1990] can serve as a test case for "cultural" as versus "genetic" models of behavioral transmission). In "virtual" behaviors like these it is impossible to reasonably understand any behavioral contingencies because the abilities called upon by these "indices" are so varied and so muddled by the method of lumping them together to form a test score. The labels are unreliable because they tell you about the interests of the designers of the test, not about what it actually measures at a more concrete level, which is never determined in detail. Conglomerate indices represent filtering at its most pernicious because information on their construction is frequently unavailable owing to proprietary concerns (many of them have considerable commercial success). Although Blinkhorn and Johnson (1990) discuss personality tests in the realm of job applicant screening, their critical characterizations of test development and use apply with equal force to the "instruments" used in human behavioral genetics today.

22.2.3 The Study of Development

There is disagreement in the communities of researchers studying behavior and genetics over how to study development. Much of the dialogue in the human behavior–genetics community about development has been co-opted by heritability modelers, who have filtered the study of genetics and behavioral development in two misleading ways. The first is to reduce the study of development to studying how heritabilities change with age (Matheny 1990; Corley and Fulker 1990; Hewitt 1990a,b; Eaves, Hewitt et al. 1990; Henderson 1990; Plomin, Owen, and McGuffin 1994). The second is to conceptualize the relationship of genotypes, phenotypes, and environments as genes specifying a set of constrained phenotypes across a set of environments called a "reaction range" (phenotype versus environment for one gentoype) or "reaction surface" (phenotype versus environment versus genotype) (Gottesman 1963, 1974; Turkheimer and Gottesman 1991; Gottesman and Goldsmith 1994; Turkheimer, Goldsmith, and Gottesman 1995).

Proponents of the "development as changing heritabilities" view suggest that heritability changes indicate differences in the extent to which genes exert "control" over behaviors during ontogeny rather than fluctuations in an unreliable estimate. Hewitt (1990b) maintains that, using heritabilities and genetic correlations (estimated from phenotypic correlations),

> we can test whether the same genes influence behavior in the same way at different ages. . . . The answers to such questions come from our ability to estimate both the genetic effects at two or more ages *and the genetic correlation between the effects at one age and those at another.* Different sets of genes acting at different ages will result in low genetic correlation. We can take our analysis of developmental continuity a step further and test whether the effects of genes, manifest at one age, persist in the organism at older ages, or whether continuity of genetic influences is a consequence of continuous de novo activity of genes. For example, do genetic influences on conduct disorder manifested early in childhood have persistent consequences for later adolescence, or are the genetic influences on later adolescent behavior unaffected by earlier expression? This test . . . is provided by the different patterns of variances and covariances predicted for longitudinal twin data by the different mechanisms that give rise to age to age correlations. The practical importance of the distinction between these two mechanisms is that in the absence of persistence, intervention to prevent the expression of genes at one age will have no benefits at later ages unless the intervention is continued. (pp. 298–9, emphasis in original)

And Scarr (1995) asserts that

> an arrogance exists among some mechanistic experimentalists in developmental research who would like to claim that theirs is the only legitimate approach to studying development. They are, however, a tiny minority among behavioral scientists. Most behavioral scientists do not conduct laboratory experiments on isolated variables, because this approach has not proved fruitful in addressing important human problems or in advancing our understanding of the development of complex human phenotypes. . . . In the normal range of human variation, however, mechanistic approaches – whether molecular genetics, molecular psychobiology, or molecular behaviorism – have not proved useful in advancing knowledge about individual variation in intelligence, personality, or mental health. By contrast, correlational studies, whether motivated by developmental behavior genetics or other population-based theories, have identified complex webs of predictors and outcomes that have proved useful in understanding and improving human lives. . . . I have found that students experience a nearly revelatory mind shift when they realize that they can think about both individual development in a mechanistic sense and about sources of individual variation in populations. (p. 157)

There seems once again to be a fundamental confusion between measures of population properties (such as heritability and genetic correlation), estimated in a linear model by lumping together subjects from a large number of genetically very different lineages and parsing out average "effects," and measures of properties shown by individuals (the developmental and ongoing dynamic action of genes). One can always imagine a sequence of developmental events involving genes in an individual which (if all individuals in the population were homogeneous for this set of genes, and genotype led to phenotype in the same way in all individuals) would produce the observed pattern of heritability and genetic correlational change as a function of age. The problem is that there are a huge number of different scenarios that could also be consistent with these events, and thus consistency is not a good measure for evaluating the usefulness of this procedure. There is in fact no way of evaluating which of the many possibilities is true without uncovering the actual developmental processes and examining the action of genes that impact on them. What do we gain by calculating heritability and genetic correlation as a function of age? Some may say that this exercise is useful because we rule out a few hypotheses even though we still have a universe of others. But this is only true if what is called "genetic" in the model that parses phenotypic variation is mediated via the action of identifiable genes. I have argued earlier in this chapter that this is not necessarily the case.

The phrasing of potential applications of these methods by Hewitt (and the claims of their superiority by Scarr) are also absurd. Is it a good idea to tell the public that longitudinal heritability analyses will lead to the identity of a discrete set of genes "influencing" something as diffusely defined as conduct disorders and that an appropriate intervention to forestall them would be to "prevent the expression of genes"? Although these analyses may provoke someone to search for such genes using other methods, heritability analyses will not directly contribute to finding any.

It is true that gene products do not only participate in developmental processes, for they also participate in the ongoing dynamics of keeping an organism running. The former, more "organizational" effects of genes are mostly over by the time an organism like a human is born. The latter functions of genes continue throughout life. Dynamically regulated genes are part of an organism's response to coping with a slowly changing body and a world that has important change and variety simultaneously at many different time scales. Separating "organizational" from "dynamic" effects of genes is not always possible because a gene product expressed at one time can serve both functions. We do not understand the relationship of temporal variation in gene expression to temporal variation in behavioral phenotypes in any meaningful way except to note that gene products are typically produced well in advance of when they actually "function" in a developing organism. One thing heritability analyses will never do is to tell you when to turn particular genes on or off to get particular phenotypic effects.

The second misconception about development, that of the "reaction range," is based on a concept derived from an empirical result in biology, which made its appearance in the same year as the "gene" (Woltereck 1909). Organisms that have the same genotype do not have the same phenotype when raised in environments that are quantitatively different in some measurable parameter. An empirical plot of phenotypes associated with different environments for single genotypes is called a "norm of reaction" (Falk, this volume). In biology, this empirical plot has been used in two different ways: to emphasize the unpredictability of the phenotype from the genotype at the level of the individual (and vice versa) and to study how natural selection makes certain "important" features of the phenotype relatively independent of the environment, which is called "buffering" (Falk, this volume). A recent series of articles in the journal *Human Development* (Gottlieb 1995a,b; Turkheimer et al. 1995; Scarr 1995; Burgess and Molenaar 1995) has critically explored some of the differences between the "reaction range or surface" and the "norm of reaction" and some of the reasons heritability modelers find these concepts to be of value. What is missing from this dialogue and from the idea of reaction ranges or surfaces is an interface with biological reality.

In biology a norm of reaction can be an empirical fact. One can put genetically identical organisms into a series of carefully defined and controlled environments and measure the phenotypes obtained and their variance. Although we could proceed to measure this in all environments we can imagine, we have absolutely no theory that would tell us what a phenotype would look like if we put our chosen genotype in a new environment. This is what we would need if we would ever want to talk cogently of predicting anything about phenotypes from genotypes, especially in humans, for whom it is not possible to measure norms of reaction empirically. Norms of reaction and reaction ranges are descriptors, not predictors. Heritability modelers are happy to assume that development has just those characteristics that make reaction ranges predictive. People who do not like such models will criticize them for making these assumptions. Until we understand something more about why particular genotypes give rise to particular phenotypes in particular environments, this will be a sterile exercise.

22.3 Can the Filters Be Deconvoluted?

As genetic techniques become more powerful and behavioral sophistication is increasingly left behind, we have to face the possibility that much of what will be discovered in the next decade about "specific" relationships between genotypes and behavioral phenotypes using current approaches for characterizing behavior will not even be wrong: it will simply be a ridiculous dead end. There is still considerable societal shock value (and attendant publicity with all of the goodies this can bring to individual researchers) in talking about genetic variation underlying intelligence, personality, fearfulness, sexual orientation, or social success. It is much more difficult to actually say what we are measuring when we presume to study such phenotypes. The labels we stick onto what we study and the specificity we think that gene effects wield over these traits advance researcher's individual ambitions more than they do our knowledge of biology. If these measures are the best people can do at the present time to study complex, "higher-order" behaviors, then they should be used with extreme conservatism and caution lacking in work published to date. Better yet, those interested in studying such questions should put much more energy into finding behavioral measures more directly reflecting the characteristics they want to study.

There are findings of real value that can emerge from putting genetic techniques to work with behavioral phenotypes, but in order to realize them we must reconcile the different convolutions performed by the simplicity filters we have discussed. This will only happen when we increase our degree of sophistication in devising and interpreting behavioral measurements and when the diverse groups of researchers trying to understand how genetic variation and behavior relate to each other find some common ground.

It is difficult to study behavior in a more sophisticated fashion. The first step in this process will be achieved with a more widespread recognition of some of the pitfalls of the common simplicity filters discussed here. Some ways to do this might include the following:

1. Making an effort to be more explicit about goals and assumptions of research and attempting to evaluate bias in behavioral measures by measuring the same behavior in multiple ways and by having independent investigators critically evaluate measures for possible sources of bias;
2. Making more serious attempts to evaluate and report both the genetic and behavioral specificity of potential gene effects on behavior;
3. Gathering more valid information on contextual covariates and including these in behavioral analyses;
4. Giving more attention to the problems of quantifying environments and finding valid quantitative measures for environmental variation;
5. Applying labels to behavioral tests (and the genes that affect them) with extreme caution and making the labels correspond more closely to the quantities being measured.

A common ground for relating genes and behavior is logically provided by those processes that intervene between a genotype and a behavioral phenotype: the processes of development, especially development of the nervous system. A start to a more useful dialogue has been provided by dynamic models based on explicit developmental events such as one proposed to describe defects in heart formation by Kurnit, Layton, and Matthysse (1987) and a model of early visual system development proposed by Fraser and Perkel (1990). These models portray developmental events in which the timing of gene expression and structural changes in genes affecting cell–cell interactions could have major effects on the phenotype. But at the same time, because the models are of cellular processes, stochastic events involving the cell interactions and events external to the cells ("environmental") can affect the same developmental processes in the same ways. It would not be difficult to imagine similar models (applied to the development of the nervous system) being used to examine behavioral pathology and, with sufficient sophistication, to variation in normal behavior.

There will be a true meeting of the minds when enough of us, in spite of differences in goals, training, and methods decide it is worth our while to work together on what will surely be the "glue" for relating genotypes to behavioral phenotypes: the explication of the complexity and diversity of the developmental and functional pathways that participate in building and running nervous systems via both stochastic processes and processes linked to events in the "environment." Here the computer and mathematical expertise of the heritability modeler will surely synergize better with the biological expertise of the molecular biologist, neurobiologist, and behavioral scientist. A more neutral scientific base and an appropriate explanatory context for posing questions about the involvement of biology in the genesis of complex behavioral phenotypes may help to deconvolve the most egregious simplicity filters, which seriously hamper our understanding of the role of biology in behavioral systems.

22.4 Acknowledgments

I thank Richard Lewontin for his friendship and the many contributions he made (both before and after I actually met him) to my thinking about biology. I am grateful to Raphael Falk, Gilbert Gottlieb, and Benno Muller-Hill for helpful comments. The preparation of this manuscript was supported by the NIMH (MH47149) and by the Neurosciences Research Foundation.

REFERENCES

Aceves-Pina, E. O., Booker, R., Duerr, J. S., Livingstone, M. S., Quinn, W. G., Smith, R. F., Sziber, P. P., Temple, B. L., and Tully, T. P. (1983). Learning and memory in *Drosophila*, studied with mutants. *Cold Spring Harbor Symposia in Quantitative Biology* 48:831–40.

Arnold, S. J. (1990). Inheritance and the evolution of behavioral ontogenies. In *Developmental Behavior Genetics*, (eds.) M. E. Hahn, J. K. Hewitt, N. D. Henderson, and R. Benno, pp. 167–89. Oxford, UK: Oxford University Press.

Arnold, S. J. (1994). Multivariate inheritance and evolution: A review of concepts. In *Quantitative Studies of Behavioral Evolution*, (ed.) C. R. B. Boake, pp. 17–48. Chicago: University of Chicago Press.

Balaban, E. (1996). Reflections on Wye Woods: Crime, biology, and self-interest. *Politics and the Life Sciences* 15:86–8.

Balaban, E., Alper, J. S., and Kasamon, Y. (1996). Mean genes and the biology of aggression. A critical review of recent animal and human research. *Journal of Neurogenetics* 11:1–43.

Begley, S. (1995). The mice that roared ... and raped and killed, due to a genetic defect. *Newsweek* 126:76.

Benjamin, J., Li, L., Patterson, C., Greenberg, B. D., Murphy, D. L., and Hamer, D. H. (1996). Population and familial association between the D4 dopamine receptor gene and measures of novelty seeking. *Nature Genetics* 12:81–4.

Blinkhorn, S., and Johnson, C. (1990). The insignificance of personality testing. *Nature* 348:671–2.

Boake, C. R. B., (ed.) (1994). *Quantitative Studies of Behavioral Evolution.* Chicago: University of Chicago Press.

Bouchard, T. J., Jr. (1994). Genes, environment and personality. *Science* 264:1700–1701.

Bouchard, T. J., Jr., Lykken, D. T., McGue, M., Segal, N. L., and Tellegen, A. (1990). Sources of human psychological differences: The Minnesota study of twins reared apart. *Science* 250:223–8.

Braungart, J. M., Fulker, D. W., and Plomin, R. (1992). Genetic mediation of the home environment during infancy: A sibling adoption study of the HOME. *Developmental Psychology* 28:1048–55.

Brunner, H. G., Nelen, M., Breakefield, X. O., Ropers, H. H., and van Oost, B. A. (1993). Abnormal behavior associated with a point mutation in the structural gene for monoamine oxidase A. *Science* 262:578–80.

Brunner, H. G., Nelen, M. R., van Zandvoort, P., Abeling, N. G. G. M., van Gennip, A. H., Wolters, E. C., Kuiper, M. A., Ropers, H. H., and van Oost, B. A. (1993). X-linked borderline mental retardation with prominent behavioral disturbance: Phenotype, genetic localization, and evidence for disturbed monoamine metabolism. *American Journal of Human Genetics* 52:1032–9.

Burgess, R. L., and Molenaar, P. C. M. (1995). Commentary on "Some conceptual deficiencies in 'developmental' behavior genetics." *Human Development* 38:159–64.

Caldwell, B. H., and Bradley, R. H. (1978). *Home Observation for Measurement of the Environment.* Little Rock: University of Arkansas Press.

Chen, C., Rainnie, D. G., Greene, R. W., and Tonegawa, S. (1994). Abnormal fear response and aggressive behavior in mutant mice deficient for α-calcium-calmodulin kinase II. *Science* 266:291–4.

Chipuer, H. M., Plomin, R., Pedersen, N. L., McClearn, G. E., and Nesselroade, J. R. (1993). Genetic influence on family environment: The role of personality. *Developmental Psychology* 29:110–8.

Cloninger, C. R., Adolfsson, R., and Svrakic, N. M. (1996). Mapping genes for human personality. *Nature Genetics* 12:3–4.

Corfas, G., and Dudai, Y. (1990). Adaptation and fatigue of a mechanosensory neuron in wildtype *Drosophila* and in memory mutants. *Journal of Neuroscience* 10:491–9.

Corley, R. P., and Fulker, D. W. (1990). What can adoption studies tell us about cognitive development? In *Developmental Behavior Genetics,* (eds.) M. E. Hahn, J. K. Hewitt, N. D. Henderson, and R. Benno, pp. 236–65. Oxford, UK: Oxford University Press.

Crabbe, J. C., Belknap, J. K., and Buck, K. (1994). Genetic models of alcohol and drug abuse. *Science* 264:1715–23.

Crawley, J. N. (1996). Unusual behavioral phenotypes of inbred mouse strains. *Trends in Neuroscience* 19:181–2.

Crowe, R. R. (1993). Candidate genes in psychiatry: An epidemiological perspective. *American Journal of Medical Genetics* (*Neuropsychiatric Genetics*) 48:74–7.

Crusio, W. (1996). Gene-targeting studies: New methods, old problems. *Trends in Neuroscience* 19:186–7.

De Belle, J. S., and Heisenberg. M. (1996). Expression of *Drosophila* mushroom body mutations in alternative genetic backgrounds: A case study of the mushroom body miniature gene (*mbm*). *Proceedings of the National Academy of Sciences USA* 93:9875–80.

DeFries, J. C., and Fulker, D. W. (eds.) (1986). Symposium on multivariate behavioral genetics and development. *Behavior Genetics* 16:1–235.

Delgado, R., Hidalgo, P., Diaz, F., Latorre, R., and Labarca, P. (1991). A cyclic AMP-activated K+ channel in *Drosophila* larval muscle is persistently activated in dunce. *Proceedings of the National Academy of Sciences USA* 88:557–60.

Eaves, L. J., Hewitt, J. K., Meyer, J., and Neale, M. (1990). Approaches to the quantitative genetic modeling of development and age-related changes. In *Developmental Behavior Genetics,* (eds.) M. E. Hahn, J. K. Hewitt, N. D. Henderson, and R. Benno, pp. 266–80. Oxford, UK: Oxford University Press.

Eaves, L. J., Martin, N. G., and Heath, A. C. (1990). Religious affiliation in twins and their parents: Testing a model of cultural inheritance. *Behavior Genetics* 20:1–22.

Ebstein, R. P., Novick, O., Umansky, R., Priel, B., Osher, Y., Blaine, D., Bennett, E. R., Nemanov, L., Katz, M., and Belmaker, R. H. (1996). Dopamine D4 receptor exon III polymorphism associated with the human personality trait of novelty seeking. *Nature Genetics* 12:78–80.

Falconer, D. S. (1989). *Introduction to Quantitative Genetics,* 3d edition. New York: John Wiley & Sons.

Flint, J., Corley, R., DeFries, J. C., Fulker, D. W., Gray, J. A., Miller, S., and Collins, A. C. (1995). A simple genetic basis for a complex psychological trait in laboratory mice. *Science* 269:1432–35.

Fraser, S. E., and Perkel, D. H. (1990). Competitive and positional cues in the patterning of nerve connections. *Journal of Neurobiology* 21:51–72.

Gerlai, R. (1996a). Gene-targeting studies in mammalian behavior: Is it the mutation or the background genotype? *Trends in Neuroscience* 19:177–81.

Gerlai, R. (1996b). Molecular genetic analysis of mammalian behavior and brain processes: Caveats and perspectives. *Seminars in the Neurosciences* 8:153–61.

Gilbert, S. (1994). *Developmental Biology,* 4th edition. Sunderland, MA: Sinauer Associates.

Gottesman, I. I. (1963). Genetic aspects of intelligent behavior. In *The Handbook of Mental Deficiency: Psychological Theory and Research,* (ed.) N. Ellis, pp. 253–96. New York: McGraw-Hill.

Gottesman, I. I. (1974). Developmental genetics and ontogenetic psychology: Overdue detente and propositions from a matchmaker. In *Minnesota Symposium on Child Psychology* (*Vol. 6*), (ed.) A. Pick, pp. 55–80. Minneapolis: University of Minnesota Press.

Gottesman, I. I., and Goldsmith, H. H. (1994). Developmental psychopathology of antisocial behavior: Inserting genes into its ontogenesis and epigenesis. In *Threats to Optimal Development: Integrating Biological, Psychological, and Social Risk Factors,* (ed.) C. A. Nelson, pp. 69–104. Hillsdale, NJ: Lawrence Erlbaum Associates.

Gottlieb, G. (1995a). Some conceptual deficiencies in "developmental" behavior genetics. *Human Development* 38:131–41.

Gottlieb, G. (1995b). Reply to commentaries on "Some conceptual deficiencies in 'developmental' behavior genetics." *Human Development* 38:165–9.

Griffin, D. R. (1950). Foreword. In *Bees: Their Vision, Chemical Senses, and Language,* (ed.) pp. vii–xiii. Ithaca: Cornell University Press.

Griffin, D. R. (1976). *The Question of Animal Awareness: Evolutionary Continuity of Mental Experience.* New York: The Rockefeller University Press.

Hall, C. S. (1934). Emotional behavior in the rat: I. Defecation and urination as measures of individual differences in emotionality. *Journal of Comparative Psychology* 18:385–403.

Hall, J. C. (1994). The mating of a fly. *Science* 264:1702–14.

Hall, J. C. (1995). Pleiotropy of behavioral genes. In *Flexibility and Constraint in Behavioral Systems,* (eds.) R. J. Greenspan and C. P. Kyriacou, pp. 15–28. New York: John Wiley and Sons.

Hamer, D. H., Hu, S., Magnuson, V. L., Hu, N., and Pattatucci, A. M. L. (1993). A linkage between DNA marker on the X chromosome and male sexual orientation. *Science* 261:321–7.

Hen, R. (1996). Mean genes. *Neuron* 16, 17–21.

Henderson, N. D. (1990). Quantitative genetic analysis of neurobehavioral phenotypes. In *Developmental Behavior Genetics,* (eds.) M. E. Hahn, J. K. Hewitt, N. D. Henderson, and R. Benno, pp. 283–97. Oxford, UK: Oxford University Press.

Hewitt, J. K. (1990a). Changes in genetic control during learning, development, and aging. In *Developmental Behavior Genetics,* (eds.) M. E. Hahn, J. K. Hewitt, N. D. Henderson, and R. Benno, pp. 217–35. Oxford, UK: Oxford University Press.

Hewitt, J. K. (1990b). Genetics as a framework for the study of behavioral development. In *Developmental Behavior Genetics,* (eds.) M. E. Hahn, J. K. Hewitt, N. D. Henderson, and R. Benno, pp. 298–303, Oxford, UK: Oxford University Press.

Hu, S., Pattatucci, A. M. L., Patterson, C., Li, L., Fulker, D. W., Cherny, S. S., Kruglyak, L., and Hamer, D. H. (1995). Linkage between sexual orientation and chromosome Xq28 in males but not in females. *Nature Genetics* 11:248–56.

Kendler, K. S., Neale, M., Kessler, R., Heath, A., and Eaves, L. (1993). A twin study of recent life events and difficulties. *Archives of General Psychiatry* 50:789–96.

Kidd, K. K. (1993). Associations of disease with genetic markers: *Déjà vu* all over again. *Am. J. Med. Genet.* (*Neuropsychiatr. Genet.*) 48:71–3.

Kretschmer, E. (1936). *Physique and Character: An Investigation of the Nature of Constitution and of the Theory of Temperament.* New York: Cooper Square Publishers, Inc.

Kurnit, D. M., Layton, W. M., and Matthysse, S. (1987). Genetics, chance and morphogenesis. *American Journal of Human Genetics* 41:979–95.

Lande, R. (1984). The genetic correlation between characters maintained by selection, linkage, and inbreeding. *Genetical Research, Cambridge* 44:309–20.

Lander, E. S., and Kruglyak, L. (1995). Genetic dissection of complex traits: Guidelines for interpreting and reporting linkage results. *Nature Genetics* 11:241–7.

Lander, E. S., and Schork, N. J. (1994). Genetic dissection of complex traits. *Science* 265:2037–48.

Lashley, K. (1929). *Brain Mechanisms and Intelligence.* Chicago: University of Chicago Press.

Lathe, R. (1996). Mice, gene targeting and behaviour: More than just genetic background. *Trends in Neuroscience* 19:183–6.

Lewontin, R. C. (1974). The analysis of variance and the analysis of causes. *American Journal of Human Genetics* 26:400–411.

Malpighi, M. (1686). *Opera Omnia et Appendix Repetitas Auctasque de Ovo Incubato.* London: Robert Scott & George Wells.

Martin, N. G., and Boomsma, D. I. (1989). Willingness to drive when drunk and personality: A twin study. *Behavior Genetics* 9:97–112.

Matheny, A. P., Jr. (1990). Developmental behavior genetics: Contributions from the Louisville twin study. In *Developmental Behavior Genetics,* (eds.) M. E. Hahn, J. K. Hewitt, N. D. Henderson, and R. Benno, pp. 25–39. Oxford, UK: Oxford University Press.

Morris, R., and Nosten-Bertrand, M. (1996). NOS and aggression. *Trends in Neuroscience* 19:277–8.

Neale, M. C., and Cardon, L. R. (1992). *Methodology for Genetic Studies of Twins and Families.* Dordrecht, The Netherlands: Kluwer Academic Publishers.

Nelson, R. J., Demas, G. E., Huang, P. L., Fishman, M. C., Dawson, V. L., Dawson, T. M., and Snyder, S. H. (1995). Behavioral abnormalities in male mice lacking neuronal nitric oxide synthase. *Nature* 378:383–6.

Plomin, R., and Bergeman, C. S. (1991). The nature of nurture: Genetic influence on "environmental" measures. *Behavioral and Brain Sciences* 14:373–427.

Plomin, R., DeFries, J. C., and McClearn, G. E. (1990). *Behavioral Genetics: A Primer,* 2d edition. New York: W. F. Freeman.

Plomin, R., McClearn, G. E., Smith, D. L., Vignetti, S., Chorney, M. J., Chorney, K., Venditti, C. P., Kasarda, S., Thompson, L. A., Detterman, D., Daniels, J., Owen, M., and McGuffin, P. (1994). DNA markers associated with high versus low IQ: The quantitative trait loci (QTL) project. *Behavior Genetics* 24:107–18.

Plomin, R., Owen, M. J., and McGuffin, P. (1994). The genetic basis of complex human behaviors. *Science* 264:1733–9.

de Réaumur, René, A. F. (1740). *Mémoires pour servir a l'histoire des insectes,* Vol. V., Paris: Imprimarie Royale.

Risch, N., and Merikangas, K. (1996). The future of genetic studies of complex human diseases. *Science* 273:1516–17.

Risch, N., and Zhang, H. (1995). Extreme discordant sib pairs for mapping quantitative trait loci in humans. *Science* 268:1584–9.

Rosenthal, R. (1966). *Experimenter Effects in Behavioral Research.* New York: Appleton–Century–Crofts.

Scarr, S. (1995). Commentary on "Some conceptual deficiencies in 'developmental' behavior genetics." *Human Development* 38:154–7.

Scarr, S., and McCartney, K. (1983). How people make their own environments: A theory of genotype–environment effects. *Child Development* 54:424–35.

Silva, A. J., Paylor, R., Wehner, J. M., and Tonegawa, S. (1992). Impaired spatial learning in α-calcium-calmodulin kinase II mutant mice. *Science* 257:206–11.

Spitzner, M. J. E. (1788). Ausfürliche Beschreibung der Korbbienenzucht im sächsischen Churkreise. Leipzig. Translated in von Frisch, K. (1967). *The Dance Language and Orientation of Bees.* Cambridge, UK: Belknap Press, p. 6.

Takahashi, J. S., Pinto, L. H., and Vitaterna, M. H. (1994). Forward and reverse genetic approaches to behavior in the mouse. *Science* 264:1724–33.

Tecott, L. H., and Barondes, S. H. (1996). Genes and aggressiveness. *Current Biology* 6:238–40.

Thorpe, W. H. (1949). Orientation and methods of communication of the honey bee and its sensitivity to the polarization of the light. *Nature* 164:11–14.

Tully, T. (1991). Physiology of mutations affecting learning and memory in *Drosophila*–The missing link between gene product and behavior. *Trends in Neuroscience* 14:163–4.

Turkheimer, E., Goldsmith, H. H., and Gottesman, I. I. (1995). Commentary on "Some conceptual deficiencies in 'developmental' behavior genetics." *Human Development* 38:142–53.

Turkheimer, E., and Gottesman, I. I. (1991). Individual differences and the canalization of human behavior. *Developmental Psychology* 27:18–22.

Unhoch, N. (1823). Anleitung zur wahren Kenntnis und zweckmäβigsten Behandlung der Bienen Munich. Translated in von Frisch, K. (1967). *The Dance Language and Orientation of Bees.* Cambridge, UK: Belknap Press, p. 7.

von Frisch, K. (1946). Die Tanz der Bienen. *Östrr. Zool. Zeit.* 1:1–48.

Wahlsten, D. (1990). Insensitivity of the analysis of variance to heredity–environment interaction. *Behavioral and Brain Sciences* 13:109–61.

Whitney, G. (1970). Timidity and fearfulness of laboratory mice: An illustration of problems in animal temperament. *Behavior Genetics* 1:77–85.

Woltereck, R. (1909). Weitere experimentelle Untersuchungen über Artveränderung, speziell über des Wesen quantitativer Artunterscheide bei Daphnien. *Verhandlungen der Deutschen Zoologischen Gesellschaft* 19:110–73.

Zhong, Y., and Wu, C.-F. (1991). Altered synaptic plasticity in *Drosophila* memory mutants with a defective cyclic AMP cascade. *Science* 251:198–201.

CHAPTER TWENTY-THREE

Identity Politics and Biology

RUTH HUBBARD

Identity politics is part of a process of liberation adopted by oppressed groups, but the significance of the boundaries (women, blacks, gays, etc.) is established in the culture at large with the result that overlaps are ignored and at least some aspects of such identities are cast in a negative light. As part of their resistance, the oppressed build internal cohesion and pride within the group and devise strategies for resistance such as forming coalitions among overlapping groups.

To be successful in their identity politics, people must acknowledge their oppression and identify the oppressor. Beyond that, following Freire (1993), the oppressed must rid themselves of the image of themselves which they internalize while growing up in an oppressive culture, for that image is designed to make them accept their oppression as just and their due. In their identity politics, oppressed groups therefore tend to assign positive value to the stigma(s) on which their oppression focuses.

Oppressed and excluded groups are almost always characterized as different *in kind* from the dominant culture. In the last two or three centuries in the West, these groups' stigmatized differences have often been turned into biological essences from which there is no escape. The oppressed, then, must either accept such assignments, while reinterpreting them as positive, or show that the naturalization is wrong and that the identity has been socially constructed or is simply false. To celebrate their identity, groups sometimes elaborate on the naturalization of their supposedly negative characteristics or else explore the attributes' social origins while assigning them positive value.

It would be futile, as always, for me to try to unravel the web of ongoing, cumulative interactions between biology and culture involved in such identity formations and to assign primacy to either. What I want to do instead is to compare four identity groupings – women, African Americans, gays and lesbians, and people with disabilities – to see how they navigate among these attributions and how outsiders classify them in trying to honor or disparage them.

The extent to which group members accept essentialist descriptions, along with their implication of intrinsic differences from the culture at large, differs among identity groupings and also within them. In the case of the groups I shall discuss, our culture often considers differences in gender or disability as grounded in nature, differences in race or sexual orientation only variably so. But irrespective of whether societal oppression and rejection are based on presumed differences in biology, to be liberating, the group itself often responds by standing the stigma on its head and celebrating the despised characteristics.

23.1 Sisterhood Is Powerful

The nineteenth-century women's (or rather, "woman's") movement by and large accepted the essentialized "woman's nature" projected on women by the male-dominated culture. Biologists and medical men waxed eloquent about woman's (always homogenized and singular) aptitude for wife- and motherhood. That this aptitude was inborn hardly needed saying initially, but it emphatically needed saying once women's rights advocates began to argue that upper-class girls should be educated the same as boys of their class and have access to the professions.

This is where the testimony of physicians like Edward Clarke (1874) acquired significance because such men could throw the full authority of their professional expertise and status (Clarke was a professor at Harvard Medical School and later a member of its Board of Overseers) behind their pronouncements. Clarke, for example, argued that, if girls were educated the same as boys, they would become sterile and thus threaten the survival of the upper classes. Eloquently summing up the threat to the way of life to which these men were accustomed, the British physician Withers Moore (1886), in an address as president of the British Medical Association, pontificated: "Women are made and meant to be, not men, but mothers of men." And, given that most women's rights advocates accepted the ideology that women's procreative functions were all-consuming, they agreed that affluent women needed to decide whether to be wives and mothers or function outside the home. This "choice," of course, was not available to poor women. Indeed, that poor women worked hard and yet had (too) many children was taken as a sign of their evolutionary inferiority.

The women's (plural) movements of the second half of the twentieth century have developed more varied and nuanced attitudes toward biological essentialism. Firestone (1972) argued that equality of the sexes could not be achieved unless technology liberated women from having to bear the next generation. This thinking also resonates in the feminist science fiction of the period such as Ursula LeGuin's *Left Hand of Darkness* and Marge Piercy's *Woman on the Edge of Time.* Now, some 25 years later, some feminists still accept, and indeed celebrate, the idea that women and men are intrinsically different, and not just in our procreative functions. They take comfort in evolutionary, sociobiological, and genetic arguments claiming that not only our reproductive organs and so-called sex hormones, but also our brains and abilities to reason and experience and express emotions differ in kind from men's. And they accept the proposition that women's nurturing qualities and greater passivity and men's aggressiveness and competitiveness are inborn.

Other feminists refuse to accept that the differences in women's and men's procreative anatomies and functions spill over into the rest of our biology or psychology and explain the way we live in society. Such feminists, myself included, stress the historical and cultural diversity among the ways women and men are expected to behave and, in consequence, do behave. Remembering Simone de Beauvoir's "One is not born a woman; one becomes a woman," we emphasize the constant, multiple interactions between biological and social experiences that mold our biology as well as our behavior. To illustrate the dialectical interactions between biology and behavior, let me quote a passage I have published before (Hubbard 1990, pp. 115–16):

> If a society puts half its children in dresses and skirts but warns them not to move in ways that reveal their underpants, while putting the other half in jeans and overalls and encouraging them to climb trees and play ball and other active outdoor games; if later, during adolescence, the half that has worn trousers is exhorted to "eat like a growing boy," while the half in skirts is warned to watch its weight and not get fat; if the half in jeans trots around in sneakers and boots, while the half in skirts totters about in spike heels, then these two groups of people will be biologically as well as socially different. Their muscles will be different, as will their reflexes, posture, arms, legs and feet, hand-eye coordination, spatial perception, and so on. They will also be biologically different if, as adults, they spend eight hours a day sitting in front of a visual display terminal or work on a construction job or in a mine. I am not saying that one is more healthful than the other, only that they will have different biological effects.

To the extent that much feminist politics is concerned with working for increased access to information and services for women's health, including reproductive health and abortion, feminism may seem to be embracing essentialist issues. But that is based on a misunderstanding of the nature and causes of illness and health. Though health ultimately is expressed in terms of biology, biology alone does not determine our state of health. Our economic status, where and how we live, and the stresses and pleasures we meet with at work and in our personal lives all affect our health and well-being. Thus, a concern with women's health need not be essentialist at all but, in fact, embraces all aspects of women's experience and therefore is intrinsic to feminist politics.

In addition to the range of attitudes toward essentializing positions that we find within the present-day feminist movement, there is another important difference between it and the movements of the nineteenth and early twentieth centuries. As the largely middle-class and Euro–American movement has become more varied and multicultural, feminists have needed to become aware of the ludicrousness of universalizing women's experiences – including our prototypically "feminine" experience (singular) of pregnancy and childbearing and of nurturing children and men. Thus, as feminists have needed to acknowledge that class, racial–ethnic backgrounds, sexual and affectional orientations, and physical abilities and disabilities affect our cultural *and* biological experiences, the movement has had to begin to integrate and validate a wider range of identities without submerging specific ones. This is a tall order but one all identity movements must fulfill if they want to avoid recreating the oppressive hierarchies in response to which they have grown up.

23.2 Black Is Beautiful

If the eighteenth and nineteenth centuries made major contributions to essentializing sex, they outdid themselves when it came to race. In the sixteenth century, Europeans described Africans as superior in wit and intelligence to the inhabitants of colder regions, arguing that the hot, dry climate "enlivened their temperament," (Schiebinger 1989, p. 165), and two centuries later Rousseau still rhapsodized about the Noble Savage. The invention of "race" and its naturalization went hand in hand with European colonization of the other continents, the slave trade, and industrialization. And so, by the nineteenth century, the Noble Savage had become a lying, thieving Indian, and Africans and enslaved African Americans were described as ugly, slow, stupid, and fit only to be slaves.

What changed were, of course, not the "Indians" and Africans. Europeans and Euro–Americans began to need such derogatory characterizations. Particularly after the invocations of *liberté, egalité, fraternité* and the rights of man, political and economic circumstances made it essential to explain why some people were born more equal than others. To the extent that European and American societies were striving to abolish hereditary privilege, the answer necessarily lay in biology. A naturalized racism became an essential component of eighteenth- and nineteenth-century European and American politics. To this end, natural scientists theorized about separate origins of different, and unequal, species of humans, or alternatively about evolutionary processes that had brought humans from the lower forms, such as Africans and Native Americans, to the highest – Europeans, and most particularly European men.

Allan Chase (1977), Stephen J. Gould (1996), and others have analyzed the flawed science that went into accomplishing this transformation. This science, however, emanated from such distinguished men as the Philadelphia physician Samuel George Morton and the Harvard biologist Louis Agassiz, as they and other scientists "proved" that blacks were by their nature destined to be slaves.

The epidemiologist Nancy Krieger (1987) has reviewed the nineteenth-century medical debates about the biology of "the Negro." The ideological underpinnings are evident in statements like one made in 1851 by Samuel A. Cartwright, a prominent Southern physician and prolific writer about the health of slaves: "The abolitionist theory that the Negro is only a lampblacked white man is the cause of all those political agitations threatening to dissolve our union." His scientific investigations led Cartwright to conclude that "the want of a sufficiency of red, vital blood . . . chains [the Negro's] mind to ignorance and barbarism when in freedom" and a defect in the "atmospherization of the blood conjoined with a deficiency of cerebral matter in the cranium . . . led to that debasement of mind which has rendered the people of Africa unable to take care of themselves" (Krieger 1987, p. 268; Burnham 1983, p. 35). Indeed,

President Andrew Jackson, in his December 1867 message to Congress, was drawing on contemporary science when he argued that blacks exhibit "less capacity for government than any other race of people.... Wherever they have been left to their own devices they have shown a constant tendency to relapse into barbarism" (Foner 1988, p. 180). In a word, slavery and disenfranchisement were the white man's gift to blacks.

As might be expected from the viciousness of the biological attack on blacks, black scientists have seen little reason to look to biology to explain black people's oppression, and have pointed to differences in their social and economic situations to explain black–white "biological" differences, including differences in health. Already in the nineteenth century, the black physician Dr. John S. Rock argued that "as long as [racist] doctors refused to examine how slavery and poverty caused disease, they would have only biological rationales to explain racial differences in behavior and health" (Krieger 1987). He then went on and ridiculed racist descriptions of blacks by standing them on their head. "When I contrast," he wrote,

> the fine, tough muscular system, the beautiful, rich color, the full broad features and the graceful frizzled hair of the Negro with the delicate physical organization, wan color, sharp features and lank hair of Caucasians, I am inclined to believe that when the white man was created, nature was pretty well exhausted. But, determined to keep up appearance, she pinched up the features and did the best she could under the circumstances (Krieger 1987, p. 271).

To legitimate the continuous attacks on blacks since Reconstruction has required an on-going supply of scientific evidence for blacks' unalterable inability to participate on equal terms in the nation's political and social structure. Racist science, therefore, did not end in the nineteenth century. It has been revived again and again, most recently by the Canadian psychologist J. Phillipe Rushton (1988) and the late Harvard psychologist Richard Herrnstein together with the conservative social critic Charles Murray (1994). But, whereas the nature–nurture question has consistently been on the agenda of some feminists, this question has aroused little interest among blacks. Only occasionally, black nationalists, such as Marcus Garvey or, currently, some members of the Nation of Islam, have put forward essentialist arguments to espouse black superiority. Meanwhile, scientists such as Richard Lewontin (1982) have clearly established that the concept of race has no biological significance, and Herrnstein and Murray's *The Bell Curve* has been refuted in detail by Fischer et al. (1996).

To understand black identity movements, we need to look at how, despite continuous physical, economic, political, and ideological assaults, African Americans have managed to construct an identity that enabled them to coalesce into the civil rights movement during the 1950s and 1960s. Clearly, black religious leaders and their churches were foci of resistance as were black abolitionists and political leaders before and during Reconstruction and black educators and scholars since. Noteworthy among the latter was W. E. B. Du Bois, who eloquently articulated both the aspirations *and the right* of blacks to achieve equality.

Long before Frantz Fanon's and Paulo Freire's emphases on the need for the Wretched of the Earth to rid themselves of the internalized oppressor, Du Bois wrote of the need for blacks to acquire "double consciousness" (Gaines 1996, p. 9). Blacks, he argued, need to be aware not only of their own thoughts and feelings, but of the negative images whites have about them in order to surmount the barriers whites regularly place in their path. The destiny of both black and white America, he argued, depends on blacks' achieving "self-conscious manhood," which to him required melding both their African and American identities. (It is unfortunate, though understandable, that, in battling his period's racism, Du Bois failed to overcome its implicit sexism.) Du Bois constantly emphasized that, beyond striving for personal awareness and advancement, educated and successful blacks needed to "uplift" the entire race.

Such advocacy of education and uplift must be credited with sowing the seeds of the contemporary black liberation movement. Thus, although the black identity that had begun to coalesce even before Reconstruction around blacks' economic and political achievements suffered major setbacks under the fierce oppression that followed Reconstruction, blacks continued to claim their rights, articulate their achievements, and organize campaigns against lynching and other forms of oppression. Among the factors that helped was the migration, beginning in the 1930s and continuing until after World War II, of large numbers of Southern blacks to the urban centers of the North. Ghettoization, despite its drawbacks, gave opportunities for forming a politicized, self-consciously black culture. Important also were the return of large numbers of black soldiers from the victories in Europe and Asia and the subsequent access of unprecedented numbers of them to higher education under the G.I. bill. By the 1960s, the number of educated, articulate, politicized blacks had increased enormously, and they began not only to have an impact on predominantly white Left politics but to initiate their own.

The liberal, largely middle-class NAACP and Urban League had for some time agitated for legal redress of racial injustices. By the 1960s, radical, young blacks joined organizations such as SNCC and CORE that initially involved also whites but later promoted the need for black-only politics under the banner of Black Power. They and the Black Panthers provided militant young blacks with a sense of identity that was not only political but stressed pride in African Americans' and Africans' accomplishments, in African roots, and in actively standing up to white oppression. Thus, the identity politics of the 1960s went well beyond "double consciousness" to celebrate African culture, which earlier generations of blacks had tried to tone down in deference to internalized white esthetic standards. But, the civil rights movement never assigned sufficient merit to the nature–nurture question to incorporate it into its intellectual or political agenda.

And yet, though officially "race" no longer has biological meaning, the U.S. government still has us define ourselves by racial categories on the decennial census, and to many people – black or white – "race" carries biological implications. Lately this impression is being reinforced by the fact that racial census categories are being used for DNA-based identifications in criminal prosecutions, which clearly has biological overtones.

In reality, "racial" and "ethnic" census data often differ dramatically from one census to the next as people change the ways they list themselves. The resulting headcounts, however, have important economic and political consequences by serving as a basis for civil rights enforcement and other government programs (Waters 1997). Nor is the fluidity of our self-identifications evenly distributed across races and ethnicities. Identities assigned to various U.S. ethnic groups, other than blacks, are fairly fluid. This includes such "racial" census groupings as Hispanic and Asian for which intermarriage with higher-status white ethnic groups "whitens" and anglicizes. Intermarriage with blacks, however, continues to "blacken," and the children do not become white or, in the case of immigrants from the Caribbean, more "American-born" (Sanjek 1994). So, although, when celebrating the historical and cultural aspects of blackness, black identity movements may take it for granted that "black" is a social construct, a quasi-biological permanence of "race" and "black" still haunts our institutions (see Krieger 1992).

23.3 Gay and Lesbian Pride

The origins of homosexuality are the subject of hot debate among gays and lesbians, as well as among their "straight" supporters and opponents. With the increasing reliance on genetic explanations for all sorts of health conditions and behaviors, homosexuality is getting its share of genes (Hamer et al. 1993). But hormones and brain structures also get invoked (LeVay 1991), "programmed," I suppose, by gay genes. Cultural, historical, and political explanations, however, continue to have vocal exponents.

Just about everyone seems to grant that same-sex erotic love has existed at all times and in all places. People's inventiveness in matters of sex guarantees that we find pleasure in all manner of sex-related activities. And, gender need not be an issue: it takes women and men to make babies, but to get pleasure from sex is quite another story. There is therefore every reason to assume that the rules societies have laid down are grounded not in biology, but in cultural and political norms.

Most historians agree with Foucault (1980) that our culture invented the polarity of homosexuality and heterosexuality in the late nineteenth century. This is also when "homosexual" was transformed from an adjective used to describe what some people did into a noun that defined them, and homosexuality became a typology and an aberration from normative heterosexuality. Shortly afterwards, "homosexuality" and "the homosexual" became biologized through the work of sexual liberators, such as Havelock Ellis, who tried to rescue homosexuals from the threat of prison in this life or hell in the next (Weeks 1977). Homosexuals, they argued, should not be punished or blamed because they were born "that way."

As the gay and lesbian movements have come together in this century, attitudes on the subject of origins have varied. The forerunners of the "out" identity movement did not seem overly preoccupied with analyzing why their affectional and sexual inclinations differed from those of the mainstream. They were busy finding places where they could come together more or less safely to find partners and cultural and political support. Beginning in the early 1900s, they built enclaves in the growing U.S. cities, mostly on the East and West coasts (Chauncey 1994, D'Emilio and Freedman 1988, Kennedy and Davis 1993). Just as for blacks, World War II was a turning point because it drew men and women, many from small towns in which they had never known anyone with needs like theirs, into large, single-sex groups where they inevitably met other gays and lesbians. And although they were forced to lead secret lives in the face of overwhelming oppression, they established the roots of the present-day gay and lesbian cultures. The Stonewall rebellion of 1969 initiated vocal, self-conscious expressions of pride, which extended beyond the urban bar culture and the communities of mainly white and affluent gay men of the 1950s and 1960s.

Since the 1970s, the essentialist–constructivist debates within feminism have sparked similar debates about how one gets to be gay or lesbian. Lesbian and straight feminists tend to fall on both sides of that question. Gay men tend to be more biodeterminist than lesbians. Yet, some of the most trenchant cultural analyses to counter biodeterminist explanations have been developed by gay, lesbian, and bisexual scholars and activists (Weeks 1977, Garber 1995).

It is basic to the constructivist argument to question the gay–straight dichotomy. As historian Steven Zeeland (1995, p. 11) points out,

> definitions of what is gay and what is straight are . . . as likely to determine, as be determined by, sexual behavior. A straight man who experiences homosexual desire knows he cannot speak of it without his heterosexual identity being questioned. . . . A gay man who feels a heterosexual desire must wonder if he is not denying the gay identity he has worked so hard to feel good about.

Scholars and activists who have developed the arguments for the cultural origins of homosexuality note a basic asymmetry in biodeterminist explanations (Hubbard and Wald 1999): scientists do not look to biology to figure out why some people are repeatedly attracted to the same type of person as long as that person is of the approved sex. Nor do they explore the genetic basis of why some heterosexuals adopt only specific sexual practices. The only question that engages scientists is why some people are not heterosexual. This makes neither biological sense nor is it conducive to developing a progressive political position on sexuality. It grows out of our society's heterosexism and homophobia.

In political terms, the thinking of present-day biological determinists resembles that of the turn-of-the-century sex liberators. They believe that naturalistic explanations will reduce discrimination. Indeed, lesbian and gay rights organizations opposed to Colorado's Amendment 2 (which sought to prevent giving legal protection to homosexuals) invited scientists who believe homosexuality is genetic to testify against the amendment. Similarly, the conservative originators of the amendment tried to get scientists opposed to biodeteminist explanations to testify in its favor. The rationale on both sides is that if people are born homosexual, they are entitled to the same civil rights as people who are born black or female. If, on the other hand, they are homosexual "by choice," the state has no obligation to protect them because they just could change.

But, the term "choice" misrepresents and trivializes affectional and sexual commitments. It is like saying that if I, a white woman, "choose" to love a black man, the state owes me no protection against racial bigots. Of course, that was precisely the situation in the Jim Crow South. Unfortunately, as regards civil rights, the present situation of gays and lesbians is not that different. The existence of such prejudices, however, is not a good reason for lesbians and gay men to hitch their wagon to biodeteminism. The claim that they cannot help loving people of their own sex is, at best, an argument for tolerance or forbearance. It does not spell sexual liberation.

Definitions of gay, straight, and bisexual have changed over time, and different ethnic groups and cultures draw lines differently. In some cultures, if a man has sex with another man, only the receptive partner is considered homosexual. And, some men who live with their wives and children do not think of themselves as gay even if they regularly have sex with men and experience their greatest satisfaction with them.

LeVay and Hamer and his colleagues feel justified in dismissing the varieties and subtleties of sexual expression that have been described by anthropologists, sociologists, psychologists, and sexologists. LeVay, in fact, knows nothing about the sexual practices of the supposedly gay and straight men whose brains he examined after they had died. Hamer and his group select for their DNA analysis men who say they have always been attracted only to men. Indeed, Hamer and his colleague Angela Pattatucci complain that they cannot do comparable analyses with lesbians because, when they ask prospective subjects something like "Are you now, have you always been and will you always be sexually attracted only to members of your own sex?," women answer "What on earth are you talking about?," whereas men say yes or no (Burr 1996, pp. 169–71).

When I try to think what the DNA sequences Hamer and his colleagues identify as "gay genes" could be genes "for," I am stumped. Do they imagine DNA "codes" for preferences for certain kinds of sex partners, erotic proclivities, or positions in sex? Or is it for the fact that some little boys are considered more "girlish" than others and get called sissies? But such little boys tend to be more comfortable playing with girls than with boys. So, at what point does the gene "for" playing with girls start to "code" for having sex with men? As far as I can see, these researchers are doing sophisticated, though by now fairly standard, molecular biology against a backdrop of utterly naïve sociology, cultural anthropology, and sexology.

23.4 Disability Rights

The disability rights movement is organized around the realization that other people's assumptions about disabled people's inability to perform tasks essential for life, not actual realities, are responsible for the barriers and oppression people with noticeable disabilities encounter. This is what fundamentally distinguishes it from charities for the disabled such as the March of Dimes.

Even more than for the other groups we have been considering, disabled people are defined not by who they are but by who they fail to be – their *dys*functions. Except for well-known icons, such as Helen Keller or Stephen Hawking, they are not expected to have talents or professional commitments or to be involved in fulfilling relationships. People with disabilities are represented as pitiful or heroic but never as just ordinary.

In becoming organized into a political movement, people with disabilities have had to reject the stereotyping to which the biomedical model exposes them in exchange for a civil rights model. This has not been easy. Given the "healthist" and "ableist" bias of American culture, people with disabilities often have difficulty themselves acknowledging and overcoming the oppressive myths about disability they have internalized. Yet, customary phrases like "suffer from" or "victim of" are inaccurate for most disabled people. Their "tragedy" usually lies not in the disability but in the extent to which disabled people's abilities to contribute to society get underestimated so that they do not have the chance to use them. As a result, unemployment rates are much higher among working-age people with disabilities than in the rest of the population, and poverty is a major problem (Asch 1989).

Yet, our society can handle potentially quite serious disabilities and could handle many more. For example, eyeglasses, lens implants, and surgical techniques can enable people with severe eye problems to see. Even for people whose vision cannot be restored, seeing-eye dogs, clickers, tapping canes, audible traffic signals, translations of visual material into audiotapes, and computers that render written material in Braille or speech have made it easier to lead active and effective lives.

Nora Groce (1985) has described how people dealt with an inherited form of deafness that was prevalent on the island of Martha's Vineyard during the nineteenth century by teaching everyone to communicate in sign language, irrespective of whether they could hear. Indeed, the ability to sign was so taken for granted that Groce's older informants often did not recall which of their neighbors had been deaf. In recent times, people fluent in American Sign Language have developed a "deaf culture," the most widely known emblem of which to outsiders is the The National Theater of the Deaf. And they are organizing opposition to forcing deaf children to learn necessarily inadequate oral speech. This and the successful demonstrations in the 1980s by students and faculty at Gallaudet University in Washington, D.C. (the only university for the deaf), to protest the appointment of a hearing president, constitute landmarks of successful activism for disability rights.

The disability rights movement insists that, though some people's bodies or minds function differently from the majority's, "disability" is socially constructed: the result of our society's failure to accommodate people's special needs. And indeed, not just visual and hearing aids and sign language, but improvements in access for people with mobility problems and other kinds of accommodations can probably obviate most people's "disabilities." The real "tragedy" is the unavailability of such resources.

To overcome their marginalization and oppression, during the 1970s grassroots groups of people with disabilities began to form state or national membership organizations. Often organized at the local level, these have provided legal services or have published newsletters that advocate for "laws guaranteeing rights of access to education, employment, government services, and community life" (Asch 1989, p. 77). The legislation embodied in the Americans with Disabilities Act of 1992 is a result of this kind of activism.

In contrast to the other identity movements we have been considering, the disability rights movement addresses differences from the "norm" that everyone is likely to experience at some time in our lives. Its politics offers a constructive model for how to conceptualize in economic, political, and social terms differences that usually get treated as signs of biological inferiority. The disability rights movement also demonstrates how identity movements can help people to think constructively about the origins of "difference" and to meet such differences not with fear, hostility, or pity but by making the societal changes needed to accommodate them.

23.5 Coda

Whether we argue from a social constructionist or a biodeterminist position, we oversimplify enormously when we hitch our identity to just one of our many intertwined characteristics. As historian Martin Duberman (1996, p. 27) points out,

> identity politics reduces and simplifies; it is a kind of prison. But it is also, paradoxically, a haven. It is at once confining *and* empowering. And in the absence of alternative havens, group identity will for many continue to be the appropriate site of resistance and the main source of comfort.

Identity politics is sometimes accused of having splintered the Left, but that is true only to the extent that the Left tries to homogenize us and assigns primacy to certain aspects of our overgeneralized experience (being male, white, heterosexual, and able-bodied to pick some not so arbitrary examples.) Meanwhile, at the same time that progressive identity groups anchor themselves in the specifics of their members' identities, they need to oppose the tendency to naturalize both our commonalities and specificities.

23.6 Acknowledgments

I want to thank Jean Hardisty, Nancy Krieger, Margaret Randall, and Hilary Rose for their thoughtful comments on an earlier version of this manuscript.

REFERENCES

Asch, A. (1989). Reproductive technology and disability. In *Reproductive Laws for the 1990s,* (eds.) S. Cohen and N. Taub, pp. 69–124. Clifton, NJ: Humana Press.

Burnham, D. (1983). Black women as producers and reproducers for profit. In *Woman's Nature: Rationalizations of Inequality,* (eds.) M. Lowe and R. Hubbard, pp. 29–38. New York: Pergamon Press.

Burr, C. (1996). *A Separate Creation: The Search for the Biological Origins of Sexual Orientation.* New York: Hyperion.

Chase, A. (1977). *The Legacy of Malthus: The Social Costs of the New Scientific Racism.* New York: Alfred Knopf.

Chauncey, G. (1994). *Gay New York: Gender, Urban Culture, and the Making of the Gay Male World, 1890–1940.* New York: Basic Books.

Clarke, E. H. (1874). *Sex in Education: or, A Fair Chance for the Girls.* Boston: James R. Osgood and Co.

D'Emilio, J., and Freedman, E. (1988). *Intimate Matters: A History of Sexuality in America.* New York: Harper and Row.

Duberman, M. (1996). Bring back the enlightenment. *The Nation,* July 1:25–7.

Firestone, S. (1972). *The Dialectic of Sex.* New York: Morrow.

Fischer, C. S., Hout, M., Jankowski, M. S., Lucas, S. R., Swidler, A. and Vass, K. (1996). *Inequality by Design: Cracking the Bell Curve Myth.* Princeton, NJ: Princeton University Press.

Foner, E. (1988). *Reconstruction: America's Unfinished Revolution, 1863–1877.* New York: Harper and Row.

Foucault, M. (1980). *The History of Sexuality.* Vol. I. New York: Vintage Books.

Freire, P. (1993). *Pedagogy of the Oppressed.* Revised edition. New York: Continuum Publishing Company.

Gaines, K. K. (1996). *Uplifting the Race: Black Leadership, Politics, and Culture in the Twentieth Century.* Chapel Hill: University of North Carolina Press.

Garber, M. (1995). *Vice Versa: Bisexuality and the Eroticism of Everyday Life.* New York: Simon and Schuster.

Gould, S. J. (1996). *The Mismeasure of Man.* Revised Edition. New York: W. W. Norton and Co.

Groce, N. E. (1985). *Everyone Here Spoke Sign Language: Hereditary Deafness on Martha's Vineyard.* Cambridge, MA: Harvard University Press.

Hamer, D. H., Hu, S., Magnuson, V. L., Hu, N., and Pattatucci, A. M. L. (1993). A linkage between DNA markers on the X chromosome and male sexual orientation. *Science* 261:321–7.

Herrnstein, R. J., and Murray, C. (1994). *The Bell Curve: The Reshaping of American Life by Differences in Intelligence.* New York: Free Press.

Hubbard, R. (1990). *The Politics of Women's Biology.* New Brunswick, NJ: Rutgers University Press.

Hubbard, R., and Wald, E. (1999). *Exploding the Gene Myth.* Revised edition. Boston: Beacon Press.

Kennedy, E. L., and Davis, M. D. (1993). *Boots of Leather, Slippers of Gold: The History of a Lesbian Community.* New York: Routledge.

Krieger, N. (1987). Shades of difference: Theoretical underpinnings of the medical controversy on black–white differences, 1830–1870. *Int. J. Health Serv.* 7:258–79.

Krieger, N. (1992). The making of public health data: Paradigms, politics, and policy. *J. Public Health Policy.* 13:412–27.

LeVay, S. (1991). A difference in hypothalamic structure between heterosexual and homosexual men. *Science* 253:1034–7.

Lewontin, R. (1982). *Human Diversity.* New York: Scientific American Books.

Moore, W. (1886). British Medical Association, Fifty-fourth Annual Meeting. *The Lancet,* August 14:314–15.

Rushton, J. P. (1988). Race differences in behavior: A review and evolutionary analysis. *Personalities and Individual Differences* 9:1009–1024.

Sanjek, R. (1994). Intermarriage and the future of races in the United States. In *Race,* (eds.) S. Gregory and R. Sanjek, pp. 103–30.

Schiebinger, L. (1989). *The Mind Has No Sex? Women and the Origins of Modern Science.* Cambridge, MA: Harvard University Press.

Waters, M. C. (1997). The social construction of race and ethnicity: Some examples from demography. In *American Diversity: A Demographic Challenge for the Twenty-First Century,* (eds.) N. Denton and S. Tolnay. Albany: State University of New York Press.

Weeks, J. (1977). *Coming Out: Homosexual Politics in Britain, from the Nineteenth Century to the Present.* London: Quartet Books.

Zeeland, S. (1995). *Sailors and Sexual Identity: Crossing the Line Between "Straight" and "Gay" in the U.S. Navy.* New York: Harrington Park Press.

CHAPTER TWENTY-FOUR

The Agroecosystem

The Modern Vision in Crisis, the Alternative Evolving

JOHN VANDERMEER

Agriculture was the second great technological revolution of *Homo sapiens*, control of fire being the first, and industry the third. All three dramatically altered the way in which resources are extracted from the Earth and apportioned among individuals. We know little of the first, a little more of the second, and quite a lot about the third. The combination of the agricultural and industrial revolutions has had an immense impact on the way humans go about extracting basic resources, such as food, fiber, and drugs. And although industrial agriculture has undeniably been cornucopian, it has also had a downside – it has produced, directly or indirectly, human catastrophes from the Irish potato famine to the Bhopal chemical calamity, and appears to carry with it, like another famous system of human organization, the seeds of its own destruction.

This downside has led to a critique. Implicit in that critique is the suggestion of an alternative, a notion that is itself eclectic with advocates ranging from pure organic farmers through so-called input substitutionists. A variety of technical alternatives have been developed, but much remains to be done, and much remains controversial. It is generally acknowledged that political changes will also be required as the alternative is sought. The exact nature of such changes has not been clearly elucidated.

In this chapter I will first describe the nature and evolution of the modern system, noting the various problems that have emerged along the way. Then I will describe my view of the state of the alternative, including key research points that need to be addressed.

24.1 The Past: Evolution of the Modern System

24.1.1 The Center

With characteristic acumen, Lewontin crystallized the defining feature of modern agriculture: "Farming is growing peanuts on the land; agriculture is making peanut butter from petroleum" (Lewontin 1982). This simple statement summarizes much of the state of modern agriculture as it has evolved since World War II. And because of this structure, farmers, those who actually have contact with the land, are caught in a great squeeze. Their metaphorical petroleum suppliers are large, monopolized corporations whose selling prices are set artificially high, and their metaphoric peanut butter makers, who buy the peanuts they produce, are large monopsonized corporations whose purchasing prices are set artificially low. The farmer is frequently caught in the middle. Powerless, both economically and politically, he or she can only hope for government handouts to make up the difference between capital outlay and gross income.

The supplier side of this equation has its origins in the last century with dramatic mechanization technologies. These technologies got an initial boost from the widespread application of stationary steam engines but found their full realization with the adoption of the internal combustion engine after the turn of the century, leading to further advances in mechanization from the combine to the cotton harvester. Not only horses, but also person power, were displaced by this massive transformation. Much of urban America as we know it today is a consequence of the mass labor migration of the 1930s and 1940s – a consequence of agricultural mechanization.

Perhaps the most significant transformation in the supply side of agriculture had to do more with chemistry than with mechanical engineering. Mainly as a consequence of war research, chemicals rapidly took over agriculture, beginning after World War I, but with unabating force after World War II. The search for control of insect-borne diseases, acknowledged as the killers of more soldiers than the battlefield enemy, led to the discovery of the modern miracle, DDT. Thus was born the self-described "chemical agriculture."

In addition to poisons to control pests, and in conjunction with explosives research, the earlier discoveries of Liebig became more important. Liebig's famous law of the minimum recognized that there would always be a single limiting nutrient in the soil because crop nutritional requirements were more or less balanced. Artificially replacing that limiting nutrient could thus restore the productivity of agriculture. The nitrate chemistry that emerged from war research made Liebig's law and its programmatic consequence practical, and direct fertilization with inorganic chemicals soon became part and parcel of modern agriculture.

Finally, plant breeders devised new methods of economic and political control over their products mainly through the development of hybrids (Berlan and Lewontin 1986). The seed industry thus became another important input component of modern agriculture.

Thus evolved the input side of modern agriculture: mechanization based on the internal combustion engine, control of pests with biocides, fertilization of soils with inorganic chemicals, and improved, mainly hybrid, seeds. The farmer had to buy machines, pesticides, fertilizers, and seeds, all of which were supplied by industries that were monopolized – at least to some extent.

The allocation of outputs had a somewhat different evolution (Morgan 1979). Long-distance trade in agricultural products had been largely restricted to luxury goods such as chocolate, tea, coffee, and sugar. The main agricultural goods that fed the Industrial Revolution, cotton and wheat, were hardly traded internationally at all before the late eighteenth century. There was not enough demand. Wool supplied the textile industry, and local production of wheat fully satisfied all needs for porridge, the staple of the lower classes, and bread, a luxury for the rich. To make bread, wheat had to be ground into flour, mixed with water and yeast, kneaded into dough, allowed to rise, and finally baked. Such a large input of labor could be afforded only by the well-to-do.

Toward the end of the eighteenth century a change in the social organization of work caused a dramatic change in this pattern – the working class began to eat bread. Improved milling technology had caused the price of bread to decline, making it more available to the masses, and more important, it was difficult for factory workers to bring gruel or porridge to work in their lunch buckets. The result was a dramatic increase in bread consumption and a concomitant increase in the demand for wheat. Local agricultural production could not keep pace. This led to an increase in the international trade in wheat which, in turn, led to the first important international wheat-trading business complex.

At the age of 17 Louis-Dreyfus left his family's farm and sold his share of the season's wheat harvest. He spent several years in the stream of wandering merchants involved in the European grain trade, during which time he amassed considerable capital, established his own grain company, and took out large loans aimed at enlarging his enterprise. With his firm and finances in place, Louis-Dreyfus arrived in Odessa in the 1860s and purchased the bulk of the grain elevators in the city. He then sent out his own agents, not to buy grain from farmers, but to sign contracts to purchase their grain in the future. At the other end of his enterprise, in Western Europe, he sold those contracts for future delivery. By the 1870s he was contracting for wheat from the Russian hinterland, shipping by rail to his storage facilities in Odessa, loading the grain on his freighters, and selling futures and grain to buyers in Hamburg, Bremen, Berlin, Mannheim, Duisburg, and Paris. The Louis-Dreyfus Company had become the first giant grain company in the world.

Similar stories could be told for four other giants, Bunge of Argentina, Continental of New York, Cargill of Minneapolis, and André of Switzerland. With Louis-Dreyfus, these five giants dominate the world grain trade much like the Seven Sisters dominated the world petroleum trade for so many years. And although this pattern is most highly developed in the grain trade, similar structures developed in other sectors of the food industry with canned vegetables, processed foods, and so on. Beginning with the Industrial Revolution and continuing with only minor variation today, the buyers of agricultural products grew large rapidly, and most financial decisions came to be made far from their source, the farmer. By the time Louis-Dreyfus arrived in Odessa, the serf farmers of Russia had little knowledge of, or control over, what happened to their grain after it left their farm. But those who purchased it were involved in high finance throughout the industrial world.

Furthermore, the monopsonized food industry was itself transformed by the post–World War II chemical revolution. With a major economic force being shelf life, preservatives became essential to the food-processing business.

With competition for consumer dollars intense, product appearance became as important as its food value, and emulsifiers and artificial colorings joined with the new high-yielding varieties to make the processing and storage of foods just as chemically intensive as their production.

24.1.2 The Periphery

Modern agriculture, like the rest of the modern capitalist economy, is really an international affair. And most important in that international structure is the relationship between the developed and underdeveloped world. The decolonizations of this century have created a different form of colonialism. The former colonies are now members of the so-called Third World and retain important remnants of their colonial structure. First, the centerpiece of their economy is frequently a few crops or natural resources that account for the bulk of export earnings. Second, they retain a significant fraction of the land not devoted to export production in the form of peasant agriculture. A consequence of this dual economy is that its disarticulated nature (the peasant sector economically independent of the export sector) virtually ensures the continual impoverishment of the small farming sector. The engine of economic growth so perceptively rediscovered by Henry Ford (that the people who produce have to be able to afford to purchase the products) is not part of the main economic structure. Cotton is grown to be shipped to textile mills in the developed world, bananas for luxury consumption in the developed world, and coffee, tea, and coca to satisfy the drug habits of the population in the developed world. It is of little consequence that local peasants do not have the purchasing power to invest in a textile mill, or buy large quantities of bananas, or get high on coke or caffeine.

Such an arrangement is hardly conducive to economic growth, nor does it promote political stability. Consequently the postwar years have seen the birth of national liberation movements, some of which have touted socialist philosophies. The response by the West has been devastating for most of the Third World. To maintain the disarticulated nature of Third World economies, the West has reverted first to subterfuge and finally to open military intervention. Although Vietnam may have signaled the end of blatant interventionism, it did not end the need to maintain the Third World in its service position, thus leading to more subtle means of control, mainly through economic warfare. In particular, the role of the World Bank and the International Monetary Fund cannot be ignored in this regard.

24.2 The Present: The Modern System in Crisis

Although the modern system has provided the world with a cornucopia of food and fiber, to say nothing of profits, it has also created a host of environmental problems. The publication of Carson's *Silent Spring* (1962) marked the beginning of a persistent critique. This critique is diverse and eclectic. However, it can probably be summarized as two general tendencies. The modern agroecosystem may be critically viewed as either a system with problems or as a problem system, depending to some extent, it seems, on whether one's focus is on farming or agriculture, as described above.

When the agroecosystem is viewed as a system with problems, probably the most cited ones are those associated with the biocides used to control pests. This aspect of the critique has become part of a more generalized popular concern over the environment. Of less visibility are the particular problems associated with the loss of soil fertility, occasionally with the loss of the soil itself. Closely related to the problems of the soil are problems with water usage, most notably the many cases in which the modern system uses water at a rate that is larger than the recharge rate. Such concerns are largely the concerns of "farming," not "agriculture."

This view that modern farming has spawned significant problems is hardly contestable (Buttel 1990, Soule et al. 1990) and has become quite general, including those whose self-interest would be obviously served through a defense of that system. For example, a recent survey in Washington state found that among both modern agricultural practitioners and their critics, there was general agreement that modern agriculture faces serious problems (Beus and Dunlap 1993). Some researchers are now referring to the 1980s as the "decade of awareness" (Par and Hornick 1994), and even deans of agricultural colleges openly admit that the modern system has spawned significant problems (Lacy 1993).

Alternatively, the modern agroecosystem may be viewed as a problem system. Because of the particular historical trajectory agriculture has taken, it has become inserted in an industrial system that drives it. As such, it will continue generating new problems as fast as old ones are solved. Consequently, searching for individual solutions to particular individual problems is seen as applying bandaids to a serious disease, and a total revamping of the system is what is called for (Rosset and Altierri 1997). Some critics even suggest that the modern agricultural system has spawned problems at the economic level that are likely to become its Achilles' heel. In the developed world such issues as commodity price supports and interest rates dominate most practicing farmer's concerns, whereas land tenure and market stability are the sorts of things a farmer in the less developed world tends to think about. Critics at this level of the analysis tend to view agriculture as a problem system (as argued above, those who are concerned with the level of "farming" tend to view it as a system with problems).

The key problem on which the post–World War II agricultural development focused was pests. The methodology for controlling them, much like the methodology in warfare (Russell 1996), was total annihilation achievable through the use of the new poisons that ultimately sprang from war research. It was not generally recognized that this could cause human health and environmental damage until the publication of *Silent Spring* in 1962, and it was not until 1978 that the general ecological and evolutionary mechanisms behind those problems were formulated so eloquently by van den Bosch (1978). This was the now well-known pesticide treadmill in which pest resistance, pest resurgence, and the formation of secondary pest outbreaks were acknowledged as the three mechanisms leading farmers into the need to spray ever greater quantities of pesticides – the pesticide treadmill.

By acknowledging the pesticide treadmill and taking some action to curb the most blatant negative effects of pesticides, the horrible environmental consequences envisioned by Carson have been partly averted. Nevertheless, much of the vision has come to pass. Some agricultural areas are largely devoid of life. A drive through the grape-producing areas of Calfornia, for example, is ominously similar to the opening sentences of *Silent Spring* with the absence of bird song, the absence of small flying creatures, and the feeling of desolation. But the casual motorist in the Central Valley does not really get the whole picture. Pesticide residues are now known to occur in most food we eat. Although the levels of their occurrence are small and usually within government safety standards, our persistent exposure to them and the fact that two or more residues, even if at concentrations known to be safe alone, may have synergistic effects on our health, constitute a time bomb. Though particular incidents may stand out in public memory because of sensational press coverage, far more ominous in the long term is the continual small dosage received each day by the majority of the population. Carcinogenic potential has long been acknowledged (Steingraber 1997), and most recently the potential problems associated with hormone-mimicking pesticides has come to the public's attention (Colbern, Dumanoski, and Myers 1996).

The problem cannot be solved by exporting the poisonous part of production to the Third World, as capitalists generally do with onerous production tasks. Since the publication of Weir and Shapiro's *The Circle of Poison* (1981), it has been difficult to ignore the human health effects of pesticides even though they may be applied far from home. Pesticide residues in tomatoes imported from Mexico are the result of exported pesticides applied by Mexican field workers. Frequently those pesticides are banned in the United States, yet we must face the residues because of the international relations that allow producers to export dangerous technology. But if political pressure is applied without an analysis of the entire political process, unexpected results may emerge. For example, because of consumer pressure in the United States, persistent pesticides were replaced by nonpersistent ones in vegetable production in the Tuliapan Valley in Mexico. But there is a general relationship between acute toxicity and environmental persistence: short persistence time is correlated with high toxicity. Thus, the less-persistent pesticides used by Mexican workers to avoid pesticide residues in the vegetables eaten by United States citizens were far more likely to poison those Mexican workers (Wright 1990).

Yet there was another problem that probably started in the early nineteenth century when Justius von Liebig first discovered that you could put inorganic chemicals on the ground to increase crop yields. The problem of soil fertility was thus, in principle, solved. But soil management proved to be far more difficult than had been expected. The Oklahoma dust bowl was for soil conservation what the Irish potato famine was for pest and disease control – something of a wake-up call. The application of inorganic fertilizers on soils as part of the overall management system carried some strong negative consequences. Although not as well understood as the mechanisms of the pesticide treadmill, it is generally understood that provisioning the soil with large amounts of inorganic ions disrupts the normal nutrient cycling characteristics of the soil, thus making it more necessary to apply inorganic fertilizer. Even though it does not seem to have been referred to this way before, it seems perfectly justifiable to refer to it as the fertilizer treadmill.

At a most general level, the extensive use of biocides and inorganic fertilizers may in the end prove to be self-defeating, for such use may be undermining the resource base on which agriculture is founded. Examples of such habitat destruction abound. Extensive use of pesticides in cotton-production in Nicaragua has made the former cotton-producing areas biological deserts and created secondary pests that threaten other crops even outside of the cotton-producing zone. It has been hypothesized that tomato production in the Sebaco Valley, some 300 km from the nearest cotton-production, faces two or three secondary pests that may have evolved their pest status from pesticide spraying in cotton and later wind dispersed into the tomato-growing areas (Rosset, personal communication).

A final ominous fact is the loss of biodiversity in the modern agricultural system. Modern agriculture has emphasized production in large monocultures as opposed to the usually highly diverse traditional systems (Altieri 1990, Vandermeer et al. 1998). The consequences of this trend are yet to be fully appreciated. There are two distinct components of biodiversity. First is the biodiversity associated with the crops and livestock purposefully included in the agroecosystem by the farmer. A traditional home garden is more diverse than a modern wheat field because many crops are planted in the former and only one in the latter. This component of biodiversity is the "planned biodiversity."

Although planned biodiversity is perhaps the most visually obvious component and has received the most attention from agroecologists, anthropologists, and the like, there is an additional component that is at least as important regardless of the relative lack of attention paid it. This component is the "associated biodiversity," and includes all the soil flora and fauna; the phytophagous, carnivorous, scavenging, and fungus-feeding insects; the vertebrates; the associated plants (some of which are weeds); and more. Although a great deal of attention has been paid to the extent and function of planned biodiversity (e.g., Vandermeer 1989, Altieri 1987), relatively little has been written on the associated biodiversity (Perfecto et al. 1996, 1997; Swift et al. 1996). As modern industrial agriculture continues to expand, now most explosively into the Third World, biodiversity continues its decline. Exactly what ultimate effect this may have is not predictable, but is not likely to be salutary.

Perhaps the most severe criticism of modern agriculture comes not from observations of ecological devastation but rather from some remarkable political and economic arrangements. Despite their position between monopolized suppliers and monopsonized buyers, by the turn of the present century, U.S. farmers wielded considerable political clout through various farmers, organizations. As in virtually every industry in the capitalist world, the management of supply through various political arrangements has been a constant, if underreported, feature of the system. Whether by government largesse or vertical integration and monopoly control, supplies of commodities are carefully regulated so as to maintain profits at targeted levels. Agriculture was no different. Turn-of-the-century farmers turned their considerable political clout into policy, and the well-known principle of supply management became law. On the basis of prior production, each farmer was allocated a certain amount of land to put into production, thus maintaining supplies at sufficiently low levels to ensure prices adequate to meet production costs and a reasonable profit. But political power is a dynamic variable, and as the century proceeded, the power of monopolized and monopsonized agribusinesses grew exponentially. The basic facts of supply management changed accordingly to reflect the interests of the politically powerful, and farmers were encouraged to plant fence row to fence row, thus saturating markets. Commodity prices consequently fell, and farmers frequently were forced to sell below production costs, receiving direct government subsidies to make up for the shortfall. Yet the basic goals of the change in supply management were clearly met: the large monopsonized grain and canning companies were purchasing commodities at rock-bottom prices. This has led to one of the chief problems of agroecosystems today: overproduction.

Third World farmers have been especially hard hit by overproduction in the First World. First World grain, heavily subsidized by managing supply at high levels to ensure low commodity prices, now enters Third World markets, forcing many basic grain producers out of business. Here, the basic disarticulated economy of the underdeveloped world has functioned exactly as planned. Grain companies, purchasing grain at rock-bottom prices, sometimes even below the cost of production, need market outlets. The Third World has provided an excellent target for developing those markets. Frequently, with economic incentives from the developed world, Third World countries one by one have been converted into net importers of basic grain from the developed world.

Although the disarticulation of the Third World does not make for economic growth, neither does it make for political stability – at least not over the long run. To discourage the disruption caused by political instability, loans, credits, and gifts from the developed world are routinely doled out to the underdeveloped regions of the globe – most frequently for projects based on the modernization of agriculture. Such loans and credits are an important part of the huge debt built up by the Third World. Furthermore, in many cases the failure of the modern agricultural model is an implicit cause of the failure of the local governments to be able to keep up with their credit obligations.

24.3 The Future: The Alternative System Evolving

Although any historical narrative certainly must begin with traditional, premodern systems, in fact our interest is rather in the modern history defined by the emergence of an alternative to the modern industrial trajectory. Thus, the beginnings of an "alternative" to the modern trajectory is most likely to be found in some form at the beginning of that trajectory itself; it was probably in the Third World, where organic agriculture was first recognized as a distinct system. According to the Council for Agricultural Science and Technology, Sir Albert Howard, Director of the Institute of Plant Industry in India early in this century, was the first to actually describe the organic system as something distinct from the pathway that world agriculture seemed to be taking (Pesek 1990). Howard noted, I think perceptively, that von Liebig's approach to the soil was strongly contrasted to that of Darwin, as reflected in the later's treatise on earthworms. Whereas Liebig insisted on a reductionist approach, reducing the soil to its chemical constituents, Darwin correctly saw the soil as a complex biological system. Perhaps Darwin was really the first advocate of an ecological approach to the study of soil. From Howard's observations of agricultural soils in India, he developed a system of composting (called Indore composting) that effectively remains the underlying idea behind modern composting (Niggli and Lockeretz 1996, Howard 1953). The history of organic agriculture since that time is rich and complicated.

In only a slightly different context, the publication of *Silent Spring* in 1962 generated an entirely different but ultimately related movement in agriculture. Rather than accept the implied challenge of the entire modern industrial agricultural system, the vast majority of research workers, policy people, farmers, and agroindustry representatives scrambled to put as good a spin as possible on the watershed of information that Carson unleashed with the publication of her marvelous book. The notion of integrated pest management is archetypical of this tendency and has as its foundation the idea that agriculture should be changed gradually, moving toward a more sustainable or ecologically benign system. The term LISA (Low Input Sustainable Agriculture) was invented by the USDA to encompass this basic philosophy, and LISA has now been incorporated into most conventional agricultural establishments (e.g., land grant universities in the United States, farmers' organizations, agribusiness suppliers). The relationship of the LISA concept and more standard forms of organic agriculture is explored in a later section.

In most general terms, much of what is technically relevant can be conveniently codified in a simple three-part classification: (1) pest management, (2) soil management, and (3) integration. In each of these three categories there are a variety of techniques, all of which have some sort of ecological basis to them. Many are based on well-known and popular ecological assumptions; others on popular but discredited assumptions. Some have been clearly demonstrated as efficacious; others are assumed to be so without evidence. Many have been thoroughly studied from an ecological point of view; others await such serious study. My purpose here will be to review briefly the ecological bases assumed to be behind the various proposals and inquire into the state of verified ecological knowledge about them. Thus, the next three sections are organized according to these three categories – pest management, soil management, and integration.

24.3.1 Pest Management

The key problem on which the post–World War II agricultural revolution focused was that of pests. The methodology for eliminating them, much like the methodology in warfare, was total annihilation (Russel 1996) achievable through the use of the new poisons that ultimately sprang from war research. It was not generally recognized that this technique could cause human health and environmental damage until *Silent Spring* in 1962, and it was not until 1978 that the general ecological and evolutionary mechanisms behind those problems were formulated so eloquently for the lay public (van den Bosh 1978). This was the now well-known pesticide treadmill in which pest resistance, pest resurgence, and the formation of secondary pest outbreaks were acknowledged as the three mechanisms leading farmers into the need to spray ever greater quantities of pesticides: the pesticide treadmill. The response, originally proposed by van den Bosch and since adopted even by those who most vigorously supported the earlier chemical program, was integrated pest management (IPM).

24.3.1.1 The General Strategy: Prophylactic and Responsive Programs

The IPM formulation derived from a systematization of what had been a variety of issues. The key ideas, originally articulated by van den Bosch (1978), were extremely simple: (1) avoid spraying poisons unless it is necessary, and (2) manage the ecosystem in such a way that it does not become necessary to spray. These principles remain at the foundation of almost all functioning IPM programs (Levins 1986).

Vandermeer and Andow (1986) presented a simple graphical approach for the qualitative design of an IPM program based on the assumption that the farmer wishes to minimize the cost of the whole program and that the program has the two classical components, the prophylactic and the responsive. This procedure provides a practical method of deciding when it makes economic sense to develop a prophylactic program and sets the stage for the continual improvement of prophylaxis to the point that pesticide application is eventually eliminated.

24.3.1.2 Biological Control

The underlying philosophy of classical biological control is the identification of a natural enemy of a specific pest and the release of that natural enemy with the idea that doing so will form a permanent solution to the problem. An alternative form of biological control, mass release, involves the repeated release of natural enemies with no particular goal of establishing the organism permanently in the environment. Under the modern industrial model, mass release seems to be preferred because it does not solve the problem but provides a product that is continuously needed to treat the problem. From an ecological point of view, of course, mass release does not solve the problem.

Application of ecological theory to classical biological control for insect populations has been an extremely active field since the explosion of theoretical ecology in the 1970s with most theory based either on the Lotka Volterra predator–prey equations or the Nicholson–Baily model (Beddington, Free, and Lawton 1978; Murdoch, J. Chesson, and P. L. Chesson 1985; Mills and Getz 1996). Both formal and informal theories have focused on the relatively obvious idea that a successful biological control agent must not itself go extinct if it is to perform its function for more than a very short time. Thus, although practitioners seek natural enemies that are efficient at killing the pest, the emphasis from ecological theory has been on agents that form stable equilibria with their prey, the pest, but at prey densities that are below the economic threshold. This view has its detractors who note that evidence from actual case studies of biological control does not suggest that concern over the population stability of the introduced enemy and its pest has ever really been a component of a real biological control example and that subtler modes of population regulation (such as through a habitat mosaic) may provide the same practical effect as that of formal population equilibrium (Murdoch 1994). Debate continues over the interpretation of data (Murdoch 1994; Hawkins, Thomas, and Hochberg 1993).

24.3.1.3 Cultural Control

Application of ecological principles to questions of cultural control of insect pests has been equally as vigorous as, if somewhat more eclectic than, biological control. These techniques have included the management of vegetation texture, various forms of tillage, sanitation, careful planning of planting and harvesting dates, resistant varieties, fertilizer manipulations, and others (Andow and Rosset 1990). The ecological principles involved in these cultural control methods are many and varied, but many of them fall within the general category of plant apparency theory (Feeny 1976). The general idea is to reduce the plant's effective apparency with regard to the pest in question.

Vegetation texture has been one of the most studied aspects of cultural pest control. It has long been a basic assumption that pest problems are fewer when vegetation is more diverse. For example, Risch, Andow, and Altieri (1983) found that the majority of literature reports demonstrated fewer specialist insects in diversified agroecosystems than in monocultures. Others have found basically the same pattern (Perrin 1977, Dempster and Coaker 1974). Vegetation diversity seems to have little effect on generalists, but its effects on specialists are enormous. The common observation that specialist insect herbivores tend to be less abundant on their hosts when found in diverse vegetation initially seems nothing more than a consequence of Feeny's (1976) plant apparency theory.

24.3.2 Soil Management

The management of soils is a complicated affair and is conceptually far more difficult than the management of pests and diseases. As the most superficial summary of soil science will show, many processes are involved, and a range of sciences from chemistry to microbiology are required. Putting all of this together is conceptually difficult. Do we consider the chemistry of mineralization as the centerpiece on which we build, or is the general biology of the soil microorganisms the place to start? In fact we know very well that the soil is a dynamic ecosystem and must be approached from a variety of perspectives, depending on what particular aspect is to be investigated.

It is useful to think of the problem of managing soil fertility in two distinct perspectives: the point of view of process versus the point of view of ecosystem health. It is not that one way is superior to the other but rather both ways of viewing the problem are valid for slightly different purposes. Most important, the conclusions one reaches are sometimes limited by which of these two foci dominate one's perspective.

The process-oriented view is mainly concerned with pools of nutrients and their transfer. After all, the bottom line is that soils contain the pools of nutrients that plants need and all else is unimportant if the plants cannot get those nutrients. At any one point in time there are three distinct pools of nutrients: (1) readily available, (2) mobilizable, and (3) long-term storage. These three pools obviously feed into one another, but all are important to soil-management goals. The readily available pool is that which can be immediately used by the plants but also immediately leached out of the system. That which is mobilizable can be ideally fed into the readily available pool when necessary and is not subject to leaching. The long-term storage component is slowly converted into the mobilizable form and is not of much concern to the goal of maximizing production over the short term. Managerial concern with sustainability tends to concentrate on the long-term storage, whereas concern with production tends to concentrate on the readily available nutrients. Good husbandry should be concerned with both. For example, in the case of phosphorous, sustainability criteria might emphasize application of rock phosphate, whereas production criteria would emphasize direct application of phosphate fertilizer. Good husbandry would look at the whole picture and ask about stores and flows. If more long-term storage is needed for sustainability, then rock phosphorous or phosphorous-rich organic matter should be added to the soil. If exchangeable phosphorous is in short supply, the addition of phosphorous solubilizing microorganisms or increasing mycorrhizal infection might be appropriate to encourage weathering of the long-term storage material, or addition of organic matter to provide more exchange sites if that is the limiting factor. If readily available phosphorous is in short supply, addition of inorganic phosphorous is always possible, or perhaps changing the pH of the soil to alter the exchange capacity could be contemplated. In all of these possible strategies the point is to reduce the amount of leaching from the system and maximize the plant uptake.

A more integrated approach to these soil processes was provided in the model of van Noordwijk and de Willigen (1986). These authors consider inputs into the system as being transformed by four generalizable functions. An input could be a variety of things, from rock phosphate to compost, from manure to nitrate. The point is that the input has some nutrient associated with it, and our desire is to determine how the crop responds to that input. The classical agronomic approach is to treat the soil as a black box and empirically create a graph relating response of crop to input of nutrient material. At the other extreme, modern agronomists may reduce this problem to a large number of micromechanisms (e.g., Loomis and Connor 1993), providing a detailed picture of physiological and chemical processes. This latter approach in its most modern form has resulted in large computer models purporting to predict the transformation of nutrient resources into plant productivity. The popular approach of developing large computer models has been criticized from the point of view that a large computer model may provide good quantitative predictions of what will happen, but it rarely provides understanding of why. A computer model that we do not understand is not much better than a natural phenomenon we do not understand.

The van Noordwijk–de Willigen model is something of a compromise between the black box model previously so common in conventional agriculture and the hopelessly grandiose computer systems models. The former involves three transformations. First, the material applied must somehow become available to the plants in the soil. For example, application of ammonia to a field involves considerable volatile loss as anyone passing a zone of application can readily appreciate from the odor. Thus, what becomes available in the soil is less than that which was applied originally.

The next phase of the process is realized when the material available in the soil is actually taken up by the plant. This is largely the subject of root ecology. Theoretically, the plant could take up all the material available. But plants are not 100% efficient, and some material that is potentially available will not be taken up. The shape of the uptake curve depends on root biomass density, microbial actions (e.g., mycorrhizal infections or pathogenic infections), and physical and chemical factors associated with rhizosphere dynamics.

Finally, the material taken up by the plant is converted into dry matter of the plant, which is largely a subject of plant physiology. The point on the production function in which material is taken up from the soil but not converted into dry matter production (e.g., used to manufacture secondary compounds for defense against insect herbivores) is normally referred to as luxury consumption.

The strength of looking at the system from this point of view is that each output is an input to the next level. Thus, the three functions can be composed, where the composed function is the black box so commonly viewed in the classical agronomic literature. With this composition approach one can easily visualize the component functions that lead to the basic relationship between dry matter production and the application of an input. Furthermore, various classical sciences can be more or less associated with each function.

Process-level models are certainly useful, even necessary, when one focuses on a particular input – particularly nutrient inputs. An alternative, and perhaps more general, way of viewing soil management is from the point of view of ecosystem health, rather than that of view of analyzing inputs. There are four features of the soil that most soil scientists agree are crucial to good soil health: acidity, physical structure, organic matter content, and biodiversity. These four factors are obviously interconnected in complicated ways, and understanding soils requires an understanding not of each one of the subjects but rather of the way all four are interrelated. Indeed, a detailed knowledge of all four plus their interrelations with each other is ultimately what is needed for good soil management. Nevertheless, these four subjects are all measurable and all provide indices that most soil scientists would agree are correlated with the "health" of the soils: low acidity, high organic matter, high biodiversity, and good physical structure. Perhaps such characterizations are oversimplified, but they seem to represent a first step in judging the health of the soil. Further refinements, such as cation exchange capacity, soil bulk density, earthworm density, the C:N ratio of the organic matter, and many others are obviously desirable pieces of knowledge. But as a first approximation it seems perfectly reasonable to say that high organic matter, high biodiversity, low acidity, and good physical structure are the four pillars of soil ecosystem health.

24.3.3 Integration

Probably the idea most promoted by advocates of an alternative agriculture is integration of all aspects of the farming operation. This may take various forms, depending on the advocate or practitioner. One persistent idea is that an agroecosystem should mimic the functioning of nonmanaged ecosystems (Ewel 1986, Oldeman 1983, Soule and Piper 1992) with tight nutrient cycling, vertical structure, and the preservation of biodiversity. An obvious manifestation of this idea is the incorporation of trees into the agroecosystem, which is something that has been part of traditional agroecosystems in tropical regions for some time. Another manifestation has been the attempt to use ecological principles as part of the design criterion, thus replacing what had become a strictly economic decision-making process with one that includes ecological ideas also. For example, the natural biodiversity of most unmanaged ecosystems has been used as a rationale to suggest that multiple cropping is generally a laudable goal (this is discussed in Section 24.3.3.2) or that the famous diversity stability hypothesis of theoretical ecology might be related to promoting stability through diversity in agroecosystems (Vandermeer and Schultz 1990). Many of these ideas suffer from a lack of rigorous definition in the field of ecology itself, and thus their application to the design of agroecosystems is perhaps premature.

We can envision the incorporation of various levels of ecosystem function at different levels of complexity as all ecosystem functions become integrated into the agroecosystem. Thus, beginning with a stereotype conventional monoculture, we commence with the simplest integrating effect, rotations, including animals as one of the possible "crops." Rotations are in fact common even in conventional agriculture but are the core of much thinking about ecological forms of agriculture. The next higher level of integration is intercropping, in which two crops are physically associated with one another in such a way that they compete with or facilitate one another. The next level is the incorporation of more structural diversity with the addition of trees, agroforestry. This is followed by the incorporation of multiple species of trees, crops, and animals in the home–garden- type situation still so common in most of the world's tropical regions. The final integrating effect is planning at the landscape level, first at the level of the farm in which interactions both within fields and among fields is important, and second at the regional level wherein interactions among farms is also important.

24.3.3.1 Rotations

Rotations are a historical centerpiece of the modern agricultural system, for they represented the key to the new husbandry system that not only was the springboard for the modern system but also remains something of a model for modern ecological agriculture in temperate zones. The very first agricultural experiments in Europe, set up by Albrecht Thaer in the beginning of this century, were focused on the optimization of traditional crop rotations (Niggli and Lockeretz 1996).

Crop rotations serve basically two ecological functions: control of pests and diseases and rejuvenation of nutrient status of the soil. With regard to the nutrient status of the soil, the basic principles of plant competition and facilitation are relevant. A cereal mines nutrients in certain proportions that are significantly different from a legume, for example.

24.3.3.2 Intercropping

Intercropping, has often been cited as an important integrative component of the alternative agriculture agenda (Francis 1986; Willey 1979a,b; Vandermeer 1989). It is frequently observed that more traditional forms of agriculture, especially in the tropics, include some form of multiple cropping at their cores, and it is thought that the abandonment of this procedure was specifically to accommodate the methods of modern conventional agriculture (i.e., application of chemicals and mechanization). If this were true, the ecological benefits of multiple cropping would be lost in the modern system, and it would make sense to argue for their reintegration in the alternative program.

Exactly what are the ecological benefits thought to accrue from the practice of multiple cropping? The hypothesized ecological benefits have been divided into two categories (Vandermeer 1989), reduced competition (or the competitive production principle), and facilitation. In the case of the competitive production principle, it is thought that two different species occupying the same space will utilize all the necessary resources more efficiently than a single species occupying that same space, much as is sometimes believed to happen in natural ecosystems. The competitive production principle actually suggests that there is nothing special about the crop combination but simply that the intensity of interspecific competition is relatively weak compared with the intensity of intraspecific competition. The theoretical details of facilitation are substantially more complicated than those of competition (ecologists generally do not pay much attention to facilitation) and are beyond the scope of this article. Details can be found in Vandermeer (1984, 1989).

The evidence for the competitive production principle is scant. At one extreme a review of experiments with mixtures of grasses (Trenbath 1974) found little evidence that interspecific competition, on average, was smaller than intraspecific competition. On the contrary, the overall pattern of the data suggested that something more akin to the neutral competition hypothesis, in which inter- and intraspecific competition are not significantly different, was the case. In many ways this is not surprising because, for all practical purposes, plants of similar stature use similar environmental factors, which is to say they basically do live in the same niche. On the other hand, if the competitive production principle is parallel to the competitive coexistence principle, it too may be overly dependent on the underlying assumptions of linearity. Perhaps in modern industrial agriculture the environment has been simplified to such an extent that linear rules do in fact apply, and plants with the same "niche" may not be able to coexist (or demonstrate a yield advantage living together). On the other hand, in the agriculture of the future, with ecological complexity reintroduced into the equation, it could be that the nonlinearities of nature will once again dominate, making the competitive exclusion principle (and by implication the competitive production principle) invalid on first principles (Levins 1979). This is an intriguing avenue for future research.

At the other extreme, mixtures of legumes and nonlegumes frequently seem to provide evidence of intercrop advantage (Haynes 1980, Nair et al. 1979). In a literature survey, Trenbath (1976) demonstrated that legume–nonlegume combinations did on average yield better than monocultures. Yet even here it is not really clear what mechanism is involved. On the one hand it could simply be a case of the competitive production principle if the legume and nonlegume are simply tapping different pools of nitrogen and thus reducing interspecific competition for that nutrient. On the other hand, the legume could be facilitating the growth of the nonlegume by supplying it with extra nitrogen. Very few experiments have addressed this problem (Reeves 1992), and the general literature is not amenable to differentiation between these two alternatives. That the legume may simply sequester nitrogen, which would otherwise leach out of the soil and later releases it to the nonlegume, has been suggested in a few cases (Agamuthu and Broughton 1985).

The general conclusion seems to be that, although theoretically it seems reasonable to suggest that two ecologically different crops could fit into an environment more efficiently than an equal biomass of the same crop, accumulated evidence offers little support for that idea. It would seem to be equally logical, especially in the light of recent theory in plant ecology, to suggest that different species of annual crops are more or less interchangeable ecologically, and the only advantage of intercropping is either from a socioeconomic point of view, perhaps from the long-term point of view of soil conservation, or from some sort of special facilitative effect characteristic of the particular combination involved. Perhaps the competitive production principle will turn out to be the competitive neutrality principle.

24.3.3.3 Agroforestry

The next logical level of integration is the incorporation of trees into the agroecosystem – something that is seemingly universal in more traditional systems. In temperate and tropical ecosystems, before the development of the industrial system, it was common for trees to be incorporated in some way. In the temperate zone it was common, and in the tropics it was almost universal.

The actual practice of agroforestry is extremely diverse, ranging from the use of trees as boundary markers to their use for shade to the incorporation of trees and crops into the same complex mixed garden. Although traditional practices still commonly incorporate trees into the agroecosystem, efforts by agricultural researchers and extensionists to promote more systematic designs that combine crops and trees have been largely ineffective. The ecological advantages of trees on farms, especially in tropical areas, remain obvious, but socioeconomic and political forces seem to provide effective barriers to expansion in the real world. Farmers without secure titles to their land are not likely to make investment in permanent fixtures such as trees. Farmers forced to live on the margin of profitability are not likely to incorporate an investment (trees) that promises returns many years later.

24.3.3.4 High Diversity Mixed Gardens

In all tropical areas it has been a common observation that farms always have some component that is extremely diverse and frequently appears to have much of the look of a "natural" ecosystem. A typical case is a traditional farm in eastern Nicaragua, as described by Shrader (1994). The pivotal elements of this farm are the fruit trees, especially the coconuts and the citrus, both of which are sold at local markets. But in addition to the coconuts and citrus there are at least 20 different species of fruit trees, all of which provide food for the family as well as occasional products to sell at local markets. Also scattered around the farm are at least five species of timber trees, most of which were either left from when the farm was established some 30 years ago or were volunteers that were subsequently nourished by the farmer. In the understory of the tree canopy a variety of shade-tolerant crops are grown, such as cassava and taro, in addition to tree seedlings that are purposefully cared for whether they were purposefully planted or not. Throughout the farm there are relatively open areas usually caused by the recent felling of an older tree either for timber or because it was diseased. Some of the open areas are thought to be locally unsuited for trees (for obscure reasons) and are thus maintained free of trees. In the open areas is a rotational system of maize and beans followed by a fallow of *Ipomea* and *Heliconia*, providing food for the family.

Although this is a relatively large farm that is managed in this high-biodiversity fashion, most high-biodiversity systems are smaller gardens and are frequently part of a larger farming operation. They are extremely popular topics for those interested in the ecology of agroecosystems because they seem to provide the kind of structural and species diversity more normally seen in unmanaged ecosystems. Yet it is for precisely this reason that they are mainly studied by anthropologists and sociologists whose interest is less ecological and more related to the interface of human engineering with the natural world. Exactly how one would go about studying them is not clear, for their diversity is similar to a natural system and yet they also incorporate all the complexity of human intervention.

Most ecological studies of such high-diversity systems treat them as if they were natural systems, asking questions similar to those that are asked of natural systems. For example, the biodiversity patterns of these systems can be studied in the same way as unmanaged systems (e.g., Perfecto and Vandermeer 1996). Indeed, there has been a recent surge of interest in the biodiversity of complicated agroecosystems simply from the point of view of conservation (Vandermeer and Perfecto 1997, Perfecto et al. 1996). Others have looked at other ecosystem functions such as nutrient cycling in such complex systems, but the bulk of serious study of complex home–garden-type agroecosystems remains largely the domain of social scientists.

24.3.3.5 The Farm Landscape

The next level of integration is the farm level, incorporating all the other levels into the patchwork that eventually forms the farm itself. This level is usually focused on by those who plan ecosystems, whether they be the farmers themselves who are forced to do it or the proponents of the organic transformation who are trying to rationalize the planning process at this landscape level.

The farming landscape has been a popular subject for geographers and anthropologists. The raw data that they use is basically a map of the farm. The basic geographic plan of a farm derives from two major forces: the ecological landscape itself and the proximity to the central dwelling place. For example, a stream flowing on the border of the farm presents an ecological requirement that forested land remain on its banks to prevent erosion, but the position of the home garden is usually for the convenience of the family members who must more or less continuously tend to it. The general principle is that those farming activities requiring the most intensive human labor are located nearest the dwelling place within the constraints imposed by the underlying ecology.

These principles apply to farms that are not highly capitalized, especially with respect to transportation systems. If transport is mainly by means of internal combustion machinery, the locational restrictions imposed by labor are largely eliminated. On ecological farms in Europe or the United States, for example, the constraints of both ecological background and labor requirements have been largely removed, and the landscape is primarily dictated by the requirements of rotatational sequences.

24.3.3.6 Landscape in the Large

At this level we begin, of necessity, incorporating nonecological subjects. At a purely technical level, rural landscapes are thought of differently from farming landscapes. At one extreme the romantic vision of artists from the beginning of modernism has idealized the agricultural landscape. There is clearly an aesthetic component to the rural landscape, and this is one feature that deserves consideration when governments plan (whether they actively plan or clandestinely plan through other means, they in fact always do plan). For example, the postcommunist government of the Czech Republic was paying subsidies for farmers to mow their fields even though there was no incentive for production simply because the rural landscape with its mosaic of field and forest was an aesthetic desire of the population. Unfortunately, in most of the world today some form of neoliberal ideology seems to dominate, and if one were to suggest that landscapes should be planned, for whatever reason, one would risk losing credibility.

Yet large landscapes do develop as a consequence of human and nonhuman factors, and they do make a difference. As the topic of landscape ecology continues to become more popular, there seem to be two general foci developing simultaneously. First, mainly in some tropical areas of underdeveloped countries, there is a concern that agricultural expansion is coming to dominate rural landscapes, thus crowding out the sometimes romantically defined "natural areas" so dear to the hearts of First World conservationists. The response has been a generalized plan to "zone" rural areas such that a core is left as a biological preserve and surrounded by a "buffer" zone, which itself is then surrounded by anarchic agricultural development. The main concern of such landscape planning is to avoid further agricultural incursions into the natural area. This particular form of landscape planning has practical problems to say nothing of the ethics involved (Vandermeer and Perfecto 1995).

The other focus of landscape ecology is characteristic of more developed areas of the world, and usually in highly populated temperate countries. The Netherlands is a case in point in which the production of "nature" is now seen as a legitimate product for agricultural development. It is ironic that almost the reverse of the landscape problem conservationists see in the tropics is the concern of landscape ecologists in the Netherlands, which is one of the most densely populated countries in the world with one of the most chemically and energy intense agricultural systems.

We thus see two generalized approaches to even asking questions about landscape at the large level: one emphasizes how the expansion of agriculture "ruins" the landscape, and the other emphasizes how agriculture creates the rural landscape so loved by sixteenth-century Dutch painters. Where does agroecosystem ecology fit into all of this? The general framework offered by Levins and Vandermeer (1990) for landscape planning (discussed below) still seems worth consideration, although they assume a rational socioeconomic system, which is something that appears less and less attainable in the increasingly sycophantic world in which we live.

Ultimately ecological principles speak to many other integrative aspects of a potential alternative agriculture program. Topics such as multiple interacting components and complex dynamics, the problem of viewing as simple that which is complex, or the various integrative problems associated with conversion provide issues that need to be explored from an ecological point of view and for which there is much recent ecological theory that may very well be relevant. Additionally the interface of a socioeconomic framework with ecology, although always acknowledged, needs to be dealt with more explicitly.

24.4 The Future of Agriculture

24.4.1 Searching for Models, Encountering Contradictions

In searching for a general plan for the transformation of agriculture from the modern industrial system to an ecologically sound sustainable one, we most naturally look to models, both for how the new agriculture will look and for how to transform the present system. Traditional systems present us with one form of model (Altieri 1990), for by definition they were not subject to the same forces that produced the modern industrial system. Extant organic farms, isolated amidst the ocean of industrial agriculture, provide another. But in both of these cases we are presented with systems in which some of the key social variables have been canceled out of the equation. Traditional systems, by their very nature, do not exist in the context of industrial agriculture but rather in isolated pockets in which the industrial system has not yet fully penetrated. Extant organic or ecological farms are already run by people whose attitudes have been transformed and who largely are not concerned with promoting a global transformation but rather with the survival and prosperity of their individual farms. Although it is clear that much is to be learned from traditional systems and extant organic farms, I rather expect that most of that knowledge will be technical, leaving us still to ponder the vexing social and political questions yet unanswered about the transition.

But the now well-known performance of organic farms (Lockertz, Shearer, and Kohl 1981), when coupled with experimental results from side-by-side plot experiments, leads to what I refer to as the first or fundamental contradiction of alternative agriculture (Vandermeer 1995). Almost all studies that have compared conventional with organic farms (e.g., Faeth et al. 1991, NRC 1989) have come to the same conclusion. Either there is no significant difference between the organic and conventional farms, or the organic ones actually perform better. However, a variety of studies have been done comparing conventional techniques with alternative ones on side-by-side plots (e.g., Lanini et al. 1994, Liebman et al. 1993, Temple et al. 1994). In almost all cases these experiments suggest that the alternative techniques do not perform as well as the conventional ones. And herein lies the fundamental contradiction – extant organic farms do as well as or better than conventional ones, and yet attempts to demonstrate the benefits of alternative techniques invariably fail.

A variety of mechanisms could account for this contradiction. It could be that the already-existing organic farms happen to be the ones that have survived because of some special circumstances, whereas the ones that failed are never part of the sample. Thus, we would be comparing farms that exist under the special circumstances with farms that exist randomly with respect to those circumstances; the factor we are comparing could be something other than organic versus nonorganic. Alternatively, it could be that under current economic and social conditions only those farmers who are especially skillful don to enter into the organic farm movement. Thus, we would be comparing good farmers with average farmers and not really organic with nonorganic.

An alternative possibility, and the one I suspect is the truth, is that we are dealing with "syndromes of production" in the sense of Andow and Hidaka (1989). Side-by-side plots are usually set up in such a way that one or a few factors are being compared between experimental and control plots. Furthermore, they are of a duration of 1 or 2 years. Just as the conventional industrial system involves a host of factors that combine with one another to form the whole system, the organic system combines a host of factors. That is, both the organic and conventional systems represent "syndromes of production," and it would be difficult to mimic either of them outside of their entire context.

Using very simple ecological and economic functions in a one-dimensional nonlinear model, Vandermeer (1997) suggested that two general syndromes are likely to evolve. The first is characterized by low prices and stable markets and is ecologically benign with low cultivation intensity and high overall supply of agricultural products. The other syndrome is associated with high prices, market volatility, ecological decline, high cultivation intensity, and low overall supply. Furthermore, depending on the values of the parameters, either of the syndromes may be stable, or the system could unpredictably wander from one syndrome to the other in a chaotic fashion. These particular syndromes result from some very simplistic assumptions about real systems and are not likely to actually represent the conditions found in the real world. However, the demonstration that alternative syndromes exist even with such a simple model and that dynamic rules are associated with them that could even lead to an alteration between the two systems is important. It is likely that such general rules will indeed apply to the real world, although our ability to predict when one system or the other (or one of the various other syndromes that may potentially exist) will evolve, or the conditions under which syndromes will appear and disappear seemingly at random, is limited.

Although the fundamental contradiction may be explained by the notion of syndromes of production, we are still left with the fact that there is a lack of convincing models for the transition. Models for ecological forms of agriculture exist in the many traditional and organic production systems. But a model for the actual transformation from the industrial system to the sustainable alternative is elusive to say the least. Theoretical models have been proposed, and many of them look good, but no country or region has yet undergone the transition. However, a useful model may be evolving right now – Cuba. Recent changes on the world political stage have forced Cuba to strive for a different model of production and may provide us with a preliminary glimpse of the problems we will likely face in the upcoming transformation. The Cuban situation is discussed in the next section.

Despite the relative lack of models of the actual transition, what might be referred to as a political plan does seem to be emerging. A combination of the encouragement of LISA-type agriculture along with the expansion of the organic sector seems to be a rational plan. To be sure there is some criticism of this plan based on the failure of LISA to face up to some of the deeper and politically difficult issues involved. Rossett and Altieri (1997) argue that the LISA-style approach has actually allowed for the co-optation of the ecological agricultural movement. What has happened, they argue, is that the underlying structure of the conventional agricultural system has been allowed to remain with simple substitutions of biological inputs for the previous chemical ones, which is a process they refer to as "input substitution." If it is the underlying structure of the system that is the problem in the first place, focusing on input substitution is putting a Band-Aid on a cancerous tumor. Rosset and Altieri argue that the term agroecology should be reserved for that form of ecological agriculture that incorporates a radical restructuring of the overall agricultural system.

Others argue that input substitution is an important or even essential form of transition. Many farmers simply scoff at the idea of converting to more ecological forms of agriculture. Input substitution (or what is referred to more generally as LISA-type approaches) is an important transitional argument for this sector. Although LISA can clearly be co-opted (Rosset and Altieri 1997), it nevertheless can also be made to be transitional. "Don't apply unless you have to" remains an important principle of LISA and is an argument understood by any farmer in the world. The combination of promoting the expansion of ecological agriculture (or agroecology) with the promotion of LISA (input substitution) seems to offer the best overall strategy for transforming the world's agroecosystems.

24.4.2 Cuba as a Model of Transition

Cuba's agricultural system evolved along the same lines as industrialized agriculture in the rest of the world after the revolutionary war of 1959. Technological progress was based on heavy use of mechanization and synthetic chemicals, and social formations were modeled after the emerging command economies of the Soviet Union and Eastern Europe. Large state farms were managed as factories. Proletarianized farm workers enjoyed considerable advantages over their counterparts in the capitalist world but nevertheless were tied neither to pieces of land nor particular production technologies; farmers worked to receive an hourly pay, not to produce food or fiber from the land.

With the breakup of the Eastern command economies, Cuba suddenly encountered conditions of extreme scarcity of basic inputs such as fuel, pesticides, and fertilizers, virtually the definition of the technical side of the industrial model. Entering this "special period," Cubans were rapidly forced to rethink their agricultural model. What they have done thus far holds many lessons for the rest of the world (Vandermeer et al. 1993; Perfecto 1994; Rosset and Benjamin 1994; Levins 1990, 1993). Quite purposefully a new model has been advanced in which sustainability has replaced profitability as the key guiding force in planning. For pest control, pesticides have been replaced with biological and cultural control forms. For soil management, biofertilizers have replaced most chemical fertilizers. For land preparation and cultivation, animal traction has replaced the tractor.

Many of the technical substitutions are generally well known outside of Cuba. For example, vermiculture is now widespread and reportedly has made up for almost all nitrogen needs in food production. Vermiculture has been well known in organic farming, for its history traces back at least to the nineteenth century in Europe. Another example is the use of *Trichogramma* wasps in biological control. These small parasites are standard fare in IPM technology in the United States and other developed countries. Cuba's contribution here is in the rapid and widespread application of these technologies.

But Cuba has also made significant technical contributions outside of the mainstream of technical alternatives. For example, control of the sweet potato weevil is accomplished through the physical relocation of nests of the predatory ant *Pheidole megacephala.* Banana stems are laced with sugar and placed in areas designated as reserves (because they naturally harbor high densities of *Pheidole megacephala* nests). The ants, attracted to the high levels of sugar, relocate their nests inside the banana stems. Subsequently, the banana stems, with ants' nests inside, are placed along the rows of newly planted sweet potatoes. The sun's radiation heats up the stems, causing the ants to move their nests underground, where they encounter the sweet-potato weevils and prey upon them. A variety of other technical examples could be cited (see Dlott et al. 1993).

Cuba thus represents something of a success story for alternative agriculture – at least at the level of developing and implementing alternative technical procedures. In this sense Cuba can be seen as a model. On the other hand, the problems Cuba faces in implementing its program are perhaps the most interesting aspects of the entire transformation process. For example, most alternative technologies require, to some extent, the substitution of human labor for fossil fuels. Here Cuba has already been forced to face problems that have yet to be adequately formulated by the alternative agriculture movement in the United States. Foremost among these problems is labor. The alternative model requires a great deal more labor than the classical one. Both short- and long-term plans are under consideration for dealing with the problem of labor. The short-term plan is to organize brigades of city volunteers to spend short periods of time (generally 2 weeks) in agriculture. The long-term solution is considerably more ambitious. A high degree of community participation is contemplated with the reorientation of research, development, and extension from a top–down system compatible with the classical model to a bottom–up one that captures and directs local knowledge for creative alternative methodologies. Such a laudable goal requires communities from which participation can emanate and local populations from which knowledge can be gained. Yet the last 30 years of urban migration make that impossible. If this objective is indeed to be a goal, rural life must be newly created. Consequently, one component of the overall plan for attracting agricultural labor is the creation of rural facilities that will be considered attractive to Havana's residents. Accordingly, in rural areas around the country, apartment complexes serviced with health clinics, sports facilities, stores, and related conveniences are being planned with the hope that labor camp volunteers will seek permanent relocation.

24.4.3 Importance of Local Knowledge

A major impediment to the development of the alternative model in Cuba is the lack of locally tuned knowledge about particular ecological circumstances, which is a consequence of 30 years of almost religious commitment to the industrial model. This will clearly be similar in the developed world. As Levins has noted, academic knowledge is general but shallow, whereas local knowledge is specific and deep. Local knowledge is frequently "flawed" in the sense of modern ecological understanding yet is set in some other worldview that provides it with context. That a local farmer in Nicaragua "knows" that grasses "burn" the corn is not ultimately different from the scientific ecologist's "knowledge" that the corn "dries out" because of "competition" from grasses. Whether we say burn or competition really only identifies the discourse under which the actual observation of the corn's performance with grass present is communicated from one person to another. That the farmer knows the details of what happens to the corn when grasses grow around it is key to developing local alternatives that work. The fact that the ecologist uses the word *competition* to describe the phenomenon brings the rest of the experience of the world, to the extent it is cataloged, as possible knowledge of the planning process. The farmer knows what happens in his or her field. The ecologist can generalize that knowledge and compare it with what happens in other fields, effectively expanding the potential knowledge base.

The deep and local knowledge of the farmer, especially the traditional farmer, is essential to the ultimate development of the alternative model. The ecologist may help generalize and contextualize that knowledge to actually make it richer, but the local knowledge is imperative owing to the unpredictability and locality-specific nature of many ecological processes. Yet much traditional knowledge has been lost during the past 50 years because of the hegemony of the industrial model.

We are thus faced with a contradiction that is similar to the one identified by Rosa Luxumborg – it is difficult to construct an ideal system with less than ideal people. Today's world is filled with urban workers, whereas what we need are rural ecologists. Although this problem is not as universally recognized as the lack of appropriate technology, it may turn out to be a more important one.

We thus are faced with another contradiction of the move to ecological agriculture. On the one hand, we do not have a complete catalog of techniques that are tried and proven to work under all circumstances – the technical side of the contradiction. On the other hand, the destruction of rural society has taken with it the knowledge base and labor force that will be needed for the transformation – the social side of the contradiction. We are thus faced with the task of building a new system based on incomplete technical information and with a society designed to function only under the current system. The truth is that we do not know, technically, how to control, for example, the whitefly in vegetable production in Central America. Yet even if we did have a technical solution, it is not clear that the local farmers or farmworkers would be available or willing to undertake that procedure. As we strive to transform the current system, this is a principal contradiction that must be resolved.

24.4.4 Toward a Healthy Agroecosystem

What are the mechanisms of a healthy agroecosystem? Many have written on this topic (e.g., Rapport 1989, Costanza 1992), and the following features seem to emerge repeatedly as major mechanisms promoting agroecosystem health (Levins and Vandermeer 1990) in the sense of year-to-year production and over the long run:

1. Closed nutrient cycles with minimal external input;
2. Control of pests with biological or cultural practices with a minimum use of synthetic pesticides;
3. Maintenance of soil biodiversity to at least ensure that members of all functional groups of organisms remain in the system (Anderson 1988), including macro- and microfaunal and floral elements;
4. Maintenance of beneficials, such as parasitic wasps, spiders, ants, birds (the pest forms of these groups are covered in item 2 above);
5. Reasonable rewards to the practitioner (the farmer, or worker – the people who are in contact with the land in the production process);

6. Matching the needs of the consumer community (be it a local community or the international marketplace) to the capabilities of the land;
7. Maintaining the health of farmers, farm workers, and consumers;
8. Provisioning of landscape diversity with ecological interactions among patch types both in space (e.g., mixed farming, intercropping) and time (e.g., rotations).

All of these items are offered here as general and even tentative guidelines, most seem obvious and are relatively easily operationalized.

24.4.5 A Concluding Thought

As we look to the future we see clearly three needed changes in agroecosystem evolution. First, we need a change in the agenda for agroecosystem change. Agroecosystems are governed by ecological relations just as any other ecosystem. Any planned changes must take into account those relations and not, as much of industrial agriculture has done in the past, assume that human institutions can simply dominate. Just as modern chemistry changed our attitude about spinning silk into gold by understanding the laws of physics, agroecology needs to change our attitudes about how agroecosystems can be organized by understanding the laws of ecology. Second, we need major improvements in understanding those ecological laws that govern agroecosystems. Ecology is a young science, and our knowledge of ecological rules is metaphorically pre-Newtonian. Apart from our need to change attitudes of those seeking agroecosystem improvement, we need to push forward in ecological research so as to understand the ecological forces at work in agroecosystems. Third, we need a new emphasis in philosophy of change. Agroecosystems are the only ecosystems that have an inherent purpose; by definition they exist to serve humankind. Agroecosystems should be designed to benefit human beings rather than as support structures for some temporary human construction, be it the nineteenth-century dreams of Napoleon to dominate Europe or the twentieth-century dreams of the international banking industry to dominate the world. Agroecosystems emerged from the need to provide food and fiber to the human population, not to create political power for an elite.

REFERENCES

Agamuthu, P., and Broughton, W. J. (1985). Nutrient cycling within the developing oilpalm–legume ecosystem. *Agric. Ecosys. Environ.* 13:111–23.

Altieri, M. A. (1987). *Agroecology: The Scientific Basis of Alternative Agriculture.* Boulder, CO: Westview Press.

Altieri, M. A. (1990). Why study traditional agriculture? In *Agroecology,* (eds.) C. R. Carroll, J. H. Vandermeer, and P. Rosset, pp. 551–64. New York: McGraw–Hill.

Anderson, J. M. (1988). Spatiotemporal effects of invertebrates on soil processes. *Biol. Fert. Soils.* 6:216–27.

Andow, D. A., and Hidaka, K. (1989). Experimental natural history of sustainable agriculture: Syndromes of production. *Agric. Ecosyst. Environ.* 27:447–62.

Andow, D. A., and Rosset, P. M. (1990). Integrated pest management. In *Agroecology*, (eds.) C. R. Carroll, J. H. Vandermeer, and P. M. Rosset. New York: McGraw–Hill.

Beddington, J. R., Free, C. A., and Lawton, J. H. (1978). Characteristics of successful enemies in models of biological control of insect pests. *Nature* 273:513–19.

Berlan, J. P., and Lewontin, R. C. (1986). The political economy of hybrid corn. *Monthly Review* 38:35–47.

Beus, C. E., and Dunlap, R. E. (1993). Agricultural policy debates: Examining the alternative and conventional perspectives. *Am. J. Alt. Agric.* 8:98–106.

Buttel, F. H. (1990). Social relations and the growth of modern agriculture. In *Agroecology*, (eds.) C. R. Carroll, J. H. Vandermeer, and P. Rosset, pp. 113–45. New York: McGraw–Hill.

Carson, R. (1962).*Silent Spring.* New York: Houghton–Mifflin.

Colbern, T., Dumanoski, D., and Myers, J. P. (1996). *Our Stolen Future: Are We Threatening Our Fertility, Intelligence, and Survival? – A Scientific Detective Story.* New York: Dutton, Penguin Group.

Costanza, R. (1992). Toward an operational definition of health. In *Ecosystem Health: New Goals for Environmental Management.* (eds.) R. Costanza, B. G. Norton, and B. D. Haskell. Washington, DC: Island Press.

Dempster, J. P., and Coaker, T. H. (1974). Diversification of crop ecosystems as a means of controlling pests. In *Biology in Pest and Disease Control.* (eds.) D. Price Jones and M. E. Soloom, pp. 106–14. Oxford, UK: Blackwell Scientific.

Dlot, J., Perfecto, I., Rosset, P., Burkham, P., Monterrey, J., and Vandermeer, J. H. (1993). Management of insect pests and weeds. *Agric. Hum. Values* X:9–15.

Ewel, J. J. (1986). Designing agricultural ecosystems for the humid tropics. *Ann Rev. Ecol. Syst.* 17:245–71.

Faeth, P., Repetto, R., Kroll, K., Dai, Q., and Helmers, G. (1991). *Paying the Farm Bill: U.S. Agricultural Policy and the Transition to Sustainable Agriculture.* Washington, DC: World Resources Institute.

Feeny, P. P. (1976). Plant apparency and chemical defense. In *Biochemical Interaction between Plants and Insects. Rec. Adv. Phytochem.* Vol. 10. (eds.) J. Wallace and R. Mansel, pp. 1–40.

Francis, C. A. (1986). *Multiple Cropping: Practices and Potentials.* New York: Macmillan.

Hawkins, B. A., Thomas, M. B., and Hochberg, M. E. (1993). Refuge theory and biological control. *Science* 262:1429–32.

Haynes, R. J. (1980). Competitive aspects of the grass-legume association. *Adv. in Agron.* 33:227–61.

Howard, L. E. (1953). *Sir Albert Howard in India.* London: Faber and Faber.

Lacy, W. B. (1993). Can agricultural colleges meet the needs of sustainable agriculture? *Am. J. Alt. Agric.* 8:40–5.

Lanini, W. T., Zalom, F., Marois, J., and Ferris, H. (1994). Researchers find short-term insect problems, long-term weed problems. *Calif. Agric.* 48:27–33.

Levins, R. (1979). Coexistence in a variable environment. *Am. Nat.* 114:765–83.

Levins, R. (1986). Perspectives on IPM: From an industrial to an ecological model. In *Ecological Theory and Integrated Pest Management.* (ed). M. Kogan, New York: John Wiley & Sons.

Levins, R. (1990). The struggle for ecological agriculture in Cuba. *Capitalism Nature and Socialism* 5:121–41.

Levins, R. (1993). The ecological transformation of Cuba. *Agriculture and Human Values* 10:52–60.

Levins, R., and Vandermeer, J. H. (1990). The agroecosystem embedded in a complex system. In *Agroecology*, (eds.) C. R. Carroll, J. H. Vandermeer, and P. Rosset, New York: McGraw–Hill.

Lewontin, R. C. (1982). Agricultural research and the penetration of capital. *Science for the People.* January–February:12–17.

Liebman, M., Rowe, R. J., Corson, S., Marra, M. C., Honeycutt, C. W., and Murphy, B. A. (1993). Agronomic and economic performance of conventional vs. reduced input bean cropping systems. *J. Prod. Agric.* 6:369–77.

Lockeretz, W., Shearer, G., and Kohl, D. H. (1981). Organic farming in the corn belt. *Science* 211:540–47.

Loomis, R. S., and Connor, D. J. (1993). *Crop Ecology.* Cambridge, UK: Cambridge University Press.

Mills, N. J., and Getz, W. M. (1996). Modelling the biological control of insect pests: A review of host–parasitoid models. *Ecol. Modelling* 92:121–43.

Morgan, D. (1979). *Merchants of Grain.* New York: Viking.

Murdoch, W. W. (1994). Population regulation in theory and practice. *Ecology* 75: 271–87.

Murdoch, W. W., Chesson, J., and Chesson, P. L. (1985). Biological control in theory and practice. *Am. Nat.* 125:344–66.

Nair, P. K. R., Patel, U. K., Singh, R. P., and Kaushik, M. K. (1979). Evaluation of legume intercropping in conservation of fertilizer nitrogen in maize culture. *J. Agric. Sci. Cambridge* 93:189–94.

National Research Council (NRC) (1989). *Alternative Agriculture.* National Academy of Sciences, Board on Agriculture, Washington, DC: National Academy Press.

Niggli, U., and Lockeretz, W. (1996). Development of Research in Organic Agriculture. In *Fundamentals of Organic Agriculture. 11th IFOAM International Scientific Conference, Copehagen. Proceedings,* Vol. 1, (ed.) T. V. Østergaard, pp. 9–23.

Oldeman, R. D. A. (1983). The design of ecologically sound agroforests. In *Plant Research and Agroforestry,* (ed.) P. A. Huxley, pp. 173–207. Nairobi, Kenya: Int. Cntr. Res. Agroforestry (ICRAF).

Par, J. F., and Hornick, S. B. (1994). Agricultural use of organic amendments: A historical perspective. *Am. J. Altern. Agric.* 7:181–89.

Perfecto, I. (1994). The transformation of Cuban agriculture after the Cold War. *Am. J. Alt. Agric.* 9:98–108.

Perfecto, I., and Vandermeer, J. H. (1996). Physical factors and the structure of an ant community in a tropical agroecosystem. *Oecologia* 108:577–82.

Perfecto, I., Rice, R., Greenberg, R., and Van der Voort, M. A. (1996). Shade Coffee. A disappearing refuge for biodiversity. *Bioscience* 46:598–608.

Perfecto, I., Vandermeer, J., Hanson, P., and Cartin, V. (1997). Arthropod biodiversity loss and the transformation of a tropical agro-ecosystem. *Biodiversity and Conservation* 6:935–45.

Perrin, R. M. (1977). Pest management in multiple cropping systems. *Agroecosystems* 3: 93–118.

Pesek, J. (1990). Historical perspective. In *Sustainable Agricultural Systems,* (eds.) J. L. Hatfield and D. L. Karlen. Boca Raton, FL: Lewis Publications.

Rapport, D. J. (1989). Symptoms of pathology in the Gulf of Bothnia (Baltic Sea): Ecosystem response to stress from human activity. *Biol. J. Linn. Soc.* 37:33–49.

Reeves, M. (1992). Nitrogen dynamics in a maize bean intercrop in Costa Rica. Ph.D. thesis, University of Michigan.

Risch, S. J., Andow, D., and Altieri, M. (1983). Agroecosystem diversity and pest control: Data, tentative conclusions, and new research directions. *Environ. Entomol.* 12:625–29.

Rosset, P., and Altieri, M. (1997). Agroecology versus input substitution: A fundamental contradiction of sustainable agriculture. *Society and Natural Resources* (in press).

Rosset, P., and Benjamin, M. (1994). The greening of the revolution: Cuba's experiment with organic agriculture. Melborne: Ocean Press.

Russell, E. (1996). "Speaking of annihilation": Mobilizing for war against human and insect enemies, 1914–1945. *J. Am. History* 82:1505–29.

Shrader, E. (1994). Investigations of a traditional agroecosystem in the Atlantic Lowlands of Nicaragua. M.S. thesis, University of Michigan.

Soule, J. D., and Piper, J. K. (1992). *Farming in Nature's Image: An Ecological Approach to Agriculture.* Washington, DC: Island Press.

Soule, J., Carré, D., and Jackson, W. (1990). Ecological impact of modern agriculture. In *Agroecology,* (eds.) C. R. Carroll, J. H. Vandermeer, and P. Rosset, pp. 165–88. New York: McGraw–Hill.

Steingraber, S. (1997). *Living Downstream.* New York: Addison–Wesley.

Swift, M. J., Vandermeer, J. H., Ramakrishnan, P. S. Anderson, J. M., Ong, C., and Hawkins, B. (1996). Biological diversity and ecosystem processes. In *Biodiversity and Ecosystem Functioning: Ecosystem Analyses. Global Diversity Assessment,* H. A. Mooney, J. Lubchenco, R. Dirzo, and O. E. Sala, pp. 433–43. Cambridge, UK: Cambridge University Press.

Temple, S. R., Somasco, O. A., Kirk, M., and Friedman, D. (1994). Conventional, low-input and organic farming systems compared. *Calif. Agric.* 48:14–9.

Trenbath, B. R. (1974). Biomass productivity of mixtures. *Adv. Agron.* 26:177–210.

Trenbath, B. R. (1976). Plant interactions in mixed crop communities. In *Multiple Cropping,* (eds.) R. I. Papendick, A. Sanchez, and G. B. Triplett, pp. 129–70. ASA (American Society of Agronomy) Special Publication 27.

van den Bosch, R. (1978). *The Pesticide Conspiracy.* New York: Doubleday.

van Noordwijk, M., and de Willigen, P. (1986). Qualitative root ecology as element of soil fertility theory. *Netherlands J. Agric. Sci.* 34:273–82.

Vandermeer, J. H. (1984). The interpretation and design of intercrop systems involving environmental modification by one of the components: A theoretical framework. *Bio. Agr. Hort.* 2:135–1156.

Vandermeer, J. H. (1989). *The Ecology of Intercropping.* Cambridge, UK: Cambridge University Press.

Vandermeer, J. H. (1995). The ecological basis of alternative agriculture. *Ann. Rev. Ecol. Syst.* 26:201–24.

Vandermeer, J. H. (1997). Syndromes of production: An emergent property of simple agroecosystem dynamics. *J. Env. Managment* 51:59–72.

Vandermeer, J., and Andow, D. A. (1986). Prophylactic and responsive components of an integrated pest management program. *J. Econ Entomol.* 79:299–302.

Vandermeer, J. H., and Perfecto, I. (1995). *A Breakfast of Biodiversity: The Truth of Rain Forest Loss.* San Francisco: Institute for Food and Developmental Policy.

Vandermeer, J. H., and Perfecto, I. (1997). The agroecosystem: A need for the conservation biologist's lens. *Cons. Bio.* 11:591–92.

Vandermeer, J. H., Carney, J., Gesper, P., Perfecto, I., and Rosset, P. (1993). Cuba and the dilemma of modern agriculture. *Agriculture and Human Values* X:3–8.

Vandermeer, J. H., and Schultz, B. (1990). Variability, stability, and risk in intercropping. In *Agroecology: Researching the Ecological Basis for Sustainable Agriculture.* (ed.) S. R. Gliessman, pp. 205–29. New York: Springer–Verlag.

Vandermeer, J. H., van Noordwijk, M., Anderson, J., Ong, C., and Perfecto, I. (1998). Global change and multi-species agroecosystems: Concepts and issues. *Agric. Ecosyst. and Environ.* (in press).

Weir, D., and Shapiro, M. (1981). *The Circle of Poison.* San Franscico: Institution for Food and Development Policy.

Willey, R. W. (1979a). Intercropping – Its importance and its research needs. Part I. Competition and yield advantages. *Field Crop Abstracts* 32:1–10.

Willey, R. W. (1979b). Intercropping – Its importance and its research needs. Part II. Agronomic relationships. *Field Crop Abstracts* 32:73–85.

Wright, A. (1990). *The Death of Roman Gonzalez: The Modern Agricultural Dilemma.* Austin: University of Texas Press.

CHAPTER TWENTY-FIVE

Political Economy of Agricultural Genetics

JEAN-PIERRE BERLAN

Take a software diskette. The little value of this square piece of plastic stems from manufacturing and sale. But for the tag, there is no visible difference with a highly priced diskette. The price difference stems from the *nonmaterial software* embodied in the second. Let us now make a thought experiment and copy the software unto the virgin diskette. For the user, the copied software has the same "utility" as the original one. But for the software company stockholder the copied software has no "value" in the sense that it does not bring in any profit. To avoid this, our society made it illegal to use copied software by creating a protection derived from the patent and copyright system. This distinction between utility and value is the key to understanding the political economy of agricultural genetics.

Now take a grain. It has neither value nor utility. Let us make the thought experiment of sowing it. The grain becomes a seed. It germinates and the plant grows, flowers, and produces more grains. Something extraordinary happens: the original grain has reproduced and multiplied, and it will continue to do so generation after generation. This characteristic of a seed, to reproduce and multiply, is the very characteristic of life. During almost the entire human history, this phenomenon has remained *incomprehensible* in the double meaning of the term: it is beyond understanding and beyond seizure. Then, in the nineteenth century, breeders and biologists began to postulate that "something" existed between the grain and the seed and began to build it up. The scientific method, Cartesian mechanicist reductionism, made it possible to impoverish this wonder to a single dimension, that of a "genetic program" or "software," which has become comprehensible in the peculiar sense that it can be seized, manipulated, appropriated, and turned into a new source of value. Although such advancement of our comprehension (knowledge) of life is a matter of debate, for investors this reduction of life to a unidimensional source of profit is a triumph, much celebrated for this reason. Those who have capital to invest to take hold of life qua "genetic software" will become "comme maîtres et possesseurs de la Nature" (Descartes 1637).

But here the farmer is an obstacle. Farming implies, as does any biological production, *copying* and multiplying the corresponding genetic software

(Berlan and Lewontin 1986). As long as farmers can sow their harvested grain, the corresponding "genetic software" has no value.[1] Potential returns on capital invested in this genetic software creation promise to be very high, but reaping them requires first preventing farmers from sowing their harvested grain. Doing this legally was *politically* unthinkable because it amounted to confiscating a basic human right.[2] Investors had to turn to biological methods to copy-protect "their" varietal software, namely to breeding *sterile* varieties.[3] Achieving such a goal required that it remains invisible. This is the secret of agricultural genetics. Space limits us to two case studies of commodification of plant breeding and a brief discussion of the underlying pattern or model.

25.1 Cereal Breeding in Nineteenth-Century Europe

In 1831, Professor La Gasca, curator of the Royal Gardens in Madrid, visited Le Couteur's wheat fields in Jersey. To the gentleman-farmer's dismay, he pointed out that Le Couteur's fields, instead of being "tolerably pure" (Le Couteur 1836, p. 42), were a mixture of 23 sorts "of which some have been discovered, through the experimental researches made by the Author, to be three weeks later in ripening than others" (Le Couteur, p. 42). Such a mixture could not give the best yield, "for the largest quantity would be obtained when every ear produced that fine, plump, thin-skinned, coffee-like looking grain which evidently contain much meal, in a delicate, transparent, thin-coated bran, such as some Dantzic, selected from the high-mixed produces" (Le Couteur, p. 43). "No previous writer had yet called the attention of the agricultural world to the cultivation of pure sorts, originating from one single grain, or a single ear" (Le Couteur, p. 44). La Gasca and Le Couteur were the discoverers of the isolation method, which is still the basic crop improvement method. This method entails replacing a population by the best isolated plant, the seeds (i.e., the genetic software) of which are multiplied at will.

In fact, Le Couteur codified a practice based on the empirical observation that cereal varieties "bred true," that is, remained identical from one generation to the next, provided they had been multiplied from a single grain or ear. "The old Chidham wheat grown in this country from about 1800 to 1880 or later was derived from a single ear found growing in a hedge at Chidham in Sussex" (Percival 1921, p. 78). In Scotland, P. Shirreff developed his wheat Mungoswell from a plant that had survived the severe 1813 winter exceptionally well (Evershed 1884). Later,

calling upon a friend in the autumn of 1832, I was struck with an ear of wheat which had been culled from one of his fields on the farm of Drew, East Lothian, and resolved to propagate from its seeds.... The produce of the ear proved to be a new variety, which has been named Hopetoun wheat, and which was sold for the first time in 1839" (Shirreff 1841).

According to Percival, "the variety Fenton, also an excellent sort much cultivated during the nineteenth century was discovered by Mr. Hope of Fenton Barns, Scotland, in a quarry in 1835." (Percival 1921, p. 78)

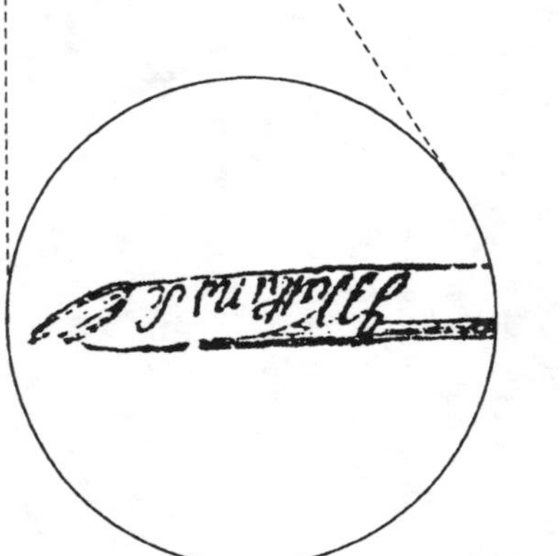

A

Figure 25.1. Advertising for Hallett's pedigree wheat, *The Agricultural Gazette*; October 31, 1881.

Grown from a Single Grain, grown in 1857, and
Re-started in each succeding year for TWENTY-FOUR YEARS (up to 1880) from the proved best Single Grain.

1878.

Hallett's Pedigree, Chevalier.

Costs less for Seed per acre, produces far more per acre, and is of finer quality than any other Barle

Apply to MAJOR HALLETT, F.L.S., BRIGHTON.

NOTE. -An Ear of the Pedigree OATS is too Large for pourtrayal in a Newspaper C

B

Figure 25.1. (*continued*)

During his long career as a breeder with a keen eye to detect "sports," Shirreff "had the good fortune to raise in his lifetime seven new varieties, which are now extensively grown in many parts of Britain" (Darwin 1868, vol. 1, p. 386). Such varieties bred true: "Mr. Shirreff, and a higher authority cannot be given . . . says, 'I have never seen seed grain which has either been improved or degenerated by cultivation, so as to convey the change to the succeeding crop' (Darwin 1868, p. 389).

This doctrine received its scientific blessing from Louis de Vilmorin. The Vilmorin method combined individual selection with progeny testing, thus anticipating the pure-line breeding technique of the end of the ninteenth century (Gayon and Zallen 1992). By growing wheat plants isolated in 1836 and 1856 for 50 years, he showed that they remained identical in all respects. "This fixity is shown not only in the characters of the ear but also in all the other characters of the plant even that of precocity, which would appear to most dependent upon climate" (Babcock and Clausen 1918, p. 257).

Whatever the theoretical and practical worth of the isolation method, in the 1860s, new approaches stressing that only *continuous* selection could improve and maintain the qualities of a variety that "deteriorated" in farmers' hands gained ground in England and Germany. These approaches rested on the continuous character of Darwin's natural selection (1859, 1868).

Like Le Couteur and Shirreff, Hallett in England started from the best single grain of the best ear of the best plant (or the best potato, although this is a vegetatively reproduced species). Unlike them, he repeated the process generation after generation, growing the selected plants in the best conditions. Each generation improved over the preceding one. In Darwin's words, "Major Hallett has gone much farther (than Le Couteur), and by continued selection of plants from the grains of the same ear, during successive generations, has made his 'Pedigree in Wheat' (and other cereals) now famous in many quarters of the world" (Darwin 1868, p. 386).

Major Hallett insisted on his scientific approach. He published articles in scientific periodicals and agricultural journals (for example 1862, 1870, 1882, 1887), ran advertisements in various journals (*The Times* in 1862 carried a two pages figure advertisement from his 1862 paper in the *Journal of the Royal Agricultural Society*) to promote his technique and sell his "pedigree" races. Because he had started in 1857, Hallett claimed that his advance made it a better business to purchase seeds from his company. The advertisement from *The Agricultural Gazette* (October 31, 1881) reproduced here (Figure 25.1) tells the story from "The scientific discovery of *The Law of Development of Cereals*" to the appropriation device by a trademark – regretably "the only means." In 1887, he added, "it is highly important to purchase fresh seed every year from Brighton where the selection is continued, and without which no 'breed' of anything can be kept up" (Halett 1887).

This advertisement not only captured almost a century of breeding but also announced the next century of genetic triumphs. A careful observer[4] can see a signature on the straw. One cannot be sure that it is F. Hallett's (see the blowup) although at the same time, Hallett ran an advertisement for his Chevalier barley with a tag attached to the ear – again a symbolic claim of his property rights. This draws our attention to the obvious fact (with hindsight) that claiming one's right over a living "material" implies that it be identifiable. Hallett's "dream" of 1881 has finally come true: the "finger-printing" of living organisms with a molecular signature becomes common – just like branding cattle with an iron.

The same Darwinian idea of imperceptible and continuous change buttressed the German breeding doctrine. It consisted in selecting the "best representatives of the varieties they wished to improve in order to be sure to retain all of its good features in the new race" (de Vries 1907, p. 47) and repeating the process year after year. Contrary to Hallett, German breeders grew their plants under ordinary field conditions, assuming that the environment had no influence upon heredity. As an improvement technique, this mass selection was probably inefficient, except in the case of rye, a cross-pollinated species. But because cultivation was bound to change the composition of the breeder's original mixture, it was probably correct to claim that such varieties deteriorated in the farmer's field. G. Liebscher, director of the Göttingen Agricultural Institute wrote, "varieties will deteriorate unless the greatest care be taken to prevent it, by always selecting the most conspicuous individuals as the mother plants for the continued propagation of the variety." (Liebscher 1897, p. 351).

Hugo de Vries (1907), the most famous biologist of the first decade of this century, understood the reason for the shift to such unstable "breeds":

> This assertion has a distinct and deep significance in agricultural practice, and has gained a great deal of influence *in the discussion of theoretical questions* as well. (...) [my italics]. It is easy to see that the gain made by the breeder of a new variety depends, for a large part, on the acceptance of this proposition. In the varieties produced by Le Couteur and Shireff, all seed is of equal value, provided that all the races are kept pure and free from admixture. Any one can multiply them with the same success as the original breeder, but on Hallett's principle all the profit of the production of reliable seed grain was given into the hands of him who kept the original pedigree. (de Vries 1907, p. 43)

This held true for the German breeders "since it kept the production of the seed-grains of his race in his own hands, at least for a long succession of years, and thereby enabled him to secure very considerable profit. On this account it is only natural that many breeders of cereals of the present time still adhere to these old convictions" (de Vries 1907, p. 66).

Before we condemn Hallett's quackery (we have known since 1903 that his method eliminated all genetic variations), one should remember that the issues about inheritance were obscure. Modern history and sociology of science has repeatedly shown that "facts" do not suffice to determine the "right" scientific theory. Many other considerations settle an issue. In the case of an applied discipline such as agricultural genetics, things are simpler. The scientific truth must also be value (profit)-creating, as de Vries suggested. Le Couteur was a gentleman-farmer interested in improving his crops, whereas Hallett was interested in the returns of his breeding enterprise. The latter had no use for true-breeding varieties. Because the economic nature of breeding had surreptitiously changed, plant reproduction had to change. This change was easy to achieve. Hallett's technique fitted well into the scientific Darwinian view of gradual accumulation of small improvements by selection. His method of growing his plants in the best agricultural conditions so that they could transmit their vigor and productivity to their offspring fitted the general view shared by Darwin (1868, Chap. XXVII) that the environment could directly affect their heredity. The best science of the times could support the replacement of a socially efficient breeding technique by an inefficient but profit-creating one – a pattern that "hybridization" repeated on an enlarged scale during the entire twentieth century.

It was only in 1892 that Haljmaar Nilsson at Svalöf, disatisfied with the German continuous technique, rediscovered the Vilmorin method (Babcock and Clausen 1918, p. 293) – the isolation technique. Interestingly, Nilsson's assignment was to improve the situation of farming in Sweden by improving yields, not breeders' profits.

25.2 "Hybrid" Corn and "Hybrid" Breeding

Biologists, geneticists, breeders, historians, sociologists, and economists have celebrated hybrid corn as the triumph of agricultural research in the twentieth century (Berlan 1987, pp. 130–8). The story of this "symbol of U.S. agriculture" (Welch 1961) unfolds from the discovery of hybrid vigor or heterosis to its vigorous implementation thanks to a daring scientific "coup d'etat." On the suggestion of his son Henry Agard Wallace, the Secretary of Agriculture Henry C. Wallace made hybridization the official U.S. breeding technique in February 1922 (Crabb 1947, pp. 93–102). One hundred or so *hybrid* breeders were recruited to work within a coordinated *hybrid* research program (Jenkins 1936). The final success almost 15 years later, although a tour de force,[5] was a matter of funding, organization, and administration. From then on, hybrids made it possible to harvest inestimable additional tons of maize all over the world. In 1946, this increased production, it was claimed, had paid for "the cost of the Manhattan project," and H. A. Wallace could compare the power of heterosis to that of the atomic bomb. Harnessing the heterosis power of other species then became the breeders' twentieth-century paradigm (Pickett and Galwey 1997). Mangelsdorf forcefully articulated this program in his forecast concluding his 1951 *Scientific American* celebration of hybrid corn by observing, "the time is rapidly approaching when the majority of our cultivated plants and domestic animals will be hybrid forms. Hybrid corn has shown the way. Man has only begun to exploit the rich 'gift of hybridity'" (Mangelsdorf 1951, p. 47).

We have to conquer in a few pages this bastion protected by decades of scientific claims, supported by powerful institutions from agricultural research to wealthy seed companies, buttressed by advertising and newspaper reports, and crowned by the scientific economic estimate of an ad aeternam return of 7 dollars to each dollar invested in research (Griliches 1958). Fortunately, we know the weak spot of this medieval *place forte*: confusing value with utility.

25.2.1 Seeding and Multiplication Rates

Breeders hired to select "hybrids" have focused on yield increase *per hectare*, that is, utility. After all, it was this task that the political power had entrusted to them. This shortsightedness is the immediate source of a great scientific mystification of this century. For what a seed company sells and what a farmer buys are not a yield gain *per hectare* but a quantity of "seed,"[6] say a quintal. Both are interested in the gain made by trading a *quintal of* "seed." Let all units be in quintals of grain, G being the total gain per quintal of "seed" traded, ΔY the yield gain per hectare, ΔC the cost difference between farmer's seed and commercial hybrid seed per quintal of seed traded, and R the seeding rate (the quantity of seed sown in quintals per hectare); G is given by:[7]

$$G = \Delta Y/R - \Delta C \tag{1}$$

The seed company and the farmer must share the gross gain per quintal of seed traded. If Π is the seed company's profit and G_1 is the farmer's net gain (a temporary innovation rent) from trading a quintal of seed (in quintals of grain), and α and $1 - \alpha$ are, respectively, the seed company's and the farmer's share, we have

$$\Pi = \alpha \Delta Y/R \tag{2}$$

Because the farmer buys the seed, his gain is

$$G_1 = (1 - \alpha)\Delta Y/R - \Delta C. \tag{3}$$

He will buy the seed if $G_1 > 0$:

$$(1 - \alpha)\Delta Y/R > \Delta C. \tag{4}$$

Dividing both members by Y, the average yield, and rearranging, we obtain

$$\Delta Y/Y > \Delta C R/Y(1 - \alpha), \tag{5}$$

where $\Delta Y/Y$ is Γ, the profitability threshold (the *computed* minimum percentage yield increase to induce a farmer to shift to hybrids) and Y/R is θ, the agronomic multiplication rate of the species. Equation (5) becomes

$$\Gamma > \Delta C/(1 - \alpha)\theta \tag{6}$$

This simple expression summarizes (almost) everything that a hybridizer should know. Because success is all the more likely, easy, and rapid when Γ is smaller, a hybridizer should do the following:

- Increase the efficiency of seed production, that is, reduce ΔC.

Table 25.1. *Estimates of the Seeding Rate for Various Crops (United States)*

	Hybrid Crops			Nonhybrid Crops		
	Maïze	Sorghum	Sunflower	Soybeans	Wheat	Barley
Time of hybrid success	mid-1930s	late 1950s	1970s	—	—	—
Yield	30–35	15				
Seeding rate	0.08	0.08	0.05			
Multiplication rate	350–400	180	300	17[1]	17	17

[1] In the 1950s. About 30–35 now.
(Source: Agricultural Statistics, various years).

- Keep α low by postponing gratification and even incurring losses until farmers have shifted to hybrids.[8]
- Pick up high multiplication rate species.

Crops fall into two categories: low (<20–25) and high (>250–300) multiplication rates (Table 25.1). The threshold will then be 10 to 20 times higher for certain crops than for others. With data from Table 25.1, and $\alpha = 0.3$ and $\Delta C = 4$ q, the computed profit threshold is 1.6% for maize (any *breeding* program, even hybridization, can achieve such a gain) and 33% for wheat, which is clearly out of reach. *Wallaces' hybrid corn breeders succeeded by chance: They were assigned a high-multiplication-rate crop.*

In the name of hybrid vigor (yield gains *for the farmer*), breeders have attempted for decades to hybridize crops, concentrating on large-market but low-multiplication-rate crops – particularly wheat, but also soybeans and barley. Breeders failed for these but succeeded for high-multiplication-rate minor species (sorghum, sunflower) (Table 25.1). In France, INRA, the public agricultural research service, claims it is close to success for rapeseed, a high-multiplication-rate species (500), whereas hybrid wheat, which "was going out the laboratory" (Rousset 1986) is still stuck, the multiplication rate of wheat in France being 40. These full-sized breeding experiments confirm that the multiplication rate is the key to the *success* of a hybridization program. Hybrid vigor or heterosis supposedly responsible for yield gains are, in practice, irrelevant. What is, then, its function?

25.2.2 Heterosis and Hybrid Vigor

25.2.2.1 Are "Hybrids" Hybrid?

Hybrid vigor, it is asserted, led to hybrid corn. It is true that agronomists at experiment stations had attempted to use "hybrid vigor" – the empirical fact that crossing unerelated materials often produced offspring displaying a higher general vigor (or some other specific characteristic) than their parents – to increase maize yield. Influenced by Darwin's work, Beal (1876, 1880) at the University of Michigan successfully tried to take advantage of the empirical hybrid vigor by crossing corn varieties "distant of at least 100 miles." This scheme was sporadically tested and offered by various experiment stations, the most thorough attempts being those of Morrow and Gardner (1893, 1894) at the University of Illinois, who suggested that farmers set up a breeding plot to produce varietal hybrids. But these agronomists' attempts to induce farmers to take advantage of an empirical hybrid vigor to increase yield by growing their own varietal hybrids have almost nothing in common with Shull's "pure-line method in corn breeding."

In fact, Shull's "pure-line method in corn breeding" (title of his second seminal article, in January 1909) extended the La Gasca–Le Couteur–Vilmorin–Nilsson isolation method to a cross-pollinated species, maize. This had a crucial consequence. When read carefully, the very first sentence of Shull's first seminal article, "The Composition of a Field of Maize" tells everything:

> While most of the newer scientific results show the theoretical importance of the isolation methods, and practical breeders have demonstrated the value of the same in the improvement of many varieties, the attempt to employ them in the breeding of Indian corn has met with peculiar difficulties, owing to the fact that self-fertilization, or even inbreeding between much wider than individual limits, results in deterioration. (1908, p. 296)

Isolation methods consist in replacing a set of plants by a single *selected* one. In the case of self-pollinated species, they lead to true-breeding varieties – the last thing investors and breeders want. But in the case of a cross-pollinated species such as maize, isolation leads to what investors crave, "deteriorating" varieties. A field of maize made up of genetically identical plants will as any field of maize *physically* cross-fertilize. But, *genetically*, this amounts to a *self-fertilization on a field scale*. The following generation will suffer from the "universally observed deterioration in self-fertilized maize" (Shull 1908, p. 297). Shull's method made it "necessary to go back each year to the original combination, instead of selecting from among the hybrid offspring the stock for continued breeding" (Shull 1908, p. 299). *He had solved the breeders' historic value problem.*

The year 1900 saw the "rediscovery" of Mendel's laws. Shull was studying inheritance on maize, a plant easy to work with because the female and male flowers are separated. In the course of his self-fertilization experiments, he realized that self-fertilizing corn would eventually lead, according to Mendel's law of segregation, to "pure-lines," to homozygous *fixed* plants. He *imagined* then that a field of maize was a population of plants, each one combining two of such "elementary species" – an "idea that was (not) new, for de Vries in his little book on "Plant Breeding" presents this view" (Shull 1908, p. 301). It led him to the "unexpected suggestion for a new method of corn breeding" (1909, p. 52), namely, "to find and maintain (i.e., to *fix*) the best hybrid combination." This method consisted first in producing (fixed) pure-lines by self-fertilization and crossing them to have a set of (*fixed*) maize plants and second in *selecting* the best (*fixed*) plant from this artificially fixed, scaled-down model of the natural population.

Shull did not address farmers but rather breeders: he presented his founding articles before the American Breeders' Association annual convention to people "who dealt with the business of heredity," whereas agronomists, including East, until he became a Harvard professor in 1909, wrote in experiment station publications for an agricultural constituency. Briefly stated, the two techniques have nothing in common, be it foundations (empiricism versus science), actors (agronomists versus geneticists), constituencies (farmers versus breeders), ends (public interest versus appropriation), mode of legitimation (yield increase versus science, namely Mendelism plus Darwinism) with the exception of one point, the necessity to *select* to increase yield. But what was an *end* for agronomists, became a *means* for geneticists and breeders.

Briefly stated, Shull's *pure-line* method has nothing to do with hybrids and hybridization. His *sterifixation* technique was a dialectical triumph: (1) the use of self-fertilization (or inbreeding) depression in the farmer's field to make *sterile* varieties, and (2) isolation of a more *fertile* (productive) plant thanks to selection–fixation program. Shull's method only required variations – the cause of which was irrelevant – and a method to fix them (i.e., the Mendelian stratagem of pure-lines). Shull could not tell the truth. For his "farmers' friends," he stressed the supposed yield-increasing virtues of his discovery. In this regard, his method was less than convincing.

FROM THE PURE-LINE METHOD IN CORN BREEDING TO HYBRIDS. Let us make the thought experiment of turning Shull's "magnificent design" into "practical reality" (Mangelsdorf 1951 p. 42). Starting from a single ear of corn with only 100 seeds and with only 100 seeds at each generation, after 6 generations of self-fertilization to get our pure-lines, we have 10^{14} pure-lines, which represents 10 times U.S. maize production. This is because we cannot select during the inbreeding process inasmuch as good lines are ones that gives a good plant and not the ones that carry any identifiable phenotypic character. Now we have to cross these lines in any order to get our fixed maize plants. After 8 years of work we have 10^{28} fixed plants, the seeds of which are carefully bagged, tagged, and stored to be tested to select the best. The sun will have finished burning when we have completed this selection program. Shull's "magnificent design" was impractical because it implied the search with no logical rule of extremely rare combinations that "nick." The only solution was to make "as many self-fertilizations as practical" (Shull 1909), hoping that by some miracle two "nicking" lines would occur. For cost reasons, the random sample of lines used to build up the scaled-down *fixed* model of the natural population from which the best plant is isolated had to be very small. The yield gain to be expected from this sterifixation scheme could only be correspondingly small. As stated recently by CIMMYT breeders, "In the preliminary phases of hybrid development, inbred lines were tested for productivity and combining ability by crossing all inbreds in all possible combinations. It was soon realized that for a few hundred inbred lines, the single cross diallel was virtually impossible because of the large number of crosses required" (McLean et al. 1997, p. 26).

Suppose now we have found two lines that "nick". Because the seed-bearing plant is an inbred pure-line that suffers from self-fertilization depression and produces little seed, we cannot produce seeds economically![9]

Finally, Shull's method was turning its back on the two basic breeding principles: "breed from the best" and "like engenders like." Here was a method that required destroying the corn plant for at least six generations in the hope that such debased parental stock might generate a vigorous offspring. Only credulous maize breeders tried his method – and failed.

In the case of true-breeding species, implementing the isolation technique did not require a theory as to why some plants were more vigorous and productive than others – why plants differed. Such was the case with Shull's maize sterifixation technique. Whatever the reason for the conspicuous variations of a corn field, the isolation method required fixing part of these variations and isolating the best fixed plant. This strength proved to be a weakness. Why should one go through the ordeal of sterifixing maize to improve it when evidence was accumulating that maize could be improved by selection? Shull had overlooked that he needed an overwhelming biological argument to support such an unlikely method, one that would make it the *only one* to improve corn. Genes of a cross-pollinated species had to have special characteristics.

At this juncture, the agronomist Edward East, who worked at the Illinois Agricultural Experiment Station from 1900 to 1905 and the Connecticut Experiment Station until 1909, offered him this very argument. East turned the correlation between the loss of heterozygosity during self-fertilization and its recovery in crosses into a causal explanation: hybridity, heterozygous genes, caused a "stimulation to development" (East 1909). This theory is known as overdominance. There was a cost: Shull's *pure-line* method became the hybrid one. Shull had to share "the monument I (Shull) could have raised to myself which could be worthy to stand with the best biological work of recent times" (Shull to East, March 3, 1908, in Jones 1945, p. 224). He brushed off the gadfly by remarking, "I care very little for the question of priority. What we are more concerned in is the *triumph* of the *truth* and especially of useful *truth*" (italics in original; Shull to East, February 1909, in Jones 1945, p. 226). This lesson in scientific ethics was enough to intimidate his younger rival: "I freely admit," answered East, "that your paper of 1908 (*The Composition of a field of maize*) gave me the first idea of inbreeding separating the biotypes and that on this hinged the whole matter" (East to Shull, February 9, 1909, in Jones 1945, p. 226).

In 1910, these leading geneticists went into a secret "gentleman's agreement" (Crabb 1947, p. 50). In 1942, after East's death, Shull described it: "He and I agreed between us that we would not enter into any personal controversy about priority, in order not to impede the progress of the hybrid corn program" (Crabb 1947, p. 51). As to East, he made the link with a skeptical agronomists' community: "This takes nothing away from Doctor East, as there can be no doubt that there would have been a still greater lag between the proposal of my 'pure-line method' and its full fruition had he and his students not jumped in so vigorously and effectively to promote it" (Shull in Crabb 1947, pp. 50–1). Why did Shull finally accept to share the monument he was planning to raise to himself?

On November 6, 1910, *Science* published a letter by a British geneticist Bruce offering a mathematical model of the decline of vigor during selfing and its recovery in crosses. A few days later, Keeble and Pellew, also British, offered the corresponding experimental evidence. Because corn is a botanical curiosity in England, British geneticists were interested in "pure science." The two powerful American geneticists were fighting bitterly over a revolutionary breeding technique for the emblematic American crop. The British theory was a lethal threat to their revolutionary sterifixation technique, for mass selection would be more efficient. At the famous symposium of the American Naturalists titled the Genotype Hypothesis (December 28–29, 1910), Shull rallied East's theory of hybrid vigor and grafted it unto his invention, devoting half of his paper to refute (rather to distort) Bruce's Mendelian model without giving the reference.

Bruce's model was "overlooked more frequently than it should have been, in view of its importance" (Richey 1945). In fact, it disappeared. It took Shull 4 years to mastermind the concept of heterosis and fill the void. Instead of the alternative between the British geneticist's Mendelian dominance model of hybrid vigor and East's non-Mendelian overdominance (the stimulating effect of hybridity), the increased vigor was attributed to "the effective differences between uniting gametes ... due to anything else than Mendelian genes, as well as the differences caused by such Mendelian genes, if the latter were not individually analysable as Mendelian genes" (Shull 1911, 1948, p. 441). Heterosis became a general biological phenomenon and a research program.

The first success of this research program was Jones' (East's student) "*pseudo*-overdominance." At last, heterosis got a not implausible Mendelian basis, although no data supported the model (Jones 1917). The following year, Jones' double-cross overcame (apparently) the problem of producing seed on a depressed and low-yielding pure-line (Jones 1918). The road was cleared. Heterosis *scientifically* imposed the implementation of Shull's expropriation technique to improve corn for the sake of general interest. Hybrid corn looked feasible. What was required was an all-out effort of the state to socialize the immense costs of searching pure-lines that "nicked." Wallaces' scientific coup d'état eliminated more efficient alternative improvement techniques. After nearly 15 years of unprecedented public research mobilization, breeders succeeded in extracting fixed plants that were consistently better than the farmers' open-pollinated corn left in its pre-1920 genetic state. In 1926, Henry A. Wallace, the future New-Deal secretary of agriculture, founded Pioneer, still the largest (hybrid) seed company. Public hybridizers, happy with their apparent task of improving corn, never realized that heterosis was, like the famed prophecy, *self-fulfilling*, nor that it was their breeding work that made it, nor that their task, the *hybrid* improvement of maize, was a means, and not the end, they thought. Quite to the contrary, they eagerly strove to repeat the same "miracle" in other species (Vassal 1993).

Briefly stated, with heterosis, a social relationship between men took the fantastic form of a relation between genes.[10]

25.3 The Enclosures of "Life"

Hybridization is *not a crop improvement* technique. It is an expropriation and appropriation scheme. Investors impose the most profitable technique, not the most useful one. It entails high social losses (Figure 25.2).

Yield loss I stems from the inefficiency of an expropriation technique at improving crops as compared with *improvement* techniques (for a given effort of R&D).

Yield loss II. ΔC is the difference in seed production costs between sterifix varieties and self-reproducing varieties. Producing (hybrid) seeds implies specific costs such as producing inbred lines, double sowing, double harvesting, detasselling, and so on. These costs are necessary to build up a copy-protected (self-destroying) genetic combination. ΔC then measures the costs of *expropriation.* The farmer (i.e., we), in effect, pays for his or her (our) own expropriation. α measures the costs of *appropriation,* namely, the overcosts of proprietary software as compared with free software: management, marketing, and sale efforts to set up a monopolistic market structure for the proprietary genetic software to provide a high profit rate – again at our expense.

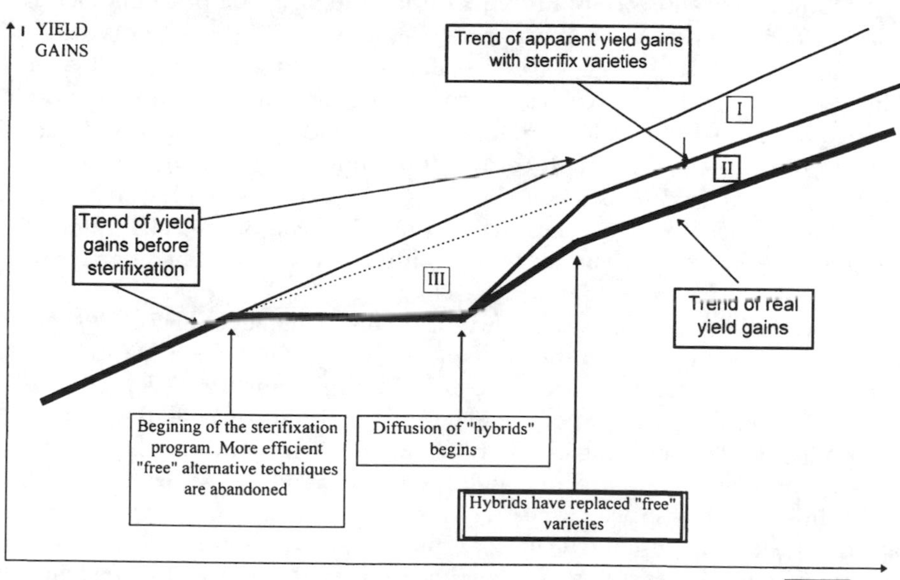

Figure 25.2. Three types of production loss due to a sterifixation program.

We can lump together our two sources of losses by writing Eq. (6) differently to distinguish *apparent* gains Γ from *real* gains $\Gamma - \Delta C/(1-\alpha)\theta$. If we consider the example of hybrid wheat breeding in France, a 12 q/ha *apparent* yield gain brought about by hybrid seed (a 15% yield jump, which a selection program will take a long, long, time to achieve) would translate into a *real* gain of only a few quintals per hectare because total overcosts are estimated at 8–10 q/ha (this does not include the increased production costs of this 15% apparent yield gain). The use of a scientific vocabulary such as sterifix instead of hybrid would have revealed immediately the mystification. Because wheat is a naturally fixed species, it is not necessary to fix it to improve it. What is left is sterility. No one will deny that this is what interests the corporate investors – the Monsantos, Cargills, Shells, Pioneer–DuPonts, Novartis, and others – that have taken over agricultural genetics during the last two decades.

Yield loss III. The seed cost difference ΔC entails a discontinuous rate of yield gain. With "free" (self-reproducing) varieties, in Eq. (6) $\Delta C = 0$. A farmer will adopt immediately any variety bringing about a gain. Improvement is *continuous* along a trend line. When a sterifixation program begins, *improvement programs must stop* to pave the way to the less-efficient sterifixation technique. Thus, in the United States, 12–15 years of exclusive hybrid selection were necessary to overcome the profitability threshold. In the meantime, there was no improvement.[11]

To be a source of value, life qua "genetic software" implies the enclosures of a common good, the faculty of life to reproduce and multiply. This socially costly process appears to take place according to the following pattern.

First, in a capitalist society, investors decide the direction of technological change. They choose what is most profitable, here sterility breeding, and not what is most useful. Second, investors' decisions must appear as nature- or biology-dictated. Science provides the corresponding evidence. Third, because science has hidden the wolf of value and profit under the sheepskin of general interest (utility), the socially costly privatization of this common good can be done at taxpayer's expense. Public breeding disguises sterile varieties as more fertile ones. Last, final success wipes out all tracks of this self-fulfilment process. This process is again under way with genetically modified organisms (GMOs).

This revisiting of our experience with agricultural genetics may offer some insights into what goes on in the twin applied-biology field, medical care. The genetic–industrial complex "Doctrine of DNA" (Lewontin 1993) is preparing minds for the new paradigm of the gene theory of disease while the Human Genome Project is socializing its cost. Predictive medicine is turning any healthy person into a potentially sick one and in effect expanding the health market to its limit, which is exactly what "hybrids" did for the "seed" market. Once again, investors are pushing the most profitable, and here destructive (no *social* security system will resist this *individual* view of health) course at the expense of our future well-being.

We cannot count on biologists to resist a course that gives them power, status, and money. But this goes deeper than crass interests. Let us again read Descartes, the founder of modern science, in the third part of the *Discours*: "Ma troisième maxime était de tâcher toujours plutôt à me vaincre que la fortune, et à changer mes désirs que l'ordre du monde" (Descartes 1637, p. 53).

25.4 Acknowledgments

This chapter is an extension of my long-time association with R. Lewontin to revisit the paradigmatic innovation of agricultural genetics, hybrid corn. A research grant from the Conseil Scientifique du Département d'Economie of INRA contributed to making it possible.

Notes

1. Even if breeder's rights are in force (Berlan 1983, Berlan and Lewontin 1986).
2. This is the crucial difference between computer and genetic software.
3. The patent granted to the U.S. Department of Agriculture for "public" research and to a private company in March 1998 for a transgenic technology to make second-generation seed *biologically* sterile confirms this theoretical analysis. The new element is that the genetic–industrial complex now feels powerful enough to state its objective. Farmers are gone, seed firms are part of powerful agrochemical and pharmaceutical corporations and "Life," reduced to strings of DNA, has lost its holy character.
4. M. Gensou of INRA drew my attention to this writing.
5. On this tour de force, see Hayes 1963.
6. The occasional use of quotation marks is to remind the reader that "seeds" have a dual nature of seed software and seed grain or seed diskette.
7. With figures corresponding to the middle 30's in the United States ($\Delta Y = 4$, $R = 0.08$, $\Delta C = 4$), the net gain from the shift is 46 quintals of grain per quintal of seed – a large gain that sets the maximum exchange value of a quintal of "hybrid" seed.
8. To grab market shares; to trap farmers into "hybrids." Once farmers have shifted to hybrids the variable setting the profit becomes the *much larger* loss incurred if they use second-generation seed (their only alternative to hybrid seed).
9. Jones' double-cross scheme (1918) combining four lines A, B, C, and D in a two-stage crossing procedure solved, it is claimed, this problem. The first step provided single crosses such as A × B and C × D (or A × C and B × D, etc.), and these single-crosses would then be combined into a double-cross AB × CD (or AC × BD etc.). The seed-bearing plant is a normal maize plant that has recovered its vigor and produces seeds economically. Although Jones' "miraculous" success may have solved the seed production problem, it made matters much worse: we have to combine four lines in any order instead of two. With only 100 lines, Shull's original single-cross scheme required testing 4,950 plants, but Jones' double-cross scheme multiplied this number 2,500 times! *The double-cross scheme is unable to improve maize.*

If it did, it is because, by chance, Jones' final cross combined in a Bealian fashion two different gene pools. This use of "genetic diversity" turned out to be the key to unlock the inefficiency of a sterifixation scheme at increasing yield (Hayes 1964, Chapter 7).

One realizes why the terms "single cross" and "double cross *hybrid*" have misled plant breeders into endless discussions about hybridity, heterosis, and the like. Biologically, a (single cross) hybrid is an *ordinary* corn plant, no more, no less "hybrid" than any corn plant; a double cross hybrid is *a population* with a reduced polymorphism since a maximum of four different alleles are possible for each gene. So that the double cross technique contradicts Shull's insight extending the isolation method – replacing a population by a *single* plant – to maize! The biological objects (an ordinary corn plant, a population with reduced polymorphism – both *reproducible by the breeder only*, thanks to the stratagem of pure-lines) are thus confused with the method (crossing pure lines in a one- or two-step procedure) to obtain them. What would one think of a chef confusing a dish with the method (steam, grilled, étouffé etc.) to cook it?

10. Not surprisingly, heterosis is still unexplained as shown by the 1997 CIMMYT (Centro International de Mejoramento de Maize y Trigo) symposium "Heterosis in crops." Symposium announcement: "we atually understand relatively little about the genetics, physiology, biochemistry or molecular bases of hybrid vigor." A number of contributors concurred.

    "Even though heterosis has been in the forefront of our thinking for many years, the phenomenon is not much better understood today than it was when Gowen's famous book on heterosis was published 45 years ago" (Phillips 1997, p. 350). "The genetic mechanisms underlying heterosis are largely unknown" (Coors 1997, p. 170). "What do we really know about the biological basis and mechanisms underlying heterosis? Very little." (Lee 1997, p. 110: "the causal factors for heterosis at the physiological, biochemical, and molecular levels are today almost as obscure as they were at the time of the Conference on Heterosis held in 1950 (Stuber 1994)" (Stuber 1997, p. 108).

11. From 1921 to 1946, a period of "hybrid" development and diffusion, wheat yield increased almost twice as rapidly as corn yield (Berlan 1987, pp. 225–32). The miracle of heterosis was for "seed" companies, not farmers nor consumers.

REFERENCES

Babcock, E. B., and Clausen, R. E. (1918). *Genetics in Relation to Agriculture.* New York: McGraw–Hill.

Beal, W. J. (1876, 1880). Report, Michigan Board of Agriculture. 1880:212–3, 1880:283, 287–8.

Berlan, J.-P. (1983). L'industrie des semences, économie et politique. *Economie Rurale* 158:18–28.

Berlan, J.-P. (1987). *Recherches sur l'économie politique d'un changement technique: les mythes du maïs hybride,* thèse d'Etat, Faculté des Sciences Economiques, Université Aix-Marseille II, 767 pp.

Berlan, J.-P., and Lewontin, R. C. (1986). Plant breeders' rights and the patenting of life forms. *Nature* 322:785–8.

Bruce, A. B. (1910). The Mendelian theory of heredity and the augmentation of vigor. *Science* 827:627–8.

Coors, J. G. (1997). Selection methodologies and heterosis. CIMMYT 1997. Book of Abstracts. *The Genetics and Exploitation of Heterosis in Crops: An International Symposium.* Mexico, D.F., Mexico.

Crabb, R. A. (1947). *The Hybrid Corn Makers, Prophets of Plently.* New Brunswick, NJ: Rutgers University Press.

Darwin, C. (1868). *The Variation of Animals and Plants under Domestication,* Vol. I and Vol. II, 1905 edition. London: John Murray.

Descartes, R. (1637). *Discours de la Methode.* Paris: Flammarion, 1966 edition.

de Vries, H. (1907). *Plant-Breeding.* Chicago: The Open Court Publishing Co.

East, E. M. (1909). The distinction between development and heredity in inbreeding. *The American Naturalist* 43:173–81.

Evershed, H. (1884). Improvement of the plants of the farm. *J. Roy. Ag. Soc.* 45:77–113.

Gayon, J., and Zallen, D. (1992). Le rôle des Vilmorin dans les recherches sur l'hybridation en France au XIXe et XXe siècles, 117e Congrès national des sociétés savantes, Clermont-Ferrand.

Griliches, Z. (1958). Research costs and social returns: Hybrid corn and related innovations. *Journal of Political Economy* 5:419–31.

Hallett, F. (1862). On "pedigree" in wheat as a means of increasing the crop. *J. Roy. Ag. Soc.* 23:371–81.

Hallett, F. (1870). On the law of development of cereal. *Report of the British Society for the Advancement of Science* 39:113–293.

Hallet, F. (1882). Food plant improvement. *Nature* May 25:91–4.

Hallett, F. (1887). *The Agricultural Gazette.* February 7, p. 144, October 31, p. 424.

Hayes, H. K. (1964). *A Professor's Story of Hybrid Corn.* Minneapolis: Burgess Publishing Co.

Jenkins, M. T. (1936). Corn improvement: Corn, the most typical American plant, yields rich results in breeding and genetics. In *Yearbook of Agriculture,* pp. 455–522, Washington, DC: U.S. Department of Agriculture.

Jones, D. F. (1917). Dominance of linked factors as a means of accounting for heterosis. *Genetics* 2:466–79.

Jones, D. F. (1918). The effect of inbreeding and crossbreeding upon development. Connecticut Agricultural Experiment Station Bulletin 207.

Jones, D. F. (1945). Biographical memoir of Edward Murray East, National Academy of Science. *Bibliographical Memoirs,* vol. XXIII, ninth memoir.

Le Couteur, J. (1836). *On the Varieties, Properties, and Classification of Wheat.* London: W. J. Johnson.

Lee, M. (1997). Towards understanding and manipulating heterosis in crops. Can Mendelian genetics help? Cimmyt. 1997. Book of Abstracts. *The Genetics and Exploitation of Heterosis in Crops: An International Symposium.* Mexico, D.F., Mexico.

Lewontin, R. C. (1993). *The Doctrine of DNA Biology as Ideology.* London: Penguin.

Liebscher, G. (1897). The principles and methods of breeding cultivated plants. USDA. *Exp. Sta. Rec.* 7:347–60.

Mangelsdorf, P. C. (1951). Hybrid corn. *Scientific American.* August 1951:39–47.

McLean, S. D., Vasal, S. K., Pandey, S., and Srinivasan, G. (1997). The use of testers to exploit heterosis in tropical maize. CIMMYT 1997. Book of Abstracts. *The Genetics and Exploitation of Heterosis in Crops: An International Symposium.* Mexico, D.F., Mexico.

Morrow, G. E., and Gardner, F. D. (1893 and 1894). Field experiments with corn. *Univ. Illinois Ag. Exp. Stat., Bull. 25 and 31.*

Percival, J. (1921). *The Wheat Plant.* London: Duchworth and Co.

Phillips, R. L. (1997). Research needs in heterosis. CIMMYT 1997. Book of Abstracts. *The Genetics and Exploitation of Heterosis in Crops: An International Symposium.* Mexico, D.F., Mexico.

Pickett, A. A., and Galwey, N. V. (1997). A further evaluation of hybrid wheat. *Plant Varieties and Seeds* 10:15–32.

Richey, F. D. (1945). Bruce's explanation of hybrid vigor. *J. Heredity* 36:243–4.

Rousset, M. (1986). Les blés hybrides sortent du laboratoire. *La Recherche* 17(173):86–9.

Shirreff, P. (1841). On the Hopetoun wheat and of comparative trials of wheat. *J. Roy. Ag. Soc.* 2:344–6.

Shull, G. (1948). What is heterosis. *Genetics* 33:439, September.

Shull, G. H. (1908). The composition of a field of maize. *American Breeder's Association* 4:296–301.

Shull, G. H. (1909). A pure-line method in corn breeding. *American Breeders Association* 5:51–59.

Shull, G. H. (1911). The genotypes of maize, Cornell Symposium, The Genotype Hypothesis. *The American Naturalist* 45:234–52.

Stuber, C. W. (1997). The biology and physiology of heterosis. CIMMYT 1997. Book of Abstracts. *The Genetics and Exploitation of Heterosis in Crops: An International Symposium.* Mexico, D.F., Mexico.

Vasal, S. K. (1993). Manifestation and genotype X environment interaction of heterosis. *Crop Improvement* 20(1):1–16.

Welch, F. J. (1961). Hybrid corn: A symbol of American agriculture. American Seed Trade Association, 16th Hybrid Corn Industry Research Conference.

CHAPTER TWENTY-SIX

The Butterfly ex Machina

RICHARD LEVINS

In 1963, MIT professor Lorenz published an article with a set of three differential equations meant to describe atmospheric conditions (Lorenz 1963). The solutions to these equations did not do what was thought to be the only decent thing for variables whose motion was described by equations, namely, either approach an equilibrium or a permanent repetitive oscillation. Instead, Lorenz's variables ended up with trajectories more like a tangled skein of yarn, and every time the equation was solved numerically on a computer the results were different. Then others began to look for similar behavior elsewhere. Robert May (1976) showed that even the simple and familiar logistic equation of discrete population growth models can show this kind of aberrant behavior for some values of the initial growth rate.

It was no wonder that these new unexpected behaviors were labeled chaotic and that chaos has caught the public imagination in a world that seems so unpredictable and in which people are so helpless. The popularization of chaos is having an impact comparable to that of the discoveries of quantum mechanics in the 1920s and 1930s, when the uncertainty principle and probabilistic transitions of atomic states became a metaphor for the uncertainty, randomness, and ultimate irrationality of life in a Europe still reeling from the unexpected horrors of World War I. (It is noteworthy that the other conspicuous aspect of quantum theory – that change occurs in leaps rather than by slow, continuous increments – was not incorporated into the popular consciousness. It seemed to come down on the wrong side of the "evolution versus revolution" debate.)

The same thing is happening now with chaos. Some authors counterpose "chaos" to "order" as if chaos showed no orderliness. Others have decided that complexity implies chaos. They use the terms almost interchangeably and conclude that the goal of prediction, and with it the possibility of having any program of change certain enough to make commitments and sacrifice for, is illusory. The physicist Peter Carruthers, on the National Public Radio program "Talk of the Nation" on January 17, 1994, said that chaos overturned the whole basis of science.

Deepak Chopra, in arguing the case for an alternative, holistic medicine based on the Ayurvedic tradition, counterposed the simple linear processes that occur out in the open on the "Newtonian table" to the mysterious world "under the table" of nonlinear, quantal, and chaotic motion.

The German socialist Peter Kruger, in an interview with Michael Hilliard (1993) stated:

> And currently you can observe a very interesting development in physics, that is the theory of chaos. I think that it's incredibly important in understanding the behavior of humankind and seeing its future. And it would be very, very fruitful for all Marxists who define themselves in the narrower sense to study the theory of chaos. They will see from this that the idea of designing an ideal society can only be a grand failure.

Chaos is very appealing to the postmodernist mood, which would deny any lawfulness in the world, reject theory as "Grand Designs," and see all theories as merely matters of discourse.

The claim that the earth's orbit is chaotic suggested to some that we may fly into the sun or off into cold space at any moment. But others have seen chaos as beneficial (e.g., in the rhythms of heart contractions) and see regularity as a risk factor. Throughout these discussions there is frequent casual reference to "the science of chaos."

Perhaps the most dramatic expression of pop chaos is Lorenz's provocative "Predictability: Does the Flap of a Butterfly's Wings in Brazil Set Off a Tornado in Texas?" (the title of an address given to the American Association for the Advancement of Science in 1979).

There is of course no "science of chaos." Chaos refers to a class of mathematical phenomena within the general subdiscipline of nonlinear dynamical systems comparable in scope to time series or local equilibria in linear analysis.

How common chaos is in nature or in social life still remains to be determined. Not all nonlinear equations are chaotic. In fact, population dynamics cannot be truly chaotic because the size of a population is always an integer, whereas chaos requires a continuum of possible values so that different initial conditions can be arbitrarily close to each other. Not all nonlinear equations can be made chaotic by an appropriate choice of parameters. The implications of particular kinds of chaos are still to be worked out. Furthermore, there are very few mathematical proofs of anything having to do with chaos, and thus most of the research still consists of setting up equations, computing numerically the trajectories of solutions starting with different initial conditions, displaying them on a computer monitor, and saying, "Doesn't this look like chaos?"

But chaotic dynamics does represent a radical departure from previously familiar behaviors and from the basic Laplacian approach implicit in the scientific agenda: if I know exactly the initial conditions and laws of motion of all the variables in a system, then I can predict the whole future course of that system.

The first modification of Laplace's notion was the recognition that we cannot know the initial conditions "exactly." Every measurement has its own confidence interval, and in many systems the act of measuring changes the system. Secondly, the laws of motion are only approximate descriptions of what really happens because no system of variables is really isolated. There is always an "outside" that imparts an additional push. This external push may be extremely small, but that small push may be enough to alter the outcome. Third, the models always include simplifying assumptions such as the lumping of variables as if they were identical except for the property of interest, or ignore friction, or treat the external as acting uniformly. Models of population genetics treat individuals as interchangeable except for their genes, whereas population ecology distinguishes ages and nutritional states but ignores genetic differences. Epidemiological models separate infective and uninfected individuals but do not usually deal with a range of susceptibilities in populations.

Therefore, the Laplacian expectation was modified: if we know the initial conditions and laws of motion "approximately," then we can know the future of the system "approximately." There were of course exceptions to this that were recognized early. Suppose that we are studying the trajectories of marbles rolling off a peaked roof. If two marbles start out on the same side of the peak near each other, then they usually will end up near each other. But if they are on opposite sides of the peak, then no matter how close together they start they will diverge to different end states. And if we make even the smallest error in locating the starting position of a marble near the peak, we may be completely wrong in predicting the outcome.

The peak of the roof is a boundary separating two domains of behavior, two basins of attraction. Whenever a system has more than one possible final outcome, depending on the initial conditions, there is correspondingly more than one basin of attraction separated by boundaries. But "most" points lie comfortably within their basins of attraction, and accurate measurement leads to accurate prediction.

But suppose that there are the equivalent of peaked roofs everywhere – that most values of the variables are near boundaries. Then, no matter how accurate our measurements, our predictions may be far from accurate, and two examples of the same model with only slightly different starting points may give quite different trajectories. This is one of the properties of chaos.

But chaos is also in the eye of the beholder. Once the initial bewilderment passes, patterns become discernible. Different kinds of chaos can be distinguished: regularities in chaotic dynamics, bounds to chaotic trajectories, correlation patterns among chaotic variables, prescriptions for detecting chaos – prone or chaos-resistant systems, and ways of intervening that suppress or enhance chaotic properties.

At first glance chaotic trajectories look like random numbers. And indeed chaotic equations can be used to generate the "pseudorandom" numbers for studying random processes. However, the randomness is only apparent, and with the right viewpoint patterns become obvious.

One task of mathematics is to make the arcane obvious and even trivial. Throughout history, changes of perspective have made it possible for quite sophisticated ideas to become part of the common sense of the public and used in everyday discourse. In medieval Europe, literate monks could add, subtract, and even multiply, but they had to resort to specialists for division. The shift from Roman to Arabic numbers was decisive for making long division a part of the culture of the educated.

The pendulum became widely used in Renaissance Europe, but the "swings of the pendulum" has already become a common metaphor for describing changes in politics or fashion. And long before systems theory, positive feedback was part of common sense as "a viscious circle."

Or consider the graphs of price trends that often appear on the front pages of newpapers. Readers now perceive at a glance and without special effort that upward on the page means greater, further to the right means more recent, and steep slopes mean rapid change. But the idea that nonspatial variables such as prices and time can be represented by spatial arrangements of points and lines on a plane is not "natural." It carries behind it a history of abstraction embodied in measure theory and the general notion of mapping.

The same applies to mathematical chaos. The unfamiliarity of these new kinds of dynamics and their recalcitrance when we apply methods appropriate to older mathematical systems encourage philosophies of despair. But with a change of viewpoint and a bit of practice their properties become obvious. In what follows we explore one kind of chaos to show how a change of perspective makes the dynamics intelligible.

26.1 Simple Discrete Chaos

Chaos can occur in continuous or discrete equations. However, much more is known for the discrete case. If the dynamics of a variable are described by an equation of the form

$$x_{n+1} = g(x_n), \tag{26.1.1}$$

then there are several properties that can be demonstrated rigorously and used as working hypotheses for other cases. Li and Yorke's famous article "Period Three Implies Chaos" (1975) showed that if Eq. (26.1.1) has a solution of period three, then it also has solutions of every other period, that there are also nonperiodic solutions that pass close to the periodic ones, and that there is "extreme sensitivity to intitial conditions." These three properties constitute a definition of chaos for the discrete equation without delays. In other situations the last property is usually the focus of attention. Li and Yorke also offered a method for demonstrating that there is a solution of period three: if you can find a sequence of consecutive points in a trajectory such that

$$x_3 < x_0 < x_1 < x_2, \tag{26.1.2}$$

then there is a solution of period three and hence chaos.

In our own research, modeling the dynamics of the growth of grass in a savanna (Grove et al. 1994) derived from the research of Tilman and Wedin (1991), we wanted a simple qualitative approach for understanding the dynamics from the shape of the curve $g(x)$. The shape had to reflect the fact that the quantity of grass present affects growth in opposing ways: By way of reproduction, the more grass now the more grass later, but through litter accumulation on the ground, old grass inhibits new growth. Therefore, the curve $g(x)$ would start at zero, rise to some peak level, and taper off asymptotically toward zero when litter completely covers the ground and suppresses growth. (The full model also took into account the decay of litter and its release of nutrients.) We considered using either of the two equations

$$x_{n+1} = Ax_n/\left[1 + (A-1)x_n^2\right] \quad (26.1.3)$$

and

$$x_{n+1} = x_n \exp[b(1 - x_n)] \quad (26.1.4)$$

They have roughly the same shape. In both, $g(0)=0$, and $g(x)$ rises to a peak and then decreases asymptotically toward 0. Both have equilibrium points at $x=1$. Yet the first equation always has a stable equilibrium and is never chaotic, whereas the second equation may have oscillatory solutions and even chaos when b is large enough. Therefore, the notion of the "shape" of the function needed refinement. In what follows we show how to find the solutions of the difference equation (26.1.1) graphically and then introduce a series of landmarks, the tools for understanding "shape."

26.2 Dynamics of the Interval Map

Figure 26.1 shows an example of one kind of equation that may be chaotic, the interval map with the curve $g(x)$. Draw the 45° diagonal. Where it intersects the curve,

$$g(x) = x, \quad (26.2.1)$$

and thus this is the positive equilibrium point. If the slope of $g(x)$ is less steep than -1 at the equilibrium, then the equilibrium is stable, whereas if the slope is steeper than -1, it is unstable. The stability of the equilibrium depends on the derivative $dg(x)/dx$ at the equilibrium only. It is a local property compatible with any "shape" in the global sense.

Start at any point x_n along the x-axis and go vertically to the curve $g(x)$. Then from the intersection with $g(x)$ draw a horizontal line until it intersects the diagonal. The x-value at that intersection is x_{n+1}. Now repeat the process: first draw a vertical line to $g(x)$ and then a horizontal line to the diagonal. This gives the next value, and so on. Depending on the shape of the curve, the sequence of steps may approach an equilibrium, enter into a periodic oscillation, or become aperiodic. If you draw curves $g(x)$ derived from any equations or from a data set, or even free hand, you can repeat the steps and get a feel for the process.

The equilibrium value x^* is important not because all processes go to equilibrium but because it is a landmark in the interval along with other landmarks that give the shape. The other landmarks are found as follows:

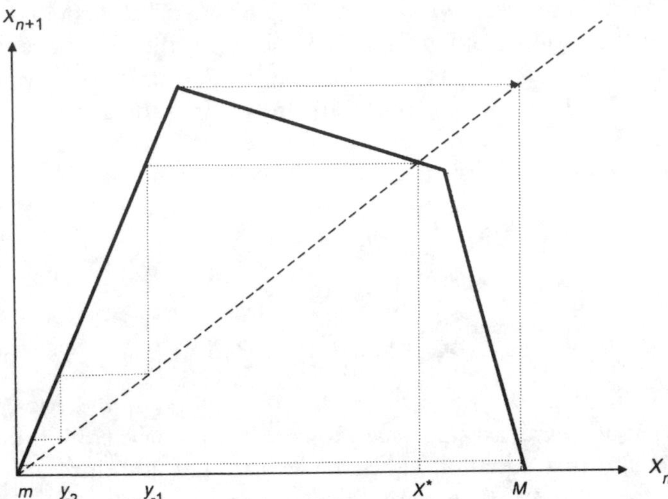

Figure 26.1. The curve $x_{n+1} = g(x_n)$, the generation of its solution and the landmarks of $g(x)$. The procedure is always to go vertically to the curve and then horizontally to the 45° bisector to find the next x. The intersection of the bisector with $g(x)$ locates the equilibrium. Start from the peak and move horizontally to the bisector to get the maximum M. Then moving vertically to $g(m)$ and horizontally to the bisector locates the minimum m. The inverse process, horizontal from equilibrium to $g(x)$, gives y_{-1}, and then moving vertically to the bisector and horizontally to $g(x)$ locates y_{-2}, and so on.

Find the peak of $g(x)$. Draw a horizontal line to the diagonal. This gives the maximum value M. Now draw the vertical down from here to $g(x)$ and once again horizontally to the diagonal. This identifies the lower bound m. The interval $[m, M]$ is the region of permanence: all trajectories eventually fall within this interval.

Next, from the equilibrium point draw a horizontal line to $g(x)$. This identifies the preimage y_{-1} of the equilibrium value, the point from which the trajectory would get to equilibrium in a single step. Now go vertically to the diagonal and horizontally to the left to the curve $g(x)$. This locates the preimage of y_{-1}, which is labeled y_{-2}. Repeat the process, horizontal to $g(x)$ and vertical to the diagonal. This is the inverse process of generating the trajectory and gives the preimage set $y_{-1}, y_{-2}, y_{-3}, \ldots$.

If y_{-2} lies within the permanent region (that is, if $y_{-2} > m$), then inequality 26.1.2 is satisfied and the equation is chaotic.

Semicycles: The negative and positive semicycles S_- and S_+ are the number of consecutive steps during which the variable is below (S_-) or above (S_+) equilibrium. Semicycles are easier to identify than periods because they do not require that a variable return to exactly the same previous value.

It can easily be proven that a trajectory starting at any initial condition gets within the permanent region in at most three semicycles. Further, the variable x_n crosses over a member of the preimage set in each step. Therefore, the maximum length of a semicycle is the number of members of the preimage set on that side of equilibrium that are within the permanent region.

Note that although chaos implies periodic solutions of every period, the lengths of the semicycles can still be bounded. The longer periods are then formed from many semicycles.

These results can be used to identify chaotic equations in several ways. If we have the functional relation $g(x)$, then a simple calculation or plotting of the landmarks suffices to determine whether y_{-2} lies within or outside the permanent region. If we have only data points from which to plot $g(x)$, we can still make the same determination although with a margin for error. If we are working from data rather than the equation we may not know the equilibrium point and therefore cannot be sure of the semicycles. But if any ascending sequence is more than three times as long as the shortest ascending sequence then the equation is chaotic. Thus even short sequences of observation may be sufficient to make a determination.

26.3 The Logistic Equation

The logistic equation has a special importance in the study of discrete chaos because of its widespread use in population genetics and ecology. Yet until May's (1976) observations, it was used mostly for studying equilibrium (stable or unstable genetic polymorphism or species coexistence).

Because the logistic equation

$$x_{n+1} = r x_n (1 - x_n) \tag{26.3.1}$$

is quadratic in x_n, it can be solved for the landmarks.

The landmarks are the following:

$x^* = (r-1)/r$	equilibrium
$M = r/4$	maximum x in permanent region
$m = r^2(4-r)/16$	minimum x in permanent region

The positive preimage set is empty. Therefore, when $x_n > x^*$, $x_{n+1} < x^*$. The trajectory can only be greater than equilibrium for one consecutive time interval. The first three members of the negative preimage set are given by

$$y_0 = x^*, \quad \text{which equals } (r-1)/r. \tag{26.3.2}$$

$$r y_{-1}(1 - y_{-1}) = x^* \tag{26.3.3}$$

so that

$$y_{-1} = 1/r$$

and

$$r y_{-2}(1 - y_{-2}) = y_{-1} \tag{26.3.4}$$

so that

$$y_{-2} = 1/2(1 - \tilde{A}[(r_{-4}^2)/r^2].$$

The same procedure can be used to find the other members of the preimage set. The number of these that are greater than m, and therefore within the permanent region, gives the length of the longest semicycle in any long-term solution (after the first three semicycles in order to be sure the trajectory in inside the permanent region).

Finally, we can ask, at what value of r does m cross above y_{-1} to preclude chaos or below y_{-2} to ensure chaos? To do this, solve numerically for y_{-1} or $y_{-2} = m$ (or, alternatively, $g(m) = x^*$ and $g[g(m)] = x^*$). Thus,

$$r^2(4 - r)/16 = 1/r \tag{26.3.5}$$

and

$$r^2(4 - r)/16 = 1/2[1 - \tilde{A}\{(r_{-4}^2)/r^2\}] \tag{26.3.6}$$

The roots are approximately 3.67 and 3.94. The equilibrium becomes unstable at $r = 3$ and oscillations appear. For $r < 3.67$ there cannot be chaos, whereas for $r > 3.94$ there is necessarily chaos. (Actually we can get a better limit by finding the value of r for which m equals the preimage of the maximum. This is approximately 3.83.)

The negative preimage set is a sequence $y_{-1}, y_{-2}, y_{-3}, y_{-k} \cdots$ that converges toward zero, a point of accumulation. As r increases there are more and more members of the preimage set within the permanent region, and they are closer and closer together. If zero is in the permanent region, then there are infinitely many members of P_-, and therefore there is no limit to the length of the negative semicycle. If the variable gets close to zero it will stay small for a very long time. In reality, a population with such dynamics would become extinct. This occurs at $r = 4$. But if $3.83 < r < 4$, then even though there is chaos, the semicycles cannot become infinite and x cannot become trapped indefinitely near zero.

There was only one solution to Eq. (26.3.6). Therefore, as r moves across this critical value, there is a single transition from periodic to chaotic motion. But in other models it is possible for r not to have any real roots, and the transition to chaos may be impossible. This may happen because a parameter analogous to r will in general affect all the landmarks. In Eq. (26.1.3), as A increases, m and the y_{-k} all get smaller, but the y's decrease faster than m and are always outside the region of permanence. Then, the equation is immune to chaos.

Or it may be that the equation equivalent to Eq. (26.3.6),

$$y_{-2}(r) = m(r), \tag{26.3.7}$$

may have several real roots. Then, as r increases, the equations may move in and out of chaos. If there is a double root, then the equation may be chaotic only for a single point value of r. Because the same parameters usually affect several landmarks, not all nonlinear equations can become chaotic just by changing the parameters, and we have no reason to make assumptions about how common chaos is in natural or social life. On the other hand, if $g(x)$ consists of two or three straight line segments, then we can manipulate the landmarks independently and design chaotic and nonchaotic systems at will.

26.4 Discussion

Some other properties of Eq. (26.1.1) can be deduced from the shape of $g(x)$. The local stability of an equilibrium depends on the slope $g'(x^*)$ at equilibrium, $-1 < g(x^*) < 1$ giving stability. But the chaotic properties depend on the relations among the landmarks. Therefore, it is possible to have a curve $g(x)$ that gives a locally stable equilibrium and yet is chaotic. This is shown in Figure 26.2a. We can also check the local stability of periodic solutions of Eq. (26.1.1).

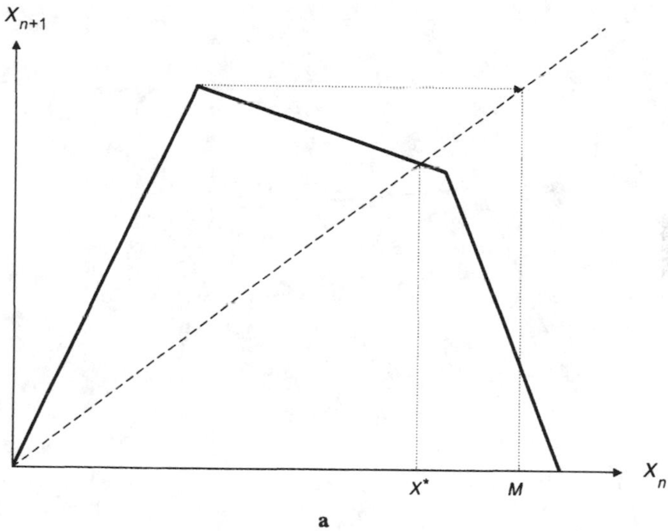

Figure 26.2. The behaviors of solutions for 2- and 3-segment $g(x)$. (**a**) The slope at equilibrium is flatter than -1. Therefore, the equilibrium is locally stable. But segment 3 is steep enough to ensure that m is less than y_{-2} and the equation is chaotic. We would observe erratic oscillations until the process is trapped by the equilibrium. (**b**) Both slopes are steeper than 1 or -1. Therefore, all periodic solutions are unstable, and we would observe a typical nonperiodic chaotic pattern. (**c**) The equilibrium is unstable because the slope is too steep at the equilibrium. Segment 2 has a flat slope, and thus periodic solutions that have points in segment 2 will be stable. These will be of cycle length 2. (**d**) The equation is chaotic with m close to 0. Orbits that include points on segment 3 move from that segment to segment 1 and will need several steps to cross equilibrium again. The product of the slopes around on orbit that has k steps in segment 1 and one step in segment 3 will be $s_1^k s_3$. Because segment 3 is very flat, this product can lie between -1 and zero if k is not too big. Thus, periodic orbits of intermediate length may be stable, but very short ones miss segment 3, and very long ones have too many steep slopes. If segment 3 is horizontal, any long orbit is stable. (**e**) In this chaotic equation, only orbits that include a point on segment 2 can be stable. But these correspond to orbits of cycle length 2. They will be of small amplitude compared with the unstable orbits and irregular oscillations.

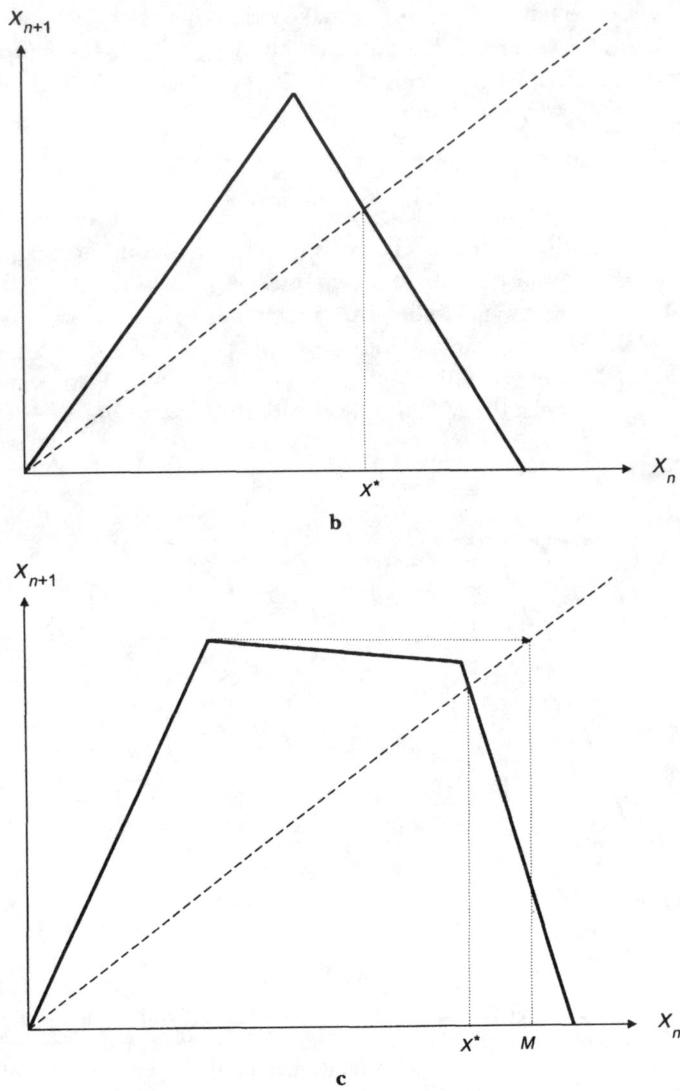

Figure 26.2. *(continued)*

The requirement for a stable periodic solution is that $-1 < \Pi g'(x_i) < 1$, where the product is taken over all points on a periodic orbit. In Figure 26.2b, we show an example of a function whose periodic orbits are all unstable; thus, we would only observe aperiodic trajectories. On the other hand, in Figure 26.2c the curve $g(x)$ is flat near the peak value but steep at equilibrium. Here any periodic solution that is stable must certainly have one point in the flat segment 2. The next point is near the maximum, and thus the stable period 2 orbits have maximum amplitude. In Figure 26.2d, any stable periodic solution must have a point on the flat segment 3 and therefore a moderately long period.

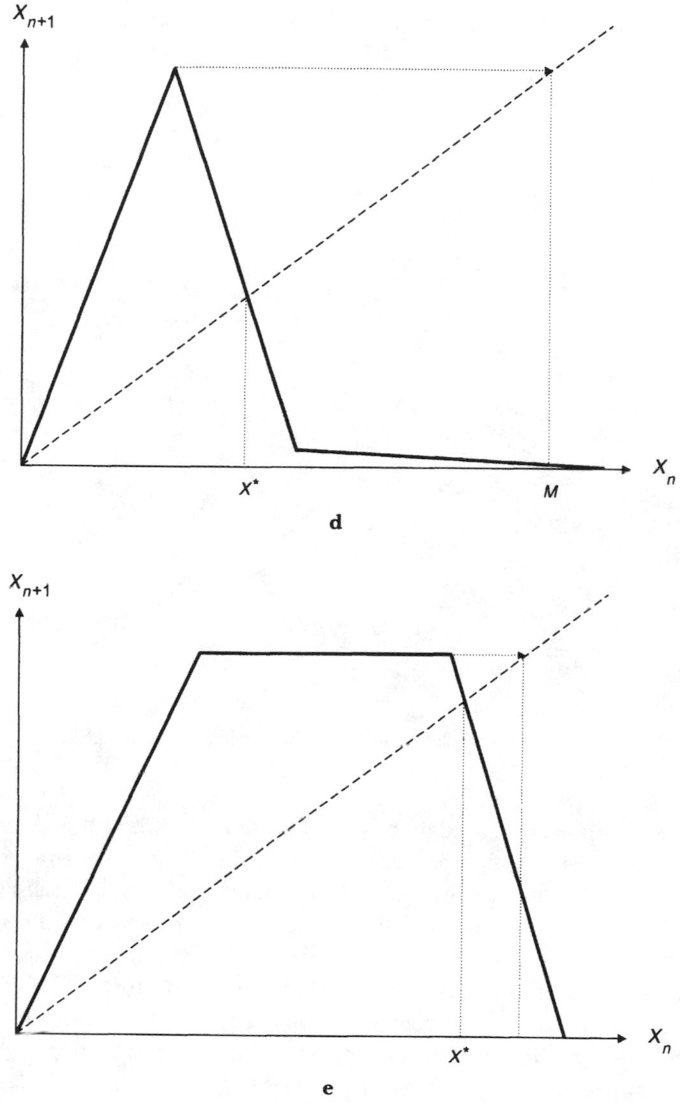

Figure 26.2. *(continued)*

But it cannot have too many points on segment 1, or the product will still be below −1, and the orbit will be unstable. Finally, in Figure 26.2e, the flat segment, segment 2, is followed by a point not too much greater than equilibrium. Therefore, there will be stable low-amplitude oscillations, whereas longer cycles will be of greater amplitude and will be unstable.

The analysis of dynamic properties from the "shape" of the curve $g(x)$ can be applied to situations in which we have reason to believe that the real situation deviates from the model in particular ways. For instance, suppose that the curve

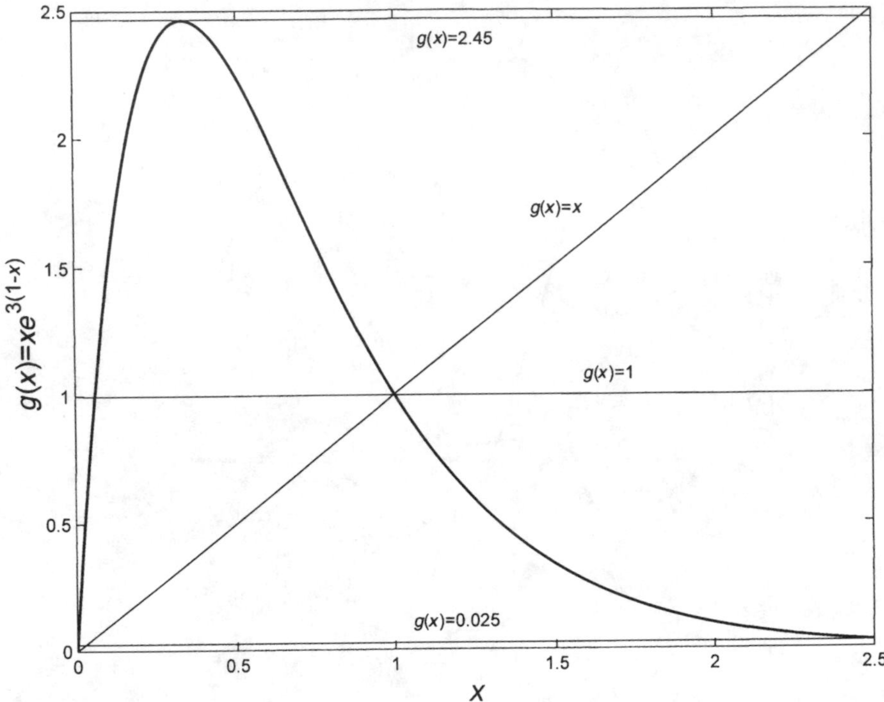

Figure 26.3. An idealized population model from which to show realistic departure.

in Figure 26.3 represents the dynamics of an epidemic when it is assumed that all susceptible individuals have equal susceptibility. We know that this is not true, although we do not know what the shape of $g(x)$ should be. The greater susceptibility of part of the population would alter $g(x)$ by making it steeper at low prevalences, where each case transmits the infection to more people than in the model, and flatter at high values of x when most of the uninfected people are more resistant. Therefore, the preimage set y_{-i} would be displaced to the left, whereas the flattening of $g(x)$ would later on move m to the right. Thus, heterogeneity of susceptibility shortens the semicycles and makes chaos less likely. Or consider a $g(x)$ for the growth of a population of herbivorous insects in a cultivated field. Suppose now that we intervene with some pesticide whenever the population exceeds some economic threshold. If that threshold is above the equilibrium, then the only effect of the intervention is to make $g(x)$ steeper to the right of the equilibrium, which itself does not change. The lower bound m is decreased, and thus the semicycle may become longer and we may be provoking chaos. An intervention when populations are small reduces the slope of $g(x)$ and may therefore make periodic solutions stable.

Or suppose that $g(x)$ applies to a pest population in a crop of beans. In the off-season the insects decline slowly in the wild vegetation. Then the next year $x_{n+1} = sg(x_n)$ where s is survival in the wild. The curve is lowered uniformly by the same fraction s. This reduces all slopes and favors stability. Now suppose that a loan from the World Bank encourages farmers to plant several crops of beans per year with less time between crops. Then s increases, steepening all slopes and making periodic and chaotic solutions more likely.

This kind of qualitative analysis is more robust than the more precise models that give equations for $g(x)$ because it makes fewer restrictive and usually unrealistic assumptions about the shape of that function.

26.5 Conclusions

The popular hyperbole about chaos and the claims that it is ubiquitous, that it is the antithesis of order, that it overthrows science, that it makes the world unintelligible and unpredictable, and that it guarantees that programs for change will be ineffective are all unjustified.

We do not know how common chaos is in the world. It is difficult to detect chaos for two opposite reasons: on the one hand, obviously irregular trajectories may be produced by external perturbations as well as by chaotic dynamics, and telling them apart is quite difficult. On the other hand, nonperiodic oscillations with constrained semicycles may look periodic. Nor does the mathematics provide an answer: adding complexity to a dynamical system does not necessarily give chaos because the parameters change the landmarks in constrained ways. Finally, in nature the dynamics of physiological and demographic processes are influenced by natural selection. The parameters evolve and may evolve toward or away from producing chaotic behavior according to the fitness consequences.

It is clear that chaotic systems are not without order. There are restrictions on the possible solutions to some range of values, there are bounds on semicycle lengths, there are indications about the stability and instability of long or short and small or large amplitude trajectories, and we can understand which features of the curve $g(x)$ give or prevent chaos. In systems of differential equations, which have to have three or more variables or delays in order for chaos to appear, the shape of the trajectories in the three- or higher-dimensional space is also not arbitrary. We can make predictions about the correlations among the variables or the correlations between values of the same variable observed at different times.

The dynamics of the difference equation without delays (26.1.1) can be grasped quickly and intuitively from the shape of the curve $g(x)$. In order to do this, we have to look at $g(x)$ in terms of its landmarks, the equilibrium value x^*, the boundaries of the permanent region M and m, and the preimage set. It is this change of perspective that makes the mysterious obvious. In the case discussed here of discrete time, nondelay difference equations, we have rigorously demonstrable results. For equations with delay, or for differential equations, there are fewer analytic results, but the qualitative conclusions above can be used as working hypotheses that can guide explorations by numerical methods and analysis.

Far from overthrowing science, the study of chaos opens up new areas for investigation and obliges a more subtle approach to the relations of prediction and understanding. Prediction has often played an important role in science, but science is not prediction. Nor is it necessary for uniform causes to give uniform results. The uniqueness of each ecological site and each individual person overthrows only the most mechanistic, reductionist kind of science and the technocratic ambition to achieve complete control. Rather, we can use our investigative tools to explain patterns of difference under seemingly uniform influences, learn to appreciate the richness of the world, and develop a strategy for coping with uncertainty (Levins 1995).

The regular properties of even chaotic equations give us a different take on that ominous butterfly who threatens to flap its wings. Historians and natural scientists have looked at the possibility of small events having big consequences in very different ways long before the arrival of chaos. Claims have been made that if Cleopatra's nose had had a different length (I do not recall now whether the author preferred long or short noses), if King George III had selected a better prime minister, if Rosa Lee Parks had not been so tired that day in Montgomery, then the course of history would have been different. Historians use this to emphasize the unpredictability of history and the absence of lawfulness. On the other hand, in physical and biological dynamics, instability is a regular part of predictable processes. The phase transitions of materials, the emergence of asymmetry in development, and the course of natural selection when the fitness of a genotype increases with its frequency all allow prediction in the large without prediction in the small. The more insignificant the precipitating factor, the more inevitable the macro level change. Our task then becomes to examine the structure of the system that makes it unstable or chaotic when its parameters reach some critical values and the processes that bring the parameters to the critical points and to make a determination of the domain of possible outcomes.

In chaotic systems, not anything can happen but only a range of alternatives within a set of constraints. It would take more than the flap of a butterfly's wing to induce monsoon rains in Finland or a drought in the Amazon or equal representation of women on the Harvard faculty. Great quantities of energy and matter are involved in particular configurations for the major events to occur. Only when a system is poised on the brink of a qualitative change can a tiny event set it off. Therefore, the task of promoting change is one of promoting the conditions under which small, local events can precipitate the desired restructuring

REFERENCES

Carruthers, P. (1994). Interview on "Talk of the Nation," National Public Radio, January 17, 1994.

Chopra, D. (1989). *Quantum Healing*. Bantam Books.

Grove, E. A., Ladas, G., Levins, R., and Puccia, C. (1994). Oscillation and stability in models of a perennial grass. *Proc. Dynam. Syst. and Applic* 1:87–91.

Hilliard, M. (1993). The future of East Germany after the GDR: Interview with Peter Kruger. *Rethinking Marxism* 6(1):115–27.

Levins, R. (1995). Preparing for uncertainty. *Ecosystem Health* I(1):48–55.

Li, T. Y., and Yorke, J. (1975). Period three implies chaos. *Amer. Math. Monthly* 2:985–99.

Lorenz (1963). Deterministic nonperiodic flows. *J. Atmos. Sci.* 20:130–41.

May, R. (1976). Simple mathematical models with very complicated dynamics. *Nature* 261:459–67.

Tilman, D., and Wedin, D. (1991). Oscillations and chaos in the dynamics of a perennial grass. *Nature* 353:653–55.

CHAPTER TWENTY-SEVEN

Evoking Transmutational Dread

Military and Civilian Uses of Nuclear and Genetic Alchemies

ROBERT H. HAYNES

27.1 Introduction

Atoms and genes first entered nineteenth-century science as hypothetical entities useful in accounting for certain quantitative features of chemical combination and the transmission of hereditary traits, respectively. Today they are known to be structurally complex but empirically significant units of matter and life. They are distinguishable, dissectible, mutable, manipulable, and recombinable physical entities. These profound discoveries led to the rise of nuclear physics and molecular genetics as important branches of science. This expanding knowledge base has been utilized to meet a variety of civilian needs in electrical power generation, industrial fermentation processes, and pharmaceutical production as well as other areas of engineering, medicine, agriculture, and industry.

As a result of the 1939 discovery of nuclear fission in Germany and the exigencies of World War II, the first major application of nuclear physics was in the creation and use of nuclear weapons. Thus, nuclear physics entered public consciousness accompanied by vivid images of mass slaughter and destruction. The subsequent testing and stockpiling of these weapons during the Cold War, coupled with justifiable fears over the carcinogenic effects of global radioactive fallout, cast a pall over the civilian development of nuclear power even though it was touted as providing a virtually limitless source of energy and other benefits for humankind.

In the early 1970s, it became possible to covalently link replicable fragments of genomic DNA from unrelated viruses, organisms, or both not known to recombine in nature. Many molecular geneticists became alarmed that certain planned experiments in animal tumor virology using these new "recombinant DNA techniques" of genetic engineering might prove to be hazardous to laboratory workers and perhaps to the general public. Others warned that the possible escape into the environment of new organisms created by transgressing some inscrutable "natural barriers" between prokaryotes and eukaryotes might trigger an ecological catastrophe. These scientists also feared that governments might seek to develop artificial pathogens which, unlike natural ones, would be genuinely useful as biological weapons (BWs). More recently, worries have been exacerbated about "old-fashioned" BW agents such as the pathogens that cause anthrax or smallpox because of the difficulty of defending civilian populations against surreptitious attacks with them.

Apart from the use of radioisotopic labeling and related techniques in biomedical research, nuclear physics and molecular genetics are essentially unconnected scientific specialities. However, they have been described metaphorically as "nuclear alchemy" and "genetic alchemy," respectively. Unfortunately, these sobriquets conjure images of black as well as white magic – of sorcerers calling up familiar spirits to effect evil ends as well as of eccentric alchemists striving to transmute base metals into gold ad libitum.

Many ethical, social, and political variables undoubtedly contribute to public anxieties about the development of nuclear power and the introduction of genetically engineered organisms into commerce and the environment. It is difficult, even for social scientists, to identify these often cryptic factors and determine their relative importance for any given individual let alone their distribution among social groups or the citizenry as a whole. However, as writers and speakers know, the ability to call up vivid, emotionally significant images or stereotypes from popular culture is an important element in the arts of communication and persuasion. Although alchemy is of interest primarily to scholars today, its indirect influence on writers and film makers in this century has been remarkable. In her study of representations of the scientist in Western literature, Roslynn Haynes (1994, p. 19) commented that "the intimate connection in the popular mind between alchemy and magic was to issue in a steady stream of works featuring some kind of magus who possessed special powers, often implicitly satanic, over nature. For the most part, such magicians are presented as evil or, at best, awesome figures."

In this rather eclectic essay I argue that a deep-rooted fear, which I call "transmutational dread," has been aroused in many sectors of society by the use of alchemical imagery in popular depictions of nuclear physics and molecular genetics. I shall begin with a few remarks on the military applications of "nuclear alchemy" and then recount some experiences as a participant in two technical reviews of the risks of the civilian nuclear power program in Canada (Sections 27.2 to 27.7). In Sections 27.8 to 27.10, I describe the scientific contexts in which alchemical and apocalyptic imagery was grafted onto nuclear physics and mutagenesis and the consequences of this in popular culture. Finally, in Sections 27.11 to 27.15, I deal with civilian and the mooted military applications of "genetic alchemy."

27.2 Nuclear and Nonnuclear Bombs

The yields of the fission bombs dropped on Hiroshima and Nagasaki in August 1945 were 0.0125 and 0.022 megatons (TNT equivalent), respectively. Fusion bombs of 20 megatons have been tested by the United States, and the former Soviet Union once set off a 59-megaton device. The fallout even from a one megaton ground burst of a "dirty" hydrogen bomb would produce serious radioactive contamination. It would be virtually impossible to decontaminate the soil and groundwater. People would continue to die of radiation-induced illnesses many years later. It has been argued that in the "worst case" a sufficiently large exchange of such weapons could lead to "nuclear winter" with devastating consequences for the biosphere as a whole.

Early opposition to nuclear weaponeering was raised by nuclear physicists who had participated in the bomb project itself such as Joseph Rotblat[1] and Leo Szilard. They (and other scientists) had become depressed and morally outraged by the fearful consequences of a future nuclear war. Thus, they abandoned weapons-related research and devoted much of their subsequent careers to promoting international awareness of the hazards of nuclear weapons and working for their abolition and peaceful cooperation among nations. This concern of the nuclear élite lent scientific credibility and moral support to the burgeoning antiwar sentiments of many young people during the 1960s.

A marked contrast exists between the fierce opposition to nuclear weapons and the relatively ineffectual protests against the massive bombing of cities with incendiaries and chemical explosives during World War II. The numbers of people killed in the largest air raids on Hamburg, Dresden, and Tokyo were comparable to the levels attained with the fission bombs exploded over Hiroshima and Nagasaki. For example, it has been estimated that 140,000 people in Hiroshima had died of injuries by the end of 1945, and yet similar numbers were killed during the fire-storm set over Dresden in February 1945. This dreadful slaughter had little strategic significance in affecting the outcome of the war even though tactically the nuclear attacks on Japan ended it (Alperovitz 1996). Since then a taboo against the use, but so far not the deployment, of nuclear weapons seems to have emerged among the most advanced military powers (Gizewski 1996).

Public opposition to nuclear weapons has always seemed to me to be more deeply visceral than that directed against conventional weapons – or indeed against war itself. If this perception is correct, what might account for these contrasting responses, especially because the most obvious effects of nuclear weapons – blast, shock, and fire – are similar to those of sufficiently large bombing attacks? One suggestion has been that nuclear explosives arrived on the world scene suddenly and by surprise, whereas people have been familiar with fire and gunpowder for centuries; such familiarity may breed fatalistic acceptance. Furthermore, the physical damage resulting from chemical explosives and fire-storm attacks can, in principle, be cleared away fairly quickly. On the other hand, the radioactivity released by nuclear explosions is something new and strange. It lingers in the environment and has the potential to produce various adverse health effects. At a deeper cultural level, religious concepts also might contribute to these reactions. Throughout the Bible, burning to death is a punishment often meted out by the Lord (e.g., Genesis 19, Leviticus 10,

Numbers 16, 2 Peter 3). The Divinely-ignited fires of purgatory are believed to be both punishing and purifying, whereas nuclear weapons, created through the inquisitiveness of fallen man, may readily be seen as tools of the devil – but more on this later in the related context of alchemical imagery.

27.3 Hazards of Nuclear Power

The prominent nuclear physicist Edward Teller served as the first chairman (1947–53) of the Advisory Committee on Reactor Safeguards (RSC) of the U.S. Atomic Energy Commission. He was one of the first to draw attention to the potential hazards of nuclear reactors. Under his leadership, the RSC was led to consider various hypothetical "maximum credible accidents," or "worst case" scenarios. In 1953, he made the sensational public statement that "a runaway reactor can be relatively more dangerous than an atomic bomb producing the same radioactivity." He went on to say that "with all the inherent safeguards that can be put into a reactor, there is still no foolproof system that couldn't be made to work wrongly by a great enough fool. The real danger occurs when a false sense of security causes a letdown of caution" (Teller 1953, P. 80). Sadly, the Chernobyl reactor accident justified his second comment, but the first was alarming only to those with little technical knowledge of power reactors.

Worst case scenarios depend on the occurrence of specific sequences of unlikely events for their realization. Thus "reactor safety" came to imply that reactors must be designed to guard against the most appalling calamities that anyone could imagine. However, when such accidents are considered seriously enough for costly design steps to be taken to guard against them, they may appear to the average person to be quite plausible indeed, however low their technically estimated probabilities may be. Not surprisingly, as the number of civilian power reactors increased, many people began to see them as infernal machines ready to explode in their backyards (Weart 1988, pp. 280–94). I happen to have participated in formal assessments of the risks of nuclear power in Canada, and so I will now recount some personal experiences in this troublesome area.

I was trained in physics, but much of my research has been in radiation and molecular genetics – in particular, studies on mutagenesis and DNA repair in microorganisms. Presumably because of this background, I was asked to serve on two independent review committees of the Canadian nuclear power program. The first was the Technical Advisory Committee (TAC) to the Nuclear Fuel Waste Management Program (NFWMP) of Atomic Energy of Canada Limited (AECL) chaired by Professor L. W. Shemilt, a chemical engineer at McMaster University. It began its work in 1979 with a remit to provide ongoing peer review and *technical* advice to AECL. The NFWMP, initiated in 1978, has developed generic plans for the permanent, geological disposal of used nuclear fuel from CANDU (Canada deuterium uranium) reactors.

The second review, known as the Hare Commission, focused on the safety of nuclear power generation (NPG) by Ontario Hydro, the provincial electrical utility, with Professor F. K. Hare of the University of Toronto as commissioner. All nuclear-generated electricity in Canada is produced using CANDU reactors fueled with natural (nonenriched) uranium dioxide (UO_2). These reactors are heavy-water moderated and cooled through separate circuits. Thus, their design differs significantly from the various light–water- or graphite-moderated reactors built for NPG in other countries. Technology assessments indicate that NPG is an environmentally benign source of electricity in contrast with the mining and burning of fossil fuels – especially coal. Nonetheless, technically sophisticated groups opposed specifically to nuclear power have been active and effective in Canada and other countries for many years.

27.4 Nuclear Fuel Waste Management

The first committee (TAC) completed its work in 1996 with the publication of its fifteenth annual report. The CANDU fuel bundles contain about 22 kg of UO_2. Some 100,000 bundles of used fuel, a highly radioactive material with long-lived components, are removed annually from reactors at Ontario's three NPG stations. As of 1997, the total accumulation of used fuel in the province consisted of about 1.1 million bundles. They are stored at the power stations themselves: initially for about 10 years of cooling in wet storage ("swimming pool" installations) after which they are transferred for dry storage in ground-level concrete canisters designed to maintain their structural integrity for 50 years.

Current plans for permanent disposal call for isolation of the used fuel behind multiple engineered and natural barriers in a specially excavated underground vault. Estimates of "time to failure" from corrosion for copper-clad fuel-bundle containers in the proposed vault are in the range of 10^4 to 10^6 years. If approved by government, the vault will be constructed at a depth of 500 to 1000 meters at a site yet to be chosen in seismically stable precambrian rock of the Canadian Shield. The incremental radiation on the ground above the vault will be less than the regional variations in natural background levels. The radiotoxicity within the vault will decay ultimately to levels found in high-grade uranium ore deposits. Thus, the material would, in effect, be returned to the rock from which it came.

This concept of deep geological disposal was later reviewed (1989–97) by a citizen panel appointed by the federal government under the auspices of the Canadian Environmental Assessment Agency (CEAA 1998). The panel consulted widely with individuals, interest groups, and other organizations, and numerous public hearings were held in Ontario and other provinces. It found that "for various reasons, there exists in many quarters *an apprehension* about nuclear power that bedevils the activities and proposals of the nuclear industry" (my emphasis). It concluded that "from a technical perspective, the safety of the AECL concept has been on balance adequately demonstrated for a conceptual stage of development, but from a social perspective it has not. As it stands, the AECL concept... has not been demonstrated to have broad public support." At present it is not clear how the government will respond to this report. Meanwhile the used fuel continues to accumulate on site at the NPG stations.

27.5 Reactor Safety in Ontario

The second review in which I participated was set up by the provincial government as a result of public concerns over the safety of Ontario's nuclear power stations in the aftermath of the April 1986 accident at the Chernobyl graphite-moderated reactor located in Ukraine (Hare 1988). About 30 people were killed immediately and some 200 others developed various degrees of radiation sickness, environmental contamination was widespread, and ultimately about 135,000 people were evacuated from an area of 300 square miles around the plant. The accident was caused by operators who foolishly violated several safety procedures while running a test on one of the reactors at the station. Furthermore, safety seems not to have been a major criterion in the design of this reactor in particular: it was not fully enclosed within a sealed containment shell, and its top was covered only by a screen containing ports to facilitate frequent refueling.

The work of the Hare Commission included the consideration of numerous submissions by antinuclear groups and individuals as well as those presented by industry representatives and government officials. We concluded that the technical risks associated with Ontario's NPG program are very small. Thus, most committee members were puzzled by the intransigent opposition mounted during the hearings by some technically astute people who urged that all NPG stations be shut down as soon as possible. These facilities normally supply up to 60% of the province's electricity. The only practical alternative, in Ontario's northern latitudes, would entail a greatly increased use of fossil fuels.

It seems not to be widely appreciated that, like the earth itself, all fossil fuels contain trace quantities of naturally occurring radionuclides that are released into the atmosphere when the fuel is burned. The committee did not compare the radiation risks of NPG with those of fossil fuel plants.[2] However, in my view at least, the *overall* health risks of NPG are less than those associated with producing an equivalent amount of electricity through the fossil fuel cycle. For example, the emissions from coal-fired plants include carcinogen-containing fly ash, the dioxides of sulfur and carbon, nitrogen oxides, and ozone-depleting halons. These pollutants contribute substantially to the accumulation of acid rain and atmospheric greenhouse gases. New and costly technologies of emission control have ameliorated these problems, but much remains to be done in this regard.

No life-threatening radiation accidents have occurred during the 30-year history of NPG in Ontario, although the industry has had its fair share of political and technical "growing pains." In 1997, a performance assessment review by outside experts concluded that Ontario's power reactors were operating at *minimally acceptable* safety levels. It was not the design or construction of the CANDU reactors that led to this conclusion. Rather, the assessors found shortcomings in management, routine operations and maintenance, employee attitudes, staff culture, and labor relations. As a result of this report, Ontario Hydro plans to shut down temporarily 7 of its 19 reactors to concentrate its resources on bringing the other 12 back to a high level of operational safety. Operational problems of this kind are of course not unique to NPG stations.

Partly as a result of unremitting public opposition, which extends to the use of radiation for the sterilization of ground beef, the nuclear industry in Canada is in stasis if not absolute decline today. Excellent science and engineering programs in AECL laboratories have been phased out, and many employees have been discharged before their normal age of retirement. Cynical observers have commented that the only seriously damaging effect of Canada's NPG program has been on the careers of its staff.

27.6 Nuclear and Nonnuclear Accidents

The world's first serious accident at an NPG station occurred in March 1979 at the Three Mile Island (TMI) installation in Pennsylvania (the Chernobyl accident was the second). The cost of clean-up at TMI was high (over one billion dollars), and the event made headlines everywhere. However, no one was killed or injured, and no radioactivity was released beyond the confines of the plant. Occasional releases of low levels of radioactivity from NPG stations are regularly featured in news reports. They seem to elicit more attention and concern than do the thousands of annual casualties resulting from motor vehicle, airline, marine, and railway crashes; the bursting of dams; disasters in mining, construction, chemical, and other industries; and many other sources of accidental death or disability.

On November 11, 1979, a speeding 106-car tanker train derailed, piled up, and caught fire at a road crossing in the large suburban city of Mississauga located immediately to the west of metropolitan Toronto. The fully loaded tank cars were carrying caustic soda, propane, chlorine, styrene, and toluene. On impact, a propane tanker caught fire. The flames were fed by styrene and toluene leaking from other damaged cars. An explosion occurred, and the resulting fireball rose to a height of 1500 m; it could be seen at distances up to 100 km from the accident site. Then, a supertanker loaded with chlorine ruptured and released most of its contents into the blaze. As a result, chlorine gas began to drift in a large, ground-level plume into the surrounding suburban area.

This accident triggered the world's largest-ever peacetime evacuation of people. Some 218,000 residents (including 1450 patients from hospitals and nursing homes) were dispersed into the surrounding metropolitan area and remained away from their homes for periods of 3 to 5 days. Amazingly, there were no casualties in all of this. The derailment stimulated much discussion about the risks of shipping deadly chemicals by rail, and revised safety regulations for such shipments were quickly put into effect. However, the accident was forgotten within 2 or 3 months, at least by the local media, and it did not inspire the kind of chronic criticism long directed against the Pickering NPG station located just east of Toronto. The short public lifetime of this event surprised me because chlorine is well known as a chemical weapon, which was used with appalling effects on Canadian and other troops during World War I.

27.7 Risk Assessment and Perception

During the public hearings of the Hare Commission, industry officials and representatives of the antinuclear groups seemed, for the most part, to argue from discordant presuppositions about the appropriate approach to risk analysis itself as well as differing personal, social, and political agendas. There were communication failures, even though everyone spoke and wrote politely and clearly enough in plain English. In a recent book, Powell and Leiss (1997) have identified seven distinguishing characteristics of "expert" and "public" languages typical of discourse in risk analysis. The most obvious of these is that expert language is grounded in probabilistic reasoning about acceptable levels of risk to the average person exposed to a wide range of hazards. On the other hand, public (and often juridical) language is intuitive and dichotomous (is it *safe or not?*) and focuses on particular hazards and events.

Public perceptions of the risks of nuclear plants relative to other industrial hazards are inconsistent with actuarial data on the incidence of death and injury associated with NPG. However, the *qualitative* perceptions of the citizenry count for much more politically than do the calculations of risk assessors (Perrow 1984). For example, in Taiwan, opposition to nuclear power actually *increased* after the government launched an extensive public education program designed to alleviate public concerns (Hamilton 1990). Expert technical analysis has little chance of putting to rest worries over the hazards of a mysterious technology, especially if people feel that they have no personal control over exposure to its risks. Furthermore, risk analysts have grave difficulties dealing quantitatively with low probability–high consequence accidents; concerned citizens do not have this problem in the case of nuclear power.

These considerations are consistent with Slovic's (1987) comment that

> there is wisdom as well as error in public attitudes and perceptions.... However, their basic conceptualization of risk is much richer than that of the experts and reflects legitimate concerns that are typically omitted from expert risk assessments. As a result, risk communication and risk management efforts are destined to fail unless they are structured as a two-way process. Each side, expert and public, has something valid to contribute. Each side must respect the insights and intelligence of the other.

Clearly, participants in such debates have a responsibility to understand the dissonant mentalities within which *intuitions* of risk arise and judgments are formed on questions for which there are no simple answers (cf. Lowrance 1976, pp. 109–14; Starr and Whipple 1980).

27.8 Atoms, Genes, and Alchemy

In 1903 Ernest Rutherford and Frederick Soddy, then at McGill University, suggested that spontaneous transformations of atomic nuclei were responsible for natural radioactivity (Soddy 1904, p. 93). Thus was born "nuclear alchemy" (cf. Soddy 1912, Redgrove 1922, Rutherford 1937). For 23 years after the initial discovery of radioactivity by Henri Becquerel in 1896, few suspected that it might be possible to stimulate such processes *artificially*. It was not until 1919 that Rutherford showed that if nitrogen was bombarded with high-energy alpha particles, it could be changed into oxygen.

Undoubtedly wishing to dramatize his first important scientific discovery for popular audiences, Soddy invoked the alchemical metaphor of transmutation for these processes even though coeval knowledge of atomic physics could scarcely support its use.[3] He concluded an article on natural radioactivity published in a widely read British intellectual magazine with the suggestion that we should "regard the planet on which we live as a storehouse stuffed with explosives, inconceivably more powerful than any we know of, and possibly only awaiting a suitable detonator to cause the earth to revert to chaos" (Soddy 1903).

In a similar vein, a reviewer of a 1903 article by Rutherford and Soddy concluded with the following anecdote:

> Professor Rutherford has playfully suggested to the writer the disquieting idea that, could a proper detonator be discovered, an explosive wave of atomic disintegration might be started through all matter which would transmute the whole mass of the globe into helium or similar gases, and, in very truth, leave not one stone upon another. Such a speculation is, of course, only a nightmare dream of the scientific imagination, but it serves to show the illimitable avenues of thought opened up by the study of radio-activity (Whetham 1904).

A similar warning was expressed as late as 1935 by Frédéric Joliot-Curie in his Nobel Prize speech. Thus, radioactivity, alchemy, and the final apocalyptic agony of Earth were conflated by some of the most prominent contributors to the nascent field of nuclear physics.

Soddy's fantasy of a planet primed to explode if properly detonated was revived decades later in the bugbear of triggering a chain reaction in the atmosphere during the first fission bomb test at Alamogordo, New Mexico (Rhodes 1986, pp. 664–5). Furthermore, some time after the first hydrogen bomb test at Eniwetok in 1952, Sir Robert Robinson, in his presidential address to the British Association for the Advancement of Science, expressed the fear that exploding *fusion* bombs might yet set off a global-scale thermonuclear catastrophe. Such remarks by a prominent scientist prompted M. H. L. Pryce, a distinguished British mathematical physicist, to write a popular article explaining why this could not happen (Pryce 1955). Indeed, as early as 1946, Hans Bethe had written a similar article in connection with *fission* bomb tests. In it he commented that

> it is necessary to examine in great detail the possible dangers of atomic bomb tests before carrying them out. *Such considerations can be made with the help of existing nuclear*

theory without making any assumptions going beyond the range of well explored phenomena. The conclusion from the calculations is that there is no danger of a nuclear explosion in any substance naturally occurring on earth with any of the atomic bombs which have been developed or been conceived on paper. (my emphasis) (Bethe 1946, p. 2).

He concluded with the following statement

There is one problem which is of really over-whelming concern to us all: the problem of international control of the bomb. While some scientists may argue that there might be a remote chance of our concepts of theoretical physics being false . . . they will agree that there is an immeasurably greater danger that the peoples of the world will not have sufficient wisdom to settle their differences before a war breaks out which will be fought with atomic weapons. To draw everybody's attention to this possible catastrophe has been, and still is, the main concern of all of us.

Spontaneous mutations in cells occur naturally without regard to their phenotypic consequences. They were regarded for many years as rare, sudden, and discrete events of unknown origin that cause genes to pass suddenly, rather like quantum jumps, from one stable state to another. No one knew whether it might be possible to induce artificially the change of one gene into another. However, in 1927 Hermann Muller showed that mutations could be produced in *Drosophila* by exposing the flies to ionizing radiation (X rays). He entitled his first article on this discovery "Artificial Transmutation of the Gene." As Carlson (1981, p. 150) commented regarding Muller's public announcement of these findings at the 1927 International Congress of Genetics in Berlin; "the paper created a sensation. The press dispatched the news around the world. Man's most precious substance, the hereditary material that he could pass on to his offspring, was now potentially in his control. . . . Like the discoveries . . . of Rutherford, Muller's tampering with a fundamental aspect of nature provoked the public awe. When Muller returned to the United States, he found, to his surprise, that he was famous." For many years after Muller's discovery, mutations could only be induced haphazardly in organisms by exposure to ionizing and ultraviolet radiations and a few chemical mutagens. However, it is now possible to create, at the single DNA base-pair level, specific mutations in specified genes, and with recombinant DNA techniques, to generate "designer organisms" à la mode. These accomplishments have been described popularly by various authors as "genetic alchemy" (cf. Krimsky 1982.)

27.9 'Transmutational Dread'

The harnessing of nuclear reactions seems uniquely frightening, even eerie, to many people. This was pointed out in the aforementioned CEAA (1998) report, and I quote the relevant paragraph in full (p. 18):

A deeply entrenched fear and mistrust of nuclear technology exists within some segments of our society. *This 'dread factor' is real and palpable.* It is an important element in decision-making processes concerning nuclear matters, as it will undoubtedly affect public confidence resulting from such processes. The dread factor stems not only from the imperceptibility, mobility and longevity of the radiation hazard and its disturbing potential health effects, but also from association with nuclear weapons and with past disasters, nuclear and other, involving human error or engineering failures. A combination of these elements led many participants to *express great anxiety over worst-case scenarios with terrible and long-lasting consequences, regardless of their low likelihood.* Although experts may challenge or debate the perception that nuclear fuel wastes pose unprecedented hazards due to their extreme toxicity and longevity, these challenges are not, by themselves, likely to materially reduce the dread factor. (my emphasis).

Alchemical imagery also might contribute to this "dread factor." Nuclear (and genetic) alchemy is seen by many as a black art. Nonscientist friends have sometimes asked me, in effect, Are not scientists exposing the world to spooky radiations that can cause weird and irreversible transmutations of matter and of life itself?

Other authors also have suggested that nuclear technology evokes images associated with the alchemists of old and the mad scientists of popular culture today. Thus, Spencer Weart has argued that those opposed to nuclear power should not be dismissed by experts as being ignorant of the facts or little more than vulgar Luddites (Weart 1988). He attributed their opposition to a mixture of antiauthoritarian sociopolitical values coupled with serious environmental concerns. However, Weart also claimed that these more obvious motivating factors may be supplemented at a deeper psychological level by the arousal of archetypal images of various sorts. He described the emotional response in people influenced by these fantasies as *nuclear fear* and concluded that nuclear energy has "become a symbolic representation for the magical transmutation of society and the individual" (p. 421). Further insights into the cultural sources of nuclear fear have come from literary studies on the several, mostly negative, stereotypes of scientists over the past century. They have been propagated relentlessly through much of our popular culture which, in the words of Roslynn Haynes (1994, p. 1) "is influenced more by images than by demonstrable facts." If public reactions to the development of genetic technologies are considered together with those aroused by nuclear power, Weart's concept of "nuclear fear" might be expanded into the general notion of "transmutational dread" – a vague fear or anxiety about scientists usurping the power of God by meddling with the sacrosanct mysteries of the universe.

27.10 Cultural Roots of Transmutational Dread

The occult arts encompass alchemy, astrology, divination, magic, and witchcraft. In both Eastern and Western cultures they have interacted strongly among one another and with basic religious beliefs. The British chemist and student of transcendentalism, H. Stanley Redgrove, argued that "the main alchemistic hypotheses were drawn from the domain of mystical theology and applied to physics and chemistry by way of analogy (Redgrove 1922, p. V." However, "in its attempt to demonstrate the applicability of the fundamental principles of Mysticism to the things of the physical realm Alchemy apparently failed and ended its days in fraud" (Redgrove 1922, p. 16).

Popular interest in alchemy quickened during the occult revival that began among poets, novelists, mystics, and pseudoscientists in the nineteenth century. It went on to gain a certain fashionable credibility in this century through C. G. Jung's theory of archetypes and the "collective unconscious." These seductive influences live today among New Age addicts and others who fear that science has gained a monopoly on "truth." For example, Baigent and Leigh (1997) recently asked (rhetorically) whether

> our culture, technology and its products constitute a form of magic circle. From within the supposed safety of this circle, we invoke powers with apocalyptically destructive potential. We pollute our world with plastics and radiation, with toxic chemicals and industrial waste... we arrogate to ourselves the power of godhood and perpetrate experiments which nature herself does not – in genetic engineering, in nuclear fission and fusion, in the development of biological and chemical weaponry. Like Dr. Frankenstein, we create monstrosities. And like the careless magician ... we all too easily lose control of the forces we conjure up – as happened, for example, at Chernobyl.... We all become victims.

Black magic and witchcraft were traditionally believed to be sinful practices worthy of damnation. *White* magic was an acceptable part of the hermetic tradition insofar as it was not employed for maleficent purposes. In its long heyday, alchemy was believed by many to partake of *both* black and white magic. For some, it was (and may still be) viewed as a pious art reflective of God's handiwork and the sacred mysteries of religion. The alchemist's fiery crucible is said to have prefigured the conflagrations and purifications of the Last Judgment. However, for others it was "an art of dubious authority that vainly presumed to imitate divine acts of making and changing proper only to God" (Roberts 1994). In early Christendom, knowledge gained outside the Church was always suspect, and at least two church fathers, Clement of Alexandria and Tertullian, wrote that fallen angels had revealed to humankind the techniques of metal-working, divination, and magic.

Though influenced strongly by Chinese and Islamic ideas, alchemical theories of matter in the Western world were based primarily on two Aristotelian notions. The first was that of the fundamental unity of matter in a *prima materia* from which the four elements (earth, water, air, and fire together with their four corresponding binary qualities, dry and cold, cold and wet, wet and hot, and hot and dry) were somehow derived. It was assumed further that all bodies were composed of these elements in different proportions and that when properly decomposed in the fiery crucible they could be transmuted into one another. The second notion was that nature strives toward perfection. Gold was considered to be the perfect metal, and thus base metals must somehow be striving to attain the perfection of gold. If we read "debased humanity tainted with original sin," for "base metals lacking the perfection of gold," strong metaphorical similarities between alchemical transmutation and apocalyptic visions of death and resurrection come into view, and the latter is not a wholly joyful prospect for many comfortable Christians today (see Revelation 21).

Alchemists sought to assist nature in this progress toward perfection with the aid of the "philosopher's stone" (Read 1936, Roberts 1994). This protean substance was widely thought to be capable of catalyzing these and other marvels. Thus, other avatars of the stone were the *elixir,* which was thought to restore youth or grant immortality, the *panacea,* which could cure every disease, and the *alkahest,* a universal solvent. The philosopher's stone was taken by some to be symbolic of the redemptive power of Jesus Christ, and from this point of view it is easy to understand how it acquired its miraculous powers.

Yet another alchemical dream was the artificial creation of a homunculus, a tiny perfect man, or manikin. According to one recipe, this little beast would emerge from a mixture of semen and menstrual blood without recourse to a uterus. Roslynn Haynes (1994, p. 13) has argued that "this was also perceived as a threat to the divine basis of life expounded by the Church, and much the same furor attended the attempts as obtains today over in vitro fertilization or genetic engineering."

The surviving alchemical literature is vast, complex, deliberately obscure, and utterly fascinating – even to skeptics like me. Obviously it was not subject to peer review. Its bizarre but numinous iconography transports one into a surreal world of portentous symbols and mythic drama (Roob 1997). The more I have browsed through collections of beautiful illustrations from alchemical manuscripts and viewed paintings and engravings of scenes from the Christian apocalypse, the more difficult it becomes to fathom the mentality of my religious and intellectual forebears as well as the New Age folk of today.

Mary Shelley's novel *Frankenstein, or the Modern Prometheus* is based on the gruesome alchemical theme of creating an artificial man from dismembered body parts (see also Roob, 1997, p. 211). Unfortunately, Frankenstein's idealistic endeavor produced a monster endowed with supernatural strength that inspired loathing in anyone who gazed upon it. Attacks by weird creatures have been featured in films such as *The Andromeda Strain* (Panavision 1971) and *Twelve Monkeys* (Universal Pictures 1995). There is at least one film, *The Omega Man* (Panavision 1971) whose theme is based upon organisms created by genetic engineering. Thus Roslynn Haynes (1994, p. 3) has described the alchemist as one of the recurring stereotypes of the obsessed scientist, "driven to pursue an arcane intellectual goal that carries suggestions of ideological evil . . . and *reincarnated recently as the sinister biologist producing new species through the quasi-magical processes of genetic engineering*" (my emphasis).

27.11 Radioactivity and Recombinant DNA: Actual and Improbable Hazards

In 1895 Wilhelm Röntgen discovered X rays for which he was awarded the first (1901) Nobel Prize in physics. Recognition of the value of ionizing radiations as a tool in many areas of scientific research and medical practice grew rapidly in the ensuing decades, and most of their basic biological effects were discovered before World War II. By 1925, the dangers of excessive human exposures to radiation had become obvious to radiologists, and in 1928 the International Commission on Radiological Protection (ICRP) was established. The main tasks of the ICRP, in cooperation with national regulatory authorities and scientific academies, were to establish, for various categories of people, maximum permissible dose levels for whole-body exposures to external sources of radiation and maximum body burdens for ingested radionuclides. Thus, by the time that civilian nuclear power appeared on the scene, the fact that radiation poses a health hazard was well known.

In the late 1960s, various enzymes were discovered that made it possible to cut and splice specific segments of DNA and thereby create in vitro recombinant DNA molecules from different "parental" sources; these chimeric molecules might then be cloned in bacteria or other living cells. In 1970, Paul Berg of Stanford University proposed to insert the genome of an animal tumor virus (SV40) into the human colonic bacterium *E. coli*, as a result of which alarm over the *possible* carcinogenic hazards of such experiments arose among a few of his colleagues. Some of these individuals quickly spun out various "worst case" scenarios of worldwide health disasters and claimed that molecular genetics might be in a "pre-Hiroshima situation." As a result of this fuss, Berg postponed the experiment.

At the 1973 Gordon Conference on Nucleic Acids, the hypothetical hazards of recombinant DNA experiments were discussed hurriedly on the last day of the meeting, and a letter was sent to the presidents of the National Academy of Sciences and the Institute of Medicine asking them to establish a committee to study the issues. In addition this letter was published in *Science* (Krimsky 1982). The Committee on Recombinant DNA Molecules was duly appointed by the Academy with Berg as chairman. In July 1974 the committee published a letter in leading scientific journals calling for a voluntary moratorium on two types of recombinant DNA experiments (Watson and Tooze 1981, p. 11).

As a followup to the moratorium letter, an international meeting to consider the risks of several generic classes of recombinant DNA experiments was organized by Berg at Asilomar, California, in February 1975; a report was published 4 months later (Berg et al. 1975). In the report it was stated that "accurate estimates of the risks associated with various types of experiments are difficult to obtain because of our ignorance of the probability that the anticipated dangers will manifest themselves." In the meantime, the National Institutes of Health (NIH) had established the Recombinant DNA Molecule Program Advisory Committee (RAC for short) in October 1974. The RAC issued a set of very stringent containment rules that took effect in June 1976. Several other countries doing recombinant DNA research issued their own biohazard guidelines which slowed down, or increased the cost, or both, of research in this area. However, as experience was gained as a result of many experiments carried out under such guidelines, the "worst case" possibilities that first caused the alarm began to be challenged with observations and numerical estimates of the probabilities of deleterious consequences. After much scientific and political debate (see Watson and Tooze 1981), revised U.S. guidelines came into force in January 1979, and similarly relaxed controls were adopted in other countries.

In Canada national guidelines are still officially in effect, but responsibility for biosafety has devolved largely to the institutions under whose auspices the experiments are being carried out and ultimately to the common sense of the scientists themselves. The criteria for good laboratory practice are much as they were before 1976, and most experiments with nonpathogenic organisms no longer require strict containment. However, I notice that my colleagues now are more sensitive to biosafety issues than they were previously. The international brouhaha over recombinant DNA seems to have been a word to the wise.

27.12 Biological Weapons: Natural and Novel Pathogens[4]

Biological and nuclear weapons are often lumped together by strategists as tools of potentially mass slaughter. In the case of BWs the justification for this view is persuasive but largely theoretical because they have been so rarely used in war. On the other hand, the enfeeblement of armies and the breakdown of societies by epidemic diseases have been well-documented historically (Bray 1996). Residues from BW attacks could be difficult to eradicate, and their natural rates of inactivation might be outstripped by their proliferation in some ecological niches. It is difficult to make sound predictions about the ecological consequences of releasing large quantities of natural BW agents. However, there is no reason to think they would necessarily be trivial, short-lived, or remain localized in the target zone. For example, anthrax spores, released by the United Kingdom in tests carried out in 1941–42 on Gruinard Island, were found to persist for decades in the soil (Manchee et al. 1981; Carter and Pearson 1998).

Nature has provided bioweaponeers with a wide choice of natural pathogens that might be used as BWs, but it is difficult for anyone outside official BW establishments to prepare an accurate world list of those that actually have been weaponized or studied with hostile intent. According to Franz et al. (1997), we have reason to be concerned about some 10 types of bacteria, viruses, and biotoxins. Burck (1990) presented a much longer list of agents that have been studied for offensive and defensive purposes in U.S. military establishments since the end of World War II. Of these, the microorganisms that cause anthrax, brucellosis, tularemia, and Q-fever and the viruses for Venezuelan equine encephalitis (VEE) and yellow fever apparently were standardized and once stockpiled by the U.S. armed forces. In addition, 11 biotoxins have been studied of which 4 were weaponized (botulin, cobrotoxin, saxitoxin and *Staphylococcus* enterotoxin). However, in 1969 President Nixon renounced biological warfare, and over the next few years U.S. stockpiles of biological and toxin weapons were destroyed.

It would appear that the world's largest BW program was built up by the former Soviet Union, components of which may still be operational in facilities run by the Russian Ministry of Defense. If defecting scientists are to be believed, Soviet BWs included "improved" variants of smallpox, anthrax, and plague, and antibiotic-resistant variants of plague and a smallpox–VEE chimeric microbe were produced using genetic engineering techniques (Preston 1998). By 1995, some 17 countries (most in the Third World) were suspected by U.S. authorities as possessing some BW capabilities (Cole 1996, p. 62).

A list of criteria for the military effectiveness of BWs has been assembled from publicly available military documents (Geissler 1986). None of the weaponized agents named above would appear to meet all of these criteria in the fullest degree, especially with regard to their levels of contagion and lethality as estimated on the basis of civilian medical data. However, infections by any of them will produce severely debilitating physical and psychological symptoms, and often death, especially in otherwise weakened or untreated victims (Franz et al. 1997; Holloway et al. 1997).

Geissler's list can be summarized as follows: The agent(s) should be highly contagious and lead to disease or death with high probability at low inoculating doses for the normal routes of entry into humans. The incubation time from initial infection to the onset of symptoms should be predictable and short. The pathogenic potency of the agent should be environmentally stable and not undergo extensive decay during production, storage, and transport. For efficient dissemination, agents that infect by inhalation should be capable of being incorporated into aerosols and remain viable in the target area for tactically effective periods of time. The target population(s) should have little or no natural or acquired immunity, nor other forms of protection, against the agents used. The agent(s) should be difficult to identify, and little or no treatment for the infections should be available in the target area or enemy country. Those using the weapons should have the means to immunize, or appropriately protect their own military personnel and, if practicable, their civilian populations.

Rosenberg and Burck (1990, p. 307) have given some "worst case" examples of how natural pathogens might be altered using the techniques of genetic engineering to make them more effective as weapons. These include the combination of pathogenicity with drug resistance, altered antigenicity, enhanced stability, the incorporation of additional pathogenic factors, extended host ranges or tissue specificities, increased invasiveness, and other properties that would compromise detection or correct identification. Whether any such strains and their foreign DNA sequences would survive and spread in any particular environment is hard to predict accurately; indeed ecologists and microbiologists appear to be divided on this issue (Campbell 1991, Levin 1991).

"Genetic alchemy" is not known to have been utilized successfully by the bioweaponeers of any country even though suspicions are life. Robert Sinsheimer and a few other prominent molecular geneticists were among the first to warn publicly about this possibility (Wright and Sinsheimer 1983, Geissler 1984). However, like some others I remain skeptical that novel BW agents would prove to be any more useful as weapons than existing pathogens (cf. Kaplan 1983). Moreover, this view reflects only my personal intution and that of some other interested scientists who have neither worked in BW research laboratories nor are known to have been privy to the relevant classified information. Furthermore, the effectiveness of *natural* pathogens has not been subject to test in combat situations. Thus, there exists no "standard," or base-line data against which the effectiveness of novel BWs may be compared with natural BWs, and any such tests would be immoral in the extreme. If it should be possible to develop novel BWs by "genetic alchemy," their deployment could excite images of a "biological Armageddon," a world wasted by highly infectious, incurable diseases. For connoisseurs of eschatology, the millennium would arrive not with a nuclear bang but a germy whimper.

27.13 Reactions against Field Tests of Bioengineered Microorganisms

In 1983 permission was given by the NIH for field tests, to be carried out in California, of a soil microorganism (*Pseudomonas syringae*) from whose genome a specific DNA fragment was deleted. This fragment encodes a protein that provides a nucleation point for ice on the bacteria. These so-called ice-minus strains were designed to reduce the temperature at which frost forms on potato and strawberry plants and other crops. The activist Jeremy Rifkin brought national attention to the tests by successfully suing NIH on the ground that approval was given before a study of their environmental consequences had been carried out. One California resident wrote to a local newspaper that "the new technology conjures up images of a science fiction thriller where the creation of mad scientists threatens the delicate balance of nature and our ecosystem." Some farmers in California feared that the public might associate the release of genetically engineered organisms with the nuclear reactor accidents at Chernobyl and Three Mile Island and decide to boycott their farm produce. Official statements regarding the safety of the tests did not calm local concerns.

What can we presently say with confidence about the possible hazards of releasing novel microorganisms into the environment? Unfortunately, the answer is "not much" (Fowle 1987). Some introduced species are known to have caused serious ecological disruptions. Ecologists who have appraised the possible effects of releasing (nonpathogenic) organisms have concluded that risk assessments in this area should be carried out on a case-by-case basis (Lenski 1987). Given the variety and complexity of ecological systems, it may be that few general principles exist upon which the consequences of the release of specific recombinant microorganisms or viruses can be predicted (Krimsky 1987). On the other hand, there is no a priori scientific reason to think that catastrophic consequences would ensue (Miller 1991).

27.14 International Prohibitions against Biological Weapons[5]

Deliberate attempts to expose one's enemies to infectious diseases have occurred on remarkably few occasions in the long and tragic history of organized warfare (Christopher et al. 1997; Geissler and van Courtland Moon 1998). The Geneva Protocol of 1925 prohibited the *use* of biological and chemical weapons. Nonetheless, research, development, testing, and stockpiling of BWs, and contingency plans for their use, did go on in several countries during World War II, and these agents were employed by Japanese troops in China in the early 1940s (Harris 1998). After the war, the Biological and Toxin Weapons Convention (BWC) was signed in 1972 by three co-depositary countries, the United States, the United Kingdom, and the Soviet Union. It came into force in 1975. As of 1997, it had been ratified or acceded to by 140 States Parties. The BWC has been hailed as the world's first significant disarmament treaty (Gizewski 1987). It is a landmark in arms control because it prohibits not only the development, production, and stockpiling but also the *possession* of an existing class of weapons. Unfortunately, some countries maintained secret BW programs long after adhering publicly to the BWC.

Certain weaknesses in the BWC came to light soon after it came into force in 1975. Considerable effort has been devoted to rectifying them (cf. Geissler and Haynes 1991). However, it is difficult to ensure compliance even with the most explicit and unambiguous legal prohibitions against BWs. In this area it is notoriously easy to cheat. BWs have been called the "poor man's atomic bomb." Unlike nuclear weapons, BW agents can be grown inconspicuously in small fermentation plants. Even if located and destroyed, such facilities can be readily replaced at a later date. Still, it is necessary to strengthen the existing BWC despite the possibility of violation. Arms control agreements are prerequisites for establishing international verification régimes and providing legal grounds for sanctions against violators. Indeed, negotiation of the strongest possible BWC is itself an important confidence-building measure among nations. Recent reviews of this endeavor may be found in Pearson and Dando (1996), Thränert (1996), and Kadlec, Zelicoff, and Vrtis (1997). For discussions of the current threat of BWs as perceived by academic strategists see, for example, Tucker 1997, Betts 1998, Gavaghan 1998, Pearson 1998, and Steinbruner 1998.

Offensive and defensive BW programs have much in common, especially at the level of laboratory research. Honestly defensive work on means to protect civilians and military personnel against BWs can be misinterpreted as being carried out in preparation for attacks on others. However, ready access to vaccines and antitoxins against BW agents would benefit everyone, especially civilian populations that might be exposed to terrorist attacks. It would be wise therefore to develop BW defenses on an internationally collaborative basis. Any such program should include the exchange of working scientists among countries which have the relevant expertise and BW research facilities, and be open to inspection by civilian experts (cf., NAS 1997).

27.15 Problematic Aspects of Novel Biological Weapons

Many reasons have been advanced to explain the unpopularity of biological weapons. It has been said that they are singularly immoral, impractical, or ungentlemanly tools of war – in curious contrast to conventional weapons. However, scientists and experts in the arms control community have expressed anxiety that governments might try to develop novel BW agents using recombinant DNA techniques (Geissler 1986, Dando 1994). The concerns that have arisen over novel BWs indicate that considerations of morality and international law are inadequate to ensure that governments would eschew development of BWs if they promised to be tactically effective in battle or strategically advantageous in diplomacy. What, then, is there to restrain governments from trying to develop more effective biological weapons using genetic engineering and other modern microbiological techniques?

I think there are at least two reasons, but neither of these should make anyone complacent about the need for a strengthened BWC. First, it is by no means clear that novel BWs would prove to be any more useful than those employing natural pathogens (Novick and Shulman 1990). Second, the civilian biotechnology industry could suffer much the same fate as the nuclear industry if genetic engineering techniques were used successfully to develop novel BWs, and a widespread belief has arisen that small accidental releases of such recombinant pathogens could trigger pandemics or ecological catastrophe. Under such circumstances the cry to "Ban the Bomb" could quickly change to "Ban the Bugs" and perhaps to a ban on biotechnology altogether.

A more robust international control régime against all biological and toxin weapons is essential and should be concluded expeditiously. Rigorous international verification mechanisms must obviously be included in it (Kadlec et al. 1997). Provisions for unannounced "challenge inspections" and sanctions against violating nations are obviously necessary. Cant about the sanctity of national sovereignty in this area of arms control must be rejected (cf. *Nature* 1998). Unfortunately, because of the relative ease of surreptitiously growing small quantities of pathogenic bacteria and viruses, I fear it will remain difficult, if not impossible, to eliminate entirely the threat of terrorist attacks with biological and toxin weapons.

27.16 Final Comments

The development of nuclear power was promoted initially with promises of inexhaustible sources of cheap energy and other marvels. Unfortunately, people will continue to associate nuclear power with "the bomb" and its potential to transmute Earth into a planet fit only for mutants: hopeful promises cannot overcome deep-rooted sources of human dread.

Weapons of mass slaughter are morally repugnant to all but sociopaths, and their proliferation among poor countries comes at the expense of social and economic development. The acquisition of such weapons feeds only the hubris of leaders and the jingoism of the ignorant, and worse, it may well exacerbate regional political and religious instabilities.

The sensational imagery of nuclear and genetic alchemy may have contributed significantly to public opposition to the civilian applications of nuclear physics and molecular genetics. If it is plausible to invoke the concept of "transmutational dread" to anticipate public responses to alarming stories about novel biological weapons, then the future of civilian biotechnology could become as problematic as that of nuclear power today.

Scientists owe it to themselves to eschew both overblown promises and apocalyptic images of death and destruction when communicating seriously with others. On these occasions they have a unique responsibility to emphasize the distinction between real science and science fiction for the simple reason that science and technology are seen as deeply threatening by many people today. Proliferation of "worst case" scenarios in the public domain contributes only to antiscience sentiment, and "best case" scenarios that fail to materialize provoke backlash and cynicism. This bodes ill for everyone, for it undermines public understanding of the spiritual values of science that have freed so many from superstition and the fear of ancient gods.

27.17 Acknowledgments

It has been a pleasure to contribute to this Festschrift in honor of Dick Lewontin. My intellectual horizons have been broadened through conversations with him and by reading his wonderful books, essays, and reviews. He is not only a great geneticist and evolutionary biologist but also a natural philosopher and social critic nonpareil.

I am much indebted to Professor Erhard Geissler of the Max Delbrück Center for Molecular Medicine, Berlin–Buch, for first alerting me to the possible use of genetic engineering in BW research, and for many instructive discussions on this problem. I thank Iris Hunger for referring me to recent publications on efforts to strengthen the BWC. I also thank K. W. Dormuth, M. Gascoyne, P. Gizewski, F. K. Hare, T. F. Kempe, M. S. Mahdy, R. G. E. Murray, L. W. Shemilt, and G. Vachon for their help with this chapter. Finally, I am grateful to my wife, Jane Banfield, for her patience, good humor, and editorial help as this essay evolved slowly through more drafts than she or I care to recall.

This work was supported by the Natural Sciences and Engineering Research Council of Canada.

Notes

1. Professor Rotblat was my postdoctoral research supervisor (1957–58) in the Physics Department of St. Bartholomew's Hospital Medical College. In December 1944, he left the bomb project at Los Alamos for essentially moral reasons. On July 9, 1955,

he chaired the public meeting at Caxton Hall in London where Bertrand Russell launched the Russell–Einstein Manifesto for the abolition of nuclear weapons. This meeting led to the establishment in 1957 of the Pugwash Conferences on Science and World Affairs (Rotblat 1972). In 1995, Rotblat and the Pugwash organization were awarded the Nobel Prize for Peace.

2. Recent (1993) dose assessments indicate that the radiological impact on individuals living in the vicinity of an Ontario coal-fired electrical generating plant, rated at 4000 MWe, is smaller than the dose for an equivalent-sized nuclear plant by a factor of five (24 versus 120 FSv per year). The normal radiation dose from natural background sources is 3000 FSv annually.
3. The word *transmutation* has been widely used since the fourteenth century to describe many types of change other than those sought by alchemists. For example, Lamarck, Darwin, Huxley, and other biologists often used the word transmutation to describe the general notion of species changing over time. It was Soddy who first brought the metaphysical baggage of transmutation into popular scientific writing.
4. See the August 6, 1997, issue (volume 278, number 5) of the *Journal of the American Medical Association* for an excellent series of review papers and editorial comment on the major scientific, clinical, and policy aspects of biological warfare.
5. The full texts of the 1925 Geneva Protocol, the 1972 BWC, and the list of States Parties to these agreements, together with other relevant documents, may be found in Wright (1990). The "rolling text" of the current state of the negotiation of the Protocol with a view to strengthening the 1972 BWC, as of 4 August 1997, drafted by the *Ad Hoc Group of the States Parties to the Convention on the Prohibition of the Development, Production and Stockpiling of Bacteriological (Biological) and Toxin Weapons and on their Destruction* is contained in the United Nations, Geneva, document BWC/AD HOC Group/36, seventh session.

REFERENCES

Alperovitz, G. (1996). *The Decision to Use the Atomic Bomb.* New York: Vintage Books.

Baigent, M., and Leigh, R. (1997). *The Elixir and the Stone.* London: Viking.

Berg, P., Baltimore, D., Brenner, S., Roblin, R. O., and Singer, M. (1975). Asilomar Conference on recombinant DNA molecules. *Science* 188:991–4.

Bethe, H. A. (1946). Can air or water be exploded? *Bull. Atomic Scientists* 1:2.

Betts, R. K. (1998). The new threat of mass destruction. *Foreign Affairs* 77:26–41.

Bray, R. S. (1996). *Armies of Pestilence: The Effects of Pandemics on History.* Cambridge, UK: Lutterworth Press.

Burck, G. M. (1990). Biological, chemical, and toxin warfare agents. In Wright (1990), pp. 352–67.

Campbell, A. M. (1991). Microbes: The laboratory and the field. In *The Genetic Revolution: Scientific Prospects and Public Perceptions,* (ed.) B. D. Davis, pp. 28–44. Baltimore: The Johns Hopkins University Press.

Carlson, E. A. (1981). *Genes, Radiation and Society: The Life and Work of H. J. Muller.* Ithaca and London: Cornell University Press.

Carter, G. B., and Pearson, G. S. (1998). British biological warfare and biological defence: 1925–1945. In Geissler and van Courtland Moon (1998), chapter 9.

Christopher, G. W., Cieslak, T. J., Pavlin, J. A., and Eitzen, E. M. (1997). Biological warfare: A historical perspective. *J. Amer. Med. Assoc.* 278:412–7.

Cole, L. A. (1996). The spectre of biological weapons. *Scientific American* 275(12):60–65.

Dando, M. (1994). *Biological Warfare in the 21st Century: Biotechnology and the Proliferation of Biological Weapons.* London and New York: Brassey's.

Fowle, J. R. (1987). *Applications of Biotechnology: Environmental and Policy Issues.* Boulder, CO: Westview Press.

Franz, D. R., Jahrling, P. B., Friedlander, A. M., McClain, D. J., Hoover, D. L., Bryne, W. R., Pavlin, J. A., Christopher, G. W., and Eitzen, E. M. (1997). Clinical recognition and management of patients exposed to biological warfare agents. *J. Amer. Med. Assoc.* 278:399–411.

Gavaghan, H. (1998). Arms control enters the biology lab. *Science* 281:29–30.

Geissler, E. (1984). Implications of genetic engineering for chemical and biological warfare. In *World Armaments and Disarmament, SIPRI Yearbook,* 1984, pp. 421–54. Philadelphia: Taylor & Francis.

Geissler, E. (ed.) (1986). *Biological and Toxin Weapons Today.* Oxford: Oxford University Press.

Geissler, E., and Haynes, R. H. (eds.) (1991). *Prevention of a Biological and Toxin Arms Race and the Responsibility of Scientists.* Berlin: Akademie Verlag.

Geissler, E., and van Courtland Moon, J. E. (eds.) (1998). *Biological and Toxin Weapons Research, Development and Use from the Middle Ages to 1945.* SIPRI Chemical & Biological Warfare Studies No. 17. Oxford, UK: Oxford University Press.

Gizewski, P. J. (1987). *Biological Weapons Control,* Issues Brief Number 5. Ottawa: Canadian Centre for Arms Control and Disarmament.

Gizewski, P. J. (1996). From winning weapon to destroyer of worlds: The nuclear taboo in international politics. *Internat. J.* 51:397–421.

Hamilton, D. P. (1990). Fear and loathing of nuclear power. *Science* 250:28.

Hare, F. K. (1988). *The Safety of Ontario's Nuclear Power Reactors,* vol. 1. A Royal Society of Canada Report. Toronto: Government of Ontario Ministry of Energy.

Harris, S. (1998). Japanese biological warfare programme: An overview. In Geissler and van Courtland Moon (1998), chapter 7.

Haynes, Roslynn D. (1994). *From Faust to Strangelove: Representations of the Scientist in Western Literature.* Baltimore: Johns Hopkins University Press.

Holloway, H. C., Norwood, A. E., Fullerton, C. S., Engel, C. C., and Ursano, R. J. (1997). The threat of biological weapons: Prophylaxis and mitigation of psychological and social consequences. *J. Amer. Med. Assoc.* 278:425–7.

Kadlec, R. P., Zelicoff, A. P., and Vrtis, A. M. (1997). Biological weapons control: Prospects and implications for the future. *J. Amer. Med. Assoc.* 278:351–6.

Kaplan, M. M. (1983). Another view. *Bull. Atomic Scientists* 39(9):27.

Krimsky, S. (1982). *Genetic Alchemy. The Social History of the Recombinant DNA Controversy.* Cambridge, MA: MIT Press.

Krimsky, S. (1987). Gene splicing enters the environment: The socio-historical context of the debate over deliberate release. In Fowle (1987), pp. 27–53.

Lenski, R. E. (1987). The infectious spread of engineered genes. In Fowle (1987), pp. 99–124.

Levin, S. A. (1991). An ecological perspective. In *The Genetic Revolution: Scientific Prospects and Public Perceptions,* (ed.) B. D. Davis, pp. 45–59. Baltimore: The Johns Hopkins University Press.

Lowrance, W. W. (1976). *Of Acceptable Risk: Science and the Determination of Safety.* Los Altos, CA: William Kaufmann, Inc.

Manchee, R. J., Broster, M. G., Melling, J., Henstridge, R. M., and Stagg, A. J. (1981). *Bacillus anthracis* on Gruinard Island. *Nature* 294:254–5.

Miller, H. I. (1991). Regulation. In *The Genetic Revolution: Scientific Prospects and Public Perceptions*, (ed.) B. D. Davis, pp. 196–211. Baltimore: The Johns Hopkins University Press.

NAS (1997). *Controlling Dangerous Pathogens: A Blueprint for U.S.–Russian Collaboration.* Washington, DC: National Academy of Sciences.

Nature (1998). Time to accept realities of bioweapon control. *Lead Editorial*, (26 February) 391:823.

Novick, R., and Shulman, S. (1990). New forms of biological warfare. In Wright (1990), pp. 103–19.

Pearson, G. S. (1998). How to make microbes safer. *Nature* 394:217–8.

Pearson, G. S., and Dando, M. R. (eds.) (1996). *Strengthening the Biological Weapons Convention. Key Points for the Fourth Review Conference.* Geneva: Quaker United Nations Office.

Perrow, C. (1984). *Normal Accidents.* New York: Basic Books.

Powell, D., and Leiss, W. (1997). *Mad Cows and Mother's Milk: The Perils of Poor Risk Communication.* Montreal & Kingston: McGill-Queen's University Press.

Preston, R. (1998). The bioweaponeers. *The New Yorker*, March 9 issue, pp. 52–65.

Pryce, M. H. L. (1955). Global thermonuclear explosions are impossible. *Discovery* 16(12):495–7.

Read, J. (1936). *Prelude to Chemistry: An Outline of Alchemy, Its Literature and Relationships.* London: G. Bell and Sons.

Redgrove, H. S. (1922). *Alchemy: Ancient and Modern.* London: William Rider & Son.

Rhodes, R. (1986). *The Making of the Atomic Bomb.* New York: Simon and Schuster.

Roberts, G. (1994). *The Mirror of Alchemy.* London: The British Library.

Roob, A. (1997). *The Hermetic Museum: Alchemy and Mysticism.* Köln: Taschen Verlag.

Rosenberg, B. H., and Burck, G. (1990). Verification of compliance with the biological weapons convention. In Wright (1990), chapter 14.

Rotblat, J. (1972). *Scientists in the Quest for Peace – A History of the Pugwash Conferences.* Cambridge, MA: MIT Press.

Rutherford, E. (1937). *The Newer Alchemy.* Cambridge, UK: Cambridge University Press.

Slovic, P. (1987). Perception of risk. *Science* 236:280–5.

Soddy, F. (1903). Some recent advances in radioactivity. *Contemporary Rev.* 83:708–20.

Soddy, F. (1904). *Radioactivity.* London: "The Electrician" Printing and Publishing Company.

Soddy, F. (1912). Transmutation: The vital problem of the future. Reprinted in Kauffman, G. B. (ed.) (1986). *Frederick Soddy (1877–1956).* Dordrecht, The Netherlands: D. Reidel.

Starr, C., and Whipple, C. (1980). Risks of risk decisions. *Science* 208:1114–19.

Steinburner, J. D. (1998). Biological weapons: A plague upon all houses. *Foreign Policy*, No. 109:85–96.

Teller, E. (1953). News report of Teller's remarks. *Nucleonics* 11(11):80.

Thränert, O. (ed.) (1996). *Enhancing the Biological Weapons Convention.* Bonn: Dietz.

Tucker, J. B. (1997). The biological weapons threat. *Current History* 96(609): 167–72.

Watson, J. D., and Tooze, J. (1981). *The DNA Story: A Documentary History of Gene Cloning.* San Francisco: W. H. Freeman.

Weart, S. R. (1988). *Nuclear Fear.* Cambridge, MA: Harvard University Press.

Whetham, W. C. D. (1904). Matter and electricity. *Quarterly Rev.* 199:100–126.

Wright, S., and Sinsheimer, R. L. (1983). Recombinant DNA and biological warfare. *Bull. Atomic Scientists* 39(9):20–6.

Wright, S. (ed.) (1990). *Preventing a Biological Arms Race.* Cambridge, MA: MIT Press.

EDITORS' NOTE

We are sad to report that Dr. Robert Haynes (1931-1998) died on December 22, 1998 during the final publication of his chapter in this volume. He worked continuously on his manuscript and submitted numerous revised versions of it right up to his death. He looked forward to its publication. Robert Haynes, born in London, Ontario (August 27, 1931), and educated at the University of Western Ontario, was dedicated to the promotion and popularization of science. Trained as a biophysicist, he became a yeast geneticist and pioneered research into the repair of DNA damage. He became a member of the Department of Biophysics at the University of Chicago in 1958, Associate Professor of Biophysics at the University of California at Berkeley in 1964, and Professor and Chair of the Department of Biology at York University in 1968. From the beginning his interest in the philosophical basis of science and its implications to society led him to contribute significantly to these aspects of scientific research and policy. Through his influence as president of the Genetics Society of Canada and Royal Society of Canada and as member of the National Research Council of Canada, he made substantial contributions to the progress of science in Canada. His many honors and awards indicate his deep commitment to using science for improving human conditions. He will be deeply missed.

The editorial assistance of Dr. Richard Morton and Dr. Jane Banfield with chapter proofs is greatly appreciated.

Editors

CHAPTER TWENTY-EIGHT

What Causes Cancer? A Political History of Recent Debates

ROBERT N. PROCTOR

For every Ph.D. there is an equal and opposite Ph.D.

Gibson's Law

28.1 Good News and Bad News

There is good news from the cancer front: Age-adjusted mortality from the industrial world's second largest killer is down in the United States from 135/100,000 per year in 1990 to 130,000 per year in 1995, which represents a drop of about 3%. That is the first clear-cut decline in the past 100 years and probably the first for which we have records anywhere on the globe. More Americans are still dying of cancer – more than half a million in 2000 – but the rate at which they are dying, adjusted for age, of course, is somewhat lower than a few years ago. The decline seems to stem from the steady drop in smoking rates over the past 35 years and a consequent decline in lung cancer rates, which is by far the largest cause of cancer death in most industrial nations.[1]

The good news raises disturbing questions. If this is the first true evidence of a decline, what are we to make of the glowing assurances of progress we have heard since Richard Nixon declared war on cancer in 1971? Does this mean that all those previous assertions of gains were disingenuous? (The mortality rate is still well above what it was three decades ago.) And if indeed the decline is real – which I believe it is – who or what should get the credit? Improved therapeutics? Public health education? Antitobacco activism? Science journalism? Exercise fashions? High-speed X-ray film?

The causes of cancer's decline are no doubt as complex as the causes of cancer itself. Benefits are rarely distributed equally, of course, and in this case improvements in the United States are dwarfed by the skyrocketing global cancer rates – mainly due to the worldwide increase in cigarette smoking. Cancer and AIDS are now the two most rapidly growing diseases on a planetary scale. Poverty and ignorance are important causes of the latter, but wealth and corporate marketing are important causes of the former. American tobacco corporations now sell only about 5% of their cigarettes in the United States; the overwhelming bulk of their sales goes to Africa, Asia, and Eastern Europe, where cigarette consumption is increasing by several percentage points per year (British companies have largely claimed the African market, and the Americans have claimed the Asian market).

The point I want to explore here, though, is that there is much more to cancer causation than physical agents. Cancer is caused by tobacco, asbestos, and countless petrochemical exudates, but it is also caused by regulatory myopia, corporate neglect, campaign contributions, and long-standing cultural traditions. There are proximal and distal causes of cancer – causes that lie close to molecular processes and causes that begin in corporate boardrooms. Which of these we tend to emphasize – where we put our research dollars, for example – has much to do with how we view the politics of cancer. The fashion in most national cancer campaigns has been to look for causes in the realm of biochemical mechanisms, but there are many other ways to approach the problem, many other points at which the chains that lead to cancer might be broken.

The relation between knowledge and ignorance in these matters is complex. It is not just that we do not know what causes cancer; it is not even that we do not act when we do know. The problem is partly that ignorance can be manufactured, controversy can be engineered. The problem is also that cancer is conceived as a research problem – when research can actually block the pursuit of intelligent policies.

28.2 Ignorance Is Epidemic?

The historian of science controversy is obviously in for a feast when it comes to cancer. There is controversy over what causes cancer, and there is controversy over how to identify and quantify hazards. There is controversy over what should be done to remediate carcinogens and who should pay, and there are questions about what to do in the face of ignorance, such as, What does it mean to "err on the side of caution" in making a risk estimate? Is it worse to *overstate* a hazard, risking public fears and costly but perhaps unnecessary cleanups, or to *underestimate* a hazard, risking public complacency and perhaps also public health? What does it mean to be "conservative" in estimating risks – to conserve the body of reliable knowledge or to conserve public and environmental health?

The manufacture of controversy serves a number of different social functions. The news media profit from fast-breaking reports of novel hazards – the so-called "carcinogen of the week" syndrome – but trade associations also depend upon the manufacture of doubt to neutralize public fears. The Chlorine Institute defends chlorinated hydrocarbons, the Styrene Information and Research Center defends the styrene monomer, the Lawn Institute defends pesticides, and so on. There are hundreds of such institutes – often with considerable political muscle – defending everything from asbestos to zinc. Trade association business is in fact the second largest industry in Washington, DC (second only to tourism), and the political muscle that gets flexed is considerable. Trade associations played an important role in helping trash the 1978 NCI–OSHA "Califano Report" cautioning that workplace exposures are responsible for a sizeable fraction of all U.S. cancers; trade associations have kept saccharine on the market years after it was shown to be a carcinogen (diabetics were mobilized for the campaign led by the Calorie Control Council representing diet food and drug interests).

Trade associations have become masters of the art of insinuating doubt, recognizing the force of the public relations principle that "for every Ph.D. there is an equal and opposite Ph.D." (also known as "Gibson's Law"). The manufacture of doubt has rescued many an embattled product from consumer product safety scrutiny. Maintaining controversy is often the name of the game: controversy can help stave off regulation as can the perennial call for ever "more research" to eliminate residual doubts about the magnitude of a particular hazard. This is the corporate version of the idea that knowledge is always tentative, that facts always need rechecking, that every question answered raises two in its place.

The net effect in public space is an unceasing barrage of affirmations and denials about what causes cancer and to what degree. Do electromagnetic fields cause cancer? What about cell phones or high-altitude flying or the mercury amalgam used in so-called silver dental fillings? What about the chlorine used in swimming pools or the fiberglass we use to insulate our homes? What about vasectomies or abortions or tight-fitting, wire-supported bras? Are diesel fumes more carcinogenic than gasoline exhausts? Does a high cholesterol level increase one's risk? Could it really be true, as John Gofman maintains, that a sizable fraction of breast cancers are caused by mammograms and other kinds of chest X rays? (Healy 1993, Singer and Grismaijer 1995, Gordon 1996, Gofman 1995).

Similar questions can be asked about many of the purported preventives we hear about. Does aspirin prevent cancer (Giovannucci 1995)? What about vitamin C, or broccoli, or beta-carotene,[2] or RU-486 (Brown 1990) or, for that matter, pesticides? Bruce Ames, the Friedmanite libertarian of Berkeley (and the 23rd most-cited scientist in the world), claims that pesticides can actually *prevent* cancer by lowering the cost of fresh fruits and vegetables. The more pesticides a nation sprays, he suggests, the cheaper will be its produce, leading people to eat more cancer-fighting fruits and vegetables. Ames does not deny that high-dose exposures are hazardous; he simply claims that the hazards of low-dose exposures have been exaggerated. Even more radical are the arguments put forward by advocates of the so-called *hormesis thesis*, the idea that substances carcinogenic at high doses may actually help prevent cancer at low dosages. The argument has been made that this is the case for many forms of radiation and even dioxin.[3]

These are not idle questions. Bruce Ames says we should dismantle the EPA and the FDA, sell them off to the highest bidder, and let the market regulate itself. Tens of thousands of Germans have had their dental fillings pulled, and the American Dental Association has launched an inquiry into whether the mercury abraded from such fillings may cause cancer. Steps are also being taken to monitor and limit exposure to cosmic radiation during high-altitude flight. Studies financed by European pilot associations have suggested that pilots have a tenfold increase in chromosomal abnormalities, elevated leukemia rates (purportedly higher than any other professional group), and brain cancer rates several times above normal. Studies financed by flight attendant unions have indicated a doubling of breast cancer among female flight attendants and bone cancer rates fifteen times higher than expected (Pukkala et al. 1995).

Are such studies to be taken seriously? Some people apparently think so. The Concorde already has an alarm that goes off whenever radiation levels exceed 100 millirem per hour, and British and German pilot associations now require all of their members to wear radiation monitors. The Federal Aviation Administration has published instructions for airflight crews on how to handle high radiation situations (FAA 1994). By the year 2000 *all* European airlines will be required to outfit their planes with radiation monitors and to limit radiation exposure for crew members (Hunt 1996).[4]

What should be done if it turns out that more people are killed from aviation-induced cancer than from plane crashes? Should planes be forced to increase shielding? Will window seats become less popular? Should flights be limited over the North Pole, where cosmic radiation is most intense? Will airlines be willing to alert travelers to the possibility of a hazard when sunspot activity is high or there is evidence of solar flares?

And what about electromagnetic fields (EMF)? A 1996 National Academy of Sciences study found "no conclusive evidence" of a hazard, but is this better interpreted as an absence of evidence or the absence of a hazard? A *New York Times* editorial in November 1996 proclaimed that the question had "finally been put to rest" (Park 1996), but the Academy's text itself was not so sanguine. Childhood leukemias were found to be slightly higher near power lines (the difference is statistically significant), though the explanation offered was that the children living near power lines are usually poor and subjected to other kinds of carcinogens. Stronger correlations were reported for occupational exposures (e.g., among railway linemen), but media attention has tended not to focus on this finding. The EPA has decided to move ahead with its own appraisal, and I would not be surprised if a somewhat different conclusion is reached – partly because there may be some "citizen advocates" on the investigating committee.

I do not want to leave the impression that the hazards I have mentioned are or are not real. They may or may not be, and it is often difficult to tell. The sources of possible bias are immense and fascinating, and seldom the object of well-funded scholarly scrutiny. There is also the problem that epidemiology is not a very sensitive tool. Financial constraints often dictate the size and significance of a study – and exposed populations without access to resources simply do not get studied. Even under the best of circumstances, an epidemiologic study is not going to tell you anything of significance if the risk is small, the health effects are heterogeneous, or the numbers affected are not particularly large. Under the most common conditions of epidemiologic investigation, it is virtually impossible to identify a hazard that increases one's risk of cancer by less than about 30%. Cigarette smoking and occupational asbestos typically confer increased risks of many hundreds of a percent, but most carcinogens are nowhere near this strong.

How many carcinogens are there that confer a .1- or .01% increased lifetime risk of cancer? How does one detect them, and what does one do about them? Hazards of such a magnitude are "epidemiologically invisible" owing not just to financial constraints but also to the heterogeneity of most low-dose exposures. Cigarettes are an epidemiologist's dream come true because the dosages are so uniform, the habit so widespread, and the production and consumption of the carcinogenic agent is so easy to monitor (it's legal, taxed, highly regulated, generally public, visible, etc.). A toxic waste dump presents the opposite situation. Rates and even routes of exposure are often uncertain, the toxic agents may be diverse or unknown, the population affected relatively small and heterogeneous, mixing young and old, male and female, and so forth. Health effects are difficult to quantify in such environments – especially in a court of law, where push often comes to shove.

28.3 Much, Though, Is Known

It is also important to keep in mind, of course, that *theories* of what causes cancer have changed dramatically over time. In the eighteenth century, nuns were said to suffer high rates of breast cancer because they never had sexual intercourse – this is arguably the first postulate of a cancer of occupational origins (Ramazzini 1964). In the nineteenth century, tomatoes were commonly blamed for cancer, as were broken teeth, "trauma," sexual abstinence, and diverse "cancer germs." Cancer-causing viruses have come and gone several times (they are now back in fashion in the form of oncogenes), and we should not forget poor Johannes Fibiger of Denmark – winner of the world's first Nobel prize for cancer research – whose work on nematode-induced rat cancer never could be replicated. Fibiger's may well be the only Nobel prize ever awarded for work that turned out to be completely wrong (economics excepted, of course); the will to believe is very strong when it comes to cancer.

Many of these claims seem quaint today, as does the turn-of-the-century effort to bring the U.S. government into the campaign against cancer – an effort that relied upon the theory that cancer was most likely caused by drinking water from streams containing trout (Proctor 1995). I do not want to leave the impression, though, that little is known about what causes cancer. Much is known, and reliably so. The Tobacco Institute wants us to believe that cancer is a great mystery, but in large part it is not. Most reliable experts (I will not define the term) agree that cancers are caused by active and passive smoking, asbestos, household radon, petrochemicals in our air, food and water, diverse bacteria and viruses – like *Helicobacter pylori*, linked to stomach cancer, and hepatitis B, linked to cancer of the liver – and a few other well-characterized agents. There is really no question that beta-naphthylamine, nickel carbonyl, vinyl chloride, and plutonium cause cancers of the bladder, nose, liver, and lung, and it is pretty clear that early onset of menarche increases one's risk for breast cancer (Henderson 1991). We know that Japanese living in Japan have breast cancer rates four or five times lower than U.S. rates, but we also know that Japanese who migrate to the U.S. end up with cancer rates quite similar to those of this country. We know that bladder cancers in the so-called Third World are often caused by bilharzia and that bladder cancers in industrial nations are more often caused by work with aniline dyes or smoking.

Much of what we know comes from occupational studies. Vinyl chloride cancers were first discovered in an occupational context (in the 1970s) as were chimney soot cancers (in the eighteenth century), arsenic cancers (in the 1920s and 1930s among vintners using arsenic insecticides), uranium mine lung cancers (in Schneeberg and Joachimsthal in the 1870s, though radiation was not suspected until the 1920s), asbestos lung tumors and mesotheliomas (1930s), and many others. Workplace exposures are often high and homogeneous; cause and effect are therefore clearly, and sometimes criminally, displayed for convenient study. We know about aniline dye factories, for example, in which *every single worker* developed bladder cancer: 18 of 19 died of bladder cancer in the most notorious case; the one remaining worker died in an automobile accident and was diagnosed with bladder cancer at autopsy (Doll 1977).

But we also know a lot about other kinds of causes. We know that cancer is caused by bad habits, bad working conditions, bad government, and bad luck, including the luck of your genetic draw and the culture into which you are born. Thanks to Stanton Glantz, we have a pretty good idea of how political contributions from tobacco firms influence voting behavior of legislators (Glantz and Begay 1996); it's not such a stretch to claim that cancer can be caused by elections, or interest rates, or World Bank policies that encourage the use of asbestos in Third World building projects, or trade policies that promote tobacco exports.

28.4 Troubling Trends

The World Health Organization in 1964 declared that more than half of all cancers were of environmental origins and preventable, and there is little we have learned in recent years to change that view. What is very often in question is the relative contribution of particular agents and what should be done to combat them. The differences of opinion can be truly remarkable. There is no well-defined consensus on whether stress and depression play a major role (there is much more sympathy toward this view in Europe); there is even some disagreement about whether overall rates are on the rise or falling. We really do not know much about what cancer rates were like prior to the twentieth century, and statistical methods have changed so much even in recent decades that it is sometimes difficult to distinguish real changes from "trends" due to changing recording conventions and medical practices (especially novel diagnostic techniques and increasing access to hospitals).

Even when we do have good data, it is often difficult to say why some cancers are on the rise while others are on the decline. Stomach cancer is probably the most famous example. Here we have the great success story of the twentieth century – one of the few major cancers to show a clear and steady decline. Stomach cancer was the most common cancer killer of the first half of the century, accounting for about half of all cancers in many parts of the United Sates, Japan, and Europe. I am always puzzled by how little attention has been given to its decline: in the United States, the age-adjusted mortality today is about a sixth of what it was in the 1920s. The crucial fact may be that turn-of-the-century diets were full of exotic chemical preservatives and colorants, though bacterial contamination may also have played a role. Food laws were weak and often unenforced, and cancer was of little concern in any event. Arsenic was sprayed on wine grapes and dimethyl-amino-azobenzene ("butter yellow") was mixed to color margarine. Most of these were not banned until the so-called Delaney Clause of the 1950s. (The Nazis, interestingly, were the first to ban butter yellow.)

The high-salt diet of the time may also have played a role in the genesis of stomach cancer. Salt was used to preserve many meats, as were nitrates, nitrites, and several even more powerful picklings. Preservation must often have been imperfect, and food may have contained considerably more carcinogenic molds and fungi than we today are used to. (Many molds produce cancer-causing aflatoxins.) Food spoilage may have contributed to the incidence of stomach cancer, and the popular consumption of smoked meats surely didn't help. We today eat a lot of fresh fruits and vegetables, but a century ago the unavailability of fresh produce for many months of the year may have been yet another cause of stomach cancer. (This might also help explain the prevalence of the disease in the northern United States.) If so, then the development of refrigerated transport may well have been the pivotal cause of the decline. Medical improvements in any event seem to have played only a minor role, for incidence rates have fallen almost as rapidly as mortality.

Lung cancer presents a more straightforward story. It remains the most common cause of cancer death in most parts of the Western world; in the United States, the age-adjusted incidence among men has increased by about a factor of 25 since 1900, and because survival rates are still low – roughly 10% for 5 years after diagnosis – mortality has increased nearly as fast. A small fraction of the increase is probably due to improved diagnostics: it was almost impossible to diagnose lung cancer before the development of bronchoscopy and X rays; lung cancer was difficult to distinguish from tuberculosis or silicosis, and diagnoses were generally made at autopsy. Much of what passed as "consumption" may have been lung cancer or some other chronic obstructive disease, such as silicosis.

We now have excellent evidence, though, that tobacco is chiefly responsible for the twentieth century's lung cancer epidemic. Tobacco began to be suspected in the 1920s, when skyrocketing lung cancer rates first began to be noticed. By the end of the 1930s we had the first experimental epidemiologic evidence – using case-control methods – and not just that tobacco played a role, but that tobacco was indeed "the major cause" of the upsurge in lung cancer (Müller 1939). That conclusion was confirmed in the 1940s by work at Jena's Institute for Tobacco Hazards Research (Hitler provided the funding directly from his Reichskanzlei) and again in the 1950s, with the better-known articles of Richard Doll, A. Bradford Hill, and Ernest Wynder and others.

Tobacco is surely the major culprit, but what role do industrial pollutants or household radon play? Why do 20,000 nonsmokers die of lung cancer in the United States every year? Does radon cause 7,000 to 30,000 American lung cancer deaths, as the EPA maintains? Are there strong radon–smoking synergies? And what may be causing increases in cancers that seem to have nothing to do with smoking, such as brain cancer among the elderly or breast cancer? Does the single-minded focus on tobacco distract from efforts to address other causes? Or does the focus on nontobacco agents distract from tobacco, the undisputed king of carcinogenesis?

28.5 Complex Interests

It is not hard to understand why confusion reigns in the science of carcinogenesis and the politics of what to do about it. There are powerful political and professional interests that come into play – more so, I think, than in the case of, say, heart disease or even AIDS. But not all politics belongs to industry of course. Environmental organizations have begun to exercise political clout in recent years, and with this has come a (perhaps) predictable abuse of power. The most dramatic case was the Alar scare of 1989 following a CBS *60 Minutes* report accusing the ripening agent daminozide – used on apples to ensure uniform ripening and a brighter red color – of being a major carcinogen. The public response bordered on hysteria: one worried mother called the state police to stop her child's school bus to remove an apple from her child's lunch pail; another called the EPA to ask whether it was OK to pour apple juice down the drain (she was worried it might poison the sewer). Antienvironmentalists have feasted on the story (e.g., Fumento 1993), but it does show how cancer fears can be manipulated and that environmentalists can have biases.

Radon is a case in which "industry interests" are not as simple as many of us might like to think. It used to be possible to argue that radon had no natural political defenders; there was never a "Radon Institute" with deep pockets comparable to those of the Chemical Manufacturers' Association or the American Industrial Health Council, which is probably why efforts to attribute lung cancer to radon encountered so little resistance in the early 1980s. Some of the strongest resistance came from the environmental Left, which feared that radon was being used to divert attention from manmade environmental hazards. And in fact the nuclear industry did use radon, for a time, to argue that conservation posed a far greater cancer threat than nuclear power (Proctor 1995). The argument was that by sealing ourselves inside our weather-tightened homes, we were actually *increasing* our exposure to natural radon, which posed a far greater hazard than one would ever confront with nuclear power!

The politics of radon are complicated by the fact that radon mitigation is now a multibillion dollar business, and however real the danger, the radon remediators have sometimes been less than candid about ambiguities in our estimates of the magnitude of the hazard. The same is true of asbestos remediation, another multibillion dollar industry known to have exaggerated the risks of nonfraying asbestos insulation.

28.6 Thresholds

Most people will agree that Alar, dioxin, formaldehyde, and styrene can all cause cancer when inhaled or ingested in sufficient quantities. But what about very low doses? Is the likelihood of contracting cancer a linear function of exposure? The question of "how much" turns out to be much harder than "whether." The NCI and OSHA in the late 1970s held that thousands of people are killed every year by exposure to benzene; the EPA under Reagan said the true figure was closer to three. The NCI in the late 1970s said that between 50,000 and 100,000 people are killed every year by asbestos; today we hear that the true figure is less by an order of magnitude. Who is right?[5]

Animal studies and bacterial bioassays are two of the primary ways industrial laboratories and federal agencies test for carcinogenicity; both, however, have come under attack from scientists who question whether high-dose animal tests have any relevance at all to actual human hazards. We know that different animals metabolize toxins in very different ways; that is probably why cows never get cancer of the udder and horses never get cancer of the prostate. There is evidence (to prejudice the case) that mutagenicity is not necessarily an indication of carcinogenicity and that bacterial bioassays may therefore not be very reliable (Bruce Ames is now a champion of this view). What should be done about this?

Part of the problem is that we don't know much about how the human body responds to very low doses of toxins. We don't know much about DNA repair or how easily those repair mechanisms may be damaged. We don't know much about how fair it is to extrapolate from high-dose studies to low-dose effects, and we often don't know whether there are *thresholds* of exposure (to radiation or chemical toxins, for example) below which a given carcinogen is completely safe. We know that thresholds are the rule for some carcinogens. We know that a high-salt diet, for example, is associated with certain kinds of cancer, but no one claims that the salt in a *low*-salt diet still causes a little bit of cancer. There are other examples: we know that selenium is an essential nutrient in trace amounts, but we also know that it is carcinogenic at higher doses. We know that it is better to get a little bit of sun over a few days before you expose yourself heavily all at once.

How typical, though, are thresholds? Is dioxin just like salt in this respect? You might think it would be hard to argue that lead or plutonium evidence thresholds, but there are scientists who have made such claims: one reason tetraethyl lead was not banned in the 1920s, despite a strong public health campaign opposing the antiknock compound, was that oil company scientists successfully argued that ingestion of the metal was safe so long as the rate of excretion was equal to the rate of ingestion.[6] (This theory makes about as much sense, I would suggest, as the argument that it is alright for bullets to pass through your body so long as the number going out is the same as the number going in.) Plutonium for a time was suspected of having cancer-curative capacities, which helps us understand some of the now-notorious radiation experiments of the 1950s.

The whole question of thresholds has become highly politicized – so much so that it is possible to talk about a "political morphology of dose response curves." The linearity of dose response has long been a mainstay of radiation risk assessment, but there are those who maintain that dose-response curves may be supra- or sublinear – concave or convex about the origin (see Figure 28.1). It is relatively easy to map someone's attitude toward nuclear power onto such

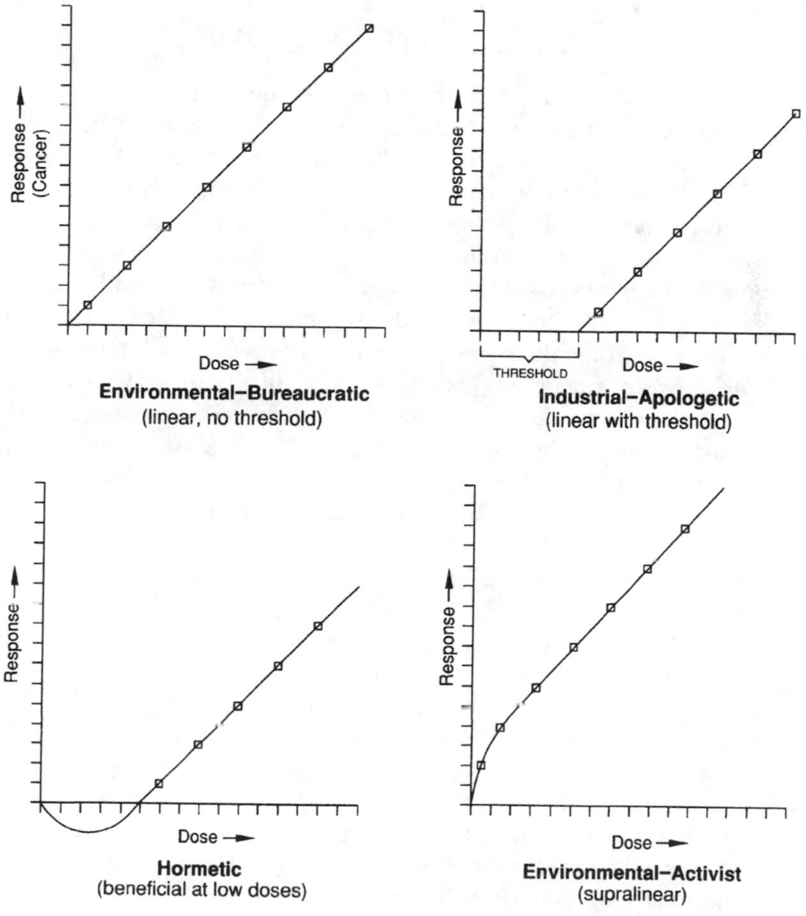

Figure 28.1. The political morphology of dose-response curves. It is very difficult to know how exposures to minute amounts of a carcinogen will affect the body. Given such uncertainty, and given the difficulty of extrapolating from high-dose animal studies, political ideologies often intervene to fill the empirical vacuum. These are some of the possibilities.

curves. The differences are also expressed in different images of the human body, suggesting different views of how well we are physically able to tolerate or resist environmental insults. The linear or supralinear model is generally followed by those who uphold a "body victimology," whereas thresholds and hormesis models are generally upheld by subscribers to what I have called a "body machismo." The former tends to emphasize the inability of the body to respond to toxic insults; the latter maintains that the body is strong and resilient in the face of toxic shocks. The models reflect deep-seated prejudices concerning the body's ability to withstand toxic insults – the strength of DNA repair capabilities, for example, or the effectiveness of epithelial shedding mechanisms in destroying early cancers, and so forth (Proctor 1995). Body history can inform or deform risk analysis.

28.7 Cultural and Political Causation

I want to end by drawing attention to the fact that discussions of cancer causation are too often focused on material agents rather than larger cultural and political processes. The problem is what might be called the "100% fallacy," the notion that the causes of cancer can be neatly totaled up to 100%, from which we derive the conclusion that attention should be focused first and foremost on the larger fractions on the list and perhaps later on the smaller. This was essentially the method followed by Richard Doll and Richard Peto in their 1981 *Causes of Cancer* commissioned by the (now defunct) Office of Technology Assessment to determine the fraction of cancers caused by occupational exposures. Doll and Peto concluded that about a third of all cancers were caused by diet, another third by smoking, roughly 10% by infections and another 7% for sexual behavior, followed by 4% for occupation, 3% each for alcohol and "geophysical factors" (presumably radon), 2% for pollution, and another percent or so each for medicines (including X rays) and food additives (Doll and Peto 1981, Proctor 1995).

The political use of Doll and Peto's book is an interesting topic in its own right; here I will simply point to the methodological narrowness of the text insofar as it failed to explore causes of cancer that are *more than physical agents.* Cancer, after all, is caused by many things, and not just physical agents like asbestos, dioxin, or tobacco. Cancer, again, is caused by bad habits, bad government, bad working conditions, and bad luck. Cancer is caused by advertisers who offer cigarettes to children, and by employers who fail to clean up their factories or force a sedentary work style. Cancer is caused by policies that encourage the export of tobacco, including tax-breaks for advertising but also strong-arm threats against governments (like those in Hong Kong or Thailand) that attempt to limit the import tobacco products.[7]

What we need, in other words, is a broader concept of causality, one that includes what might be called the "causes of causes." Smoking causes cancer, but what causes smoking? Poverty? Peer pressure? Youthful illusions of immortality? The cinematic equation of sex and smoke? The cigarette is not the beginning of cancer causation but a midpoint along that path: why not say that cancer is caused by tobacco advertising, or tobacco lawyers, or the failure to tax, or the inability to maintain taxes in the face of cross-border smuggling?[8] We should also not forget that tobacco is not just consumed but produced. Cancers must therefore be caused by agricultural policies that encourage tobacco farms, by price-supports and subsidies, and much else besides.

Of course, tobacco is not the only cause with causes; the same is true for radiation, asbestos, bilharzia, and every other carcinogen one can name. Cancers were clearly caused by the regulatory myopia that made the Atomic Energy Commission responsible for both the promotion of nuclear power and the supervision of nuclear health and safety – a situation critics characterized as "the fox guarding the henhouse." Cancers were caused by the secrecy stressed early in the atomic age, when even the word *plutonium* was classified and the science designed to study radiogenic cancer ("health physics") was phrased in such a way as to disguise the fact that radiation was the focus. Cancers were caused when Wilhelm Hueper at the NCI was barred from warning uranium miners on the Colorado Plateau that lung cancers were in store for them; the failure to publicize the danger and to ventilate the mines cost thousands of lives in the United States – and even more abroad.

I mention this need for a broader conception of causality because I think it is important to realize that the chain of causality can be broken at many different points. We have to follow this chain wherever it leads, and we have to be creative about finding the weakest links. A list can be made of how many cancers could be prevented by eliminating specific agents; this is the kernel of truth in Doll and Peto's list, since it may well be true that 30% of all cancers could be eliminated by eliminating smoking, that cancer mortality would drop by 3% without alcohol, and so forth. What we also have to recognize, however, is that many other lists could be constructed. Bad government doesn't feature in Doll and Peto's list, nor does bad science journalism, trade association obfuscation, or the failure to fund prevention-oriented research and activism. Some fraction of tobacco cancers must be caused by governmental policies because governments have always played a role in either the suppression or encouragement of the tobacco trade. After World War II, for example, the United States shipped more than ninety thousand tons of tobacco to Germany free of charge as part of the Marshall Plan to rebuild Europe (Proctor 1996); tobacco exports have been supported under the Food for Peace program and by diverse tax advantages. The United States today is collaborating in the largest increase in cancer the world has ever seen – I mean the spread of tobacco into Asia and Africa, where smoking is increasing by leaps and bounds. The United States exports three times more tobacco than any other nation on Earth, though British corporations also have their finger in the pie (the United States dominates the Asian market; the British control Africa). The World Health Organization estimates that tobacco causes two million deaths per year in developing nations and expects this figure to grow to seven million per year by the year 2025 (WHO 1997). Even here at home, tobacco remains less regulated than meat or cars or many other consumer products. You cannot manufacture a toy that might kill one in a million children, but no one will stop you from manufacturing tobacco products that kill one in ten when used according to the manufacturer's instructions.

Recall for a moment why we talk about causes in the first place. We talk about causes because we want to know how the world works but also because we want to know how the world might be different. We generally don't spend much time talking about causes that we cannot do anything about: life, after all, is (trivially) a cause of *all* cancers; there would be no cancer without life. The same is true of oxygen, gravity, mitosis, sexual intercourse, and much else as well; there would be no cancer without these things. This is uninteresting because we want to know about causes we might be able to do something about. The *significance* of a given causal agent is therefore not necessarily the same as the *percentage* of cancers caused by that agent. Thirty-five percent of all cancers may be caused by diet, but if only half of these are avoidable, then the significance of that particular agent is diminished. This is important to keep in mind when we hear that half of all cancers are caused by p53 dysfunction or that oncogene expression is involved in nearly every cancer.

The great hope, of course, is that a better understanding of genetic mutation, angiogenesis inhibition, or cell-surface properties will allow us to devise better treatments. That may or may not be the case – I hope it is. Cancer is unfortunately not like smallpox, but perhaps we will see some effective therapies evolve from knowledge of genetic mechanisms. What is unfortunate, however, is how little attention is given to policies and practices we already know will work. We already know that half of all cancers are preventable, and we already have the knowledge required to avert those cancers. Cancer campaigns have neglected prevention for several reasons, having to do with the prestige of basic research and the reductionist conception of causes in terms of mechanisms but also from the political fact that prevention generally requires stepping on a lot of powerful political toes. Until this is realized, and acted upon, I fear we will be in a position something akin to that of a Cassandra – blessed with ever finer knowledge of our fate but with little ability to change it.

Notes

1. The decline is most pronounced in men; female cancer mortality has been holding pretty much steady, the consequence of a decade or so lag between the peak of male smoking (1964) and that of women (early 1970s).
2. Marilyn Menkes of Johns Hopkins School of Medicine in a 1986 study found that people with high beta-carotene in their blood were 2.2 times less likely to contract lung cancer than people with low levels; she estimated that eating one carrot a day could prevent as many as 15,000 to 20,000 lung cancer deaths per year (Menkes et al. 1986). Subsequent studies found that beta-carotene actually increased the risk of lung cancer among smokers (De Luca 1996).
3. The conservative publishers of the journal *BELLE* (Biological Effects of Low Level Exposures) are dedicated to demonstrating hormesis; the organization includes scientists from the U.S. Departments of Energy and Defense plus representatives

from the EPA, the tobacco industry's Center for Indoor Air Research (dedicated to assuaging fears of passive smoke hazards), and various universities and private corporations.

4. A standard health physics text dismisses the hazard, reassuring readers that "even in the case of a giant flare there would be ample warning and ample time for evasive action" (Faw and Shultis 1993).
5. In World War II, when German and British governments estimated the number of planes shot down in a bombing raid, some said that a close approximation to the truth could be had by adding the opposing estimates and dividing by two. Today, epidemiologists use a similar technique to calculate cancer hazards. In what is usually called "meta-analysis," you essentially combine all the published studies you can get your hands on and then analyze these as if they were one single study. The wartime averaging process worked pretty well because it was generally safe to assume that both German and British estimates were inflated by roughly equal amounts, and both had equal input into the estimate. Meta-analysis presents a different story because the studies contributing data are not necessarily equal and opposite in their biases. If two-thirds of all EMF hazard studies are financed by a power utility consortium, just to take a purely hypothetical case, it might well turn out that the meta-analysis will simply magnify a systematic bias in the overall data set.
6. Poland was one of the few countries to ban leaded gas in the 1920s. U.S. production was stopped for 9 months in the mid-1920s – New York managed a 3-year ban – but production resumed shortly thereafter. A nationwide U.S. ban was not enacted until the 1970s (Rosner and Markowitz 1987).
7. Senator Robert Dole in the mid-1980s helped crush Hong Kong's ambitious campaign against tobacco imports; similar campaigns elsewhere in Asia have been met with fierce U.S. resistance (MacKay 1996).
8. The reference is to Canada's tobacco tax campaign that was ruined by illegal tobacco imports smuggled from the United States into Canada. The $3.72 per pack tax cut Canadian cigarette consumption by nearly 50 percent; many of these gains were lost when the government was forced to lower the tax in February 1994 (Cunningham 1996).

REFERENCES

Brown, P. (1990). Abortion pill may help fight breast cancer. *New Scientist* 128 (November 3):21.

Cunningham, R. (1996) *Smoke and Mirrors: The Canadian Tobacco War.* Ottawa: International Development Research Centre.

De Luca, L. M. (1996). Beta-carotene increases lung cancer incidence in cigarette smokers. *Nutrition Reviews* 54:178–80.

Doll, R. (1977). Strategy for detection of cancer hazards to man. *Nature* 265:589–96.

Doll, R., and Peto, R. (1981). *Causes of Cancer.* New York: Oxford University Press.

FAA. (1994). *Crewmember Training on In-flight Radiation Exposure.* Washington, DC: FAA.

Faw, R. E., and Shultis, J. K. (1993). *Radiological Assessment: Sources and Exposures.* Englewood Cliffs, NJ: Prentice–Hall.

Fumento, M. (1993). *Science Under Siege: Balancing Technology and the Environment.* New York: William Morrow.

Giovannucci, E. (1995). Aspirin and the risk of colorectal cancer in women. *New England Journal of Medicine* 333:609–14.

Glantz, S. A., and Begay, M. E. (1996). Tobacco industry campaign contributions are affecting tobacco control policymaking in California. *JAMA* 272:1176–82.

Gofman, J. W. (1995). *Preventing Breast Cancer.* San Francisco: Committee for Nuclear Responsibility, Inc.

Gordon, D. B. (1996). Complex cancer-cholesterol link (letter). *Science News* 149:307.

Healy, B. (1993). Does vasectomy cause prostate cancer? *JAMA* 269:2620.

Henderson, B. E. (1991). Toward the primary prevention of cancer. *Science* 254:1131–8.

Hunt, L. (1996). Planes, trains and automobiles: Cancer fear for airline crews. *The Independent,* August 27, 1996.

MacKay, J. L. (1996). An advocate in Asia. In *The Doctor-Activist: Physicians Fighting for Social Change,* (ed.) E. L. Bassuk, pp. 29–56. New York: Plenum.

Menkes, M. S., et al. (1986). Serum beta-carotene, vitamins A and E, selenium, and the risk of lung cancer. *New England Journal of Medicine* 315:1250–4.

Müller, F. H. (1939). Tabakmissbrauch and Lungencarcinom. *Zeitschrift für Krebsforschung* 49:57–85.

Park, R. L. (1996). Power line paranoia. *New York Times,* November 13, 1996.

Proctor, R. N. (1995). *Cancer Wars: How Politics Shapes What We Know and Don't Know About Cancer.* New York: Basic Books.

Proctor, R. N. (1996). The anti-tobacco campaign of the Nazis: A little known aspect of public health in Germany, 1933–1945. *British Medical Journal* 313:1450–3.

Pukkala, E., et al. (1995). Incidence of cancer among Finnish airline attendants, 1967–1992. *British Medical Journal* 311:649–52.

Ramazzini, B. (1964). *Diseases of Workers* (1713) trans. W. C. Wright. New York: Hafner Pub. Co.

Rosner, D., and Markowitz, G. (1987). "A Gift of God"?: The public health controversy over leaded gasoline during the 1920s. In *Dying for Work: Workers' Safety and Health in Twentieth-Century America,* (ed.) D. Rosner and G. Markowitz, pp. 121–39. Bloomington, IN: Indiana University Press.

Singer, S. R., and Grismaijer, S. (1995). *Dressed to Kill: The Link Between Breast Cancer and Bras.* Garden City Park, NY: Avery Publishing.

World Health Organization (WHO) (1997). *Tobacco or Health: A Global Status Report.* Geneva: WHO.

Publications of R. C. Lewontin

1952. An elementary text on evolution. Review of *A Textbook of Evolution* by E. O. Dodson. *Evolution* **6(2)**:247–248.

1953. The effect of compensation on populations subject to natural selection. *Am. Nat.* **87**:375–381.

1954. Review of *Problems of Life. Am. J. Sci.* **252**:123–124.

1954. Review of *Biology and Language. Am. J. Sci.* **252**:124–126.

1954. Familial occurrence of migraine headache: A study of heredity. *A.M.A. Arch. Neurol. Psych.* **72**:325–334 (with H. Goodell and H. G. Wolff).

1955. The effects of population density and composition on viability in *Drosophila melanogaster. Evolution* **9**:27–41.

1956. Estimation of the number of different classes in a population. *Biometrics* **12**:211–223 (with T. Prout).

1956. Studies on homeostasis and heterozygosity. I. General considerations. Abdominal bristle number in second chromosome homozygotes of *Drosophila melanogaster. Am. Nat.* **90**:237–255.

1956. A reply to Professor Dempster's comments on homeostasis (Letters to the Editors). *Am. Nat.* **90**:386–388.

1957. The adaptations of populations to varying environments. *Cold Spring Harbor Symp. Quant. Biol.* **22**:395–408.

1958. A general method for investigating the equilibrium of gene frequency in a population. *Genetics* **43**:419–434.

1948. Studies of heterozygosity and homeostasis. II. Loss of heterosis in a constant environment. *Evolution* **12**:494–503.

1959. On the anomalous response of *Drosophila pseudoobscura* to light. *Am. Nat.* **93**:321–328.

1959. The goodness-of-fit test for detecting natural selection. *Evolution* **13**:561–564 (with C. C. Cockerham).

1960. *Quantitative Zoology*, 2nd ed. New York: Harcourt Brace (with G. G. Simpson and A. Roe).

1960. Interaction between inversion polymorphism of two chromosome pairs in the grasshopper *Moraba scurra. Evolution* **14**:116–129.

1960. The evolutionary dynamics of a polymorphism in the house mouse. *Genetics* **45**:705–722 (with L. C. Dunn).

1960. The evolutionary dynamics of complex polymorphisms. *Evolution* **14**:458–472 (with K. Kojima).

1960. Review of *Introduction to Quantitative Genetics* by D. S. Falconer. *Am. Sci.* **48**:274A–276A.

1961. Evolution and the theory of games. *J. Theor. Biol.* **1**:382–403.

1961. Review of *Biochemical Genetics. Am. Sci.* **49**:190A.

1962. Review of *Introduction to the Mathematical Theory of Genetic Linkage* by N. T. J. Bailey. *Am. Sci.* **50**:320A–322A.

1962. Review of *An Outline of Chemical Genetics* by B. Strauss. *Hum. Biol.* 235–236.

1962. Interdeme selection controlling a polymorphism in the house mouse. *Am. Nat.* **96**:65–78.

1963. Interaction of genotypes determining viability in *Drosophila busckii. Proc. Natl. Acad. Sci.* **49**:270–278 (with Y. Matsuo).

1963. Relative fitness of geographic races of *Drosophila serrata. Evolution* **17**:72–83 (with L. C. Birch, T. Dobzhansky, and P. O. Elliott).

1963. Models, mathematics, and metaphors. *Synthese* **15**:222–244.

1963. Cytogenetics of the grasshopper *Moraba scurra.* VII. Geographic variation of adaptive properties of inversions. *Evolution* **17**:147–162 (with M. J. D. White and L. E. Andrew).

1964. The interaction of selection and linkage. I. General considerations: Heterotic models. *Genetics* **49**:49–67.

1964. A molecular messiah: The new gospel of genetics? Essay review of *The Mechanics of Inheritance* by F. Stahl. *Science* **145**:525.

1964. The role of linkage in natural selection. Proceedings of the XI International Congress of Genetics, The Hague, September 1964. *Genet. Today* 517–525.

1964. The interaction of selection and linkage. II. Optimum models. *Genetics* **50**:757–782.

1964. Review of *Elizabethan Acting. The Seventeenth Century News.*

1964. The capacity for increase in chromosomally polymorphic and monomorphic populations of *Drosophila pseudoobscura. Heredity* **19**:597–614 (with T. Dobzhansky and O. Pavlovsky).

1965. Selection in and of populations. In *Ideas in Modern Biology*, (Proceedings of the XVI International Congress of Zoology, vol. 6) J. A. Moore, ed., pp. 299–311. Garden City, NY: National History Press.

1965. The robustness of homogeneity tests in 2 X N tables. *Biometrics* **21**:19–33 (with J. Felsenstein).

1965. Selection for colonizing ability. In *The Genetics of Colonizing Species*, Herbert Baker, ed., pp. 77–94 New York: Academic Press.

1965. Review of *The Effects of Inbreeding on Japanese Children. Science* **150**:332–333.

1965. Review of *Stochastic Models in Medicine and Biology. Am. Sci.* **53**:254A–255A.

1966. Adaptation and natural selection (essay review). *Science* **152**:338–339.

1966. Is nature probable or capricious? *Bio Science* **16**:25–27.

1966. Differences in bristle-making abilities in scute and wild-type *Drosophila melanogaster. Genet. Res.* **7**:295–301 (with S. S. Y. Young).

1966. On the measurement of relative variability. *Syst. Zool.* **15**:141–142.

1966. Stable equilibria under optimizing selection. *Proc. Natl. Acad. Sci.* **56**:1345–1348 (with M. Singh).

1966. A molecular approach to the study of genic heterozygosity in natural populations. I. The number of alleles at different loci in *Drosophila pseudoobscura. Genetics* **54**:577–594 (with J. L. Hubby).

1966. A molecular approach to the study of genic heterozygosity in natural populations. II. Amount of variation and degree of heterozygosity in the natural populations of *Drosophila pseudoobscura. Genetics* **54**:595–609 (with J. L. Hubby).

1966. Hybridization as a source of variation for adaptation to new environments. *Evolution* **20**:315–336 (with L. C. Birch).

1966. Review of *The Theory of Inbreeding. Science* **150**:1800–1801.

1967. The genetics of complex systems. *Proceedings of the 5th Berkeley Symposium on Mathematical Statistics and Probability*, Vol. IV. Berkeley, CA: U. of Calif. Press, pp. 439–455.

1967. The interaction of selection and linkage. III. Synergistic effect of blocks of genes. *Der Zuchter* **37**:93–98 (with P. Hull).

1967. The principle of historicity in evolution. In P. S. Moorhead and M. M. Kaplan, *Mathematical Challenges of the Neo-Darwinian Theory of Evolution.* Wistar Symposium Monograph No. 5, pp. 81–94.

1967. An estimate of average heterozygosity in man. *Am. J. Human. Genet.* **19**:681–685.

1967. Population genetics. *Ann. Rev. Genet.* **1**:37–70.

1968. A molecular approach to the study of genic heterozygosity in natural populations. III. Direct evidence of coadaptation in gene arrangements of Drosophila. *Proc. Natl. Acad. Sci.* **59**:398–405 (with S. Prakash).

1968. A note on evolution and changes in the quantity of genetic information. In *Towards a Theoretical Biology*, I, pp. 109–110. Aldin (with C. H. Waddington).

1968. Essay review of *Phage and the Origins of Molecular Biology*, J. Cairns et al., eds., (Cold Spring Harbor, New York: Cold Spring Harbor Laboratory of Quantative Biology, 1966, xii + 340). *J. Hist. Biol.* **1**(1):155–161.

1968. The concept of evolution. In "Evolution," *International Encyclopedia of the Social Sciences*, pp. 202–210. Macmillan and The Free Press.

1968. The effect of differential viability on the population dynamics of *t* alleles in the house mouse. *Evolution* **22**:262–273.

1968. Selective mating, assortative mating, and inbreeding: Definitions and implications. *Eugen. Quar.* **15**:141–143 (with D. Kirk and J. Crow).

1969. The bases of conflict in biological explanation. *J. Hist. Biol.* **2**(1):35–45.

1969. On population growth in a randomly varying environment. *Proc. Natl. Acad. Sci.* **62**(4):1056–1060 (with D. Cohen).

1969. A molecular approach to the study of genic heterozygosity in natural populations. IV. Patterns of genic variation in central, marginal and isolated populations of *Drosophila pseudoobscura. Genetics* **61**:841–858 (with S. Prakash and J. L. Hubby).

1969. The meaning of stability. (Reprinted from *Diversity and Stability in Ecological Systems*) *Brookhave Symp. Biol.* **22**:13–24.

1970. On the irrelevance of genes. In *Towards a Theoretical Biology*, 3: Drafts. Edinburgh, UK: Edinburgh U. Press, pp. 63–72.

1970. Race and intelligence. *Bull. Atom. Sci.* **26**(Mar):2–8.

1970. Further remarks on race and the genetics of intelligence. *Bull. Atom. Sci.* **26**(May):23–25.

1970. Genetic variation in the horseshoe crab (*Limulus polyphemus*), a phylogenetic "relic." *Evolution* **24**:402–414 (with R. K. Selander, S. Y. Yang, and W. E. Johnson).

1970. Is the gene the unit of selection? *Genetics* **65**:707–734 (with I. Franklin).

1970. The units of selection. *Ann. Rev. Ecol. Syst.* **1**:1–18.

1971. Genes in populations – end of the beginning. Review of *An Introduction to Population Genetics Theory* by J. F. Crow and M. Kimura. *Quart. Rev. Biol.* **46**:66–67.

1971. Evolutionary significance of linkage and epistasis. In *Biomathematics Vol. I: Mathematical Topics in Population Genetics*, pp. 367–388 (with K. Kojima).
1971. The effect of genetic linkage on the mean fitness of a population. *Proc. Natl. Acad. Sci. USA* **68**:984–986.
1971. The Yahoos ride again. *Evolution* **25**:442.
1971. Science and ethics. *BioScience* **21**:799.
1971. A molecular approach to the study of genic heterozygosity in natural populations. V. Further direct evidence of coadaptation in inversions of *Drosophila. Genetics* **69**:405–408 (with S. Prakash).
1972. Testing the theory of natural selection. *Nature* **236**:181–182.
1972. The apportionment of human diversity. *Evol. Biol.* **6**:381–398.
1972. Comparative evolution at the levels of molecules, organisms and populations. *Proceedings of the VI Berkeley Symposium on Mathematical Statistics and Probability* **5**:23–42 (with G. L. Stebbins).
1973. Distribution of gene frequency as a test of the theory of the selective neutrality of polymorphism. *Genetics* **74**:175–195 (with J. Krakauer).
1974. Population genetics. *Ann. Rev. Gen.* **7**:1–17.
1974. Molecular heterosis for heat-sensitive enzyme alleles. *Proc. Natl. Acad. Sci. USA* **71**:1808–1810 (with R. S. Singh and J. L. Hubby).
1974. *The Genetic Basis of Evolutionary Change.* New York: Columbia U. Press.
1974. Annotation: The analysis of variance and the analysis of causes. *Am. J. Hum. Gen.* **26**:400–411.
1974. Darwin and Mendel – the materialist revolution. In *The Heritage of Copernicus: Theories "More Pleasing to the Mind,"* J. Neyman, ed., pp. 166–183. Cambridge, MA: MIT Press.
1975. The problem of genetic diversity. *Harvey Lec. Ser.* **70**:1–20.
1975. Selection in complex genetic systems. III. An effect of allele multiplicity with two loci. *Genetics* **79**:333–347 (with M. W. Feldman, I. R. Franklin, and F. B. Christiansen).
1975. Review of *The Modern Concept of Nature* by H. J. Muller. *Soc. Biol.* **22**:96–98.
1975. The heritability hang-up. *Science* **190**:1163–1168 (with M. W. Feldman).
1975. Genetic aspects of intelligence. *Ann. Rev. Genet.* **9**:382–405.
1976. Review of *Race Differences in Intelligence* by J. C. Loehlin, G. Lindzey, and J. N. Spuhler. *Am. J. Hum. Genet.* **28**:92–97.
1976. Adattamento genetico. *Enciclopedia del Novecento* **1**:61–68.
1976. Genetic heterogeneity within electrophoretic "alleles" of xanthine dehydrogenase in *Drosophila pseudoobscura. Genetics* **84**:609–629 (with R. S. Singh and A. A. Felton).
1976. The problem of Lysenkoism. In *The Radicalisation of Science* by H. Rose and S. Rose, eds., pp. 32–64. London: MacMillan (with R. Levins).
1976. The fallacy of biological determinism. *The Sciences* **16**:6–10.
1976. Sociobiology – a caricature of Darwinism. *Phil. Sci. Assoc.* **2**:22–31.
1977. The relevance of molecular biology to plant and animal breeding. *Proceedings of the International Conference on Quantitative Genetics*, Ames, IA: Iowa State U. Press, pp. 55–62.
1977. Population genetics. *Proceedings of the 5th International Congress of Human Genetics, Excerpta Medica International Congress* Series No. 441, pp. 13–18.
1977. Adattamento. *Enciclopedia Einaudi* **1**:198–214.
1977. Biological determinism as a social weapon. In *Biology as a Social Weapon*, pp. 6–18. Minneapolis, MN: Burgess.

1977. Caricature of Darwinism. Book review of *The Selfish Gene* by R. Dawkins. *Nature* **266**:283–284.

1978. Heterosis as an explanation for large amounts of genic polymorphism. *Genetics* **88**:149–170 (with L. R. Ginzburg and S. D. Tuljapurkar).

1978. Fitness, survival and optimality. In *Analysis of Ecological Systems* by D. J. Horn et al., eds., Columbia, OH: Ohio State U. Press.

1978. The extent of genetic variation at a highly polymorphic esterase locus in *Drosophila pseudoobscura. Proc. Natl. Acad. Sci. USA* **75**:5090 (with J. Coyne and A. A. Felton).

1978. Evoluzione. *Enciclopedia Einaudi* **5**:995–1051.

1978. Adaptation. *Sci. Am.* **239**(3):212–228.

1979. Sociobiology as an adaptationist program. *Behav. Sci.* **24**:5–14.

1979. Theodosius Dobzhansky. (Biographical article) In *International Encyclopedia of the Social Sciences.* New York: The Free Press, MacMillan.

1979. Single- and multiple-locus measures of genetic distance between groups. *Am. Nat.* **112**(988):1138–1139.

1979. The genetics of electrophoretic variation. *Genetics* **92**:353–361 (with J. A. Coyne and W. F. Eanes).

1979. The spandrels of San Marco and the Panglossian paradigm: a critique of the adaptationist programme. *Proc. Royal Soc. Lond. B.* **205**:581–598 (with S. J. Gould).

1979. The sensitivity of gel electrophoresis as a detector of genetic variation. *Genetics* **93**:1019–1037 (with J. A. M. Ramshaw and J. A. Coyne).

1979. *Mutazione/selezione. Enciclopedia Einaudi* **9**:647–695.

1980. Dialectics and reductionism in ecology. *Synthese* **43**:47–78 (with R. Levins).

1980. Sociobiology: Another biological determinism. *Int. J. Health Sci.* **10**(3):347–363.

1980. Economics down on the farm. Review of *Farm and Food Policy: Issues of the 1980's* by D. Paarlberg. *Nature* **287**:661–662.

1980. The political economy of food and agriculture (World Agricultural Research Project). *Int. J. Health Sci.* **10**:161–170.

1981. *An Introduction to Genetic Analysis,* 2nd ed., San Francisco: W. H. Freeman (with D. T. Suzuki and A. J. F. Griffiths).

1981. Evolution/Creation debate: A time for truth. *BioScience* **31**:559.

1981. Sleight of hand. Review of *Genes, Mind and Culture* by C. J. Lumsden and E. O. Wilson. *The Sciences* **21**:23–26.

1981. Review of *The Mismeasure of Man* by S. J. Gould. *NY Rev. Books.*

1981. Gene flow and the geographical distribution of a molecular polymorphism in *Drosophila pseudoobscura. Genetics* **98**:157–178 (with J. S. Jones, S. H. Bryant, J. A. Moore, and T. Prout).

1981. Theoretical population genetics in the evolutionary synthesis. In *The Evolutionary Synthesis,* E. Mayr and W. Province, eds. Cambridge, MA: Harvard U. Press.

1981. *Dobzhansky's Genetics of Natural Populations I-XLIII.* New York: Columbia U. Press (with J. A. Moore, W. B. Provine, and B. Wallace, ed.).

1981. L'Evolution. *La Pensee* **223**:16–24.

1982. Artifact, cause and genic selection. *Phil. Sci.* **49**:157–180 (with E. Sober).

1982. A study of reaction norms in natural populations of *Drosophila pseudoobscura. Evolution* **36**:934–938 (with A. Gupta).

1982. Review of *Matter, Life and Generation* by S. A. Roe. *The Sciences,* December.

1982. Prospectives, perspectives, and retrospectives. Review of *Perspectives on Evolution* by R. Milkman. *Paleobiology* **8**(3):309–313.

1982. Organism and environment. In *Learning, Development and Culture: Essays in Evolutionary Epistemology*, H. Plotkin, ed., Chichester, England, UK: Wiley.

1982. Review of *Evolution and the Theory of Games* by J. M. Smith, Cambridge U. Press. *Nature* **300**:113–114.

1982. *Human Diversity*. Sci. Am., Redding, CT: W. H. Freeman.

1982. Elementary errors about evolution. Review of *Intentional systems in cognitive ethology:* The '*Panglossian paradigm' defined* by D. Dennett. *Behav. Brain Sci.* **6**:367.

1983. The corpse in the elevator. Reviews of *Against Biological Determinism and Toward a Liberating Biology* by The Dialectics of Biology Group. *NY Rev. Books.*

1983. Biological determinism. In *The Tanner Lectures on Human Values* Vol. IV, p. 147–183. Salt Lake City, UT: U. Utah Press.

1983. Science as a social weapon. In *Occasional Papers I*, pp. 13–29. Amherst, MA: Inst. for Advanced Study in the Humanities, U. Massachusetts.

1983. Review of books by J. Miller & B. Van Loon, J. M. Smith, B. G. Gale, J. Gribben & J. Cherfas, N. Eldrige, I. Tattersall, D. Futuyma, P. Kitcher, and M. Ruse. *NY Rev. Books.*

1983. The organism as the subject and object of evolution. *Scientia* **188**:65–82.

1983. Gene, organism and environment. In *Evolution from Molecules to Men*, D. S. Bendall, ed., pp. 273–285. Cambridge: Cambridge U. Press.

1983. Introduction. In *Scientists Confront Creationism*, L. R. Godfrey and J. R. Coles, eds., W. W. Norton.

1983. Discussion: Reply to Rosenberg on genic selectionism. *Phil. Sci.* **50**:648–650 (with E. Sober).

1984. Detecting population differences in quantitative characters as opposed to gene frequencies. *Am. Nat.* **123**(1):115–124.

1984. *Not In Our Genes: Biology, Ideology and Human Nature.* New York: Pantheon (with S. Rose and L. Kamin).

1984. Review of *Women in Science: Portraits of a World in Transition* by V. Gornick. New York: *NY Rev. Books.*

1984. Le determinisme biologique comme arme social. In *Les Enjeux du Progress*, A. Cambrosio and R. Duchesne, eds., pp. 233–251. Québec, PQ, Canada: Presses de l'U. du Québec.

1985. Nearly identical allelic distributions of xanthine dehydrogenase in two populations of *Drosophila pseudoobscura. Mol. Biol. Evol.* **2**:206–216 (with T. P. Keith, L. D. Brooks, J. C. Martinez-Cruzado, and D. L. Rigbny).

1985. Population genetics. *Ann. Rev. Genet.* **19**:81–102.

1985. Population genetics. In *Evolution, Essays in Honor of John Maynard Smith*, J. Greenwood and M. Slatkins, eds., pp. 3–18. Cambridge, England, UK: Cambridge U. Press.

1985. *The Dialectical Biologist.* Cambridge, MA: Harvard University Press (with R. Levins).

1985. Review of books by R. W. Clark, C. Darwin, V. Orel, L. J. Jordanova, and J. P. Changeeuex. *NY Rev. Books* **32**(15).

1985. In *Current Contents.* This week's citation classic. October (with J. L. Hubby).

1986. Technology, research, and the penetration of capital: The case of agriculture. *Monthly Rev.* **38**:21–34 (with J. P. Berlan).

1986. The political economy of hybrid corn. *Monthly Rev.* **38**:35–47 (with J. P. Berlan).

1986. Review of *In the Name of Eugenics* by D. J. Kevles. *Rev. Symp. Isis* **77**(2):314–317.

1986. *Education and Class.* Oxford England, UK: Oxford U. Press (with M. Schiff).

1986. A comment on the comments of Rogers and Felsenstein. *Am. Nat.* **127**(5):733–734.

1986. An Introduction to *Genetic Analysis*, 3rd. ed., by D. T. Suzuki, A. J. F. Griffiths, J. H. Miller and R. C. Lewontin. New York: W. H. Freeman and Co.

1986. Breeder's rights and patenting of life forms. *Nature* **322**:785–788 (with J. P. Berlan).

1986. How important is genetics for an understanding of evolution? In *Science as a Way of Knowing*, Vol III, p. 811–820. American Society of Zoologists.

1987. The shape of optimality. In *The Latest on the Best: Essays on Evolution and Optimality*, J. Dupre, ed., pp. 151–159. Cambridge, MA: MIT Press.

1987. Sequence of the structural gene for xanthine dehydrogenase (*rosy* locus) in *Drosophila melanogaster. Genetics* **116**:64–73 (with T. P. Keith, M. A. Riley, M. Kreitman, D. Curtis, and G. Chambers).

1987. Polymorphism and heterosis: Old wine in new bottles and *vice versa. J. Hist. Biol.* **20**:337–349.

1988. A general asymptotic property of two-locus selection models. *Theor. Popul. Biol.* **34**(2):177–193 (with M. W. Feldman).

1988. On measures of gametic disequilibrium. *Genetics* **120**:849–852.

1988. Aspects of wholes and parts in population biology. In *Evolution of Social Behavior and Integrative Levels*, G. Greenberg and E. Tobach, eds., pp. 31–52. NJ: Erlbaum (with R. Levins).

1988. La paradoja de la adaptacion biologica. In *Polemicas contemporáneas en Evolución*, A. O. Franco, ed. pp. 57–65. SA, Mexico: AGT Editor.

1989. Inferring the number of evolutionary events from DNA coding sequence differences. *Mol. Biol. Evol.* **6**(1):15–32.

1989. Review of *Controlling Life: Jacques Loeb and the Engineering Ideal in Biology* by P. Pauly, and *Topobiology: An Introduction to Molecular Embryology* by G. Edelman. *NY Rev. Books.*

1989. DNA sequence polymorphism. In *Essays in Honor of Alan Robertson*, W. G. Hill, and T. F. C. Mackey eds., pp. 33–37. Edinburgh, Scotland, UK: U. Edinburgh Press.

1989. On the characterization of density and resource availability, *Am. Nat.* **134**:513–524 (with R. Levins).

1989. Distinguishing the forces controlling genetic variation at the *Xdh* locus in *Drosophila pseudoobscura*, *Genetics* **123**:359–369 (with M. A. Riley and M. E. Hallas).

1989. Review of *Evolutionary Genetics* by J. Maynard Smith. *Nature* **339**:107 (May 11).

1990. The political economy of agricultural research: The case of hybrid corn. In *Agroecology*, C. R. Carroll, J. H. Vandermeer, and P. Rosset, eds., pp. 613–626. New York: McGraw-Hill.

1990. Review of *Wonderful Life: The Burgess Shale and the Nature of History* by S. J. Gould, Norton. *NY Rev. Books.*

1990. The evolution of cognition. In *Thinking, an Invitation to Cognitive Sciences*, Vol 3., D. N. Osherson and E. E. Smith, eds., pp. 229–240. Cambridge, MA: MIT Press.

1991. *How much did the brain have to change for speech?* (Pinker and Bloom Commentary) San Diego, CA: Academic Press.

1991. Review of *The Structure and Confirmation of Evolution Theory* by E. A. Lloyd, Greenwood Press, New York, 1988. *Biol. and Phil.* **6**:461–466.

1991. Foreword. In *Organism and the Origins of Self*, A. I. Tauber, ed., pp. xiii–xix. Boston: Kluwer Academic.

1991. Facts and the factitious in natural science. *Crit. Inq.* **18**:140–153.

1991. *Biology as Ideology: The Doctrine of DNA*. Ontario, Canada: Stoddart.

1991. Population genetics. *Encyl. Hum. Biol.* **6**:107–115.

1991. Perspectives: 25 years ago in Genetics: Electrophoresis in the development of evolutionary genetics: Milestone or Millstone? *Genetics* **128**:657–662.

1991. Population genetic problems in forensic DNA typing. *Science* **254**:1745–1750 (with D. L. Hartl).

1992. Genotype and phenotype. In *Keywords in Evolutionary Biology*, E. F. Keller and E. A. Lloyd, eds. pp. 137–144. Cambridge, MA: Harvard U. Press.

1992. Biology. In *Academic Press Dictionary of Science and Technology*, C. Morris, ed., p. 261. San Diego, CA: Academic Press.

1992. The dream of the human genome. *NY Rev. Books* **39**(10):31–40.

1992. Polemiche sul genoma umano, I. *La Riv. dei Libri*, Oct. 7–10.

1992. Polemiche sul genoma umano, II. *La Riv. dei Libri*, Nov. 6–9.

1992. The dimensions of selection. *J. Phil. Sci.* **60**:373–395 (with P. Godfrey-Smith).

1992. Open peer commentary: Gene talk on target. *Soc. Epist.* **6**(2):179–181.

1992. Letter to the Editor: Which population? *Am. J. Hum. Genet.* **52**:205.

1992. Inside and Outside: Genetics, environment and organism. *Heinz Werner Lecture Series*, Vol. XX. Worcester, MA: Clark U. Press.

1992. DNA data banking and the public interest. In *DNA on Trial*, P. R. Billings, ed., pp. 141–149. Cold Spring Harbor Press. (with N. L. Wilker, S. Stawksi, and P. R. Billings).

1993. *Biology as Ideology: The Doctrine of DNA*. New York: Harper Collins.

1993. *The Doctrine of DNA. Biology as Ideology*. Penguin Books.

1993. Correlation between relatives for colorectal cancer mortality in familial adenomatous polyposis. *Ann. Hum. Genet.* **57**:105–115 (with S. Presciuttini, L. Bertario, P. Saia, and C. Rossetti).

1993. Letters *Science* **260**:473–474 (with D. L. Hartl).

1993. Risposta a Sgaramella (Reply to Sgaramella). *La Riv. dei Libri.*

1993. Biology as Ideology. In *The Dancer and the Dance*, E. Sagarra and M. Sagarra, eds., pp. 54–64. Trinity Jameson Quartercentenary Symposium.

1994. Comment: The use of DNA profiles in forensic context. In *DNA Fingerprinting: A Review of the Controversy*, by K. Roeder. *Stat. Sci.* **9**(2):259–262.

1994. *Biologia come Ideologia: La Dottrina del DNA*. Italy: Bollati Boringhieri.

1994. Women *versus* the biologists. *NY Rev. Books* **41**(7):31–35.

1994. Forensic DNA typing dispute. *Nature* (Correspondence) **372**:398.

1994. DNA fingerprinting. *Science* (Letters) **266**:201 (with D. L. Hartl).

1994. Here we go again. Neri e bianchi per mi pari sono. *La Republica* (Milan), Oct. 25, pp. 24–25.

1994. Response to Goldberg and Hardy. *NY Rev. Books.*

1994. Holism and reductionism in ecology. *CNS* **5**(4):33–40 (with R. Levins).

1994. Facts and the factitious in natural sciences. In *Questions of Evidence*, J. Chandler, A. I. Davidson, and H. Harootunian, eds., pp. 478–491. Chicago: U. Chicago Press.

1994. A rejoinder to William Wimsatt. In *Questions of Evidence*, J. Chandler, A. I. Davidson, and H. Harootunian, eds., pp. 504–509. Chicago: U. Chicago Press.

1994. Response to Lander and Budowle. *Nature* (Letter to the Editor) **372**:398.

1994. *De DNA Doctrine*. Amsterdam, The Netherlands: Uitgevery Bert Bakker.

1995. *Human Diversity*. New York: Scientific American Library.

1995. Promesses, promesses. *Genetique et Evolution* **11**:II–V.

1995. Genes, environment and organisms. In *Hidden Histories of Science,* R. B. Silvers, ed., pp. 115–139. *NY Rev. Books.*

1995. The detection of linkage disequilibrium in molecular sequence data. *Genetics* **140**:377–388.

1995. A la recherche du temps perdu. *Configurations* **2**:257–265.

1995. Sex, lies and social science. *NY Rev. Books* **42**(7):24–29.

1995. Il potere del progetto (The power of the project). *SFERA Magazine* **43**:11–34. Rome and Milan.

1995. Theodosius Dobzhansky – A theoretician without tools. In *Genetics of Natural Populations: The Continuing Importance of Theodocius Dobzhansky,* L. Levine, ed., pp. 87–101. New York: Columbia U. Press.

1995. The dream of the human genome. In *Politics and the Human Body,* J. B. Elshtain and J. T. Lloyd, eds., pp. 41–66. Nashville, TN: Vanderbilt U. Press.

1995. IV. Primate models of human traits. In *Aping Science* by B. P. Reines, pp. 17–35.

1996. What does electrophoretic variation tell us about protein variation? *Mol. Biol. Evol.* **13**(2):427–432. (with A. Barbadilla and L. M. King).

1996. La recherche du temps perdue. In *Science Wars,* ed., Durham, NC: Duke U. Press.

1996. *Of genes and genitals.* Cambridge, MA: Transition Publishers.

1996. The end of natural history? *CNS* **7**(1):1–4 (with R. Levins).

1996. Pitfalls of genetic testing. *NEJM* (Sounding Board) **334**(18):1192–1194 (with R. Hubbard).

1996. Authors' Reply. *NEJM* (Correspondence) **335**:1236–1237 (with R. Hubbard).

1996. In defense of science. *Society* **33**(4):30–31.

1996. Evolution as engineering. In *Integrative Approaches to Molecular Biology.* J. Collado, T. Smith, and B. Magasnik, eds., pp. 1–10. Cambridge, MA: MIT Press.

1996. Primate models of human traits. In *Monkeying with Public Health.* R. Reines and S. Kaufman, eds., pp. 17–35. WI: One Voice.

1996. Indiana Jones meets King Kong. Review of three books by M. Crichton. *NY Rev. Books* (February 29).

1996. Letter. (Zola biography review) *NY Rev. Books* (May '96).

1996. *Evolution and Religion.* Am. Jewish Congress.

1996. Letter. (On Horton's essay on genetics and homosexuality.) *NY Rev. Books.*

1996. Population genetic issues in the forensic use of DNA. In *The West Companion to Scientific Evidence.* D. Faigman, ed., pp. 673–696.

1996. The evolution of cognition: questions we will never answer. In *An Invitation to Cognitive Science—Methods, Models, and Conceptual Issues,* Vol. 4, D. Scarborough and S. Sternberg, eds., pp. 132–197, 2nd ed. Cambridge, MA: MIT Press

1996. Detecting heterogeneity of substitutions along DNA and protein sequences. P. J. E. Goss and R. C. Lewontin, eds., *Genetics* **143**:589–602.

1996. False dichotomies. *CNS* **7**(3):27–30 (with R. Levins).

1996. The return of old diseases and the appearance of new ones. *CNS* **7**(2):103–107 (with R. Levins).

1997. Billions and billions of demons. Review of *The Demon Haunted World: Science as a Candle in the Dark* by C. Sagan. *NY Rev. Books.*

1997. Nucleotide variation and conservation at the dpp locus, a gene controlling early development. *Genetics* **145**(2):311–323 (with B. Richter, M. Long, E. Nitasaka).

1997. A scientist meets a visionary. In *Bucky's 100.* (In press.)

1997. Population genetics. In *Encyclopedia of Human Biology,* 2nd ed. New York: Academic Press.

1997. Genetics, plant breeding and patents: Conceptual contradictions and practical problems in protecting biological innovation. *Plant Genet. Res. Newsletter* **112**:1–8 (with M. de Miranda Santos).

1997. The evolution of cognition: Questions we will never answer. In *Invitation to Cognitive Science,* Vol. 4., D. N. Osherson, D. Scarborough, and S. Sternberg, eds., pp. 107–132. Cambridge, MA: MIT Press.

1997. Dobzhansky's "Genetics and the Origin of Species": Is it still relevant? *Genetics* **147**:351–355.

1997. The Cold War and the transformation of the Academy. In *The Cold War and the University,* Vol. 1., pp. 1–34. New York: The New Press.

1997. The biological and the social. *CNS* **8**(3):89–92 (with R. Levins).

1997. Chance and Necessity. *CNS* **8**(2):95–98 (with R. Levins).

1997. Confusing over cloning. Review of: *Cloning Human Beings: Report and Recommendations of the National Bioethics Advisory Commission* (June, 1997). *NY Rev. Books* **44**(16).

1997. Biologie as Ideologie: Ursache und Wirkung bei der Tuberkulose, den menschlichen Genen und in der Landwirtschaft. *Streitbarer Materialismus* **21**:111–128.

1997. A question of biology: Are the races different? In *Beyond Heroes and Holidays,* E. Lee, D. Menkart, and M. Okazawa-Rey, eds., Network of Educators on the Americas.

1998. *Gene, organismo, ambiente.* Rome: Laterza. (In press.)

1998. The maturing of capitalist agriculture: The farmer as proletarian. *Monthly Rev. Press.* (In press.)

1998. The economics of "hybrid" corn. *Am. Econom. Rev.* (with J. P. Berlan).

1998. Forward. In *The Ontogeny of Information,* S. Oyama, R. Gray, and P. Griffiths, eds., Duke U. Press. (In press.)

1998. Review of *"Unto Others": The Evolution and Psychology of Unselfish Behavior.* by E. Sober and D. S. Wilson. Cambridge, MA: Harvard U. Press. *NY Rev. Books.*

1998. *Gene, Organismo, e Ambiente.* Rome: Guis. Laterza and Figli Spa.

1998. The maturing of capitalist agriculture: Farmer as Proletarian. *Monthly Rev.* **50**(3):72–85.

1998. Survival of the Nicest? *NY Rev. Books* **45**(16):59–63.

1998. Foreword. In *Building a New Biocultural Synthesis,* A. H. Goodman and T. L. Leatherman, eds., pp. xi–xv (with R. Levins).

1998. The confusion over cloning. In *Flesh of My Flesh. The Ethics of Cloning Humans,* G. E. Pense, ed., pp. 129–139. Lanham, MD: Rowman and Littlefield.

1998. How different are natural and social science? *CNS* **9**(1):85–89 (with R. Levins).

1998. Does anything new ever happen? *CNS* **9**(2):53–56 (with R. Levins).

1998. Life on other worlds. *CNS* **9**(4):39–42 (with R. Levins).

1999. Menace of the genetic-industrial complex. *The Guardian Weekly.*

1999. *Modern Genetic Analysis.* New York: W. H. Freeman (with A. J. F. Griffiths, W. M. Gelbart, and J. H. Miller).

1999. Foreword. In *The Ontogeny of Information,* S. Oyama, R. Gray, and P. Griffiths, eds., Duke U. Press. (In press.)

1999. *Dream of the Human Genome – and Other Confusions.* R. Silvers, ed. *NY Rev. Books.* (In press.)

1999. *Gene, Organism and Environment.* Cambridge, MA: Harvard U. Press. (In press.)

1999. Genotype and phenotype. In *International Encyclopedia of the Social and Behavioral,* N. J. Smelser and P. D. Baltes, eds. Elsevier Science, Pergamon. (In preparation.)

1999. Are we programmed? *CNS* **10**(2):71–75 (with R. Levins).

1999. *Does Culture Evolve?* (with J. Fracchia) (In preparation.)

1999. Does Culture Evolve? *History and Theory: Studies in the Philosophy of History* **38**:52–78. The Return of Science: Evolutionary Ideas and History. Wesleyan University, CT.

1999. Locating regions of differential variability in DNA and protein sequences. *Genetics* **153**:485–495 (with H. Tang).

2000. Let the numbers speak. *CNS* **11**(2):63–67 (with R. Levins).

2000. The politics of averages. *CNS* **11**(2):111–114 (with R. Levins).

2000. Computing the organism. End Paper. *Natural History* **4**:94–95.

2000. What do population geneticists know and how do they know it? In *Biology and Epistemology*, R. Creath and J. Maienschen, eds., pp. 191–214. Cambridge, UK: Cambridge University Press.

2000. The problems of population genetic. In *Evolutionary Genetics from Molecules to Morphology*, R. S. Singh and C. B. Krimbas, eds., pp. 5–23. Cambridge, UK: Cambridge University Press.

2000. *It Ain't Necessarily So: The Dream of the Human Genome and Other Illusions.* The New York Review of Books, NY.

2000. *The Triple Helix: Gene, Organism, and Environment.* Cambridge, MA: Harvard University Press.

2000. *An Introduction to Genetic Analysis*, 7th ed. (with A. J. F. Griffiths, J. H. Miller, D. T. Suzuki, and W. M. Gelbart). New York: W. H. Freeman.

2000. Dobzhansky, Theodosious. In *Encyclopedia of Genetics*, S. Brenner and J. H. Miller, eds. New York: Academic Press.

2000. Natural history and formalism in evolutionary genetics. In *Thinking About Evolution: Historical, Philosophical, and Political Perspectives*, R. S. Singh et al., eds. Cambridge, UK: Cambridge University Press.

Index